Student Solutions Manual for Stewart's

SINGLE VARIABLE
CALCULUS
Early Transcendentals

FOURTH EDITION

DANIEL ANDERSON
University of Iowa

JEFFERY A. COLE
Anoka-Ramsey Community College

DANIEL DRUCKER
Wayne State University

BROOKS/COLE PUBLISHING COMPANY
I(T)P® An International Thomson Publishing Company

Pacific Grove · Albany · Belmont · Bonn · Boston · Cincinnati · Detroit
Johannesburg · London · Madrid · Melbourne · Mexico City · New York
Paris · Singapore · Tokyo · Toronto · Washington

 A GARY W. OSTEDT BOOK

Assistant Editor: *Carol Ann Benedict*   Typesetting: *Andrew Bulman-Fleming*
Marketing Team: *Caroline Croley, Debra Johnston*   Cover Illustration: *dan clegg*
Ancillaries Coordinator: *Dorothy Bell*   Printing and Binding: *Webcom Limited*

For more information, contact:

BROOKS/COLE PUBLISHING COMPANY
511 Forest Lodge Road
Pacific Grove, CA 93950
USA

International Thomson Editores
Seneca 53
Col. Polanco
11560 México, D. F., México

International Thomson Publishing Europe
Berkshire House 168-173
High Holborn
London WC1V 7AA
England

International Thomson Publishing GmbH
Königswinterer Strasse 418
53227 Bonn
Germany

Thomas Nelson Australia
102 Dodds Street
South Melbourne, 3205
Victoria, Australia

International Thomson Publishing Asia
60 Albert Street
#15-01 Albert Complex
Singapore 189969

Nelson Canada
1120 Birchmount Road
Scarborough, Ontario
Canada M1K 5G4

International Thomson Publishing Japan
Palaceside Building, 5F
1-1-1 Hitotsubashi
Chiyoda-ku, Tokyo 100-0003
Japan

Printed in Canada

10 9 8 7 6 5 4 3 2

ISBN 0-534-36301-6

# Preface

This *Student Solutions Manual* contains strategies for solving and solutions to selected exercises in the text *Single Variable Calculus: Early Transcendentals, Fourth Edition*, by James Stewart. It contains solutions to the odd-numbered exercises in each section, the review sections, the True-False Quizzes, and the Problem Solving sections, as well as solutions to all the exercises in the Concept Checks.

We use some non-standard notation in order to save space. If you see a symbol which you don't recognize, refer to the Table of Abbreviations and Symbols on page iv.

This manual is a text supplement and should be read along *with* the text. You should read all exercise solutions in this manual because many concept explanations are given and then used in subsequent solutions. All concepts necessary to solve a particular problem are not reviewed for every exercise. If you are having difficulty with a previously covered concept, refer back to the section where it was covered for more complete help.

A significant number of today's students are involved in various outside activities, and find it difficult, if not impossible, to attend all class sessions; this manual should help meet the needs of these students. In addition, it is our hope that this manual's solutions will enhance the understanding of all readers of the material and provide insights to solving other exercises.

We appreciate feedback concerning errors, solution correctness or style, and manual style. Any comments may be sent directly to jcole@an.cc.mn.us, or in care of the publisher: Brooks/Cole Publishing Company, 511 Forest Lodge Road, Pacific Grove, CA 93950.

We would like to thank Andrew Bulman-Fleming, for typesetting the manuscript; Brian Betsill, Stephanie Kuhns, and Kathi Townes, of TECH-arts, for their production services; and Carol Ann Benedict, of Brooks/Cole Publishing Company, for her patience and support. All of these people have provided invaluable help in creating this manual.

Jeffery A. Cole
Anoka-Ramsey Community College

James Stewart
McMaster University

Daniel Drucker
Wayne State University

Daniel Anderson
University of Iowa

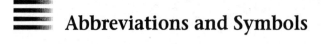

Abbreviations and Symbols

CD	concave downward
CU	concave upward
D	the domain of f
FDT	First Derivative Test
HA	horizontal asymptote(s)
I	interval of convergence
IP	inflection point(s)
R	radius of convergence
VA	vertical asymptote(s)
$\overset{H}{=}$	indicates the use of l'Hospital's Rule.
$\overset{j}{=}$	indicates the use of Formula j in the Table of Integrals in the back endpapers.
$\overset{s}{=}$	indicates the use of the substitution $\{u = \sin x, \, du = \cos x \, dx\}$.
$\overset{c}{=}$	indicates the use of the substitution $\{u = \cos x, \, du = -\sin x \, dx\}$.

Contents

▤ Infinite Sequences and Series   389

▤ Appendixes   441

Functions and Models

1.1 Four Ways to Represent a Function

In exercises requiring estimations or approximations, your answers may vary slightly from the answers given here.

1. (a) The point $(-1, -2)$ is on the graph of f, so $f(-1) = -2$.

 (b) When $x = 2$, y is about 2.8, so $f(2) \approx 2.8$.

 (c) $f(x) = 2$ is equivalent to $y = 2$. When $y = 2$, we have $x = -3$ and $x = 1$.

 (d) Reasonable estimates for x when $y = 0$ are $x = -2.5$ and $x = 0.3$.

 (e) The domain of f consists of all x-values on the graph of f. For this function, the domain is $-3 \le x \le 3$. The range of f consists of all y-values on the graph of f. For this function, the range is $-2 \le y \le 3$.

 (f) As x increases from -1 to 3, y increases from -2 to 3. Thus, f is increasing on the interval $[-1, 3]$.

3. From Figure 1 in the text, the lowest point occurs at about $(t, a) = (12, -85)$. The highest point occurs at about $(17, 115)$. Thus, the range of the vertical ground acceleration is $-85 \le a \le 115$. In Figure 11, the range of the north-south acceleration is approximately $-325 \le a \le 485$. In Figure 12, the range of the east-west acceleration is approximately $-210 \le a \le 200$.

5. Yes, the curve is the graph of a function because it passes the Vertical Line Test. The domain is $[-3, 2]$ and the range is $[-2, 2]$.

7. No, the curve is not the graph of a function since for $x = -1$ there are infinitely many points on the curve.

9. The person's weight increased to about 160 pounds at age 20 and stayed fairly steady for 10 years. The person's weight dropped to about 120 pounds for the next 5 years, then increased rapidly to about 170 pounds. The next 30 years saw a gradual increase to 190 pounds. Possible reasons for the drop in weight at 30 years of age: diet, exercise, health problems.

11. The water will cool down almost to freezing as the ice melts. Then, when the ice has melted, the water will slowly warm up to room temperature.

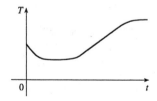

13. Of course, this graph depends strongly on the geographical location!

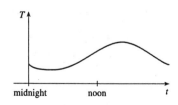

15.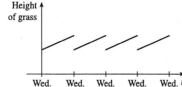

17. (a)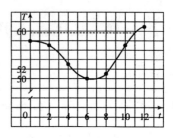

(b) $T(11) \approx 59°F$

19. $f(x) = 2x^2 + 3x - 4$, so $f(0) = 2(0)^2 + 3(0) - 4 = -4$,

$f(2) = 2(2)^2 + 3(2) - 4 = 10$, $f(\sqrt{2}) = 2(\sqrt{2})^2 + 3(\sqrt{2}) - 4 = 3\sqrt{2}$,

$f(1 + \sqrt{2}) = 2(1 + \sqrt{2})^2 + 3(1 + \sqrt{2}) - 4 = 2\left(1 + 2 + 2\sqrt{2}\right) + 3 + 3\sqrt{2} - 4 = 5 + 7\sqrt{2}$,

$f(-x) = 2(-x)^2 + 3(-x) - 4 = 2x^2 - 3x - 4$,

$f(x + 1) = 2(x + 1)^2 + 3(x + 1) - 4 = 2(x^2 + 2x + 1) + 3x + 3 - 4 = 2x^2 + 7x + 1$,

$2f(x) = 2(2x^2 + 3x - 4) = 4x^2 + 6x - 8$, and

$f(2x) = 2(2x)^2 + 3(2x) - 4 = 2(4x^2) + 6x - 4 = 8x^2 + 6x - 4$.

21. $f(x) = x - x^2$, so $f(2 + h) = 2 + h - (2 + h)^2 = 2 + h - 4 - 4h - h^2 = -\left(h^2 + 3h + 2\right)$,

$f(x + h) = x + h - (x + h)^2 = x + h - x^2 - 2xh - h^2$, and

$\dfrac{f(x + h) - f(x)}{h} = \dfrac{x + h - x^2 - 2xh - h^2 - x + x^2}{h} = \dfrac{h - 2xh - h^2}{h} = 1 - 2x - h$.

23. $f(x) = \dfrac{x + 2}{x^2 - 1}$ is defined for all x except when $x^2 - 1 = 0 \iff x = 1$ or $x = -1$, so the domain is $\{x \mid x \neq \pm 1\}$.

25. $g(x) = \sqrt[4]{x^2 - 6x}$ is defined when $0 \leq x^2 - 6x = x(x - 6) \iff x \geq 6$ or $x \leq 0$, so the domain is $(-\infty, 0] \cup [6, \infty)$.

27. $f(t) = \sqrt[3]{t - 1}$ is defined for every t, since every real number has a cube root. The domain is the set of all real numbers, $\mathbb{R}$.

29. $f(x) = 3 - 2x$. Domain is $\mathbb{R}$.

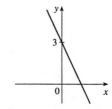

31. $g(x) = \sqrt{x-5}$ is defined when $x - 5 \geq 0$ or $x \geq 5$, so the domain is

$[5, \infty)$. Since $y = \sqrt{x-5} \Rightarrow y^2 = x - 5 \Rightarrow x = y^2 + 5$,

we see that g is the top half of a parabola.

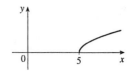

33. $G(x) = |x| + x$. Since $|x| = \begin{cases} x & \text{if } x \geq 0 \\ -x & \text{if } x < 0 \end{cases}$ we have

$$G(x) = \begin{cases} x + x & \text{if } x \geq 0 \\ -x + x & \text{if } x < 0 \end{cases} = \begin{cases} 2x & \text{if } x \geq 0 \\ 0 & \text{if } x < 0 \end{cases}$$

Domain is $\mathbb{R}$. Note that the negative x-axis is part of the graph of G.

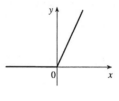

35. $f(x) = \dfrac{x}{|x|} = \begin{cases} x/x & \text{if } x > 0 \\ x/(-x) & \text{if } x < 0 \end{cases} = \begin{cases} 1 & \text{if } x > 0 \\ -1 & \text{if } x < 0 \end{cases}$

Note that we did not use $x \geq 0$, because $x \neq 0$. Hence, the domain of f is
$\{x \mid x \neq 0\}$.

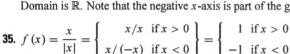

37. $f(x) = \begin{cases} x & \text{if } x \leq 0 \\ x + 1 & \text{if } x > 0 \end{cases}$

Domain is $\mathbb{R}$.

39. $f(x) = \begin{cases} x + 2 & \text{if } x \leq -1 \\ x^2 & \text{if } x > -1 \end{cases}$

Domain is $\mathbb{R}$.

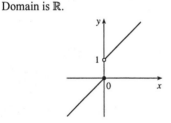

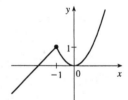

41. Recall that the slope m of a line between the two points (x_1, y_1) and (x_2, y_2) is $m = \dfrac{y_2 - y_1}{x_2 - x_1}$ and an equation of the

line connecting those two points is $y - y_1 = m(x - x_1)$. The slope of this line segment is $\dfrac{-6 - 1}{4 - (-2)} = -\dfrac{7}{6}$, so an

equation is $y - 1 = -\frac{7}{6}(x + 2)$. The function is $f(x) = -\frac{7}{6}x - \frac{4}{3}$, $-2 \leq x \leq 4$.

43. We need to solve the given equation for y. $x + (y-1)^2 = 0 \Rightarrow (y-1)^2 = -x \Rightarrow y - 1 = \pm\sqrt{-x} \Rightarrow$
$y = 1 \pm \sqrt{-x}$. The expression with the positive radical represents the top half of the parabola, and the one with the
negative radical represents the bottom half. Hence, we want $f(x) = 1 - \sqrt{-x}$, $x \leq 0$.

45. For $-1 \leq x \leq 2$, the graph is the line with slope 1 and y-intercept 1, that is, the line $y = x + 1$. For $2 < x \leq 4$, the
graph is the line with slope $-\frac{3}{2}$ and x-intercept 4, so $y = -\frac{3}{2}(x - 4) = -\frac{3}{2}x + 6$. So the function is

$$f(x) = \begin{cases} x + 1 & \text{if } -1 \leq x \leq 2 \\ -\frac{3}{2}x + 6 & \text{if } 2 < x \leq 4 \end{cases}$$

47. Let the length and width of the rectangle be L and W. Then the perimeter is $2L + 2W = 20$ and the area is

$A = LW$. Solving the first equation for W in terms of L gives $W = \dfrac{20 - 2L}{2} = 10 - L$. Thus,

$A(L) = L(10 - L) = 10L - L^2$. Since lengths are positive, the domain of A is $0 < L < 10$. If we further restrict
L to be larger than W, then $5 < L < 10$ would be the domain.

49. Let the length of a side of the equilateral triangle be x. Then by the Pythagorean Theorem, the height y of the triangle satisfies $y^2 + \left(\frac{1}{2}x\right)^2 = x^2$, so that $y = \frac{\sqrt{3}}{2}x$. Using the formula for the area A of a triangle, $A = \frac{1}{2}$ (base) (height), we obtain $A(x) = \frac{1}{2}(x)\left(\frac{\sqrt{3}}{2}x\right) = \frac{\sqrt{3}}{4}x^2$, with domain $x > 0$.

51. Let each side of the base of the box have length x, and let the height of the box be h. Since the volume is 2, we know that $2 = hx^2$, so that $h = 2/x^2$, and the surface area is $S = x^2 + 4xh$. Thus, $S(x) = x^2 + 4x\left(2/x^2\right) = x^2 + 8/x$, with domain $x > 0$.

53. The height of the box is x and the length and width are $L = 20 - 2x$, $W = 12 - 2x$. Then $V = LWx$ and so

$$V(x) = (20 - 2x)(12 - 2x)(x) = 4(10 - x)(6 - x)(x) = 4x(60 - 16x + x^2)$$
$$= 4x^3 - 64x^2 + 240x$$

The sides L, W, and x must be positive. Thus, $L > 0 \iff 20 - 2x > 0 \iff x < 10$; $w > 0 \iff 12 - 2x > 0 \iff x < 6$; and $x > 0$. Combining these restrictions gives us the domain $0 < x < 6$.

55. (a) 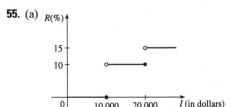 (c)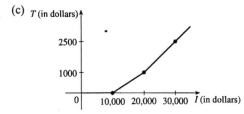

(b) On \$14,000, tax is assessed on \$4000, and 10% (\$4000) = \$400.

On \$26,000, tax is assessed on \$16,000, and 10% (\$10,000) + 15% (\$6000) = \$1000 + \$900 = \$1900.

57. (a) Because an even function is symmetric with respect to the y-axis, and the point $(5, 3)$ is on the graph of this even function, the point $(-5, 3)$ must also be on its graph.

(b) Because an odd function is symmetric with respect to the origin, and the point $(5, 3)$ is on the graph of this odd function, the point $(-5, -3)$ must also be on its graph.

59. $f(-x) = (-x)^{-2} = \dfrac{1}{(-x)^2} = \dfrac{1}{x^2}$

$= x^{-2} = f(x)$

so f is an even function.

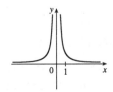

61. $f(-x) = (-x)^2 + (-x) = x^2 - x$. Since this is neither $f(x)$ nor $-f(x)$, the function f is neither even nor odd.

63. $f(-x) = (-x)^3 - (-x) = -x^3 + x$

$= -\left(x^3 - x\right) = -f(x)$

so f is odd.

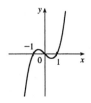

1.2 Mathematical Models

1. (a) $f(x) = \sqrt[5]{x}$ is a root function.

(b) $g(x) = \sqrt{1 - x^2}$ is an algebraic function because it is a root of a polynomial.

(c) $h(x) = x^9 + x^4$ is a polynomial of degree 9.

(d) $r(x) = \dfrac{x^2 + 1}{x^3 + x}$ is a rational function because it is a ratio of polynomials.

(e) $s(x) = \tan 2x$ is a trigonometric function.

(f) $t(x) = \log_{10} x$ is a logarithmic function.

3. We notice from the figure that g and h are even functions (symmetric with respect to the y-axis) and that f is an odd function (symmetric with respect to the origin). So (b) $[y = x^5]$ must be f. Since g is flatter than h near the origin, we must have (c) $[y = x^8]$ matched with g and (a) $[y = x^2]$ matched with h.

5. (a) An equation for the family of linear functions with slope 2 is

$y = f(x) = 2x + b$, where b is the y-intercept.

(b) $f(2) = 1$ means that the point $(2, 1)$ is on the graph of f. We can use the point-slope form of a line to obtain an equation for the family of linear functions through the point $(2, 1)$. $y - 1 = m(x - 2)$, which is equivalent to $y = mx + (1 - 2m)$ in slope-intercept form.

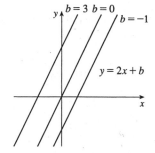

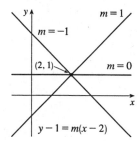

(c) The slope m must equal 2, so the equation in part (b), $y = mx + (1 - 2m)$, becomes $y = 2x - 3$. It is the *only* function that belongs to both families.

7. (a)

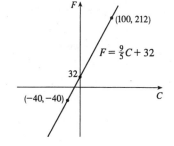

(b) The slope of $\frac{9}{5}$ means that F increases $\frac{9}{5}$ degrees for each increase of $1°C$. (Equivalently, F increases by 9 when C increases by 5 and F decreases by 9 when C decreases by 5.) The F-intercept of 32 is the Fahrenheit temperature corresponding to a Celsius temperature of 0.

9. (a) Using N in place of x and T in place of y, we find the slope to be $\dfrac{T_2 - T_1}{N_2 - N_1} = \dfrac{80 - 70}{173 - 113} = \dfrac{10}{60} = \dfrac{1}{6}$. So a

linear equation is $T - 80 = \frac{1}{6}(N - 173)$ ⟺ $T - 80 = \frac{1}{6}N - \frac{173}{6}$ ⟺ $T = \frac{1}{6}N + \frac{307}{6}$ [$\frac{307}{6} = 51.1\overline{6}$].

(b) The slope of $\frac{1}{6}$ means that the temperature in Fahrenheit degrees increases one-sixth as rapidly as the number of cricket chirps per minute. Said differently, each increase of 6 cricket chirps per minute corresponds to an increase of 1°F.

(c) When $N = 150$, the temperature is given approximately by $T = \frac{1}{6}(150) + \frac{307}{6} = 76.1\overline{6}°\text{F} \approx 76°\text{F}$.

11. (a) We are given $\dfrac{\text{change in pressure}}{10 \text{ feet change in depth}} = \dfrac{4.34}{10} = 0.434$. Using P for pressure and d for depth with the point
$(d, P) = (0, 15)$, we have $P - 15 = 0.434(d - 0)$ ⟺ $P = 0.434d + 15$.

(b) When $P = 100$, then $100 = 0.434d + 15$ ⟺ $0.434d = 85$ ⟺ $d \approx 195.85$ feet. Thus, the pressure is 100 lb/in^2 at a depth of approximately 196 feet.

13. (a) The data appear to be periodic and a sine or cosine function would make the best model. A model of the form
$f(x) = a\cos(bx) + c$ seems appropriate.

(b) The data appear to be decreasing in a linear fashion. A model of the form $f(x) = mx + b$ seems appropriate.

Some values are given to many decimal places. These are the results given by several computer algebra systems — rounding is left to the reader.

15. (a)

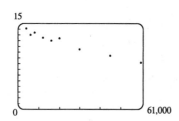

A linear model does seem appropriate.

(b) Using the points $(4000, 14.1)$ and $(60{,}000, 8.2)$, we obtain

$$y - 14.1 = \frac{8.2 - 14.1}{60{,}000 - 4000}(x - 4000) \text{ or, equivalently,}$$

$$y \approx -0.000105357x + 14.521429.$$

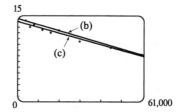

(c) Using a computing device, we obtain the least squares regression line $y = -0.0000997855x + 13.950764$.

(d) When $x = 25{,}000$, $y \approx 11.456$; or about 11.5 per 100 population.

(e) When $x = 80{,}000$, $y \approx 5.968$; or about a 6% chance.

(f) When $x = 200{,}000$, y is negative, so the model does not apply.

17. (a)

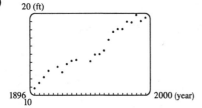

A linear model does seem appropriate.

(b)

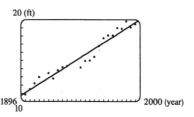

Using a computing device, we obtain the least squares regression line $y = -158.2403249x + 0.089119747$, where x is the year and y is the height in feet.

(c) When $x = 2000$, $y \approx 20.00$ ft.

(d) When $x = 2100$, $y \approx 28.91$ ft. This would be an increase of 9.49 ft from 1996 to 2100. Even though there was an increase of 8.59 ft from 1900 to 1996, it is unlikely that a similar increase will occur over the next 100 years.

19.

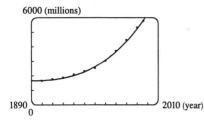

Using a computing device, we obtain the cubic function $y = ax^3 + bx^2 + cx + d$ with $a = 0.00232567051876$, $b = -13.064877957628$, $c = 24{,}463.10846422$, and $d = -15{,}265{,}793.872507$. When $x = 1925$, $y \approx 1922$ (millions).

1.3 New Functions from Old Functions

1. (a) If the graph of f is shifted 3 units upward, its equation becomes $y = f(x) + 3$.

(b) If the graph of f is shifted 3 units downward, its equation becomes $y = f(x) - 3$.

(c) If the graph of f is shifted 3 units to the right, its equation becomes $y = f(x - 3)$.

(d) If the graph of f is shifted 3 units to the left, its equation becomes $y = f(x + 3)$.

(e) If the graph of f is reflected about the x-axis, its equation becomes $y = -f(x)$.

(f) If the graph of f is reflected about the y-axis, its equation becomes $y = f(-x)$.

(g) If the graph of f is stretched vertically by a factor of 3, its equation becomes $y = 3f(x)$.

(h) If the graph of f is shrunk vertically by a factor of 3, its equation becomes $y = \frac{1}{3}f(x)$.

3. (a) (graph 3) The graph of f is shifted 4 units to the right and has equation $y = f(x - 4)$.

(b) (graph 1) The graph of f is shifted 3 units upward and has equation $y = f(x) + 3$.

(c) (graph 4) The graph of f is shrunk vertically by a factor of 3 and has equation $y = \frac{1}{3}f(x)$.

(d) (graph 5) The graph of f is shifted 4 units to the left and reflected about the x-axis. Its equation is $y = -f(x + 4)$.

(e) (graph 2) The graph of f is shifted 6 units to the left and stretched vertically by a factor of 2. Its equation is $y = 2f(x + 6)$.

5. (a) To graph $y = f(2x)$ we shrink the graph of f horizontally by a factor of 2.

(b) To graph $y = f\left(\frac{1}{2}x\right)$ we stretch the graph of f horizontally by a factor of 2.

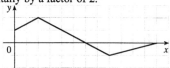

(c) To graph $y = f(-x)$ we reflect the graph of f about the y-axis.

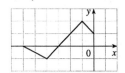

(d) To graph $y = -f(-x)$ we reflect the graph of f about the y-axis, then about the x-axis.

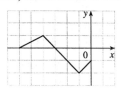

7. The graph of $y = f(x) = \sqrt{3x - x^2}$ has been shifted 4 units to the left, reflected about the x-axis, and shifted downward 1 unit. Thus, a function describing the graph is

$$y = \underbrace{-1 \cdot}_{\substack{\text{reflect} \\ \text{about} \\ x\text{-axis}}} \underbrace{f(x+4)}_{\substack{\text{shift} \\ 4 \text{ units} \\ \text{left}}} \underbrace{-1}_{\substack{\text{shift} \\ 1 \text{ unit} \\ \text{down}}}$$

This function can be written as

$$y = -f(x+4) - 1 = -\sqrt{3(x+4) - (x+4)^2} - 1$$

$$= -\sqrt{3x + 12 - (x^2 + 8x + 16)} - 1 = -\sqrt{-x^2 - 5x - 4} - 1$$

9. $y = -1/x$: Start with the graph of $y = 1/x$ and reflect about the x-axis.

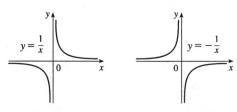

11. $y = \tan 2x$: Start with the graph of $y = \tan x$ and compress horizontally by a factor of 2.

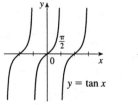

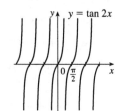

13. $y = \cos(x/2)$: Start with the graph of $y = \cos x$ and stretch horizontally by a factor of 2.

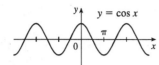
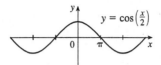

15. $y = \dfrac{1}{x-3}$: Start with the graph of $y = 1/x$ and shift 3 units to the right.

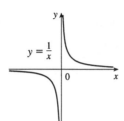

17. $y = \frac{1}{3}\sin\left(x - \frac{\pi}{6}\right)$: Start with the graph of $y = \sin x$, shift $\frac{\pi}{6}$ units to the right, and then compress vertically by a factor of 3.

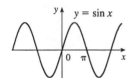
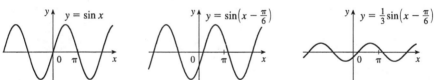
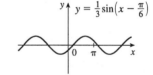

19. $y = 1 + 2x - x^2 = -x^2 + 2x + 1 = -\left(x^2 - 2x + 1\right) + 1 + 1 = -(x-1)^2 + 2$: Start with the graph of $y = x^2$, shift 1 unit right, reflect about the x-axis, and then shift 2 units upward.

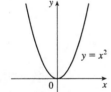
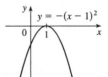
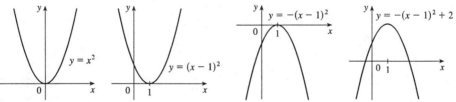
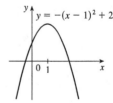

21. $y = 2 - \sqrt{x+1}$: Start with the graph of $y = \sqrt{x}$, reflect about the x-axis, shift 1 unit to the left, and then shift 2 units upward.

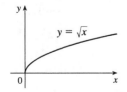
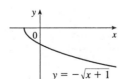
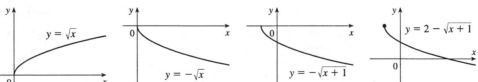
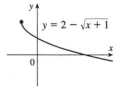

23. $y = ||x| - 1|$: Start with the graph of $y = |x|$, shift 1 unit downward, and then reflect the part of the graph from $x = -1$ to $x = 1$ about the x-axis.

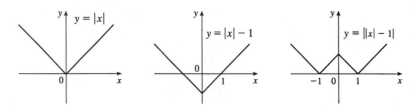

25. This is just like the solution to Example 4 except the amplitude of the curve is $14 - 12 = 2$. So the function is $L(t) = 12 + 2\sin\left[\frac{2\pi}{365}(t - 80)\right]$. March 31 is the 90th day of the year, so the model gives $L(90) \approx 12.34$ h. The daylight time (5:51 A.M. to 6:18 P.M.) is 12 hours and 27 minutes, or 12.45 h. The model value differs from the actual value by $\frac{12.45 - 12.34}{12.45} \approx 0.009$, less than 1%.

27. (a) To obtain $y = f(|x|)$, the portion of $y = f(x)$ to the right of the y-axis is reflected about the y-axis.

(b) $y = \sin|x|$

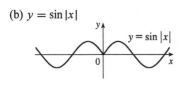

(c) $y = \sqrt{|x|}$

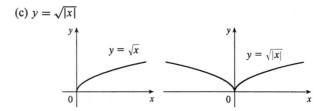

29. Assuming that successive horizontal and vertical gridlines are a unit apart, we can make a table of approximate values as follows.

x	0	1	2	3	4	5	6
$f(x)$	2	1.7	1.3	1.0	0.7	0.3	0
$g(x)$	2	2.7	3	2.8	2.4	1.7	0
$f(x) + g(x)$	4	4.4	4.3	3.8	3.1	2.0	0

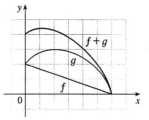

Connecting the points $(x, f(x) + g(x))$ with a smooth curve gives an approximation to the graph of $f + g$. Extra points can be plotted between those listed above if necessary.

31. $f(x) = x^3 + 2x^2$; $g(x) = 3x^2 - 1$. $D = \mathbb{R}$ for both f and g.
$(f + g)(x) = x^3 + 2x^2 + 3x^2 - 1 = x^3 + 5x^2 - 1$, $D = \mathbb{R}$.
$(f - g)(x) = x^3 + 2x^2 - (3x^2 - 1) = x^3 - x^2 + 1$, $D = \mathbb{R}$.
$(fg)(x) = (x^3 + 2x^2)(3x^2 - 1) = 3x^5 + 6x^4 - x^3 - 2x^2$, $D = \mathbb{R}$.
$\left(\frac{f}{g}\right)(x) = \frac{x^3 + 2x^2}{3x^2 - 1}$, $D = \left\{x \mid x \neq \pm\frac{1}{\sqrt{3}}\right\}$ since $3x^2 - 1 \neq 0$.

33. $f(x) = x$, $g(x) = 1/x$

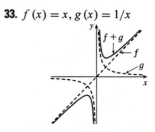

35. $f(x) = 2x^2 - x$; $g(x) = 3x + 2$. $D = \mathbb{R}$ for both f and g, and hence for their composites.

$(f \circ g)(x) = f(g(x)) = f(3x + 2) = 2(3x + 2)^2 - (3x + 2) = 18x^2 + 21x + 6$.

$(g \circ f)(x) = g(f(x)) = g(2x^2 - x) = 3(2x^2 - x) + 2 = 6x^2 - 3x + 2$.

$(f \circ f)(x) = f(f(x)) = f(2x^2 - x) = 2(2x^2 - x)^2 - (2x^2 - x) = 8x^4 - 8x^3 + x$.

$(g \circ g)(x) = g(g(x)) = g(3x + 2) = 3(3x + 2) + 2 = 9x + 8$.

37. $f(x) = 1/x$, $D = \{x \mid x \neq 0\}$; $g(x) = x^3 + 2x$, $D = \mathbb{R}$.

$(f \circ g)(x) = f(g(x)) = f(x^3 + 2x) = 1/(x^3 + 2x)$, $D = \{x \mid x^3 + 2x \neq 0\} = \{x \mid x \neq 0\}$.

$(g \circ f)(x) = g(f(x)) = g(1/x) = 1/x^3 + 2/x$, $D = \{x \mid x \neq 0\}$.

$(f \circ f)(x) = f(f(x)) = f(1/x) = \dfrac{1}{1/x} = x$, $D = \{x \mid x \neq 0\}$.

$(g \circ g)(x) = g(g(x)) = g(x^3 + 2x) = (x^3 + 2x)^3 + 2(x^3 + 2x) = x^9 + 6x^7 + 12x^5 + 10x^3 + 4x$, $D = \mathbb{R}$.

39. $f(x) = \sin x$, $D = \mathbb{R}$; $g(x) = 1 - \sqrt{x}$, $D = [0, \infty)$.

$(f \circ g)(x) = f(g(x)) = f(1 - \sqrt{x}) = \sin(1 - \sqrt{x})$, $D = [0, \infty)$.

$(g \circ f)(x) = g(f(x)) = g(\sin x) = 1 - \sqrt{\sin x}$. For $\sqrt{\sin x}$ to be defined, we must have

$\sin x \geq 0 \Leftrightarrow x \in [0, \pi], [2\pi, 3\pi], [-2\pi, -\pi], [4\pi, 5\pi], [-4\pi, -3\pi], \ldots$, so

$D = \{x \mid x \in [2n\pi, \pi + 2n\pi]$, where n is an integer$\}$.

$(f \circ f)(x) = f(f(x)) = f(\sin x) = \sin(\sin x)$, $D = \mathbb{R}$.

$(g \circ g)(x) = g(g(x)) = g(1 - \sqrt{x}) = 1 - \sqrt{1 - \sqrt{x}}$, $D = \{x \geq 0 \mid 1 - \sqrt{x} \geq 0\} = [0, 1]$.

41. $(f \circ g \circ h)(x) = f(g(h(x))) = f(g(x - 1)) = f(\sqrt{x - 1}) = \sqrt{x - 1} - 1$

43. $(f \circ g \circ h)(x) = f(g(h(x))) = f(g(\sqrt{x})) = f(\sqrt{x} - 5) = (\sqrt{x} - 5)^4 + 1$

45. Let $g(x) = x - 9$ and $f(x) = x^5$. Then $(f \circ g)(x) = (x - 9)^5 = F(x)$.

47. Let $g(x) = x^2$ and $f(x) = \dfrac{x}{x + 4}$. Then $(f \circ g)(x) = \dfrac{x^2}{x^2 + 4} = G(x)$.

49. Let $g(t) = \cos t$ and $f(t) = \sqrt{t}$. Then $(f \circ g)(t) = \sqrt{\cos t} = u(t)$.

51. Let $h(x) = x^2$, $g(x) = 3^x$, and $f(x) = 1 - x$. Then $(f \circ g \circ h)(x) = 1 - 3^{x^2} = H(x)$.

53. Let $h(x) = \sqrt{x}$, $g(x) = \sec x$, and $f(x) = x^4$. Then $(f \circ g \circ h)(x) = (\sec \sqrt{x})^4 = \sec^4(\sqrt{x}) = H(x)$.

55. (a) $g(2) = 5$, because the point $(2, 5)$ is on the graph of g. Thus, $f(g(2)) = f(5) = 4$, because the point $(5, 4)$ is on the graph of f.

(b) $g(f(0)) = g(0) = 3$

(c) $(f \circ g)(0) = f(g(0)) = f(3) = 0$

(d) $(g \circ f)(6) = g(f(6)) = g(6)$. This value is not defined, because there is no point on the graph of g that has x-coordinate 6.

(e) $(g \circ g)(-2) = g(g(-2)) = g(1) = 4$

(f) $(f \circ f)(4) = f(f(4)) = f(2) = -2$

57. (a) Using the relationship *distance* = *rate* · *time* with the radius r as the distance, we have $r(t) = 60t$.

(b) $A = \pi r^2 \Rightarrow (A \circ r)(t) = A(r(t)) = \pi(60t)^2 = 3600\pi t^2$. This formula gives us the extent of the rippled area (in cm^2) at any time t.

59. (a)

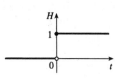

(b)

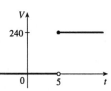

$$V(t) = \begin{cases} 0 & \text{if } t < 0 \\ 120 & \text{if } t \geq 0 \end{cases}$$

so $V(t) = 120H(t)$.

(c)

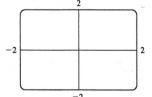

Starting with the formula in part (b), we replace 120 with 240 to reflect the different voltage. Also, because we are starting 5 units to the right of $t = 0$, we replace t with $t - 5$. Thus, the formula is $V(t) = 240H(t - 5)$.

61. (a) By examining the variable terms in g and h, we deduce that we must square g to get the terms $4x^2$ and $4x$ in h. If we let $f(x) = x^2 + c$, then $(f \circ g)(x) = f(g(x)) = f(2x + 1) = (2x + 1)^2 + c = 4x^2 + 4x + (1 + c)$. Since $h(x) = 4x^2 + 4x + 7$, we must have $1 + c = 7$. So $c = 6$ and $g(x) = x^2 + 6$.

(b) We need a function g so that
$$f(g(x)) = 3(g(x)) + 5 = h(x) = 3x^2 + 3x + 2 = 3(x^2 + x) + 2 = 3(x^2 + x - 1) + 5. \text{ So we see that}$$
$g(x) = x^2 + x - 1$.

63. We need to examine $h(-x)$.

$$h(-x) = (f \circ g)(-x) = f(g(-x)) = f(g(x)) \quad \text{[because } g \text{ is even]} \quad = h(x)$$

Because $h(-x) = h(x)$, h is an even function.

65. (a) $P = (a, g(a))$ and $Q = (g(a), g(a))$ because Q has the same y-value as P and it is on the line $y = x$.

(b) The x-value of Q is $g(a)$; this is also the x-value of R. The y-value of R is therefore $f(x\text{-value})$, that is, $f(g(a))$. Hence, $R = (g(a), f(g(a)))$.

(c) The coordinates of S are $(a, f(g(a)))$ or, equivalently, $(a, h(a))$.

(d)

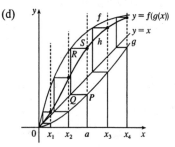

1.4 Graphing Calculators and Computers

1. $f(x) = x^4 + 2$

(a) $[-2, 2]$ by $[-2, 2]$

(b) $[0, 4]$ by $[0, 4]$

(c) $[-4, 4]$ by $[-4, 4]$

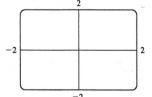

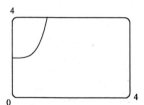

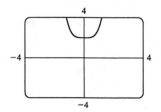

(d)　　　[−8, 8] by [−4, 40]　　　(e)　　　[−40, 40] by [−80, 800]

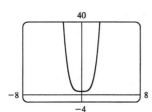

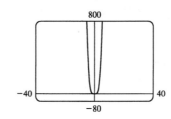

The most appropriate graph is produced in viewing rectangle (d).

3. $f(x) = 10 + 25x - x^3$

(a)　　　[−4, 4] by [−4, 4]　　　　　　(b)　　　[−10, 10] by [−10, 10]

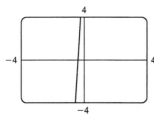

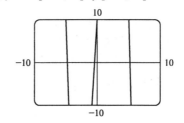

(c)　　　[−20, 20] by [−100, 100]　　　(d)　　　[−100, 100] by [−200, 200]

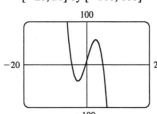

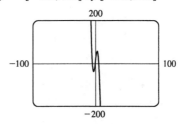

The most appropriate graph is produced in viewing rectangle (c) because the maximum and minimum points are fairly easy to see and estimate.

5. Since the graph of $f(x) = 5 + 20x - x^2$ is a parabola opening downward, an appropriate viewing rectangle should include the maximum point.

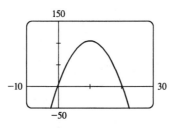

7. $f(x) = \sqrt[4]{256 - x^2}$. To find an appropriate viewing rectangle, we calculate f's domain and range: $256 - x^2 \geq 0 \iff x^2 \leq 256 \iff |x| \leq 16 \iff -16 \leq x \leq 16$, so the domain is $[-16, 16]$. Also, $0 \leq \sqrt[4]{256 - x^2} \leq \sqrt[4]{256} = 4$, so the range is $[0, 4]$. Thus, we choose the viewing rectangle to be $[-20, 20]$ by $[-2, 6]$.

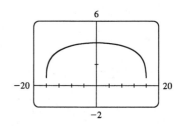

9. $f(x) = 0.01x^3 - x^2 + 5$. Graphing f in a standard viewing rectangle, $[-10, 10]$ by $[-10, 10]$, shows us what appears to be a parabola. But since this is a cubic polynomial, we know that a larger viewing rectangle will reveal a minimum point as well as the maximum point. After some trial and error, we choose the viewing rectangle $[-50, 150]$ by $[-2000, 2000]$.

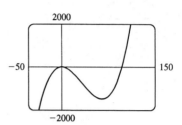

11. $y = \dfrac{1}{x^2 + 25}$

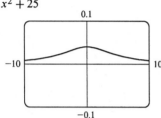

13. $y = x^4 - 4x^3$

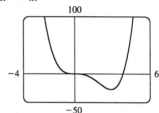

15. $y = \dfrac{2x - 1}{x + 3}$

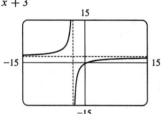

17. $f(x) = \cos(100x)$

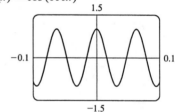

19. $f(x) = \sin(x/40)$

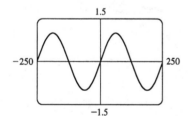

21. $y = 3^{\cos(x^2)}$

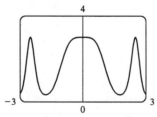

23. We must solve the given equation for y to obtain equations for the upper and lower halves of the ellipse. $4x^2 + 2y^2 = 1$

$$\Leftrightarrow \quad 2y^2 = 1 - 4x^2 \quad \Leftrightarrow \quad y^2 = \frac{1 - 4x^2}{2} \quad \Leftrightarrow$$

$$y = \pm\sqrt{\frac{1 - 4x^2}{2}}$$

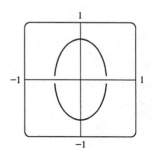

25. Graphing $f(x) = 3x^3 + x^2 + x - 2$ in a standard viewing rectangle, $[-10, 10]$ by $[-10, 10]$, reveals one real root between 0 and 1. The second figure shows a close-up of this region. By using a root finder or by zooming in, we find the value of the root to be approximately 0.67.

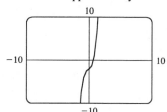   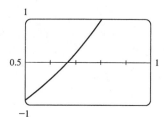

27. From the graph of $f(x) = 2 \sin x$ and $g(x) = x$, we see that there are three points of intersection. The intersection point $(0, 0)$ is obvious and due to the symmetry of the graphs (both functions are odd), we only need to find one of the other two points of intersection. Using an intersection finder or zooming in, we find the x-value of the intersection to be approximately 1.90. Hence, the solutions are $x = 0$ and $x \approx \pm 1.90$.

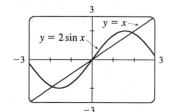   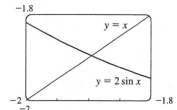

29. $g(x) = x^3/10$ is larger than $f(x) = 10x^2$ whenever $x > 100$.

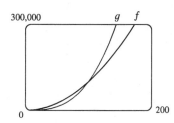

31. (a) (i) $[0, 5]$ by $[0, 20]$ **(ii)** $[0, 25]$ by $[0, 10^7]$ **(iii)** $[0, 50]$ by $[0, 10^8]$

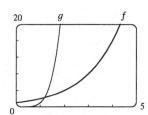   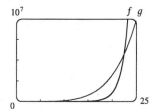   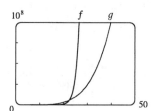

As x gets large, $f(x) = 2^x$ grows more rapidly than $g(x) = x^5$.

(b) From the graphs in part (a), it appears that the two solutions are $x \approx 1.2$ and 22.4.

33.

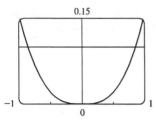

We see from the graphs of $y = |\sin x - x|$ and $y = 0.1$ that there are two solutions to the equation $|\sin x - x| = 0.1$: $x \approx -0.85$ and $x \approx 0.85$. The condition $|\sin x - x| < 0.1$ holds for any x lying between these two values.

35. (a) The root functions $y = \sqrt{x}$, $y = \sqrt[4]{x}$ and $y = \sqrt[6]{x}$

(b) The root functions $y = x$, $y = \sqrt[3]{x}$ and $y = \sqrt[5]{x}$

(c) The root functions $y = \sqrt{x}$, $y = \sqrt[3]{x}$, $y = \sqrt[4]{x}$ and $y = \sqrt[5]{x}$

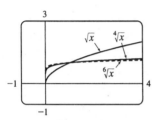

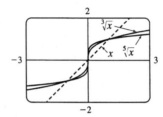

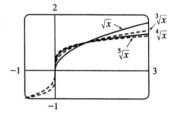

(d) • For any n, the nth root of 0 is 0 and the nth root of 1 is 1; that is, all nth root functions pass through the points $(0, 0)$ and $(1, 1)$.

• For odd n, the domain of the nth root function is $\mathbb{R}$, while for even n, it is $\{x \in \mathbb{R} \mid x \geq 0\}$.

• Graphs of even root functions look similar to that of $\sqrt{x}$, while those of odd root functions resemble that of $\sqrt[3]{x}$.

• As n increases, the graph of $\sqrt[n]{x}$ becomes steeper near 0 and flatter for $x > 1$.

37. $f(x) = x^4 + cx^2 + x$. If $c < 0$, there are three humps: two minimum points and a maximum point. These humps get flatter as c increases, until at $c = 0$ two of the humps disappear and there is only one minimum point. This single hump then moves to the right and approaches the origin as c increases.

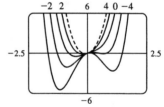

39. $y = x^n 2^{-x}$. As n increases, the maximum of the function moves further from the origin, and gets larger. Note, however, that regardless of n, the function approaches 0 as $x \to \infty$.

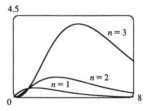

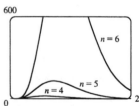

41. $y^2 = cx^3 + x^2$

If $c < 0$, the loop is to the right of the origin, and if c is positive, it is to the left. In both cases, the closer c is to 0, the larger the loop is. (In the limiting case, $c = 0$, the loop is "infinite", that is, it doesn't close.) Also, the larger $|c|$ is, the steeper the slope is on the loopless side of the origin.

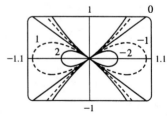

1.5 Exponential Functions

1. (a) $f(x) = a^x, a > 0$ (b) $\mathbb{R}$

   (c) $(0, \infty)$ (d) See Figures 4(c), 4(b), and 4(a), respectively.

3. All of these graphs approach 0 as $x \to -\infty$, all of them pass through the point $(0, 1)$, and all of them are increasing and approach ∞ as $x \to \infty$. The larger the base, the faster the function increases for $x > 0$, and the faster it approaches 0 as $x \to -\infty$.

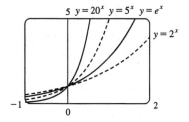

5. The functions with bases greater than 1 (3^x and 10^x) are increasing, while those with bases less than 1 $\left[\left(\frac{1}{3}\right)^x \text{ and } \left(\frac{1}{10}\right)^x\right]$ are decreasing. The graph of $\left(\frac{1}{3}\right)^x$ is the reflection of that of 3^x about the y-axis, and the graph of $\left(\frac{1}{10}\right)^x$ is the reflection of that of 10^x about the y-axis. The graph of 10^x increases more quickly than that of 3^x for $x > 0$, and approaches 0 faster as $x \to -\infty$.

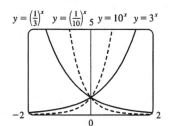

7. We start with the graph of $y = 2^x$ and shift 1 unit upward.

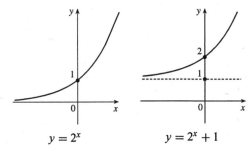

$$y = 2^x \qquad\qquad y = 2^x + 1$$

9. We start with the graph of $y = 3^x$ and reflect it about the y-axis.

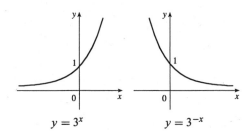

$$y = 3^x \qquad\qquad y = 3^{-x}$$

11. We start with the graph of $y = 3^{-x}$ (from Exercise 9) and reflect it about the x-axis.

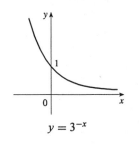

$$y = 3^{-x}$$

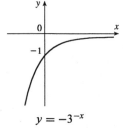

$$y = -3^{-x}$$

13. We start with the graph of $y = e^x$ (Figure 13), reflect it about the x-axis, and then shift 3 units upward. Note the horizontal asymptote at $y = 3$.

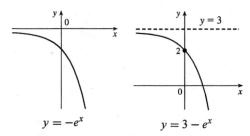

$$y = -e^x \qquad\qquad y = 3 - e^x$$

15. (a) To find the equation of the graph that results from shifting the graph of $y = e^x$ 2 units downward, we subtract 2 from the original function to get $y = e^x - 2$.

(b) To find the equation of the graph that results from shifting the graph of $y = e^x$ 2 units to the right, we replace x with $x - 2$ in the original function to get $y = e^{(x-2)}$.

(c) To find the equation of the graph that results from reflecting the graph of $y = e^x$ about the x-axis, we multiply the original function by -1 to get $y = -e^x$.

(d) To find the equation of the graph that results from reflecting the graph of $y = e^x$ about the y-axis, we replace x with $-x$ in the original function to get $y = e^{-x}$.

(e) To find the equation of the graph that results from reflecting the graph of $y = e^x$ about the x-axis and then about the y-axis, we first multiply the original function by -1 (to get $y = -e^x$) and then replace x with $-x$ in this equation to get $y = -e^{-x}$.

17. Use $y = Ca^x$ with the points $(1, 6)$ and $(3, 24)$. $6 = Ca^1$ and $24 = Ca^3$ $\Rightarrow$ $24 = \left(\dfrac{6}{a}\right)a^3$ $\Rightarrow$ $4 = a^2$ $\Rightarrow$ $a = 2$ (since $a > 0$) and $C = 3$. The function is $f(x) = 3 \cdot 2^x$.

19. 2 ft $= 24$ in, $f(24) = 24^2$ in $= 576$ in $= 48$ ft. $g(24) = 2^{24}$ in $= 2^{24}/(12 \cdot 5280)$ mi ≈ 265 mi

21. The graph of g finally surpasses that of f at $x \approx 35.8$.

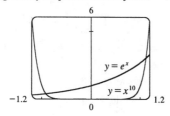

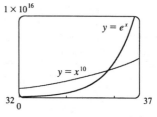

23. (a) Fifteen hours represents 5 doubling periods.

$100 \cdot 2^5 = 3200$

(b) In t hours, there will be $t/3$ doubling periods. The initial population is 100, so the population y at time t is $y = 100 \cdot 2^{t/3}$.

(c) $t = 20$ $\Rightarrow$ $y = 100 \cdot 2^{20/3} \approx 10{,}159$

(d) We graph $y_1 = 100 \cdot 2^{x/3}$ and $y_2 = 50{,}000$. The two curves intersect at $x \approx 26.9$.

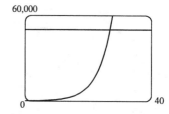

25. An exponential model is $y(x) = a \cdot b^x$, where

$a = 8.5688194337975 \times 10^{-13}$ and
$b = 1.0184365505945$. This model gives
$y(1993) \approx 5563$ million and
$y(2010) \approx 7590$ million.

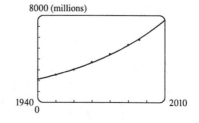

1.6 Inverse Functions and Logarithms

1. (a) See Definition 1.

(b) It must pass the Horizontal Line Test.

3. f is not one-to-one because $2 \neq 6$, but $f(2) = f(6)$.

5. The diagram shows that there is a horizontal line which intersects the graph more than once, so the function is not one-to-one.

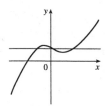

7. The function is one-to-one because no horizontal line intersects the graph more than once.

9. $x_1 \neq x_2 \ \Rightarrow \ 7x_1 \neq 7x_2 \ \Rightarrow \ 7x_1 - 3 \neq 7x_2 - 3 \ \Rightarrow \ f(x_1) \neq f(x_2)$, so f is 1-1.

11. $g(x) = |x| \ \Rightarrow \ g(-1) = 1 = g(1)$, so g is not one-to-one.

13. A football will attain every height h up to its maximum height twice: once on the way up, and again on the way down. Thus, even if t_1 does not equal t_2, $f(t_1)$ may equal $f(t_2)$, so f is not 1-1.

15. f does not pass the Horizontal Line Test, so f is not 1-1.

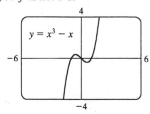

17. Since $f(2) = 9$ and f is 1-1, we know that $f^{-1}(9) = 2$. Remember, if the point $(2, 9)$ is on the graph of f, then the point $(9, 2)$ is on the graph of f^{-1}.

19. First, we must determine x such that $g(x) = 4$. By inspection, we see that if $x = 0$, then $g(x) = 4$. Since g is 1-1 (g is an increasing function), it has an inverse, and $g^{-1}(4) = 0$.

21. We solve $C = \frac{5}{9}(F - 32)$ for F: $\frac{9}{5}C = F - 32 \ \Rightarrow \ F = \frac{9}{5}C + 32$. This gives us the Fahrenheit temperature F as a function of the Celsius temperature C. $F \geq -459.67 \ \Rightarrow \ \frac{9}{5}C + 32 \geq -459.67 \ \Rightarrow \ \frac{9}{5}C \geq -491.67 \ \Rightarrow \ C \geq -273.15$, the domain of the inverse function.

23. $y = f(x) = \dfrac{1 + 3x}{5 - 2x} \ \Rightarrow \ 5y - 2xy = 1 + 3x \ \Rightarrow \ 5y - 1 = 3x + 2xy \ \Rightarrow \ x(3 + 2y) = 5y - 1 \ \Rightarrow$ $x = \dfrac{5y - 1}{2y + 3}$. Interchange x and y: $y = \dfrac{5x - 1}{2x + 3}$. So $f^{-1}(x) = \dfrac{5x - 1}{2x + 3}$.

25. $y = f(x) = \sqrt{2 + 5x} \ \Rightarrow \ y^2 = 2 + 5x$ and $y \geq 0 \ \Rightarrow \ 5x = y^2 - 2 \ \Rightarrow \ x = \dfrac{y^2 - 2}{5}, y \geq 0$. Interchange x and y: $y = \dfrac{x^2 - 2}{5}, x \geq 0$. So $f^{-1}(x) = \dfrac{x^2 - 2}{5}, x \geq 0$.

27. $y = \ln(x + 3) \ \Rightarrow \ x + 3 = e^y \ \Rightarrow \ x = e^y - 3$. Interchange x and y: $y = e^x - 3$. So $f^{-1}(x) = e^x - 3$.

29. $y = f(x) = 1 - \dfrac{2}{x^2} \Rightarrow 1 - y = \dfrac{2}{x^2} \Rightarrow x^2 = \dfrac{2}{1-y} \Rightarrow$

$x = \sqrt{\dfrac{2}{1-y}}$, since $x > 0$. Interchange x and y: $y = \sqrt{\dfrac{2}{1-x}}$. So

$f^{-1}(x) = \sqrt{\dfrac{2}{1-x}}$.

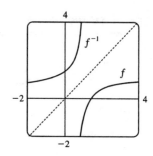

31. The function f is one-to-one, so its inverse exists and the graph of its inverse can be obtained by reflecting the graph of f through the line $y = x$.

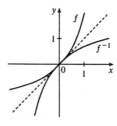

33. (a) It is defined as the inverse of the exponential function with base a, that is, $\log_a x = y \iff a^y = x$.

(b) $(0, \infty)$ (c) $\mathbb{R}$ (d) See Figure 11.

35. (a) $\log_2 64 = 6$ since $2^6 = 64$.   (b) $\log_6 \frac{1}{36} = -2$ since $6^{-2} = \frac{1}{36}$.

37. (a) $\log_{10} 1.25 + \log_{10} 80 = \log_{10}(1.25 \cdot 80) = \log_{10} 100 = \log_{10} 10^2 = 2$

(b) $\log_5 10 + \log_5 20 - 3\log_5 2 = \log_5(10 \cdot 20) - \log_5 2^3 = \log_5 \frac{200}{8} = \log_5 25 = \log_5 5^2 = 2$

39. $2\ln 4 - \ln 2 = \ln 4^2 - \ln 2 = \ln 16 - \ln 2 = \ln \frac{16}{2} = \ln 8$

41. (a) $\log_2 5 = \dfrac{\ln 5}{\ln 2} \approx 2.321928$ (b) $\log_5 26.05 = \dfrac{\ln 26.05}{\ln 5} \approx 2.025563$

43. To graph these functions, we use $\log_{1.5} x = \dfrac{\ln x}{\ln 1.5}$ and

$\log_{50} x = \dfrac{\ln x}{\ln 50}$. These graphs all approach $-\infty$ as $x \to 0^+$, and

they all pass through the point $(1, 0)$. Also, they are all increasing,

and all approach ∞ as $x \to \infty$. The functions with larger bases

increase extremely slowly, and the ones with smaller bases do so

somewhat more quickly. The functions with large bases approach the

y-axis more closely as $x \to 0^+$.

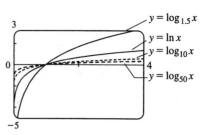

45. $3 \text{ ft} = 36 \text{ in}$, so we need x such that $\log_2 x = 36 \iff x = 2^{36} = 68{,}719{,}476{,}736$. In miles, this is

$68{,}719{,}476{,}736 \text{ in} \cdot \dfrac{1 \text{ ft}}{12 \text{ in}} \cdot \dfrac{1 \text{ mi}}{5280 \text{ ft}} \approx 1{,}084{,}587.7 \text{ mi}$.

47. (a) $y = \log_{10} x$ $\qquad$ $y = \log_{10}(x+5)$ $\qquad$ (b) $\qquad$ $y = \ln x$ $\qquad$ $y = -\ln x$

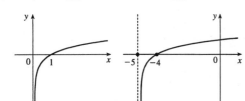
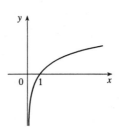
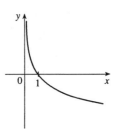

49. (a) $e^x = 16 \iff \ln e^x = \ln 16 \iff x = \ln 16 = \ln 2^4 = 4\ln 2$

$\quad$ (b) $\ln x = -1 \iff e^{\ln x} = e^{-1} \iff x = 1/e$

51. (a) $2^{x-5} = 3 \iff \log_2 3 = x - 5 \iff x = 5 + \log_2 3$.

$\quad$ *Or:* $2^{x-5} = 3 \iff \ln(2^{x-5}) = \ln 3 \iff (x-5)\ln 2 = \ln 3 \iff x - 5 = \dfrac{\ln 3}{\ln 2} \iff x = 5 + \dfrac{\ln 3}{\ln 2}$

$\quad$ (b) $\ln x + \ln(x-1) = \ln(x(x-1)) = 1 \iff x(x-1) = e^1 \iff x^2 - x - e = 0$. The quadratic formula

$\quad$ gives $x = \frac{1}{2}\left(1 \pm \sqrt{1+4e}\right)$, but we reject the negative root since the natural logarithm is not defined for $x < 0$.

$\quad$ So $x = \frac{1}{2}\left(1 + \sqrt{1+4e}\right)$.

53. We see that the graph of $y = f(x) = \sqrt{x^3 + x^2 + x + 1}$ is increasing, so

$\quad$ f is 1-1. Enter $x = \sqrt{y^3 + y^2 + y + 1}$ and use your CAS to solve the

$\quad$ equation for y. Using Derive, we get two (irrelevant) solutions involving

$\quad$ imaginary expressions, as well as one which can be simplified to the

$\quad$ following:

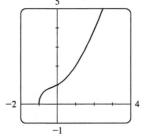

$$y = f^{-1}(x) = -\tfrac{\sqrt[3]{4}}{6}\left(\sqrt[3]{D - 27x^2 + 20} - \sqrt[3]{D + 27x^2 - 20} + \sqrt[3]{2}\right)$$

where $D = 3\sqrt{3}\sqrt{27x^4 - 40x^2 + 16}$. Maple and Mathematica each give two complex expressions and one real

expression, and the real expression is equivalent to that given by Derive. For example, Maple's expression

simplifies to $\dfrac{1}{6}\dfrac{M^{2/3} - 8 - 2M^{1/3}}{2M^{1/3}}$, where $M = 108x^2 + 12\sqrt{48 - 120x^2 + 81x^4} - 80$.

55. (a) $n = 100 \cdot 2^{t/3} \implies \dfrac{n}{100} = 2^{t/3} \implies \log_2\left(\dfrac{n}{100}\right) = \dfrac{t}{3} \implies t = 3\log_2\left(\dfrac{n}{100}\right)$. This function tells us how

$\quad$ long it will take to obtain n bacteria (given the number n).

$\quad$ (b) $n = 50{,}000 \implies t = 3\log_2 \dfrac{50{,}000}{100} = 3\log_2 500 = 3\left(\dfrac{\ln 500}{\ln 2}\right) \approx 26.9$ hours

57. (a) To find the equation of the graph that results from shifting the graph of $y = \ln x$ 3 units upward, we add 3 to the

$\quad$ original function to get $y = \ln x + 3$.

$\quad$ (b) To find the equation of the graph that results from shifting the graph of $y = \ln x$ 3 units to the left, we replace x

$\quad$ with $x + 3$ in the original function to get $y = \ln(x+3)$.

$\quad$ (c) To find the equation of the graph that results from reflecting the graph of $y = \ln x$ about the x-axis, we multiply

$\quad$ the original equation by -1 to get $y = -\ln x$.

$\quad$ (d) To find the equation of the graph that results from reflecting the graph of $y = \ln x$ about the y-axis, we replace

$\quad$ x with $-x$ in the original equation to get $y = \ln(-x)$.

(e) To find the equation of the graph that results from reflecting the graph of $y = \ln x$ about the line $y = x$, we interchange x and y in the original equation to get $x = \ln y$ $\Leftrightarrow$ $y = e^x$.

(f) To find the equation of the graph that results from reflecting the graph of $y = \ln x$ about the x-axis and then about the line $y = x$, we first multiply the original equation by -1 [to get $y = -\ln x$] and then interchange x and y in this equation to get $x = -\ln y$ $\Leftrightarrow$ $\ln y = -x$ $\Leftrightarrow$ $y = e^{-x}$.

(g) To find the equation of the graph that results from reflecting the graph of $y = \ln x$ about the y-axis and then about the line $y = x$, we first replace x with $-x$ in the original equation [to get $y = \ln(-x)$] and then interchange x and y to get $x = \ln(-y)$ $\Leftrightarrow$ $-y = e^x$ $\Leftrightarrow$ $y = -e^x$.

(h) To find the equation of the graph that results from shifting the graph of $y = \ln x$ 3 units to the left and then reflecting it about the line $y = x$, we first replace x with $x + 3$ in the original equation [to get $y = \ln(x + 3)$] and then interchange x and y in this equation to get $x = \ln(y + 3)$ $\Leftrightarrow$ $y + 3 = e^x$ $\Leftrightarrow$ $y = e^x - 3$.

Review

CONCEPT CHECK

1. (a) A **function** f is a rule that assigns to each element x in a set A exactly one element, called $f(x)$, in a set B. The set A is called the **domain** of the function. The **range** of f is the set of all possible values of $f(x)$ as x varies throughout the domain.

(b) If f is a function with domain A, then its **graph** is the set of ordered pairs $\{(x, f(x)) \mid x \in A\}$.

(c) Use the Vertical Line Test on page 17.

2. The four ways to represent a function are: verbally, numerically, visually, and algebraically. An example of each is given below.
Verbally: An assignment of students to chairs in a classroom (a description in words)
Numerically: A tax table that assigns an amount of tax to an income (a table of values)
Visually: A graphical history of the Dow Jones average (a graph)
Algebraically: A relationship between distance, rate, and time: $d = rt$ (an explicit formula)

3. (a) An **even function** f satisfies $f(-x) = f(x)$ for every number x in its domain. It is symmetric with respect to the y-axis.

(b) An **odd function** g satisfies $g(-x) = -g(x)$ for every number x in its domain. It is symmetric with respect to the origin.

4. A function f is called **increasing** on an interval I if $f(x_1) < f(x_2)$ whenever $x_1 < x_2$ in I.

5. A **mathematical model** is a mathematical description (often by means of a function or an equation) of a real-world phenomenon.

6. (a) Linear function: $f(x) = 2x + 1$, $f(x) = ax + b$

(b) Power function: $f(x) = x^2$, $f(x) = x^a$

(c) Exponential function: $f(x) = 2^x$, $f(x) = a^x$

(d) Quadratic function: $f(x) = x^2 + x + 1$,

$f(x) = ax^2 + bx + c$

(e) Polynomial of degree 5: $f(x) = x^5 + 2$

(f) Rational function: $f(x) = \dfrac{x}{x+2}$, $f(x) = \dfrac{P(x)}{Q(x)}$ where

$P(x)$ and $Q(x)$ are polynomials

7.

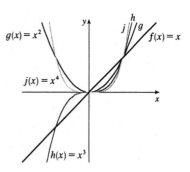

8. (a)

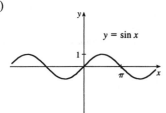

(b)

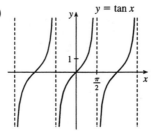

(c)

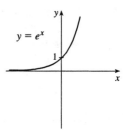

(d)

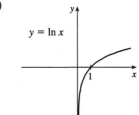

(e)

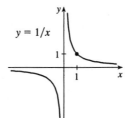

(f)

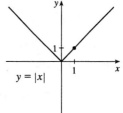

(g)

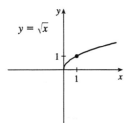

9. (a) The domain of $f + g$ is the intersection of the domain of f and the domain of g; that is, $A \cap B$.

(b) The domain of fg is also $A \cap B$.

(c) The domain of f/g must exclude values of x that make g equal to 0; that is, $\{x \in A \cap B \mid g(x) \neq 0\}$.

10. Given two functions f and g, the **composite** function $f \circ g$ is defined by $(f \circ g)(x) = f(g(x))$. The domain of $f \circ g$ is the set of all x in the domain of g such that $g(x)$ is in the domain of f.

11. (a) If the graph of f is shifted 2 units upward, its equation becomes $y = f(x) + 2$.

(b) If the graph of f is shifted 2 units downward, its equation becomes $y = f(x) - 2$.

(c) If the graph of f is shifted 2 units to the right, its equation becomes $y = f(x - 2)$.

(d) If the graph of f is shifted 2 units to the left, its equation becomes $y = f(x + 2)$.

(e) If the graph of f is reflected about the x-axis, its equation becomes $y = -f(x)$.

(f) If the graph of f is reflected about the y-axis, its equation becomes $y = f(-x)$.

(g) If the graph of f is stretched vertically by a factor of 2, its equation becomes $y = 2f(x)$.

(h) If the graph of f is shrunk vertically by a factor of 2, its equation becomes $y = \frac{1}{2}f(x)$.

(i) If the graph of f is stretched horizontally by a factor of 2, its equation becomes $y = f\left(\frac{1}{2}x\right)$.

(j) If the graph of f is shrunk horizontally by a factor of 2, its equation becomes $y = f(2x)$.

TRUE-FALSE QUIZ

1. False. Let $f(x) = x^2$, $s = -1$, and $t = 1$. Then $f(s + t) = (-1 + 1)^2 = 0^2 = 0$, but
$$f(s) + f(t) = (-1)^2 + 1^2 = 2 \neq 0 = f(s + t).$$

3. False. Let $f(x) = x^2$. Then $f(3x) = (3x)^2 = 9x^2$ and $3f(x) = 3x^2$. So $f(3x) \neq 3f(x)$.

5. True. See the Vertical Line Test.

7. False. Let $f(x) = x^3$. Then f is one-to-one and $f^{-1}(x) = \sqrt[3]{x}$. But $1/f(x) = 1/x^3$, which is not equal to $f^{-1}(x)$.

9. True. The function $\ln x$ is an increasing function on $(0, \infty)$.

EXERCISES

1. (a) When $x = 2$, $y \approx 2.7$. Thus, $f(2) \approx 2.7$.

(b) $f(x) = 3 \Rightarrow x \approx 2.3, 5.6$

(c) The domain of f is $-6 \leq x \leq 6$, or $[-6, 6]$.

(d) The range of f is $-4 \leq y \leq 4$, or $[-4, 4]$.

(e) f is increasing on $(-4, 4)$.

(f) f is not one-to-one since it fails the Horizontal Line Test.

(g) f is odd since its graph is symmetric with respect to the origin.

3. (a)

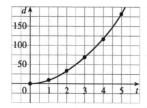

(b) From the graph, we see that the distance is slightly less than 150 feet.

5. $f(x) = \sqrt{4 - 3x^2}$. Domain: $4 - 3x^2 \geq 0 \implies 3x^2 \leq 4 \implies x^2 \leq \frac{4}{3} \implies |x| \leq \frac{2}{\sqrt{3}}$. Range: $y \geq 0$ and $y \leq \sqrt{4} \implies 0 \leq y \leq 2$.

7. $y = 1 + \sin x$. Domain: $\mathbb{R}$. Range: $-1 \leq \sin x \leq 1 \implies 0 \leq 1 + \sin x \leq 2 \implies 0 \leq y \leq 2$.

9. (a) To obtain the graph of $y = f(x) + 8$, we shift the graph of $y = f(x)$ up 8 units.

(b) To obtain the graph of $y = f(x + 8)$, we shift the graph of $y = f(x)$ left 8 units.

(c) To obtain the graph of $y = 1 + 2f(x)$, we stretch the graph of $y = f(x)$ vertically by a factor of 2, and then shift the resulting graph 1 unit upward.

(d) To obtain the graph of $y = f(x - 2) - 2$, we shift the graph of $y = f(x)$ right 2 units, and then shift the resulting graph 2 units downward.

(e) To obtain the graph of $y = -f(x)$, we reflect the graph of $y = f(x)$ about the x-axis.

(f) To obtain the graph of $y = f^{-1}(x)$, we reflect the graph of $y = f(x)$ about the line $y = x$ (assuming f is one-to-one).

11. To sketch the graph of $y = 1 + \sqrt{x + 2}$, we shift the graph of $y = \sqrt{x}$ left 2 units and up 1 unit.

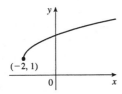

$(-2, 1)$

13. To sketch the graph of $y = \cos 3x$, we compress the graph of $y = \cos x$ horizontally by a factor of 3.

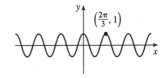

$\left(\frac{2\pi}{3}, 1\right)$

15. To sketch the graph of $y = -e^x$, we reflect the graph of $y = e^x$ about the x-axis.

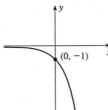

$(0, -1)$

17. (a) The terms of f are a mixture of odd and even powers of x, so f is neither even nor odd.

(b) The terms of f are all odd powers of x, so f is odd.

(c) $f(-x) = e^{-(-x)^2} = e^{-x^2} = f(x)$, so f is even.

(d) $f(-x) = 1 + \sin(-x) = 1 - \sin x$. Now $f(-x) \neq f(x)$ and $f(-x) \neq -f(x)$, so f is neither even nor odd.

19. $(f \circ g)(x) = f(g(x)) = f(x^2 - 9) = \ln(x^2 - 9)$.

Domain: $x^2 - 9 > 0 \implies x^2 > 9 \implies |x| > 3 \implies x \in (-\infty, -3) \cup (3, \infty)$

$(g \circ f)(x) = g(f(x)) = g(\ln x) = (\ln x)^2 - 9$. Domain: $x > 0$, or $(0, \infty)$

$(f \circ f)(x) = f(f(x)) = f(\ln x) = \ln(\ln x)$. Domain: $\ln x > 0 \implies x > e^0 = 1$, or $(1, \infty)$

$(g \circ g)(x) = g(g(x)) = g(x^2 - 9) = (x^2 - 9)^2 - 9$. Domain: $x \in \mathbb{R}$, or $(-\infty, \infty)$

21.

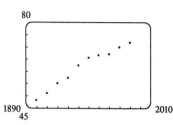

Many models appear to be plausible. Your choice depends on whether you think medical advances will keep increasing life expectancy, or if there is bound to be a natural leveling-off of life expectancy. A linear model, $y = 0.263x - 450.034$, gives us an estimate of 76.0 years for the year 2000.

23. We need to know the value of x such that $f(x) = 2$. Since $x = 1$ gives us $y = 2$, $f^{-1}(2) = 1$.

25. (a) $e^{2\ln 3} = \left(e^{\ln 3}\right)^2 = 3^2 = 9$

(b) $\log_{10} 25 + \log_{10} 4 = \log_{10}(25 \cdot 4) = \log_{10} 100 = \log_{10} 10^2 = 2$

27. (a) After 4 days, $\frac{1}{2}$ gram remains; after 8 days, $\frac{1}{4}$ g; after 12 days, $\frac{1}{8}$ g; after 16 days, $\frac{1}{16}$ g.

(b) $m(4) = \dfrac{1}{2}$, $m(8) = \dfrac{1}{2^2}$, $m(12) = \dfrac{1}{2^3}$, $m(16) = \dfrac{1}{2^4}$. From the pattern, we see that $m(t) = \dfrac{1}{2^{t/4}}$, or $2^{-t/4}$.

(c) $m = 2^{-t/4} \implies \log_2 m = -t/4 \implies t = -4\log_2 m$; this is the time elapsed when there are m grams of ^{100}Pd.

(d) $m = 0.01 \implies t = -4\log_2 0.01 = -4\left(\dfrac{\ln 0.01}{\ln 2}\right) \approx 26.6$ days

29. $f(x) = \ln(x^2 - c)$. If $c < 0$, the domain of f is $\mathbb{R}$. If $c = 0$, the domain of f is $(-\infty, 0) \cup (0, \infty)$. If $c > 0$, the domain of f is $\left(-\infty, -\sqrt{c}\right) \cup \left(\sqrt{c}, \infty\right)$. As c increases, the dip at $x = 0$ becomes deeper. For $c \geq 0$, the graph has asymptotes at $x = \pm\sqrt{c}$.

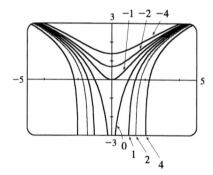

Principles of Problem Solving

1.

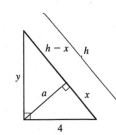

By using the area formula for a triangle, $\frac{1}{2}$ (base) (height), in two ways, we see that $\frac{1}{2}\,(4)\,(y) = \frac{1}{2}\,(h)\,(a)$, so $a = \dfrac{4y}{h}$. Since $4^2 + y^2 = h^2$,

$$y = \sqrt{h^2 - 16}, \text{ and } a = \frac{4\sqrt{h^2 - 16}}{h}.$$

3. $|2x - 1| = \begin{cases} 1 - 2x & \text{if } x < \frac{1}{2} \\ 2x - 1 & \text{if } x \geq \frac{1}{2} \end{cases}$ and $|x + 5| = \begin{cases} -x - 5 & \text{if } x < -5 \\ x + 5 & \text{if } x \geq -5 \end{cases}$

Therefore, we consider the three cases $x < -5$, $-5 \leq x < \frac{1}{2}$, and $x \geq \frac{1}{2}$.

If $x < -5$, we must have $1 - 2x - (-x - 5) = 3 \iff x = 3$, which is false, since we are considering $x < -5$.

If $-5 \leq x < \frac{1}{2}$, we must have $1 - 2x - (x + 5) = 3 \iff x = -\frac{7}{3}$.

If $x \geq \frac{1}{2}$, we must have $2x - 1 - (x + 5) = 3 \iff x = 9$.

So the two solutions of the equation are $x = -\frac{7}{3}$ and $x = 9$.

5. $f(x) = |x^2 - 4|x| + 3|$. If $x \geq 0$, then $f(x) = |x^2 - 4x + 3| = |(x - 1)(x - 3)|$.

              Case (i):   If $0 < x \leq 1$, then $f(x) = x^2 - 4x + 3$.

              Case (ii):   If $1 < x \leq 3$, then $f(x) = -(x^2 - 4x + 3) = -x^2 + 4x - 3$.

              Case (iii):   If $x > 3$, then $f(x) = x^2 - 4x + 3$.

This enables us to sketch the graph for $x \geq 0$. Then we use the fact that f is an even function to reflect this part of the graph about the y-axis to obtain the entire graph. Or, we could consider also the cases $x < -3$, $-3 \leq x < -1$, and $-1 \leq x < 0$.

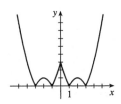

7. $|x| + |y| = 1 + |xy| \iff |xy| - |x| - |y| + 1 = 0 \iff$

$|x|\,|y| - |x| - |y| + 1 = 0 \iff (|x| - 1)(|y| - 1) = 0 \iff x = \pm 1$ or $y = \pm 1$.

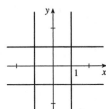

9. $|x| + |y| \leq 1$. The boundary of the region has equation $|x| + |y| = 1$. In quadrants I, II, III, and IV, this becomes the lines $x + y = 1$, $-x + y = 1$, $-x - y = 1$, and $x - y = 1$ respectively.

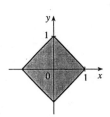

11. $(\log_2 3)(\log_3 4)(\log_4 5)\cdots(\log_{31} 32) = \left(\dfrac{\ln 3}{\ln 2}\right)\left(\dfrac{\ln 4}{\ln 3}\right)\left(\dfrac{\ln 5}{\ln 4}\right)\cdots\left(\dfrac{\ln 32}{\ln 31}\right) = \dfrac{\ln 32}{\ln 2} = \dfrac{\ln 2^5}{\ln 2} = \dfrac{5\ln 2}{\ln 2} = 5$

13. $\ln\left(x^2 - 2x - 2\right) \le 0 \;\Rightarrow\; x^2 - 2x - 2 \le e^0 = 1 \;\Rightarrow\; x^2 - 2x - 3 \le 0 \;\Rightarrow\; (x-3)(x+1) \le 0 \;\Rightarrow$

$x \in [-1, 3]$. Since the argument must be positive, $x^2 - 2x - 2 > 0 \;\Rightarrow\; \left[x - \left(1 - \sqrt{3}\right)\right]\left[x - \left(1 + \sqrt{3}\right)\right] > 0$

$\Rightarrow\; x \in \left(-\infty, 1 - \sqrt{3}\right) \cup \left(1 + \sqrt{3}, \infty\right)$. The intersection of these intervals is $\left[-1, 1 - \sqrt{3}\right) \cup \left(1 + \sqrt{3}, 3\right]$.

15. Let d be the distance traveled on each half of the trip. Let t_1 and t_2 be the times taken for the first and second halves of the trip.

For the first half of the trip we have $t_1 = d/30$ and for the second half we have $t_2 = d/60$. Thus, the average speed

for the entire trip is $\dfrac{\text{total distance}}{\text{total time}} = \dfrac{2d}{t_1 + t_2} = \dfrac{2d}{\dfrac{d}{30} + \dfrac{d}{60}} \cdot \dfrac{60}{60} = \dfrac{120d}{2d + d} = \dfrac{120d}{3d} = 40$. The average speed for

the entire trip is 40 mi/h.

17. Let S_n be the statement that $7^n - 1$ is divisible by 6.

- S_1 is true because $7^1 - 1 = 6$ is divisible by 6.
- Assume S_k is true, that is, $7^k - 1$ is divisible by 6. In other words, $7^k - 1 = 6m$ for some positive integer m. Then $7^{k+1} - 1 = 7^k \cdot 7 - 1 = (6m + 1) \cdot 7 - 1 = 42m + 6 = 6(7m + 1)$, which is divisible by 6, so S_{k+1} is true.
- Therefore, by mathematical induction, $7^n - 1$ is divisible by 6 for every positive integer n.

19. $f_0(x) = x^2$ and $f_{n+1}(x) = f_0(f_n(x))$ for $n = 0, 1, 2, \ldots$.

$f_1(x) = f_0(f_0(x)) = f_0\left(x^2\right) = \left(x^2\right)^2 = x^4$, $f_2(x) = f_0(f_1(x)) = f_0\left(x^4\right) = \left(x^4\right)^2 = x^8$,

$f_3(x) = f_0(f_2(x)) = f_0\left(x^8\right) = \left(x^8\right)^2 = x^{16}, \ldots$. Thus, a general formula is $f_n(x) = x^{2^{n+1}}$.

Limits and Derivatives

2.1 The Tangent and Velocity Problems

1. (a) Using $P(15, 250)$, we construct the following table:

t	Q	slope $= m_{PQ}$
5	$(5, 694)$	$\frac{694-250}{5-15} = -\frac{444}{10} = -44.4$
10	$(10, 444)$	$\frac{444-250}{10-15} = -\frac{194}{5} = -38.8$
20	$(20, 111)$	$\frac{111-250}{20-15} = -\frac{139}{5} = -27.8$
25	$(25, 28)$	$\frac{28-250}{25-15} = -\frac{222}{10} = -22.2$
30	$(30, 0)$	$\frac{0-250}{30-15} = -\frac{250}{15} = -16.\overline{6}$

(b) Using the values of t that correspond to the points closest to P ($t = 10$ and $t = 20$), we have

$$\frac{-38.8 + (-27.8)}{2} = -33.3.$$

(c) From the graph, we can estimate the slope of the tangent line at P to be $\frac{-300}{9} = -33.\overline{3}$.

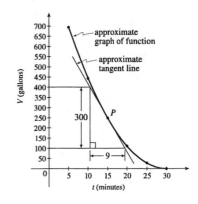

3. For the curve $y = \sqrt{x}$ and the point $P(4, 2)$:

(a)

	x	Q	m_{PQ}
(i)	5	$(5, 2.236068)$	0.236068
(ii)	4.5	$(4.5, 2.121320)$	0.242641
(iii)	4.1	$(4.1, 2.024846)$	0.248457
(iv)	4.01	$(4.01, 2.002498)$	0.249844
(v)	4.001	$(4.001, 2.000250)$	0.249984
(vi)	3	$(3, 1.732051)$	0.267949
(vii)	3.5	$(3.5, 1.870829)$	0.258343
(viii)	3.9	$(3.9, 1.974842)$	0.251582
(ix)	3.99	$(3.99, 1.997498)$	0.250156
(x)	3.999	$(3.999, 1.999750)$	0.250016

(b) The slope appears to be $\frac{1}{4}$.

(c) $y - 2 = \frac{1}{4}(x - 4)$ or

$y = \frac{1}{4}x + 1.$

5. (a) At $t = 2$, $y = 40\,(2) - 16\,(2)^2 = 16$. The average velocity between times 2 and $2 + h$ is

$$\frac{40\,(2+h) - 16\,(2+h)^2 - 16}{h} = \frac{-24h - 16h^2}{h} = -24 - 16h, \text{ if } h \neq 0.$$

(i) $h = 0.5$, -32 ft/s   (ii) $h = 0.1$, -25.6 ft/s

(iii) $h = 0.05$, -24.8 ft/s   (iv) $h = 0.01$, -24.16 ft/s

(b) The instantaneous velocity when $t = 2$ is -24 ft/s.

7. Average velocity between times 1 and $1 + h$ is

$$\frac{s\,(1+h) - s\,(1)}{h} = \frac{(1+h)^3/6 - 1/6}{h} = \frac{h^3 + 3h^2 + 3h}{6h} = \frac{h^2 + 3h + 3}{6} \text{ if } h \neq 0$$

(a) (i) [1, 3]: $h = 2$, $\frac{13}{6}$ ft/s   (ii) [1, 2]: $h = 1$, $\frac{7}{6}$ ft/s

(iii) [1, 1.5]: $h = 0.5$, $\frac{19}{24}$ ft/s   (iv) [1, 1.1]: $h = 0.1$, $\frac{331}{600}$ ft/s

(b) As h approaches 0, the velocity approaches $\frac{1}{2}$ ft/s.

(c)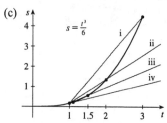

(d)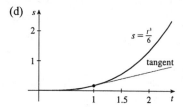

9. For the curve $y = \sin(10\pi/x)$ and the point $P\,(1, 0)$:

(a)

x	Q	m_{PQ}
2	$(2, 0)$	0
1.5	$(1.5, 0.8660)$	1.7321
1.4	$(1.4, -0.4339)$	-1.0847
1.3	$(1.3, -0.8230)$	-2.7433
1.2	$(1.2, 0.8660)$	4.3301
1.1	$(1.1, -0.2817)$	-2.8173

x	Q	m_{PQ}
0.5	$(0.5, 0)$	0
0.6	$(0.6, 0.8660)$	-2.1651
0.7	$(0.7, 0.7818)$	-2.6061
0.8	$(0.8, 1)$	-5
0.9	$(0.9, -0.3420)$	3.4202
0.99	$(0.99, 0.3120)$	-31.2033

As x approaches 1, the slopes do not appear to be approaching any particular value.

(b)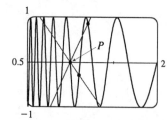

We see that problems with estimation are caused by the frequent oscillations of the graph. The tangent is so steep at P that we need to take x-values much closer to 1 in order to get accurate estimates of its slope.

(c) If we choose $x = 1.001$, then the point Q is $(1.001, -0.0314)$ and $m_{PQ} \approx -31.3794$. If $x = 0.999$, then Q is $(0.999, 0.0314)$ and $m_{PQ} = -31.4422$. Averaging these two slopes gives us the estimate -31.4108.

2.2 The Limit of a Function

1. As x approaches 2, $f(x)$ approaches 5. [Or, the values of $f(x)$ can be made as close to 5 as we like by taking x sufficiently close to 2 (but $x \neq 2$).] Yes, the graph could have a hole at $(2, 5)$ and be defined such that $f(2) = 3$.

3. (a) $\lim\limits_{x \to -3} f(x) = \infty$ means that the values of $f(x)$ can be made arbitrarily large (as large as we please) by taking x sufficiently close to -3 (but not equal to -3).

(b) $\lim\limits_{x \to 4^+} f(x) = -\infty$ means that the values of $f(x)$ can be made arbitrarily large negative by taking x sufficiently close to 4 through values larger than 4.

5. (a) $\lim\limits_{x \to 1} f(x) = 3$ **(b)** $\lim\limits_{x \to 3^-} f(x) = 2$ **(c)** $\lim\limits_{x \to 3^+} f(x) = -2$

(d) $\lim\limits_{x \to 3} f(x)$ doesn't exist because the limits in part (b) and part (c) are not equal.

(e) $f(3) = 1$ **(f)** $\lim\limits_{x \to -2^-} f(x) = -1$ **(g)** $\lim\limits_{x \to -2^+} f(x) = -1$

(h) $\lim\limits_{x \to -2} f(x) = -1$ **(i)** $f(-2) = -3$

7. (a) $\lim\limits_{x \to 3} f(x) = 2$ **(b)** $\lim\limits_{x \to 1} f(x) = -1$ **(c)** $\lim\limits_{x \to -3} f(x) = 1$

(d) $\lim\limits_{x \to 2^-} f(x) = 1$ **(e)** $\lim\limits_{x \to 2^+} f(x) = 2$

(f) $\lim\limits_{x \to 2} f(x)$ doesn't exist because the limits in part (d) and part (e) are not equal.

9. (a) $\lim\limits_{x \to 3} f(x) = \infty$ **(b)** $\lim\limits_{x \to 7} f(x) = -\infty$ **(c)** $\lim\limits_{x \to -4} f(x) = -\infty$

(d) $\lim\limits_{x \to -9^-} f(x) = \infty$ **(e)** $\lim\limits_{x \to -9^+} f(x) = -\infty$

(f) The equations of the vertical asymptotes: $x = -9, x = -4, x = 3, x = 7$

11.

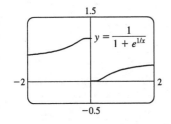

(a) $\lim\limits_{x \to 0^-} f(x) = 1$

(b) $\lim\limits_{x \to 0^+} f(x) = 0$

(c) $\lim\limits_{x \to 0} f(x) = 0$ does not exist because the limits in part (a) and part (b) are not equal.

13.

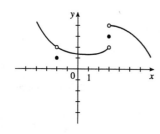

15. For $g(x) = \dfrac{x-1}{x^3-1}$:

x	$g(x)$	x	$g(x)$
0.2	0.806452	1.8	0.165563
0.4	0.641026	1.6	0.193798
0.6	0.510204	1.4	0.229358
0.8	0.409836	1.2	0.274725
0.9	0.369004	1.1	0.302115
0.99	0.336689	1.01	0.330022

It appears that $\lim\limits_{x\to 1}\dfrac{x-1}{x^3-1} = 0.\overline{3} = \frac{1}{3}$.

17. For $F(x) = \dfrac{(1/\sqrt{x}) - \frac{1}{5}}{x - 25}$:

x	$F(x)$	x	$F(x)$
26	−0.003884	24	−0.004124
25.5	−0.003941	24.5	−0.004061
25.1	−0.003988	24.9	−0.004012
25.05	−0.003994	24.95	−0.004006
25.01	−0.003999	24.99	−0.004001

It appears that $\lim\limits_{x\to 25} F(x) = -0.004$.

19. For $f(x) = \dfrac{1 - \cos x}{x^2}$:

x	$f(x)$
1	0.459698
0.5	0.489670
0.4	0.493369
0.3	0.496261
0.2	0.498336
0.1	0.499583
0.05	0.499896
0.01	0.499996

It appears that $\lim\limits_{x\to 0}\dfrac{1 - \cos x}{x^2} = 0.5$.

21. $\lim\limits_{x\to 5^+}\dfrac{6}{x-5} = \infty$ since $(x-5) \to 0$ as $x \to 5^+$ and $\dfrac{6}{x-5} > 0$ for $x > 5$.

23. $\lim\limits_{x\to 3}\dfrac{1}{(x-3)^8} = \infty$ since $(x-3) \to 0$ as $x \to 3$ and $\dfrac{1}{(x-3)^8} > 0$.

25. $\lim\limits_{x\to -2^+}\dfrac{x-1}{x^2(x+2)} = -\infty$ since $(x+2) \to 0$ as $x \to 2^+$ and $\dfrac{x-1}{x^2(x+2)} < 0$ for $-2 < x < 0$.

27. $\lim\limits_{x\to (-\pi/2)^-}\sec x = \lim\limits_{x\to (-\pi/2)^-}(1/\cos x) = -\infty$ since $\cos x \to 0$ as $x \to (-\pi/2)^-$ and $\cos x < 0$ for $-\pi < x < -\pi/2$.

29. (a)

x	$f(x)$
0.5	-1.14
0.9	-3.69
0.99	-33.7
0.999	-333.7
0.9999	-3333.7
0.99999	$-33,333.7$

x	$f(x)$
1.5	0.42
1.1	3.02
1.01	33.0
1.001	333.0
1.0001	3333.0
1.00001	33,333.3

From these calculations, it seems that $\lim\limits_{x \to 1^-} f(x) = -\infty$ and $\lim\limits_{x \to 1^+} f(x) = \infty$.

(b) If x is slightly smaller than 1, then $x^3 - 1$ will be a negative number close to 0, and the reciprocal of $x^3 - 1$, that is, $f(x)$, will be a negative number with large absolute value. So $\lim\limits_{x \to 1^-} f(x) = -\infty$.

If x is slightly larger than 1, then $x^3 - 1$ will be a small positive number, and its reciprocal, $f(x)$, will be a large positive number. So $\lim\limits_{x \to 1^+} f(x) = \infty$.

(c) It appears from the graph of f that $\lim\limits_{x \to 1^-} f(x) = -\infty$ and

$\lim\limits_{x \to 1^+} f(x) = \infty$.

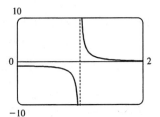

31. (a) Let $h(x) = (1+x)^{1/x}$.

(b)

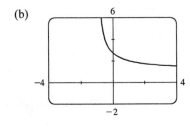

x	$h(x)$
-0.001	2.71964
-0.0001	2.71842
-0.00001	2.71830
-0.000001	2.71828
0.000001	2.71828
0.00001	2.71827
0.0001	2.71815
0.001	2.71692

It appears that $\lim\limits_{x \to 0} (1+x)^{1/x} \approx 2.71828$, which is approximately e.

In Section 3.7 we'll see that the value of the limit is exactly e.

33. (a) From the following graphs, it seems that $\lim\limits_{x \to 0} \dfrac{\tan(4x)}{x} = 4.$

(b)

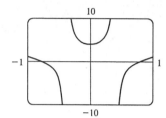

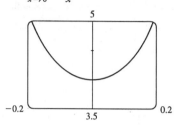

x	$f(x)$
±0.1	4.227932
±0.01	4.002135
±0.001	4.000021
±0.0001	4.000000

35. For $f(x) = x^2 - (2^x/1000)$:

(a)

x	$f(x)$
1	0.998000
0.8	0.638259
0.6	0.358484
0.4	0.158680
0.2	0.038851
0.1	0.008928
0.05	0.001465

It appears that $\lim\limits_{x \to 0} f(x) = 0.$

(b)

x	$f(x)$
0.04	0.000572
0.02	−0.000614
0.01	−0.000907
0.005	−0.000978
0.003	−0.000993
0.001	−0.001000

It appears that $\lim\limits_{x \to 0} f(x) = -0.001.$

37. No matter how many times we zoom in towards the origin, the graphs appear to consist of almost-vertical lines. This indicates more and more frequent oscillations as $x \to 0.$

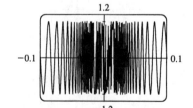

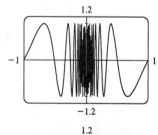

39.

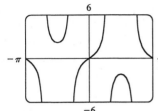

There appear to be vertical asymptotes at $x \approx \pm 0.90$ and $x \approx \pm 2.24$. To find the exact equations of these asymptotes, we note that the graph of the tangent function has vertical asymptotes at $x = \frac{\pi}{2} + \pi n$. Thus, we must have $2 \sin x = \frac{\pi}{2} + \pi n$, or equivalently, $\sin x = \frac{\pi}{4} + \frac{\pi}{2} n$. Since $-1 \le \sin x \le 1$, we must have $\sin x = \pm \frac{\pi}{4}$ and so $x = \pm \sin^{-1} \frac{\pi}{4}$ (corresponding to $x \approx \pm 0.90$).

Just as 150° is the reference angle for 30°, $\pi - \sin^{-1} \frac{\pi}{4}$ is the reference angle for $\sin^{-1} \frac{\pi}{4}$. So $x = \pm \left(\pi - \sin^{-1} \frac{\pi}{4} \right)$ are also equations of the vertical asymptotes (corresponding to $x \approx \pm 2.24$).

2.3 Calculating Limits Using the Limit Laws

1. (a) $\lim\limits_{x \to a} [f(x) + h(x)] = \lim\limits_{x \to a} f(x) + \lim\limits_{x \to a} h(x) = -3 + 8 = 5$

(b) $\lim\limits_{x \to a} [f(x)]^2 = \left[\lim\limits_{x \to a} f(x) \right]^2 = (-3)^2 = 9$

(c) $\lim\limits_{x \to a} \sqrt[3]{h(x)} = \sqrt[3]{\lim\limits_{x \to a} h(x)} = \sqrt[3]{8} = 2$

(d) $\lim\limits_{x \to a} \dfrac{1}{f(x)} = \dfrac{1}{\lim\limits_{x \to a} f(x)} = \dfrac{1}{-3} = -\dfrac{1}{3}$

(e) $\lim\limits_{x \to a} \dfrac{f(x)}{h(x)} = \dfrac{\lim\limits_{x \to a} f(x)}{\lim\limits_{x \to a} h(x)} = \dfrac{-3}{8} = -\dfrac{3}{8}$

(f) $\lim\limits_{x \to a} \dfrac{g(x)}{f(x)} = \dfrac{\lim\limits_{x \to a} g(x)}{\lim\limits_{x \to a} f(x)} = \dfrac{0}{-3} = 0$

(g) The limit does not exist, since $\lim\limits_{x \to a} g(x) = 0$ but $\lim\limits_{x \to a} f(x) \ne 0$.

(h) $\lim\limits_{x \to a} \dfrac{2f(x)}{h(x) - f(x)} = \dfrac{2 \lim\limits_{x \to a} f(x)}{\lim\limits_{x \to a} h(x) - \lim\limits_{x \to a} f(x)} = \dfrac{2(-3)}{8 - (-3)} = -\dfrac{6}{11}$

3. $\lim\limits_{x \to 4} (5x^2 - 2x + 3) = \lim\limits_{x \to 4} 5x^2 - \lim\limits_{x \to 4} 2x + \lim\limits_{x \to 4} 3$ (Limit Laws 2 & 1)

$\qquad\qquad\qquad\qquad = 5 \lim\limits_{x \to 4} x^2 - 2 \lim\limits_{x \to 4} x + 3$ (3 & 7)

$\qquad\qquad\qquad\qquad = 5 (4)^2 - 2 (4) + 3 = 75$ (9 & 8)

5. $\lim\limits_{x \to -1} \dfrac{x - 2}{x^2 + 4x - 3} = \dfrac{\lim\limits_{x \to -1} (x - 2)}{\lim\limits_{x \to -1} (x^2 + 4x - 3)}$ (5)

$\qquad\qquad\qquad\qquad = \dfrac{\lim\limits_{x \to -1} x - \lim\limits_{x \to -1} 2}{\lim\limits_{x \to -1} x^2 + 4 \lim\limits_{x \to -1} x - \lim\limits_{x \to -1} 3}$ (2, 1 & 3)

$\qquad\qquad\qquad\qquad = \dfrac{(-1) - 2}{(-1)^2 + 4(-1) - 3} = \dfrac{1}{2}$ (8, 7 & 9)

7. $\lim\limits_{t\to-2} (t+1)^9 (t^2-1) = \lim\limits_{t\to-2} (t+1)^9 \lim\limits_{t\to-2} (t^2-1)$ (4)

$$= \left[\lim\limits_{t\to-2} (t+1)\right]^9 \lim\limits_{t\to-2} (t^2-1)$$ (6)

$$= \left[\lim\limits_{t\to-2} t + \lim\limits_{t\to-2} 1\right]^9 \left[\lim\limits_{t\to-2} t^2 - \lim\limits_{t\to-2} 1\right]$$ (1 & 2)

$$= [(-2)+1]^9 \left[(-2)^2-1\right] = -3$$ (8, 7 & 9)

9. $\lim\limits_{x\to4^-} \sqrt{16-x^2} = \sqrt{\lim\limits_{x\to4^-} (16-x^2)}$ (11)

$$= \sqrt{\lim\limits_{x\to4^-} 16 - \lim\limits_{x\to4^-} x^2}$$ (2)

$$= \sqrt{16-(4)^2} = 0$$ (7 & 9)

11. $\lim\limits_{x\to-3} \dfrac{x^2-x+12}{x+3}$ does not exist since $x+3\to0$ but $x^2-x+12\to24$ as $x\to-3$.

13. $\lim\limits_{x\to-2} \dfrac{x+2}{x^2-x-6} = \lim\limits_{x\to-2} \dfrac{x+2}{(x-3)(x+2)} = \lim\limits_{x\to-2} \dfrac{1}{x-3} = -\dfrac{1}{5}$

15. $\lim\limits_{h\to0} \dfrac{(h-5)^2-25}{h} = \lim\limits_{h\to0} \dfrac{(h^2-10h+25)-25}{h} = \lim\limits_{h\to0} \dfrac{h^2-10h}{h} = \lim\limits_{h\to0} (h-10) = -10$

17. $\lim\limits_{h\to0} \dfrac{(1+h)^4-1}{h} = \lim\limits_{h\to0} \dfrac{(1+4h+6h^2+4h^3+h^4)-1}{h} = \lim\limits_{h\to0} \dfrac{4h+6h^2+4h^3+h^4}{h}$

$$= \lim\limits_{h\to0} (4+6h+4h^2+h^3) = 4$$

19. $\lim\limits_{t\to9} \dfrac{9-t}{3-\sqrt{t}} = \lim\limits_{t\to9} \dfrac{(3+\sqrt{t})(3-\sqrt{t})}{3-\sqrt{t}} = \lim\limits_{t\to9} (3+\sqrt{t}) = 3+\sqrt{9} = 6$

21. $\lim\limits_{t\to0} \dfrac{\sqrt{2-t}-\sqrt{2}}{t} = \lim\limits_{t\to0} \dfrac{\sqrt{2-t}-\sqrt{2}}{t} \cdot \dfrac{\sqrt{2-t}+\sqrt{2}}{\sqrt{2-t}+\sqrt{2}} = \lim\limits_{t\to0} \dfrac{-t}{t\left(\sqrt{2-t}+\sqrt{2}\right)} = \lim\limits_{t\to0} \dfrac{-1}{\sqrt{2-t}+\sqrt{2}}$

$$= -\dfrac{1}{2\sqrt{2}} = -\dfrac{\sqrt{2}}{4}$$

23. $\lim\limits_{x\to9} \dfrac{x^2-81}{\sqrt{x}-3} = \lim\limits_{x\to9} \dfrac{(x-9)(x+9)}{\sqrt{x}-3} = \lim\limits_{x\to9} \dfrac{(\sqrt{x}-3)(\sqrt{x}+3)(x+9)}{\sqrt{x}-3}$

$$= \lim\limits_{x\to9} (\sqrt{x}+3)(x+9) = \lim\limits_{x\to9} (\sqrt{x}+3) \lim\limits_{x\to9} (x+9) = \left(\sqrt{9}+3\right)(9+9) = 108$$

25. $\lim\limits_{t\to0} \left(\dfrac{1}{t\sqrt{1+t}} - \dfrac{1}{t}\right) = \lim\limits_{t\to0} \dfrac{1-\sqrt{1+t}}{t\sqrt{1+t}} = \lim\limits_{t\to0} \dfrac{(1-\sqrt{1+t})(1+\sqrt{1+t})}{t\sqrt{t+1}(1+\sqrt{1+t})} = \lim\limits_{t\to0} \dfrac{-t}{t\sqrt{1+t}(1+\sqrt{1+t})}$

$$= \lim\limits_{t\to0} \dfrac{-1}{\sqrt{1+t}(1+\sqrt{1+t})} = \dfrac{-1}{\sqrt{1+0}(1+\sqrt{1+0})} = -\dfrac{1}{2}$$

27. $\lim\limits_{x\to2} \dfrac{1/x-\frac{1}{2}}{x-2} = \lim\limits_{x\to2} \dfrac{2-x}{2x(x-2)} = \lim\limits_{x\to2} \dfrac{-1}{2x} = -\dfrac{1}{4}$

29. (a)

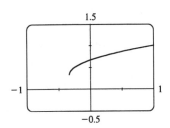

$$\lim_{x\to 0}\frac{x}{\sqrt{1+3x}-1}\approx\frac{2}{3}$$

(b)

x	$f(x)$
-0.001	0.6661663
-0.0001	0.6666167
-0.00001	0.6666617
-0.000001	0.6666662
0.000001	0.6666672
0.00001	0.6666717
0.0001	0.6667167
0.001	0.6671663

The limit appears to be $\frac{2}{3}$.

(c) $\displaystyle\lim_{x\to 0}\left(\frac{x}{\sqrt{1+3x}-1}\cdot\frac{\sqrt{1+3x}+1}{\sqrt{1+3x}+1}\right)=\lim_{x\to 0}\frac{x\left(\sqrt{1+3x}+1\right)}{(1+3x)-1}=\lim_{x\to 0}\frac{x\left(\sqrt{1+3x}+1\right)}{3x}$

$\qquad\qquad = \frac{1}{3}\lim_{x\to 0}\left(\sqrt{1+3x}+1\right)$ (Limit Law 3)

$\qquad\qquad = \frac{1}{3}\left[\sqrt{\lim_{x\to 0}(1+3x)}+\lim_{x\to 0}1\right]$ (1 & 11)

$\qquad\qquad = \frac{1}{3}\left(\sqrt{\lim_{x\to 0}1+3\lim_{x\to 0}x}+1\right)$ (1, 3 & 7)

$\qquad\qquad = \frac{1}{3}\left(\sqrt{1+3\cdot 0}+1\right)$ (7 & 8)

$\qquad\qquad = \frac{2}{3}$

31. Let $f(x)=-x^2$, $g(x)=x^2\cos 20\pi x$ and $h(x)=x^2$. Then

$-1\le\cos 20\pi x\le 1\ \Rightarrow\ -x^2\le x^2\cos 20\pi x\le x^2\ \Rightarrow$

$f(x)\le g(x)\le h(x)$. So since $\lim_{x\to 0}f(x)=\lim_{x\to 0}h(x)=0$, by the

Squeeze Theorem we have $\lim_{x\to 0}g(x)=0$.

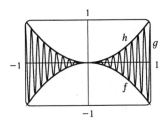

33. $1\le f(x)\le x^2+2x+2$ for all x. Now $\lim_{x\to -1}1=1$ and

$\lim_{x\to -1}(x^2+2x+2)=\lim_{x\to -1}x^2+2\lim_{x\to -1}x+\lim_{x\to -1}2=(-1)^2+2(-1)+2=1$. Therefore, by the Squeeze

Theorem, $\lim_{x\to -1}f(x)=1$.

35. $-1\le\cos(2/x)\le 1\ \Rightarrow\ -x^4\le x^4\cos(2/x)\le x^4$. Since $\lim_{x\to 0}(-x^4)=0$ and $\lim_{x\to 0}x^4=0$, we have

$\lim_{x\to 0}\left[x^4\cos(2/x)\right]=0$ by the Squeeze Theorem.

37. If $x>-4$, then $|x+4|=x+4$, so $\lim_{x\to -4^+}|x+4|=\lim_{x\to -4^+}(x+4)=-4+4=0$.

If $x<-4$, then $|x+4|=-(x+4)$, so $\lim_{x\to -4^-}|x+4|=\lim_{x\to -4^-}-(x+4)=-(-4+4)=0$.

Since the right and left limits are equal, $\lim_{x\to -4}|x+4|=0$.

39. If $x > 2$, then $|x - 2| = x - 2$, so $\lim\limits_{x \to 2^+} \dfrac{|x-2|}{x-2} = \lim\limits_{x \to 2^+} \dfrac{x-2}{x-2} = \lim\limits_{x \to 2^+} 1 = 1$. If $x < 2$, then

$|x - 2| = -(x - 2)$, so $\lim\limits_{x \to 2^-} \dfrac{|x-2|}{x-2} = \lim\limits_{x \to 2^-} \dfrac{-(x-2)}{x-2} = \lim\limits_{x \to 2^-} -1 = -1$. The right and left limits are

different, so $\lim\limits_{x \to 2} \dfrac{|x-2|}{x-2}$ does not exist.

41. Since $|x| = -x$ for $x < 0$, we have $\lim\limits_{x \to 0^-} \left(\dfrac{1}{x} - \dfrac{1}{|x|} \right) = \lim\limits_{x \to 0^-} \left(\dfrac{1}{x} - \dfrac{1}{-x} \right) = \lim\limits_{x \to 0^-} \dfrac{2}{x}$, which does not exist since

the denominator approaches 0 and the numerator does not.

43. (a)

(b) (i) Since $\operatorname{sgn} x = 1$ for $x > 0$, $\lim\limits_{x \to 0^+} \operatorname{sgn} x = \lim\limits_{x \to 0^+} 1 = 1$.

(ii) Since $\operatorname{sgn} x = -1$ for $x < 0$, $\lim\limits_{x \to 0^-} \operatorname{sgn} x = \lim\limits_{x \to 0^-} -1 = -1$.

(iii) Since $\lim\limits_{x \to 0^-} \operatorname{sgn} x \neq \lim\limits_{x \to 0^+} \operatorname{sgn} x$, $\lim\limits_{x \to 0} \operatorname{sgn} x$ does not exist.

(iv) Since $|\operatorname{sgn} x| = 1$ for $x \neq 0$, $\lim\limits_{x \to 0} |\operatorname{sgn} x| = \lim\limits_{x \to 0} 1 = 1$.

45. (a) (i) $\lim\limits_{x \to 1^+} \dfrac{x^2-1}{|x-1|} = \lim\limits_{x \to 1^+} \dfrac{x^2-1}{x-1} = \lim\limits_{x \to 1^+} (x+1) = 2$

(c)

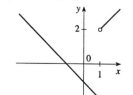

(ii) $\lim\limits_{x \to 1^-} \dfrac{x^2-1}{|x-1|} = \lim\limits_{x \to 1^-} \dfrac{x^2-1}{-(x-1)} = \lim\limits_{x \to 1^-} -(x+1) = -2$

(b) No, $\lim\limits_{x \to 1} F(x)$ does not exist since $\lim\limits_{x \to 1^+} F(x) \neq \lim\limits_{x \to 1^-} F(x)$.

47. (a) (i) $[\![x]\!] = -2$ for $-2 \leq x < -1$, so $\lim\limits_{x \to -2^+} [\![x]\!] = \lim\limits_{x \to -2^+} (-2) = -2$

(ii) $[\![x]\!] = -3$ for $-3 \leq x < -2$, so $\lim\limits_{x \to -2^-} [\![x]\!] = \lim\limits_{x \to -2^-} (-3) = -3$. The right and left limits are different, so

$\lim\limits_{x \to -2} [\![x]\!]$ does not exist.

(iii) $[\![x]\!] = -3$ for $-3 \leq x < -2$, so $\lim\limits_{x \to -2.4} [\![x]\!] = \lim\limits_{x \to -2.4} (-3) = -3$.

(b) (i) $[\![x]\!] = n - 1$ for $n - 1 \leq x < n$, so $\lim\limits_{x \to n^-} [\![x]\!] = \lim\limits_{x \to n^-} (n-1) = n - 1$.

(ii) $[\![x]\!] = n$ for $n \leq x < n + 1$, so $\lim\limits_{x \to n^+} [\![x]\!] = \lim\limits_{x \to n^+} n = n$.

(c) $\lim\limits_{x \to a} [\![x]\!]$ exists $\Leftrightarrow$ a is not an integer.

49. The graph of $f(x) = [\![x]\!] + [\![-x]\!]$ is the same as the graph of $g(x) = -1$ with holes at each integer, since

$f(a) = 0$ for any integer a. Thus, $\lim\limits_{x \to 2^-} f(x) = -1$ and $\lim\limits_{x \to 2^+} f(x) = -1$, so $\lim\limits_{x \to 2} f(x) = -1$.

$f(2) = [\![2]\!] + [\![-2]\!] = 2 + (-2) = 0$.

51. Since $p(x)$ is a polynomial, $p(x) = a_0 + a_1 x + a_2 x^2 + \cdots + a_n x^n$. Thus, by the Limit Laws,

$$\lim\limits_{x \to a} p(x) = \lim\limits_{x \to a} \left(a_0 + a_1 x + a_2 x^2 + \cdots + a_n x^n \right)$$

$$= a_0 + a_1 \lim\limits_{x \to a} x + a_2 \lim\limits_{x \to a} x^2 + \cdots + a_n \lim\limits_{x \to a} x^n$$

$$= a_0 + a_1 a + a_2 a^2 + \cdots + a_n a^n = p(a)$$

Thus, for any polynomial p, $\lim\limits_{x \to a} p(x) = p(a)$.

53. Observe that $0 \leq f(x) \leq x^2$ for all x, and $\lim\limits_{x \to 0} 0 = 0 = \lim\limits_{x \to 0} x^2$. So, by the Squeeze Theorem, $\lim\limits_{x \to 0} f(x) = 0$.

55. Let $f(x) = H(x)$ and $g(x) = 1 - H(x)$, where H is the Heaviside function defined in Exercise 1.3.59. Thus, either f or g is 0 for any value of x. Then $\lim\limits_{x \to 0} f(x)$ and $\lim\limits_{x \to 0} g(x)$ do not exist, but

$$\lim_{x \to 0} [f(x) g(x)] = \lim_{x \to 0} 0 = 0.$$

57. Since the denominator approaches 0 as $x \to -2$, the limit will exist only if the numerator also approaches 0 as $x \to -2$. In order for this to happen, we need $\lim\limits_{x \to -2} \left(3x^2 + ax + a + 3\right) = 0 \iff$

$3(-2)^2 + a(-2) + a + 3 = 0 \iff 12 - 2a + a + 3 = 0 \iff a = 15$. With $a = 15$, the limit becomes

$$\lim_{x \to -2} \frac{3x^2 + 15x + 18}{x^2 + x - 2} = \lim_{x \to -2} \frac{3(x+2)(x+3)}{(x-1)(x+2)} = \frac{3(-2+3)}{-2-1} = -1.$$

2.4 The Precise Definition of a Limit

1. (a) To have $5x + 3$ within a distance of 0.1 of 13, we must have $12.9 \le 5x + 3 \le 13.1 \implies 9.9 \le 5x \le 10.1$
$\implies 1.98 \le x \le 2.02$. Thus, x must be within 0.02 units of 2 so that $5x + 3$ is within 0.1 of 13.

(b) Use 0.01 in place of 0.1 in part (a) to obtain 0.002.

3. On the left side, we need $|x - 2| < \left|\frac{10}{7} - 2\right| = \frac{4}{7}$. On the right side, we need $|x - 2| < \left|\frac{10}{3} - 2\right| = \frac{4}{3}$. For both of these conditions to be satisfied at once, we need the more restrictive of the two to hold, that is, $|x - 2| < \frac{4}{7}$. So we can choose $\delta = \frac{4}{7}$, or any smaller positive number.

5. The leftmost question mark is the solution of $\sqrt{x} = 1.6$ and the rightmost, $\sqrt{x} = 2.4$. So the values are $1.6^2 = 2.56$ and $2.4^2 = 5.76$. On the left side, we need $|x - 4| < |2.56 - 4| = 1.44$. On the right side, we need $|x - 4| < |5.76 - 4| = 1.76$. To satisfy both conditions, we need the more restrictive condition to hold — namely, $|x - 4| < 1.44$. Thus, we can choose $\delta = 1.44$, or any smaller positive number.

7. $\left|\sqrt{4x + 1} - 3\right| < 0.5 \iff 2.5 < \sqrt{4x + 1} < 3.5$. We plot the three parts of this inequality on the same screen and identify the x-coordinates of the points of intersection using the cursor. It appears that the inequality holds for $1.32 \le x \le 2.81$. Since $|2 - 1.32| = 0.68$ and $|2 - 2.81| = 0.81$, we choose $0 < \delta \le \min\{0.68, 0.81\} = 0.68$.

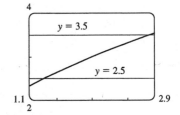

9. For $\varepsilon = 1$, the definition of a limit requires that we find δ such that $\left|(4 + x - 3x^3) - 2\right| < 1$ $\Leftrightarrow$
$1 < 4 + x - 3x^3 < 3$ whenever $|x - 1| < \delta$. If we plot the graphs of $y = 1$, $y = 4 + x - 3x^3$ and $y = 3$ on the
same screen, we see that we need $0.86 \leq x \leq 1.11$. So since $|1 - 0.86| = 0.14$ and $|1 - 1.11| = 0.11$, we choose
$\delta = 0.11$ (or any smaller positive number). For $\varepsilon = 0.1$, we must find δ such that $\left|(4 + x - 3x^3) - 2\right| < 0.1$ $\Leftrightarrow$
$1.9 < 4 + x - 3x^3 < 2.1$ whenever $|x - 1| < \delta$. From the graph, we see that we need $0.988 \leq x \leq 1.012$. So since
$|1 - 0.988| = 0.012$ and $|1 - 1.012| = 0.012$, we must choose $\delta = 0.012$ (or any smaller positive number) for the
inequality to hold.

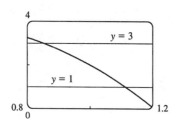

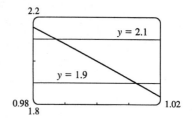

11. From the graph, we see that $\dfrac{x}{(x^2 + 1)(x - 1)^2} > 100$ whenever

$0.93 \leq x \leq 1.07$. So since $|1 - 0.93| = 0.07$ and $|1 - 1.07| = 0.07$,
we can take $\delta = 0.07$ (or any smaller positive number).

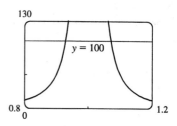

13. (a) $A = \pi r^2$ and $A = 1000$ cm^2 $\Rightarrow$ $\pi r^2 = 1000$ $\Rightarrow$ $r^2 = \dfrac{1000}{\pi}$ $\Rightarrow$ $r = \sqrt{\dfrac{1000}{\pi}}$ $(r > 0)$ ≈ 17.8412 cm.

(b) $|A - 1000| \leq 5$ $\Rightarrow$ $1000 - 5 \leq \pi r^2 \leq 1000 + 5$ $\Rightarrow$ $\sqrt{\dfrac{995}{\pi}} \leq r \leq \sqrt{\dfrac{1005}{\pi}}$ $\Rightarrow$

$17.7966 \leq r \leq 17.8858$. $\sqrt{\dfrac{1000}{\pi}} - \sqrt{\dfrac{995}{\pi}} \approx 0.04466$ and $\sqrt{\dfrac{1005}{\pi}} - \sqrt{\dfrac{1000}{\pi}} \approx 0.04455$. So if the machinist gets
the radius within 0.0445 cm of 17.8412, the area will be within 5 cm^2 of 1000.

(c) x is the radius, $f(x)$ is the area, a is the target radius given in part (a), L is the target area (1000), ε is the
tolerance in the area (5), and δ is the tolerance in the radius given in part (b).

15. Given $\varepsilon > 0$, we need $\delta > 0$ such that if $|x - 2| < \delta$, then
$|(3x - 2) - 4| < \varepsilon$ $\Leftrightarrow$ $|3x - 6| < \varepsilon$ $\Leftrightarrow$ $3|x - 2| < \varepsilon$ $\Leftrightarrow$
$|x - 2| < \varepsilon/3$. So if we choose $\delta = \varepsilon/3$, then $|x - 2| < \delta$ $\Rightarrow$
$|(3x - 2) - 4| < \varepsilon$. Thus, $\lim\limits_{x \to 2} (3x - 2) = 4$ by the definition of a
limit.

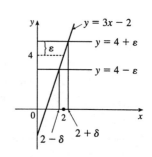

17. Given $\varepsilon > 0$, we need $\delta > 0$ such that if $|x - (-1)| < \delta$, then

$|(5x + 8) - 3| < \varepsilon$ ⇔ $|5x + 5| < \varepsilon$ ⇔ $5|x + 1| < \varepsilon$ ⇔

$|x - (-1)| < \varepsilon/5$. So if we choose $\delta = \varepsilon/5$, then $|x - (-1)| < \delta$

⇒ $|(5x + 8) - 3| < \varepsilon$. Thus, $\lim\limits_{x \to -1} (5x + 8) = 3$ by the definition

of a limit.

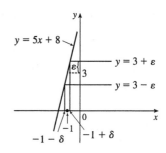

19. Given $\varepsilon > 0$, we need $\delta > 0$ such that if $0 < |x - 3| < \delta$, then $\left|\dfrac{x}{5} - \dfrac{3}{5}\right| < \varepsilon$ ⇔ $\frac{1}{5}|x - 3| < \varepsilon$ ⇔

$|x - 3| < 5\varepsilon$. So choose $\delta = 5\varepsilon$. Then $0 < |x - 3| < \delta$ ⇒ $|x - 3| < 5\varepsilon$ ⇒ $\dfrac{|x - 3|}{5} < \varepsilon$ ⇒

$\left|\dfrac{x}{5} - \dfrac{3}{5}\right| < \varepsilon$. By the definition of a limit, $\lim\limits_{x \to 3} \dfrac{x}{5} = \dfrac{3}{5}$.

21. Given $\varepsilon > 0$, we need $\delta > 0$ such that if $|x - (-5)| < \delta$ then $\left|\left(4 - \frac{3}{5}x\right) - 7\right| < \varepsilon$ ⇔ $\frac{3}{5}|x + 5| < \varepsilon$ ⇔

$|x - (-5)| < \frac{5}{3}\varepsilon$. So take $\delta = \frac{5}{3}\varepsilon$. Then $|x - (-5)| < \delta$ ⇒ $\left|\left(4 - \frac{3}{5}x\right) - 7\right| < \varepsilon$. Thus, $\lim\limits_{x \to -5} \left(4 - \frac{3}{5}x\right) = 7$

by the definition of a limit.

23. Given $\varepsilon > 0$, we need $\delta > 0$ such that if $|x - a| < \delta$ then $|x - a| < \varepsilon$. So $\delta = \varepsilon$ will work.

25. Given $\varepsilon > 0$, we need $\delta > 0$ such that if $|x| < \delta$ then $\left|x^2 - 0\right| < \varepsilon$ ⇔ $x^2 < \varepsilon$ ⇔ $|x| < \sqrt{\varepsilon}$. Take $\delta = \sqrt{\varepsilon}$.

Then $|x - 0| < \delta$ ⇒ $\left|x^2 - 0\right| < \varepsilon$. Thus, $\lim\limits_{x \to 0} x^2 = 0$ by the definition of a limit.

27. Given $\varepsilon > 0$, we need $\delta > 0$ such that if $|x - 0| < \delta$ then $||x| - 0| < \varepsilon$. But $||x|| = |x|$. So this is true if we pick

$\delta = \varepsilon$.

29. Given $\varepsilon > 0$, we need $\delta > 0$ such that if $|x - 2| < \delta$, then $\left|(x^2 - 4x + 5) - 1\right| < \varepsilon$ ⇔ $\left|x^2 - 4x + 4\right| < \varepsilon$ ⇔

$\left|(x - 2)^2\right| < \varepsilon$. So take $\delta = \sqrt{\varepsilon}$. Then $|x - 2| < \delta$ ⇔ $|x - 2| < \sqrt{\varepsilon}$ ⇔ $\left|(x - 2)^2\right| < \varepsilon$. So

$\lim\limits_{x \to 2} (x^2 - 4x + 5) = 1$ by the definition of a limit.

31. Given $\varepsilon > 0$, we need $\delta > 0$ such that if $|x - (-2)| < \delta$ then $\left|(x^2 - 1) - 3\right| < \varepsilon$ or upon simplifying we need

$\left|x^2 - 4\right| < \varepsilon$ whenever $|x + 2| < \delta$. Notice that if $|x + 2| < 1$, then $-1 < x + 2 < 1$ ⇒ $-5 < x - 2 < -3$

⇒ $|x - 2| < 5$. So take $\delta = \min\{\varepsilon/5, 1\}$. Then $|x - 2| < 5$ and $|x + 2| < \varepsilon/5$, so

$\left|(x^2 - 1) - 3\right| = |(x + 2)(x - 2)| = |x + 2||x - 2| < (\varepsilon/5)(5) = \varepsilon$. Therefore, by the definition of a limit,

$\lim\limits_{x \to -2} (x^2 - 1) = 3$.

33. Given $\varepsilon > 0$, we let $\delta = \min\left\{2, \frac{\varepsilon}{8}\right\}$. If $0 < |x - 3| < \delta$ then $|x - 3| < 2$ ⇒ $1 < x < 5$ ⇒ $|x + 3| < 8$.

Also $|x - 3| < \frac{\varepsilon}{8}$, so $\left|x^2 - 9\right| = |x + 3||x - 3| < 8 \cdot \frac{\varepsilon}{8} = \varepsilon$. Thus, $\lim\limits_{x \to 3} x^2 = 9$.

35. *1. Guessing a value for δ*   Given $\varepsilon > 0$, we must find $\delta > 0$ such that $\left| \sqrt{x} - \sqrt{a} \right| < \varepsilon$ whenever

$0 < |x - a| < \delta$. But $\left| \sqrt{x} - \sqrt{a} \right| = \dfrac{|x - a|}{\sqrt{x} + \sqrt{a}} < \varepsilon$ (from the hint). Now if we can find a positive constant C such

that $\sqrt{x} + \sqrt{a} > C$ then $\dfrac{|x - a|}{\sqrt{x} + \sqrt{a}} < \dfrac{|x - a|}{C} < \varepsilon$, and we take $|x - a| < C\varepsilon$. We can find this number by

restricting x to lie in some interval centered at a. If $|x - a| < \frac{1}{2}a$, then $\frac{1}{2}a < x < \frac{3}{2}a \ \Rightarrow$

$\sqrt{x} + \sqrt{a} > \sqrt{\frac{1}{2}a} + \sqrt{a}$, and so $C = \sqrt{\frac{1}{2}a} + \sqrt{a}$ is a suitable choice for the constant. So

$|x - a| < \left(\sqrt{\frac{1}{2}a} + \sqrt{a} \right) \varepsilon$. This suggests that we let $\delta = \min\left\{ \frac{1}{2}a, \left(\sqrt{\frac{1}{2}a} + \sqrt{a} \right) \varepsilon \right\}$.

2. Showing that δ works   Given $\varepsilon > 0$, we let $\delta = \min\left\{ \frac{1}{2}a, \left(\sqrt{\frac{1}{2}a} + \sqrt{a} \right) \varepsilon \right\}$. If $0 < |x - a| < \delta$, then

$|x - a| < \frac{1}{2}a \ \Rightarrow \ \sqrt{x} + \sqrt{a} > \sqrt{\frac{1}{2}a} + \sqrt{a}$ (as in part 1). Also $|x - a| < \left(\sqrt{\frac{1}{2}a} + \sqrt{a} \right) \varepsilon$, so

$\left| \sqrt{x} + \sqrt{a} \right| = \dfrac{|x - a|}{\sqrt{x} + \sqrt{a}} < \dfrac{\left(\sqrt{a/2} + \sqrt{a} \right) \varepsilon}{\left(\sqrt{a/2} + \sqrt{a} \right)} = \varepsilon$. Therefore, $\lim\limits_{x \to a} \sqrt{x} = \sqrt{a}$ by the definition of a limit.

37. Suppose that $\lim\limits_{x \to 0} f(x) = L$. Given $\varepsilon = \frac{1}{2}$, there exists $\delta > 0$ such that $0 < |x| < \delta \ \Rightarrow \ |f(x) - L| < \frac{1}{2}$. Take

any rational number r with $0 < |r| < \delta$. Then $f(r) = 0$, so $|0 - L| < \frac{1}{2}$, so $L \leq |L| < \frac{1}{2}$. Now take any irrational

number s with $0 < |s| < \delta$. Then $f(s) = 1$, so $|1 - L| < \frac{1}{2}$. Hence, $1 - L < \frac{1}{2}$, so $L > \frac{1}{2}$. This contradicts

$L < \frac{1}{2}$, so $\lim\limits_{x \to 0} f(x)$ does not exist.

39. $\dfrac{1}{(x + 3)^4} > 10{,}000 \ \Leftrightarrow \ (x + 3)^4 < \dfrac{1}{10{,}000} \ \Leftrightarrow \ |x - (-3)| = |x + 3| < \dfrac{1}{10}$

41. Given $M < 0$ we need $\delta > 0$ so that $\ln x < M$ whenever $0 < x < \delta$; that is, $x = e^{\ln x} < e^M$ whenever $0 < x < \delta$.
This suggests that we take $\delta = e^M$. If $0 < x < e^M$, then $\ln x < \ln e^M = M$. By the definition of a limit,
$\lim\limits_{x \to 0^+} \ln x = -\infty$.

2.5 Continuity

1. From Equation 1, $\lim\limits_{x \to 4} f(x) = f(4)$.

3. (a) The following are the numbers at which f is discontinuous and the type of discontinuity at that number:
   -5 (jump), -3 (infinite), -1 (undefined), 3 (removable), 5 (infinite), 8 (jump), 10 (undefined).

   (b) f is continuous from the left at -5 and -3, and continuous from the right at 8. It is continuous from neither
   side at -1, 3, 5, and 10.

5.

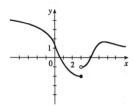

7. (a)

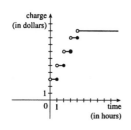

(b) There are discontinuities at $t = 1, 2, 3$, and 4. A person parking in the lot would want to keep in mind that the charge will jump at the beginning of each hour.

9. Since f and g are continuous functions,

$$\lim_{x \to 3} [2f(x) - g(x)] = 2 \lim_{x \to 3} f(x) - \lim_{x \to 3} g(x) \text{ (by Limit Laws 2 \& 3)}$$

$$= 2f(3) - g(3) \text{ (by continuity of } f \text{ and } g \text{ at } x = 3)$$

$$= 2 \cdot 5 - g(3) = 10 - g(3)$$

Since it is given that $\lim_{x \to 3} [2f(x) - g(x)] = 4$, we have $10 - g(3) = 4$, or $g(3) = 6$.

11. $\lim_{x \to -1} f(x) = \lim_{x \to -1} (x + 2x^3)^4 = \left(\lim_{x \to -1} x + 2 \lim_{x \to -1} x^3 \right)^4 = [-1 + 2(-1)^3]^4 = (-3)^4 = 81 = f(-1)$. By the definition of continuity, f is continuous at $a = -1$.

13. For $-4 < a < 4$ we have $\lim_{x \to a} f(x) = \lim_{x \to a} x\sqrt{16 - x^2} = \lim_{x \to a} x \sqrt{\lim_{x \to a} 16 - \lim_{x \to a} x^2} = a\sqrt{16 - a^2} = f(a)$, so f is continuous on $(-4, 4)$. Similarly, we get $\lim_{x \to 4^-} f(x) = 0 = f(4)$ and $\lim_{x \to -4^+} f(x) = 0 = f(-4)$, so f is continuous from the left at 4 and from the right at -4. Thus, f is continuous on $[-4, 4]$.

15. $f(x) = \ln|x - 2|$ is discontinuous at 2 since $f(2) = \ln 0$ is not defined.

17. $f(x) = \dfrac{x^2 - 1}{x + 1}$ is discontinuous at -1 because $f(-1)$ is not defined.

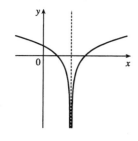

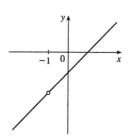

19. Since $f(x) = \dfrac{x^2 - 2x - 8}{x - 4}$ if $x \neq 4$,

$$\lim_{x \to 4} f(x) = \lim_{x \to 4} \frac{x^2 - 2x - 8}{x - 4} = \lim_{x \to 4} \frac{(x - 4)(x + 2)}{x - 4}$$

$$= \lim_{x \to 4} (x + 2) = 4 + 2 = 6.$$

But $f(4) = 3$, so $\lim_{x \to 4} f(x) \neq f(4)$. Therefore, f is discontinuous at 4.

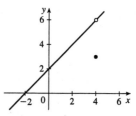

21. $F(x) = \dfrac{x}{x^2 + 5x + 6}$ is a rational function. So by Theorem 5 (or Theorem 7), F is continuous at every number in its domain, $\{x \mid x^2 + 5x + 6 \neq 0\} = \{x \mid x \neq -2, -3\}$ or $(-\infty, -3) \cup (-3, -2) \cup (-2, \infty)$.

23. $g(x) = x - 1$ and $G(x) = x^2 - 2$ are both polynomials, so by Theorem 5 they are continuous. Also $f(x) = \sqrt[5]{x}$ is continuous by Theorem 6, so $f(g(x)) = \sqrt[5]{x - 1}$ is continuous on $\mathbb{R}$ by Theorem 8. Thus, the product $h(x) = \sqrt[5]{x - 1}\,(x^2 - 2)$ is continuous on $\mathbb{R}$ by Theorem 4 #4.

25. By Theorem 5, the polynomial $5x$ is continuous on $(-\infty, \infty)$. By Theorems 9 and 7, $\sin 5x$ is continuous on $(-\infty, \infty)$. By Theorem 7, e^x is continuous on $(-\infty, \infty)$. By Theorem 4 #4, the product of e^x and $\sin 5x$ is continuous at all numbers which are in both of their domains, that is, on $(-\infty, \infty)$.

27. By Theorem 5, the polynomial $t^4 - 1$ is continuous $(-\infty, \infty)$. By Theorem 7, $\ln x$ is continuous on its domain, $(0, \infty)$. By Theorem 9, $\ln(t^4 - 1)$ is continuous on its domain, which is
$$\{t \mid t^4 - 1 > 0\} = \{t \mid |t| > 1\} = (-\infty, 1) \cup (1, \infty).$$

29. The function $y = 1/(1 + e^{1/x})$ is discontinuous at $x = 0$ because the left- and right-hand limits at $x = 0$ are different.

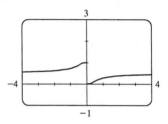

31. Because we are dealing with root functions, $5 + \sqrt{x}$ is continuous on $[0, \infty)$, $\sqrt{x + 5}$ is continuous on $[-5, \infty)$, so the quotient $f(x) = \dfrac{5 + \sqrt{x}}{\sqrt{5 + x}}$ is continuous on $[0, \infty)$. Since f is continuous at $x = 4$, $\lim\limits_{x \to 4} f(x) = f(4) = \frac{7}{3}$.

33. Because $x^2 - x$ is continuous on $\mathbb{R}$, the composite function $f(x) = e^{x^2 - x}$ is continuous on $\mathbb{R}$, so
$$\lim_{x \to 1} f(x) = f(1) = e^{1 - 1} = e^0 = 1.$$

35. f is continuous on $(-\infty, 3)$ and $(3, \infty)$ since on each of these intervals it is a polynomial.
Also $\lim\limits_{x \to 3^+} f(x) = \lim\limits_{x \to 3^+} (5 - x) = 2$ and $\lim\limits_{x \to 3^-} f(x) = \lim\limits_{x \to 3^-} (x - 1) = 2$, so $\lim\limits_{x \to 3} f(x) = 2$.
Since $f(3) = 5 - 3 = 2$, f is also continuous at 3. Thus, f is continuous on $(-\infty, \infty)$.

37. f is continuous on $(-\infty, 0)$ and $(0, \infty)$ since on each of these intervals it is a polynomial. Now $\lim\limits_{x \to 0^-} f(x) = \lim\limits_{x \to 0^-} (x - 1)^3 = -1$ and
$\lim\limits_{x \to 0^+} f(x) = \lim\limits_{x \to 0^+} (x + 1)^3 = 1$. Thus, $\lim\limits_{x \to 0} f(x)$ does not exist, so f is discontinuous at 0. Since $f(0) = 1$, f is continuous from the right at 0.

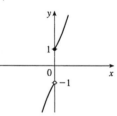

39. f is continuous on $(-\infty, 3)$ and $(3, \infty)$. Now $\lim\limits_{x \to 3^-} f(x) = \lim\limits_{x \to 3^-} (cx + 1) = 3c + 1$ and
$\lim\limits_{x \to 3^+} f(x) = \lim\limits_{x \to 3^+} (cx^2 - 1) = 9c - 1$. So f is continuous $\Leftrightarrow$ $3c + 1 = 9c - 1$ $\Leftrightarrow$ $6c = 2$ $\Leftrightarrow$ $c = \frac{1}{3}$.
Thus, for f to be continuous on $(-\infty, \infty)$, $c = \frac{1}{3}$.

41. (a) $f(x) = \dfrac{x^2 - 2x - 8}{x + 2} = \dfrac{(x - 4)(x + 2)}{x + 2}$ has a removable discontinuity at -2 because $g(x) = x - 4$ is continuous on $\mathbb{R}$ and $f(x) = g(x)$ for $x \neq -2$. [The discontinuity is removed by defining $f(-2) = -6$.]

(b) $f(x) = \dfrac{x - 7}{|x - 7|}$ $\Rightarrow$ $\lim\limits_{x \to 7^-} f(x) = -1$ and $\lim\limits_{x \to 7^+} f(x) = 1$. Thus, $\lim\limits_{x \to 7} f(x)$ does not exist, so the discontinuity is not removable. (It is a jump discontinuity.)

(c) $f(x) = \dfrac{x^3 + 64}{x + 4} = \dfrac{(x+4)\left(x^2 - 4x + 16\right)}{x + 4}$ has a removable discontinuity at -4 because

$g(x) = x^2 - 4x + 16$ is continuous on $\mathbb{R}$ and $f(x) = g(x)$ for $x \neq -4$. [The discontinuity is removed by defining $f(-4) = 48$.]

(d) $f(x) = \dfrac{3 - \sqrt{x}}{9 - x} = \dfrac{3 - \sqrt{x}}{\left(3 - \sqrt{x}\right)\left(3 + \sqrt{x}\right)}$ has a removable discontinuity at 9 because $g(x) = 1/\left(3 + \sqrt{x}\right)$ is

continuous on $\mathbb{R}$ and $f(x) = g(x)$ for $x \neq 9$. [The discontinuity is removed by defining $f(9) = \frac{1}{6}$.]

43. $f(x) = x^3 - x^2 + x$ is continuous on the interval $[2, 3]$, $f(2) = 6$, and $f(3) = 21$. Since $6 < 10 < 21$, there is a number c in $(2, 3)$ such that $f(c) = 10$ by the Intermediate Value Theorem.

45. $f(x) = x^3 - 3x + 1$ is continuous on the interval $[0, 1]$, $f(0) = 1$, and $f(1) = -1$. Since $-1 < 0 < 1$, there is a number c in $(0, 1)$ such that $f(c) = 0$ by the Intermediate Value Theorem. Thus, there is a root of the equation $x^3 - 3x + 1 = 0$ in the interval $(0, 1)$.

47. $f(x) = \cos x - x$ is continuous on the interval $[0, 1]$, $f(0) = 1$, and $f(1) = \cos 1 - 1 \approx -0.46$. Since $-0.46 < 0 < 1$, there is a number c in $(0, 1)$ such that $f(c) = 0$ by the Intermediate Value Theorem. Thus, there is a root of the equation $\cos x - x = 0$, or $\cos x = x$, in the interval $(0, 1)$.

49. (a) $f(x) = e^x + x - 2$ is continuous on the interval $[0, 1]$, $f(0) = -1$, and $f(1) = e - 1 \approx 1.72$. Since $-1 < 0 < 1.72$, there is a number c in $(0, 1)$ such that $f(c) = 0$ by the Intermediate Value Theorem. Thus, there is a root of the equation $e^x + x - 2 = 0$, or $e^x = 2 - x$, in the interval $(0, 1)$.

(b) $f(0.44) \approx -0.007$ and $f(0.45) \approx 0.018$, so there is a root between 0.44 and 0.45.

51. (a) Let $f(x) = x^5 - x^2 - 4$. Then $f(1) = 1^5 - 1^2 - 4 = -4 < 0$ and $f(2) = 2^5 - 2^2 - 4 = 24 > 0$. So by the Intermediate Value Theorem, there is a number c in $(1, 2)$ such that $c^5 - c^2 - 4 = 0$.

(b) We can see from the graphs that, correct to three decimal places, the root is $x \approx 1.434$.

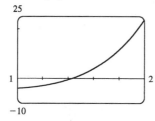
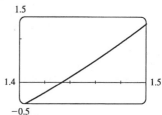

53. ($\Rightarrow$) If f is continuous at a, then by Theorem 7 with $g(h) = a + h$, we have

$$\lim_{h \to 0} f(a + h) = f\left(\lim_{h \to 0}(a + h)\right) = f(a).$$

($\Leftarrow$) Let $\varepsilon > 0$. Since $\lim_{h \to 0} f(a + h) = f(a)$, there exists $\delta > 0$ such that $|h| < \delta \Rightarrow$

$|f(a + h) - f(a)| < \varepsilon$. So if $|x - a| < \delta$, then $|f(x) - f(a)| = |f(a + (x - a)) - f(a)| < \varepsilon$. Thus, $\lim_{x \to a} f(x) = f(a)$ and so f is continuous at a.

55. As in the previous exercise, we must show that $\lim_{h \to 0} \cos(a + h) = \cos a$ to prove that the cosine function is continuous.

$$\lim_{h \to 0} \cos(a + h) = \lim_{h \to 0} (\cos a \cos h - \sin a \sin h)$$

$$= \lim_{h \to 0} (\cos a \cos h) - \lim_{h \to 0} (\sin a \sin h)$$

$$= \left(\lim_{h \to 0} \cos a \right) \left(\lim_{h \to 0} \cos h \right) - \left(\lim_{h \to 0} \sin a \right) \left(\lim_{h \to 0} \sin h \right)$$

$$= (\cos a)(1) - (\sin a)(0) = \cos a$$

57. $f(x) = \begin{cases} 0 & \text{if } x \text{ is rational} \\ 1 & \text{if } x \text{ is irrational} \end{cases}$ is continuous nowhere. For, given any number a and any $\delta > 0$, the interval $(a - \delta, a + \delta)$ contains both infinitely many rational and infinitely many irrational numbers. Since $f(a) = 0$ or 1, there are infinitely many numbers x with $|x - a| < \delta$ and $|f(x) - f(a)| = 1$. Thus, $\lim_{x \to a} f(x) \neq f(a)$. [In fact, $\lim_{x \to a} f(x)$ does not even exist.]

59. If there is such a number, it satisfies the equation $x^3 + 1 = x \Leftrightarrow x^3 - x + 1 = 0$. Let the LHS of this equation be called $f(x)$. Now $f(-2) = -5 < 0$, and $f(-1) = 1 > 0$. Note also that $f(x)$ is a polynomial, and thus continuous. So by the Intermediate Value Theorem, there is a number c between -2 and -1 such that $f(c) = 0$, so that $c = c^3 + 1$.

61. Define $u(t)$ to be the monk's distance from the monastery, as a function of time, on the first day, and define $d(t)$ to be his distance from the monastery, as a function of time, on the second day. Let D be the distance from the monastery to the top of the mountain. From the given information we know that $u(0) = 0$, $u(12) = D$, $d(0) = D$ and $d(12) = 0$. Now consider the function $u - d$, which is clearly continuous. We calculate that $(u - d)(0) = -D$ and $(u - d)(12) = D$. So by the Intermediate Value Theorem, there must be some time t_0 between 0 and 12 such that $(u - d)(t_0) = 0 \Leftrightarrow u(t_0) = d(t_0)$. So at time t_0 after 7:00 A.M., the monk will be at the same place on both days.

2.6 Limits at Infinity; Horizontal Asymptotes

1. (a) As x becomes large, the values of $f(x)$ approach 5.

(b) As x becomes large negative, the values of $f(x)$ approach 3.

3. (a) $\lim_{x \to 2} f(x) = \infty$

(b) $\lim_{x \to -1^-} f(x) = \infty$

(c) $\lim_{x \to -1^+} f(x) = -\infty$

(d) $\lim_{x \to \infty} f(x) = 1$

(e) $\lim_{x \to -\infty} f(x) = 2$

(f) Vertical: $x = -1$, $x = 2$; Horizontal: $y = 1$, $y = 2$

5.

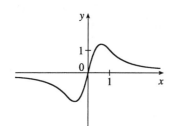

7.

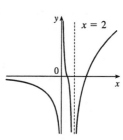

9. If $f(x) = x^2/2^x$, then a calculator gives $f(0) = 0$, $f(1) = 0.5$, $f(2) = 1$, $f(3) = 1.125$, $f(4) = 1$, $f(5) = 0.78125$, $f(6) = 0.5625$, $f(7) = 0.3828125$, $f(8) = 0.25$, $f(9) = 0.158203125$, $f(10) = 0.09765625$, $f(20) \approx 0.00038147$, $f(50) \approx 2.2204 \times 10^{-12}$, $f(100) \approx 7.8886 \times 10^{-27}$. It appears that $\lim_{x \to \infty} (x^2/2^x) = 0$.

11. $\displaystyle \lim_{x \to \infty} \frac{x+4}{x^2 - 2x + 5} = \lim_{x \to \infty} \frac{\dfrac{1}{x} + \dfrac{4}{x^2}}{1 - \dfrac{2}{x} + \dfrac{5}{x^2}} = \frac{\displaystyle\lim_{x \to \infty} \left(\dfrac{1}{x} + \dfrac{4}{x^2}\right)}{\displaystyle\lim_{x \to \infty}\left(1 - \dfrac{2}{x} + \dfrac{5}{x^2}\right)}$

$$= \frac{\displaystyle\lim_{x \to \infty} \frac{1}{x} + 4 \lim_{x \to \infty} \frac{1}{x^2}}{\displaystyle\lim_{x \to \infty} 1 - 2 \lim_{x \to \infty} \frac{1}{x} + 5 \lim_{x \to \infty} \frac{1}{x^2}} = \frac{0 + 4(0)}{1 - 2(0) + 5(0)} = 0$$

13. $\displaystyle \lim_{x \to -\infty} \frac{(1-x)(2+x)}{(1+2x)(2-3x)} = \lim_{x \to -\infty} \frac{\left[\dfrac{1}{x} - 1\right]\left[\dfrac{2}{x} + 1\right]}{\left[\dfrac{1}{x} + 2\right]\left[\dfrac{2}{x} - 3\right]} = \frac{\left[\displaystyle\lim_{x \to -\infty} \dfrac{1}{x} - 1\right]\left[\displaystyle\lim_{x \to -\infty} \dfrac{2}{x} + 1\right]}{\left[\displaystyle\lim_{x \to -\infty} \dfrac{1}{x} + 2\right]\left[\displaystyle\lim_{x \to -\infty} \dfrac{2}{x} - 3\right]}$

$$\overset{(5,4,1,2,7)}{=} \frac{(0-1)(0+1)}{(0+2)(0-3)} = \frac{1}{6}$$

15. $\displaystyle \lim_{r \to \infty} \frac{r^4 - r^2 + 1}{r^5 + r^3 - r} = \lim_{r \to \infty} \frac{\dfrac{1}{r} - \dfrac{1}{r^3} + \dfrac{1}{r^5}}{1 + \dfrac{1}{r^2} - \dfrac{1}{r^4}} = \frac{\displaystyle\lim_{r \to \infty} \frac{1}{r} - \lim_{r \to \infty} \frac{1}{r^3} + \lim_{r \to \infty} \frac{1}{r^5}}{\displaystyle\lim_{r \to \infty} 1 + \lim_{r \to \infty} \frac{1}{r^2} - \lim_{r \to \infty} \frac{1}{r^4}} = \frac{0 - 0 + 0}{1 + 0 - 0} = 0$

17. $\displaystyle \lim_{x \to \infty} \frac{\sqrt{1 + 4x^2}}{4 + x} = \lim_{x \to \infty} \frac{\sqrt{(1/x^2) + 4}}{(4/x) + 1} = \frac{\sqrt{0 + 4}}{0 + 1} = 2$

19. $\displaystyle \lim_{x \to \infty} \frac{1 - \sqrt{x}}{1 + \sqrt{x}} = \lim_{x \to \infty} \frac{(1/\sqrt{x}) - 1}{(1/\sqrt{x}) + 1} = \frac{0 - 1}{0 + 1} = -1$

21. $\displaystyle \lim_{x \to \infty} \left(\sqrt{x^2 + 1} - \sqrt{x^2 - 1}\right) = \lim_{x \to \infty} \left(\sqrt{x^2 + 1} - \sqrt{x^2 - 1}\right) \frac{\sqrt{x^2 + 1} + \sqrt{x^2 - 1}}{\sqrt{x^2 + 1} + \sqrt{x^2 - 1}}$

$$= \lim_{x \to \infty} \frac{(x^2 + 1) - (x^2 - 1)}{\sqrt{x^2 + 1} + \sqrt{x^2 - 1}} = \lim_{x \to \infty} \frac{2}{\sqrt{x^2 + 1} + \sqrt{x^2 - 1}}$$

$$= \lim_{x \to \infty} \frac{2/x}{\sqrt{1 + (1/x^2)} + \sqrt{1 - (1/x^2)}} = \frac{0}{\sqrt{1 + 0} + \sqrt{1 - 0}} = 0$$

23. $\lim\limits_{x\to\infty}\left(\sqrt{9x^2+x}-3x\right)=\lim\limits_{x\to\infty}\dfrac{\left(\sqrt{9x^2+x}-3x\right)\left(\sqrt{9x^2+x}+3x\right)}{\sqrt{9x^2+x}+3x}=\lim\limits_{x\to\infty}\dfrac{\left(\sqrt{9x^2+x}\right)^2-(3x)^2}{\sqrt{9x^2+x}+3x}$

$\qquad=\lim\limits_{x\to\infty}\dfrac{(9x^2+x)-9x^2}{\sqrt{9x^2+x}+3x}=\lim\limits_{x\to\infty}\dfrac{x/x}{\left(\sqrt{9x^2+x}+3x\right)/x}=\lim\limits_{x\to\infty}\dfrac{1}{\sqrt{9+1/x}+3}$

$\qquad=\dfrac{1}{\sqrt{9}+3}=\dfrac{1}{3+3}=\dfrac{1}{6}$

25. $\sqrt{x}$ is large when x is large, so $\lim\limits_{x\to\infty}\sqrt{x}=\infty$.

27. $\lim\limits_{x\to\infty}\left(x-\sqrt{x}\right)=\lim\limits_{x\to\infty}\sqrt{x}\left(\sqrt{x}-1\right)=\infty$ since $\sqrt{x}\to\infty$ and $\sqrt{x}-1\to\infty$ as $x\to\infty$.

29. $\lim\limits_{x\to-\infty}\left(x^3-5x^2\right)=-\infty$ since $x^3\to-\infty$ and $-5x^2\to-\infty$ as $x\to-\infty$.

$\quad$ *Or:* $\lim\limits_{x\to-\infty}\left(x^3-5x^2\right)=\lim\limits_{x\to-\infty}x^2\left(x-5\right)=-\infty$ since $x^2\to\infty$ and $x-5\to-\infty$.

31. $\lim\limits_{x\to\infty}\dfrac{x^7-1}{x^6+1}=\lim\limits_{x\to\infty}\dfrac{1-1/x^7}{(1/x)+\left(1/x^7\right)}=\infty$ since $1-\dfrac{1}{x^7}\to1$ while $\dfrac{1}{x}+\dfrac{1}{x^7}\to0^+$ as $x\to\infty$.

$\quad$ *Or:* Divide numerator and denominator by x^6 instead of x^7.

33. (a)

![graph from -100 to 0, with value near -1]

From the graph of $f(x)=\sqrt{x^2+x+1}+x$, we estimate the value of $\lim\limits_{x\to-\infty}f(x)$ to be -0.5.

(b)

x	$f(x)$
$-10{,}000$	-0.4999625
$-100{,}000$	-0.4999962
$-1{,}000{,}000$	-0.4999996

From the table, we estimate the limit to be -0.5.

(c) $\lim\limits_{x\to-\infty}\left(\sqrt{x^2+x+1}+x\right)=\lim\limits_{x\to-\infty}\left(\sqrt{x^2+x+1}+x\right)\left[\dfrac{\sqrt{x^2+x+1}-x}{\sqrt{x^2+x+1}-x}\right]=\lim\limits_{x\to-\infty}\dfrac{(x^2+x+1)-x^2}{\sqrt{x^2+x+1}-x}$

$\qquad=\lim\limits_{x\to-\infty}\dfrac{(x+1)\,(1/x)}{\left(\sqrt{x^2+x+1}-x\right)(1/x)}=\lim\limits_{x\to-\infty}\dfrac{1+(1/x)}{-\sqrt{1+(1/x)+\left(1/x^2\right)}-1}=\dfrac{1+0}{-\sqrt{1+0+0}-1}=-\dfrac{1}{2}$

Note that for $x<0$, we have $\sqrt{x^2}=|x|=-x$, so when we divide the radical by x, with $x<0$, we get

$\dfrac{1}{x}\sqrt{x^2+x+1}=-\dfrac{1}{\sqrt{x^2}}\sqrt{x^2+x+1}=-\sqrt{1+(1/x)+\left(1/x^2\right)}.$

35. $\lim\limits_{x\to\pm\infty}\dfrac{x}{x+4}=\lim\limits_{x\to\pm\infty}\dfrac{1}{1+4/x}=\dfrac{1}{1+0}=1$, so $y=1$ is a horizontal

asymptote. $\lim\limits_{x\to-4^-}\dfrac{x}{x+4}=\infty$ and $\lim\limits_{x\to-4^+}\dfrac{x}{x+4}=-\infty$, so $x=-4$ is a

vertical asymptote. The graph confirms these calculations.

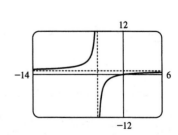

37. $\lim\limits_{x \to \pm\infty} \dfrac{x^3}{x^2 + 3x - 10} = \lim\limits_{x \to \pm\infty} \dfrac{x}{1 + (3/x) - (10/x^2)} = \pm\infty$, so there is

no horizontal asymptote.

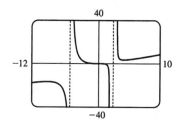

$\lim\limits_{x \to 2^+} \dfrac{x^3}{x^2 + 3x - 10} = \lim\limits_{x \to 2^+} \dfrac{x^3}{(x + 5)(x - 2)} = \infty$, since

$\dfrac{x^3}{(x + 5)(x - 2)} > 0$ for $x > 2$. Similarly, $\lim\limits_{x \to 2^-} \dfrac{x^3}{x^2 + 3x - 10} = -\infty$

and $\lim\limits_{x \to -5^-} \dfrac{x^3}{x^2 + 3x - 10} = -\infty$, $\lim\limits_{x \to -5^+} \dfrac{x^3}{x^2 + 3x - 10} = \infty$, so $x = 2$

and $x = -5$ are vertical asymptotes. The graph confirms these

calculations.

39. $\lim\limits_{x \to \infty} \dfrac{x}{\sqrt[4]{x^4 + 1}} = \lim\limits_{x \to \infty} \dfrac{1}{\sqrt[4]{1 + (1/x^4)}} = \dfrac{1}{\sqrt[4]{1 + 0}} = 1$ and

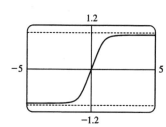

$\lim\limits_{x \to -\infty} \dfrac{x}{\sqrt[4]{x^4 + 1}} = \lim\limits_{x \to -\infty} \dfrac{1}{-\sqrt[4]{1 + (1/x^4)}} = \dfrac{1}{-\sqrt[4]{1 + 0}} = -1$, so

$y = \pm 1$ are horizontal asymptotes. There is no vertical asymptote.

41. Let's look for a rational function.

(1) $\lim\limits_{x \to \pm\infty} f(x) = 0 \Rightarrow$ degree of numerator < degree of denominator

(2) $\lim\limits_{x \to 0} f(x) = -\infty \Rightarrow$ there is a factor of x^2 in the denominator (not just x, since that would produce a

sign change at $x = 0$), and the function is negative near $x = 0$.

(3) $\lim\limits_{x \to 3^-} f(x) = \infty$ and $\lim\limits_{x \to 3^+} f(x) = -\infty \Rightarrow$ vertical asymptote at $x = 3$; there is a factor of $(x - 3)$ in

the denominator.

(4) $f(2) = 0 \Rightarrow$ 2 is an x-intercept; there is at least one factor of $(x - 2)$ in the numerator.

Combining all of this information, and putting in a negative sign to give us the desired left- and right-hand limits,

gives us $f(x) = \dfrac{2 - x}{x^2(x - 3)}$ as one possibility.

43. $y = f(x) = x^2(x - 2)(1 - x)$. The y-intercept is $f(0) = 0$, and the x-intercepts

occur when $y = 0 \Rightarrow x = 0, 1, 2$. Notice that, since x^2 is always positive, the

graph does not cross the x-axis at 0, but does cross the x-axis at 1 and 2.

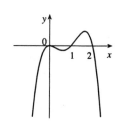

$\lim\limits_{x \to \infty} x^2(x - 2)(1 - x) = -\infty$, since the first two factors are large positive and the

third large negative when x is large positive. $\lim\limits_{x \to -\infty} x^2(x - 2)(1 - x) = -\infty$ because

the first and third factors are large positive and the second large negative as $x \to -\infty$.

45. $y = f(x) = (x+4)^5 (x-3)^4$. The y-intercept is $f(0) = 4^5 (-3)^4 = 82{,}944$.

The x-intercepts occur when $y = 0 \implies x = -4, 3$. Notice that the graph does

not cross the x-axis at 3 because $(x-3)^4$ is always positive, but does cross the

x-axis at -4. $\lim\limits_{x \to \infty} (x+4)^5 (x-3)^4 = \infty$ since both factors are large positive

when x is large positive. $\lim\limits_{x \to -\infty} (x+4)^5 (x-3)^4 = -\infty$ since the first factor is

large negative and the second factor is large positive when x is large negative.

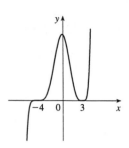

47. Since $0 \le \sin^2 x \le 1$, we have $0 \le \dfrac{\sin^2 x}{x^2} \le \dfrac{1}{x^2}$ for all $x \ne 0$. Now $\lim\limits_{x \to \infty} 0 = 0$ and $\lim\limits_{x \to \infty} \dfrac{1}{x^2} = 0$, so by the

Squeeze Theorem, $\lim\limits_{x \to \infty} \dfrac{\sin^2 x}{x^2} = 0$.

49. Divide numerator and denominator by the highest power of x in $Q(x)$.

(a) If $\deg P < \deg Q$, then numerator $\to 0$ but denominator doesn't. So $\lim\limits_{x \to \infty} [P(x)/Q(x)] = 0$.

(b) If $\deg P > \deg Q$, then numerator $\to \pm\infty$ but denominator doesn't, so $\lim\limits_{x \to \infty} [P(x)/Q(x)] = \pm\infty$ (depending

on the ratio of the leading coefficients of P and Q).

51. $\lim\limits_{x \to \infty} \dfrac{4x-1}{x} = \lim\limits_{x \to \infty} \left(4 - \dfrac{1}{x}\right) = 4$, and $\lim\limits_{x \to \infty} \dfrac{4x^2+3x}{x^2} = \lim\limits_{x \to \infty} \left(4 + \dfrac{3}{x}\right) = 4$. Therefore, by the Squeeze

Theorem, $\lim\limits_{x \to \infty} f(x) = 4$.

53. (a) $\lim\limits_{t \to \infty} v(t) = \lim\limits_{t \to \infty} v^* \left(1 - e^{-gt/v^*}\right) = v^* (1 - 0) = v^*$

(b) We graph $v(t) = 1 - e^{-9.8t}$ and $v(t) = 0.99 v^*$, or in this case,

$v(t) = 0.99$. Using an intersect feature or zooming in on the point

of intersection, we find that $t \approx 0.47$ s.

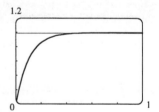

55. $\left| \dfrac{6x^2 + 5x - 3}{2x^2 - 1} - 3 \right| < 0.2 \iff 2.8 < \dfrac{6x^2 + 5x - 3}{2x^2 - 1} < 3.2$. So we

graph the three parts of this inequality on the same screen, and find

that the curve $y = \dfrac{6x^2 + 5x - 3}{2x^2 - 1}$ seems to lie between the lines

$y = 2.8$ and $y = 3.2$ whenever $x > 12.8$. So we can choose $N = 13$

(or any larger number), so that the inequality holds whenever $x \ge N$.

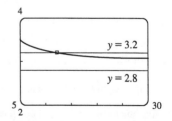

57. For $\varepsilon = 0.5$, we need to find N such that $\left| \dfrac{\sqrt{4x^2 + 1}}{x + 1} - (-2) \right| < 0.5$

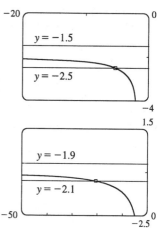

$\Leftrightarrow \ -2.5 < \dfrac{\sqrt{4x^2 + 1}}{x + 1} < -1.5$ whenever $x \leq N$. We graph the

three parts of this inequality on the same screen, and see that the
inequality holds for $x \leq -6$. So we choose $N = -6$ (or any smaller
number).

For $\varepsilon = 0.1$, we need $-2.1 < \dfrac{\sqrt{4x^2 + 1}}{x + 1} < -1.9$ whenever

$x \leq N$. From the graph, it seems that this inequality holds for
$x \leq -22$. So we choose any $N = -22$ (or any smaller number).

59. (a) $1/x^2 < 0.0001 \ \Leftrightarrow \ x^2 > 1/0.0001 = 10{,}000 \ \Leftrightarrow \ x > 100 \quad (x > 0)$

(b) If $\varepsilon > 0$ is given, then $1/x^2 < \varepsilon \ \Leftrightarrow \ x^2 > 1/\varepsilon \ \Leftrightarrow \ x > 1/\sqrt{\varepsilon}$. Let $N = 1/\sqrt{\varepsilon}$.

Then $x > N \ \Rightarrow \ x > \dfrac{1}{\sqrt{\varepsilon}} \ \Rightarrow \ \left| \dfrac{1}{x^2} - 0 \right| = \dfrac{1}{x^2} < \varepsilon$, so $\lim\limits_{x \to \infty} \dfrac{1}{x^2} = 0$.

61. For $x < 0$, $|1/x - 0| = -1/x$. If $\varepsilon > 0$ is given, then $-1/x < \varepsilon \ \Leftrightarrow \ x < -1/\varepsilon$.

Take $N = -1/\varepsilon$. Then $x < N \ \Rightarrow \ x < -1/\varepsilon \ \Rightarrow \ |(1/x) - 0| = -1/x < \varepsilon$, so $\lim\limits_{x \to -\infty} (1/x) = 0$.

63. Given $M > 0$, we need $N > 0$ such that $x > N \ \Rightarrow \ e^x > M$. Now $e^x > M \ \Leftrightarrow \ x > \ln M$, so take

$N = \max(1, \ln M)$. (This ensures that $N > 0$.) Then $x > N = \max(1, \ln M) \ \Rightarrow \ e^x > \max(e, M) \geq M$, so
$\lim\limits_{x \to \infty} e^x = \infty$.

65. Suppose that $\lim\limits_{x \to \infty} f(x) = L$. Then for every $\varepsilon > 0$ there is a corresponding positive number N such that

$|f(x) - L| < \varepsilon$ whenever $x > N$. If $t = 1/x$, then $x > N \ \Leftrightarrow \ 0 < t < 1/N$. Thus, for every $\varepsilon > 0$ there is a
corresponding $\delta > 0$ (namely $1/N$) such that $|f(1/t) - L| < \varepsilon$ whenever $t < \delta$. This proves that
$\lim\limits_{t \to 0^+} f(1/t) = L = \lim\limits_{x \to \infty} f(x)$.

Now suppose that $\lim\limits_{x \to -\infty} f(x) = L$. Then for every $\varepsilon > 0$ there is a corresponding negative number N such

that $|f(x) - L| < \varepsilon$ whenever $x < N$. If $t = 1/x$, then $x < N \ \Leftrightarrow \ 1/N < 1/x < 0$. Thus, for every $\varepsilon > 0$
there is a corresponding $\delta > 0$ (namely $-1/N$) such that $|f(1/t) - L| < \varepsilon$ whenever $-\delta < t < 0$. This proves
that $\lim\limits_{t \to 0^-} f(1/t) = L = \lim\limits_{x \to -\infty} f(x)$.

2.7 Tangents, Velocities, and Other Rates of Change

1. (a) This is just the slope of the line through two points: $m_{PQ} = \dfrac{\Delta y}{\Delta x} = \dfrac{f(x) - f(3)}{x - 3}$.

(b) This is the limit of the slope of the secant line PQ as Q approaches P: $m = \lim\limits_{x \to 3} \dfrac{f(x) - f(3)}{x - 3}$.

3. The slope at D is the largest positive slope, followed by the positive slope at E. The slope at C is zero. The slope at B is steeper than at A (both are negative). In decreasing order, we have the slopes at: D, E, C, A, B.

5. (a) (i) $m = \lim\limits_{x \to -3} \dfrac{f(x) - f(-3)}{x - (-3)} = \lim\limits_{x \to -3} \dfrac{(x^2 + 2x) - (3)}{x - (-3)} = \lim\limits_{x \to -3} \dfrac{(x + 3)(x - 1)}{x + 3} = \lim\limits_{x \to -3} (x - 1)$

$$= -4$$

(ii) $m = \lim\limits_{h \to 0} \dfrac{f(-3 + h) - f(-3)}{h} = \lim\limits_{h \to 0} \dfrac{[(-3 + h)^2 + 2(-3 + h)] - (3)}{h}$

$$= \lim\limits_{h \to 0} \dfrac{9 - 6h + h^2 - 6 + 2h - 3}{h} = \lim\limits_{h \to 0} \dfrac{h(h - 4)}{h} = \lim\limits_{h \to 0} (h - 4) = -4$$

(b) The equation of the tangent line is

$y - 3 = -4(x + 3)$ or $y = -4x - 9$.

(c)

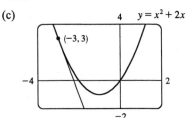

7. Using (1),

$m = \lim\limits_{x \to a} \dfrac{f(x) - f(a)}{x - a} = \lim\limits_{x \to -2} \dfrac{(1 - 2x - 3x^2) - (-7)}{x - (-2)} = \lim\limits_{x \to -2} \dfrac{-3x^2 - 2x + 8}{x + 2} = \lim\limits_{x \to -2} \dfrac{(-3x + 4)(x + 2)}{x + 2}$

$$= \lim\limits_{x \to -2} (-3x + 4) = 10$$

Thus, an equation of the tangent is $y + 7 = 10(x + 2)$ or $y = 10x + 13$.

Alternate Solution: Using (2),

$m = \lim\limits_{h \to 0} \dfrac{f(a + h) - f(a)}{h} = \lim\limits_{h \to 0} \dfrac{f(-2 + h) - f(-2)}{h} = \lim\limits_{h \to 0} \dfrac{[1 - 2(-2 + h) - 3(-2 + h)^2] - (-7)}{h}$

$$= \lim\limits_{h \to 0} \dfrac{(-3h^2 + 10h - 7) + 7}{h} = \lim\limits_{h \to 0} \dfrac{h(-3h + 10)}{h} = \lim\limits_{h \to 0} (-3h + 10) = 10$$

9. Using (1), $m = \lim\limits_{x \to -2} \dfrac{1/x^2 - \frac{1}{4}}{x - (-2)} = \lim\limits_{x \to -2} \dfrac{4 - x^2}{4x^2(x + 2)} = \lim\limits_{x \to -2} \dfrac{(2 - x)(2 + x)}{4x^2(x + 2)} = \lim\limits_{x \to -2} \dfrac{2 - x}{4x^2} = \dfrac{1}{4}$.

Thus, an equation of the tangent line is $y - \frac{1}{4} = \frac{1}{4}(x + 2) \;\Rightarrow\; y = \frac{1}{4}x + \frac{3}{4}$.

11. (a) $m = \lim\limits_{x \to a} \dfrac{2/(x + 3) - 2/(a + 3)}{x - a} = \lim\limits_{x \to a} \dfrac{2(a - x)}{(x - a)(x + 3)(a + 3)} = \lim\limits_{x \to a} \dfrac{-2}{(x + 3)(a + 3)} = \dfrac{-2}{(a + 3)^2}$

(b) (i) $a = -1 \;\Rightarrow\; m = \dfrac{-2}{(-1 + 3)^2} = -\dfrac{1}{2}$

(ii) $a = 0 \;\Rightarrow\; m = \dfrac{-2}{(0 + 3)^2} = -\dfrac{2}{9}$

(iii) $a = 1 \;\Rightarrow\; m = \dfrac{-2}{(1 + 3)^2} = -\dfrac{1}{8}$

13. (a) Using (1),

$$m = \lim_{x \to a} \frac{(x^3 - 4x + 1) - (a^3 - 4a + 1)}{x - a} = \lim_{x \to a} \frac{(x^3 - a^3) - 4(x - a)}{x - a}$$

$$= \lim_{x \to a} \frac{(x - a)(x^2 + ax + a^2) - 4(x - a)}{x - a} = \lim_{x \to a} \left(x^2 + ax + a^2 - 4\right) = 3a^2 - 4$$

(b) At $(1, -2)$: $m = 3(1)^2 - 4 = -1$, so an equation of the tangent line (c)
is $y - (-2) = -1(x - 1)$ ⇔ $y = -x - 1$. At $(2, 1)$:
$m = 3(2)^2 - 4 = 8$, so an equation of the tangent line is
$y - 1 = 8(x - 2)$ ⇔ $y = 8x - 15$.

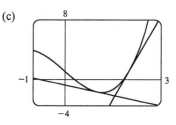

15. (a) Since the slope of the tangent at $s = 0$ is 0, the car's initial velocity was 0.

(b) The slope of the tangent is greater at C than at B, so the car was going faster at C.

(c) Near A, the tangent lines are becoming steeper as x increases, so the velocity was increasing, so the car was speeding up. Near B, the tangent lines are becoming less steep, so the car was slowing down. The steepest tangent near C is the one at C, so at C the car had just finished speeding up, and was about to start slowing down.

(d) Between D and E, the slope of the tangent is 0, so the car did not move during that time.

17. Let $s(t) = 40t - 16t^2$.

$$v(2) = \lim_{t \to 2} \frac{s(t) - s(2)}{t - 2} = \lim_{t \to 2} \frac{(40t - 16t^2) - 16}{t - 2} = \lim_{t \to 2} \frac{-8(t - 2)(2t - 1)}{t - 2}$$

$$= -8 \lim_{t \to 2} (2t - 1) = -24$$

Thus, the instantaneous velocity when $t = 2$ is -24 ft/s.

19. $v(a) = \lim_{h \to 0} \dfrac{s(a + h) - s(a)}{h} = \lim_{h \to 0} \dfrac{4(a + h)^3 + 6(a + h) + 2 - (4a^3 + 6a + 2)}{h}$

$$= \lim_{h \to 0} \frac{4a^3 + 12a^2h + 12ah^2 + 4h^3 + 6a + 6h + 2 - 4a^3 - 6a - 2}{h}$$

$$= \lim_{h \to 0} \frac{12a^2h + 12ah^2 + 4h^3 + 6h}{h} = \lim_{h \to 0} \left(12a^2 + 12ah + 4h^2 + 6\right)$$

$$= (12a^2 + 6) \text{ m/s}$$

So $v(1) = 12(1)^2 + 6 = 18$ m/s, $v(2) = 12(2)^2 + 6 = 54$ m/s, and $v(3) = 12(3)^2 + 6 = 114$ m/s.

21. The sketch shows the graph for a room temperature of $72°$ and a refrigerator temperature of $38°$. The initial rate of change is greater in magnitude than the rate of change after an hour.

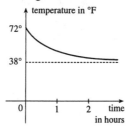

23. (a) (i) [8, 11]: $\dfrac{7.9 - 11.5}{3} = -1.2°/\text{h}$

(ii) [8, 10]: $\dfrac{9.0 - 11.5}{2} = -1.25°/\text{h}$

(iii) [8, 9]: $\dfrac{10.2 - 11.5}{1} = -1.3°/\text{h}$

(b) In the figure, we estimate A to be
$(18, 15.5)$ and B as $(23, 6)$. So the

slope is $\dfrac{6 - 15.5}{23 - 18} = -1.9°/\text{h}$ at

8:00 P.M.

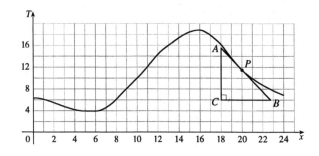

25. (a) (i) $\dfrac{\Delta C}{\Delta x} = \dfrac{C(105) - C(100)}{105 - 100} = \dfrac{6601.25 - 6500}{5} = 20.25/\text{unit}.$

(ii) $\dfrac{\Delta C}{\Delta x} = \dfrac{C(101) - C(100)}{101 - 100} = \dfrac{6520.05 - 6500}{1} = 20.05/\text{unit}.$

(b) $\dfrac{C(100 + h) - C(100)}{h} = \dfrac{\left[5000 + 10(100 + h) + 0.05(100 + h)^2\right] - 6500}{h} = \dfrac{20h + 0.05h^2}{h}$

$= 20 + 0.05h, \; h \neq 0$

So the instantaneous rate of change is $\displaystyle\lim_{h \to 0} \dfrac{C(100 + h) - C(100)}{h} = \lim_{h \to 0} (20 + 0.05h) = \$20/\text{unit}.$

2.8 Derivatives

1.

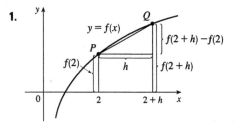

The line from P to Q is the line that has slope

$$\dfrac{f(2 + h) - f(2)}{h}.$$

3. $g'(0)$ is the only negative value. The slope at $x = 4$ is smaller than the slope at $x = 2$ and both are smaller than the slope at $x = -2$. Thus, $g'(0) < 0 < g'(4) < g'(2) < g'(-2)$.

5.

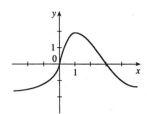

7. $f'(2) = \lim\limits_{h\to0} \dfrac{f(2+h) - f(2)}{h} = \lim\limits_{h\to0} \dfrac{[3(2+h)^2 - 5(2+h)] - [3(2)^2 - 5(2)]}{h}$

$= \lim\limits_{h\to0} \dfrac{(12 + 12h + 3h^2 - 10 - 5h) - (2)}{h} = \lim\limits_{h\to0} \dfrac{3h^2 + 7h}{h} = \lim\limits_{h\to0} (3h + 7) = 7$

So an equation of the tangent line at $(2, 2)$ is $y - 2 = 7(x - 2)$ or $y = 7x - 12$.

9. (a) $F'(1) = \lim\limits_{x\to1} \dfrac{F(x) - F(1)}{x - 1} = \lim\limits_{x\to1} \dfrac{(x^3 - 5x + 1) - (-3)}{x - 1} = \lim\limits_{x\to1} \dfrac{x^3 - 5x + 4}{x - 1}$

$= \lim\limits_{x\to1} \dfrac{(x - 1)(x^2 + x - 4)}{x - 1} = \lim\limits_{x\to1} (x^2 + x - 4) = -2$

So an equation of the tangent line at $(1, -3)$ is

$y - (-3) = -2(x - 1) \Leftrightarrow y = -2x - 1$.

Note: Instead of using Equation 3 to compute $F'(1)$, we could have used Equation 1.

(b)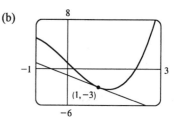

11. (a) $f'(1) = \lim\limits_{h\to0} \dfrac{f(1+h) - f(1)}{h} = \lim\limits_{h\to0} \dfrac{3^{1+h} - 3^1}{h}$.

So let $F(h) = \dfrac{3^{1+h} - 3}{h}$. We calculate:

(b)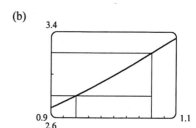

h	$F(h)$
0.1	3.484
0.01	3.314
0.001	3.298
0.0001	3.296
−0.1	3.121
−0.01	3.278
−0.001	3.294
−0.0001	3.296

From the graph, we estimate that the slope of the tangent is about

$\dfrac{3.2 - 2.8}{1.06 - 0.94} = \dfrac{0.4}{0.12} \approx 3.3.$

We estimate that $f'(1) \approx 3.296$.

13. $f'(a) = \lim\limits_{h\to0} \dfrac{f(a+h) - f(a)}{h} = \lim\limits_{h\to0} \dfrac{[1 + (a+h) - 2(a+h)^2] - (1 + a - 2a^2)}{h}$

$= \lim\limits_{h\to0} \dfrac{h - 4ah - 2h^2}{h} = \lim\limits_{h\to0} (1 - 4a - 2h) = 1 - 4a$

15. $f'(a) = \lim\limits_{h\to0} \dfrac{f(a+h) - f(a)}{h} = \lim\limits_{h\to0} \dfrac{\dfrac{a+h}{2(a+h)-1} - \dfrac{a}{2a-1}}{h}$

$= \lim\limits_{h\to0} \dfrac{(a+h)(2a-1) - a(2a+2h-1)}{h(2a+2h-1)(2a-1)} = \lim\limits_{h\to0} \dfrac{-h}{h(2a+2h-1)(2a-1)}$

$= \lim\limits_{h\to0} \dfrac{-1}{(2a+2h-1)(2a-1)} = -\dfrac{1}{(2a-1)^2}$

17. $f'(a) = \lim\limits_{h \to 0} \dfrac{f(a+h) - f(a)}{h} = \lim\limits_{h \to 0} \dfrac{\dfrac{2}{\sqrt{3 - (a+h)}} - \dfrac{2}{\sqrt{3-a}}}{h} = \lim\limits_{h \to 0} \dfrac{2\left(\sqrt{3-a} - \sqrt{3-a-h}\right)}{h\sqrt{3-a} \cdot h\sqrt{3-a}}$

$\qquad = \lim\limits_{h \to 0} \dfrac{2\left(\sqrt{3-a} - \sqrt{3-a-h}\right)}{h\sqrt{3-a} - h\sqrt{3-a}} \cdot \dfrac{\sqrt{3-a} + \sqrt{3-a-h}}{\sqrt{3-a} + \sqrt{3-a-h}}$

$\qquad = \lim\limits_{h \to 0} \dfrac{2\,[3-a-(3-a-h)]}{h\sqrt{3-a} - h\sqrt{3-a}\left(\sqrt{3-a} + \sqrt{3-a-h}\right)}$

$\qquad = \lim\limits_{h \to 0} \dfrac{2}{\sqrt{3-a} - h\sqrt{3-a}\left(\sqrt{3-a} + \sqrt{3-a-h}\right)}$

$\qquad = \dfrac{2}{\sqrt{3-a}\sqrt{3-a}\left(2\sqrt{3-a}\right)} = \dfrac{1}{(3-a)^{3/2}}$

Note that the answers to Exercises 19–23 are not unique.

19. By Equation 1, $\lim\limits_{h \to 0} \dfrac{\sqrt{1+h} - 1}{h} = f'(1)$, where $f(x) = \sqrt{x}$.

$\quad$ *Or:* $f'(0)$, where $f(x) = \sqrt{1+x}$

21. By Equation 3, $\lim\limits_{x \to 1} \dfrac{x^9 - 1}{x - 1} = f'(1)$, where $f(x) = x^9$.

23. By Equation 1, $\lim\limits_{t \to 0} \dfrac{\sin\left(\frac{\pi}{2} + t\right) - 1}{t} = f'\left(\frac{\pi}{2}\right)$, where $f(x) = \sin x$.

25. $v(2) = f'(2) = \lim\limits_{h \to 0} \dfrac{f(2+h) - f(2)}{h} = \lim\limits_{h \to 0} \dfrac{[(2+h)^2 - 6(2+h) - 5] - [2^2 - 6(2) - 5]}{h}$

$\qquad = \lim\limits_{h \to 0} \dfrac{(4 + 4h + h^2 - 12 - 6h - 5) - (-13)}{h} = \lim\limits_{h \to 0} \dfrac{h^2 - 2h}{h} = \lim\limits_{h \to 0} (h - 2) = -2 \text{ m/s}$

27. (a) $f'(x)$ is the rate of change of the production cost with respect to the number of ounces of gold produced. Its units are dollars per ounce.

$\quad$ (b) After 800 ounces of gold have been produced, the rate at which the production cost is increasing is $17/ounce. So the cost of producing the 800th (or 801st) ounce is about $17.

$\quad$ (c) In the short term, the values of $f'(x)$ will decrease because more efficient use is made of start-up costs as x increases. But eventually $f'(x)$ might increase due to large-scale operations.

29. (a) $f'(v)$ is the rate at which the fuel consumption is changing with respect to the speed. Its units are (gal/h) / (mi/h).

$\quad$ (b) The fuel consumption is decreasing by 0.05 (gal/h) / (mi/h) as the car's speed reaches 20 mi/h. So if you increase your speed to 21 mi/h, you could expect to decrease your fuel consumption by about 0.05 (gal/h) / (mi/h).

31. $C'(1980)$ is the rate of change of U.S. cash per capita in circulation with respect to time. To estimate the value of $C'(1980)$, we will average the difference quotients obtained using the years 1970 and 1990.

$\quad$ Let $A = \dfrac{C(1970) - C(1980)}{1970 - 1980} = \dfrac{265 - 571}{-10} = 30.6$ and $B = \dfrac{C(1990) - C(1980)}{1990 - 1980} = \dfrac{1063 - 571}{10} = 49.2$.

$\quad$ Then $C'(1980) = \lim\limits_{t \to 1980} \dfrac{C(t) - C(1980)}{t - 1980} \approx \dfrac{A + B}{2} = 39.9$ dollars per year.

33. Since $f(x) = x \sin(1/x)$ when $x \neq 0$ and $f(0) = 0$, we have

$$f'(0) = \lim_{h \to 0} \frac{f(0+h) - f(0)}{h} = \lim_{h \to 0} \frac{h \sin(1/h) - 0}{h} = \lim_{h \to 0} \sin(1/h).$$ This limit does not exist since $\sin(1/h)$

takes the values -1 and 1 on any interval containing 0. (Compare with Example 4 in Section 2.2.)

2.9 The Derivative as a Function

1. *Note:* Your answers may vary depending on your estimates. From the graph of f, it appears that

(a) $f'(1) \approx -2$ (b) $f'(2) \approx 0.8$

(c) $f'(3) \approx -1$ (d) $f'(4) \approx -0.5$

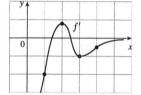

3. It appears that f is an odd function, so f' will be an even function — that is, $f'(-a) = f'(a)$.

(a) $f'(-3) \approx 1.5$ (b) $f'(-2) \approx 1$

(c) $f'(-1) \approx 0$ (d) $f'(0) \approx -4$

(e) $f'(1) \approx 0$ (f) $f'(2) \approx 1$

(g) $f'(3) \approx 1.5$

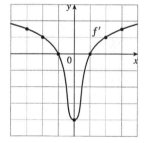

5.

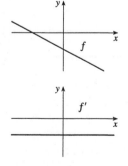

7.

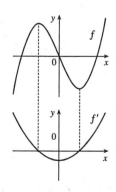

9.

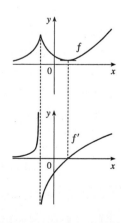

11.

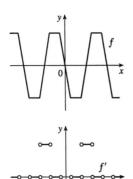

13.

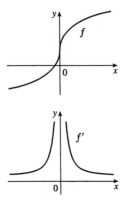

15.

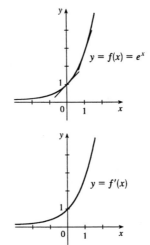

The slope at 0 appears to be 1 and the slope at 1 appears to be 2.7. As x decreases, the slope gets closer to 0. Since the graphs are so similar, we might guess that $f'(x) = e^x$.

17. (a) By zooming in, we estimate that $f'(0) = 0$, $f'\left(\frac{1}{2}\right) = 1$, $f'(1) = 2$, and $f'(2) = 4$.

    (b) By symmetry, $f'(-x) = -f'(x)$. So $f'\left(-\frac{1}{2}\right) = -1$, $f'(-1) = -2$, and $f'(-2) = -4$.

    (c) It appears that $f'(x)$ is twice the value of x, so we guess that $f'(x) = 2x$.

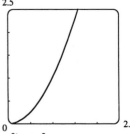

    (d) $f'(x) = \lim\limits_{h \to 0} \dfrac{f(x+h) - f(x)}{h} = \lim\limits_{h \to 0} \dfrac{(x+h)^2 - x^2}{h} = \lim\limits_{h \to 0} \dfrac{(x^2 + 2hx + h^2) - x^2}{h}$

        $= \lim\limits_{h \to 0} \dfrac{h(2x+h)}{h} = \lim\limits_{h \to 0} (2x + h) = 2x$

19. $f'(x) = \lim\limits_{h \to 0} \dfrac{f(x+h) - f(x)}{h} = \lim\limits_{h \to 0} \dfrac{[5(x+h) + 3] - (5x + 3)}{h} = \lim\limits_{h \to 0} \dfrac{5h}{h} = \lim\limits_{h \to 0} 5 = 5$

Domain of f = domain of $f' = \mathbb{R}$.

21. $f'(x) = \lim\limits_{h \to 0} \dfrac{f(x+h) - f(x)}{h} = \lim\limits_{h \to 0} \dfrac{\left[(x+h)^3 - (x+h)^2 + 2(x+h)\right] - (x^3 - x^2 + 2x)}{h}$

$= \lim\limits_{h \to 0} \dfrac{3x^2h + 3xh^2 + h^3 - 2xh - h^2 + 2h}{h} = \lim\limits_{h \to 0} \left(3x^2 + 3xh + h^2 - 2x - h + 2\right)$

$= 3x^2 - 2x + 2$

Domain of f = domain of f' = $\mathbb{R}$.

23. $g'(x) = \lim\limits_{h \to 0} \dfrac{g(x+h) - g(x)}{h} = \lim\limits_{h \to 0} \dfrac{\sqrt{1 + 2(x+h)} - \sqrt{1+2x}}{h} \left[\dfrac{\sqrt{1 + 2(x+h)} + \sqrt{1+2x}}{\sqrt{1 + 2(x+h)} + \sqrt{1+2x}} \right]$

$= \lim\limits_{h \to 0} \dfrac{(1 + 2x + 2h) - (1 + 2x)}{h\left[\sqrt{1 + 2(x+h)} + \sqrt{1+2x}\right]} = \lim\limits_{h \to 0} \dfrac{2}{\sqrt{1 + 2(x+h)} + \sqrt{1+2x}} = \dfrac{1}{\sqrt{1+2x}}$

Domain of $g = \left[-\frac{1}{2}, \infty\right)$, domain of $g' = \left(-\frac{1}{2}, \infty\right)$.

25. $G'(x) = \lim\limits_{h \to 0} \dfrac{G(x+h) - G(x)}{h} = \lim\limits_{h \to 0} \dfrac{\dfrac{4 - 3(x+h)}{2 + (x+h)} - \dfrac{4 - 3x}{2 + x}}{h}$

$= \lim\limits_{h \to 0} \dfrac{(4 - 3x - 3h)(2+x) - (4 - 3x)(2 + x + h)}{h(2 + x + h)(2 + x)} = \lim\limits_{h \to 0} \dfrac{-10h}{h(2 + x + h)(2 + x)}$

$= \lim\limits_{h \to 0} \dfrac{-10}{(2 + x + h)(2 + x)} = \dfrac{-10}{(2+x)^2}$

Domain of G = domain of G' = $\{x \mid x \neq -2\}$.

27. $f'(x) = \lim\limits_{h \to 0} \dfrac{f(x+h) - f(x)}{h} = \lim\limits_{h \to 0} \dfrac{(x+h)^4 - x^4}{h} = \lim\limits_{h \to 0} \dfrac{4x^3h + 6x^2h^2 + 4xh^3 + h^4}{h}$

$= \lim\limits_{h \to 0} \left(4x^3 + 6x^2h + 4xh^2 + h^3\right) = 4x^3$

Domain of f = domain of f' = $\mathbb{R}$.

29. (a) $f'(x) = \lim\limits_{h \to 0} \dfrac{f(x+h) - f(x)}{h} = \lim\limits_{h \to 0} \dfrac{\left[x + h - \left(\dfrac{2}{x+h}\right)\right] - \left[x - \left(\dfrac{2}{x}\right)\right]}{h}$

$= \lim\limits_{h \to 0} \left[\dfrac{h + \dfrac{2}{x} - \dfrac{2}{(x+h)}}{h}\right] = \lim\limits_{h \to 0} \left[1 + \dfrac{2(x+h) - 2x}{(h)(x)(x+h)}\right] = \lim\limits_{h \to 0} \left[1 + \dfrac{2}{x(x+h)}\right]$

$= 1 + 2x^{-2}$

(b) Notice that when f has steep tangent lines, $f'(x)$ is very large.
When f is flatter, $f'(x)$ is smaller.

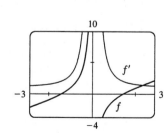

31. (a) $U'(t)$ is the rate at which the unemployment rate is changing with respect to time. Its units are percent per year.

(b) To find $U'(t)$, we use $\lim\limits_{h \to 0} \dfrac{U(t+h) - U(t)}{h} \approx \dfrac{U(t+h) - U(t)}{h}$ for small values of h.

For 1988: $U'(1988) = \dfrac{U(1989) - U(1988)}{1989 - 1988} = \dfrac{5.3 - 5.5}{1} = -0.20$

For 1989: We estimate $U'(1989)$ both using $h = -1$ and using $h = 1$, and then average the two results to obtain a final estimate.

$h = -1 \quad \Rightarrow \quad U'(1989) \approx \dfrac{U(1988) - U(1989)}{1988 - 1989} = \dfrac{5.5 - 5.3}{-1} = -0.20;$

$h = 1 \quad \Rightarrow \quad U'(1989) \approx \dfrac{U(1990) - U(1989)}{1990 - 1989} = \dfrac{5.6 - 5.3}{1} = 0.30.$

So we estimate that $U'(1989) \approx \frac{1}{2}(-0.20 + 0.30) = 0.05.$

t	1988	1989	1990	1991	1992	1993	1994	1995	1996	1997
$U'(t)$	−0.20	0.05	0.75	0.95	0.05	−0.70	−0.65	−0.35	−0.35	−0.50

33. f is not differentiable at $x = -1$ or at $x = 11$ because the graph has vertical tangents at those points; at $x = 4$, because there is a discontinuity there; and at $x = 8$, because the graph has a corner there.

35. As we zoom in toward $(-1, 0)$, the curve appears more and more like a straight line, so f is differentiable at $x = -1$. But no matter how much we zoom in toward the origin, the curve doesn't straighten out — we can't eliminate the sharp point (a cusp). So f is not differentiable at $x = 0$.

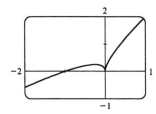

37. (a) $f'(a) = \lim\limits_{x \to a} \dfrac{f(x) - f(a)}{x - a} = \lim\limits_{x \to a} \dfrac{x^{1/3} - a^{1/3}}{x - a} = \lim\limits_{x \to a} \dfrac{x^{1/3} - a^{1/3}}{\left(x^{1/3} - a^{1/3}\right)\left(x^{2/3} + x^{1/3}a^{1/3} + a^{2/3}\right)}$

$= \lim\limits_{x \to a} \dfrac{1}{x^{2/3} + x^{1/3}a^{1/3} + a^{2/3}} = \lim\limits_{x \to a} \dfrac{1}{3x^{2/3}} = \dfrac{1}{3a^{2/3}}$

(b) $f'(0) = \lim\limits_{h \to 0} \dfrac{f(0+h) - f(0)}{h} = \lim\limits_{h \to 0} \dfrac{\sqrt[3]{h} - 0}{h} = \lim\limits_{h \to 0} \dfrac{1}{h^{2/3}}.$ This limit does not exist, and therefore $f'(0)$ does not exist.

(c) $\lim\limits_{x \to 0} |f'(x)| = \lim\limits_{x \to 0} \dfrac{1}{3x^{2/3}} = \infty$ and f is continuous at $x = 0$ (root function), so f has a vertical tangent at $x = 0$.

39. $f(x) = |x - 6| = \begin{cases} 6 - x & \text{if } x < 6 \\ x - 6 & \text{if } x \ge 6 \end{cases}$

$\lim\limits_{x \to 6^+} \dfrac{f(x) - f(6)}{x - 6} = \lim\limits_{x \to 6^+} \dfrac{|x - 6| - 0}{x - 6} = \lim\limits_{x \to 6^+} \dfrac{x - 6}{x - 6} = \lim\limits_{x \to 6^+} 1 = 1.$

But $\lim\limits_{x \to 6^-} \dfrac{f(x) - f(6)}{x - 6} = \lim\limits_{x \to 6^-} \dfrac{|x - 6| - 0}{x - 6} = \lim\limits_{x \to 6^-} \dfrac{6 - x}{x - 6} = \lim\limits_{x \to 6^-} (-1) = -1.$

So $f'(6) = \lim\limits_{x \to 6} \dfrac{f(x) - f(6)}{x - 6}$ does not exist. However, $f'(x) = \begin{cases} -1 & \text{if } x < 6 \\ 1 & \text{if } x > 6 \end{cases}$

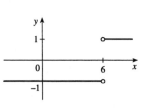

41. (a) $f(x) = x|x| = \begin{cases} x^2 & \text{if } x \geq 0 \\ -x^2 & \text{if } x < 0 \end{cases}$

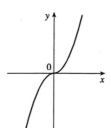

(c) From part (b), we have $f'(x) = \begin{cases} 2x & \text{if } x \geq 0 \\ -2x & \text{if } x < 0 \end{cases} = 2|x|.$

(b) Since $f(x) = x^2$ for $x \geq 0$, we have $f'(x) = 2x$ for $x > 0$. [See Exercise 2.9.17(d).] Similarly, since $f(x) = -x^2$ for $x < 0$, we have $f'(x) = -2x$ for $x < 0$. At $x = 0$, we have

$$f'(0) = \lim_{x \to 0} \frac{f(x) - f(0)}{x - 0} = \lim_{x \to 0} \frac{x|x|}{x} = \lim_{x \to 0} |x| = 0$$

So f is differentiable at 0. Thus, f is differentiable for all x.

43. (a) If f is even, then

$$f'(-x) = \lim_{h \to 0} \frac{f(-x+h) - f(-x)}{h} = \lim_{h \to 0} \frac{f(x-h) - f(x)}{h}$$

$$= -\lim_{h \to 0} \frac{f(x-h) - f(x)}{-h} \quad \text{[let } \Delta x = -h] \quad = -\lim_{\Delta x \to 0} \frac{f(x + \Delta x) - f(x)}{\Delta x} = -f'(x)$$

Therefore, f' is odd.

(b) If f is odd, then

$$f'(-x) = \lim_{h \to 0} \frac{f(-x+h) - f(-x)}{h} = \lim_{h \to 0} \frac{-f(x-h) + f(x)}{h}$$

$$= \lim_{h \to 0} \frac{f(x-h) - f(x)}{-h} \quad \text{[let } \Delta x = -h] \quad = \lim_{\Delta x \to 0} \frac{f(x + \Delta x) - f(x)}{\Delta x} = f'(x)$$

Therefore, f' is even.

45.

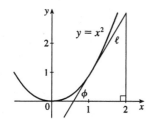

From the diagram, we see that the slope of the tangent is equal to $\tan \phi$, and also that $0 < \phi < \frac{\pi}{2}$. We know (see Exercise 17) that the derivative of $f(x) = x^2$ is $f'(x) = 2x$. So the slope of the tangent to the curve at the point $(1, 1)$ is 2. So ϕ is the angle between 0 and $\frac{\pi}{2}$ whose tangent is 2, that is, $\phi = \tan^{-1} 2 \approx 63°$.

≡2 Review

CONCEPT CHECK

1. (a) $\lim_{x \to a} f(x) = L$: See Definition 2.2.1 and Figures 1 and 2 in Section 2.2.

(b) $\lim_{x \to a^+} f(x) = L$: See the paragraph after Definition 2.2.2 and Figure 9(b) in Section 2.2.

(c) $\lim_{x \to a^-} f(x) = L$: See Definition 2.2.2 and Figure 9(a) in Section 2.2.

(d) $\lim_{x \to a} f(x) = \infty$: See Definition 2.2.4 and Figure 12 in Section 2.2.

(e) $\lim_{x \to a} f(x) = -\infty$: See Definition 2.2.5 and Figure 13 in Section 2.2.

2. (a)–(g) See the statements of Limit Laws 1–6 and 11 in Section 2.3.

3. See Theorem 3 in Section 2.3.

4. (a) See Definition 2.2.6 and Figures 12–14 in Section 2.2.

(b) See Definition 2.6.3 and Figures 2 and 3 in Section 2.6.

5. (a) $y = x^4$: No asymptote

(b) $y = \sin x$: No asymptote

(c) $y = \tan x$: Vertical asymptotes $x = \frac{\pi}{2} + \pi n$,

    n an integer

(d) $y = \tan^{-1} x$: Horizontal asymptotes $y = \pm\frac{\pi}{2}$

(e) $y = e^x$: Horizontal asymptote $y = 0$ ($\lim\limits_{x \to -\infty} e^x = 0$)

(f) $y = \ln x$: Vertical asymptote $x = 0$ ($\lim\limits_{x \to 0^+} \ln x = -\infty$)

(g) $y = 1/x$: Vertical asymptote $x = 0$, horizontal asymptote $y = 0$

(h) $y = \sqrt{x}$: No asymptote

6. (a) A function f is continuous at a number a if $f(x)$ gets closer to $f(a)$ as x gets close to a, that is,
$\lim\limits_{x \to a} f(x) = f(a)$.

(b) A function f is continuous on the interval $(-\infty, \infty)$ if f is continuous at every real number a. The graph of such a function has no breaks and every vertical line crosses it.

7. See Theorem 2.5.10.

8. See Definition 2.6.1.

9. See the paragraph containing Formula 3 in Section 2.6.

10. (a) The average rate of change of y with respect to x over the interval $[x_1, x_2]$ is $\dfrac{f(x_2) - f(x_1)}{x_2 - x_1}$.

(b) The instantaneous rate of change of y with respect to x at $x = x_1$ is $\lim\limits_{x_2 \to x_1} \dfrac{f(x_2) - f(x_1)}{x_2 - x_1}$.

11. See Definition 2.8.2. The pages following the definition discuss interpretations of $f'(a)$ as the slope of a tangent line to the graph of f at $x = a$ and as an instantaneous rate of change of $f(x)$ with respect to x when $x = a$.

12. (a) A function f is differentiable at a number a if its derivative f' exists at $x = a$, that is, if $f'(a)$ exists.

(b) See Theorem 2.9.4. This theorem also tells us that if f is *not* continuous at a, then f is *not* differentiable at a.

TRUE-FALSE QUIZ

1. False. Limit Law 2 applies only if the individual limits exist (these don't.)

3. True. Limit Law 5 applies.

5. False. Consider $\lim\limits_{x \to 5} \dfrac{x^2 - 5x}{x - 5}$ or $\lim\limits_{x \to 5} \dfrac{\sin(x - 5)}{x - 5}$. By Example 3 in Section 2.2, we know that the latter limit exists (and it is equal to 1).

7. True. A polynomial is continuous everywhere, so $\lim\limits_{x \to b} p(x)$ exists and is equal to $p(b)$.

9. False. Consider $f(x) = \begin{cases} 1/(x-1) & \text{if } x \neq 1 \\ 2 & \text{if } x = 1 \end{cases}$

11. True. Use Theorem 2.5.8 with $a = 2$, $b = 5$, and $g(x) = 4x^2 - 11$. Note that $f(4) = 3$ is not needed.

13. True, by the definition of a limit with $\varepsilon = 1$.

15. False. See the note after Theorem 4 in Section 2.9.

EXERCISES

1. (a) (i) $\lim\limits_{x \to 2^+} f(x) = 3$

(ii) $\lim\limits_{x \to -3^+} f(x) = 0$

(iii) $\lim\limits_{x \to -3} f(x)$ does not exist since the left and right limits are not equal. (The left limit is -2.)

(iv) $\lim\limits_{x \to 4} f(x) = 2$

(v) $\lim\limits_{x \to 0} f(x) = \infty$

(vi) $\lim\limits_{x \to 2^-} f(x) = -\infty$

(vii) $\lim\limits_{x \to \infty} f(x) = 4$

(viii) $\lim\limits_{x \to -\infty} f(x) = -1$

(b) The horizontal asymptotes are $y = 4$ and $y = -1$.

(c) The vertical asymptotes are $x = 0$ and $x = 2$.

(d) f is discontinuous at $x = -3, 0, 2$, and 4. The discontinuities are jump, infinite, infinite, and removable, respectively.

3. $\lim\limits_{x \to 0} \tan\left(x^2\right) = \tan 0 = 0$ because the tangent function is continuous at $x = 0$.

5. $\lim\limits_{h \to 0} \dfrac{(1+h)^2 - 1}{h} = \lim\limits_{h \to 0} \dfrac{1 + 2h + h^2 - 1}{h} = \lim\limits_{h \to 0} \dfrac{2h + h^2}{h} = \lim\limits_{h \to 0} (2 + h) = 2$

7. $\lim\limits_{x \to -1} \dfrac{x^2 - x - 2}{x^2 + 3x - 2} = \dfrac{\lim\limits_{x \to -1} \left(x^2 - x - 2\right)}{\lim\limits_{x \to -1} \left(x^2 + 3x - 2\right)} = \dfrac{(-1)^2 - (-1) - 2}{(-1)^2 + 3(-1) - 2} = \dfrac{0}{-4} = 0$

9. $\lim\limits_{t \to 6} \dfrac{17}{(t-6)^2} = \infty$ since $(t-6)^2 \to 0$ and $\dfrac{17}{(t-6)^2} > 0$.

11. $\lim\limits_{s \to 16} \dfrac{4 - \sqrt{s}}{s - 16} = \lim\limits_{s \to 16} \dfrac{4 - \sqrt{s}}{\left(\sqrt{s} + 4\right)\left(\sqrt{s} - 4\right)} = \lim\limits_{s \to 16} \dfrac{-1}{\sqrt{s} + 4} = \dfrac{-1}{\sqrt{16} + 4} = -\dfrac{1}{8}$

13. $\lim\limits_{x \to 8^-} \dfrac{|x - 8|}{x - 8} = \lim\limits_{x \to 8^-} \dfrac{-(x - 8)}{x - 8} = \lim\limits_{x \to 8^-} (-1) = -1$

15. $\lim\limits_{x \to 0} \dfrac{1 - \sqrt{1 - x^2}}{x} \cdot \dfrac{1 + \sqrt{1 - x^2}}{1 + \sqrt{1 - x^2}} = \lim\limits_{x \to 0} \dfrac{1 - \left(1 - x^2\right)}{x\left(1 + \sqrt{1 - x^2}\right)} = \lim\limits_{x \to 0} \dfrac{x^2}{x\left(1 + \sqrt{1 - x^2}\right)} = \lim\limits_{x \to 0} \dfrac{x}{1 + \sqrt{1 - x^2}} = 0$

17. $\lim\limits_{x \to \infty} \dfrac{1 + 2x - x^2}{1 - x + 2x^2} = \lim\limits_{x \to \infty} \dfrac{\left(1 + 2x - x^2\right)/x^2}{\left(1 - x + 2x^2\right)/x^2} = \lim\limits_{x \to \infty} \dfrac{1/x^2 + 2/x - 1}{1/x^2 - 1/x + 2} = \dfrac{0 + 0 - 1}{0 - 0 + 2} = -\dfrac{1}{2}$

19. Since x is positive, $\sqrt{x^2} = |x| = x$. $\lim\limits_{x \to \infty} \dfrac{\sqrt{x^2 - 9}}{2x - 6} = \lim\limits_{x \to \infty} \dfrac{\sqrt{x^2 - 9}/\sqrt{x^2}}{(2x - 6)/x} = \lim\limits_{x \to \infty} \dfrac{\sqrt{1 - 9/x^2}}{2 - 6/x} = \dfrac{\sqrt{1 - 0}}{2 - 0} = \dfrac{1}{2}$

21. If $y = -3x$, then as $x \to \infty$, $y \to -\infty$. $\lim\limits_{x \to \infty} e^{-3x} = \lim\limits_{y \to -\infty} e^y = 0$ by (2.6.6).

23. From the graph of $y = \left(\cos^2 x\right) / x^2$, it appears that $y = 0$ is the horizontal

asymptote and $x = 0$ is the vertical asymptote. Now $0 \leq (\cos x)^2 \leq 1 \Rightarrow$

$$\frac{0}{x^2} \leq \frac{\cos^2 x}{x^2} \leq \frac{1}{x^2} \Rightarrow 0 \leq \frac{\cos^2 x}{x^2} \leq \frac{1}{x^2}. \text{ But } \lim_{x \to \pm\infty} 0 = 0 \text{ and}$$

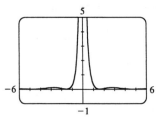

$$\lim_{x \to \pm\infty} \frac{1}{x^2} = 0, \text{ so by the Squeeze Theorem,}$$

$$\lim_{x \to \pm\infty} \frac{\cos^2 x}{x^2} = 0. \text{ Thus, } y = 0 \text{ is the horizontal asymptote. } \lim_{x \to 0} \frac{\cos^2 x}{x^2} = \infty \text{ because } \cos^2 x \to 1 \text{ and } x^2 \to 0 \text{ as}$$

$x \to 0$, so $x = 0$ is the vertical asymptote.

25. Since $2x - 1 \leq f(x) \leq x^2$ for $0 < x < 3$ and $\lim_{x \to 1} (2x - 1) = 1 = \lim_{x \to 1} x^2$, we have $\lim_{x \to 1} f(x) = 1$ by the Squeeze

Theorem.

27. Given $\varepsilon > 0$, we need $\delta > 0$ so that if $|x - 5| < \delta$ then $|(7x - 27) - 8| < \varepsilon \Leftrightarrow |7x - 35| < \varepsilon \Leftrightarrow$

$|x - 5| < \varepsilon/7$. So take $\delta = \varepsilon/7$. Then $|x - 5| < \delta \Rightarrow |(7x - 27) - 8| < \varepsilon$. Thus, $\lim_{x \to 5} (7x - 27) = 8$ by the

definition of a limit.

29. Given $\varepsilon > 0$, we need $\delta > 0$ so that if $|x - 2| < \delta$ then $\left|x^2 - 3x - (-2)\right| < \varepsilon$. First, note that if $|x - 2| < 1$, then

$-1 < x - 2 < 1$, so $0 < x - 1 < 2 \Rightarrow |x - 1| < 2$. Now let $\delta = \min\{\varepsilon/2, 1\}$. Then $|x - 2| < \delta \Rightarrow$

$\left|x^2 - 3x - (-2)\right| = |(x - 2)(x - 1)| = |x - 2| \, |x - 1| < (\varepsilon/2)(2) = \varepsilon$.

Thus, $\lim_{x \to 2} \left(x^2 - 3x\right) = -2$ by the definition of a limit.

31. (a) $f(x) = \sqrt{-x}$ if $x < 0$, $f(x) = 3 - x$ if $0 \leq x < 3$, $f(x) = (x - 3)^2$ if $x > 3$.

   (i) $\lim_{x \to 0^+} f(x) = \lim_{x \to 0^+} (3 - x) = 3$ (ii) $\lim_{x \to 0^-} f(x) = \lim_{x \to 0^-} \sqrt{-x} = 0$

   (iii) Because of (i) and (ii), $\lim_{x \to 0} f(x)$ does not exist.   (iv) $\lim_{x \to 3^-} f(x) = \lim_{x \to 3^-} (3 - x) = 0$

   (v) $\lim_{x \to 3^+} f(x) = \lim_{x \to 3^+} (x - 3)^2 = 0$ (vi) Because of (iv) and (v), $\lim_{x \to 3} f(x) = 0$.

   (b) f is discontinuous at 0 since $\lim_{x \to 0} f(x)$ does not exist. f is   (c)

   discontinuous at 3 since $f(3)$ does not exist.

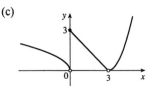

33. $\sin x$ is continuous on $\mathbb{R}$ by Theorem 7 in Section 2.4. Since e^x is continuous on $\mathbb{R}$, $e^{\sin x}$ is continuous on $\mathbb{R}$ by

Theorem 9 in Section 2.4. Lastly, x is continuous on $\mathbb{R}$ since it's a polynomial and the product $xe^{\sin x}$ is continuous

on its domain $\mathbb{R}$ by Theorem 4 in Section 2.4.

35. $f(x) = 2x^3 + x^2 + 2$ is a polynomial, so it is continuous on $[-2, -1]$ and $f(-2) = -10 < 0 < 1 = f(-1)$. So

by the Intermediate Value Theorem there is a number c in $(-2, -1)$ such that $f(c) = 0$, that is, the equation

$2x^3 + x^2 + 2 = 0$ has a root in $(-2, -1)$.

37. (a) The slope of the tangent line at $(2, 1)$ is

$$\lim_{x \to 2} \frac{f(x) - f(2)}{x - 2} = \lim_{x \to 2} \frac{9 - 2x^2 - 1}{x - 2} = \lim_{x \to 2} \frac{8 - 2x^2}{x - 2} = \lim_{x \to 2} \frac{-2\left(x^2 - 4\right)}{x - 2} = \lim_{x \to 2} \frac{-2(x - 2)(x + 2)}{x - 2}$$

$$= \lim_{x \to 2} -2(x + 2) = -8$$

(b) An equation of this tangent line is $y - 1 = -8(x - 2)$ or $y = -8x + 17$.

39. (a) $s = 1 + 2t + t^2/4$. The average velocity over the time interval $[1, 1 + h]$ is

$$\frac{s(1+h) - s(1)}{h} = \frac{1 + 2(1+h) + (1+h)^2/4 - 13/4}{h} = \frac{10h + h^2}{4h} = \frac{10+h}{4}. \text{ So for the following}$$

intervals the average velocities are:

(i) $[1, 3]$: $(10 + 2)/4 = 3$ m/s

(ii) $[1, 2]$: $(10 + 1)/4 = 2.75$ m/s

(iii) $[1, 1.5]$: $(10 + 0.5)/4 = 2.625$ m/s

(iv) $[1, 1.1]$: $(10 + 0.1)/4 = 2.525$ m/s

(b) When $t = 1$ the velocity is $\displaystyle\lim_{h \to 0} \frac{s(1+h) - s(1)}{h} = \lim_{h \to 0} \frac{10 + h}{4} = 2.5$ m/s.

41. Estimating the slopes of the tangent lines at $x = 2, 3$, and 5, we obtain approximate values $0.4, 2$, and 0.1. Since the graph is concave downward at $x = 5$, $f''(5)$ is negative. Arranging the numbers in increasing order, we have: $f''(5), 0, f'(5), f'(2), 1, f'(3)$.

43. (a) Estimating $f'(1)$ from the triangle in the graph,

we get $\dfrac{\Delta y}{\Delta x} \approx \dfrac{-0.37}{0.50} = -0.74.$

To estimate $f'(1)$ numerically, we have

$$f'(1) = \lim_{h \to 0} \frac{f(1+h) - f(1)}{h}$$

$$= \lim_{h \to 0} \frac{e^{-(1+h)^2} - e^{-1}}{h} = y$$

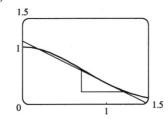

h	y
0.01	-0.732
0.001	-0.735
0.0001	-0.736
-0.01	-0.739
-0.001	-0.736
-0.0001	-0.736

From the table, we have $f'(1) \approx -0.736$.

(b) $y - e^{-1} \approx -0.736(x - 1)$ or $y \approx -0.736x + 1.104$

(c) See the graph in part (a).

45. (a) $f'(r)$ is the rate at which the total cost changes with respect to the interest rate. Its units are dollars/ (percent per year).

(b) The total cost of paying off the loan is increasing by \$1200/ (percent per year) as the interest rate reaches 10%. So if the interest rate goes up from 10% to 11%, the cost goes up approximately \$1200.

(c) As r increases, C increases. So $f'(r)$ will always be positive.

47.

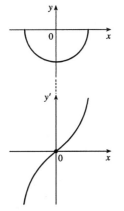

49. (a) $f'(x) = \lim\limits_{h \to 0} \dfrac{f(x+h) - f(x)}{h} = \lim\limits_{h \to 0} \dfrac{\sqrt{3 - 5(x+h)} - \sqrt{3 - 5x}}{h} \dfrac{\left(\sqrt{3 - 5(x+h)} + \sqrt{3 - 5x}\right)}{\left(\sqrt{3 - 5(x+h)} + \sqrt{3 - 5x}\right)}$

$= \lim\limits_{h \to 0} \dfrac{[3 - 5(x+h)] - (3 - 5x)}{h\left(\sqrt{3 - 5(x+h)} + \sqrt{3 - 5x}\right)} = \lim\limits_{h \to 0} \dfrac{-5}{\sqrt{3 - 5(x+h)} + \sqrt{3 - 5x}} = \dfrac{-5}{2\sqrt{3 - 5x}}$

(b) Domain of f: $3 - 5x \geq 0 \;\Rightarrow\; 5x \leq 3 \;\Rightarrow\; x \in \left(-\infty, \frac{3}{5}\right]$

Domain of f': exclude $\frac{3}{5}$; $x \in \left(-\infty, \frac{3}{5}\right)$

(c) Our answer to part (a) is reasonable because $f'(x)$ is always negative and f is always decreasing.

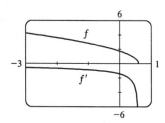

51. f is not differentiable: at $x = -4$ because f is not continuous, at $x = -1$ because f has a corner, at $x = 2$ because f is not continuous, and at $x = 5$ because f has a vertical tangent.

53. The inequality $\left|\dfrac{x+1}{x-1} - 3\right| < 0.2$ is equivalent to the double

inequality $2.8 < \dfrac{x+1}{x-1} < 3.2$. Graphing the functions $y = 2.8$,

$y = |(x+1)/(x-1)|$ and $y = 3.2$ on the interval $[1.9, 2.15]$, we

see that the inequality holds whenever $1.91 < x < 2.11$

(approximately). So since $|2 - 1.91| = 0.09$ and $|2 - 2.15| = 0.15$,

any positive $\delta \leq 0.09$ will do.

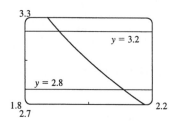

55. $|f(x)| \leq g(x) \;\Leftrightarrow\; -g(x) \leq f(x) \leq g(x)$ and $\lim\limits_{x \to a} g(x) = 0 = \lim\limits_{x \to a} -g(x)$. Thus, by the Squeeze Theorem,

$\lim\limits_{x \to a} f(x) = 0$.

Problems Plus

1. Let $t = \sqrt[6]{x}$, so $x = t^6$. Then $t \to 1$ as $x \to 1$, so

$$\lim_{x \to 1} \frac{\sqrt[3]{x} - 1}{\sqrt{x} - 1} = \lim_{t \to 1} \frac{t^2 - 1}{t^3 - 1} = \lim_{t \to 1} \frac{(t - 1)(t + 1)}{(t - 1)(t^2 + t + 1)} = \lim_{t \to 1} \frac{t + 1}{t^2 + t + 1} = \frac{1 + 1}{1^2 + 1 + 1} = \frac{2}{3}.$$

Another Method: Multiply both numerator and denominator by $(\sqrt{x} + 1)\left(\sqrt[3]{x^2} + \sqrt[3]{x} + 1\right)$.

3. For $-\frac{1}{2} < x < \frac{1}{2}$, we have $2x - 1 < 0$ and $2x + 1 > 0$, so $|2x - 1| = -(2x - 1)$ and $|2x + 1| = 2x + 1$.

Therefore, $\lim_{x \to 0} \frac{|2x - 1| - |2x + 1|}{x} = \lim_{x \to 0} \frac{-(2x - 1) - (2x + 1)}{x} = \lim_{x \to 0} \frac{-4x}{x} = \lim_{x \to 0} (-4) = -4.$

5. Since $[\![x]\!] \le x < [\![x]\!] + 1$, we have $1 \le \dfrac{x}{[\![x]\!]} \le 1 + \dfrac{1}{[\![x]\!]}$ for $x > 0$. As $x \to \infty$, $[\![x]\!] \to \infty$, so $\dfrac{1}{[\![x]\!]} \to 0$ and

$1 + \dfrac{1}{[\![x]\!]} \to 1$. Thus, $\lim_{x \to \infty} \dfrac{x}{[\![x]\!]} = 1$ by the Squeeze Theorem.

7. f is continuous on $(-\infty, a)$ and (a, ∞). To make f continuous on $\mathbb{R}$, we must have continuity at a. Thus,

$$\lim_{x \to a^+} f(x) = \lim_{x \to a^-} f(x) \ \Rightarrow \ \lim_{x \to a^+} x^2 = \lim_{x \to a^-} (x + 1) \ \Rightarrow \ a^2 = a + 1 \ \Rightarrow \ a = \left(1 \pm \sqrt{5}\right) / 2 \approx 1.618$$

or -0.618.

9. $\lim_{x \to a} f(x) = \lim_{x \to a} \left(\frac{1}{2} [f(x) + g(x)] + \frac{1}{2} [f(x) - g(x)]\right)$

$\qquad = \frac{1}{2} \lim_{x \to a} [f(x) + g(x)] + \frac{1}{2} \lim_{x \to a} [f(x) - g(x)]$

$\qquad = \frac{1}{2} \cdot 2 + \frac{1}{2} \cdot 1 = \frac{3}{2}$, and

$\lim_{x \to a} g(x) = \lim_{x \to a} ([f(x) + g(x)] - f(x)) = \lim_{x \to a} [f(x) + g(x)] - \lim_{x \to a} f(x) = 2 - \frac{3}{2} = \frac{1}{2}.$

So $\lim_{x \to a} [f(x) g(x)] = \left[\lim_{x \to a} f(x)\right]\left[\lim_{x \to a} g(x)\right] = \frac{3}{2} \cdot \frac{1}{2} = \frac{3}{4}.$

Another Solution: Since $\lim_{x \to a} [f(x) + g(x)]$ and $\lim_{x \to a} [f(x) - g(x)]$ exist, we must have

$\lim_{x \to a} [f(x) + g(x)]^2 = \left(\lim_{x \to a} [f(x) + g(x)]\right)^2$ and $\lim_{x \to a} [f(x) - g(x)]^2 = \left(\lim_{x \to a} [f(x) - g(x)]\right)^2$, so

$\lim_{x \to a} [f(x) g(x)] = \lim_{x \to a} \frac{1}{4} \left([f(x) + g(x)]^2 - [f(x) - g(x)]^2\right)$ (because all of the f^2 and g^2 cancel)

$\qquad = \frac{1}{4} \left(\lim_{x \to a} [f(x) + g(x)]^2 - \lim_{x \to a} [f(x) - g(x)]^2\right) = \frac{1}{4} (2^2 - 1^2) = \frac{3}{4}.$

11. (a) Consider $G(x) = T(x + 180°) - T(x)$. Fix any number a. If $G(a) = 0$, we are
done: Temperature at $a =$ Temperature at $a + 180°$. If $G(a) > 0$, then
$G(a + 180°) = T(a + 360°) - T(a + 180°) = T(a) - T(a + 180°) = -G(a) < 0.$ Also, G is continuous
since temperature varies continuously. So, by the Intermediate Value Theorem, G has a zero on the interval
$[a, a + 180°]$. If $G(a) < 0$, then a similar argument applies.

(b) Yes. The same argument applies.

(c) The same argument applies for quantities that vary continuously, such as barometric pressure. But one could
argue that altitude above sea level is sometimes discontinuous, so the result might not always hold for that
quantity.

13. (a) Put $x = 0$ and $y = 0$ in the equation: $f(0) = f(0 + 0) = f(0) + f(0) + 0^2 \cdot 0 + 0 \cdot 0^2 = 2f(0)$.
Subtracting $f(0)$ from each side of this equation gives $f(0) = 0$.

(b) $f'(0) = \lim\limits_{h \to 0} \dfrac{f(0 + h) - f(0)}{h} = \lim\limits_{h \to 0} \dfrac{\left[f(0) + f(h) + 0^2 h + 0h^2\right] - f(0)}{h}$

$\qquad = \lim\limits_{h \to 0} \dfrac{f(h)}{h} = \lim\limits_{x \to 0} \dfrac{f(x)}{x} = 1$

(c) $f'(x) = \lim\limits_{h \to 0} \dfrac{f(x + h) - f(x)}{h} = \lim\limits_{h \to 0} \dfrac{\left[f(x) + f(h) + x^2 h + xh^2\right] - f(x)}{h}$

$\qquad = \lim\limits_{h \to 0} \dfrac{f(h) + x^2 h + xh^2}{h} = \lim\limits_{h \to 0} \left[\dfrac{f(h)}{h} + x^2 + xh\right] = 1 + x^2$

3 Differentiation Rules

3.1 Derivatives of Polynomials and Exponential Functions

1. (a) e is the number such that $\displaystyle\lim_{h\to 0}\frac{e^h - 1}{h} = 1$.

(b)

x	$(2.7^x - 1)/x$
-0.001	0.9928
-0.0001	0.9932
0.001	0.9937
0.0001	0.9933

x	$(2.8^x - 1)/x$
-0.001	1.0291
-0.0001	1.0296
0.001	1.0301
0.0001	1.0297

From the tables (to two decimal places), $\displaystyle\lim_{h\to 0}\frac{2.7^h - 1}{h} = 0.99$ and $\displaystyle\lim_{h\to 0}\frac{2.8^h - 1}{h} = 1.03$. Since $0.99 < 1 < 1.03$, $2.7 < e < 2.8$.

3. $f(x) = 5x - 1 \;\Rightarrow\; f'(x) = 5 - 0 = 5$

5. $f(x) = x^2 + 3x - 4 \;\Rightarrow\; f'(x) = 2x^{2-1} + 3 - 0 = 2x + 3$

7. $y = x^{-2/5} \;\Rightarrow\; y' = -\frac{2}{5}x^{(-2/5)-1} = -\frac{2}{5}x^{-7/5} = -\dfrac{2}{5x^{7/5}}$

9. $V(r) = \frac{4}{3}\pi r^3 \;\Rightarrow\; V'(r) = \frac{4}{3}\pi\left(3r^2\right) = 4\pi r^2$

11. $Y(t) = 6t^{-9} \;\Rightarrow\; Y'(t) = 6(-9)t^{-10} = -54t^{-10}$

13. $F(x) = (16x)^3 = 4096x^3 \;\Rightarrow\; F'(x) = 4096\left(3x^2\right) = 12{,}288x^2$

15. $g(x) = x^2 + \dfrac{1}{x^2} = x^2 + x^{-2} \;\Rightarrow\; g'(x) = 2x + (-2)x^{-3} = 2x - \dfrac{2}{x^3}$

17. $y = \dfrac{x^2 + 4x + 3}{\sqrt{x}} = x^{3/2} + 4x^{1/2} + 3x^{-1/2} \;\Rightarrow$

$y' = \frac{3}{2}x^{1/2} + 4\left(\frac{1}{2}\right)x^{-1/2} + 3\left(-\frac{1}{2}\right)x^{-3/2} = \frac{3}{2}\sqrt{x} + \dfrac{2}{\sqrt{x}} - \dfrac{3}{2x\sqrt{x}}$

19. $y = 3x + 2e^x \;\Rightarrow\; y' = 3 + 2e^x$

21. $y = 4\pi^2 \;\Rightarrow\; y' = 0$ since $4\pi^2$ is a constant.

23. $y = ax^2 + bx + c \;\Rightarrow\; y' = 2ax + b$

25. $y = x + \sqrt[5]{x^2} = x + x^{2/5} \;\Rightarrow\; y' = 1 + \frac{2}{5}x^{-3/5} = 1 + \dfrac{2}{5\sqrt[5]{x^3}}$

27. $v = x\sqrt{x} + \dfrac{1}{x^2\sqrt{x}} = x^{3/2} + x^{-5/2} \;\Rightarrow\; v' = \frac{3}{2}x^{1/2} - \frac{5}{2}x^{-7/2} = \frac{3}{2}\sqrt{x} - \dfrac{5}{2x^3\sqrt{x}}$

29. $f(x) = 2x^2 - x^4 \Rightarrow f'(x) = 4x - 4x^3$.

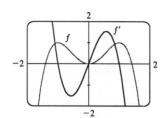

Notice that $f'(x) = 0$ when f has a horizontal tangent and that f' is an odd function while f is an even function.

31. $f(x) = 3x^{15} - 5x^3 + 3 \Rightarrow$
$f'(x) = 45x^{14} - 15x^2$.

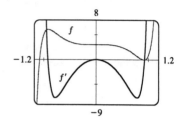

Notice that $f'(x) = 0$ when f has a horizontal tangent.

33. $f(x) = x - 3x^{1/3} \Rightarrow f'(x) = 1 - x^{-2/3} = 1 - 1/x^{2/3}$.

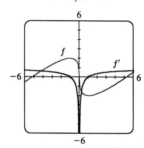

Note that $f'(x) = 0$ when f has a horizontal tangent, f' is positive when f is increasing, and f' is negative when f is decreasing.

35. (a)

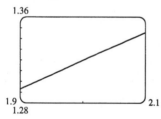

The endpoints of f in this graph are about $(1.9, 1.2927)$ and $(2.1, 1.3455)$. An estimate of $f'(2)$ is
$$\frac{1.3455 - 1.2927}{2.1 - 1.9} = \frac{0.0528}{0.2} = 0.264.$$
(b) $f(x) = x^{2/5} \Rightarrow f'(x) = \frac{2}{5}x^{-3/5} = 2/(5x^{3/5})$.
$f'(2) = 2/(5 \cdot 2^{3/5}) \approx 0.263902$.

37. $y = f(x) = x + \dfrac{4}{x} \Rightarrow f'(x) = 1 - \dfrac{4}{x^2}$. So the slope of the tangent line at $(2, 4)$ is $f'(2) = 0$ and its equation is $y - 4 = 0$ or $y = 4$.

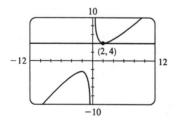

39. $y = f(x) = x + \sqrt{x} \Rightarrow f'(x) = 1 + \frac{1}{2}x^{-1/2}$. So the slope of the tangent line at $(1, 2)$ is
$f'(1) = 1 + \frac{1}{2}(1) = \frac{3}{2}$ and its equation is
$y - 2 = \frac{3}{2}(x - 1)$ or $y = \frac{3}{2}x + \frac{1}{2}$.

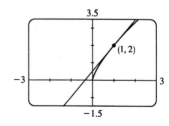

41. (a)

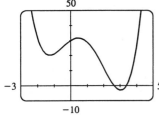

(b)

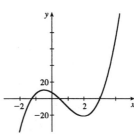

From the graph in part (a), it appears that f' is zero at $x_1 \approx -1.25$, $x_2 \approx 0.5$, and $x_3 \approx 3$. The slopes are negative (so f' is negative) on $(-\infty, x_1)$ and (x_2, x_3). The slopes are positive (so f' is positive) on (x_1, x_2) and (x_3, ∞).

(c) $f(x) = x^4 - 3x^3 - 6x^2 + 7x + 30 \Rightarrow$
$f'(x) = 4x^3 - 9x^2 - 12x + 7$

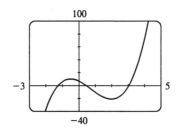

43. $y = x^3 - x^2 - x + 1$ has a horizontal tangent when $y' = 3x^2 - 2x - 1 = 0$. $(3x + 1)(x - 1) = 0 \Leftrightarrow x = 1$ or $-\frac{1}{3}$. Therefore, the points are $(1, 0)$ and $\left(-\frac{1}{3}, \frac{32}{27}\right)$.

45. $y = 6x^3 + 5x - 3 \Rightarrow m = y' = 18x^2 + 5$, but $x^2 \geq 0$ for all x, so $m \geq 5$ for all x.

47.

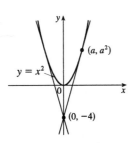

Let $\left(a, a^2\right)$ be a point on the parabola at which the tangent line passes through the point $(0, -4)$. The tangent line has slope $2a$ and equation $y - (-4) = 2a(x - 0) \Rightarrow y = 2ax - 4$. Since $\left(a, a^2\right)$ also lies on the line, $a^2 = 2a(a) - 4$, or $a^2 = 4$. So $a = \pm 2$ and the points are $(2, 4)$ and $(-2, 4)$.

49. $y = f(x) = 1 - x^2 \Rightarrow f'(x) = -2x$, so the tangent line at $(2, -3)$ has slope $f'(2) = -4$. The normal line has slope $-\frac{1}{-4} = \frac{1}{4}$ and equation $y + 3 = \frac{1}{4}(x - 2)$ or $y = \frac{1}{4}x - \frac{7}{2}$.

51. $f'(x) = \lim\limits_{h \to 0} \dfrac{f(x + h) - f(x)}{h} = \lim\limits_{h \to 0} \dfrac{\dfrac{1}{x + h} - \dfrac{1}{x}}{h} = \lim\limits_{h \to 0} \dfrac{x - (x + h)}{hx(x + h)}$

$$= \lim\limits_{h \to 0} \dfrac{-h}{hx(x + h)} = \lim\limits_{h \to 0} \dfrac{-1}{x(x + h)} = -\dfrac{1}{x^2}$$

53. $f(x) = 2 - x$ if $x \leq 1$ and $f(x) = x^2 - 2x + 2$ if $x > 1$. Now we compute
the right- and left-hand derivatives defined in Exercise 2.9.42:

$$f'_-(1) = \lim_{h \to 0^-} \frac{f(1+h) - f(1)}{h} = \lim_{h \to 0^-} \frac{2 - (1+h) - 1}{h} = \lim_{h \to 0^-} \frac{-h}{h} = \lim_{h \to 0^-} -1 = -1 \text{ and}$$

$$f'_+(1) = \lim_{h \to 0^+} \frac{f(1+h) - f(1)}{h} = \lim_{h \to 0^+} \frac{(1+h)^2 - 2(1+h) + 2 - 1}{h} = \lim_{h \to 0^+} \frac{h^2}{h} = \lim_{h \to 0^+} h = 0.$$

Thus, $f'(1)$ does not exist since $f'_-(1) \neq f'_+(1)$,
so f is not differentiable at 1. But $f'(x) = -1$
for $x < 1$ and $f'(x) = 2x - 2$ if $x > 1$.

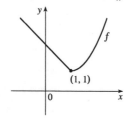

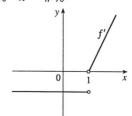

55. (a) Note that $x^2 - 9 < 0$ for $x^2 < 9 \Leftrightarrow |x| < 3 \Leftrightarrow -3 < x < 3$. So

$$f(x) = \begin{cases} x^2 - 9 & \text{if } x \leq -3 \\ -x^2 + 9 & \text{if } -3 < x < 3 \\ x^2 - 9 & \text{if } x \geq 3 \end{cases} \Rightarrow f'(x) = \begin{cases} 2x & \text{if } x < -3 \\ -2x & \text{if } -3 < x < 3 \\ 2x & \text{if } x > 3 \end{cases}$$

To show that $f'(3)$ does not exist we investigate $\lim_{h \to 0} \frac{f(3+h) - f(3)}{h}$ by computing the left- and right-hand
derivatives defined in Exercise 2.9.42.

$$f'_-(3) = \lim_{h \to 0^-} \frac{f(3+h) - f(3)}{h} = \lim_{h \to 0^-} \frac{(-(3+h)^2 + 9) - 0}{h} = \lim_{h \to 0^-} (-6 + h) = -6 \text{ and}$$

$$f'_+(3) = \lim_{h \to 0^+} \frac{f(3+h) - f(3)}{h} = \lim_{h \to 0^+} \frac{[(3+h)^2 + 9] - 0}{h} = \lim_{h \to 0^+} \frac{6h + h^2}{h} = \lim_{h \to 0^+} (6 + h) = 6.$$

Since the left and right limits are different,

$$\lim_{h \to 0} \frac{f(3+h) - f(3)}{h} \text{ does not exist, that is,}$$

$f'(3)$ does not exist. Similarly, $f'(-3)$ does not
exist. Therefore, f is not differentiable at 3 or at
-3.

(b)

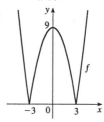

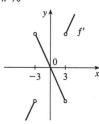

57. $y = f(x) = ax^2 \Rightarrow f'(x) = 2ax$. So the slope of the tangent to the parabola at $x = 2$ is $m = 2a(2) = 4a$.
The slope of the given line is seen to be -2, so we must have $4a = -2 \Leftrightarrow a = -\frac{1}{2}$. So the point in question
has y-coordinate $-\frac{1}{2} \cdot 2^2 = -2$. Now we simply require that the given line, whose equation is $2x + y = b$, pass
through the point $(2, -2)$: $2(2) + (-2) = b \Leftrightarrow b = 2$. So we must have $a = -\frac{1}{2}$ and $b = 2$.

59. $y = f(x) = ax^3 + bx^2 + cx + d \Rightarrow f'(x) = 3ax^2 + 2bx + c$. $f(-2) = 6 \Rightarrow$
$-8a + 4b - 2c + d = 6$ **(1)**. $f(2) = 0 \Rightarrow 8a + 4b + 2c + d = 0$ **(2)**. Since there are horizontal tangents at
$(-2, 6)$ and $(2, 0)$, $f'(\pm 2) = 0$. $f'(-2) = 0 \Rightarrow 12a - 4b + c = 0$ **(3)** and $f'(2) = 0 \Rightarrow$
$12a + 4b + c = 0$ **(4)**. Subtracting equation **(3)** from **(4)** gives $8b = 0 \Rightarrow b = 0$. Adding **(1)** and **(2)** gives
$8b + 2d = 6$, so $d = 3$ since $b = 0$. From **(3)** we have $c = -12a$, so **(2)** becomes $8a + 4(0) + 2(-12a) + 3 = 0$
$\Rightarrow 3 = 16a \Rightarrow a = \frac{3}{16}$. Now $c = -12a = -12\left(\frac{3}{16}\right) = -\frac{9}{4}$ and the desired cubic function is
$y = \frac{3}{16}x^3 - \frac{9}{4}x + 3.$

61. *Solution 1:* Let $f(x) = x^{1000}$. Then, by the definition of derivative,

$$f'(1) = \lim_{x \to 1} \frac{f(x) - f(1)}{x - 1} = \lim_{x \to 1} \frac{x^{1000} - 1}{x - 1}.$$ But this is just the limit we want to find, and we know (from the

Power Rule) that $f'(x) = 1000x^{999}$, so $f'(1) = 1000\,(1)^{999} = 1000$. So $\lim_{x \to 1} \dfrac{x^{1000} - 1}{x - 1} = 1000$.

Solution 2: Note that $\left(x^{1000} - 1\right) = (x - 1)\left(x^{999} + x^{998} + x^{997} + \cdots + x^2 + x + 1\right)$. So

$$\lim_{x \to 1} \frac{x^{1000} - 1}{x - 1} = \lim_{x \to 1} \frac{(x - 1)\left(x^{999} + x^{998} + x^{997} + \cdots + x^2 + x + 1\right)}{x - 1}$$

$$= \lim_{x \to 1} \left(x^{999} + x^{998} + x^{997} + \cdots + x^2 + x + 1\right) = \underbrace{1 + 1 + 1 + \cdots + 1 + 1 + 1}_{1000 \text{ ones}}$$

$$= 1000, \text{ as above.}$$

3.2 The Product and Quotient Rules

1. Product Rule: $y = \left(x^2 + 1\right)\left(x^3 + 1\right) \Rightarrow$

$y' = \left(x^2 + 1\right)\left(3x^2\right) + \left(x^3 + 1\right)(2x) = 3x^4 + 3x^2 + 2x^4 + 2x = 5x^4 + 3x^2 + 2x.$

Multiplying first: $y = \left(x^2 + 1\right)\left(x^3 + 1\right) = x^5 + x^3 + x^2 + 1 \Rightarrow y' = 5x^4 + 3x^2 + 2x$ (equivalent)

3. By the Product Rule, $f(x) = x^2 e^x \Rightarrow f'(x) = x^2 \dfrac{d}{dx}\left(e^x\right) + e^x \dfrac{d}{dx}\left(x^2\right) = x^2 e^x + e^x\,(2x) = xe^x\,(x + 2).$

5. By the Quotient Rule, $y = \dfrac{e^x}{x^2} \Rightarrow$

$$y' = \frac{x^2 \dfrac{d}{dx}\left(e^x\right) - e^x \dfrac{d}{dx}\left(x^2\right)}{\left(x^2\right)^2} = \frac{x^2\left(e^x\right) - e^x\,(2x)}{x^4} = \frac{xe^x\,(x - 2)}{x^4} = \frac{e^x\,(x - 2)}{x^3}.$$

7. $h(x) = \dfrac{x + 2}{x - 1} \Rightarrow h'(x) = \dfrac{(x - 1)\,(1) - (x + 2)\,(1)}{(x - 1)^2} = \dfrac{x - 1 - x - 2}{(x - 1)^2} = \dfrac{-3}{(x - 1)^2}$

9. $G(s) = \left(s^2 + s + 1\right)\left(s^2 + 2\right) \Rightarrow$

$G'(s) = \left(s^2 + s + 1\right)(2s) + \left(s^2 + 2\right)(2s + 1) = 2s^3 + 2s^2 + 2s + 2s^3 + s^2 + 4s + 2 = 4s^3 + 3s^2 + 6s + 2$

11. $H(x) = \left(x^3 - x + 1\right)\left(x^{-2} + 2x^{-3}\right) = \left(x^3 - x + 1\right)\left(x^{-2}\right) + \left(x^3 - x + 1\right)\left(2x^{-3}\right)$

$ = x - x^{-1} + x^{-2} + 2 - 2x^{-2} + 2x^{-3} = 2 + x - x^{-1} - x^{-2} + 2x^{-3} \Rightarrow$

$H'(x) = 1 + x^{-2} + 2x^{-3} - 6x^{-4}$

Another Method: Use the Product Rule.

13. $y = \dfrac{3t - 7}{t^2 + 5t - 4} \Rightarrow y' = \dfrac{\left(t^2 + 5t - 4\right)(3) - (3t - 7)(2t + 5)}{\left(t^2 + 5t - 4\right)^2} = \dfrac{-3t^2 + 14t + 23}{\left(t^2 + 5t - 4\right)^2}$

15. $y = \dfrac{x^2 + 4x + 3}{\sqrt{x}} = x^{3/2} + 4x^{1/2} + 3x^{-1/2} \Rightarrow$

$y' = \frac{3}{2}x^{1/2} + 4\left(\frac{1}{2}\right)x^{-1/2} + 3\left(-\frac{1}{2}\right)x^{-3/2} = \frac{3}{2}\sqrt{x} + \dfrac{2}{\sqrt{x}} - \dfrac{3}{2x\sqrt{x}}$

Another Method: Use the Quotient Rule.

17. $y = \left(r^2 - 2r\right)e^r \Rightarrow y' = \left(r^2 - 2r\right)\left(e^r\right) + e^r\,(2r - 2) = e^r\left(r^2 - 2r + 2r - 2\right) = e^r\left(r^2 - 2\right)$

19. $y = \dfrac{1}{x^4+x^2+1} \;\Rightarrow\; y' = \dfrac{\left(x^4+x^2+1\right)(0)-1\left(4x^3+2x\right)}{\left(x^4+x^2+1\right)^2} = -\dfrac{2x\left(2x^2+1\right)}{\left(x^4+x^2+1\right)^2}$

21. $f(x) = \dfrac{x}{x+c/x} \;\Rightarrow\; f'(x) = \dfrac{(x+c/x)(1)-x\left(1-c/x^2\right)}{(x+c/x)^2} = \dfrac{x+c/x-x+c/x}{\left(\dfrac{x^2+c}{x}\right)^2}\cdot\dfrac{x^2}{x^2} = \dfrac{2cx}{\left(x^2+c\right)^2}$

23. $y = \dfrac{2x}{x+1} \;\Rightarrow\; y' = \dfrac{(x+1)(2)-(2x)(1)}{(x+1)^2} = \dfrac{2}{(x+1)^2}$. At $(1,1)$, $y' = \frac{1}{2}$, and an equation of the tangent line is $y-1 = \frac{1}{2}(x-1)$, or $y = \frac{1}{2}x + \frac{1}{2}$.

25. $y = 2xe^x \;\Rightarrow\; y' = 2\left(x\cdot e^x+e^x\cdot 1\right) = 2e^x(x+1)$. At $(0,0)$, $y' = 2$, and an equation of the tangent line is $y-0 = 2(x-0)$, or $y = 2x$.

27. (a) $y = f(x) = \dfrac{1}{1+x^2} \;\Rightarrow\; f'(x) = \dfrac{-2x}{\left(1+x^2\right)^2}$. So the slope of the   (b)

tangent line at the point $\left(-1,\frac{1}{2}\right)$ is $f'(-1) = \frac{2}{2^2} = \frac{1}{2}$ and its

equation is $y-\frac{1}{2} = \frac{1}{2}(x+1)$ or $y = \frac{1}{2}x + 1$.

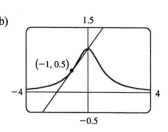

29. (a) $f(x) = \dfrac{e^x}{x^3} \;\Rightarrow\; f'(x) = \dfrac{x^3\left(e^x\right)-e^x\left(3x^2\right)}{\left(x^3\right)^2} = \dfrac{x^2 e^x(x-3)}{x^6} = \dfrac{e^x(x-3)}{x^4}$

(b)

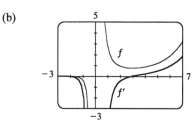

$f' = 0$ when f has a horizontal tangent line, f' is negative when f is decreasing, and f' is positive when f is increasing.

31. (a) $(fg)'(5) = f(5)g'(5)+g(5)f'(5) = (1)(2)+(-3)(6) = 2-18 = -16$

(b) $\left(\dfrac{f}{g}\right)'(5) = \dfrac{g(5)f'(5)-f(5)g'(5)}{[g(5)]^2} = \dfrac{(-3)(6)-(1)(2)}{(-3)^2} = -\dfrac{20}{9}$

(c) $\left(\dfrac{g}{f}\right)'(5) = \dfrac{f(5)g'(5)-g(5)f'(5)}{[f(5)]^2} = \dfrac{(1)(2)-(-3)(6)}{(1)^2} = 20$

33. $f(x) = e^x g(x) \;\Rightarrow\; f'(x) = e^x g'(x)+g(x)e^x = e^x\left[g'(x)+g(x)\right]$.
$f'(0) = e^0\left[g'(0)+g(0)\right] = 1(5+2) = 7$

35. (a) $u(x) = f(x) g(x)$, so $u'(1) = f(1) g'(1) + g(1) f'(1) = 2 \cdot (-1) + 1 \cdot 2 = 0$

(b) $v(x) = f(x)/g(x)$, so $v'(5) = \dfrac{g(5) f'(5) - f(5) g'(5)}{[g(5)]^2} = \dfrac{2\left(-\frac{1}{3}\right) - 3 \cdot \frac{2}{3}}{2^2} = -\dfrac{2}{3}$

37. Let $P(t)$ be the population and let $A(t)$ be the average annual income at time t, where t is measured in years and $t = 0$ corresponds to July 1993. Then the total personal income is given by $T(t) = P(t) A(t)$. We wish to find $T'(0)$. $T'(t) = P(t) A'(t) + A(t) P'(t)$. The term $P(t) A'(t)$ represents the portion of the rate of change of total income due to the existing population's increasing income. The term $A(t) P'(t)$ represents the portion of the rate of change of total income due to the increasing population.
$T'(0) = P(0) A'(0) + A(0) P'(0) \approx (3{,}354{,}000)(1900) + (21{,}107)(45{,}000) = 7{,}322{,}415{,}000$. So the total personal income was rising at a rate of about \$7.322 billion per year.

39. If $y = f(x) = \dfrac{x}{x+1}$ then $f'(x) = \dfrac{(x+1)(1) - x(1)}{(x+1)^2} = \dfrac{1}{(x+1)^2}$. When $x = a$, the equation of the tangent

line is $y - \dfrac{a}{a+1} = \dfrac{1}{(a+1)^2}(x-a)$. This line passes through $(1, 2)$ when

$2 - \dfrac{a}{a+1} = \dfrac{1}{(a+1)^2}(1-a) \quad \Leftrightarrow$

$2(a+1)^2 = a(a+1) + (1-a) = a^2 + 1 \quad \Leftrightarrow \quad a^2 + 4a + 1 = 0$. The

quadratic formula gives the roots of this equation as $-2 \pm \sqrt{3}$, so there are

two such tangent lines, which touch the curve at

$A\left(-2 + \sqrt{3}, \frac{1-\sqrt{3}}{2}\right) \approx (-0.27, -0.37)$ and

$B\left(-2 - \sqrt{3}, \frac{1+\sqrt{3}}{2}\right) \approx (-3.73, 1.37)$.

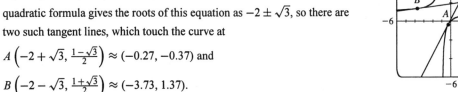

We will sometimes use the form $f'g + fg'$ rather than the form $fg' + gf'$ for the Product Rule.

41. (a) $(fgh)' = [(fg)h]' = (fg)'h + (fg)h' = (f'g + fg')h + (fg)h' = f'gh + fg'h + fgh'$

(b) Putting $f = g = h$ in part (a), we have

$$\frac{d}{dx}[f(x)]^3 = (fff)' = f'ff + ff'f + fff' = 3fff' = 3[f(x)]^2 f'(x).$$

(c) $\dfrac{d}{dx}(e^{3x}) = \dfrac{d}{dx}(e^x)^3 = 3(e^x)^2 e^x = 3e^{2x}e^x = 3e^{3x}$

43. $\dfrac{d}{dx}(x^{-n}) = \dfrac{d}{dx}\left(\dfrac{1}{x^n}\right) = -\dfrac{nx^{n-1}}{(x^n)^2} = -nx^{n-1-2n} = -nx^{-n-1}$

3.3 Rates of Change in the Natural and Social Sciences

1. (a) $s = f(t) = t^2 - 10t + 12 \implies v(t) = f'(t) = (2t - 10)$ft/s

(b) $v(3) = 2(3) - 10 = -4$ ft/s

(c) The particle is at rest when $v(t) = 0 \iff 2t - 10 = 0 \iff t = 5$ s.

(d) The particle is moving in the positive direction when $v(t) > 0 \iff 2t - 10 > 0 \iff 2t > 10 \iff$ $t > 5$.

(e) Since the particle is moving in the positive direction and in the negative direction, we need to calculate the distance traveled in the intervals $[0, 5]$ and $[5, 8]$ separately.
$|f(5) - f(0)| = |-13 - 12| = 25$ ft and
$|f(8) - f(5)| = |-4 - (-13)| = 9$ ft. The total distance traveled during the first 8 s is
$25 + 9 = 34$ ft.

(f)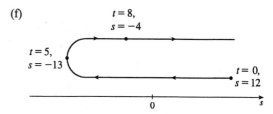

$t = 8,$
$s = -4$

$t = 5,$
$s = -13$

$t = 0,$
$s = 12$

0

s

3. (a) $s = f(t) = t^3 - 12t^2 + 36t \implies v(t) = f'(t) = 3t^2 - 24t + 36$

(b) $v(3) = 27 - 72 + 36 = -9$ ft/s

(c) The particle is at rest when $v(t) = 0$. $3t^2 - 24t + 36 = 0 \implies 3(t - 2)(t - 6) = 0 \implies t = 2, 6$.

(d) The particle is moving in the positive direction when $v(t) > 0$. $3(t - 2)(t - 6) > 0 \iff 0 \le t < 2$ or $t > 6$.

(e) Since the particle is moving in the positive direction and in the negative direction, we need to calculate the distance traveled in the intervals $[0, 2]$, $[2, 6]$, and $[6, 8]$ separately.
$|f(2) - f(0)| = |32 - 0| = 32$. $|f(6) - f(2)| = |0 - 32| = 32$.
$|f(8) - f(6)| = |32 - 0| = 32$. The total distance is
$32 + 32 + 32 = 96$ ft.

(f)

$t = 8,$
$s = 32$

$t = 6,$
$s = 0$

$t = 2,$
$s = 32$

$t = 0,$
$s = 0$

0

s

5. (a) $s = \dfrac{t}{t^2 + 1} \implies v(t) = s'(t) = \dfrac{(t^2 + 1)(1) - t(2t)}{(t^2 + 1)^2} = \dfrac{1 - t^2}{(t^2 + 1)^2}$

(b) $v(3) = \dfrac{1 - (3)^2}{(3^2 + 1)^2} = -\dfrac{2}{25}$ ft/s

(c) It is at rest when $v = 0 \iff 1 - t^2 = 0 \iff t = 1$.

(d) It moves in the positive direction when $v > 0 \iff 1 - t^2 > 0 \iff t^2 < 1 \iff 0 \le t < 1$.

(e) Distance in positive direction $= |s(1) - s(0)| = \left| \frac{1}{2} - 0 \right| = \frac{1}{2}$ ft

Distance in negative direction $= |s(8) - s(1)| = \left| \frac{8}{65} - \frac{1}{2} \right| = \frac{49}{130}$ ft

Total distance traveled $= \frac{1}{2} + \frac{49}{130} = \frac{57}{65}$ ft

(f)

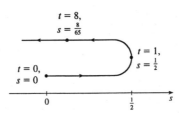

7. $s(t) = t^3 - 4.5t^2 - 7t \Rightarrow v(t) = s'(t) = 3t^2 - 9t - 7 = 5 \Leftrightarrow 3t^2 - 9t - 12 = 0 \Leftrightarrow$
$3(t-4)(t+1) = 0 \Leftrightarrow t = 4$ or -1. Since $t \geq 0$, the particle reaches a velocity of 5 m/s at $t = 4$ s.

9. (a) $A(x) = x^2 \Rightarrow A'(x) = 2x$. $A'(15) = 30$ mm^2/mm is the rate at which the area is increasing with respect
to the side length as x reaches 15 mm.

(b) The perimeter is $P(x) = 4x$, so

$A'(x) = 2x = \frac{1}{2}(4x) = \frac{1}{2}P(x)$. The figure suggests that if
Δx is small, then the change in the area of the square is
approximately half of its perimeter (2 of the 4 sides) times
Δx. From the figure, $\Delta A = 2x(\Delta x) + (\Delta x)^2$. If Δx is
small, then $\Delta A \approx 2x(\Delta x)$ and so $\Delta A/\Delta x \approx 2x$.

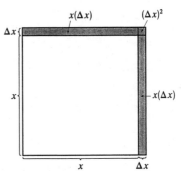

11. (a) $A(r) = \pi r^2$, so the average rate of change is:

(i) $\dfrac{A(3) - A(2)}{3 - 2} = \dfrac{9\pi - 4\pi}{1} = 5\pi$ (ii) $\dfrac{A(2.5) - A(2)}{2.5 - 2} = \dfrac{6.25\pi - 4\pi}{0.5} = 4.5\pi$

(iii) $\dfrac{A(2.1) - A(2)}{2.1 - 2} = \dfrac{4.41\pi - 4\pi}{0.1} = 4.1\pi$

(b) $A'(r) = 2\pi r$, so $A'(2) = 4\pi$.

(c) The circumference is $C(r) = 2\pi r = A'(r)$. The figure suggests that if Δr is
small, then the change in the area of the circle (a ring around the outside) is
approximately equal to its circumference times Δr. Straightening out this ring
gives us a shape that is approximately rectangular with length $2\pi r$ and width
Δr, so $\Delta A \approx 2\pi r(\Delta r)$. Algebraically,
$\Delta A = A(r + \Delta r) - A(r) = \pi(r + \Delta r)^2 - \pi r^2 = 2\pi r(\Delta r) + \pi(\Delta r)^2$. So
we see that if Δr is small, then $\Delta A \approx 2\pi r(\Delta r)$ and therefore, $\Delta A/\Delta r \approx 2\pi r$.

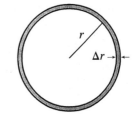

13. $S(r) = 4\pi r^2 \Rightarrow S'(r) = 8\pi r \Rightarrow$

(a) $S'(1) = 8\pi$ ft^2/ft (b) $S'(2) = 16\pi$ ft^2/ft (c) $S'(3) = 24\pi$ ft^2/ft

As the radius increases, the surface area grows at an increasing rate. In fact, the rate of change is linear with respect
to the radius.

15. $f(x) = 3x^2$, so the linear density at x is $\rho(x) = f'(x) = 6x$.

(a) $\rho(1) = 6$ kg/m (b) $\rho(2) = 12$ kg/m (c) $\rho(3) = 18$ kg/m

Since ρ is an increasing function, the density will be the highest at the right end of the rod and lowest at the left
end.

17. $Q(t) = t^3 - 2t^2 + 6t + 2$, so the current is $Q'(t) = 3t^2 - 4t + 6$.

(a) $Q'(0.5) = 3(0.5)^2 - 4(0.5) + 6 = 4.75$ A

(b) $Q'(1) = 3(1)^2 - 4(1) + 6 = 5$ A

The current is lowest when Q' has a minimum. $Q''(t) = 6t - 4 < 0$ when $t < \frac{2}{3}$. So the current decreases when $t < \frac{2}{3}$ and increases when $t > \frac{2}{3}$. Thus, the current is lowest at $\frac{2}{3}$ s.

19. (a) $PV = C \Rightarrow V = \dfrac{C}{P} \Rightarrow \dfrac{dV}{dP} = -\dfrac{C}{P^2}$

(b) From the formula for dV/dP in part (a), we see that as P increases, the absolute value of dV/dP decreases. Thus, the volume is decreasing more rapidly at the beginning.

(c) $\beta = -\dfrac{1}{V}\dfrac{dV}{dP} = -\dfrac{1}{V}\left(-\dfrac{C}{P^2}\right) = \dfrac{C}{(PV)P} = \dfrac{C}{CP} = \dfrac{1}{P}$

21. (a) **1920:** $m_1 = \frac{1860-1750}{1920-1910} = \frac{110}{10} = 11$, $m_2 = \frac{2070-1860}{1930-1920} = \frac{210}{10} = 21$,

$(m_1 + m_2)/2 = (11 + 21)/2 = 16$ million/year

1980: $m_1 = \frac{4450-3700}{1980-1970} = \frac{750}{10} = 75$, $m_2 = \frac{5300-4450}{1990-1980} = \frac{850}{10} = 85$,

$(m_1 + m_2)/2 = (75 + 85)/2 = 80$ million/year

(b) $P(t) = at^3 + bt^2 + ct + d$ where $a = 2325.67$, $b = -1.306488 \times 10^7$, $c = 2.44631 \times 10^{10}$, and $d = -1.52658 \times 10^{13}$.

(c) $P(t) = at^3 + bt^2 + ct + d \Rightarrow P'(t) = 3at^2 + 2bt + c$

(d) $P'(1920) = 3(2325.67)(1920)^2 + 2(-1.306448 \times 10^7)(1920) + 2.44631 \times 10^{10}$

$= 14{,}010{,}464$/year [smaller than the answer in part (a), but close to it]

$P'(1980) = 78{,}845{,}204$/year (smaller, but close)

(e) $P'(1985) = 86{,}515{,}627.25$/year, so the rate of growth in 1985 was about 86.5 million/year.

23. (a) $[\text{C}] = \dfrac{a^2 kt}{akt + 1} \Rightarrow$

rate of reaction $= \dfrac{d[\text{C}]}{dt} = \dfrac{(akt+1)(a^2 k) - (a^2 kt)(ak)}{(akt+1)^2} = \dfrac{a^2 k(akt + 1 - akt)}{(akt+1)^2} = \dfrac{a^2 k}{(akt+1)^2}$

(b) If $x = [\text{C}]$, then $a - x = a - \dfrac{a^2 kt}{akt + 1} = \dfrac{a^2 kt + a - a^2 kt}{akt + 1} = \dfrac{a}{akt + 1}$.

So $k(a-x)^2 = k\left(\dfrac{a}{akt + 1}\right)^2 = \dfrac{a^2 k}{(akt+1)^2} = \dfrac{d[\text{C}]}{dt}$ [from part (a)] $= \dfrac{dx}{dt}$.

(c) As $t \to \infty$, $[\text{C}] = \dfrac{a^2 kt}{akt + 1} = \dfrac{a^2 k}{ak + (1/t)} \to \dfrac{a^2 k}{ak} = a$ moles/L

(d) As $t \to \infty$, $\dfrac{d[\text{C}]}{dt} = \dfrac{a^2 k}{(akt+1)^2} \to 0$.

(e) As t increases, nearly all of the reactants A and B are converted into product C. In practical terms, the reaction virtually stops.

25. (a) Using $v = \dfrac{P}{4\eta l}(R^2 - r^2)$ with $R = 0.01$, $l = 3$, $P = 3000$, and $\eta = 0.027$, we have v as a function of r:

$v(r) = \dfrac{3000}{4(0.027)3}(0.01^2 - r^2)$. $v(0) = 0.9\overline{25}$ cm/s, $v(0.005) = 0.69\overline{4}$ cm/s, $v(0.01) = 0$.

(b) $v(r) = \dfrac{P}{4\eta l}(R^2 - r^2) \Rightarrow v'(r) = \dfrac{P}{4\eta l}(-2r) = -\dfrac{Pr}{2\eta l}$. When $l = 3$, $P = 3000$, and $\eta = 0.027$, we have

$v'(r) = -\dfrac{3000r}{2(0.027)3}$. $v'(0) = 0$, $v'(0.005) = -92.\overline{592}$ (cm/s) /cm, and $v'(0.01) = -185.\overline{185}$ (cm/s) /cm.

(c) The velocity is greatest where $r = 0$ (at the center) and the velocity is changing most where $r = R = 0.01$ cm (at the edge).

27. (a) $C(x) = 2000 + 3x + 0.01x^2 + 0.0002x^3 \Rightarrow C'(x) = 3 + 0.02x + 0.0006x^2$

(b) $C'(100) = 3 + 0.02(100) + 0.0006(10,000) = 3 + 2 + 6 = \$11/\text{yard}$. $C'(100)$ is the rate at which costs are increasing as the 100th yard is produced. It predicts the cost of the 101st yard.

(c) The cost of manufacturing the 101st yard is

$$C(101) - C(100) = (2000 + 303 + 102.01 + 206.0602) - (2000 + 300 + 100 + 200)$$
$$= 11.0702 \approx \$11.07/\text{yard}$$

29. (a) $A(x) = \dfrac{p(x)}{x} \Rightarrow A'(x) = \dfrac{xp'(x) - p(x) \cdot 1}{x^2}$. $A'(x) > 0 \Rightarrow A(x)$ is increasing; that is, the average productivity increases as the size of the workforce increases.

(b) Suppose $p'(x) > A(x)$. Then $p'(x) > \dfrac{p(x)}{x} \Rightarrow xp'(x) > p(x) \Rightarrow xp'(x) - p(x) > 0 \Rightarrow$

$\dfrac{xp'(x) - p(x)}{x^2} > 0 \Rightarrow A'(x) > 0$.

31. $PV = nRT \Rightarrow T = \dfrac{PV}{nR} = \dfrac{PV}{(10)(0.0821)} = \dfrac{1}{0.821}(PV)$.

$\dfrac{dT}{dt} = \dfrac{1}{0.821}[P(t)V'(t) + V(t)P'(t)] = \dfrac{1}{0.821}[(8)(-0.15) + (10)(0.10)] \approx -0.2436 \text{ K/min}$

33. (a) $\dfrac{dC}{dt} = 0$ and $\dfrac{dW}{dt} = 0$.

(b) The caribou go extinct $\Leftrightarrow C = 0$.

(c) We have **(1)** $0.05C - 0.001CW = 0$ and **(2)** $-0.05W + 0.0001CW = 0$. Adding 10 times **(2)** to **(1)** gives us $0.05C - 0.5W = 0 \Rightarrow C = 10W$. Substituting $C = 10W$ into **(1)** results in $W = 0$ or 50 and hence, $C = 0$ or 500. The pairs are $(0, 0)$ and $(500, 50)$. So it is possible for the two species to live in harmony.

3.4 Derivatives of Trigonometric Functions

1. $f(x) = x - 3\sin x \Rightarrow f'(x) = 1 - 3\cos x$

3. $y = \sin x + \cos x \Rightarrow dy/dx = \cos x - \sin x$

5. $g(t) = t^3 \cos t \Rightarrow g'(t) = t^3(-\sin t) + (\cos t) \cdot 3t^2 = 3t^2 \cos t - t^3 \sin t$

7. $h(\theta) = \csc\theta + e^\theta \cot\theta \Rightarrow$
$h'(\theta) = -\csc\theta\cot\theta + e^\theta(-\csc^2\theta) + (\cot\theta)e^\theta = -\csc\theta\cot\theta + e^\theta(\cot\theta - \csc^2\theta)$

9. $y = \dfrac{\tan x}{x} \Rightarrow \dfrac{dy}{dx} = \dfrac{x\sec^2 x - \tan x}{x^2}$

11. $y = \dfrac{x}{\sin x + \cos x}$ $\Rightarrow$

$$\dfrac{dy}{dx} = \dfrac{(\sin x + \cos x) - x\,(\cos x - \sin x)}{(\sin x + \cos x)^2} = \dfrac{(1 + x)\sin x + (1 - x)\cos x}{\sin^2 x + \cos^2 x + 2\sin x \cos x}$$

$$= \dfrac{(1 + x)\sin x + (1 - x)\cos x}{1 + \sin 2x}$$

13. $y = \dfrac{\sin x}{x^2}$ $\Rightarrow$ $y' = \dfrac{x^2 \cos x - (\sin x)\,(2x)}{\left(x^2\right)^2} = \dfrac{x\,(x \cos x - 2\sin x)}{x^4} = \dfrac{x \cos x - 2\sin x}{x^3}$

15. $y = \csc x \cot x$ $\Rightarrow$ $dy/dx = (-\csc x \ \cot x)\cot x + \csc x\left(-\csc^2 x\right) = -\csc x\left(\cot^2 x + \csc^2 x\right)$

17. $\dfrac{d}{dx}\left(\csc x\right) = \dfrac{d}{dx}\left(\dfrac{1}{\sin x}\right) = \dfrac{(\sin x)\,(0) - 1\,(\cos x)}{\sin^2 x} = \dfrac{-\cos x}{\sin^2 x} = -\dfrac{1}{\sin x} \cdot \dfrac{\cos x}{\sin x} = -\csc x \cot x$

19. $\dfrac{d}{dx}\left(\cot x\right) = \dfrac{d}{dx}\left(\dfrac{\cos x}{\sin x}\right) = \dfrac{(\sin x)\,(-\sin x) - (\cos x)\,(\cos x)}{\sin^2 x} = -\dfrac{\sin^2 x + \cos^2 x}{\sin^2 x} = -\dfrac{1}{\sin^2 x}$

$$= -\csc^2 x$$

21. $y = \tan x$ $\Rightarrow$ $y' = \sec^2 x \Rightarrow$ the slope of the tangent line at $\left(\frac{\pi}{4}, 1\right)$ is $\sec^2 \frac{\pi}{4} = \left(\sqrt{2}\right)^2 = 2$ and an equation is
$y - 1 = 2\left(x - \frac{\pi}{4}\right)$ or $y = 2x + 1 - \frac{\pi}{2}$.

23. $y = x + \cos x$ $\Rightarrow$ $y' = 1 - \sin x$. At $(0, 1)$, $y' = 1$, and an equation of the tangent line is $y - 1 = 1\,(x - 0)$, or
$y = x + 1$.

25. (a) $y = x \cos x$ $\Rightarrow$ $y' = x\,(-\sin x) + \cos x\,(1) = \cos x - x \sin x$. So
the slope of the tangent at the point $(\pi, -\pi)$ is
$\cos \pi - \pi \sin \pi = -1 - \pi\,(0) = -1$, and its equation is
$y + \pi = -(x - \pi)$ $\Leftrightarrow$ $y = -x$.

(b)

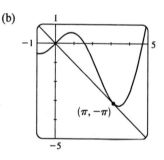

$(\pi, -\pi)$

27. (a) $f\,(x) = 2x + \cot x$ $\Rightarrow$
$f'\,(x) = 2 - \csc^2 x$

(b)

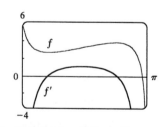

Notice that $f'\,(x) = 0$ when f has a
horizontal tangent. Also, $f'\,(x)$ is large
negative when the graph of f is steep.

29. $y = x + 2\sin x$ has a horizontal tangent when
$y' = 1 + 2\cos x = 0$ $\Leftrightarrow$ $\cos x = -\frac{1}{2}$ $\Leftrightarrow$
$x = \frac{2\pi}{3} + 2\pi n$ or $\frac{4\pi}{3} + 2\pi n$ or, equivalently,
$(2n + 1)\,\pi \pm \frac{\pi}{3}$, n an integer.

31. (a) $x(t) = 8\sin t \implies v(t) = x'(t) = 8\cos t$

(b) $x\left(\frac{2\pi}{3}\right) = 8\left(\frac{\sqrt{3}}{2}\right) = 4\sqrt{3}$, $v\left(\frac{2\pi}{3}\right) = 8\left(-\frac{1}{2}\right) = -4$. Since $v\left(\frac{2\pi}{3}\right) < 0$, the particle is moving to the left.

33.

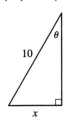

From the diagram we can see that $\sin\theta = x/10 \iff x = 10\sin\theta$. But we want to find the rate of change of x with respect to θ, that is, $dx/d\theta$. Taking the derivative of the above expression, $dx/d\theta = 10(\cos\theta)$. So when $\theta = \frac{\pi}{3}$,

$$dx/d\theta = 10\cos\frac{\pi}{3} = 10\left(\frac{1}{2}\right) = 5 \text{ ft/rad.}$$

35. $\lim\limits_{t\to 0} \dfrac{\sin 5t}{t} = \lim\limits_{t\to 0} \dfrac{5\sin 5t}{5t} = 5\lim\limits_{t\to 0} \dfrac{\sin 5t}{5t} = 5\cdot 1 = 5$

37. $\lim\limits_{\theta\to 0} \dfrac{\sin(\cos\theta)}{\sec\theta} = \dfrac{\sin\left(\lim\limits_{\theta\to 0}\cos\theta\right)}{\lim\limits_{\theta\to 0}\sec\theta} = \dfrac{\sin 1}{1} = \sin 1$

39. $\lim\limits_{\theta\to 0} \dfrac{\sin^2\theta}{\theta} = \lim\limits_{\theta\to 0}\left(\dfrac{\sin\theta}{\theta}\right)\sin\theta = \lim\limits_{\theta\to 0}\dfrac{\sin\theta}{\theta}\lim\limits_{\theta\to 0}\sin\theta = 1\cdot 0 = 0$

41. $\lim\limits_{x\to 0} \dfrac{\cot 2x}{\csc x} = \lim\limits_{x\to 0} \dfrac{\cos 2x\sin x}{\sin 2x} = \lim\limits_{x\to 0}\cos 2x\left[\dfrac{(\sin x)/x}{(\sin 2x)/x}\right] = \lim\limits_{x\to 0}\cos 2x\left[\dfrac{\lim\limits_{x\to 0}[(\sin x)/x]}{2\lim\limits_{x\to 0}[(\sin 2x)/2x]}\right]$

$$= 1\cdot\dfrac{1}{2\cdot 1} = \dfrac{1}{2}$$

43. Divide numerator and denominator by θ. ($\sin\theta$ also works.)

$$\lim\limits_{\theta\to 0} \dfrac{\sin\theta}{\theta+\tan\theta} = \lim\limits_{\theta\to 0}\dfrac{\dfrac{\sin\theta}{\theta}}{1+\dfrac{\sin\theta}{\theta}\cdot\dfrac{1}{\cos\theta}} = \dfrac{\lim\limits_{\theta\to 0}\dfrac{\sin\theta}{\theta}}{1+\lim\limits_{\theta\to 0}\dfrac{\sin\theta}{\theta}\lim\limits_{\theta\to 0}\dfrac{1}{\cos\theta}} = \dfrac{1}{1+1\cdot 1} = \dfrac{1}{2}$$

45. (a) $\dfrac{d}{dx}\tan x = \dfrac{d}{dx}\dfrac{\sin x}{\cos x} \implies \sec^2 x = \dfrac{\cos x\cos x - \sin x(-\sin x)}{\cos^2 x} = \dfrac{\cos^2 x + \sin^2 x}{\cos^2 x}$. So $\sec^2 x = \dfrac{1}{\cos^2 x}$.

(b) $\dfrac{d}{dx}\sec x = \dfrac{d}{dx}\dfrac{1}{\cos x} \implies \sec x\tan x = \dfrac{(\cos x)(0) - 1(-\sin x)}{\cos^2 x}$. So $\sec x\tan x = \dfrac{\sin x}{\cos^2 x}$.

(c) $\dfrac{d}{dx}(\sin x + \cos x) = \dfrac{d}{dx}\dfrac{1+\cot x}{\csc x} \implies$

$$\cos x - \sin x = \dfrac{\csc x\left(-\csc^2 x\right) - (1+\cot x)(-\csc x\cot x)}{\csc^2 x} = \dfrac{-\csc^2 x + \cot^2 x + \cot x}{\csc x}$$

So $\cos x - \sin x = \dfrac{\cot x - 1}{\csc x}$.

47. By the definition of radian measure, $s = r\theta$, where r is the radius of the circle.

By drawing the bisector of the angle θ, we can see that $\sin\dfrac{\theta}{2} = \dfrac{d/2}{r} \implies d = 2r\sin\dfrac{\theta}{2}$.

So $\lim\limits_{\theta\to 0^+}\dfrac{s}{d} = \lim\limits_{\theta\to 0^+}\dfrac{r\theta}{2r\sin(\theta/2)} = \lim\limits_{\theta\to 0^+}\dfrac{2\cdot(\theta/2)}{2\sin(\theta/2)} = \lim\limits_{\theta\to 0}\dfrac{\theta/2}{\sin(\theta/2)} = 1$. [This is just the reciprocal of the limit

$\lim\limits_{x\to 0}\dfrac{\sin x}{x} = 1$ combined with the fact that as $\theta\to 0$, $\dfrac{\theta}{2}\to 0$ also.]

3.5 The Chain Rule

1. Let $u = g(x) = x^2 + 4x + 6$ and $y = f(u) = u^5$.

Then $\dfrac{dy}{dx} = \dfrac{dy}{du}\dfrac{du}{dx} = (5u^4)(2x + 4) = 5(x^2 + 4x + 6)^4(2x + 4) = 10(x^2 + 4x + 6)^4(x + 2)$.

3. Let $u = g(x) = \tan x$ and $y = f(u) = \cos u$.

Then $\dfrac{dy}{dx} = \dfrac{dy}{du}\dfrac{du}{dx} = (-\sin u)(\sec^2 x) = -\sin(\tan x)\sec^2 x$.

5. Let $u = g(x) = \sqrt{x}$ and $y = f(u) = e^u$.

Then $\dfrac{dy}{dx} = \dfrac{dy}{du}\dfrac{du}{dx} = \frac{1}{2}x^{-1/2}e^u = \frac{1}{2}x^{-1/2}e^{\sqrt{x}} = \dfrac{e^{\sqrt{x}}}{2\sqrt{x}}$.

7. $F(x) = (x^3 + 4x)^7 \Rightarrow F'(x) = 7(x^3 + 4x)^6(3x^2 + 4)$ [or $7x^6(x^2 + 4)^6(3x^2 + 4)$]

9. $g(x) = \sqrt{x^2 - 7x} = (x^2 - 7x)^{1/2} \Rightarrow g'(x) = \frac{1}{2}(x^2 - 7x)^{-1/2}(2x - 7) = \dfrac{2x - 7}{2\sqrt{x^2 - 7x}}$

11. $h(t) = (t - 1/t)^{3/2} \Rightarrow h'(t) = \frac{3}{2}(t - 1/t)^{1/2}(1 + 1/t^2)$

13. $y = \cos(a^3 + x^3) \Rightarrow y' = -\sin(a^3 + x^3) \cdot 3x^2 = -3x^2\sin(a^3 + x^3)$

15. $y = e^{-mx} \Rightarrow y' = e^{-mx}(-m) = -me^{-mx}$

17. $G(x) = (3x - 2)^{10}(5x^2 - x + 1)^{12} \Rightarrow$

$\quad G'(x) = (3x - 2)^{10}(12)(5x^2 - x + 1)^{11}(10x - 1) + 10(3x - 2)^9(3)(5x^2 - x + 1)^{12}$

$\qquad = 6(3x - 2)^9(5x^2 - x + 1)^{11}[2(3x - 2)(10x - 1) + 5(5x^2 - x + 1)]$

$\qquad = 6(3x - 2)^9(5x^2 - x + 1)^{11}(85x^2 - 51x + 9)$

19. $y = (2x - 5)^4(8x^2 - 5)^{-3} \Rightarrow$

$\quad y' = 4(2x - 5)^3(2)(8x^2 - 5)^{-3} + (2x - 5)^4(-3)(8x^2 - 5)^{-4}(16x)$

$\qquad = 8(2x - 5)^3(8x^2 - 5)^{-3} - 48x(2x - 5)^4(8x^2 - 5)^{-4}$

$\quad$ [This simplifies to $8(2x - 5)^3(8x^2 - 5)^{-4}(-4x^2 + 30x - 5)$.]

21. $f(x) = xe^{-x^2} \Rightarrow f'(x) = e^{-x^2} + xe^{-x^2}(-2x) = e^{-x^2}(1 - 2x^2)$

23. $F(y) = \left(\dfrac{y - 6}{y + 7}\right)^3 \Rightarrow F'(y) = 3\left(\dfrac{y - 6}{y + 7}\right)^2 \dfrac{(y + 7)(1) - (y - 6)(1)}{(y + 7)^2} = 3\left(\dfrac{y - 6}{y + 7}\right)^2 \dfrac{13}{(y + 7)^2} = \dfrac{39(y - 6)^2}{(y + 7)^4}$

25. $f(z) = (2z - 1)^{-1/5} \Rightarrow f'(z) = -\frac{1}{5}(2z - 1)^{-6/5}(2) = -\frac{2}{5}(2z - 1)^{-6/5}$

27. $y = \tan(\cos x) \Rightarrow y' = \sec^2(\cos x) \cdot (-\sin x) = -\sin x\sec^2(\cos x)$

29. Using Formula 5 and the Chain Rule, $y = 5^{-1/x} \Rightarrow y' = 5^{-1/x}(\ln 5)[-1 \cdot (-x^{-2})] = 5^{-1/x}(\ln 5)/x^2$

31. $y = \sin^3 x + \cos^3 x \Rightarrow y' = 3\sin^2 x\cos x + 3\cos^2 x(-\sin x) = 3\sin x\cos x(\sin x - \cos x)$

33. $y = (1 + \cos^2 x)^6 \Rightarrow y' = 6(1 + \cos^2 x)^5 2\cos x(-\sin x) = -12\cos x\sin x(1 + \cos^2 x)^5$

35. $y = \dfrac{e^{3x}}{1 + e^x} \Rightarrow y' = \dfrac{3e^{3x}(1 + e^x) - e^{3x}(e^x)}{(1 + e^x)^2} = \dfrac{3e^{3x} + 3e^{4x} - e^{4x}}{(1 + e^x)^2} = \dfrac{3e^{3x} + 2e^{4x}}{(1 + e^x)^2}$

37. $y = e^{x\cos x} \Rightarrow y' = e^{x\cos x}(\cos x - x\sin x)$

39. $y = \sqrt{x + \sqrt{x}} \Rightarrow y' = \frac{1}{2}(x + \sqrt{x})^{-1/2}\left(1 + \frac{1}{2}x^{-1/2}\right) = \dfrac{1}{2\sqrt{x + \sqrt{x}}}\left(1 + \dfrac{1}{2\sqrt{x}}\right)$

41. $y = \sin\left(\tan\sqrt{\sin x}\right) \;\Rightarrow\; y' = \cos\left(\tan\sqrt{\sin x}\right)\left(\sec^2\sqrt{\sin x}\right)\left(\dfrac{1}{2\sqrt{\sin x}}\right)(\cos x)$

43. $y = f(x) = \dfrac{8}{\sqrt{4+3x}} = 8(4+3x)^{-1/2} \;\Rightarrow\; f'(x) = 8\left(-\frac{1}{2}\right)(4+3x)^{-3/2}(3) = -12(4+3x)^{-3/2}$. The

slope of the tangent at $(4, 2)$ is $f'(4) = -\frac{12}{64} = -\frac{3}{16}$ and its equation is $y - 2 = -\frac{3}{16}(x - 4)$ or $y = -\frac{3}{16}x + \frac{11}{4}$.

45. $y = \sin(\sin x) \;\Rightarrow\; y' = \cos(\sin x)\cdot\cos x$. At $(\pi, 0)$, $y' = \cos(\sin\pi)\cdot\cos\pi = \cos(0)\cdot(-1) = 1(-1) = -1$,
and an equation of the tangent line is $y - 0 = -1(x - \pi)$, or $y = -x + \pi$.

47. (a) $y = \dfrac{2}{1+e^{-x}} \;\Rightarrow$

$y' = \dfrac{(1+e^{-x})(0) - 2(-e^{-x})}{(1+e^{-x})^2} = \dfrac{2e^{-x}}{(1+e^{-x})^2}$. At $(0, 1)$,

$y' = \dfrac{2}{2^2} = \dfrac{1}{2}$.

So an equation of the tangent line is $y - 1 = \frac{1}{2}(x - 0)$ or

$y = \frac{1}{2}x + 1$.

(b)
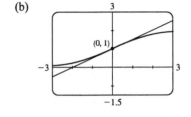

49. (a) $f(x) = \dfrac{\sqrt{1-x^2}}{x} \;\Rightarrow$

$f'(x) = \dfrac{x\cdot\frac{1}{2}(1-x^2)^{-1/2}(-2x) - \sqrt{1-x^2}}{x^2}\cdot\dfrac{\sqrt{1-x^2}}{\sqrt{1-x^2}}$

$= \dfrac{-x^2 - (1-x^2)}{x^2\sqrt{1-x^2}} = \dfrac{-1}{x^2\sqrt{1-x^2}}$

(b)
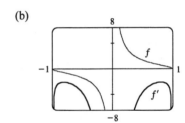

Notice that all tangents to the
graph of f have negative slopes
and $f'(x) < 0$ always.

51. For the tangent line to be horizontal $f'(x) = 0$. $f(x) = 2\sin x + \sin^2 x \;\Rightarrow$
$f'(x) = 2\cos x + 2\sin x\cos x = 0 \;\Leftrightarrow\; 2\cos x(1 + \sin x) = 0 \;\Leftrightarrow\; \cos x = 0$ or $\sin x = -1$, so
$x = \left(n + \frac{1}{2}\right)\pi$ or $\left(2n + \frac{3}{2}\right)\pi$ where n is any integer. So the points on the curve with a horizontal tangent are
$\left(\left(2n + \frac{1}{2}\right)\pi, 3\right)$ and $\left(\left(2n + \frac{3}{2}\right)\pi, -1\right)$ where n is any integer.

53. $F(x) = f(g(x)) \;\Rightarrow$
$F'(x) = f'(g(x))g'(x)$, so $F'(3) = f'(g(3))g'(3) = f'(6)g'(3) = 7\cdot 4 = 28$.

55. (a) $h(x) = f(g(x)) \;\Rightarrow\; h'(x) = f'(g(x))\cdot g'(x)$, so $h'(1) = f'(g(1))\cdot g'(1) = f'(2)\cdot 6 = 5\cdot 6 = 30$.
   (b) $H(x) = g(f(x)) \;\Rightarrow\; H'(x) = g'(f(x))\cdot f'(x)$, so $H'(1) = g'(f(1))\cdot f'(1) = g'(3)\cdot 4 = 9\cdot 4 = 36$.

57. (a) $u(x) = f(g(x)) \;\Rightarrow\; u'(x) = f'(g(x))g'(x)$.
   So $u'(1) = f'(g(1))g'(1) = f'(3)g'(1) = \left(-\frac{1}{4}\right)(-3) = \frac{3}{4}$.
   (b) $v(x) = g(f(x)) \;\Rightarrow\; v'(x) = g'(f(x))f'(x)$. So $v'(1) = g'(f(1))f'(1) = g'(2)f'(1)$, which does
   not exist since $g'(2)$ does not exist.
   (c) $w(x) = g(g(x)) \;\Rightarrow\; w'(x) = g'(g(x))g'(x)$.
   So $w'(1) = g'(g(1))g'(1) = g'(3)g'(1) = \left(\frac{2}{3}\right)(-3) = -2$.

59. $h(x) = f(g(x)) \Rightarrow h'(x) = f'(g(x))g'(x)$. So $h'(0.5) = f'(g(0.5))g'(0.5) = f'(0.1)g'(0.5)$. We can estimate the derivatives by taking the average of two secant slopes.

For $f'(0.1)$: $m_1 = \dfrac{14.8 - 12.6}{0.1 - 0} = 22$, $m_2 = \dfrac{18.4 - 14.8}{0.2 - 0.1} = 36$. So $f'(0.1) \approx \dfrac{m_1 + m_2}{2} = 29$.

For $g'(0.5)$: $m_1 = \dfrac{0.10 - 0.17}{0.5 - 0.4} = -0.7$, $m_2 = \dfrac{0.05 - 0.10}{0.6 - 0.5} = -0.5$.

So $g'(0.5) \approx (m_1 + m_2)/2 = -0.6$. Hence, $h'(0.5) \approx (29)(-0.6) = -17.4$.

61. (a) $F(x) = f(e^x) \Rightarrow F'(x) = f'(e^x)\dfrac{d}{dx}(e^x) = f'(e^x)e^x$

(b) $G(x) = e^{f(x)} \Rightarrow G'(x) = e^{f(x)}\dfrac{d}{dx}f(x) = e^{f(x)}f'(x)$

63. $s(t) = 10 + \frac{1}{4}\sin(10\pi t) \Rightarrow$ the velocity after t seconds is
$v(t) = s'(t) = \frac{1}{4}\cos(10\pi t)(10\pi) = \frac{5\pi}{2}\cos(10\pi t)$ cm/s.

65. (a) $B(t) = 4.0 + 0.35\sin\dfrac{2\pi t}{5.4} \Rightarrow \dfrac{dB}{dt} = \left(0.35\cos\dfrac{2\pi t}{5.4}\right)\left(\dfrac{2\pi}{5.4}\right) = \dfrac{7\pi}{54}\cos\dfrac{2\pi t}{5.4}$

(b) At $t = 1$, $\dfrac{dB}{dt} = \dfrac{7\pi}{54}\cos\dfrac{2\pi}{5.4} \approx 0.16$.

67. $s(t) = 2e^{-1.5t}\sin 2\pi t \Rightarrow$
$v(t) = s'(t) = 2\left[e^{-1.5t}(\cos 2\pi t)(2\pi) + (\sin 2\pi t)e^{-1.5t}(-1.5)\right] = 2e^{-1.5t}(2\pi\cos 2\pi t - 1.5\sin 2\pi t)$

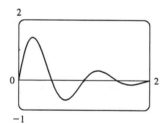
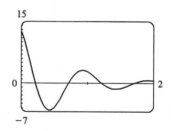

69. (a) Using a calculator or CAS, we obtain the model $Q = ab^t$ with $a = 100.0124369$ and $b = 0.000045145933$. We can change this model to one with base e and exponent $\ln b$: $Q = ae^{t\ln b} = 100.0124369e^{-10.005553063t}$.

(b) $Q'(t) = ab^t\ln b$. $Q'(0.04) \approx -670.63$ μA. The result of Example 2 in Section 2.1 was -670 μA.

71. (a) Derive gives $g'(t) = \dfrac{45(t-2)^8}{(2t+1)^{10}}$ without simplifying. With either Maple or Mathematica, we first get

$g'(t) = 9\dfrac{(t-2)^8}{(2t+1)^9} - 18\dfrac{(t-2)^9}{(2t+1)^{10}}$, and the simplification command results in the above expression.

(b) Derive gives $y' = 2(x^3 - x + 1)^3(2x+1)^4(17x^3 + 6x^2 - 9x + 3)$ without simplifying.
With either Maple or Mathematica, we first get
$y' = 10(2x+1)^4(x^3 - x + 1)^4 + 4(2x+1)^5(x^3 - x + 1)^3(3x^2 - 1)$. If we use Mathematica's Factor or Simplify, or Maple's factor, we get the above expression, but Maple's simplify gives the polynomial expansion instead. For locating horizontal tangents, the factored form is the most helpful.

73. (a) If f is even, then $f(x) = f(-x)$. Using the Chain Rule to differentiate this equation, we get

$$f'(x) = f'(-x) \frac{d}{dx}(-x) = -f'(-x). \text{ Thus, } f'(-x) = -f'(x), \text{ so } f' \text{ is odd.}$$

(b) If f is odd, then $f(x) = -f(-x)$. Differentiating this equation, we get $f'(x) = -f'(-x)(-1) = f'(-x)$, so f' is even.

75. (a) $\dfrac{d}{dx}\left(\sin^n x \cos nx\right) = n \sin^{n-1} x \cos x \cos nx + \sin^n x \, (-n \sin nx)$

$$= n \sin^{n-1} x \, (\cos nx \cos x - \sin nx \sin x) = n \sin^{n-1} x \cos(nx + x)$$

$$= n \sin^{n-1} x \cos[(n+1)x]$$

(b) $\dfrac{d}{dx}\left(\cos^n x \cos nx\right) = n \cos^{n-1} x \, (-\sin x) \cos nx + \cos^n x \, (-n \sin nx)$

$$= -n \cos^{n-1} x \, (\cos nx \sin x + \sin nx \cos x) = -n \cos^{n-1} x \sin(nx + x)$$

$$= -n \cos^{n-1} x \sin[(n+1)x]$$

77. Since $\theta° = \left(\frac{\pi}{180}\right)\theta$ rad, we have $\dfrac{d}{d\theta}(\sin \theta°) = \dfrac{d}{d\theta}\left(\sin \frac{\pi}{180}\theta\right) = \frac{\pi}{180} \cos \frac{\pi}{180}\theta = \frac{\pi}{180} \cos \theta°.$

3.6 Implicit Differentiation

1. (a) $\dfrac{d}{dx}\left(xy + 2x + 3x^2\right) = \dfrac{d}{dx}(4) \ \Rightarrow \ (x \cdot y' + y \cdot 1) + 2 + 6x = 0 \ \Rightarrow \ xy' = -y - 2 - 6x \ \Rightarrow$

$y' = \dfrac{-y - 2 - 6x}{x}$ or $y' = -6 - \dfrac{y+2}{x}.$

(b) $xy + 2x + 3x^2 = 4 \ \Rightarrow \ xy = 4 - 2x - 3x^2 \ \Rightarrow \ y = \dfrac{4 - 2x - 3x^2}{x} = \dfrac{4}{x} - 2 - 3x,$ so $y' = -\dfrac{4}{x^2} - 3.$

(c) From part (a), $y' = \dfrac{-y - 2 - 6x}{x} = \dfrac{-(4/x - 2 - 3x) - 2 - 6x}{x} = \dfrac{-4/x - 3x}{x} = -\dfrac{4}{x^2} - 3.$

3. (a) $\dfrac{d}{dx}\left(\dfrac{1}{x} + \dfrac{1}{y}\right) = \dfrac{d}{dx}(1) \ \Rightarrow \ -\dfrac{1}{x^2} - \dfrac{1}{y^2}y' = 0 \ \Rightarrow \ -\dfrac{1}{y^2}y' = \dfrac{1}{x^2} \ \Rightarrow \ y' = -\dfrac{y^2}{x^2}$

(b) $\dfrac{1}{x} + \dfrac{1}{y} = 1 \ \Rightarrow \ \dfrac{1}{y} = 1 - \dfrac{1}{x} = \dfrac{x-1}{x} \ \Rightarrow \ y = \dfrac{x}{x-1},$ so $y' = \dfrac{(x-1)(1) - (x)(1)}{(x-1)^2} = \dfrac{-1}{(x-1)^2}.$

(c) $y' = -\dfrac{y^2}{x^2} = -\dfrac{[x/(x-1)]^2}{x^2} = -\dfrac{x^2}{x^2(x-1)^2} = -\dfrac{1}{(x-1)^2}$

5. $\dfrac{d}{dx}\left(x^2 + y^2\right) = \dfrac{d}{dx}(1) \ \Rightarrow \ 2x + 2yy' = 0 \ \Rightarrow \ 2yy' = -2x \ \Rightarrow \ y' = -\dfrac{x}{y}$

7. $\dfrac{d}{dx}\left(x^3 + x^2 y + 4y^2\right) = \dfrac{d}{dx}(6) \ \Rightarrow \ 3x^2 + (x^2 y' + y \cdot 2x) + 8yy' = 0 \ \Rightarrow \ x^2 y' + 8yy' = -3x^2 - 2xy \ \Rightarrow$

$\left(x^2 + 8y\right)y' = -3x^2 - 2xy \ \Rightarrow \ y' = -\dfrac{3x^2 + 2xy}{x^2 + 8y}$

9. $\dfrac{d}{dx}\left(x^2 y + xy^2\right) = \dfrac{d}{dx}(3x) \ \Rightarrow \ (x^2 y' + y \cdot 2x) + (x \cdot 2yy' + y^2 \cdot 1) = 3 \ \Rightarrow \ x^2 y' + 2xyy' = 3 - 2xy - y^2$

$\Rightarrow \ y'\left(x^2 + 2xy\right) = 3 - 2xy - y^2 \ \Rightarrow \ y' = \dfrac{3 - 2xy - y^2}{x^2 + 2xy}$

11. $\dfrac{y}{x-y} = x^2+1 \;\Rightarrow\; 2x = \dfrac{(x-y)\,y' - y\,(1-y')}{(x-y)^2} = \dfrac{xy'-y}{(x-y)^2} \;\Rightarrow\; y' = \dfrac{y}{x} + 2\,(x-y)^2$

Another Method: Write the equation as $y = (x-y)\,(x^2+1) = x^3 + x - yx^2 - y$. Then $y' = \dfrac{3x^2+1-2xy}{x^2+2}$.

13. $\sqrt{xy} = 1 + x^2 y \;\Rightarrow\; \tfrac{1}{2}\,(xy)^{-1/2}\,(xy'+y\cdot 1) = 0 + x^2 y' + y\cdot 2x \;\Rightarrow\; \dfrac{x}{2\sqrt{xy}}\,y' + \dfrac{y}{2\sqrt{xy}} = x^2 y' + 2xy \;\Rightarrow\;$

$y'\left(\dfrac{x}{2\sqrt{xy}} - x^2\right) = 2xy - \dfrac{y}{2\sqrt{xy}} \;\Rightarrow\; y'\left(\dfrac{x-2x^2\sqrt{xy}}{2\sqrt{xy}}\right) = \dfrac{4xy\sqrt{xy}-y}{2\sqrt{xy}} \;\Rightarrow\; y' = \dfrac{4xy\sqrt{xy}-y}{x-2x^2\sqrt{xy}}$

15. $4\cos x \sin y = 1 \;\Rightarrow\; 4\cos x \cos y \cdot y' + 4\sin y\,(-\sin x) = 0 \;\Rightarrow\; y' = \dfrac{4\sin x \sin y}{4\cos x \cos y} = \tan x \tan y$

17. $\cos(x-y) = y\sin x \;\Rightarrow\; -\sin(x-y)\,(1-y') = y'\sin x + y\cos x \;\Rightarrow\; y' = \dfrac{\sin(x-y)+y\cos x}{\sin(x-y)-\sin x}$

19. $xy = \cot(xy) \;\Rightarrow\; y + xy' = -\csc^2(xy)\,(y+xy') \;\Rightarrow\; (y+xy')\left[1+\csc^2(xy)\right] = 0 \;\Rightarrow\;$

$y + xy' = 0$ [since $1 + \csc^2(xy) > 0$] $\;\Rightarrow\; y' = -y/x$

21. $x\,[f(x)]^3 + xf(x) = 6 \;\Rightarrow\; [f(x)]^3 + 3x\,[f(x)]^2\,f'(x) + f(x) + xf'(x) = 0 \;\Rightarrow\;$

$f'(x) = -\dfrac{[f(x)]^3 + f(x)}{3x\,[f(x)]^2 + x} \;\Rightarrow\; f'(3) = -\dfrac{(1)^3+1}{3\,(3)\,(1)^2+3} = -\dfrac{1}{6}$

23. $y^4 + x^2 y^2 + yx^4 = y+1 \;\Rightarrow\; 4y^3 + 2x\dfrac{dx}{dy}y^2 + 2x^2 y + x^4 + 4yx^3\dfrac{dx}{dy} = 1 \;\Rightarrow\;$

$\dfrac{dx}{dy} = \dfrac{1-4y^3-2x^2y-x^4}{2xy^2+4yx^3}$

25. $\dfrac{x^2}{16} - \dfrac{y^2}{9} = 1 \;\Rightarrow\; \dfrac{x}{8} - \dfrac{2yy'}{9} = 0 \;\Rightarrow\; y' = \dfrac{9x}{16y}$. When $x = -5$ and $y = \tfrac{9}{4}$ we have $y' = \dfrac{9\,(-5)}{16\,(9/4)} = -\dfrac{5}{4}$ so

an equation of the tangent is $y - \tfrac{9}{4} = -\tfrac{5}{4}\,(x+5)$ or $y = -\tfrac{5}{4}x - 4$.

27. $y^2 = x^3\,(2-x) = 2x^3 - x^4 \;\Rightarrow\; 2yy' = 6x^2 - 4x^3 \;\Rightarrow\; y' = \dfrac{3x^2-2x^3}{y}$. When $x = y = 1$,

$y' = \dfrac{3\,(1)^2 - 2\,(1)^3}{1} = 1$, so an equation of the tangent line is $y - 1 = 1\,(x-1)$ or $y = x$.

29. $2\,(x^2+y^2)^2 = 25\,(x^2-y^2) \;\Rightarrow\; 4\,(x^2+y^2)\,(2x+2yy') = 25\,(2x-2yy') \;\Rightarrow\;$

$4yy'\,(x^2+y^2) + 25yy' = 25x - 4x\,(x^2+y^2) \;\Rightarrow\; y' = \dfrac{25x - 4x\,(x^2+y^2)}{25y + 4y\,(x^2+y^2)}$. When $x = 3$ and $y = 1$,

$y' = \dfrac{75-120}{25+40} = -\dfrac{9}{13}$ so an equation of the tangent is $y - 1 = -\dfrac{9}{13}\,(x-3)$ or $y = -\dfrac{9}{13}x + \dfrac{40}{13}$.

31. (a) $y^2 = 5x^4 - x^2 \;\Rightarrow\; 2yy' = 5\,(4x^3) - 2x \;\Rightarrow\; y' = \dfrac{10x^3 - x}{y}$. \hspace{1em} (b)

So at the point $(1, 2)$ we have $y' = \dfrac{10\,(1)^3 - 1}{2} = \dfrac{9}{2}$, and an equation

of the tangent line is $y - 2 = \tfrac{9}{2}\,(x-1)$ or $y = \tfrac{9}{2}x - \tfrac{5}{2}$.

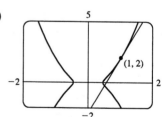

33. (a)

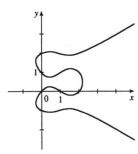

There are eight points with horizontal tangents:
four at $x \approx 1.57735$ and four at $x \approx 0.42265$.

(b) $y' = \dfrac{3x^2 - 6x + 2}{2\left(2y^3 - 3y^2 - y + 1\right)} \quad \Rightarrow \quad y' = -1$ at

$(0, 1)$ and $y' = \frac{1}{3}$ at $(0, 2)$.

Equations of the tangent lines are $y = -x + 1$

and $y = \frac{1}{3}x + 2$.

(c) $y' = 0 \quad \Rightarrow \quad 3x^2 - 6x + 2 = 0 \quad \Rightarrow$

$x = 1 \pm \frac{1}{3}\sqrt{3}$

(d) By multiplying the right side of the equation by $x - 3$, we
obtain the first graph.
By modifying the equation in other ways, we can generate
the other graphs.

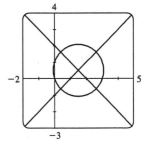

$$y\left(y^2 - 1\right)(y - 2)$$
$$= x\,(x - 1)\,(x - 2)\,(x - 3)$$

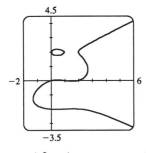

$$y\left(y^2 - 4\right)(y - 2)$$
$$= x\,(x - 1)\,(x - 2)$$

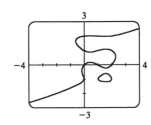

$$y\,(y + 1)\left(y^2 - 1\right)(y - 2)$$
$$= x\,(x - 1)\,(x - 2)$$

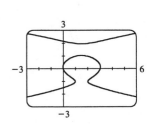

$$(y + 1)\left(y^2 - 1\right)(y - 2)$$
$$= (x - 1)\,(x - 2)$$

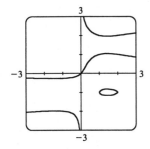

$$x\,(y + 1)\left(y^2 - 1\right)(y - 2)$$
$$= y\,(x - 1)\,(x - 2)$$

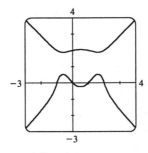

$$y\left(y^2 + 1\right)(y - 2)$$
$$= x\left(x^2 - 1\right)(x - 2)$$

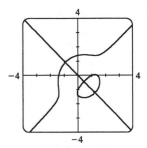

$$y\,(y + 1)\left(y^2 - 2\right)$$
$$= x\,(x - 1)\left(x^2 - 2\right)$$

35. From Exercise 29, a tangent to the lemniscate will be horizontal $\Rightarrow$ $y' = 0$ $\Rightarrow$ $25x - 4x\left(x^2 + y^2\right) = 0$ $\Rightarrow$ $x^2 + y^2 = \frac{25}{4}$. (Note that $x = 0$ $\Rightarrow$ $y = 0$ and there is no horizontal tangent at the origin.) Putting this in the equation of the lemniscate, we get $x^2 - y^2 = \frac{25}{8}$. Solving these two equations we have $x^2 = \frac{75}{16}$ and $y^2 = \frac{25}{16}$, so the four points are $\left(\pm\frac{5\sqrt{3}}{4}, \pm\frac{5}{4}\right)$.

37. $\dfrac{x^2}{a^2} - \dfrac{y^2}{b^2} = 1$ $\Rightarrow$ $\dfrac{2x}{a^2} - \dfrac{2yy'}{b^2} = 0$ $\Rightarrow$ $y' = \dfrac{b^2x}{a^2y}$ $\Rightarrow$ the equation of the tangent at (x_0, y_0) is

$y - y_0 = \dfrac{b^2x_0}{a^2y_0}(x - x_0)$. Multiplying both sides by $\dfrac{y_0}{b^2}$ gives $\dfrac{y_0y}{b^2} - \dfrac{y_0^2}{b^2} = \dfrac{x_0x}{a^2} - \dfrac{x_0^2}{a^2}$. Since (x_0, y_0) lies on the

hyperbola, we have $\dfrac{x_0x}{a^2} - \dfrac{y_0y}{b^2} = \dfrac{x_0^2}{a^2} - \dfrac{y_0^2}{b^2} = 1$.

39. If the circle has radius r, its equation is $x^2 + y^2 = r^2$ $\Rightarrow$ $2x + 2yy' = 0$ $\Rightarrow$ $y' = -\dfrac{x}{y}$, so the slope of the

tangent line at $P\,(x_0, y_0)$ is $-\dfrac{x_0}{y_0}$. The slope of OP is $\dfrac{y_0}{x_0} = \dfrac{-1}{-x_0/y_0}$, so the tangent is perpendicular to OP.

41. $y = \sin^{-1}\left(x^2\right)$ $\Rightarrow$ $y' = \dfrac{1}{\sqrt{1 - \left(x^2\right)^2}}\dfrac{d}{dx}\left(x^2\right) = \dfrac{2x}{\sqrt{1 - x^4}}$

43. $y = \tan^{-1}\left(e^x\right)$ $\Rightarrow$ $y' = \dfrac{1}{1 + \left(e^x\right)^2}\dfrac{d}{dx}\left(e^x\right) = \dfrac{e^x}{1 + e^{2x}}$

45. $H\,(x) = \left(1 + x^2\right)\arctan x$ $\Rightarrow$ $H'\,(x) = (2x)\arctan x + \left(1 + x^2\right)\dfrac{1}{1 + x^2} = 1 + 2x\arctan x$

47. $g\,(t) = \sin^{-1}\left(\dfrac{4}{t}\right)$ $\Rightarrow$ $g'\,(t) = \dfrac{1}{\sqrt{1 - (4/t)^2}}\left(-\dfrac{4}{t^2}\right) = -\dfrac{4}{\sqrt{t^4 - 16t^2}}$

49. $y = x^2\cot^{-1}(3x)$ $\Rightarrow$ $y' = 2x\cot^{-1}(3x) + x^2\left[-\dfrac{1}{1 + (3x)^2}\right](3) = 2x\cot^{-1}(3x) - \dfrac{3x^2}{1 + 9x^2}$

51. $f\,(x) = e^x - x^2\arctan x$ $\Rightarrow$

$f'\,(x) = e^x - \left[x^2\left(\dfrac{1}{1 + x^2}\right) + 2x\arctan x\right]$

$= e^x - \dfrac{x^2}{1 + x^2} - 2x\arctan x$

This is reasonable because the graphs show that f is increasing when $f'\,(x)$ is positive.

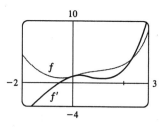

53. Let $y = \cos^{-1}x$. Then $\cos y = x$ and $0 \le y \le \pi$ $\Rightarrow$ $-\sin y\dfrac{dy}{dx} = 1$ $\Rightarrow$

$\dfrac{dy}{dx} = -\dfrac{1}{\sin y} = -\dfrac{1}{\sqrt{1 - \cos^2 y}} = -\dfrac{1}{\sqrt{1 - x^2}}$ (Note that $\sin y \ge 0$ for $0 \le y \le \pi$.)

55. $2x^2 + y^2 = 3$ and $x = y^2$ intersect when $2x^2 + x - 3 = (2x + 3)(x - 1) = 0$ $\Leftrightarrow$ $x = -\frac{3}{2}$ or 1, but $-\frac{3}{2}$ is extraneous. $2x^2 + y^2 = 3$ $\Rightarrow$ $4x + 2yy' = 0$ $\Rightarrow$ $y' = -2x/y$, and $x = y^2$ $\Rightarrow$ $1 = 2yy'$ $\Rightarrow$ $y' = 1/(2y)$. At $(1, 1)$ the slopes are $m_1 = -2$ and $m_2 = \frac{1}{2}$, so the curves are orthogonal there. By symmetry they are also orthogonal at $(1, -1)$.

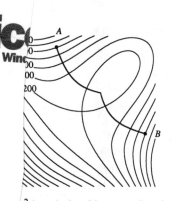

r^2 is a circle with center O and $ax + by = 0$ is a line through O.

$r^2 \Rightarrow 2x + 2yy' = 0 \Rightarrow y' = -x/y$, so the slope of the tangent (x_0, y_0) is $-x_0/y_0$. The slope of the line OP is y_0/x_0, which is the negative reciprocal of $-x_0/y_0$. Hence, the curves are orthogonal.

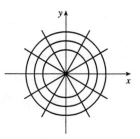

61. $y = cx^2 \Rightarrow y' = 2cx$ and $x^2 + 2y^2 = k \Rightarrow 2x + 4yy' = 0 \Rightarrow$

$y' = -\dfrac{x}{2y} = -\dfrac{x}{2cx^2} = -\dfrac{1}{2cx}$, so the curves are orthogonal.

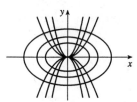

63. $y = 0 \Rightarrow x^2 - x(0) + 0^2 = 3 \Leftrightarrow x = \pm\sqrt{3}$. So the graph of the ellipse crosses the x-axis at the points $\left(\pm\sqrt{3}, 0\right)$. Using implicit differentiation to find y', we get $2x - xy' - y + 2yy' = 0 \Rightarrow y'(2y - x) = y - 2x$

$\Leftrightarrow y' = \dfrac{y - 2x}{2y - x}$. So $y'\left(\sqrt{3}, 0\right) = \dfrac{0 - 2\sqrt{3}}{2(0) - \sqrt{3}} = 2$, and $y'\left(-\sqrt{3}, 0\right) = \dfrac{0 + 2\sqrt{3}}{2(0) + \sqrt{3}} = 2 = y'\left(\sqrt{3}, 0\right)$. So the tangent lines at these points are parallel.

65. $x^2y^2 + xy = 2 \Rightarrow 2xy^2 + 2x^2yy' + y + xy' = 0 \Leftrightarrow y'(2x^2y + x) = -2xy^2 - y \Leftrightarrow y' = -\dfrac{2xy^2 + y}{2x^2y + x}$.

So $-\dfrac{2xy^2 + y}{2x^2y + x} = -1 \Leftrightarrow 2xy^2 + y = 2x^2y + x \Leftrightarrow y(2xy + 1) = x(2xy + 1) \Leftrightarrow$

$(2xy + 1)(y - x) = 0 \Leftrightarrow y = x$ or $xy = -\frac{1}{2}$. But $xy = -\frac{1}{2} \Rightarrow x^2y^2 + xy = \frac{1}{4} - \frac{1}{2} \neq 2$ so we must have $x = y$. Then $x^2y^2 + xy = 2 \Rightarrow x^4 + x^2 = 2 \Leftrightarrow x^4 + x^2 - 2 = 0 \Leftrightarrow (x^2 + 2)(x^2 - 1) = 0$. So $x^2 = -2$, which is impossible, or $x^2 = 1 \Leftrightarrow x = \pm 1$. So the points on the curve where the tangent line has a slope of -1 are $(-1, -1)$ and $(1, 1)$.

67. (a) If $y = f^{-1}(x)$, then $f(y) = x$. Differentiating implicitly with respect to x and remembering that y is a function of x, we get $f'(y)\dfrac{dy}{dx} = 1$, so $\dfrac{dy}{dx} = \dfrac{1}{f'(y)} = \dfrac{1}{f'\left(f^{-1}(x)\right)}$.

(b) $f(4) = 5 \Rightarrow f^{-1}(5) = 4$. $\left(f^{-1}\right)'(5) = 1/f'\left(f^{-1}(5)\right) = 1/f'(4) = 1\Big/\left(\frac{2}{3}\right) = \frac{3}{2}$.

69. $x^2 + 4y^2 = 5 \Rightarrow 2x + 4(2yy') = 0 \Rightarrow y' = -\dfrac{x}{4y}$. Now let h be the height of the lamp, and let (a, b) be

the point of tangency of the line passing through the points $(3, h)$ and $(-5, 0)$. This line has slope

$(h - 0)/[3 - (-5)] = \frac{1}{8}h$. But the slope of the tangent line through the point (a, b) can be expressed as

$y' = -\dfrac{a}{4b}$, or as $\dfrac{b - 0}{a - (-5)} = \dfrac{b}{a + 5}$ [since the line passes through $(-5, 0)$ and (a, b)], so $-\dfrac{a}{4b} = \dfrac{b}{a + 5}$ $\Leftrightarrow$

$4b^2 = -a^2 - 5a \Leftrightarrow a^2 + 4b^2 = -5a$. But $a^2 + 4b^2 = 5$, since (a, b) is on the ellipse, so $5 = -5a \Leftrightarrow$

$a = -1$. Then $4b^2 = -1 - 5(-1) = 4 \Rightarrow b = 1$, since the point is on the top half of the ellipse. So

$\dfrac{h}{8} = \dfrac{b}{a + 5} = \dfrac{1}{-1 + 5} = \dfrac{1}{4} \Rightarrow h = 2$. So the lamp is located 2 units above the x-axis.

3.7 Higher Derivatives

1. $a = f$, $b = f'$, $c = f''$. We can see this because where a has a horizontal tangent, $b = 0$, and where b has a
horizontal tangent, $c = 0$. We can immediately see that c can be neither f nor f', since at the points where c has a
horizontal tangent, neither a nor b is equal to 0.

3. We can immediately see that a is the graph of the acceleration function, since at the points where a has a horizontal
tangent, neither c nor b is equal to 0. Next, we note that $a = 0$ at the point where b has a horizontal tangent, so b
must be the graph of the velocity function, so that $b' = a$. We conclude that c is the graph of the position function.

5. $f(x) = x^5 + 6x^2 - 7x \Rightarrow f'(x) = 5x^4 + 12x - 7 \Rightarrow f''(x) = 20x^3 + 12$

7. $y = \cos 2\theta \Rightarrow y' = -2 \sin 2\theta \Rightarrow y'' = -4 \cos 2\theta$

9. $h(x) = \sqrt{x^2 + 1} \Rightarrow h'(x) = \dfrac{1}{2}(x^2 + 1)^{-1/2}(2x) = \dfrac{x}{\sqrt{x^2 + 1}} \Rightarrow$

$h''(x) = \dfrac{\sqrt{x^2 + 1} - x\left(x/\sqrt{x^2 + 1}\right)}{x^2 + 1} = \dfrac{x^2 + 1 - x^2}{(x^2 + 1)^{3/2}} = \dfrac{1}{(x^2 + 1)^{3/2}}$

11. $F(s) = (3s + 5)^8 \Rightarrow F'(s) = 8(3s + 5)^7 (3) = 24(3s + 5)^7 \Rightarrow$

$F''(s) = 168(3s + 5)^6 (3) = 504(3s + 5)^6$

13. $y = \dfrac{x}{1 - x} \Rightarrow y' = \dfrac{1(1 - x) - x(-1)}{(1 - x)^2} = \dfrac{1}{(1 - x)^2} \Rightarrow y'' = -2(1 - x)^{-3}(-1) = \dfrac{2}{(1 - x)^3}$

15. $y = (1 - x^2)^{3/4} \Rightarrow y' = \frac{3}{4}(1 - x^2)^{-1/4}(-2x) = -\frac{3}{2}x(1 - x^2)^{-1/4} \Rightarrow$

$y'' = -\frac{3}{2}(1 - x^2)^{-1/4} - \frac{3}{2}x\left(-\frac{1}{4}\right)(1 - x^2)^{-5/4}(-2x) = -\frac{3}{2}(1 - x^2)^{-1/4} - \frac{3}{4}x^2(1 - x^2)^{-5/4}$

$= \frac{3}{4}(1 - x^2)^{-5/4}(x^2 - 2)$

17. $H(t) = \tan 3t \Rightarrow H'(t) = 3 \sec^2 3t \Rightarrow$

$H''(t) = 2 \cdot 3 \sec 3t \dfrac{d}{dt}(\sec 3t) = 6 \sec 3t (3 \sec 3t \tan 3t) = 18 \sec^2 3t \tan 3t$

19. $g(t) = t^3 e^{5t} \Rightarrow g'(t) = t^3 e^{5t} \cdot 5 + e^{5t} \cdot 3t^2 = t^2 e^{5t}(5t + 3) \Rightarrow$

$g''(t) = (2t) e^{5t}(5t + 3) + t^2 (5e^{5t})(5t + 3) + t^2 e^{5t}(5)$

$= te^{5t}[2(5t + 3) + 5t(5t + 3) + 5t] = te^{5t}(25t^2 + 30t + 6)$

21. (a) $f(x) = 2\cos x + \sin^2 x \Rightarrow f'(x) = 2(-\sin x) + 2\sin x (\cos x) = \sin 2x - 2\sin x \Rightarrow$

$f''(x) = 2\cos 2x - 2\cos x = 2(\cos 2x - \cos x)$

(b)

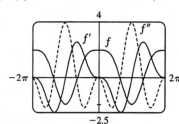

We can see that our answers are plausible, since f has horizontal tangents where $f'(x) = 0$, and f' has horizontal tangents where $f''(x) = 0$.

23. $y = \sqrt{2x+3} = (2x+3)^{1/2} \Rightarrow y' = \frac{1}{2}(2x+3)^{-1/2} \cdot 2 = (2x+3)^{-1/2} \Rightarrow$

$y'' = -\frac{1}{2}(2x+3)^{-3/2} \cdot 2 = -(2x+3)^{-3/2} \Rightarrow y''' = \frac{3}{2}(2x+3)^{-5/2} \cdot 2 = 3(2x+3)^{-5/2}$

25. $f(x) = (2-3x)^{-1/2} \Rightarrow f(0) = 2^{-1/2} = \frac{1}{\sqrt{2}}$

$f'(x) = -\frac{1}{2}(2-3x)^{-3/2}(-3) = \frac{3}{2}(2-3x)^{-3/2} \Rightarrow f'(0) = \frac{3}{2}(2)^{-3/2} = \frac{3}{4\sqrt{2}}$

$f''(x) = -\frac{9}{4}(2-3x)^{-5/2}(-3) = \frac{27}{4}(2-3x)^{-5/2} \Rightarrow f''(0) = \frac{27}{4}(2)^{-5/2} = \frac{27}{16\sqrt{2}}$

$f'''(x) = \frac{405}{8}(2-3x)^{-7/2} \Rightarrow f'''(0) = \frac{405}{8}(2)^{-7/2} = \frac{405}{64\sqrt{2}}$

27. $f(\theta) = \cot\theta \Rightarrow f'(\theta) = -\csc^2\theta \Rightarrow f''(\theta) = -2\csc\theta(-\csc\theta\cot\theta) = 2\csc^2\theta\cot\theta \Rightarrow$

$f'''(\theta) = 2(-2\csc^2\theta\cot\theta)\cot\theta + 2\csc^2\theta(-\csc^2\theta) = -2\csc^2\theta(2\cot^2\theta + \csc^2\theta) \Rightarrow$

$f'''\left(\frac{\pi}{6}\right) = -2(2)^2\left[2\left(\sqrt{3}\right)^2 + (2)^2\right] = -80$

29. $x^3 + y^3 = 1 \Rightarrow 3x^2 + 3y^2 y' = 0 \Rightarrow y' = -\frac{x^2}{y^2} \Rightarrow$

$y'' = -\frac{2xy^2 - 2x^2 yy'}{y^4} = -\frac{2xy^2 - 2x^2 y\left(-x^2/y^2\right)}{y^4} = -\frac{2xy^3 + 2x^4}{y^5} = -\frac{2x\left(y^3 + x^3\right)}{y^5} = -\frac{2x}{y^5}$, since x and y

must satisfy the original equation, $x^3 + y^3 = 1$.

31. $x^2 + xy + y^2 = 1 \Rightarrow 2x + xy' + y + 2yy' = 0 \Rightarrow y'(x+2y) = -2x - y \Rightarrow y' = -\frac{2x+y}{x+2y} \Rightarrow$

$y'' = -\frac{(x+2y)(2+y') - (2x+y)(1+2y')}{(x+2y)^2} = -\frac{(2x+xy'+4y+2yy') - (2x+4xy'+y+2yy')}{(x+2y)^2}$

$= -\frac{-3xy'+3y}{(x+2y)^2} = -\frac{-3x\left(-\dfrac{2x+y}{x+2y}\right)+3y}{(x+2y)^2} = -\frac{\dfrac{3x(2x+y)+3y(x+2y)}{x+2y}}{(x+2y)^2}$

$= -\frac{6x^2 + 3xy + 3xy + 6y^2}{(x+2y)^3} = -\frac{6\left(x^2 + xy + y^2\right)}{(x+2y)^3}$

$= -\frac{6}{(x+2y)^3}$, since x and y must satisfy the original equation, $x^2 + xy + y^2 = 1$.

33. $f(x) = x^n \Rightarrow f'(x) = nx^{n-1} \Rightarrow f''(x) = n(n-1)x^{n-2} \Rightarrow \cdots \Rightarrow$

$f^{(n)}(x) = n(n-1)(n-2)\cdots 2 \cdot 1 x^{n-n} = n!$

35. $f(x) = e^{2x} \Rightarrow f'(x) = 2e^{2x} \Rightarrow f''(x) = 2 \cdot 2e^{2x} = 2^2 e^{2x} \Rightarrow f'''(x) = 2^2 \cdot 2e^{2x} = 2^3 e^{2x} \Rightarrow \cdots$

$\Rightarrow f^{(n)}(x) = 2^n e^{2x}$

37. $f(x) = 1/(3x^3) = \frac{1}{3}x^{-3} \Rightarrow f'(x) = \frac{1}{3}(-3)x^{-4} \Rightarrow$

$f''(x) = \frac{1}{3}(-3)(-4)x^{-5} \Rightarrow f'''(x) = \frac{1}{3}(-3)(-4)(-5)x^{-6} \Rightarrow \cdots \Rightarrow$

$f^{(n)}(x) = \frac{1}{3}(-3)(-4)\cdots[-(n+2)]x^{-(n+3)} = \dfrac{(-1)^n \cdot 3 \cdot 4 \cdot 5 \cdots \cdot (n+2)}{3x^{n+3}} = \dfrac{(-1)^n (n+2)!}{6x^{n+3}}$

39. In general, $Df(2x) = 2f'(2x)$, $D^2 f(2x) = 4f''(2x)$, ..., $D^n f(2x) = 2^n f^{(n)}(2x)$. Specifically, since $f(x) = \cos x$ and $50 = 4(12) + 2$, we have $f^{(50)}(x) = f^{(2)}(x) = -\cos x$, so $D^{50} \cos 2x = -2^{50} \cos 2x$.

41. By measuring the slope of the graph of $s = f(t)$ at $t = 0, 1, 2, 3, 4$, and 5, and using the method of Example 1 in Section 2.9, we plot the graph of the velocity function $v = f'(t)$ in the first figure. The acceleration when $t = 2$ s is $a = f''(2)$, the slope of the tangent line to the graph of f' when $t = 2$. We estimate the slope of this tangent line to be $a(2) = f''(2) = v'(2) \approx \frac{27}{3} = 9$ ft/s^2. Similar measurements enable us to graph the acceleration function in the second figure.

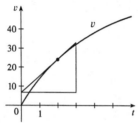
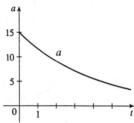

43. (a) $s = t^3 - 3t \Rightarrow v(t) = s'(t) = 3t^2 - 3 \Rightarrow a(t) = v'(t) = 6t$

(b) $a(1) = 6(1) = 6$ m/s^2

(c) $v(t) = 3t^2 - 3 = 0$ when $t^2 = 1$, that is, $t = 1$ and $a(1) = 6$ m/s^2.

45. (a) $s = \sin 2\pi t \Rightarrow v(t) = s'(t) = 2\pi \cos 2\pi t \Rightarrow a(t) = v'(t) = -4\pi^2 \sin 2\pi t$

(b) $a(1) = -4\pi^2 \sin 2\pi(1) = -4\pi^2(0) = 0$ m/s^2

(c) $v(t) = 2\pi \cos 2\pi t = 0$ when $2\pi t$ is an odd multiple of $\frac{\pi}{2} \Leftrightarrow t$ is an odd multiple of $\frac{1}{4} \Leftrightarrow$

$t \in \left\{ \pm\frac{1}{4}, \pm\frac{3}{4}, \pm\frac{5}{4}, \ldots \right\}$. When $t \in \left\{ \ldots, -\frac{7}{4}, -\frac{3}{4}, \frac{1}{4}, \frac{5}{4}, \ldots \right\}$, $\sin 2\pi t = 1$ and $a(t) = -4\pi^2$ m/s^2. When

$t \in \left\{ \ldots, -\frac{5}{4}, -\frac{1}{4}, \frac{3}{4}, \frac{7}{4}, \ldots \right\}$, $\sin 2\pi t = -1$ and $a(t) = 4\pi^2$ m/s^2.

47. (a) $s(t) = t^4 - 4t^3 + 2 \Rightarrow v(t) = s'(t) = 4t^3 - 12t^2 \Rightarrow a(t) = v'(t) = 12t^2 - 24t = 12t(t-2) = 0$ when $t = 0$ or 2.

(b) $s(0) = 2$ m, $v(0) = 0$ m/s, $s(2) = -14$ m, $v(2) = -16$ m/s

49. (a) $s = f(t) = t^3 - 12t^2 + 36t \Rightarrow v(t) = s'(t) = 3t^2 - 24t + 36 \Rightarrow a(t) = v'(t) = 6t - 24$.

$a(3) = -6$ (m/s)/s or m/s^2

(b)

(c) First note that $v(t) = 0$ when $t = 2$ and when $t = 6$, and $a(t) = 0$ when $t = 4$. The particle is speeding up when v and a have the same sign. This occurs when $2 < t < 4$ and when $t > 6$. It is slowing down when v and a have opposite signs; that is, when $0 \le t < 2$ and $4 < t < 6$.

51. (a) $y(t) = A \sin \omega t \Rightarrow v(t) = y'(t) = A\omega \cos \omega t \Rightarrow a(t) = v'(t) = -A\omega^2 \sin \omega t$

(b) $a(t) = -A\omega^2 \sin \omega t = -\omega^2 y(t)$

(c) $|v(t)| = A\omega |\cos \omega t|$ is a maximum when $\cos \omega t = \pm 1 \Leftrightarrow \sin \omega t = 0 \Leftrightarrow a(t) = -A\omega^2 \sin^2 \omega t = 0$.

53. Let $P(x) = ax^2 + bx + c$. Then $P'(x) = 2ax + b$ and $P''(x) = 2a$.

$P''(2) = 2 \Rightarrow 2a = 2 \Rightarrow a = 1$. $P'(2) = 3 \Rightarrow 4a + b = 4 + b = 3 \Rightarrow b = -1$.

$P(2) = 5 \Rightarrow 2^2 - 2 + c = 5 \Rightarrow c = 3$. So $P(x) = x^2 - x + 3$.

55. $y = A\sin x + B\cos x \Rightarrow y' = A\cos x - B\sin x \Rightarrow y'' = -A\sin x - B\cos x$. Substituting into

$y'' + y' - 2y = \sin x$ gives us $(-3A - B)\sin x + (A - 3B)\cos x = 1\sin x$, so we must have $-3A - B = 1$ and

$A - 3B = 0$. Solving for A and B, we add the first equation to three times the second to get $B = -\frac{1}{10}$ and

$A = -\frac{3}{10}$.

57. $y = e^{rx} \Rightarrow y' = re^{rx} \Rightarrow y'' = r^2 e^{rx}$, so

$y'' + 5y' - 6y = r^2 e^{rx} + 5re^{rx} - 6e^{rx} = e^{rx}(r^2 + 5r - 6) = e^{rx}(r + 6)(r - 1) = 0 \Rightarrow (r + 6)(r - 1) = 0$

$\Rightarrow r = 1$ or -6.

59. $f(x) = xg(x^2) \Rightarrow f'(x) = g(x^2) + xg'(x^2)\,2x = g(x^2) + 2x^2 g'(x^2) \Rightarrow$

$f''(x) = 2xg'(x^2) + 4xg'(x^2) + 4x^3 g''(x^2) = 6xg'(x^2) + 4x^3 g''(x^2)$

61. $f(x) = g(\sqrt{x}) \Rightarrow f'(x) = \dfrac{g'(\sqrt{x})}{2\sqrt{x}} \Rightarrow f''(x) = \dfrac{\dfrac{g''(\sqrt{x})}{2\sqrt{x}} \cdot 2\sqrt{x} - \dfrac{g'(\sqrt{x})}{\sqrt{x}}}{4x} = \dfrac{\sqrt{x}\,g''(\sqrt{x}) - g'(\sqrt{x})}{4x\sqrt{x}}$

63. (a) $f(x) = \dfrac{1}{x^2 + x} \Rightarrow f'(x) = \dfrac{-(2x + 1)}{(x^2 + x)^2} \Rightarrow$

$f''(x) = \dfrac{(x^2 + x)^2(-2) + (2x + 1)(2)(x^2 + x)(2x + 1)}{(x^2 + x)^4} = \dfrac{2(3x^2 + 3x + 1)}{(x^2 + x)^3} \Rightarrow$

$f'''(x) = \dfrac{(x^2 + x)^3(2)(6x + 3) - 2(3x^2 + 3x + 1)(3)(x^2 + x)^2(2x + 1)}{(x^2 + x)^6}$

$= \dfrac{-6(4x^3 + 6x^2 + 4x + 1)}{(x^2 + x)^4} \Rightarrow$

$f^{(4)}(x) = \dfrac{(x^2 + x)^4(-6)(12x^2 + 12x + 4) + 6(4x^3 + 6x^2 + 4x + 1)(4)(x^2 + x)^3(2x + 1)}{(x^2 + x)^8}$

$= \dfrac{24(5x^4 + 10x^3 + 10x^2 + 5x + 1)}{(x^2 + x)^5}$

$f^{(5)}(x) = ?$

(b) $f(x) = \dfrac{1}{x(x + 1)} = \dfrac{1}{x} - \dfrac{1}{x + 1} \Rightarrow f'(x) = -x^{-2} + (x + 1)^{-2} \Rightarrow$

$f''(x) = 2x^{-3} - 2(x + 1)^{-3} \Rightarrow f'''(x) = (-3)(2)x^{-4} + (3)(2)(x + 1)^{-4} \Rightarrow \cdots \Rightarrow$

$f^{(n)}(x) = (-1)^n n! \left[x^{-(n+1)} - (x + 1)^{-(n+1)} \right]$

65. The Chain Rule says that $\dfrac{dy}{dx} = \dfrac{dy}{du}\dfrac{du}{dx}$, so

$\dfrac{d^2y}{dx^2} = \dfrac{d}{dx}\left(\dfrac{dy}{dx}\right) = \dfrac{d}{dx}\left(\dfrac{dy}{du}\dfrac{du}{dx}\right) = \left[\dfrac{d}{dx}\left(\dfrac{dy}{du}\right)\right]\dfrac{du}{dx} + \dfrac{dy}{du}\dfrac{d}{dx}\left(\dfrac{du}{dx}\right)$ (Product Rule)

$= \left[\dfrac{d}{du}\left(\dfrac{dy}{du}\right)\dfrac{du}{dx}\right]\dfrac{du}{dx} + \dfrac{dy}{du}\dfrac{d^2u}{dx^2} = \dfrac{d^2y}{du^2}\left(\dfrac{du}{dx}\right)^2 + \dfrac{dy}{du}\dfrac{d^2u}{dx^2}$

3.8 Derivatives of Logarithmic Functions

1. The differentiation formula for logarithmic functions, $\dfrac{d}{dx}\left(\log_a x\right) = \dfrac{1}{x \ln a}$, is simplest when $a = e$ because $\ln e = 1$.

3. $f(\theta) = \ln(\cos\theta) \Rightarrow f'(\theta) = \dfrac{1}{\cos\theta}\dfrac{d}{d\theta}(\cos\theta) = \dfrac{-\sin\theta}{\cos\theta} = -\tan\theta$

5. $f(x) = \log_3\left(x^2 - 4\right) \Rightarrow f'(x) = \dfrac{1}{\left(x^2 - 4\right)\ln 3}(2x) = \dfrac{2x}{\left(x^2 - 4\right)\ln 3}$

7. $F(x) = \ln\sqrt{x} = \ln x^{1/2} = \frac{1}{2}\ln x \Rightarrow F'(x) = \dfrac{1}{2}\left(\dfrac{1}{x}\right) = \dfrac{1}{2x}$

9. $f(x) = \sqrt{x}\ln x \Rightarrow f'(x) = \dfrac{1}{2\sqrt{x}}\ln x + \sqrt{x}\left(\dfrac{1}{x}\right) = \dfrac{\ln x}{2\sqrt{x}} + \dfrac{1}{\sqrt{x}} = \dfrac{\ln x + 2}{2\sqrt{x}}$

11. $g(x) = \ln\dfrac{a-x}{a+x} = \ln(a-x) - \ln(a+x) \Rightarrow$

$g'(x) = \dfrac{1}{a-x}(-1) - \dfrac{1}{a+x} = \dfrac{-(a+x) - (a-x)}{(a-x)(a+x)} = \dfrac{-2a}{a^2 - x^2}$

13. $F(x) = e^x \ln x \Rightarrow F'(x) = e^x \ln x + e^x\left(\dfrac{1}{x}\right) = e^x\left(\ln x + \dfrac{1}{x}\right)$

15. $y = \dfrac{\ln x}{1+x} \Rightarrow y' = \dfrac{(1+x)(1/x) - (\ln x)(1)}{(1+x)^2} = \dfrac{\dfrac{1+x}{x} - \dfrac{x\ln x}{x}}{(1+x)^2} = \dfrac{1+x - x\ln x}{x(1+x)^2}$

17. $y = \ln\left|x^3 - x^2\right| \Rightarrow y' = \dfrac{1}{x^3 - x^2}\left(3x^2 - 2x\right) = \dfrac{x(3x - 2)}{x^2(x-1)} = \dfrac{3x - 2}{x(x - 1)}$

19. $y = \ln\left(e^{-x}(1+x)\right) = \ln\left(e^{-x}\right) + \ln(1+x) = -x + \ln(1+x) \Rightarrow y' = -1 + \dfrac{1}{1+x} = -\dfrac{x}{1+x}$

21. $y = x\ln x \Rightarrow y' = \ln x + x(1/x) = \ln x + 1 \Rightarrow y'' = 1/x$

23. $y = \log_{10} x \Rightarrow y' = \dfrac{1}{x \ln 10} = \dfrac{1}{\ln 10}\left(\dfrac{1}{x}\right) \Rightarrow y'' = \dfrac{1}{\ln 10}\left(-\dfrac{1}{x^2}\right) = -\dfrac{1}{x^2 \ln 10}$

25. $f(x) = \ln(2x + 1) \Rightarrow f'(x) = \dfrac{1}{2x + 1}\cdot 2 = \dfrac{2}{2x + 1}$. $\text{Dom}(f) = \{x \mid 2x + 1 > 0\} = \left(-\frac{1}{2}, \infty\right)$.

27. $f(x) = x^2 \ln\left(1 - x^2\right) \Rightarrow f'(x) = 2x\ln\left(1 - x^2\right) + \dfrac{x^2(-2x)}{1 - x^2} = 2x\ln\left(1 - x^2\right) - \dfrac{2x^3}{1 - x^2}$.

$\text{Dom}(f) = \{x \mid 1 - x^2 > 0\} = \{x \mid |x| < 1\} = (-1, 1)$.

29. $f(x) = \dfrac{x}{\ln x}$ $\Rightarrow$ $f'(x) = \dfrac{\ln x - x(1/x)}{(\ln x)^2} = \dfrac{\ln x - 1}{(\ln x)^2}$ $\Rightarrow$ $f'(e) = \dfrac{1-1}{1^2} = 0$

31. $y = f(x) = \ln \ln x$ $\Rightarrow$ $f'(x) = \dfrac{1}{\ln x}\left(\dfrac{1}{x}\right)$ $\Rightarrow$ $f'(e) = \dfrac{1}{e}$, so an equation of the tangent line at $(e, 0)$ is

$y - 0 = \dfrac{1}{e}(x - e)$, or $y = \dfrac{1}{e}x - 1$, or $x - ey = e$.

33. $f(x) = \sin x + \ln x$ $\Rightarrow$ $f'(x) = \cos x + 1/x$. This is reasonable, because the graph shows that f increases when $f'(x)$ is positive, and $f'(x) = 0$ when f has a horizontal tangent.

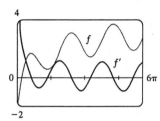

35. $y = (2x+1)^5 (x^4 - 3)^6$ $\Rightarrow$ $\ln y = \ln\left((2x+1)^5 (x^4 - 3)^6\right)$ $\Rightarrow$ $\ln y = 5\ln(2x+1) + 6\ln(x^4 - 3)$ $\Rightarrow$

$\dfrac{1}{y}y' = 5 \cdot \dfrac{1}{2x+1} \cdot 2 + 6 \cdot \dfrac{1}{x^4 - 3} \cdot 4x^3$ $\Rightarrow$

$y' = y\left(\dfrac{10}{2x+1} + \dfrac{24x^3}{x^4 - 3}\right) = y \cdot \dfrac{10(x^4 - 3) + 24x^3(2x+1)}{(2x+1)(x^4 - 3)} = (2x+1)^5 (x^4 - 3)^6 \cdot \dfrac{58x^4 + 24x^3 - 30}{(2x+1)(x^4 - 3)}$

$= 2(2x+1)^4 (x^4 - 3)^5 (29x^4 + 12x^3 - 15)$

37. $y = \dfrac{\sin^2 x \tan^4 x}{(x^2 + 1)^2}$ $\Rightarrow$ $\ln y = \ln(\sin^2 x \tan^4 x) - \ln(x^2 + 1)^2 = 2\ln \sin x + 4\ln \tan x - 2\ln(x^2 + 1)$ $\Rightarrow$

$\dfrac{1}{y}y' = 2 \cdot \dfrac{1}{\sin x} \cdot \cos x + 4 \cdot \dfrac{1}{\tan x} \cdot \sec^2 x - 2 \cdot \dfrac{1}{x^2 + 1} \cdot 2x$ $\Rightarrow$

$y' = \dfrac{\sin^2 x \tan^4 x}{(x^2 + 1)^2}\left(2\cot x + \dfrac{4\sec^2 x}{\tan x} - \dfrac{4x}{x^2 + 1}\right)$

39. $y = x^x$ $\Rightarrow$ $\ln y = x \ln x$ $\Rightarrow$ $y'/y = \ln x + x(1/x)$ $\Rightarrow$ $y' = x^x(\ln x + 1)$

41. $y = x^{\sin x}$ $\Rightarrow$ $\ln y = \sin x \ln x$ $\Rightarrow$ $\dfrac{y'}{y} = \cos x \ln x + \dfrac{\sin x}{x}$ $\Rightarrow$ $y' = x^{\sin x}\left(\cos x \ln x + \dfrac{\sin x}{x}\right)$

43. $y = (\ln x)^x$ $\Rightarrow$ $\ln y = x \ln \ln x$ $\Rightarrow$ $\dfrac{y'}{y} = \ln \ln x + x \cdot \dfrac{1}{\ln x} \cdot \dfrac{1}{x}$ $\Rightarrow$ $y' = (\ln x)^x\left(\ln \ln x + \dfrac{1}{\ln x}\right)$

45. $y = x^{e^x}$ $\Rightarrow$ $\ln y = e^x \ln x$ $\Rightarrow$ $\dfrac{y'}{y} = e^x \ln x + \dfrac{e^x}{x}$ $\Rightarrow$ $y' = x^{e^x} e^x\left(\ln x + \dfrac{1}{x}\right)$

47. $y = \ln(x^2 + y^2)$ $\Rightarrow$ $y' = \dfrac{2x + 2yy'}{x^2 + y^2}$ $\Rightarrow$ $x^2 y' + y^2 y' = 2x + 2yy'$ $\Rightarrow$ $y' = \dfrac{2x}{x^2 + y^2 - 2y}$

49. $f(x) = \ln(x - 1)$ $\Rightarrow$ $f'(x) = 1/(x - 1) = (x - 1)^{-1}$ $\Rightarrow$ $f''(x) = -(x - 1)^{-2}$

$\Rightarrow$ $f'''(x) = 2(x - 1)^{-3}$ $\Rightarrow$ $f^{(4)}(x) = -2 \cdot 3(x - 1)^{-4}$ $\Rightarrow$ $\cdots$ $\Rightarrow$

$f^{(n)}(x) = (-1)^{n-1} \cdot 2 \cdot 3 \cdot 4 \cdots \cdots (n - 1)(x - 1)^{-n} = (-1)^{n-1}\dfrac{(n-1)!}{(x-1)^n}$

51. If $f(x) = \ln(1 + x)$, then $f'(x) = \dfrac{1}{1 + x}$, so $f'(0) = 1$.

Thus, $\displaystyle\lim_{x \to 0} \dfrac{\ln(1 + x)}{x} = \lim_{x \to 0} \dfrac{f(x)}{x} = \lim_{x \to 0} \dfrac{f(x) - f(0)}{x - 0} = f'(0) = 1$.

◼3.9 Hyperbolic Functions

1. (a) $\sinh 0 = \frac{1}{2}\left(e^0 - e^0\right) = 0$ (b) $\cosh 0 = \frac{1}{2}\left(e^0 + e^0\right) = \frac{1}{2}(1+1) = 1$

3. (a) $\sinh(\ln 2) = \dfrac{e^{\ln 2} - e^{-\ln 2}}{2} = \dfrac{2 - \frac{1}{2}}{2} = \dfrac{3}{4}$ (b) $\sinh 2 = \frac{1}{2}\left(e^2 - e^{-2}\right) \approx 3.62686$

5. (a) $\operatorname{sech} 0 = \dfrac{1}{\cosh 0} = \dfrac{1}{1} = 1$ (b) $\cosh^{-1} 1 = 0$ because $\cosh 0 = 1$.

7. $\sinh(-x) = \frac{1}{2}\left[e^{-x} - e^{-(-x)}\right] = \frac{1}{2}\left(e^{-x} - e^x\right) = -\frac{1}{2}\left(e^x - e^{-x}\right) = -\sinh x$

9. $\cosh x + \sinh x = \frac{1}{2}\left(e^x + e^{-x}\right) + \frac{1}{2}\left(e^x - e^{-x}\right) = \frac{1}{2}\left(2e^x\right) = e^x$

11. $\sinh x \cosh y + \cosh x \sinh y = \left[\frac{1}{2}\left(e^x - e^{-x}\right)\right]\left[\frac{1}{2}\left(e^y + e^{-y}\right)\right] + \left[\frac{1}{2}\left(e^x + e^{-x}\right)\right]\left[\frac{1}{2}\left(e^y - e^{-y}\right)\right]$

$\qquad = \frac{1}{4}\left[\left(e^{x+y} + e^{x-y} - e^{-x+y} - e^{-x-y}\right) + \left(e^{x+y} - e^{x-y} + e^{-x+y} - e^{-x-y}\right)\right]$

$\qquad = \frac{1}{4}\left(2e^{x+y} - 2e^{-x-y}\right) = \frac{1}{2}\left[e^{x+y} - e^{-(x+y)}\right] = \sinh(x+y)$

13. Divide both sides of the identity $\cosh^2 x - \sinh^2 x = 1$ by $\sinh^2 x$:

$\dfrac{\cosh^2 x}{\sinh^2 x} - 1 = \dfrac{1}{\sinh^2 x} \quad\Leftrightarrow\quad \coth^2 x - 1 = \operatorname{csch}^2 x.$

15. By Exercise 11, $\sinh 2x = \sinh(x+x) = \sinh x \cosh x + \cosh x \sinh x = 2\sinh x \cosh x.$

17. $\tanh(\ln x) = \dfrac{\sinh(\ln x)}{\cosh(\ln x)} = \dfrac{\left(e^{\ln x} - e^{-\ln x}\right)/2}{\left(e^{\ln x} + e^{-\ln x}\right)/2} = \dfrac{x - 1/x}{x + 1/x} = \dfrac{x^2 - 1}{x^2 + 1}$

19. By Exercise 9, $(\cosh x + \sinh x)^n = (e^x)^n = e^{nx} = \cosh nx + \sinh nx.$

21. $\tanh x = \frac{4}{5} > 0$, so $x > 0$. $\coth x = 1/\tanh x = \frac{5}{4}$, $\operatorname{sech}^2 x = 1 - \tanh^2 x = 1 - \left(\frac{4}{5}\right)^2 = \frac{9}{25} \;\Rightarrow\; \operatorname{sech} x = \frac{3}{5}$

$\qquad$ (since $\operatorname{sech} x > 0$), $\cosh x = 1/\operatorname{sech} x = \frac{5}{3}$, $\sinh x = \tanh x \cosh x = \frac{4}{5}\cdot\frac{5}{3} = \frac{4}{3}$, and $\operatorname{csch} x = 1/\sinh x = \frac{3}{4}.$

23. (a) $\displaystyle\lim_{x\to\infty}\tanh x = \lim_{x\to\infty}\dfrac{e^x - e^{-x}}{e^x + e^{-x}} = \lim_{x\to\infty}\dfrac{1 - e^{-2x}}{1 + e^{-2x}} = \dfrac{1-0}{1+0} = 1$

$\qquad$ (b) $\displaystyle\lim_{x\to-\infty}\tanh x = \lim_{x\to-\infty}\dfrac{e^x - e^{-x}}{e^x + e^{-x}} = \lim_{x\to-\infty}\dfrac{e^{2x} - 1}{e^{2x} + 1} = \dfrac{0-1}{0+1} = -1$

$\qquad$ (c) $\displaystyle\lim_{x\to\infty}\sinh x = \lim_{x\to\infty}\dfrac{e^x - e^{-x}}{2} = \infty$

$\qquad$ (d) $\displaystyle\lim_{x\to-\infty}\sinh x = \lim_{x\to-\infty}\dfrac{e^x - e^{-x}}{2} = -\infty$

$\qquad$ (e) $\displaystyle\lim_{x\to\infty}\operatorname{sech} x = \lim_{x\to\infty}\dfrac{2}{e^x + e^{-x}} = 0$

$\qquad$ (f) $\displaystyle\lim_{x\to\infty}\coth x = \lim_{x\to\infty}\dfrac{e^x + e^{-x}}{e^x - e^{-x}} = \lim_{x\to\infty}\dfrac{1 + e^{-2x}}{1 - e^{-2x}} = \dfrac{1+0}{1-0} = 1$ [*Or:* Use part (a)]

$\qquad$ (g) $\displaystyle\lim_{x\to0^+}\coth x = \lim_{x\to0^+}\dfrac{\cosh x}{\sinh x} = \infty$, since $\sinh x \to 0$ and $\coth x > 0$.

$\qquad$ (h) $\displaystyle\lim_{x\to0^-}\coth x = \lim_{x\to0^-}\dfrac{\cosh x}{\sinh x} = -\infty$, since $\sinh x \to 0$ and $\coth x < 0$.

$\qquad$ (i) $\displaystyle\lim_{x\to-\infty}\operatorname{csch} x = \lim_{x\to-\infty}\dfrac{2}{e^x - e^{-x}} = 0$

25. Let $y = \sinh^{-1} x$. Then $\sinh y = x$ and, by Example 1(a), $\cosh y = \sqrt{1 + \sinh^2 y} = \sqrt{1 + x^2}$. So by Exercise 9,

$$e^y = \sinh y + \cosh y = x + \sqrt{1 + x^2} \quad \Rightarrow \quad y = \ln\left(x + \sqrt{1 + x^2}\right).$$

27. (a) Let $y = \tanh^{-1} x$. Then $x = \tanh y = \dfrac{e^y - e^{-y}}{e^y + e^{-y}} = \dfrac{e^{2y} - 1}{e^{2y} + 1} \quad \Rightarrow \quad xe^{2y} + x = e^{2y} - 1 \quad \Rightarrow \quad e^{2y} = \dfrac{1 + x}{1 - x}$

$$\Rightarrow \quad 2y = \ln\left(\frac{1 + x}{1 - x}\right) \quad \Rightarrow \quad y = \tfrac{1}{2} \ln\left(\frac{1 + x}{1 - x}\right).$$

(b) Let $y = \tanh^{-1} x$. Then $x = \tanh y$, so from Exercise 18 we have $e^{2y} = \dfrac{1 + \tanh y}{1 - \tanh y} = \dfrac{1 + x}{1 - x} \quad \Rightarrow$

$$2y = \ln\left(\frac{1 + x}{1 - x}\right) \quad \Rightarrow \quad y = \tfrac{1}{2} \ln\left(\frac{1 + x}{1 - x}\right).$$

29. (a) Let $y = \cosh^{-1} x$. Then $\cosh y = x$ and $y \geq 0 \quad \Rightarrow \quad \sinh y \dfrac{dy}{dx} = 1 \quad \Rightarrow$

$$\frac{dy}{dx} = \frac{1}{\sinh y} = \frac{1}{\sqrt{\cosh^2 y - 1}} = \frac{1}{\sqrt{x^2 - 1}} \quad \text{(since } \sinh y \geq 0 \text{ for } y \geq 0\text{)}. \quad Or: \text{ Use Formula 4.}$$

(b) Let $y = \tanh^{-1} x$. Then $\tanh y = x \quad \Rightarrow \quad \text{sech}^2 y \dfrac{dy}{dx} = 1 \quad \Rightarrow \quad \dfrac{dy}{dx} = \dfrac{1}{\text{sech}^2 y} = \dfrac{1}{1 - \tanh^2 y} = \dfrac{1}{1 - x^2}.$

$Or:$ Use Formula 5.

(c) Let $y = \text{csch}^{-1} x$. Then $\text{csch } y = x \quad \Rightarrow \quad -\text{csch } y \coth y \dfrac{dy}{dx} = 1 \quad \Rightarrow \quad \dfrac{dy}{dx} = -\dfrac{1}{\text{csch } y \coth y}.$

By Exercise 13, $\coth y = \pm\sqrt{\text{csch}^2 y + 1} = \pm\sqrt{x^2 + 1}$. If $x > 0$, then $\coth y > 0$, so $\coth y = \sqrt{x^2 + 1}$.

If $x < 0$, then $\coth y < 0$, so $\coth y = -\sqrt{x^2 + 1}$. In either case we have

$$\frac{dy}{dx} = -\frac{1}{\text{csch } y \coth y} = -\frac{1}{|x|\sqrt{x^2 + 1}}.$$

(d) Let $y = \text{sech}^{-1} x$. Then $\text{sech } y = x \quad \Rightarrow \quad -\text{sech } y \tanh y \dfrac{dy}{dx} = 1 \quad \Rightarrow$

$$\frac{dy}{dx} = -\frac{1}{\text{sech } y \tanh y} = -\frac{1}{\text{sech } y\sqrt{1 - \text{sech}^2 y}} = -\frac{1}{x\sqrt{1 - x^2}}. \quad \text{(Note that } y > 0 \text{ and so } \tanh y > 0.\text{)}$$

(e) Let $y = \coth^{-1} x$. Then $\coth y = x \quad \Rightarrow \quad -\text{csch}^2 y \dfrac{dy}{dx} = 1 \quad \Rightarrow$

$$\frac{dy}{dx} = -\frac{1}{\text{csch}^2 y} = \frac{1}{1 - \coth^2 y} = \frac{1}{1 - x^2} \quad \text{by Exercise 13.}$$

31. $f(x) = x \cosh x \quad \Rightarrow \quad f'(x) = x(\cosh x)' + (\cosh x)(x)' = x \sinh x + \cosh x$

33. $h(x) = \sinh(x^2) \quad \Rightarrow \quad h'(x) = \cosh(x^2) \cdot 2x = 2x \cosh(x^2)$

35. $G(x) = \dfrac{1 - \cosh x}{1 + \cosh x} \quad \Rightarrow$

$$G'(x) = \frac{(1 + \cosh x)(-\sinh x) - (1 - \cosh x)(\sinh x)}{(1 + \cosh x)^2} = \frac{-\sinh x - \sinh x \cosh x - \sinh x + \sinh x \cosh x}{(1 + \cosh x)^2}$$

$$= \frac{-2 \sinh x}{(1 + \cosh x)^2}$$

37. $h(t) = \coth\sqrt{1 + t^2} \quad \Rightarrow \quad h'(t) = -\text{csch}^2\sqrt{1 + t^2} \cdot \tfrac{1}{2}(1 + t^2)^{-1/2}(2t) = -\dfrac{t \, \text{csch}^2\sqrt{1 + t^2}}{\sqrt{1 + t^2}}$

39. $H(t) = \tanh(e^t) \quad \Rightarrow \quad H'(t) = \text{sech}^2(e^t) \cdot e^t = e^t \text{sech}^2(e^t)$

41. $y = e^{\cosh 3x} \quad \Rightarrow \quad y' = e^{\cosh 3x} \cdot \sinh 3x \cdot 3 = 3e^{\cosh 3x} \sinh 3x$

43. $y = \tanh^{-1} \sqrt{x} \quad \Rightarrow \quad y' = \dfrac{1}{1 - (\sqrt{x})^2} \cdot \tfrac{1}{2} x^{-1/2} = \dfrac{1}{2\sqrt{x}\,(1-x)}$

45. $y = x\sinh^{-1}(x/3) - \sqrt{9+x^2} \quad \Rightarrow$

$y' = \sinh^{-1}\left(\dfrac{x}{3}\right) + x\dfrac{1/3}{\sqrt{1+(x/3)^2}} - \dfrac{2x}{2\sqrt{9+x^2}} = \sinh^{-1}\left(\dfrac{x}{3}\right) + \dfrac{x}{\sqrt{9+x^2}} - \dfrac{x}{\sqrt{9+x^2}} = \sinh^{-1}\left(\dfrac{x}{3}\right)$

47. $y = \coth^{-1}\sqrt{x^2+1} \quad \Rightarrow \quad y' = \dfrac{1}{1-(x^2+1)}\dfrac{2x}{2\sqrt{x^2+1}} = -\dfrac{1}{x\sqrt{x^2+1}}$

49. (a) $y = 20\cosh(x/20) - 15 \quad \Rightarrow \quad y' = 20\sinh(x/20) \cdot \frac{1}{20} = \sinh(x/20)$. Since the right pole is positioned at $x = 7$, we have $y' = \sinh\frac{7}{20} \approx 0.3572$.

(b) If θ is the angle between the tangent line and the x-axis, then $\tan\theta = $ slope of the line $= \sinh\frac{7}{20}$, so $\theta = \tan^{-1}\left(\sinh\frac{7}{20}\right) \approx 0.343$ rad $\approx 19.66°$. Thus, the angle between the line and the pole is about $90° - 19.66° = 70.34°$.

51. (a) $y = A\sinh mx + B\cosh mx \quad \Rightarrow \quad y' = mA\cosh mx + mB\sinh mx \quad \Rightarrow$
$y'' = m^2 A\sinh mx + m^2 B\cosh mx = m^2 y$

(b) From part (a), a solution of $y'' = 9y$ is $y(x) = A\sinh 3x + B\cosh 3x$. So
$-4 = y(0) = A\sinh 0 + B\cosh 0 = B$, so $B = -4$. Now $y'(x) = 3A\cosh 3x - 12\sinh 3x \quad \Rightarrow$
$6 = y'(0) = 3A \quad \Rightarrow \quad A = 2$, so $y = 2\sinh 3x - 4\cosh 3x$.

53. The tangent to $y = \cosh x$ has slope 1 when $y' = \sinh x = 1 \quad \Rightarrow \quad x = \sinh^{-1} 1 = \ln\left(1+\sqrt{2}\right)$, by Equation 3. Since $\sinh x = 1$ and $y = \cosh x = \sqrt{1+\sinh^2 x}$, we have $\cosh x = \sqrt{2}$. The point is $\left(\ln\left(1+\sqrt{2}\right), \sqrt{2}\right)$.

3.10 Related Rates

1. $V = x^3 \quad \Rightarrow \quad \dfrac{dV}{dt} = \dfrac{dV}{dx}\dfrac{dx}{dt} = 3x^2\dfrac{dx}{dt}$

3. $y = x^3 + 2x \quad \Rightarrow \quad \dfrac{dy}{dt} = \dfrac{dy}{dx}\dfrac{dx}{dt} = (3x^2+2)(5) = 5(3x^2+2)$. When $x = 2$, $\dfrac{dy}{dt} = 5(14) = 70$.

5. (a) Given: a plane flying horizontally at an altitude of 1 mi and a speed of 500 mi/h passes directly over a radar station. If we let t be time (in hours) and x be the horizontal distance traveled by the plane (in mi), then we are given that $dx/dt = 500$ mi/h.

(b) Unknown: the rate at which the distance from the plane to the station is increasing when it is 2 mi from the station. If we let y be the distance from the plane to the station, then we want to find dy/dt when $y = 2$ mi.

(c)

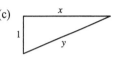

(d) By the Pythagorean Theorem, $y^2 = x^2 + 1 \;\Rightarrow\; 2y\,(dy/dt) = 2x\,(dx/dt)$.

(e) $\dfrac{dy}{dt} = \dfrac{x}{y}\dfrac{dx}{dt} = 500\dfrac{x}{y}$. When $y = 2$, $x = \sqrt{3}$, so $\dfrac{dy}{dt} = 500\left(\dfrac{\sqrt{3}}{2}\right) = 250\sqrt{3} \approx 433$ mi/h.

7. (a) Given: a man 6 ft tall walks away from a street light mounted on a 15-ft-tall pole at a rate of 5 ft/s. If we let t be time (in s) and x be the distance from the pole to the man (in ft), then we are given that $dx/dt = 5$ ft/s.

(b) Unknown: the rate at which the tip of his shadow is moving when he is 40 ft from the pole. If we let y be the distance from the man to the tip of his shadow (in ft), then we want to find $\dfrac{d}{dt}(x + y)$ when $x = 40$ ft.

(c)

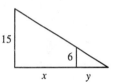

(d) By similar triangles, $\dfrac{15}{6} = \dfrac{x+y}{y} \;\Rightarrow\; 15y = 6x + 6y \;\Rightarrow\; 9y = 6x \;\Rightarrow\; y = \tfrac{2}{3}x$.

(e) The tip of the shadow moves at a rate of $\dfrac{d}{dt}(x + y) = \dfrac{d}{dt}\left(x + \tfrac{2}{3}x\right) = \dfrac{5}{3}\dfrac{dx}{dt} = \tfrac{5}{3}\,(5) = \tfrac{25}{3}$ ft/s.

9.

We are given that $\dfrac{dx}{dt} = 60$ mi/h and $\dfrac{dy}{dt} = 25$ mi/h. $z^2 = x^2 + y^2 \;\Rightarrow$

$2z\dfrac{dz}{dt} = 2x\dfrac{dx}{dt} + 2y\dfrac{dy}{dt}$. After 2 hours, $x = 2\,(60) = 120$ and $y = 2\,(25) = 50$

$\Rightarrow\; z = \sqrt{120^2 + 50^2} = 130$, so

$\dfrac{dz}{dt} = \dfrac{1}{z}\left(x\dfrac{dx}{dt} + y\dfrac{dy}{dt}\right) = \dfrac{120\,(60) + 50\,(25)}{130} = 65$ mi/h.

11.

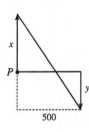

We are given that $\dfrac{dx}{dt} = 4$ ft/s and $\dfrac{dy}{dt} = 5$ ft/s. $z^2 = (x + y)^2 + 500^2 \;\Rightarrow$

$2z\dfrac{dz}{dt} = 2\,(x + y)\left(\dfrac{dx}{dt} + \dfrac{dy}{dt}\right)$. 15 minutes after the woman starts, we have

$x = (4\text{ ft/s})\,(20\text{ min})\,(60\text{ s/min}) = 4800$ ft and $y = 5 \cdot 15 \cdot 60 = 4500 \;\Rightarrow$

$z = \sqrt{(4800 + 4500)^2 + 500^2}$, so

$\dfrac{dz}{dt} = \dfrac{x + y}{z}\left(\dfrac{dx}{dt} + \dfrac{dy}{dt}\right) = \dfrac{4800 + 4500}{\sqrt{86{,}740{,}000}}\,(5 + 4) = \dfrac{837}{\sqrt{8674}} \approx 8.99$ ft/s.

13. $A = \tfrac{1}{2}bh$, where b is the base and h is the altitude. We are given that $\dfrac{dh}{dt} = 1$ cm/min and $\dfrac{dA}{dt} = 2$ cm^2/min.

Using the Product Rule, we have $\dfrac{dA}{dt} = \dfrac{1}{2}\left(b\dfrac{dh}{dt} + h\dfrac{db}{dt}\right)$. When $h = 10$ and $A = 100$, we have $b = 20$, so

$2 = \dfrac{1}{2}\left(20 \cdot 1 + 10\dfrac{db}{dt}\right) \;\Rightarrow\; 4 = 20 + 10\dfrac{db}{dt} \;\Rightarrow\; \dfrac{db}{dt} = \dfrac{4 - 20}{10} = -1.6$ cm/min.

15.

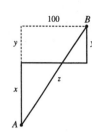

We are given that $\dfrac{dx}{dt} = 35$ km/h and $\dfrac{dy}{dt} = 25$ km/h. $z^2 = (x+y)^2 + 100^2$

$\Rightarrow 2z\dfrac{dz}{dt} = 2(x+y)\left(\dfrac{dx}{dt} + \dfrac{dy}{dt}\right)$. At 4:00 P.M., $x = 140$ and $y = 100$

$\Rightarrow z = 260$, so

$\dfrac{dz}{dt} = \dfrac{x+y}{z}\left(\dfrac{dx}{dt} + \dfrac{dy}{dt}\right) = \dfrac{140 + 100}{260}(35 + 25) = \dfrac{720}{13} \approx 55.4$ km/h.

17.

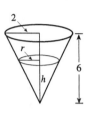

If $C =$ the rate at which water is pumped in, then $\dfrac{dV}{dt} = C - 10{,}000$, where

$V = \frac{1}{3}\pi r^2 h$ is the volume at time t. By similar triangles, $\dfrac{r}{2} = \dfrac{h}{6}$ $\Rightarrow$

$r = \frac{1}{3}h$ $\Rightarrow$ $V = \frac{1}{3}\pi\left(\frac{1}{3}h\right)^2 h = \frac{\pi}{27}h^3$ $\Rightarrow$ $\dfrac{dV}{dt} = \frac{\pi}{9}h^2\dfrac{dh}{dt}$. When

$h = 200$, $\dfrac{dh}{dt} = 20$, so $C - 10{,}000 = \frac{\pi}{9}(200)^2(20)$ $\Rightarrow$

$C = 10{,}000 + \frac{800{,}000}{9}\pi \approx 289{,}253$ cm^3/min.

19.

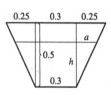

The figure is labeled in meters. The area A of a trapezoid is

$\frac{1}{2}$ (base$_1$ + base$_2$) (height), and the volume V of the trough is $10A$. Thus,

$V = \frac{1}{2}[0.3 + (0.3 + 2a)]\,h\,(10)$, where $\dfrac{a}{h} = \dfrac{0.25}{0.5} = \dfrac{1}{2}$ so $2a = h$ $\Rightarrow$

$V = 5(0.6 + h)h = 3h + 5h^2$ $\Rightarrow$ $0.2 = \dfrac{dV}{dt} = (3 + 10h)\dfrac{dh}{dt}$ $\Rightarrow$

$\dfrac{dh}{dt} = \dfrac{0.2}{3 + 10h}$. When $h = 0.3$, $\dfrac{dh}{dt} = \dfrac{0.2}{3 + 10(0.3)} = \dfrac{0.2}{6}$ m/min $= \dfrac{10}{3}$ cm/min.

21.

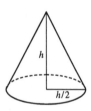

We are given that $\dfrac{dV}{dt} = 30$ ft^3/min. $V = \frac{1}{3}\pi r^2 h = \frac{1}{3}\pi\left(\dfrac{h}{2}\right)^2 h = \dfrac{\pi h^3}{12}$ $\Rightarrow$

$30 = \dfrac{dV}{dt} = \dfrac{\pi h^2}{4}\dfrac{dh}{dt}$ $\Rightarrow$ $\dfrac{dh}{dt} = \dfrac{120}{\pi h^2}$. When $h = 10$ ft,

$\dfrac{dh}{dt} = \dfrac{120}{10^2 \pi} = \dfrac{6}{5\pi} \approx 0.38$ ft/min.

23.

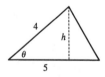

$A = \frac{1}{2}bh$, but $b = 5$ m and $h = 4\sin\theta$ so $A = 10\sin\theta$. We are given

$\dfrac{d\theta}{dt} = 0.06$ rad/s. $\dfrac{dA}{dt} = 10\cos\theta\dfrac{d\theta}{dt} = 0.6\cos\theta$. When $\theta = \frac{\pi}{3}$,

$\dfrac{dA}{dt} = 0.6\left(\cos\frac{\pi}{3}\right) = (0.6)\left(\frac{1}{2}\right) = 0.3$ m^2/s.

25. Differentiating both sides of $PV = C$ with respect to t and using the Product Rule gives us $P\dfrac{dV}{dt} + V\dfrac{dP}{dt} = 0$

$\Rightarrow \dfrac{dV}{dt} = -\dfrac{V}{P}\dfrac{dP}{dt}$. When $V = 600$, $P = 150$ and $\dfrac{dP}{dt} = 20$, we have $\dfrac{dV}{dt} = -\dfrac{600}{150}(20) = -80$, so the volume

is decreasing at a rate of 80 cm^3/min.

27. With $R_1 = 80$ and $R_2 = 100$, $\dfrac{1}{R} = \dfrac{1}{R_1} + \dfrac{1}{R_2} = \dfrac{1}{80} + \dfrac{1}{100} = \dfrac{180}{8000} = \dfrac{9}{400}$, so $R = \dfrac{400}{9}$. Differentiating

$\dfrac{1}{R} = \dfrac{1}{R_1} + \dfrac{1}{R_2}$ with respect to t, we have $-\dfrac{1}{R^2}\dfrac{dR}{dt} = -\dfrac{1}{R_1^2}\dfrac{dR_1}{dt} - \dfrac{1}{R_2^2}\dfrac{dR_2}{dt}$ $\Rightarrow$

$\dfrac{dR}{dt} = R^2 \left(\dfrac{1}{R_1^2}\dfrac{dR_1}{dt} + \dfrac{1}{R_2^2}\dfrac{dR_2}{dt} \right)$. When $R_1 = 80$ and $R_2 = 100$,

$\dfrac{dR}{dt} = \dfrac{400^2}{9^2}\left[\dfrac{1}{80^2}(0.3) + \dfrac{1}{100^2}(0.2) \right] = \dfrac{107}{810} \approx 0.132\ \Omega/\text{s}.$

29.

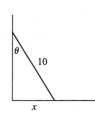

We are given that $\dfrac{dx}{dt} = 2$ ft/s. $x = 10\sin\theta$ $\Rightarrow$ $\dfrac{dx}{dt} = 10\cos\theta\dfrac{d\theta}{dt}$. When

$\theta = \frac{\pi}{4}$, $\dfrac{d\theta}{dt} = \dfrac{2}{10\left(1/\sqrt{2}\right)} = \dfrac{\sqrt{2}}{5}$ rad/s.

31. (a)

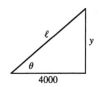

By the Pythagorean Theorem, $4000^2 + y^2 = \ell^2$. Differentiating with respect to

t, we obtain $2y\dfrac{dy}{dt} = 2\ell\dfrac{d\ell}{dt}$. We know that $\dfrac{dy}{dt} = 600$, so when $y = 3000$ and

$\ell = 5000$, $\dfrac{d\ell}{dt} = \dfrac{y\,(dy/dt)}{\ell} = \dfrac{3000\,(600)}{5000} = \dfrac{1800}{5} = 360$ ft/s.

(b) Here $\tan\theta = y/4000$, so $\sec^2\theta\dfrac{d\theta}{dt} = \dfrac{1}{4000}\dfrac{dy}{dt}$ $\Rightarrow$ $\dfrac{d\theta}{dt} = \dfrac{\cos^2\theta}{4000}\dfrac{dy}{dt}$. When $y = 3000$, $\dfrac{dy}{dt} = 600$,

$\ell = 5000$ and $\cos\theta = \dfrac{4000}{\ell} = \dfrac{4000}{5000} = \dfrac{4}{5}$, so $\dfrac{d\theta}{dt} = \dfrac{(4/5)^2}{4000}(600) = 0.096$ rad/s.

33.

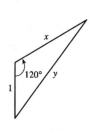

We are given that $\dfrac{dx}{dt} = 300$ km/h. By the Law of Cosines,

$y^2 = x^2 + 1 - 2\,(1)\,x\cos 120° = x^2 + 1 - 2x\left(-\frac{1}{2}\right) = x^2 + x + 1$, so

$2y\dfrac{dy}{dt} = 2x\dfrac{dx}{dt} + \dfrac{dx}{dt}$ $\Rightarrow$ $\dfrac{dy}{dt} = \dfrac{2x+1}{2y}\dfrac{dx}{dt}$. After 1 minute, $x = \frac{300}{60} = 5$

$\Rightarrow$ $y = \sqrt{5^2 + 5 + 1} = \sqrt{31}$ $\Rightarrow$

$\dfrac{dy}{dt} = \dfrac{2\,(5)+1}{2\sqrt{31}}(300) = \dfrac{1650}{\sqrt{31}} \approx 296$ km/h.

35.

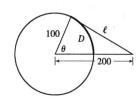

Let the distance between the runner and the friend be ℓ. Then by the Law of Cosines,

$$\ell^2 = 200^2 + 100^2 - 2 \cdot 200 \cdot 100 \cdot \cos\theta = 50{,}000 - 40{,}000\cos\theta \ (\bigstar).$$

Differentiating implicitly with respect to t, we obtain

$$2\ell\frac{d\ell}{dt} = -40{,}000\,(-\sin\theta)\,\frac{d\theta}{dt}.$$ Now if D is the distance run when

the angle is θ radians, then by the formula for the length of an arc on a circle, $s = r\theta$, we have $D = 100\theta$, so

$$\theta = \frac{1}{100}D \ \Rightarrow\ \frac{d\theta}{dt} = \frac{1}{100}\frac{dD}{dt} = \frac{7}{100}.$$ To substitute into the expression for $\frac{d\ell}{dt}$, we must know $\sin\theta$ at the time

when $\ell = 200$, which we find from ($\bigstar$): $200^2 = 50{,}000 - 40{,}000\cos\theta \ \Leftrightarrow\ \cos\theta = \frac{1}{4} \ \Rightarrow$

$$\sin\theta = \sqrt{1 - \left(\tfrac{1}{4}\right)^2} = \frac{\sqrt{15}}{4}.$$ Substituting, we get $2\,(200)\,\dfrac{d\ell}{dt} = 40{,}000\dfrac{\sqrt{15}}{4}\left(\tfrac{7}{100}\right) \ \Rightarrow$

$d\ell/dt = \frac{7\sqrt{15}}{4} \approx 6.78$ m/s. Whether the distance between them is increasing or decreasing depends on the direction in which the runner is running.

3.11 Linear Approximations and Differentials

1. As in Example 1, $T\,(0) = 185$, $T\,(10) = 172$, $T\,(20) = 160$, and

$$T'\,(20) \approx \frac{T\,(10) - T\,(20)}{10 - 20} = \frac{172 - 160}{-10} = -1.2°\,\text{F/min}.$$

$T\,(30) \approx T\,(20) + T'\,(20)\,(30 - 20) \approx 160 - 1.2\,(10) = 148°$ F. We would expect the temperature of the turkey to get closer to $75°$ F as time increases. Since the temperature decreased $13°$ F in the first 10 minutes and $12°$ F in the second 10 minutes, we can assume that the slopes of the tangent line are increasing through negative

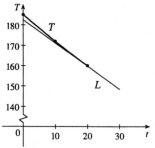

values: $-1.3, -1.2, \ldots$. Hence, the tangent lines are under the curve and $148°$ F is an underestimate. From the figure, we estimate the slope of the tangent line at $t = 20$ to be $\frac{184 - 147}{0 - 30} = -\frac{37}{30}$. Then the linear approximation becomes $T\,(30) \approx T\,(20) + T'\,(20) \cdot 10 \approx 160 - \frac{37}{30}\,(10) = 147\frac{2}{3} \approx 147.7$.

3. If $C\,(t)$ represents the cash per capita in circulation in year t, then we estimate $C\,(2000)$ using a linear approximation based on the tangent line to the graph of C at

$$(1990, C\,(1990)) = (1990, 1063).\ C'\,(1990) \approx \frac{C\,(1980) - C\,(1990)}{1980 - 1990} = \frac{571 - 1063}{-10} = 49.2\,\frac{\text{dollars}}{\text{year}}.$$

$C\,(2000) \approx C\,(1990) + C'\,(1990)\,(2000 - 1990) \approx 1063 + 49.2\,(10) = 1555.$ So our estimate of cash per capita in circulation in the year 2000 is \$1555. For the given data, C' is increasing, so the graph of C is concave upward and the tangent line approximations are below the curve, indicating that our prediction is an underestimate.

5. $f\,(x) = x^3 \ \Rightarrow\ f'\,(x) = 3x^2$ so $f\,(1) = 1$ and $f'\,(1) = 3$.
Thus, $L\,(x) = f\,(1) + f'\,(1)\,(x - 1) = 1 + 3\,(x - 1) = 3x - 2$.

7. $f\,(x) = e^{-2x} \ \Rightarrow\ f'\,(x) = -2e^{-2x}$ so $f\,(0) = 1$ and $f'\,(0) = -2$.
Thus, $L\,(x) = f\,(0) + f'\,(0)\,(x - 0) = 1 + (-2)\,(x - 0) = 1 - 2x$.

9. $f(x) = \sqrt{1-x} \Rightarrow f'(x) = \dfrac{-1}{2\sqrt{1-x}}$ so $f(0) = 1$ and $f'(0) = -\frac{1}{2}$.

Therefore,

$$\sqrt{1-x} = f(x) \approx f(0) + f'(0)(x-0)$$

$$= 1 + \left(-\frac{1}{2}\right)(x-0) = 1 - \frac{1}{2}x$$

So $\sqrt{0.9} = \sqrt{1-0.1} \approx 1 - \frac{1}{2}(0.1) = 0.95$ and

$\sqrt{0.99} = \sqrt{1-0.01} \approx 1 - \frac{1}{2}(0.01) = 0.995$.

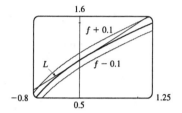

11. $f(x) = \sqrt{1+x} \Rightarrow f'(x) = \dfrac{1}{2\sqrt{1+x}}$ so $f(0) = 1$ and $f'(0) = \frac{1}{2}$.

Thus, $f(x) \approx f(0) + f'(0)(x-0) = 1 + \frac{1}{2}(x-0) = 1 + \frac{1}{2}x$.

We need $\sqrt{1+x} - 0.1 < 1 + \frac{1}{2}x < \sqrt{1+x} + 0.1$. By zooming in or

using an intersect feature, we see that this is true when $-0.69 < x < 1.09$.

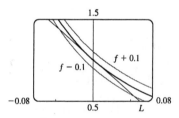

13. $f(x) = \dfrac{1}{(1+2x)^4} \Rightarrow f'(x) = \dfrac{-8}{(1+2x)^5}$ so $f(0) = 1$ and

$f'(0) = -8$.

Thus, $f(x) \approx f(0) + f'(0)(x-0) = 1 + (-8)(x-0) = 1 - 8x$.

We need $1/(1+2x)^4 - 0.1 < 1 - 8x < 1/(1+2x)^4 + 0.1$, which is true

when $-0.045 < x < 0.055$.

15. If $y = f(x)$, then the differential dy is equal to $f'(x)\,dx$. $y = x^4 + 5x \Rightarrow dy = \left(4x^3 + 5\right) dx$.

17. $y = x \ln x \Rightarrow dy = \left(x \cdot \dfrac{1}{x} + \ln x \cdot 1\right) dx = (\ln x + 1)\,dx$

19. $y = \dfrac{u+1}{u-1} \Rightarrow dy = \dfrac{(u-1)(1) - (u+1)(1)}{(u-1)^2}\,du = \dfrac{-2}{(u-1)^2}\,du$

21. (a) $y = x^2 + 2x \Rightarrow dy = (2x+2)\,dx$

(b) When $x = 3$ and $dx = \frac{1}{2}$, $dy = [2(3) + 2]\left(\frac{1}{2}\right) = 4$.

23. (a) $y = \left(x^2 + 5\right)^3 \Rightarrow dy = 3\left(x^2 + 5\right)^2 2x\,dx = 6x\left(x^2 + 5\right)^2 dx$

(b) When $x = 1$ and $dx = 0.05$, $dy = 6(1)\left(1^2 + 5\right)^2 (0.05) = 10.8$.

25. (a) $y = \cos x \Rightarrow dy = -\sin x\,dx$

(b) When $x = \frac{\pi}{6}$ and $dx = 0.05$, $dy = -\frac{1}{2}(0.05) = -0.025$.

27. $y = x^2, x = 1, \Delta x = 0.5 \Rightarrow$
$\Delta y = (1.5)^2 - 1^2 = 1.25.$
$dy = 2x\, dx = 2\,(1)\,(0.5) = 1$

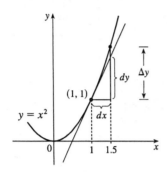

29. $y = 6 - x^2, x = -2, \Delta x = 0.4 \Rightarrow$
$\Delta y = \left(6 - (-1.6)^2\right) - \left(6 - (-2)^2\right) = 1.44$
$dy = -2x\, dx = -2\,(-2)\,(0.4) = 1.6$

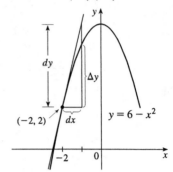

31. $y = f(x) = \sqrt{x} \Rightarrow dy = \dfrac{1}{2\sqrt{x}}\, dx.$ When $x = 36$ and $dx = 0.1$, $dy = \dfrac{1}{2\sqrt{36}}\,(0.1) = \dfrac{1}{120}$, so
$\sqrt{36.1} = f(36.1) \approx f(36) + dy = \sqrt{36} + \dfrac{1}{120} \approx 6.0083.$

33. $y = f(x) = 1/x \Rightarrow dy = \left(-1/x^2\right) dx.$ When $x = 10$ and $dx = 0.1$, $dy = \left(-\dfrac{1}{100}\right)(0.1) = -0.001$, so
$\dfrac{1}{10.1} = f(10.1) \approx f(10) + dy = 0.1 - 0.001 = 0.099.$

35. $y = f(x) = \sin x \Rightarrow dy = \cos x\, dx.$ When $x = \dfrac{\pi}{3}$ and $dx = -\dfrac{\pi}{180}$, $dy = \cos\dfrac{\pi}{3}\left(-\dfrac{\pi}{180}\right) = -\dfrac{\pi}{360}$, so
$\sin 59° = f\left(\dfrac{59}{180}\pi\right) \approx f\left(\dfrac{\pi}{3}\right) + dy = \dfrac{\sqrt{3}}{2} - \dfrac{\pi}{360} \approx 0.857.$

37. We can see from a graph of $y = \sec x$ that the tangent line approximation at $(0, 1)$ is the horizontal line $y = 1$. Since 0.08 is close to 0, approximating $\sec 0.08$ with 1 is reasonable.

39. (a) If x is the edge length, then $V = x^3 \Rightarrow dV = 3x^2\, dx.$ When $x = 30$ and $dx = 0.1$,
$dV = 3\,(30)^2\,(0.1) = 270$, so the maximum error is about 270 cm^3.

(b) $S = 6x^2 \Rightarrow dS = 12x\, dx.$ When $x = 30$ and $dx = 0.1$, $dS = 12\,(30)\,(0.1) = 36$, so the maximum error is about 36 cm^2.

41. (a) For a sphere of radius r, the circumference is $C = 2\pi r$ and the surface area is $S = 4\pi r^2$, so $r = C/(2\pi) \Rightarrow$
$S = 4\pi\,(C/2\pi)^2 = C^2/\pi \Rightarrow dS = (2/\pi)\,C\, dC.$ When $C = 84$ and $dC = 0.5$, $dS = \dfrac{2}{\pi}\,(84)\,(0.5) = \dfrac{84}{\pi}$,
so the maximum error is about $\dfrac{84}{\pi} \approx 27$ cm^2. Relative error $\approx \dfrac{dS}{S} = \dfrac{84/\pi}{84^2/\pi} = \dfrac{1}{84} \approx 0.012$

(b) $V = \dfrac{4}{3}\pi r^3 = \dfrac{4}{3}\pi\left(\dfrac{C}{2\pi}\right)^3 = \dfrac{C^3}{6\pi^2} \Rightarrow dV = \dfrac{1}{2\pi^2}C^2\, dC.$ When $C = 84$ and $dC = 0.5$,
$dV = \dfrac{1}{2\pi^2}\,(84)^2\,(0.5) = \dfrac{1764}{\pi^2}$, so the maximum error is about $\dfrac{1764}{\pi^2} \approx 179$ cm^3. The relative error is
approximately $\dfrac{dV}{V} = \dfrac{1764/\pi^2}{(84)^3/(6\pi^2)} = \dfrac{1}{56} \approx 0.018.$

43. (a) $V = \pi r^2 h \Rightarrow \Delta V \approx dV = 2\pi r h\, dr = 2\pi r h\, \Delta r$

(b) $\Delta V = \pi\,(r + \Delta r)^2 h - \pi r^2 h$, so the error is $\Delta V - dv = \pi\,(r + \Delta r)^2 h - \pi r^2 h - 2\pi r h\, \Delta r = \pi\,(\Delta r)^2 h$

45. (a) $dc = \dfrac{dc}{dx}\, dx = 0\, dx = 0$

(b) $d\,(cu) = \dfrac{d}{dx}\,(cu)\, dx = c\dfrac{du}{dx}\, dx = c\, du$

(c) $d\,(u+v) = \dfrac{d}{dx}\,(u+v)\, dx = \left(\dfrac{du}{dx} + \dfrac{dv}{dx}\right) dx = \dfrac{du}{dx}\, dx + \dfrac{dv}{dx}\, dx = du + dv$

(d) $d\,(uv) = \dfrac{d}{dx}\,(uv)\, dx = \left(u\dfrac{dv}{dx} + v\dfrac{du}{dx}\right) dx = u\dfrac{dv}{dx}\, dx + v\dfrac{du}{dx}\, dx = u\, dv + v\, du$

(e) $d\left(\dfrac{u}{v}\right) = \dfrac{d}{dx}\left(\dfrac{u}{v}\right) dx = \dfrac{v\dfrac{du}{dx} - u\dfrac{dv}{dx}}{v^2}\, dx = \dfrac{v\dfrac{du}{dx}\, dx - u\dfrac{dv}{dx}\, dx}{v^2} = \dfrac{v\, du - u\, dv}{v^2}$

(f) $d\,(x^n) = \dfrac{d}{dx}\,(x^n)\, dx = nx^{n-1}\, dx$

47. (a) The graph shows that $f'\,(1) = 2$, so $L\,(x) = f\,(1) + f'\,(1)\,(x-1) = 5 + 2\,(x-1) = 2x + 3$.
$f\,(0.9) \approx L\,(0.9) = 4.8$ and $f\,(1.1) \approx L\,(1.1) = 5.2$.

(b) From the graph, we see that $f'\,(x)$ is positive and decreasing. This means that the slopes of the tangent lines are positive, but the tangents are becoming less steep. So the tangent lines lie *above* the curve. Thus, the estimates in part (a) are too large.

Review

CONCEPT CHECK

1. (a) The Power Rule: If n is any real number, then $\dfrac{d}{dx}\,(x^n) = nx^{n-1}$. The derivative of a variable base raised to a constant power is the power times the base raised to the power minus one.

(b) The Constant Multiple Rule: If c is a constant and f is a differentiable function, then $\dfrac{d}{dx}\,[cf\,(x)] = c\dfrac{d}{dx}f\,(x)$.
The derivative of a constant times a function is the constant times the derivative of the function.

(c) The Sum Rule: If f and g are both differentiable, then $\dfrac{d}{dx}\,[f\,(x) + g\,(x)] = \dfrac{d}{dx}f\,(x) + \dfrac{d}{dx}g\,(x)$. The derivative of a sum of functions is the sum of the derivatives.

(d) The Difference Rule: If f and g are both differentiable, then $\dfrac{d}{dx}\,[f\,(x) - g\,(x)] = \dfrac{d}{dx}f\,(x) - \dfrac{d}{dx}g\,(x)$. The derivative of a difference of functions is the difference of the derivatives.

(e) The Product Rule: If f and g are both differentiable, then $\dfrac{d}{dx}\,[f\,(x)\,g\,(x)] = f\,(x)\dfrac{d}{dx}g\,(x) + g\,(x)\dfrac{d}{dx}f\,(x)$.
The derivative of a product of two functions is the first function times the derivative of the second function plus the second function times the derivative of the first function.

(f) The Quotient Rule: If f and g are both differentiable, then $\dfrac{d}{dx}\left[\dfrac{f\,(x)}{g\,(x)}\right] = \dfrac{g\,(x)\dfrac{d}{dx}f\,(x) - f\,(x)\dfrac{d}{dx}g\,(x)}{[g\,(x)]^2}$.
The derivative of a quotient of functions is the denominator times the derivative of the numerator minus the numerator times the derivative of the denominator, all divided by the square of the denominator.

(g) The Chain Rule: If f and g are both differentiable and $F = f \circ g$ is the composite function defined by $F\,(x) = f\,(g\,(x))$, then F is differentiable and F' is given by the product $F'\,(x) = f'\,(g\,(x))\,g'\,(x)$. The derivative of a composite function is the derivative of the outer function evaluated at the inner function times the derivative of the inner function.

2. (a) $y = x^n \implies y' = nx^{n-1}$

(b) $y = e^x \implies y' = e^x$

(c) $y = a^x \implies y' = a^x \ln a$

(d) $y = \ln x \implies y' = 1/x$

(e) $y = \log_a x \implies y' = 1/(x \ln a)$

(f) $y = \sin x \implies y' = \cos x$

(g) $y = \cos x \implies y' = -\sin x$

(h) $y = \tan x \implies y' = \sec^2 x$

(i) $y = \csc x \implies y' = -\csc x \cot x$

(j) $y = \sec x \implies y' = \sec x \tan x$

(k) $y = \cot x \implies y' = -\csc^2 x$

(l) $y = \sin^{-1} x \implies y' = 1/\sqrt{1 - x^2}$

(m) $y = \cos^{-1} x \implies y' = -1/\sqrt{1 - x^2}$

(n) $y = \tan^{-1} x \implies y' = 1/(1 + x^2)$

(o) $y = \sinh x \implies y' = \cosh x$

(p) $y = \cosh x \implies y' = \sinh x$

(q) $y = \tanh x \implies y' = \operatorname{sech}^2 x$

(r) $y = \sinh^{-1} x \implies y' = 1/\sqrt{1 + x^2}$

(s) $y = \cosh^{-1} x \implies y' = 1/\sqrt{x^2 - 1}$

(t) $y = \tanh^{-1} x \implies y' = 1/(1 - x^2)$

3. (a) e is the number such that $\lim\limits_{h \to 0} \dfrac{e^h - 1}{h} = 1$.

(b) $e = \lim\limits_{x \to 0} (1 + x)^{1/x}$

(c) The differentiation formula for $y = a^x$ $[y' = a^x \ln a]$ is simplest when $a = e$ because $\ln e = 1$.

(d) The differentiation formula for $y = \log_a x$ $[y' = 1/(x \ln a)]$ is simplest when $a = e$ because $\ln e = 1$.

4. (a) Implicit differentiation consists of differentiating both sides of an equation involving x and y with respect to x, and then solving the resulting equation for y'.

(b) Logarithmic differentiation consists of taking natural logarithms of both sides of an equation $y = f(x)$, simplifying, differentiating implicitly with respect to x, and then solving the resulting equation for y'.

5. The second derivative of a function f is the rate of change of the first derivative f'. The third derivative is the derivative (rate of change) of the second derivative. If f is the position function of an object, f' is its velocity function, f'' is its acceleration function, and f''' is its jerk function.

6. (a) The linearization L of f at $x = a$ is $L(x) = f(a) + f'(a)(x - a)$.

(b) If $y = f(x)$, then the differential dy is given by $dy = f'(x)\, dx$.

TRUE-FALSE QUIZ

1. True.   This is the Sum Rule.

3. True.   This is the Chain Rule.

5. False. $\dfrac{d}{dx} f(\sqrt{x}) = \dfrac{f'(\sqrt{x})}{2\sqrt{x}}$ by the Chain Rule.

7. False. $\dfrac{d}{dx} 10^x = 10^x \ln 10$

9. True. $D(\tan^2 x) = 2\tan x \sec^2 x$, and $D(\sec^2 x) = 2\sec x (\sec x \tan x) = 2\tan x \sec^2 x$. We can also show this by differentiating the identity $\tan^2 x + 1 = \sec^2 x$: we get $\dfrac{d}{dx}(\tan^2 x + 1) = \dfrac{d}{dx}\tan^2 x = \dfrac{d}{dx}\sec^2 x$.

11. True. $g(x) = x^5 \implies g'(x) = 5x^4 \implies g'(2) = 5(2)^4 = 80$, and by the definition of the derivative,

$$\lim_{x \to 2} \frac{g(x) - g(2)}{x - 2} = g'(2) = 80.$$

13. False. A tangent to the parabola has slope $dy/dx = 2x$, so at $(-2, 4)$ the slope of the tangent is $2(-2) = -4$ and the equation is $y - 4 = -4(x + 2)$. [The equation $y - 4 = 2x(x + 2)$ is not even linear!]

EXERCISES

1. $y = (x + 2)^8 (x + 3)^6 \implies$

$y' = (x + 2)^8 6(x + 3)^5 + (x + 3)^6 8(x + 2)^7$

$= 2(x + 2)^7 (x + 3)^5 [3(x + 2) + 4(x + 3)] = 2(x + 2)^7 (x + 3)^5 (7x + 18)$

3. $y = \dfrac{x}{\sqrt{9 - 4x}} \implies y' = \dfrac{\sqrt{9 - 4x} - x[-4/(2\sqrt{9 - 4x})]}{9 - 4x} \cdot \dfrac{\sqrt{9 - 4x}}{\sqrt{9 - 4x}} = \dfrac{(9 - 4x) + 2x}{(9 - 4x)^{3/2}} = \dfrac{9 - 2x}{(9 - 4x)^{3/2}}$

5. $y = \sin(\cos x) \implies y' = \cos(\cos x)(-\sin x) = -\sin x \cos(\cos x)$

7. $y = xe^{-1/x} \implies y' = e^{-1/x} + xe^{-1/x}(1/x^2) = e^{-1/x}(1 + 1/x)$

9. $y = \tan\sqrt{1 - x} \implies y' = \left(\sec^2\sqrt{1 - x}\right)\left(\dfrac{1}{2\sqrt{1 - x}}\right)(-1) = -\dfrac{\sec^2\sqrt{1 - x}}{2\sqrt{1 - x}}$

11. $y = \dfrac{x}{8 - 3x} \implies y' = \dfrac{(8 - 3x) - x(-3)}{(8 - 3x)^2} = \dfrac{8}{(8 - 3x)^2}$

13. $y = \sec 2\theta \implies y' = 2\sec 2\theta \tan 2\theta$

15. $y = (1 - x^{-1})^{-1} \implies y' = -(1 - x^{-1})^{-2} x^{-2} = -(x - 1)^{-2}$

17. $y = e^{cx}(c\sin x - \cos x) \implies$

$y' = e^{cx}(c\cos x + \sin x) + ce^{cx}(c\sin x - \cos x)$

$= e^{cx}(c^2\sin x - c\cos x + c\cos x + \sin x) = e^{cx}\sin x(c^2 + 1)$

19. $y = e^{e^x} \implies y' = e^{e^x}e^x = e^{x + e^x}$

21. $x^2 y^3 + 3y^2 = x - 4y \implies 2xy^3 + 3x^2 y^2 y' + 6yy' = 1 - 4y' \implies y' = \dfrac{1 - 2xy^3}{3x^2 y^2 + 6y + 4}$

23. $y = (x\tan x)^{1/5} \implies y' = \frac{1}{5}(x\tan x)^{-4/5}(\tan x + x\sec^2 x)$

25. $x^2 = y(y + 1) = y^2 + y \implies 2x = 2yy' + y' \implies y' = 2x/(2y + 1)$

27. $y = \dfrac{(x - 1)(x - 4)}{(x - 2)(x - 3)} = \dfrac{x^2 - 5x + 4}{x^2 - 5x + 6} \implies$

$y' = \dfrac{(x^2 - 5x + 6)(2x - 5) - (x^2 - 5x + 4)(2x - 5)}{(x^2 - 5x + 6)^2} = \dfrac{2(2x - 5)}{(x - 2)^2 (x - 3)^2}$

29. $y = \log_{10}(x^2 - x) \implies y' = \dfrac{1}{(x^2 - x)\ln 10}(2x - 1) = \dfrac{2x - 1}{(\ln 10)(x^2 - x)}$

31. $y = \ln\sin x - \frac{1}{2}\sin^2 x \implies y' = \dfrac{\cos x}{\sin x} - \sin x \cos x = \cot x - \sin x \cos x$

33. $y = \sin\left(\tan\sqrt{1 + x^3}\right) \implies y' = \cos\left(\tan\sqrt{1 + x^3}\right)\left(\sec^2\sqrt{1 + x^3}\right)\left[3x^2 / \left(2\sqrt{1 + x^3}\right)\right]$

35. $y = \cot\left(3x^2 + 5\right) \;\Rightarrow\; y' = -\csc^2\left(3x^2 + 5\right)(6x) = -6x\csc^2\left(3x^2 + 5\right)$

37. $y = \cos^2\left(\tan x\right) \;\Rightarrow\; y' = 2\cos\left(\tan x\right)\left[-\sin\left(\tan x\right)\right]\sec^2 x = -\sin\left(2\tan x\right)\sec^2 x$

39. $y = \dfrac{\sqrt{x+1}\,(2-x)^5}{(x+3)^7} \;\Rightarrow\; \ln|y| = \frac{1}{2}\ln(x+1) + 5\ln|2-x| - 7\ln(x+3) \;\Rightarrow$

$\dfrac{y'}{y} = \dfrac{1}{2(x+1)} + \dfrac{-5}{2-x} - \dfrac{7}{x+3} \;\Rightarrow\; y' = \dfrac{\sqrt{x+1}\,(2-x)^5}{(x+3)^7}\left[\dfrac{1}{2(x+1)} - \dfrac{5}{2-x} - \dfrac{7}{x+3}\right]$ or

$y' = \dfrac{(2-x)^4\left(3x^2 - 55x - 52\right)}{2\sqrt{x+1}\,(x+3)^8}.$

41. $y = x\sinh\left(x^2\right) \;\Rightarrow\; y' = x\cosh\left(x^2\right)\cdot 2x + \sinh\left(x^2\right)\cdot 1 = 2x^2\cosh\left(x^2\right) + \sinh\left(x^2\right)$

43. $y = \ln\left(\cosh 3x\right) \;\Rightarrow\; y' = (1/\cosh 3x)(\sinh 3x)(3) = 3\tanh 3x$

45. $y = \cosh^{-1}\left(\sinh x\right) \;\Rightarrow\; y' = (\cosh x)/\sqrt{\sinh^2 x - 1}$

47. $f(x) = (2x-1)^{-5} \;\Rightarrow\; f'(x) = -5(2x-1)^{-6}(2) = -10(2x-1)^{-6} \;\Rightarrow$

$f''(x) = 60(2x-1)^{-7}(2) = 120(2x-1)^{-7}.\; f''(0) = 120(-1)^{-7} = -120$

49. $x^6 + y^6 = 1 \;\Rightarrow\; 6x^5 + 6y^5 y' = 0 \;\Rightarrow\; y' = -x^5/y^5 \;\Rightarrow$

$y'' = -\dfrac{y^5\left(5x^4\right) - x^5\left(5y^4 y'\right)}{\left(y^5\right)^2} = -\dfrac{5x^4 y^4\left[y - x\left(-x^5/y^5\right)\right]}{y^{10}} = -\dfrac{5x^4\left[\left(y^6 + x^6\right)/y^5\right]}{y^6} = -\dfrac{5x^4}{y^{11}}$

51. We first show it is true for $n = 1$: $f'(x) = e^x + xe^x = (x+1)e^x$. We now assume it is true for $n = k$:

$f^{(k)}(x) = (x+k)e^x$. With this assumption, we must show it is true for $n = k+1$:

$f^{(k+1)}(x) = \dfrac{d}{dx}\left[f^{(k)}(x)\right] = \dfrac{d}{dx}\left[(x+k)e^x\right] = e^x + (x+k)e^x = [x+(k+1)]e^x$. Therefore,

$f^{(n)}(x) = (x+n)e^x$ by mathematical induction.

53. $y = \dfrac{x}{x^2 - 2} \;\Rightarrow\; y' = \dfrac{\left(x^2 - 2\right) - x(2x)}{\left(x^2 - 2\right)^2} = \dfrac{-x^2 - 2}{\left(x^2 - 2\right)^2}$. When $x = 2$, $y' = \dfrac{-2^2 - 2}{\left(2^2 - 2\right)^2} = -\dfrac{3}{2}$, so an equation of

the tangent line at $(2, 1)$ is $y - 1 = -\frac{3}{2}(x - 2)$ or $y = -\frac{3}{2}x + 4$.

55. $y = \tan x \;\Rightarrow\; y' = \sec^2 x$. When $x = \frac{\pi}{3}$, $y' = 2^2 = 4$, so an equation of the tangent line at $\left(\frac{\pi}{3}, \sqrt{3}\right)$ is

$y - \sqrt{3} = 4\left(x - \frac{\pi}{3}\right)$ or $y = 4x + \sqrt{3} - \frac{4}{3}\pi$.

57. $y = f(x) = \ln\left(e^x + e^{2x}\right) \;\Rightarrow\; f'(x) = \dfrac{e^x + 2e^{2x}}{e^x + e^{2x}} \;\Rightarrow\; f'(0) = \frac{3}{2}$, so the tangent line at $(0, \ln 2)$ is

$y - \ln 2 = \frac{3}{2}x$ or $3x - 2y + \ln 4 = 0$.

59. (a) $f(x) = x\sqrt{5-x}$ $\Rightarrow$ $f'(x) = x\left[\frac{1}{2}(5-x)^{-1/2}(-1)\right] + \sqrt{5-x} = \dfrac{-x}{2\sqrt{5-x}} + \dfrac{2(5-x)}{2\sqrt{5-x}} = \dfrac{10-3x}{2\sqrt{5-x}}$

(b) At $(1,2)$: $f'(1) = \frac{7}{4}$. So an equation of the tangent is $y - 2 = \frac{7}{4}(x-1)$ or $y = \frac{7}{4}x + \frac{1}{4}$.

At $(4,4)$: $f'(4) = -\frac{2}{2} = -1$. So an equation of the tangent is $y - 4 = -(x-4)$ or $y = -x + 8$.

(c)

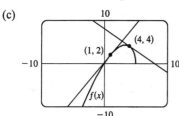

(d)

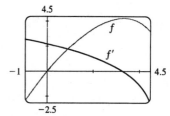

The graphs look reasonable, since f' is positive where f has tangents with positive slope, and f' is negative where f has tangents with negative slope.

61. $y = \sin x + \cos x$ $\Rightarrow$ $y' = \cos x - \sin x = 0$ $\Leftrightarrow$ $\cos x = \sin x$ and $0 \le x \le 2\pi$ $\Leftrightarrow$ $x = \frac{\pi}{4}$ or $\frac{5\pi}{4}$, so the points are $\left(\frac{\pi}{4}, \sqrt{2}\right)$ and $\left(\frac{5\pi}{4}, -\sqrt{2}\right)$.

63. $f(x) = (x-a)(x-b)(x-c)$ $\Rightarrow$ $f'(x) = (x-b)(x-c) + (x-a)(x-c) + (x-a)(x-b)$. So

$\dfrac{f'(x)}{f(x)} = \dfrac{(x-b)(x-c) + (x-a)(x-c) + (x-a)(x-b)}{(x-a)(x-b)(x-c)} = \dfrac{1}{x-a} + \dfrac{1}{x-b} + \dfrac{1}{x-c}$.

Or: $f(x) = (x-a)(x-b)(x-c)$ $\Rightarrow$ $\ln|f(x)| = \ln|x-a| + \ln|x-b| + \ln|x-c|$ $\Rightarrow$

$\dfrac{f'(x)}{f(x)} = \dfrac{1}{x-a} + \dfrac{1}{x-b} + \dfrac{1}{x-c}$

65. (a) $h(x) = f(x)g(x)$ $\Rightarrow$ $h'(x) = f'(x)g(x) + f(x)g'(x)$ $\Rightarrow$

$h'(2) = f'(2)g(2) + f(2)g'(2) = (-2)(5) + (3)(4) = 2$

(b) $F(x) = f(g(x))$ $\Rightarrow$ $F'(x) = f'(g(x))g'(x)$ $\Rightarrow$

$F'(2) = f'(g(2))g'(2) = f'(5)(4) = 11 \cdot 4 = 44$

67. $f(x) = x^2 g(x)$ $\Rightarrow$ $f'(x) = 2xg(x) + x^2 g'(x) = x\left[2g(x) + xg'(x)\right]$

69. $f(x) = [g(x)]^2$ $\Rightarrow$ $f'(x) = 2g(x)g'(x)$

71. $f(x) = g(e^x)$ $\Rightarrow$ $f'(x) = g'(e^x)e^x$

73. $f(x) = \ln|g(x)|$ $\Rightarrow$ $f'(x) = \dfrac{1}{g(x)}g'(x) = \dfrac{g'(x)}{g(x)}$

75. $h(x) = \dfrac{f(x)g(x)}{f(x) + g(x)}$ $\Rightarrow$

$h'(x) = \dfrac{[f(x) + g(x)]\left[f(x)g'(x) + g(x)f'(x)\right] - f(x)g(x)\left[f'(x) + g'(x)\right]}{[f(x) + g(x)]^2}$

$= \dfrac{[f(x)]^2 g'(x) + f(x)g(x)f'(x) + f(x)g(x)g'(x) + [g(x)]^2 f'(x) - f(x)g(x)f'(x) - f(x)g(x)g'(x)}{[f(x) + g(x)]^2}$

$= \dfrac{f'(x)[g(x)]^2 + g'(x)[f(x)]^2}{[f(x) + g(x)]^2}$

77. Using the Chain Rule repeatedly, $h(x) = f(g(\sin 4x)) \Rightarrow$

$$h'(x) = f'(g(\sin 4x)) \cdot \frac{d}{dx}(g(\sin 4x)) = f'(g(\sin 4x)) \cdot g'(\sin 4x) \cdot \frac{d}{dx}(\sin 4x)$$

$$= f'(g(\sin 4x)) g'(\sin 4x)(\cos 4x)(4)$$

79. $y = [\ln(x+4)]^2 \Rightarrow y' = 2\dfrac{\ln(x+4)}{x+4}$ and $y' = 0 \iff \ln(x+4) = 0 \iff x+4 = 1 \iff x = -3$, so the tangent is horizontal at $(-3, 0)$.

81. $y = f(x) = ax^2 + bx + c \Rightarrow f'(x) = 2ax + b$. We know that $f'(-1) = 6$ and $f'(5) = -2$, so $-2a + b = 6$ and $10a + b = -2$. Subtracting the first equation from the second gives $12a = -8 \Rightarrow a = -\frac{2}{3}$. Substituting $-\frac{2}{3}$ for a in the first equation gives $b = \frac{14}{3}$. Now $f(1) = 4 \Rightarrow 4 = a + b + c$, so $c = 4 + \frac{2}{3} - \frac{14}{3} = 0$ and hence, $f(x) = -\frac{2}{3}x^2 + \frac{14}{3}x$.

83. $s(t) = Ae^{-ct}\cos(\omega t + \delta) \Rightarrow$

$$v(t) = s'(t) = -cAe^{-ct}\cos(\omega t + \delta) + Ae^{-ct}[-\omega\sin(\omega t + \delta)]$$

$$= -Ae^{-ct}[c\cos(\omega t + \delta) + \omega\sin(\omega t + \delta)] \Rightarrow$$

$$a(t) = v'(t) = cAe^{-ct}[c\cos(\omega t + \delta) + \omega\sin(\omega t + \delta)] + (-Ae^{-ct})[-\omega c\sin(\omega t + \delta) + \omega^2\cos(\omega t + \delta)]$$

$$= Ae^{-ct}[(c^2 - \omega^2)\cos(\omega t + \delta) + 2c\omega\sin(\omega t + \delta)]$$

85. (a) $y = t^3 - 12t + 3 \Rightarrow v(t) = y' = 3t^2 - 12 \Rightarrow a(t) = v'(t) = 6t$

(b) $v(t) = 3(t^2 - 4) > 0$ when $t > 2$, so it moves upward when $t > 2$ and downward when $0 \le t < 2$.

(c) Distance upward $= y(3) - y(2) = -6 - (-13) = 7$,

Distance downward $= y(0) - y(2) = 3 - (-13) = 16$. Total distance $= 7 + 16 = 23$

87. The linear density ρ is the rate of change of mass m with respect to length x. $m = x(1 + \sqrt{x}) = x + x^{3/2} \Rightarrow$ $\rho = dm/dx = 1 + \frac{3}{2}\sqrt{x}$, so the linear density when $x = 4$ is $1 + \frac{3}{2}\sqrt{4} = 4$ kg/m.

89. If $x =$ edge length, then $V = x^3 \Rightarrow dV/dt = 3x^2dx/dt = 10 \Rightarrow dx/dt = 10/(3x^2)$ and $S = 6x^2 \Rightarrow$ $dS/dt = (12x)\,dx/dt = 12x[10/(3x^2)] = 40/x$. When $x = 30$, $dS/dt = \frac{40}{30} = \frac{4}{3}$ cm^2/min.

91. Given $dh/dt = 5$ and $dx/dt = 15$, find dz/dt. $z^2 = x^2 + h^2 \Rightarrow$

$$2z\frac{dz}{dt} = 2x\frac{dx}{dt} + 2h\frac{dh}{dt} \Rightarrow \frac{dz}{dt} = \frac{1}{z}(15x + 5h).$$ When $t = 3$,

$h = 45 + 3(5) = 60$ and $x = 15(3) = 45 \Rightarrow z = 75$, so

$$\frac{dz}{dt} = \frac{1}{75}[15(45) + 5(60)] = 13 \text{ ft/s}.$$

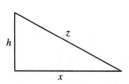

93. We are given $d\theta/dt = -0.25$ rad/h. $x = 400\cot\theta \Rightarrow$

$$\frac{dx}{dt} = -400\csc^2\theta\frac{d\theta}{dt}. \text{ When } \theta = \frac{\pi}{6},$$

$$\frac{dx}{dt} = -400(2)^2(-0.25) = 400 \text{ ft/h}.$$

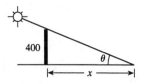

95. (a) $f(x) = \sqrt[3]{1+3x} = (1+3x)^{1/3}$ ⇒ $f'(x) = (1+3x)^{-2/3}$ so

$L(x) = f(0) + f'(0)(x-0) = 1^{1/3} + 1^{-2/3}x = 1 + x$. Thus, $\sqrt[3]{1+3x} \approx 1 + x$ ⇒

$\sqrt[3]{1.03} = \sqrt[3]{1 + 3(0.01)} \approx 1 + (0.01) = 1.01$.

(b) The linear approximation is $\sqrt[3]{1+3x} \approx 1 + x$, so for the required

accuracy we want $\sqrt[3]{1+3x} - 0.1 < 1 + x < \sqrt[3]{1+3x} + 0.1$. From

the graph, it appears that this is true when $-0.23 < x < 0.40$.

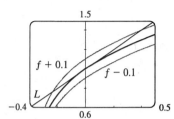

97. $A = x^2 + \frac{1}{2}\pi \left(\frac{1}{2}x\right)^2 = \left(1 + \frac{\pi}{8}\right)x^2$ ⇒ $dA = \left(2 + \frac{\pi}{4}\right)x\,dx$. When $x = 60$ and

$dx = 0.1$, $dA = \left(2 + \frac{\pi}{4}\right)60\,(0.1) = 12 + \frac{3\pi}{2}$, so the maximum error is

approximately $12 + \frac{3\pi}{2} \approx 16.7$ cm^2.

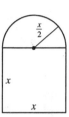

99. $\lim\limits_{h \to 0} \dfrac{\sqrt[4]{16+h} - 2}{h} = \left[\dfrac{d}{dx}\sqrt[4]{x}\right]_{x=16} = \left.\tfrac{1}{4}x^{-3/4}\right|_{x=16} = \dfrac{1}{4\left(\sqrt[4]{16}\right)^3} = \dfrac{1}{32}$

101. $\lim\limits_{x \to 0} \dfrac{\sqrt{1+\tan x} - \sqrt{1+\sin x}}{x^3} = \lim\limits_{x \to 0} \dfrac{\left(\sqrt{1+\tan x} - \sqrt{1+\sin x}\right)\left(\sqrt{1+\tan x} + \sqrt{1+\sin x}\right)}{x^3\left(\sqrt{1+\tan x} + \sqrt{1+\sin x}\right)}$

$= \lim\limits_{x \to 0} \dfrac{(1+\tan x) - (1+\sin x)}{x^3\left(\sqrt{1+\tan x} + \sqrt{1+\sin x}\right)} = \lim\limits_{x \to 0} \dfrac{\sin x\,(1/\cos x - 1)\cos x}{x^3\left(\sqrt{1+\tan x} + \sqrt{1+\sin x}\right)\cos x}$

$= \lim\limits_{x \to 0} \dfrac{\sin x\,(1 - \cos x)(1 + \cos x)}{x^3\left(\sqrt{1+\tan x} + \sqrt{1+\sin x}\right)\cos x\,(1+\cos x)}$

$= \lim\limits_{x \to 0} \dfrac{\sin x \cdot \sin^2 x}{x^3\left(\sqrt{1+\tan x} + \sqrt{1+\sin x}\right)\cos x\,(1+\cos x)}$

$= \left(\lim\limits_{x \to 0} \dfrac{\sin x}{x}\right)^3 \lim\limits_{x \to 0} \dfrac{1}{\left(\sqrt{1+\tan x} + \sqrt{1+\sin x}\right)\cos x\,(1+\cos x)}$

$= 1^3 \cdot \dfrac{1}{\left(\sqrt{1} + \sqrt{1}\right) \cdot 1 \cdot (1+1)} = \dfrac{1}{4}$

103. $\dfrac{d}{dx}[f(2x)] = x^2$ ⇒ $f'(2x) \cdot 2 = x^2$ ⇒ $f'(2x) = \tfrac{1}{2}x^2$. Let $t = 2x$. Then $f'(t) = \tfrac{1}{2}\left(\tfrac{1}{2}t\right)^2 = \tfrac{1}{8}t^2$,

so $f'(x) = \tfrac{1}{8}x^2$.

Problems Plus

1. Let a be the x-coordinate of Q. Since the derivative of $y = 1 - x^2$ is $y' = -2x$, the slope at Q is $-2a$. But since the triangle is equilateral, $\overline{AO}/\overline{OC} = \sqrt{3}/1$, so the slope at Q is $-\sqrt{3}$. Therefore, we must have that $-2a = -\sqrt{3}$ $\Rightarrow$ $a = \frac{\sqrt{3}}{2}$. Thus, the point Q has coordinates $\left(\frac{\sqrt{3}}{2}, 1 - \left(\frac{\sqrt{3}}{2}\right)^2\right) = \left(\frac{\sqrt{3}}{2}, \frac{1}{4}\right)$ and by symmetry, P has coordinates $\left(-\frac{\sqrt{3}}{2}, \frac{1}{4}\right)$.

3. Let $y = \tan^{-1} x$. Then $\tan y = x$, so from the triangle we see that

$$\sin\left(\tan^{-1} x\right) = \sin y = \frac{x}{\sqrt{1 + x^2}}.$$ Using this fact we have that

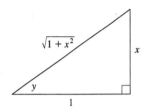

$$\sin\left(\tan^{-1}(\sinh x)\right) = \frac{\sinh x}{\sqrt{1 + \sinh^2 x}} = \frac{\sinh x}{\cosh x} = \tanh x.$$ Hence,

$$\sin^{-1}(\tanh x) = \sin^{-1}\left(\sin\left(\tan^{-1}(\sinh x)\right)\right) = \tan^{-1}(\sinh x).$$

5. We use mathematical induction. Let S_n be the statement that $\dfrac{d^n}{dx^n}\left(\sin^4 x + \cos^4 x\right) = 4^{n-1}\cos(4x + n\pi/2)$. S_1 is true because

$$\frac{d}{dx}\left(\sin^4 x + \cos^4 x\right) = 4\sin^3 x \cos x - 4\cos^3 x \sin x = 4\sin x \cos x \left(\sin^2 x - \cos^2 x\right)$$

$$= -4\sin x \cos x \cos 2x = -2\sin 2x \cos 2x = -\sin 4x$$

$$= \cos\left(\tfrac{\pi}{2} - (-4x)\right) = \cos\left(\tfrac{\pi}{2} + 4x\right) = 4^{n-1}\cos\left(4x + n\tfrac{\pi}{2}\right) \text{ when } n = 1$$

Now assume S_k is true, that is, $\dfrac{d^k}{dx^k}\left(\sin^4 x + \cos^4 x\right) = 4^{k-1}\cos\left(4x + k\tfrac{\pi}{2}\right)$. Then

$$\frac{d^{k+1}}{dx^{k+1}}\left(\sin^4 x + \cos^4 x\right) = \frac{d}{dx}\left[\frac{d^k}{dx^k}\left(\sin^4 x + \cos^4 x\right)\right] = \frac{d}{dx}\left[4^{k-1}\cos\left(4x + k\tfrac{\pi}{2}\right)\right]$$

$$= -4^{k-1}\sin\left(4x + k\tfrac{\pi}{2}\right) \cdot \frac{d}{dx}\left(4x + k\tfrac{\pi}{2}\right) = -4^k \sin\left(4x + k\tfrac{\pi}{2}\right)$$

$$= 4^k \sin\left(-4x - k\tfrac{\pi}{2}\right) = 4^k \cos\left(\tfrac{\pi}{2} - \left(-4x - k\tfrac{\pi}{2}\right)\right)$$

$$= 4^k \cos\left(4x + (k + 1)\tfrac{\pi}{2}\right)$$

which shows that S_{k+1} is true.

Therefore, $\dfrac{d^n}{dx^n}\left(\sin^4 x + \cos^4 x\right) = 4^{n-1}\cos\left(4x + n\tfrac{\pi}{2}\right)$ for every positive integer n, by mathematical induction.

Another Proof: First write

$$\sin^4 x + \cos^4 x = \left(\sin^2 x + \cos^2 x\right)^2 - 2\sin^2 x \cos^2 x = 1 - \tfrac{1}{2}\sin^2 2x = 1 - \tfrac{1}{4}(1 - \cos 4x) = \tfrac{3}{4} + \tfrac{1}{4}\cos 4x.$$

Then we have $\dfrac{d^n}{dx^n}\left(\sin^4 x + \cos^4 x\right) = \dfrac{d^n}{dx^n}\left(\tfrac{3}{4} + \tfrac{1}{4}\cos 4x\right) = \tfrac{1}{4} \cdot 4^n \cos\left(4x + n\tfrac{\pi}{2}\right) = 4^{n-1}\cos\left(4x + n\tfrac{\pi}{2}\right).$

7. We must find a value x_0 such that the normal lines to the parabola $y = x^2$ at $x = \pm x_0$ intersect at a point one unit

from the points $(\pm x_0, x_0^2)$. The normals to $y = x^2$ at $x = \pm x_0$ have slopes $-\dfrac{1}{\pm 2x_0}$ and pass through $(\pm x_0, x_0^2)$

respectively, so the normals have the equations $y - x_0^2 = -\dfrac{1}{2x_0}(x - x_0)$ and $y - x_0^2 = \dfrac{1}{2x_0}(x + x_0)$. The

common y-intercept is $x_0^2 + \frac{1}{2}$. We want to find the value of x_0 for which the distance from $\left(0, x_0^2 + \frac{1}{2}\right)$ to (x_0, x_0^2)

equals 1. The square of the distance is $(x_0 - 0)^2 + \left[x_0^2 - \left(x_0^2 + \frac{1}{2}\right)\right]^2 = x_0^2 + \frac{1}{4} = 1 \quad \Leftrightarrow \quad x_0 = \pm\frac{\sqrt{3}}{2}$. For these

values of x_0, the y-intercept is $x_0^2 + \frac{1}{2} = \frac{5}{4}$, so the center of the circle is at $\left(0, \frac{5}{4}\right)$.

Another Solution: Let the center of the circle be $(0, a)$. Then the equation of the circle is $x^2 + (y - a)^2 = 1$.

Solving with the equation of the parabola, $y = x^2$, we get $x^2 + \left(x^2 - a\right)^2 = 1 \quad \Leftrightarrow \quad x^2 + x^4 - 2ax^2 + a^2 = 1$

$\Leftrightarrow \quad x^4 + (1 - 2a)x^2 + a^2 - 1 = 0$. The parabola and the circle will be tangent to each other when this quadratic

equation in x^2 has equal roots, that is, when the discriminant is 0. Thus, $(1 - 2a)^2 - 4\left(a^2 - 1\right) = 0 \quad \Leftrightarrow$

$1 - 4a + 4a^2 - 4a^2 + 4 = 0 \quad \Leftrightarrow \quad 4a = 5$, so $a = \frac{5}{4}$. The center of the circle is $\left(0, \frac{5}{4}\right)$.

9. We can assume without loss of generality that $\theta = 0$ at time $t = 0$, so that $\theta = 12\pi t$ rad. [The angular velocity of

the wheel is 360 rpm $= 360 \cdot (2\pi \text{ rad}) / (60 \text{ s}) = 12\pi$ rad/s.] Then the position of A as a function of time is

$A = (40 \cos \theta, 40 \sin \theta) = (40 \cos 12\pi t, 40 \sin 12\pi t)$, so $\sin \alpha = \dfrac{40 \sin \theta}{120} = \dfrac{\sin \theta}{3} = \frac{1}{3} \sin 12\pi t$.

(a) Differentiating the expression for $\sin \alpha$, we get $\cos \alpha \cdot \dfrac{d\alpha}{dt} = \frac{1}{3} \cdot 12\pi \cdot \cos 12\pi t = 4\pi \cos \theta$. When

$\theta = \frac{\pi}{3}$, we have $\sin \alpha = \frac{1}{3} \sin \theta = \frac{\sqrt{3}}{6}$, so $\cos \alpha = \sqrt{1 - \left(\frac{\sqrt{3}}{6}\right)^2} = \sqrt{\frac{11}{12}}$ and

$\dfrac{d\alpha}{dt} = \dfrac{4\pi \cos \frac{\pi}{3}}{\cos \alpha} = \dfrac{2\pi}{\sqrt{11/12}} = \dfrac{4\pi\sqrt{3}}{\sqrt{11}} \approx 6.56$ rad/s.

(b) By the Law of Cosines, $|AP|^2 = |OA|^2 + |OP|^2 - 2|OA||OP|\cos\theta \quad \Rightarrow$

$\quad 120^2 = 40^2 + |OP|^2 - 2 \cdot 40|OP|\cos\theta \quad \Rightarrow \quad |OP|^2 - (80\cos\theta)|OP| - 12{,}800 = 0 \quad \Rightarrow$

$$|OP| = \frac{1}{2}\left(80\cos\theta \pm \sqrt{6400\cos^2\theta + 51{,}200}\right) = 40\cos\theta \pm 40\sqrt{\cos^2\theta + 8}$$

$$= 40\left(\cos\theta + \sqrt{8 + \cos^2\theta}\right) \text{ cm (since } |OP| > 0)$$

As a check, note that $|OP| = 160$ cm when $\theta = 0$ and $|OP| = 80\sqrt{2}$ cm when $\theta = \frac{\pi}{2}$.

(c) By part (b), the x-coordinate of P is given by $x = 40\left(\cos\theta + \sqrt{8 + \cos^2\theta}\right)$, so

$\dfrac{dx}{dt} = \dfrac{dx}{d\theta}\dfrac{d\theta}{dt} = 40\left(-\sin\theta - \dfrac{2\cos\theta\sin\theta}{2\sqrt{8+\cos^2\theta}}\right) \cdot 12\pi = -480\pi\sin\theta\left(1 + \dfrac{\cos\theta}{\sqrt{8+\cos^2\theta}}\right)$ cm/s.

In particular, $dx/dt = 0$ cm/s when $\theta = 0$ and $dx/dt = -480\pi$ cm/s when $\theta = \frac{\pi}{2}$.

11. Consider the statement that $\dfrac{d^n}{dx^n}(e^{ax}\sin bx) = r^n e^{ax}\sin(bx + n\theta)$. For $n = 1$,

$$\frac{d}{dx}(e^{ax}\sin bx) = ae^{ax}\sin bx + be^{ax}\cos bx, \text{ and}$$

$$re^{ax}\sin(bx + \theta) = re^{ax}[\sin bx\cos\theta + \cos bx\sin\theta] = re^{ax}\left(\frac{a}{r}\sin bx + \frac{b}{r}\cos bx\right)$$

$$= ae^{ax}\sin bx + be^{ax}\cos bx$$

since $\tan\theta = \dfrac{b}{a} \;\Rightarrow\; \sin\theta = \dfrac{b}{r}$ and $\cos\theta = \dfrac{a}{r}$.

So the statement is true for $n = 1$. Assume it is true for $n = k$. Then

$$\frac{d^{k+1}}{dx^{k+1}}(e^{ax}\sin bx) = \frac{d}{dx}\left[r^k e^{ax}\sin(bx + k\theta)\right] = r^k a e^{ax}\sin(bx + k\theta) + r^k e^{ax} b\cos(bx + k\theta)$$

$$= r^k e^{ax}[a\sin(bx + k\theta) + b\cos(bx + k\theta)]$$

But

$$\sin[bx + (k + 1)\theta] = \sin[(bx + k\theta) + \theta] = \sin(bx + k\theta)\cos\theta + \sin\theta\cos(bx + k\theta)$$

$$= \frac{a}{r}\sin(bx + k\theta) + \frac{b}{r}\cos(bx + k\theta)$$

Hence, $a\sin(bx + k\theta) + b\cos(bx + k\theta) = r\sin[bx + (k + 1)\theta]$. So

$$\frac{d^{k+1}}{dx^{k+1}}(e^{ax}\sin bx) = r^k e^{ax}[a\sin(bx + k\theta) + b\cos(bx + k\theta)] = r^k e^{ax}[r\sin(bx + (k + 1)\theta)]$$

$$= r^{k+1}e^{ax}[\sin(bx + (k + 1)\theta)]$$

Therefore, the statement is true for all n by mathematical induction.

13. It seems from the figure that as P approaches the point $(0, 2)$ from the right, $x_T \to \infty$ and $y_T \to 2^+$. As P approaches the point $(3, 0)$ from the left, it appears that $x_T \to 3^+$ and $y_T \to \infty$. So we guess that $x_T \in (3, \infty)$ and $y_T \in (2, \infty)$. It is more difficult to estimate the range of values for x_N and y_N. We might perhaps guess that $x_N \in (0, 3)$, and $y_N \in (-\infty, 0)$ or $(-2, 0)$.

In order to actually solve the problem, we implicitly differentiate the equation of the ellipse to find the equation of the tangent line: $\dfrac{x^2}{9} + \dfrac{y^2}{4} = 1 \implies \dfrac{2x}{9} + \dfrac{2y}{4}y' = 0$, so $y' = -\dfrac{4}{9}\dfrac{x}{y}$. So at the point (x_0, y_0) on the ellipse, an

equation of the tangent line is $y - y_0 = -\dfrac{4}{9}\dfrac{x_0}{y_0}(x - x_0)$ or $4x_0 x + 9y_0 y = 4x_0^2 + 9y_0^2$. This can be written as

$\dfrac{x_0 x}{9} + \dfrac{y_0 y}{4} = \dfrac{x_0^2}{9} + \dfrac{y_0^2}{4} = 1$, because (x_0, y_0) lies on the ellipse. So an equation of the tangent line is

$\dfrac{x_0 x}{9} + \dfrac{y_0 y}{4} = 1$.

Therefore, the x-intercept x_T for the tangent line is given by $\dfrac{x_0 x_T}{9} = 1 \iff x_T = \dfrac{9}{x_0}$, and the y-intercept y_T

is given by $\dfrac{y_0 y_T}{4} = 1 \iff y_T = \dfrac{4}{y_0}$.

So as x_0 takes on all values in $(0, 3)$, x_T takes on all values in $(3, \infty)$, and as y_0 takes on all values in $(0, 2)$, y_T takes on all values in $(2, \infty)$.

At the point (x_0, y_0) on the ellipse, the slope of the normal line is $-\dfrac{1}{y'(x_0, y_0)} = \dfrac{9}{4}\dfrac{y_0}{x_0}$, and its equation is

$y - y_0 = \dfrac{9}{4}\dfrac{y_0}{x_0}(x - x_0)$. So the x-intercept x_N for the normal line is given by $0 - y_0 = \dfrac{9}{4}\dfrac{y_0}{x_0}(x_N - x_0) \implies$

$x_N = -\dfrac{4x_0}{9} + x_0 = \dfrac{5x_0}{9}$, and the y-intercept y_N is given by $y_N - y_0 = \dfrac{9}{4}\dfrac{y_0}{x_0}(0 - x_0) \implies$

$y_N = -\dfrac{9y_0}{4} + y_0 = -\dfrac{5y_0}{4}$.

So as x_0 takes on all values in $(0, 3)$, x_N takes on all values in $\left(0, \frac{5}{3}\right)$, and as y_0 takes on all values in $(0, 2)$, y_N

takes on all values in $\left(-\frac{5}{2}, 0\right)$.

15. (a)

If the two lines L_1 and L_2 have slopes m_1 and m_2 and angles of inclination ϕ_1 and ϕ_2, then $m_1 = \tan \phi_1$ and $m_2 = \tan \phi_2$. The triangle in the figure shows that $\phi_1 + \alpha + (180° - \phi_2) = 180°$ and so $\alpha = \phi_2 - \phi_1$. Therefore, using the identity for $\tan(x - y)$, we have

$$\tan \alpha = \tan(\phi_2 - \phi_1) = \dfrac{\tan \phi_2 - \tan \phi_1}{1 + \tan \phi_2 \tan \phi_1} \text{ and so } \tan \alpha = \dfrac{m_2 - m_1}{1 + m_1 m_2}.$$

(b) (i) The parabolas intersect when $x^2 = (x - 2)^2 \implies x = 1$. If $y = x^2$, then $y' = 2x$, so the slope of the tangent to $y = x^2$ at $(1, 1)$ is $m_1 = 2(1) = 2$. If $y = (x - 2)^2$, then $y' = 2(x - 2)$, so the slope of the tangent to $y = (x - 2)^2$ at $(1, 1)$ is $m_2 = 2(1 - 2) = -2$. Therefore,

$$\tan \alpha = \dfrac{m_2 - m_1}{1 + m_1 m_2} = \dfrac{-2 - 2}{1 + 2(-2)} = \tfrac{4}{3} \text{ and so } \alpha = \tan^{-1}\tfrac{4}{3} \approx 53°.$$

(ii) $x^2 - y^2 = 3$ and $x^2 - 4x + y^2 + 3 = 0$ intersect when $x^2 - 4x + (x^2 - 3) + 3 = 0$ $\Leftrightarrow$ $2x(x - 2) = 0$ $\Rightarrow$ $x = 0$ or 2, but 0 is extraneous. If $x = 2$, then $y = \pm 1$. If $x^2 - y^2 = 3$ then $2x - 2yy' = 0$ $\Rightarrow$ $y' = x/y$ and $x^2 - 4x + y^2 + 3 = 0$ $\Rightarrow$ $2x - 4 + 2yy' = 0$ $\Rightarrow$ $y' = \dfrac{2 - x}{y}$. At (2, 1) the slopes are $m_1 = 2$ and $m_2 = 0$, so $\tan \alpha = \dfrac{0 - 2}{1 + 2 \cdot 0} = -2$ $\Rightarrow$ $\alpha \approx 117°$. At (2, -1) the slopes are $m_1 = -2$ and $m_2 = 0$, so $\tan \alpha = \dfrac{0 - (-2)}{1 + (-2)(0)} = 2$ $\Rightarrow$ $\alpha \approx 63°$.

17. Since $\angle ROQ = \angle OQP = \theta$, the triangle QOR is isosceles, so $|QR| = |RO| = x$. By the Law of Cosines, $x^2 = x^2 + r^2 - 2rx \cos \theta$. Hence,

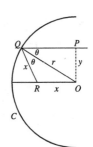

$2rx \cos \theta = r^2$, so $x = \dfrac{r^2}{2r \cos \theta} = \dfrac{r}{2 \cos \theta}$. Note that as $y \to 0^+, \theta \to 0^+$ (since $\sin \theta = y/r$), and hence $x \to \dfrac{r}{2 \cos 0} = \dfrac{r}{2}$. Thus, as P is taken closer and closer to the x-axis, the point R approaches the midpoint of the radius AO.

19. $\displaystyle\lim_{x \to 0} \dfrac{\sin(a + 2x) - 2\sin(a + x) + \sin a}{x^2}$

$= \displaystyle\lim_{x \to 0} \dfrac{\sin a \cos 2x + \cos a \sin 2x - 2 \sin a \cos x - 2 \cos a \sin x + \sin a}{x^2}$

$= \displaystyle\lim_{x \to 0} \dfrac{\sin a (\cos 2x - 2\cos x + 1) + \cos a (\sin 2x - 2\sin x)}{x^2}$

$= \displaystyle\lim_{x \to 0} \dfrac{\sin a (2\cos^2 x - 1 - 2\cos x + 1) + \cos a (2\sin x \cos x - 2\sin x)}{x^2}$

$= \displaystyle\lim_{x \to 0} \dfrac{\sin a (2\cos x)(\cos x - 1) + \cos a (2\sin x)(\cos x - 1)}{x^2}$

$= \displaystyle\lim_{x \to 0} \dfrac{2(\cos x - 1)[\sin a \cos x + \cos a \sin x](\cos x + 1)}{x^2 (\cos x + 1)}$

$= \displaystyle\lim_{x \to 0} \dfrac{-2\sin^2 x [\sin(a + x)]}{x^2 (\cos x + 1)} = -2 \lim_{x \to 0} \left(\dfrac{\sin x}{x}\right)^2 \cdot \dfrac{\sin(a + x)}{\cos x + 1} = -2(1)^2 \dfrac{\sin(a + 0)}{\cos 0 + 1} = -\sin a$

21. $y = \dfrac{x}{\sqrt{a^2 - 1}} - \dfrac{2}{\sqrt{a^2 - 1}} \arctan \dfrac{\sin x}{a + \sqrt{a^2 - 1} + \cos x}$. Let $k = a + \sqrt{a^2 - 1}$. Then

$y' = \dfrac{1}{\sqrt{a^2 - 1}} - \dfrac{2}{\sqrt{a^2 - 1}} \cdot \dfrac{1}{1 + \sin^2 x / (k + \cos x)^2} \cdot \dfrac{\cos x (k + \cos x) + \sin^2 x}{(k + \cos x)^2}$

$= \dfrac{1}{\sqrt{a^2 - 1}} - \dfrac{2}{\sqrt{a^2 - 1}} \cdot \dfrac{k \cos x + \cos^2 x + \sin^2 x}{(k + \cos x)^2 + \sin^2 x} = \dfrac{1}{\sqrt{a^2 - 1}} - \dfrac{2}{\sqrt{a^2 - 1}} \cdot \dfrac{k \cos x + 1}{k^2 + 2k \cos x + 1}$

$= \dfrac{k^2 + 2k \cos x + 1 - 2k \cos x - 2}{\sqrt{a^2 - 1} (k^2 + 2k \cos x + 1)} = \dfrac{k^2 - 1}{\sqrt{a^2 - 1} (k^2 + 2k \cos x + 1)}$

But $k^2 = 2a^2 + 2a\sqrt{a^2 - 1} - 1 = 2a \left(a + \sqrt{a^2 - 1}\right) - 1 = 2ak - 1$, so $k^2 + 1 = 2ak$, and $k^2 - 1 = 2(ak - 1)$.

So $y' = \dfrac{2(ak - 1)}{\sqrt{a^2 - 1}(2ak + 2k \cos x)} = \dfrac{ak - 1}{\sqrt{a^2 - 1}k(a + \cos x)}$. But $ak - 1 = a^2 + a\sqrt{a^2 - 1} - 1 = k\sqrt{a^2 - 1}$, so $y' = 1/(a + \cos x)$.

23. $y = x^4 - 2x^2 - x \Rightarrow y' = 4x^3 - 4x - 1$. The equation of the tangent line at $x = a$ is
$y - (a^4 - 2a^2 - a) = (4a^3 - 4a - 1)(x - a)$ or $y = (4a^3 - 4a - 1)x + (-3a^4 + 2a^2)$ and similarly for $x = b$.
So if at $x = a$ and $x = b$ we have the same tangent line, then $4a^3 - 4a - 1 = 4b^3 - 4b - 1$ and
$-3a^4 + 2a^2 = -3b^4 + 2b^2$. The first equation gives $a^3 - b^3 = a - b \Rightarrow (a - b)(a^2 + ab + b^2) = (a - b)$.
Assuming $a \neq b$, we have $1 = a^2 + ab + b^2$. The second equation gives $3(a^4 - b^4) = 2(a^2 - b^2) \Rightarrow$
$3(a^2 - b^2)(a^2 + b^2) = 2(a^2 - b^2)$ which is true if $a = -b$. Substituting into $1 = a^2 + ab + b^2$ gives
$1 = a^2 - a^2 + a^2 \Rightarrow a = \pm 1$ so that $a = 1$ and $b = -1$ or vice versa. Thus, the points $(1, -2)$ and $(-1, 0)$
have a common tangent line.

As long as there are only two such points, we are done. So we show that these are in fact the only two such
points. Suppose that $a^2 - b^2 \neq 0$. Then $3(a^2 - b^2)(a^2 + b^2) = 2(a^2 - b^2)$ gives $3(a^2 + b^2) = 2$ or
$a^2 + b^2 = \frac{2}{3}$. Thus, $ab = (a^2 + ab + b^2) - (a^2 + b^2) = 1 - \frac{2}{3} = \frac{1}{3}$, so $b = \dfrac{1}{3a}$. Hence, $a^2 + \dfrac{1}{9a^2} = \dfrac{2}{3}$, so
$9a^4 + 1 = 6a^2 \Rightarrow 0 = 9a^4 - 6a^2 + 1 = (3a^2 - 1)^2$. So $3a^2 - 1 = 0 \Rightarrow a^2 = \frac{1}{3} \Rightarrow$
$b^2 = \dfrac{1}{9a^2} = \frac{1}{3} = a^2$, contradicting our assumption that $a^2 \neq b^2$.

25.

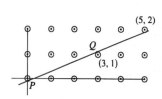
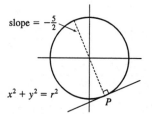

Because of the periodic nature of the lattice points, it suffices to consider the points in the 5×2 grid shown. We
can see that the minimum value of r occurs when there is a line with slope $\frac{2}{5}$ which touches the circle centered at
$(3, 1)$ and the circles centered at $(0, 0)$ and $(5, 2)$. To find P, the point at which the line is tangent to the circle at
$(0, 0)$, we simultaneously solve $x^2 + y^2 = r^2$ and $y = -\frac{5}{2}x \Rightarrow x^2 + \frac{25}{4}x^2 = r^2 \Rightarrow x^2 = \frac{4}{29}r^2 \Rightarrow$
$x = \frac{2}{\sqrt{29}}r, y = -\frac{5}{\sqrt{29}}r$. To find Q, we either use symmetry or solve $(x - 3)^2 + (y - 1)^2 = r^2$ and
$y - 1 = -\frac{5}{2}(x - 3)$. As above, we get $x = 3 - \frac{2}{\sqrt{29}}r, y = 1 + \frac{5}{\sqrt{29}}r$. Now the slope of the line PQ is $\frac{2}{5}$, so

$$m_{PQ} = \frac{1 + \frac{5}{\sqrt{29}}r - \left(-\frac{5}{\sqrt{29}}r\right)}{3 - \frac{2}{\sqrt{29}}r - \frac{2}{\sqrt{29}}r} = \frac{1 + \frac{10}{\sqrt{29}}r}{3 - \frac{4}{\sqrt{29}}r} = \frac{\sqrt{29} + 10r}{3\sqrt{29} - 4r} = \frac{2}{5}$$

$$\Rightarrow \quad 5\sqrt{29} + 50r = 6\sqrt{29} - 8r$$

$$\Leftrightarrow \quad 58r = \sqrt{29}$$

$$\Leftrightarrow \quad r = \frac{\sqrt{29}}{58}$$

So the minimum value of r for which any line with slope $\frac{2}{5}$ intersects circles with radius r centered at the lattice
points on the plane is $r = \frac{\sqrt{29}}{58} \approx 0.093$.

27.

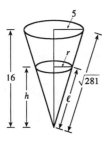

By similar triangles, $\dfrac{r}{5} = \dfrac{h}{16}$ $\Rightarrow$ $r = \dfrac{5h}{16}$. The volume of the cone is

$V = \frac{1}{3}\pi r^2 h = \frac{1}{3}\pi \left(\dfrac{5h}{16}\right)^2 h = \dfrac{25\pi}{768}h^3$, so $\dfrac{dV}{dt} = \dfrac{25\pi}{256}h^2\dfrac{dh}{dt}$. Now the rate of

change of the volume is also equal to the difference of what is being added

(2 cm^3/min) and what is oozing out ($k\pi rl$, where πrl is the area of the cone and k

is a proportionality constant). Thus, $\dfrac{dV}{dt} = 2 - k\pi rl$.

Equating the two expressions for $\dfrac{dV}{dt}$ and substituting $h = 10$, $\dfrac{dh}{dt} = -0.3$, $r = \dfrac{5(10)}{16} = \dfrac{25}{8}$, and $\dfrac{l}{\sqrt{281}} = \dfrac{10}{16}$

$\Leftrightarrow$ $l = \frac{5}{8}\sqrt{281}$, we get $\frac{25\pi}{256}(10)^2(-0.3) = 2 - k\pi\frac{25}{8} \cdot \frac{5}{8}\sqrt{281}$ $\Leftrightarrow$ $\dfrac{125k\pi\sqrt{281}}{64} = 2 + \dfrac{750\pi}{256}$. Solving for k

gives us $k = \dfrac{256 + 375\pi}{250\pi\sqrt{281}}$. To maintain a certain height, the rate of oozing, $k\pi rl$, must equal the rate of the liquid

being poured in, that is, $dV/dt = 0$. $k\pi rl = \dfrac{256 + 375\pi}{250\pi\sqrt{281}} \cdot \pi \cdot \dfrac{25}{8} \cdot \dfrac{5\sqrt{281}}{8} = \dfrac{256 + 375\pi}{128} \approx 11.204$ cm^3/min.

4 Applications of Differentiation

4.1 Maximum and Minimum Values

1. A function f has an **absolute minimum** at $x = c$ if $f(c)$ is the smallest function value on the entire domain of f, whereas f has a **local minimum** at c if $f(c)$ is the smallest function value when x is near c.

3. Absolute maximum at b; absolute minimum at d; local maxima at b, e; local minima at d, s; neither a maximum nor a minimum at a, c, r, and t.

5. Absolute maximum value is $f(4) = 4$; absolute minimum value is $f(7) = 0$; local maximum values are $f(4) = 4$ and $f(6) = 3$; local minimum values are $f(2) = 1$ and $f(5) = 2$.

7.

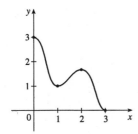

9.

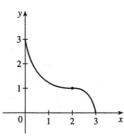

11. (a)

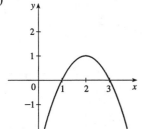

(b)

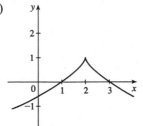

(c)

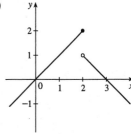

13. (a) *Note:* By the Extreme Value Theorem, f must *not* be continuous; because if it were, it would attain an absolute minimum.

(b)

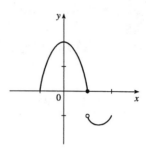

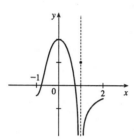

15. $f(x) = 8 - 3x$, $x \geq 1$. Absolute maximum $f(1) = 5$; no local maximum. No absolute or local minimum.

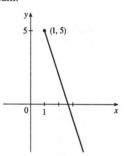

17. $f(x) = x^2$, $0 < x < 2$. No absolute or local maximum or minimum value.

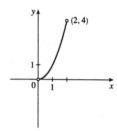

19. $f(x) = x^2$, $0 \leq x < 2$. Absolute minimum $f(0) = 0$; no local minimum. No absolute or local maximum.

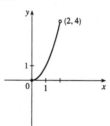

21. $f(x) = x^2$, $-3 \leq x \leq 2$. Absolute maximum $f(-3) = 9$. No local maximum. Absolute and local minimum $f(0) = 0$.

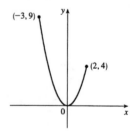

23. $f(t) = 1/t$, $0 < t < 1$. No maximum or minimum.

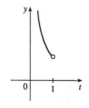

25. $f(\theta) = \sin\theta$, $-2\pi \leq \theta \leq 2\pi$. Absolute and local maxima $f\left(-\frac{3\pi}{2}\right) = f\left(\frac{\pi}{2}\right) = 1$. Absolute and local minima $f\left(-\frac{\pi}{2}\right) = f\left(\frac{3\pi}{2}\right) = -1$.

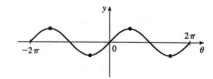

27. $f(x) = x^5$. No maximum or minimum.

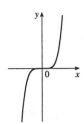

29. $f(x) = 1 - e^{-x}$, $x \geq 0$. Absolute minimum $f(0) = 0$; no local minimum. No absolute or local maximum.

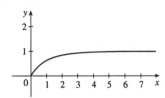

31. $f(x) = 5x^2 + 4x$ $\Rightarrow$ $f'(x) = 10x + 4$. $f'(x) = 0$ $\Rightarrow$ $x = -\frac{2}{5}$, so $-\frac{2}{5}$ is the only critical number.

33. $f(t) = 2t^3 + 3t^2 + 6t + 4$ $\Rightarrow$ $f'(t) = 6t^2 + 6t + 6$. But $t^2 + t + 1 = 0$ has no real solution since $b^2 - 4ac = 1 - 4(1)(1) = -3 < 0$. No critical number.

35. $s(t) = 2t^3 + 3t^2 - 6t + 4$ $\Rightarrow$ $s'(t) = 6t^2 + 6t - 6 = 6(t^2 + t - 1)$. By the quadratic formula, the critical numbers are $t = \left(-1 \pm \sqrt{5}\right)/2$.

37. $f(r) = \dfrac{r}{r^2 + 1}$ $\Rightarrow$ $f'(r) = \dfrac{(r^2 + 1)\,1 - r\,(2r)}{(r^2 + 1)^2} = \dfrac{-r^2 + 1}{(r^2 + 1)^2} = 0$ $\Leftrightarrow$ $r^2 = 1$ $\Leftrightarrow$ $r = \pm 1$, so these are the critical numbers. Note that $f'(x)$ always exists since $r^2 + 1 \neq 0$.

39. $g(x) = |2x + 3| = \begin{cases} 2x + 3 & \text{if } 2x + 3 \geq 0 \\ -(2x + 3) & \text{if } 2x + 3 < 0 \end{cases}$ $\Rightarrow$ $g'(x) = \begin{cases} 2 & \text{if } x > -\frac{3}{2} \\ -2 & \text{if } x < -\frac{3}{2} \end{cases}$

$g'(x)$ is never 0, but $g'(x)$ does not exist for $x = -\frac{3}{2}$, so $-\frac{3}{2}$ is the only critical number.

41. $g(t) = 5t^{2/3} + t^{5/3}$ $\Rightarrow$ $g'(t) = \frac{10}{3}t^{-1/3} + \frac{5}{3}t^{2/3}$. $g'(0)$ does not exist, so $t = 0$ is a critical number. $g'(t) = \frac{5}{3}t^{-1/3}(2 + t) = 0$ $\Leftrightarrow$ $t = -2$, so $t = -2$ is also a critical number.

43. $F(x) = x^{4/5}(x - 4)^2$ $\Rightarrow$
$F'(x) = \frac{4}{5}x^{-1/5}(x - 4)^2 + 2x^{4/5}(x - 4) = \frac{1}{5}x^{-1/5}(x - 4)[4(x - 4) + 10x]$
$= \dfrac{(x - 4)(14x - 16)}{5x^{1/5}} = \dfrac{2(x - 4)(7x - 8)}{5x^{1/5}} = 0$ when $x = 4$, $\frac{8}{7}$; and $F'(0)$ does not exist.
Critical numbers are 0, $\frac{8}{7}$, 4.

45. $f(\theta) = \sin^2(2\theta)$ $\Rightarrow$ $f'(\theta) = 2\sin(2\theta)\cos(2\theta)(2) = 2(2\sin 2\theta \cos 2\theta) = 2[\sin(2 \cdot 2\theta)] = 2\sin 4\theta = 0$ $\Leftrightarrow$ $\sin 4\theta = 0$ $\Leftrightarrow$ $4\theta = n\pi$, n an integer. So $\theta = n\pi/4$ are the critical numbers.

47. $f(x) = x\ln x$ $\Rightarrow$ $f'(x) = x(1/x) + \ln x = \ln x + 1 = 0$ $\Leftrightarrow$ $\ln x = -1$ $\Leftrightarrow$ $x = e^{-1} = 1/e$. Therefore, the only critical number is $x = 1/e$.

49. $f(x) = 3x^2 - 12x + 5$, $[0, 3]$. $f'(x) = 6x - 12 = 0$ $\Leftrightarrow$ $x = 2$. $f(0) = 5$, $f(2) = -7$, $f(3) = -4$. So $f(0) = 5$ is the absolute maximum and $f(2) = -7$ is the absolute minimum.

51. $f(x) = 2x^3 + 3x^2 + 4$, $[-2, 1]$. $f'(x) = 6x^2 + 6x = 6x(x + 1) = 0$ $\Leftrightarrow$ $x = -1, 0$. $f(-2) = 0$, $f(-1) = 5$, $f(0) = 4$, $f(1) = 9$. So $f(1) = 9$ is the absolute maximum and $f(-2) = 0$ is the absolute minimum.

53. $f(x) = x^4 - 4x^2 + 2$, $[-3, 2]$. $f'(x) = 4x^3 - 8x = 4x(x^2 - 2) = 0$ $\Leftrightarrow$ $x = 0, \pm\sqrt{2}$. $f(-3) = 47$, $f\left(-\sqrt{2}\right) = -2$, $f(0) = 2$, $f\left(\sqrt{2}\right) = -2$, $f(2) = 2$, so $f\left(\pm\sqrt{2}\right) = -2$ is the absolute minimum and $f(-3) = 47$ is the absolute maximum.

55. $f(x) = x^2 + \dfrac{2}{x}$, $\left[\frac{1}{2}, 2\right]$. $f'(x) = 2x - \dfrac{2}{x^2} = 2\dfrac{x^3 - 1}{x^2} = 0 \iff x^3 - 1 = 0 \iff (x - 1)(x^2 + x + 1) = 0$,

but $x^2 + x + 1 \neq 0$, so $x = 1$. The denominator is 0 at $x = 0$, but not in the desired interval. $f\left(\frac{1}{2}\right) = \frac{17}{4}$,

$f(1) = 3$, $f(2) = 5$. So $f(1) = 3$ is the absolute minimum and $f(2) = 5$ is the absolute maximum.

57. $f(x) = \dfrac{x}{x^2 + 1}$, $[0, 2]$. $f'(x) = \dfrac{(x^2 + 1) - x(2x)}{(x^2 + 1)^2} = \dfrac{1 - x^2}{(x^2 + 1)^2} = 0 \iff x = \pm 1$, but -1 is not in $[0, 2]$.

$f(0) = 0$, $f(1) = \frac{1}{2}$, $f(2) = \frac{2}{5}$. So $f(1) = \frac{1}{2}$ is the absolute maximum and $f(0) = 0$ is the absolute minimum.

59. $f(x) = \sin x + \cos x$, $\left[0, \frac{\pi}{3}\right]$. $f'(x) = \cos x - \sin x = 0 \iff \sin x = \cos x \implies \dfrac{\sin x}{\cos x} = 1 \implies \tan x = 1$

$\implies x = \frac{\pi}{4}$. $f(0) = 1$, $f\left(\frac{\pi}{4}\right) = \sqrt{2} \approx 1.41$, $f\left(\frac{\pi}{3}\right) = \frac{\sqrt{3}+1}{2} \approx 1.37$. So $f(0) = 1$ is the absolute minimum and

$f\left(\frac{\pi}{4}\right) = \sqrt{2}$ is the absolute maximum.

61. $f(x) = xe^{-x}$, $[0, 2]$. $f'(x) = x\left(-e^{-x}\right) + e^{-x} = e^{-x}(1 - x) = 0 \iff x = 1$. Now $f(0) = 0$,

$f(1) = e^{-1} = 1/e$, and $f(2) = 2/e^2 \approx 0.27$, so $f(0) = 0$ is the absolute minimum and $f(1) = 1/e$ is the

absolute maximum.

63. $f(x) = x - 3\ln x$, $[1, 4]$. $f'(x) = 1 - \dfrac{3}{x} = \dfrac{x - 3}{x} = 0 \iff x = 3$. f' does not exist for $x = 0$, but 0 is not in

the domain of f. $f(1) = 1$, $f(3) = 3 - 3\ln 3 \approx -0.296$, $f(4) = 4 - 3\ln 4 \approx -0.159$. So $f(1) = 1$ is the

absolute maximum and $f(3) \approx -0.296$ is the absolute minimum.

65.

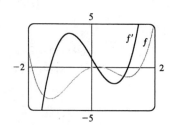

We see that $f'(x) = 0$ at about $x = -1.3, 0.2$, and 1.1. Since f' exists everywhere, these are the only critical

numbers.

67. (a)

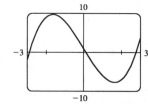

From the graph, it appears that the absolute maximum value is about

$f(-1.63) = 9.71$, and the absolute minimum value is about

$f(1.63) = -7.71$. These values make sense because the graph is

symmetric about the point $(0, 1)$. ($y = x^3 - 8x$ is symmetric about

the origin.)

(b) $f(x) = x^3 - 8x + 1 \implies f'(x) = 3x^2 - 8$. So $f'(x) = 0 \implies x = \pm\sqrt{\frac{8}{3}}$.

$f\left(\pm\sqrt{\frac{8}{3}}\right) = \left(\pm\sqrt{\frac{8}{3}}\right)^3 - 8\left(\pm\sqrt{\frac{8}{3}}\right) + 1 = \pm\frac{8}{3}\sqrt{\frac{8}{3}} \mp 8\sqrt{\frac{8}{3}} + 1$

$= -\frac{16}{3}\sqrt{\frac{8}{3}} + 1 = 1 - \frac{32\sqrt{6}}{9}$ (minimum) or $\frac{16}{3}\sqrt{\frac{8}{3}} + 1 = 1 + \frac{32\sqrt{6}}{9}$ (maximum)

(From the graph, we see that the extreme values do not occur at the endpoints.)

69. (a)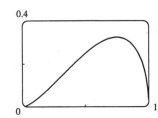

From the graph, it appears that the absolute maximum value is about $f(0.75) = 0.32$, and the absolute minimum value is $f(0) = f(1) = 0$, that is, at both endpoints.

(b) $f(x) = x\sqrt{x - x^2}$ $\Rightarrow$ $f'(x) = x \cdot \dfrac{1 - 2x}{2\sqrt{x - x^2}} + \sqrt{x - x^2} = \dfrac{(x - 2x^2) + (2x - 2x^2)}{2\sqrt{x - x^2}} = \dfrac{3x - 4x^2}{2\sqrt{x - x^2}}$.

So $f'(x) = 0$ $\Rightarrow$ $3x - 4x^2 = 0$ $\Rightarrow$ $x = 0$ or $\frac{3}{4}$. $f(0) = f(1) = 0$ (minima), and

$f\left(\frac{3}{4}\right) = \frac{3}{4}\sqrt{\frac{3}{4} - \left(\frac{3}{4}\right)^2} = \frac{3\sqrt{3}}{16}$ (maximum).

71. The density is defined as $\rho = \dfrac{\text{mass}}{\text{volume}} = \dfrac{1000}{V(T)}$ (in g/cm^3). But a critical point of ρ will also be a critical

point of V $\left[\text{since } \dfrac{d\rho}{dT} = -1000V^{-2}\dfrac{dV}{dT} \text{ and } V \text{ is never } 0\right]$, and V is easier to

differentiate than ρ. $V(T) = 999.87 - 0.06426T + 0.0085043T^2 - 0.0000679T^3$ $\Rightarrow$

$V'(T) = -0.06426 + 0.0170086T - 0.0002037T^2$. Setting this equal to 0 and using the quadratic formula to find

T, we get $T = \dfrac{-0.0170086 \pm \sqrt{0.0170086^2 - 4 \cdot 0.0002037 \cdot 0.06426}}{2(-0.0002037)} \approx 3.9665°$ or $79.5318°$. Since we are

only interested in the region $0° \le T \le 30°$, we check the density ρ at the endpoints and at $3.9665°$:

$\rho(0) \approx \dfrac{1000}{999.87} \approx 1.00013$; $\rho(30) \approx \dfrac{1000}{1003.7641} \approx 0.99625$; $\rho(3.9665) \approx \dfrac{1000}{999.7447} \approx 1.000255$. So water has

its maximum density at about $3.9665°$C.

73. We apply the Closed Interval Method to the continuous function I on $[0, 10]$. Its derivative is

$I'(t) = 0.00045225t^4 + 0.005752t^3 - 0.19683tss + 0.9196t - 0.6270$. Since I' exists for all t, the only critical

numbers of I occur when $I'(t) = 0$. We use a root-finder on a computer algebra system (or a graphing device) to

find that $I'(t) = 0$ when $t \approx -29.7186$, 0.8231, 5.1309, or 11.0459, but only the second and third roots lie in the

interval $[0, 10]$. The values of I at these critical numbers are $I(0.8231) \approx 99.09$ and $I(5.1309) \approx 100.67$. The

values of I at the endpoints of the interval are $I(0) = 99.33$ and $I(10) \approx 96.86$. Comparing these four numbers,

we see that food was most expensive at $t \approx 5.1309$ (corresponding roughly to August, 1989) and cheapest at

$t = 10$ (midyear 1994).

75. (a) $v(r) = k(r_0 - r)r^2 = kr_0 r^2 - kr^3$ $\Rightarrow$

$v'(r) = 2kr_0 r - 3kr^2$. $v'(r) = 0$ $\Rightarrow$

$kr(2r_0 - 3r) = 0$ $\Rightarrow$ $r = 0$ or $\frac{2}{3}r_0$ (but 0 is not in

the interval). Evaluating v at $\frac{1}{2}r_0$, $\frac{2}{3}r_0$, and r_0, we get

$v\left(\frac{1}{2}r_0\right) = \frac{1}{8}kr_0^3$, $v\left(\frac{2}{3}r_0\right) = \frac{4}{27}kr_0^3$, and $v(r_0) = 0$.

Since $\frac{4}{27} > \frac{1}{8}$, v attains its maximum value at $r = \frac{2}{3}r_0$.

This supports the statement in the text.

(b) From part (a), the maximum value of v is $\frac{4}{27}kr_0^3$.

(c)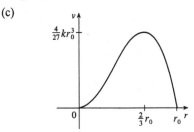

77. $f(x) = x^{101} + x^{51} + x + 1 \implies f'(x) = 101x^{100} + 51x^{50} + 1 \geq 1$ for all x, so $f'(x) = 0$ has no solution. Thus, $f(x)$ has no critical number, so $f(x)$ can have no local maximum or minimum.

79. If f has a local minimum at c, then $g(x) = -f(x)$ has a local maximum at c, so $g'(c) = 0$ by the case of Fermat's Theorem proved in the text. Thus, $f'(c) = -g'(c) = 0$.

4.2 The Mean Value Theorem

1. $f(x) = x^2 - 4x + 1$, $[0, 4]$. Since f is a polynomial, it is continuous and differentiable on $\mathbb{R}$, so it is continuous on $[0, 4]$ and differentiable on $(0, 4)$. Also, $f(0) = 1 = f(4)$. $f'(c) = 0 \iff 2c - 4 = 0 \iff c = 2$, which is in the open interval $(0, 4)$, so $c = 2$ satisfies the conclusion of Rolle's Theorem.

3. $f(x) = \sin 2\pi x$, $[-1, 1]$. f, being the composite of the sine function and the polynomial $2\pi x$, is continuous and differentiable on $\mathbb{R}$, so it is continuous on $[-1, 1]$ and differentiable on $(-1, 1)$. Also, $f(-1) = 0 = f(1)$.
$f'(c) = 0 \iff 2\pi \cos 2\pi c = 0 \iff \cos 2\pi c = 0 \iff 2\pi c = \pm\frac{\pi}{2} + 2\pi n \iff c = \pm\frac{1}{4} + n$. If $n = 0$ or ± 1, then $c = \pm\frac{1}{4}, \pm\frac{3}{4}$ is in $(-1, 1)$.

5. $f(x) = 1 - x^{2/3}$. $f(-1) = 1 - (-1)^{2/3} = 1 - 1 = 0 = f(1)$. $f'(x) = -\frac{2}{3}x^{-1/3}$, so $f'(c) = 0$ has no solution. This does not contradict Rolle's Theorem, since $f'(0)$ does not exist, and so f is not differentiable on $[-1, 1]$.

7. $\dfrac{f(8) - f(0)}{8 - 0} = \dfrac{6 - 4}{8} = \dfrac{1}{4}$. The values of c which satisfy $f'(c) = \frac{1}{4}$ seem to be about $c = 0.8, 3.2, 4.4$, and 6.1.

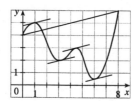

9. (a), (b) The equation of the secant line is
$$y - 5 = \frac{8.5 - 5}{8 - 1}(x - 1) \iff$$
$$y = \tfrac{1}{2}x + \tfrac{9}{2}.$$

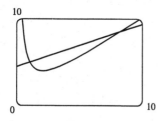

(c) $f(x) = x + 4/x \implies f'(x) = 1 - 4/x^2$. So $f'(c) = \frac{1}{2} \implies c^2 = 8 \implies c = 2\sqrt{2}$, and $f(c) = 2\sqrt{2} + \frac{4}{2\sqrt{2}} = 3\sqrt{2}$. Thus, an equation of the tangent line is $y - 3\sqrt{2} = \frac{1}{2}\left(x - 2\sqrt{2}\right)$
$\iff y = \frac{1}{2}x + 2\sqrt{2}$.

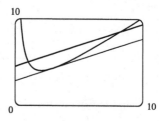

11. $f(x) = 3x^2 + 2x + 5$, $[-1, 1]$. f is continuous on $[-1, 1]$ and differentiable on $(-1, 1)$ since polynomials are
continuous and differentiable on $\mathbb{R}$. $f'(c) = \dfrac{f(b) - f(a)}{b - a} \iff 6c + 2 = \dfrac{f(1) - f(-1)}{1 - (-1)} = \dfrac{10 - 6}{2} = 2 \iff$
$6c = 0 \iff c = 0$, which is in $(-1, 1)$.

13. $f(x) = e^{-2x}$, $[0, 3]$. f is continuous and differentiable on $\mathbb{R}$, so it is continuous on $[0, 3]$ and differentiable on
$(0, 3)$. $f'(c) = \dfrac{f(b) - f(a)}{b - a} \iff -2e^{-2c} = \dfrac{e^{-6} - e^0}{3 - 0} \iff e^{-2c} = \dfrac{1 - e^{-6}}{6} \iff -2c = \ln\left(\dfrac{1 - e^{-6}}{6}\right)$
$\iff c = -\dfrac{1}{2}\ln\left(\dfrac{1 - e^{-6}}{6}\right) \approx 0.897$, which is in $(0, 3)$.

15. $f(x) = |x - 1|$. $f(3) - f(0) = |3 - 1| - |0 - 1| = 1$. Since $f'(c) = -1$ if $c < 1$ and $f'(c) = 1$ if $c > 1$,
$f'(c)(3 - 0) = \pm 3$ and so is never equal to 1. This does not contradict the Mean Value Theorem since $f'(1)$ does
not exist.

17. $f(x) = x^5 + 10x + 3$. Since f is continuous and $f(-1) = -8$ and $f(0) = 3$, the equation $f(x) = 0$ has at least
one root in $(-1, 0)$ by the Intermediate Value Theorem. Suppose that the equation has more than one root; say a
and b are both roots with $a < b$. Then $f(a) = 0 = f(b)$ so by Rolle's Theorem $f'(x) = 5x^4 + 10 = 0$ has a root
in (a, b). But this is impossible since clearly $f'(x) \geq 10 > 0$ for all real x.

19. $f(x) = x^5 - 6x + c$. Suppose that $f(x) = 0$ has two roots a and b with $-1 \leq a < b \leq 1$. Then
$f(a) = 0 = f(b)$, so by Rolle's Theorem there is a number d in (a, b) with $f'(d) = 0$. Now
$0 = f'(d) = 5d^4 - 6 \implies d = \pm\sqrt[4]{\frac{6}{5}}$, which are both outside $[-1, 1]$ and hence outside (a, b). Thus, $f(x)$ can
have at most one root in $[-1, 1]$.

21. (a) Suppose that a cubic polynomial $P(x)$ has roots $a_1 < a_2 < a_3 < a_4$, so $P(a_1) = P(a_2) = P(a_3) = P(a_4)$.
By Rolle's Theorem there are numbers c_1, c_2, c_3 with $a_1 < c_1 < a_2, a_2 < c_2 < a_3$ and $a_3 < c_3 < a_4$ and
$P'(c_1) = P'(c_2) = P'(c_3) = 0$. Thus, the second-degree polynomial $P'(x)$ has three distinct real roots,
which is impossible.

(b) We prove by induction that a polynomial of degree n has at most n real roots. This is certainly true for $n = 1$.
Suppose that the result is true for all polynomials of degree n and let $P(x)$ be a polynomial of degree $n + 1$.
Suppose that $P(x)$ has more than $n + 1$ real roots, say $a_1 < a_2 < a_3 < \cdots < a_{n+1} < a_{n+2}$. Then
$P(a_1) = P(a_2) = \cdots = P(a_{n+2}) = 0$. By Rolle's Theorem there are real numbers $c_1, \ldots, c_{n+1}$ with
$a_1 < c_1 < a_2, \ldots, \implies a_{n+1} < c_{n+1} < a_{n+2}$ and $P'(c_1) = \cdots = P'(c_{n+1}) = 0$. Thus, the nth degree
polynomial $P'(x)$ has at least $n + 1$ roots. This contradiction shows that $P(x)$ has at most $n + 1$ real roots.

23. By the Mean Value Theorem, $f(4) - f(1) = f'(c)(4 - 1)$ for some $c \in (1, 4)$. But for every $c \in (1, 4)$ we have
$f'(c) \geq 2$. Putting $f'(c) \geq 2$ into the above equation and substituting $f(1) = 10$, we get
$f(4) = f(1) + f'(c)(4 - 1) = 10 + 3f'(c) \geq 10 + 3 \cdot 2 = 16$. So the smallest possible value of $f(4)$ is 16.

25. Suppose that such a function f exists. By the Mean Value Theorem there is a number $0 < c < 2$ with
$f'(c) = \dfrac{f(2) - f(0)}{2 - 0} = \dfrac{5}{2}$. But this is impossible since $f'(x) \leq 2 < \frac{5}{2}$ for all x, so no such function can exist.

27. We use Exercise 26 with $f(x) = \sqrt{1 + x}$, $g(x) = 1 + \frac{1}{2}x$, and $a = 0$. Notice that $f(0) = 1 = g(0)$ and
$f'(x) = \dfrac{1}{2\sqrt{1 + x}} < \dfrac{1}{2} = g'(x)$ for $x > 0$. So by Exercise 26, $f(b) < g(b) \implies \sqrt{1 + b} < 1 + \frac{1}{2}b$ for $b > 0$.
Another Method: Apply the Mean Value Theorem directly to either $f(x) = 1 + \frac{1}{2}x - \sqrt{1 + x}$ or $g(x) = \sqrt{1 + x}$
on $[0, b]$.

29. Let $f(x) = \sin x$ and let $b < a$. Then $f(x)$ is continuous on $[b, a]$ and differentiable on (b, a). By the Mean Value Theorem, there is a number $c \in (b, a)$ with $\sin a - \sin b = f(a) - f(b) = f'(c)(a - b) = (\cos c)(a - b)$. Thus, $|\sin a - \sin b| \le |\cos c| |b - a| \le |a - b|$. If $a < b$, then $|\sin a - \sin b| = |\sin b - \sin a| \le |b - a| = |a - b|$. If $a = b$, both sides of the inequality are 0.

31. For $x > 0$, $f(x) = g(x)$, so $f'(x) = g'(x)$. For $x < 0$, $f'(x) = (1/x)' = -1/x^2$ and $g'(x) = (1 + 1/x)' = -1/x^2$, so again $f'(x) = g'(x)$. However, the domain of $g(x)$ is not an interval [it is $(-\infty, 0) \cup (0, \infty)$] so we cannot conclude that $f - g$ is constant (in fact it is not).

33. Let $f(x) = \arcsin\left(\dfrac{x-1}{x+1}\right) - 2\arctan\sqrt{x} + \dfrac{\pi}{2}$. Note that the domain of f is $[0, \infty)$. Thus,

$$f'(x) = \frac{1}{\sqrt{1 - \left(\dfrac{x-1}{x+1}\right)^2}} \cdot \frac{(x+1) - (x-1)}{(x+1)^2} - \frac{2}{1+x} \cdot \frac{1}{2\sqrt{x}} = \frac{1}{\sqrt{x}\,(x+1)} - \frac{1}{\sqrt{x}\,(x+1)} = 0. \text{ Then}$$

$f(x) = C$. To find C, we let $x = 0 \Rightarrow \arcsin(-1) - 2\arctan(0) + \dfrac{\pi}{2} = C \Rightarrow -\dfrac{\pi}{2} - 0 + \dfrac{\pi}{2} = 0 = C$.

Thus, $f(x) = 0 \Rightarrow \arcsin\left(\dfrac{x-1}{x+1}\right) = 2\arctan\sqrt{x} - \dfrac{\pi}{2}$.

35. Let $g(t)$ and $h(t)$ be the position functions of the two runners and let $f(t) = g(t) - h(t)$. By hypothesis $f(0) = g(0) - h(0) = 0$ and $f(b) = g(b) - h(b) = 0$ where b is the finishing time. Then by Rolle's Theorem, there is a time $0 < c < b$ with $0 = f'(c) = g'(c) - h'(c)$. Hence, $g'(c) = h'(c)$, so at time c, both runners have the same velocity $g'(c) = h'(c)$.

4.3 How Derivatives Affect the Shape of a Graph

1. (a) f is increasing on $(0, 6)$ and $(8, 9)$.

(b) f is decreasing on $(6, 8)$.

(c) f is concave upward on $(2, 4)$ and $(7, 9)$.

(d) f is concave downward on $(0, 2)$ and $(4, 7)$.

(e) The points of inflection are $(2, 3)$, $(4, 4.5)$ and $(7, 4)$ (where the concavity changes).

3. (a) Use the Increasing/Decreasing (I/D) Test.

(b) Use the Concavity Test.

(c) At any value of x where the concavity changes, we have an inflection point at $(x, f(x))$.

5. (a) Since $f'(x) > 0$ on $(-\infty, 0)$ and $(3, \infty)$, f is increasing on the same intervals. $f'(x) < 0$ and f is decreasing on $(0, 3)$.

(b) Since $f'(x) = 0$ at $x = 0$ and f' changes from positive to negative there, f changes from increasing to decreasing and has a local maximum at $x = 0$. Since $f'(x) = 0$ at $x = 3$ and changes from negative to positive there, f changes from decreasing to increasing and has a local minimum at $x = 3$.

7. There is an inflection point at $x = 1$ because $f''(x)$ changes from negative to positive there, and one at $x = 7$ because $f''(x)$ changes from positive to negative there.

9. The function must be always decreasing and concave downward.

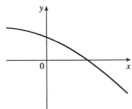

11. (a) $f(x) = x^3 - 12x + 1 \Rightarrow f'(x) = 3x^2 - 12 = 3(x+2)(x-2)$. So $f'(x) > 0 \Leftrightarrow x > 2$ or $x < -2$ and $f'(x) < 0 \Leftrightarrow -2 < x < 2$. So f is increasing on $(-\infty, -2)$ and $(2, \infty)$ and decreasing on $(-2, 2)$.

(b) f changes from increasing to decreasing at $x = -2$ and from decreasing to increasing at $x = 2$. Thus, $f(-2) = 17$ is a local maximum and $f(2) = -15$ is a local minimum.

(c) $f''(x) = 6x$. $f''(x) > 0 \Leftrightarrow x > 0$ and $f''(x) < 0 \Leftrightarrow x < 0$. Thus, f is concave upward on $(0, \infty)$ and concave downward on $(-\infty, 0)$. There is an inflection point where the concavity changes, at $(0, f(0)) = (0, 1)$.

13. (a) $f(x) = x^6 + 192x + 17 \Rightarrow f'(x) = 6x^5 + 192 = 6(x^5 + 32)$. So $f'(x) > 0 \Leftrightarrow x^5 > -32 \Leftrightarrow x > -2$ and $f'(x) < 0 \Leftrightarrow x < -2$. So f is increasing on $(-2, \infty)$ and decreasing on $(-\infty, -2)$.

(b) f changes from decreasing to increasing at its only critical number, $x = -2$. Thus, $f(-2) = -303$ is a local minimum.

(c) $f''(x) = 30x^4 \geq 0$ for all x, so the concavity of f doesn't change and there is no inflection point. f is concave upward on $(-\infty, \infty)$.

15. (a) $f(x) = x - 2\sin x$ on $(0, 3\pi) \Rightarrow f'(x) = 1 - 2\cos x$. $f'(x) > 0 \Leftrightarrow 1 - 2\cos x > 0 \Leftrightarrow \cos x < \frac{1}{2}$ $\Leftrightarrow \frac{\pi}{3} < x < \frac{5\pi}{3}$ or $\frac{7\pi}{3} < x < 3\pi$. $f'(x) < 0 \Leftrightarrow \cos x > \frac{1}{2} \Leftrightarrow 0 < x < \frac{\pi}{3}$ or $\frac{5\pi}{3} < x < \frac{7\pi}{3}$. So f is increasing on $\left(\frac{\pi}{3}, \frac{5\pi}{3}\right)$ and $\left(\frac{7\pi}{3}, 3\pi\right)$, and f is decreasing on $\left(0, \frac{\pi}{3}\right)$ and $\left(\frac{5\pi}{3}, \frac{7\pi}{3}\right)$.

(b) f changes from increasing to decreasing at $x = \frac{5\pi}{3}$, and from decreasing to increasing at $x = \frac{\pi}{3}$ and at $x = \frac{7\pi}{3}$. Thus, $f\left(\frac{5\pi}{3}\right) = \frac{5\pi}{3} + \sqrt{3} \approx 6.97$ is a local maximum and $f\left(\frac{\pi}{3}\right) = \frac{\pi}{3} - \sqrt{3} \approx -0.68$ and $f\left(\frac{7\pi}{3}\right) = \frac{7\pi}{3} - \sqrt{3} \approx 5.60$ are local minima.

(c) $f''(x) = 2\sin x > 0 \Leftrightarrow 0 < x < \pi$ and $2\pi < x < 3\pi$, $f''(x) < 0 \Leftrightarrow \pi < x < 2\pi$. Thus, f is concave upward on $(0, \pi)$ and $(2\pi, 3\pi)$, and f is concave downward on $(\pi, 2\pi)$. There are inflection points at (π, π) and $(2\pi, 2\pi)$.

17. (a) $y = f(x) = xe^x \Rightarrow f'(x) = xe^x + e^x = e^x(x+1)$. So $f'(x) > 0 \Leftrightarrow x + 1 > 0 \Leftrightarrow x > -1$. Thus, f is increasing on $(-1, \infty)$ and decreasing on $(-\infty, -1)$.

(b) f changes from decreasing to increasing at its only critical number, $x = -1$. Thus, $f(-1) = -e^{-1}$ is a local minimum.

(c) $f'(x) = e^x(x+1) \Rightarrow f''(x) = e^x(1) + (x+1)e^x = e^x(x+2)$. So $f''(x) > 0 \Leftrightarrow x + 2 > 0 \Leftrightarrow x > -2$. Thus, f is concave upward on $(-2, \infty)$ and concave downward on $(-\infty, -2)$. Since the concavity changes direction at $x = -2$, the point $(-2, -2e^{-2})$ is an inflection point.

19. (a) $y = f(x) = \dfrac{\ln x}{\sqrt{x}}$. (Note that f is only defined for $x > 0$.)

$$f'(x) = \frac{\sqrt{x}\,(1/x) - \ln x\left(\frac{1}{2}x^{-1/2}\right)}{x} = \frac{\dfrac{1}{\sqrt{x}} - \dfrac{\ln x}{2\sqrt{x}}}{x} \cdot \frac{2\sqrt{x}}{2\sqrt{x}} = \frac{2 - \ln x}{2x^{3/2}} > 0 \;\Leftrightarrow\; \ln x < 2 \;\Leftrightarrow\; x < e^2.$$

Therefore f is increasing on $\left(0, e^2\right)$ and decreasing on $\left(e^2, \infty\right)$.

(b) f changes from increasing to decreasing at $x = e^2$, so $f\left(e^2\right) = \dfrac{\ln e^2}{\sqrt{e^2}} = \dfrac{2}{e}$ is a local maximum.

(c) $f''(x) = \dfrac{2x^{3/2}\,(-1/x) - (2 - \ln x)\left(3x^{1/2}\right)}{\left(2x^{3/2}\right)^2} = \dfrac{-2x^{1/2} + 3x^{1/2}\,(\ln x - 2)}{4x^3}$

$$= \frac{x^{1/2}(-2 + 3\ln x - 6)}{4x^3} = \frac{3\ln x - 8}{4x^{5/2}}$$

$f''(x) = 0 \;\Leftrightarrow\; \ln x = \frac{8}{3} \;\Leftrightarrow\; x = e^{8/3}$. $f''(x) > 0 \;\Leftrightarrow\; x > e^{8/3}$, so f is concave upward on $\left(e^{8/3}, \infty\right)$

and concave downward on $\left(0, e^{8/3}\right)$. There is an inflection point at $\left(e^{8/3}, 8/\left(3e^{4/3}\right)\right) \approx (14.39, 0.70)$.

21. $f(x) = x^5 - 5x + 3 \;\Rightarrow\; f'(x) = 5x^4 - 5 = 5\left(x^2 + 1\right)(x + 1)(x - 1)$.

First Derivative Test: $f'(x) < 0 \;\Rightarrow\; -1 < x < 1$ and $f'(x) > 0 \;\Rightarrow\; x > 1$ or $x < -1$. Since f' changes

from positive to negative at $x = -1$, $f(-1) = 7$ is a local maximum; and since f' changes from negative to

positive at $x = 1$, $f(1) = -1$ is a local minimum.

Second Derivative Test: $f''(x) = 20x^3$. $f'(x) = 0 \;\Leftrightarrow\; x = \pm 1$. $f''(-1) = -20 < 0 \;\Rightarrow\; f(-1) = 7$ is

a local maximum. $f''(1) = 20 > 0 \;\Rightarrow\; f(1) = -1$ is a local minimum.

Preference: For this function, the two tests are equally easy.

23. $f(x) = x + \sqrt{1 - x} \;\Rightarrow\; f'(x) = 1 + \frac{1}{2}(1 - x)^{-1/2}(-1) = 1 - \dfrac{1}{2\sqrt{1 - x}}$. Note that f is defined for

$1 - x \geq 0$, that is, for $x \leq 1$. $f'(x) = 0 \;\Rightarrow\; 2\sqrt{1 - x} = 1 \;\Rightarrow\; \sqrt{1 - x} = \frac{1}{2} \;\Rightarrow\; 1 - x = \frac{1}{4} \;\Rightarrow\; x = \frac{3}{4}$.

f' does not exist at $x = 1$, but we can't have a local maximum or minimum at an endpoint.

First Derivative Test: $f'(x) > 0 \;\Rightarrow\; x < \frac{3}{4}$ and $f'(x) < 0 \;\Rightarrow\; \frac{3}{4} < x < 1$. Since f' changes from

positive to negative at $x = \frac{3}{4}$, $f\left(\frac{3}{4}\right) = \frac{5}{4}$ is a local maximum.

Second Derivative Test: $f''(x) = -\frac{1}{2}\left(-\frac{1}{2}\right)(1 - x)^{-3/2}(-1) = -\dfrac{1}{4\left(\sqrt{1 - x}\right)^3}$. $f''\left(\frac{3}{4}\right) = -2 < 0 \;\Rightarrow\;$

$f\left(\frac{3}{4}\right) = \frac{5}{4}$ is a local maximum.

Preference: The First Derivative Test may be slightly easier to apply in this case.

25. $f(-1) = 4$ and $f(1) = 0$ gives us two points to start with.

$f'(-1) = f'(1) = 0 \;\Rightarrow\;$ horizontal tangents at $x = \pm 1$. $f'(x) < 0$ if

$|x| < 1 \;\Rightarrow\; f$ is decreasing on $(-1, 1)$. $f'(x) > 0$ if $|x| > 1 \;\Rightarrow\; f$ is

increasing on $(-\infty, -1)$ and $(1, \infty)$. $f''(x) < 0$ if $x < 0 \;\Rightarrow\; f$ is concave

downward on $(-\infty, 0)$. $f''(x) > 0$ if $x > 0 \;\Rightarrow\; f$ is concave upward on

$(0, \infty)$ and there is an inflection point at $x = 0$.

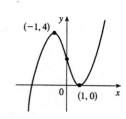

27. First we plot the points which are known to be on the graph: $(2, -1)$ and $(0, 0)$. We can also draw a short line segment of slope 0 at $x = 2$, since we are given that $f'(2) = 0$. Now we know that $f'(x) < 0$ (that is, the function is decreasing) on $(0, 2)$, and that $f''(x) < 0$ on $(0, 1)$ and $f''(x) > 0$ on $(1, 2)$. So we must join the points $(0, 0)$ and $(2, -1)$ in such a way that the curve is concave down on $(0, 1)$ and concave up on $(1, 2)$. The curve must be concave up and increasing on $(2, 4)$ and concave down and increasing on $(4, \infty)$. Now we just need to reflect the curve in the y-axis, since we are given that f is an even function.

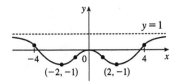

29. (a) f is increasing where f' is positive, that is, on $(0, 2)$, $(4, 6)$, and $(8, \infty)$; and decreasing where f' is negative, that is, on $(2, 4)$ and $(6, 8)$.

(b) f has local maxima where f' changes from positive to negative, at $x = 2$ and at $x = 6$, and local minima where f' changes from negative to positive, at $x = 4$ and at $x = 8$.

(c) f is concave upward where f' is increasing, that is, on $(3, 6)$ and $(6, \infty)$, and concave downward where f' is decreasing, that is, on $(0, 3)$.

(e)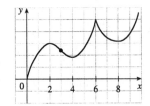

(d) There is a point of inflection where f changes from being CD to being CU, that is, at $x = 3$.

31. (a) $f(x) = 2x^3 - 3x^2 - 12x$ $\Rightarrow$ $f'(x) = 6x^2 - 6x - 12 = 6(x^2 - x - 2) = 6(x - 2)(x + 1)$. $f'(x) > 0$ $\Leftrightarrow$ $x < -1$ or $x > 2$ and $f'(x) < 0$ $\Leftrightarrow$ $-1 < x < 2$. So f is increasing on $(-\infty, -1)$ and $(2, \infty)$, and f is decreasing on $(-1, 2)$.

(b) Since f changes from increasing to decreasing at $x = -1$, $f(-1) = 7$ is a local maximum value. Since f changes from decreasing to increasing at $x = 2$, $f(2) = -20$ is a local minimum value.

(c) $f''(x) = 6(2x - 1)$ $\Rightarrow$ $f''(x) > 0$ on $\left(\frac{1}{2}, \infty\right)$ and $f''(x) < 0$ on $\left(-\infty, \frac{1}{2}\right)$. So f is concave upward on $\left(\frac{1}{2}, \infty\right)$ and concave downward on $\left(-\infty, \frac{1}{2}\right)$. There is a change in concavity at $x = \frac{1}{2}$, and we have an inflection point at $\left(\frac{1}{2}, -\frac{13}{2}\right)$.

(d)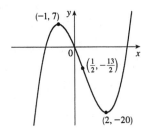

33. (a) $f(x) = x^4 - 6x^2$ $\Rightarrow$ $f'(x) = 4x^3 - 12x = 4x(x^2 - 3) = 0$ when $x = 0, \pm\sqrt{3}$.

Interval	$4x$	$x^2 - 3$	$f'(x)$	f
$x < -\sqrt{3}$	$-$	$+$	$-$	decreasing on $\left(-\infty, -\sqrt{3}\right)$
$-\sqrt{3} < x < 0$	$-$	$-$	$+$	increasing on $\left(-\sqrt{3}, 0\right)$
$0 < x < \sqrt{3}$	$+$	$-$	$-$	decreasing on $\left(0, \sqrt{3}\right)$
$x > \sqrt{3}$	$+$	$+$	$+$	increasing on $\left(\sqrt{3}, \infty\right)$

(b) Local minima $f\left(\pm\sqrt{3}\right) = -9$, local maximum $f(0) = 0$ (d)

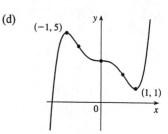

(c) $f''(x) = 12x^2 - 12 = 12\left(x^2 - 1\right) > 0 \iff x^2 > 1 \iff |x| > 1$
$\iff x > 1$ or $x < -1$, so f is CU on $(-\infty, -1)$, $(1, \infty)$ and CD on
$(-1, 1)$. Inflection points at $(\pm 1, -5)$

35. (a) $h(x) = 3x^5 - 5x^3 + 3 \implies h'(x) = 15x^4 - 15x^2 = 15x^2\left(x^2 - 1\right) = 0$ when $x = 0, \pm 1$. $h'(x) > 0 \iff$
$x^2 > 1 \iff x > 1$ or $x < -1$, so h is increasing on $(-\infty, -1)$ and $(1, \infty)$ and decreasing on $(-1, 1)$.

(b) Local maximum $h(-1) = 5$, local minimum $h(1) = 1$ (d)

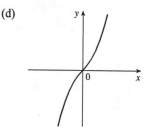

(c) $h''(x) = 60x^3 - 30x = 30x\left(2x^2 - 1\right)$
$= 60x\left(x + \frac{1}{\sqrt{2}}\right)\left(x - \frac{1}{\sqrt{2}}\right) \implies$

$h''(x) > 0$ when $x > \frac{1}{\sqrt{2}}$ or $-\frac{1}{\sqrt{2}} < x < 0$, so h is CU on $\left(-\frac{1}{\sqrt{2}}, 0\right)$

and $\left(\frac{1}{\sqrt{2}}, \infty\right)$ and CD on $\left(-\infty, -\frac{1}{\sqrt{2}}\right)$ and $\left(0, \frac{1}{\sqrt{2}}\right)$. Inflection

points at $\left(\pm\frac{1}{\sqrt{2}}, 3 \pm \frac{7}{8}\sqrt{2}\right)$ and $(0, 3)$

37. (a) $P(x) = x\sqrt{x^2 + 1} \implies$ (d)

$P'(x) = \sqrt{x^2 + 1} + \frac{x^2}{\sqrt{x^2 + 1}} = \frac{2x^2 + 1}{\sqrt{x^2 + 1}} > 0$, so P is increasing on $\mathbb{R}$.

(b) No maximum or minimum

(c) $P''(x) = \dfrac{4x\sqrt{x^2 + 1} - \left(2x^2 + 1\right)\dfrac{x}{\sqrt{x^2 + 1}}}{x^2 + 1} = \dfrac{x\left(2x^2 + 3\right)}{\left(x^2 + 1\right)^{3/2}} > 0 \iff$

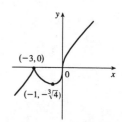

$x > 0$ so P is CU on $(0, \infty)$ and CD on $(-\infty, 0)$. IP at $(0, 0)$

39. (a) $Q(x) = x^{1/3}(x + 3)^{2/3} \implies Q'(x) = \frac{1}{3}x^{-2/3}(x + 3)^{2/3} + x^{1/3}\left(\frac{2}{3}\right)(x + 3)^{-1/3} = \dfrac{x + 1}{x^{2/3}(x + 3)^{1/3}}$. The

critical numbers are $-3, -1$, and 0. Note that $x^{2/3} \geq 0$ for all x. So $Q'(x) > 0$ when $x < -3$ or $x > -1$ and
$Q'(x) < 0$ when $-3 < x < -1 \implies Q$ is increasing on $(-\infty, -3)$ and $(-1, \infty)$ and decreasing on
$(-3, -1)$.

(b) $Q(-3) = 0$ is a local maximum and (d)
$Q(-1) = -4^{1/3} \approx -1.6$ is a local minimum.

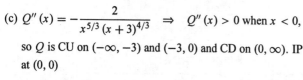

(c) $Q''(x) = -\dfrac{2}{x^{5/3}(x + 3)^{4/3}} \implies Q''(x) > 0$ when $x < 0$,

so Q is CU on $(-\infty, -3)$ and $(-3, 0)$ and CD on $(0, \infty)$. IP
at $(0, 0)$

41. (a) $f(\theta) = \sin^2 \theta \;\Rightarrow\; f'(\theta) = 2\sin\theta\cos\theta = \sin 2\theta > 0 \;\Leftrightarrow\; 2\theta \in (0, \pi) \cup (2\pi, 3\pi) \;\Leftrightarrow$
$\theta \in \left(0, \frac{\pi}{2}\right) \cup \left(\pi, \frac{3\pi}{2}\right)$. So f is increasing on $\left(0, \frac{\pi}{2}\right)$ and $\left(\pi, \frac{3\pi}{2}\right)$, and decreasing on $\left(\frac{\pi}{2}, \pi\right)$ and $\left(\frac{3\pi}{2}, 2\pi\right)$.

(b) Local minimum $f(\pi) = 0$, local maxima $f\left(\frac{\pi}{2}\right) = f\left(\frac{3\pi}{2}\right) = 1$

(c) $f''(\theta) = 2\cos 2\theta > 0 \;\Leftrightarrow$ (d)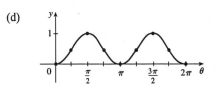
$2\theta \in \left(0, \frac{\pi}{2}\right) \cup \left(\frac{3\pi}{2}, \frac{5\pi}{2}\right) \cup \left(\frac{7\pi}{2}, 4\pi\right) \;\Leftrightarrow$
$\theta \in \left(0, \frac{\pi}{4}\right) \cup \left(\frac{3\pi}{4}, \frac{5\pi}{4}\right) \cup \left(\frac{7\pi}{4}, 2\pi\right)$, so f is CU on these
intervals and CD on $\left(\frac{\pi}{4}, \frac{3\pi}{4}\right)$ and $\left(\frac{5\pi}{4}, \frac{7\pi}{4}\right)$. IP at $\left(\frac{n\pi}{4}, \frac{1}{2}\right)$,
$n = 1, 3, 5, 7$

43. (a) $\lim\limits_{x \to \pm\infty} \dfrac{1+x^2}{1-x^2} = \lim\limits_{x \to \pm\infty} \dfrac{(1/x^2)+1}{(1/x^2)-1} = -1$, so $y = -1$ is a HA. $\lim\limits_{x \to 1^-} \dfrac{1+x^2}{1-x^2} = \infty$, $\lim\limits_{x \to 1^+} \dfrac{1+x^2}{1-x^2} = -\infty$,
$\lim\limits_{x \to -1^-} \dfrac{1+x^2}{1-x^2} = -\infty$, $\lim\limits_{x \to -1^+} \dfrac{1+x^2}{1-x^2} = \infty$. So $x = 1$ and $x = -1$ are VA.

(b) $f(x) = \dfrac{1+x^2}{1-x^2} = -1 + \dfrac{2}{1-x^2} \;\Rightarrow\; f'(x) = \dfrac{4x}{(1-x^2)^2} > 0 \;\Leftrightarrow\; x > 0 \,(x \ne 1)$, so f increases on
$(0, 1)$, $(1, \infty)$ and decreases on $(-\infty, -1)$, $(-1, 0)$.

(c) $f(0) = 1$ is a local minimum.

(d) $f''(x) = \dfrac{4(1-x^2)^2 - 4x \cdot 2(1-x^2)(-2x)}{(1-x^2)^4} = \dfrac{4(1+3x^2)}{(1-x^2)^3}$. (e)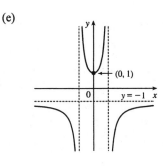
Since the numerator is always positive, the sign of $f''(x)$ is the
same as the sign of $1 - x^2$. Thus, $f''(x) > 0 \;\Leftrightarrow\; 1 - x^2 > 0$
$\Leftrightarrow\; x^2 < 1 \;\Leftrightarrow\; -1 < x < 1$, so f is CU on $(-1, 1)$ and CD on
$(-\infty, -1)$ and $(1, \infty)$. There is no IP since $x = \pm 1$ are not in the
domain of f.

45. (a) $\lim\limits_{x \to -\infty} \left(\sqrt{x^2+1} - x\right) = \infty$ and

$\lim\limits_{x \to \infty} \left(\sqrt{x^2+1} - x\right) = \lim\limits_{x \to \infty} \left(\sqrt{x^2+1} - x\right) \dfrac{\sqrt{x^2+1}+x}{\sqrt{x^2+1}+x} = \lim\limits_{x \to \infty} \dfrac{1}{\sqrt{x^2+1}+x} = 0$, so $y = 0$ is a HA.

(b) $f(x) = \sqrt{x^2+1} - x \;\Rightarrow\; f'(x) = \dfrac{x}{\sqrt{x^2+1}} - 1$. Since $\dfrac{x}{\sqrt{x^2+1}} < 1$ for all x, $f'(x) < 0$, so f is

decreasing on $\mathbb{R}$.

(c) No minimum or maximum

(e)

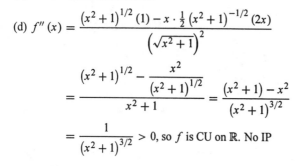

(d) $f''(x) = \dfrac{\left(x^2+1\right)^{1/2}(1) - x \cdot \frac{1}{2}\left(x^2+1\right)^{-1/2}(2x)}{\left(\sqrt{x^2+1}\right)^2}$

$= \dfrac{\left(x^2+1\right)^{1/2} - \dfrac{x^2}{\left(x^2+1\right)^{1/2}}}{x^2+1} = \dfrac{\left(x^2+1\right) - x^2}{\left(x^2+1\right)^{3/2}}$

$= \dfrac{1}{\left(x^2+1\right)^{3/2}} > 0$, so f is CU on $\mathbb{R}$. No IP

47. $f(x) = \ln(1 - \ln x)$ is defined when $x > 0$ (so that $\ln x$ is defined) and $1 - \ln x > 0$ (so that $\ln(1 - \ln x)$ is defined). The second condition is equivalent to $1 > \ln x \;\Leftrightarrow\; x < e$, so f has domain $(0, e)$.

(a) As $x \to 0^+$, $\ln x \to -\infty$, so $1 - \ln x \to \infty$ and $f(x) \to \infty$. As $x \to e^-$, $\ln x \to 1^-$, so $1 - \ln x \to 0^+$ and $f(x) \to -\infty$. Thus, $x = 0$ and $x = e$ are vertical asymptotes. There is no horizontal asymptote.

(b) $f'(x) = \dfrac{1}{1 - \ln x}\left(-\dfrac{1}{x}\right) = -\dfrac{1}{x(1 - \ln x)} < 0$ on $(0, e)$. Thus, f is decreasing on its domain, $(0, e)$.

(c) $f'(x) \neq 0$ on $(0, e)$, so f has no local maximum or minimum value.

(d) $f''(x) = -\dfrac{-[x(1-\ln x)]'}{[x(1-\ln x)]^2} = \dfrac{x(-1/x) + (1-\ln x)}{x^2(1-\ln x)^2}$

(e)

$= -\dfrac{\ln x}{x^2(1-\ln x)^2}$

so $f''(x) > 0 \;\Leftrightarrow\; \ln x < 0 \;\Leftrightarrow\; 0 < x < 1$. Thus, f is CU on
$(0, 1)$ and CD on $(1, e)$. There is an inflection point at $(1, 0)$.

49. (a) $\lim\limits_{x \to \pm\infty} e^{-1/(x+1)} = 1$ since $-1/(x+1) \to 0$, so $y = 1$ is a HA. $\lim\limits_{x \to -1^+} e^{-1/(x+1)} = 0$ since

$-1/(x+1) \to -\infty$, $\lim\limits_{x \to -1^-} e^{-1/(x+1)} = \infty$ since $-1/(x+1) \to \infty$, so $x = -1$ is a VA.

(b) $f(x) = e^{-1/(x+1)} \;\Rightarrow\; f'(x) = e^{-1/(x+1)}/(x+1)^2 \;\Rightarrow\; f'(x) > 0$ for all x except -1, so f is
increasing on $(-\infty, -1)$ and $(-1, \infty)$.

(c) No local maximum or minimum

(d) $f''(x) = \dfrac{(x+1)^2 e^{-1/(x+1)}\left[1/(x+1)^2\right] - e^{-1/(x+1)}\left[2(x+1)\right]}{\left[(x+1)^2\right]^2} = \dfrac{e^{-1/(x+1)}\left[1-(2x+2)\right]}{(x+1)^4}$

$$= -\dfrac{e^{-1/(x+1)}(2x+1)}{(x+1)^4} \quad\Rightarrow$$

$f''(x) > 0 \iff 2x+1 < 0 \iff x < -\frac{1}{2}$, so f is CU on $(-\infty, -1)$ and $\left(-1, -\frac{1}{2}\right)$, and CD on $\left(-\frac{1}{2}, \infty\right)$.

f has an IP at $\left(-\frac{1}{2}, e^{-2}\right)$.

(e)

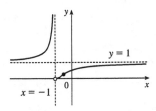

51. (a)

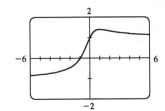

From the graph, we get us an estimate of $f(1) \approx 1.41$ as a local maximum, and no local minimum. $f(x) = \dfrac{x+1}{\sqrt{x^2+1}} \quad\Rightarrow$

$f'(x) = \dfrac{1-x}{(x^2+1)^{3/2}}$. $f'(x) = 0 \iff x = 1$. $f(1) = \dfrac{2}{\sqrt{2}} = \sqrt{2}$ is the exact value.

(b) From the graph in part (a), f increases most rapidly somewhere between $x = -\frac{1}{2}$ and $x = -\frac{1}{4}$. To find the exact value, we need to find the maximum value of f', which we can do by finding the critical numbers of f'.

$f''(x) = \dfrac{2x^2 - 3x - 1}{(x^2+1)^{5/2}} = 0 \iff x = \dfrac{3 \pm \sqrt{17}}{4}$. $x = \dfrac{3 + \sqrt{17}}{4}$ corresponds to the *minimum* value of f'.

The maximum value of f' is at $\left(\dfrac{3-\sqrt{17}}{4}, \sqrt{\dfrac{7}{6} - \dfrac{\sqrt{17}}{6}}\right) \approx (-0.28, 0.69)$.

53. (a) From the graphs of

$f(x) = 3x^5 - 40x^3 + 30x^2$, it seems that f is concave upward on $(-2, 0.25)$ and $(2, \infty)$, and concave downward on $(-\infty, -2)$ and $(0.25, 2)$, with inflection points at about $(-2, 350)$, $(0.25, 1)$, and $(2, -100)$.

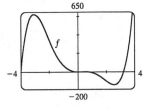

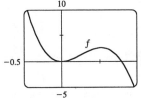

(b)

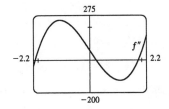

From the graph of $f''(x) = 60x^3 - 240x + 60$, it seems that f is CU on $(-2.1, 0.25)$ and $(1.9, \infty)$, and CD on $(-\infty, -2.1)$ and $(0.25, 2)$, with inflection points at about $(-2.1, 386)$, $(0.25, 1.3)$ and $(1.9, -87)$. (We have to check back on the graph of f to find the y-coordinates of the inflection points.)

55. In Maple, we define f and then use the command
`plot(diff(diff(f,x),x),x=-3..3);`. In Mathematica, we
define f and then use `Plot[Dt[Dt[f,x],x],{x,-3,3}]`. We
see that $f'' > 0$ for $x > 0.1$ and $f'' < 0$ for $x < 0.1$. So f is concave
up on $(0.1, \infty)$ and concave down on $(-\infty, 0.1)$.

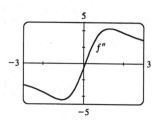

57. Most students learn more in the third hour of studying than in the eighth hour, so $K(3) - K(2)$ is larger than
$K(8) - K(7)$. In other words, as you begin studying for a test, the rate of knowledge gain is large and then starts
to taper off, so $K'(t)$ decreases and the graph of K is concave downward.

59. From the graph, we estimate that the most rapid increase in the
number of VCRs occurs at about $t = 7$. To maximize the first
derivative, we need to determine the values for which the second
derivative is 0. $V(t) = \dfrac{75}{1 + 74e^{-0.6t}}$ $\Rightarrow$

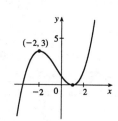

$$V'(t) = -\frac{75\left[74e^{-0.6t}(-0.6)\right]}{\left(1 + 74e^{-0.6t}\right)^2} = \frac{3330e^{-0.6t}}{\left(1 + 74e^{-0.6t}\right)^2} \quad \Rightarrow$$

$$V''(t) = \frac{\left(1 + 74e^{-0.6t}\right)^2 \left[3330e^{-0.6t}(-0.6)\right] - \left(3330e^{-0.6t}\right) 2\left(1 + 74e^{-0.6t}\right)\left[74e^{-0.6t}(-0.6)\right]}{\left[\left(1 + 74e^{-0.6t}\right)^2\right]^2}$$

$$= \frac{\left(1 + 74e^{-0.6t}\right)\left[3330e^{-0.6t}(-0.6)\right]\left[\left(1 + 74e^{-0.6t}\right) - 2\left(74e^{-0.6t}\right)\right]}{\left(1 + 74e^{-0.6t}\right)^4} = \frac{-1998e^{-0.6t}\left(1 - 74e^{-0.6t}\right)}{\left(1 + 74e^{-0.6t}\right)^3}.$$

$V''(t) = 0 \Leftrightarrow 1 = 74e^{-0.6t} \Leftrightarrow e^{0.6t} = 74 \Leftrightarrow 0.6t = \ln 74 \Leftrightarrow t = \frac{5}{3}\ln 74 \approx 7.173$ years, which
corresponds to early September 1987.

61. $f(x) = ax^3 + bx^2 + cx + d \Rightarrow f(1) = a + b + c + d = 0$ and
$f(-2) = -8a + 4b - 2c + d = 3$. Also $f'(1) = 3a + 2b + c = 0$ and
$f'(-2) = 12a - 4b + c = 0$ by Fermat's Theorem. Solving these four
equations, we get $a = \frac{2}{9}, b = \frac{1}{3}, c = -\frac{4}{3}, d = \frac{7}{9}$, so the function is
$f(x) = \frac{1}{9}\left(2x^3 + 3x^2 - 12x + 7\right)$.

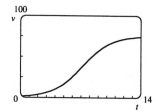

63. We will make use of the converse of the Concavity Test; that is, if f is concave upward on I, then $f'' > 0$ on I.
If f and g are CU on I, then $f'' > 0$ and $g'' > 0$ on I, so $(f + g)'' = f'' + g'' > 0$ on $I \Rightarrow f + g$ is CU on I.

65. Since f and g are positive, increasing, and CU on I, we have $f > 0, f' > 0, f'' > 0, g > 0, g' > 0, g'' > 0$ on I.
Then $(fg)' = f'g + fg' \Rightarrow (fg)'' = f''g + 2f'g' + fg'' > 0 \Rightarrow fg$ is CU on I.

67. $f(x) = \tan x - x \Rightarrow f'(x) = \sec^2 x - 1 > 0$ for $0 < x < \frac{\pi}{2}$ since $\sec^2 x > 1$ for $0 < x < \frac{\pi}{2}$. So f is
increasing on $\left(0, \frac{\pi}{2}\right)$. Thus, $f(x) > f(0) = 0$ for $0 < x < \frac{\pi}{2} \Rightarrow \tan x - x > 0 \Rightarrow \tan x > x$ for
$0 < x < \frac{\pi}{2}$.

69. Let the cubic function be $f(x) = ax^3 + bx^2 + cx + d \Rightarrow f'(x) = 3ax^2 + 2bx + c \Rightarrow$
$f''(x) = 6ax + 2b$. So f is CU when $6ax + 2b > 0 \Leftrightarrow x > -b/(3a)$, and CD when
$x < -b/(3a)$, and so the only point of inflection occurs when $x = -b/(3a)$. If the graph
has three x-intercepts x_1, x_2 and x_3, then the equation of $f(x)$ must factor as
$f(x) = a(x - x_1)(x - x_2)(x - x_3) = a[x^3 - (x_1 + x_2 + x_3)x^2 + (x_1 x_2 + x_1 x_3 + x_2 x_3)x - x_1 x_2 x_3]$.
So $b = -a(x_1 + x_2 + x_3)$. Hence, the x-coordinate of the point of inflection is
$$-\frac{b}{3a} = -\frac{-a(x_1 + x_2 + x_3)}{3a} = \frac{x_1 + x_2 + x_3}{3}.$$

71. By hypothesis $g = f'$ is differentiable on an open interval containing c. Since $(c, f(c))$ is a point of inflection, the
concavity changes at $x = c$, so $f'(x)$ changes signs at $x = c$. Hence, by the First Derivative Test, f' has a local
extremum at $x = c$. Thus, by Fermat's Theorem $f''(c) = 0$.

73. Using the fact that $|x| = \sqrt{x^2}$, we have that $g(x) = x\sqrt{x^2} \Rightarrow g'(x) = \sqrt{x^2} + \sqrt{x^2} = 2\sqrt{x^2} = 2|x| \Rightarrow$
$g''(x) = 2x(x^2)^{-1/2} = \dfrac{2x}{|x|} < 0$ for $x < 0$ and $g''(x) > 0$ for $x > 0$, so $(0, 0)$ is an inflection point. But $g''(0)$
does not exist.

4.4 Indeterminate Forms and L'Hospital's Rule

1. (a) $\lim\limits_{x \to a} \dfrac{f(x)}{g(x)}$ is an indeterminate form of type $\dfrac{0}{0}$.

(b) $\lim\limits_{x \to a} \dfrac{f(x)}{p(x)} = 0$ because the numerator approaches 0 while the denominator becomes large.

(c) $\lim\limits_{x \to a} \dfrac{h(x)}{p(x)} = 0$ because the numerator approaches a finite number while the denominator becomes large.

(d) If $\lim\limits_{x \to a} p(x) = \infty$ and $f(x) \to 0$ through positive values, then $\lim\limits_{x \to a} \dfrac{p(x)}{f(x)} = \infty$. [For example, take $a = 0$,

$p(x) = 1/x^2$, and $f(x) = x^2$.] If $f(x) \to 0$ through negative values, then $\lim\limits_{x \to a} \dfrac{p(x)}{f(x)} = -\infty$. [For example,

take $a = 0$, $p(x) = 1/x^2$, and $f(x) = -x^2$.] If $f(x) \to 0$ through both positive and negative values, then the
limit might not exist. [For example, take $a = 0$, $p(x) = 1/x^2$, and $f(x) = x$.] It is not possible to evaluate this
limit.

(e) $\lim\limits_{x \to a} \dfrac{p(x)}{q(x)}$ is an indeterminate form of type $\dfrac{\infty}{\infty}$.

3. (a) When x is near a, $f(x)$ is near 0 and $p(x)$ is large, so $f(x) - p(x)$ is large negative. Thus,
$\lim\limits_{x \to a} [f(x) - p(x)] = -\infty$.

(b) $\lim\limits_{x \to a} [p(x) - q(x)]$ is an indeterminate form of type $\infty - \infty$.

(c) When x is near a, $p(x)$ and $q(x)$ are both large, so $p(x)q(x)$ is large. Thus, $\lim\limits_{x \to a} [p(x) + q(x)] = \infty$.

5. $\lim\limits_{x \to -1} \dfrac{x^2 - 1}{x + 1} = \lim\limits_{x \to -1} \dfrac{(x + 1)(x - 1)}{x + 1} = \lim\limits_{x \to -1} (x - 1) = -2$

7. $\lim\limits_{x\to 1}\dfrac{x^9-1}{x^5-1} = \lim\limits_{x\to 1}\dfrac{(x-1)\left(x^8+x^7+x^6+x^5+x^4+x^3+x^2+x+1\right)}{(x-1)\left(x^4+x^3+x^2+x+1\right)}$

$$= \lim_{x\to 1}\frac{x^8+x^7+x^6+x^5+x^4+x^3+x^2+x+1}{x^4+x^3+x^2+x+1} = \frac{9}{5}$$

9. $\lim\limits_{x\to 0}\dfrac{e^x-1}{\sin x} \overset{\text{H}}{=} \lim\limits_{x\to 0}\dfrac{e^x}{\cos x} = \dfrac{1}{1} = 1$

11. $\lim\limits_{x\to 0}\dfrac{\sin x}{x^3} \overset{\text{H}}{=} \lim\limits_{x\to 0}\dfrac{\cos x}{3x^2} = \infty$

13. $\lim\limits_{x\to 0}\dfrac{\tan px}{\tan qx} \overset{\text{H}}{=} \lim\limits_{x\to 0}\dfrac{p\sec^2 px}{q\sec^2 qx} = \dfrac{p\,(1)^2}{q\,(1)^2} = \dfrac{p}{q}$

15. $\lim\limits_{x\to\infty}\dfrac{\ln x}{x} \overset{\text{H}}{=} \lim\limits_{x\to\infty}\dfrac{1/x}{1} = 0$

17. $\lim\limits_{x\to 0^+}\left[(\ln x)/x\right] = -\infty$ since $\ln x \to -\infty$ as $x\to 0^+$ and dividing by small values of x just increases the magnitude of the quotient $(\ln x)/x$. L'Hospital's Rule does not apply.

19. $\lim\limits_{t\to 0}\dfrac{5^t-3^t}{t} \overset{\text{H}}{=} \lim\limits_{t\to 0}\dfrac{5^t\ln 5 - 3^t\ln 3}{1} = \ln 5 - \ln 3 = \ln\tfrac{5}{3}$

21. $\lim\limits_{x\to 0}\dfrac{e^x-1-x}{x^2} \overset{\text{H}}{=} \lim\limits_{x\to 0}\dfrac{e^x-1}{2x} \overset{\text{H}}{=} \lim\limits_{x\to 0}\dfrac{e^x}{2} = \dfrac{1}{2}$

23. $\lim\limits_{x\to\infty}\dfrac{e^x}{x^3} \overset{\text{H}}{=} \lim\limits_{x\to\infty}\dfrac{e^x}{3x^2} \overset{\text{H}}{=} \lim\limits_{x\to\infty}\dfrac{e^x}{6x} \overset{\text{H}}{=} \lim\limits_{x\to\infty}\dfrac{e^x}{6} = \infty$

25. $\lim\limits_{x\to 0}\dfrac{\sin^{-1}x}{x} \overset{\text{H}}{=} \lim\limits_{x\to 0}\dfrac{1/\sqrt{1-x^2}}{1} = \lim\limits_{x\to 0}\dfrac{1}{\sqrt{1-x^2}} = \dfrac{1}{1} = 1$

27. $\lim\limits_{x\to 0}\dfrac{1-\cos x}{x^2} \overset{\text{H}}{=} \lim\limits_{x\to 0}\dfrac{\sin x}{2x} \overset{\text{H}}{=} \lim\limits_{x\to 0}\dfrac{\cos x}{2} = \dfrac{1}{2}$

29. $\lim\limits_{x\to 0}\dfrac{\sin x}{e^x} = \dfrac{0}{1} = 0$. L'Hospital's Rule does not apply.

31. $\lim\limits_{x\to 0}\dfrac{\tan\alpha x}{x} \overset{\text{H}}{=} \lim\limits_{x\to 0}\dfrac{\alpha\sec^2\alpha x}{1} = \alpha$

33. $\lim\limits_{x\to\infty}\dfrac{x}{\ln\left(1+2e^x\right)} \overset{\text{H}}{=} \lim\limits_{x\to\infty}\dfrac{1}{\dfrac{1}{1+2e^x}\cdot 2e^x} = \lim\limits_{x\to\infty}\dfrac{1+2e^x}{2e^x} \overset{\text{H}}{=} \lim\limits_{x\to\infty}\dfrac{2e^x}{2e^x} = 1$

35. $\lim\limits_{x\to 0}\dfrac{\tan 2x}{\tanh 3x} \overset{\text{H}}{=} \lim\limits_{x\to 0}\dfrac{2\sec^2 2x}{3\,\mathrm{sech}^2\,3x} = \dfrac{2}{3}$

37. $\lim\limits_{x\to 0}\dfrac{2x-\sin^{-1}x}{2x+\cos^{-1}x} = \dfrac{2\,(0)-0}{2\,(0)+\pi/2} = 0$. L'Hospital's Rule does not apply.

39. $\lim\limits_{x\to 0^+}\sqrt{x}\ln x = \lim\limits_{x\to 0^+}\dfrac{\ln x}{x^{-1/2}} \overset{\text{H}}{=} \lim\limits_{x\to 0^+}\dfrac{1/x}{-\frac{1}{2}x^{-3/2}} = \lim\limits_{x\to 0^+}\left(-2\sqrt{x}\right) = 0$

41. $\lim\limits_{x\to\infty}e^{-x}\ln x = \lim\limits_{x\to\infty}\dfrac{\ln x}{e^x} \overset{\text{H}}{=} \lim\limits_{x\to\infty}\dfrac{1/x}{e^x} = \lim\limits_{x\to\infty}\dfrac{1}{xe^x} = 0$

43. $\lim\limits_{x\to\infty}x^3e^{-x^2} = \lim\limits_{x\to\infty}\dfrac{x^3}{e^{x^2}} \overset{\text{H}}{=} \lim\limits_{x\to\infty}\dfrac{3x^2}{2xe^{x^2}} = \lim\limits_{x\to\infty}\dfrac{3x}{2e^{x^2}} \overset{\text{H}}{=} \lim\limits_{x\to\infty}\dfrac{3}{4xe^{x^2}} = 0$

45. $\lim\limits_{x\to\pi}(x-\pi)\cot x = \lim\limits_{x\to\pi}\dfrac{x-\pi}{\tan x} \overset{\text{H}}{=} \lim\limits_{x\to\pi}\dfrac{1}{\sec^2 x} = \dfrac{1}{(-1)^2} = 1$

47. $\lim\limits_{x\to 0}\left(\dfrac{1}{x^4}-\dfrac{1}{x^2}\right)=\lim\limits_{x\to 0}\dfrac{1-x^2}{x^4}=\infty$

49. $\lim\limits_{x\to 0}\left(\dfrac{1}{x}-\csc x\right)=\lim\limits_{x\to 0}\left(\dfrac{1}{x}-\dfrac{1}{\sin x}\right)=\lim\limits_{x\to 0}\dfrac{\sin x-x}{x\sin x}$

$\overset{H}{=}\lim\limits_{x\to 0}\dfrac{\cos x-1}{\sin x+x\cos x}\overset{H}{=}\lim\limits_{x\to 0}\dfrac{-\sin x}{2\cos x-x\sin x}=\dfrac{0}{2}=0$

51. $\lim\limits_{x\to\infty}\left(x-\sqrt{x^2-1}\right)=\lim\limits_{x\to\infty}\left(x-\sqrt{x^2-1}\right)\dfrac{x+\sqrt{x^2-1}}{x+\sqrt{x^2-1}}=\lim\limits_{x\to\infty}\dfrac{x^2-(x^2-1)}{x+\sqrt{x^2-1}}=\lim\limits_{x\to\infty}\dfrac{1}{x+\sqrt{x^2-1}}=0$

53. $\lim\limits_{x\to\infty}\left(\dfrac{1}{x}-\dfrac{1}{e^x-1}\right)=\lim\limits_{x\to\infty}\dfrac{1}{x}-\lim\limits_{x\to\infty}\dfrac{1}{e^x-1}$ (since both limits exist) $=0-0=0$

55. $y=x^{\sin x}\ \Rightarrow\ \ln y=\sin x\ln x$, so

$\lim\limits_{x\to 0^+}\ln y=\lim\limits_{x\to 0^+}\sin x\ln x=\lim\limits_{x\to 0^+}\dfrac{\ln x}{\csc x}\overset{H}{=}\lim\limits_{x\to 0^+}\dfrac{1/x}{-\csc x\cot x}=-\left(\lim\limits_{x\to 0^+}\dfrac{\sin x}{x}\right)\left(\lim\limits_{x\to 0^+}\tan x\right)$

$=-1\cdot 0=0\ \Rightarrow$

$\lim\limits_{x\to 0^+}x^{\sin x}=\lim\limits_{x\to 0^+}e^{\ln y}=e^0=1.$

57. $y=(1-2x)^{1/x}\ \Rightarrow\ \ln y=\dfrac{1}{x}\ln(1-2x)$, so $\lim\limits_{x\to 0}\ln y=\lim\limits_{x\to 0}\dfrac{\ln(1-2x)}{x}\overset{H}{=}\lim\limits_{x\to 0}\dfrac{-2/(1-2x)}{1}=-2\ \Rightarrow$

$\lim\limits_{x\to 0}(1-2x)^{1/x}=\lim\limits_{x\to 0}e^{\ln y}=e^{-2}.$

59. $y=\left(1+\dfrac{3}{x}+\dfrac{5}{x^2}\right)^x\ \Rightarrow\ \ln y=x\ln\left(1+\dfrac{3}{x}+\dfrac{5}{x^2}\right)\ \Rightarrow$

$\lim\limits_{x\to\infty}\ln y=\lim\limits_{x\to\infty}\dfrac{\ln\left(1+\dfrac{3}{x}+\dfrac{5}{x^2}\right)}{1/x}\overset{H}{=}\lim\limits_{x\to\infty}\dfrac{\left(-\dfrac{3}{x^2}-\dfrac{10}{x^3}\right)\Big/\left(1+\dfrac{3}{x}+\dfrac{5}{x^2}\right)}{-1/x^2}=\lim\limits_{x\to\infty}\dfrac{3+10/x}{1+3/x+5/x^2}=3$, so

$\lim\limits_{x\to\infty}\left(1+\dfrac{3}{x}+\dfrac{5}{x^2}\right)^x=\lim\limits_{x\to\infty}e^{\ln y}=e^3.$

61. $y=x^{1/x}\ \Rightarrow\ \ln y=(1/x)\ln x\ \Rightarrow\ \lim\limits_{x\to\infty}\ln y=\lim\limits_{x\to\infty}\dfrac{\ln x}{x}\overset{H}{=}\lim\limits_{x\to\infty}\dfrac{1/x}{1}=0\ \Rightarrow$

$\lim\limits_{x\to\infty}x^{1/x}=\lim\limits_{x\to\infty}e^{\ln y}=e^0=1$

63. $y=\left(\dfrac{x}{x+1}\right)^x\ \Rightarrow\ \ln y=x\ln\left(\dfrac{x}{x+1}\right)\ \Rightarrow$

$\lim\limits_{x\to\infty}\ln y=\lim\limits_{x\to\infty}x\ln\left(\dfrac{x}{x+1}\right)=\lim\limits_{x\to\infty}\dfrac{\ln x-\ln(x+1)}{1/x}\overset{H}{=}\lim\limits_{x\to\infty}\dfrac{1/x-1/(x+1)}{-1/x^2}$

$=\lim\limits_{x\to\infty}\left(-x+\dfrac{x^2}{x+1}\right)=\lim\limits_{x\to\infty}\dfrac{-x}{x+1}=-1$

so $\lim\limits_{x\to\infty}\left(\dfrac{x}{x+1}\right)^x=\lim\limits_{x\to\infty}e^{\ln y}=e^{-1}$

Or: $\lim\limits_{x\to\infty}\left(\dfrac{x}{x+1}\right)^x=\lim\limits_{x\to\infty}\left[\left(\dfrac{x+1}{x}\right)^{-1}\right]^x=\left[\lim\limits_{x\to\infty}\left(1+\dfrac{1}{x}\right)^x\right]^{-1}=e^{-1}$

65. $y = (-\ln x)^x \implies \ln y = x \ln(-\ln x)$, so

$$\lim_{x \to 0^+} \ln y = \lim_{x \to 0^+} x \ln(-\ln x) = \lim_{x \to 0^+} \frac{\ln(-\ln x)}{1/x} \overset{H}{=} \lim_{x \to 0^+} \frac{(1/-\ln x)(-1/x)}{-1/x^2} = \lim_{x \to 0^+} \frac{-x}{\ln x} = 0 \implies$$

$$\lim_{x \to 0^+} (-\ln x)^x = e^0 = 1.$$

67.

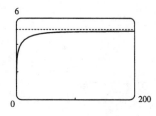

From the graph, it appears that

$$\lim_{x \to \infty} x \left[\ln(x+5) - \ln x\right] = 5. \text{ Now}$$

$$\lim_{x \to \infty} x \left[\ln(x+5) - \ln x\right] = \lim_{x \to \infty} \frac{\ln(x+5) - \ln x}{1/x}$$

$$\overset{H}{=} \lim_{x \to \infty} \frac{1/(x+5) - 1/x}{-1/x^2} = \lim_{x \to \infty} \frac{5x^2}{x(x+5)} = 5$$

69.

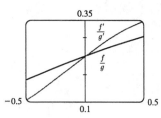

From the graph, it appears that

$$\lim_{x \to 0} \frac{f(x)}{g(x)} = \lim_{x \to 0} \frac{f'(x)}{g'(x)} = 0.25. \text{ We calculate}$$

$$\lim_{x \to 0} \frac{f(x)}{g(x)} = \lim_{x \to 0} \frac{e^x - 1}{x^3 + 4x} \overset{H}{=} \lim_{x \to 0} \frac{e^x}{3x^2 + 4} = \frac{1}{4}.$$

71. $\displaystyle\lim_{x \to \infty} \frac{e^x}{x^n} \overset{H}{=} \lim_{x \to \infty} \frac{e^x}{nx^{n-1}} \overset{H}{=} \lim_{x \to \infty} \frac{e^x}{n(n-1)x^{n-2}} \overset{H}{=} \cdots \overset{H}{=} \lim_{x \to \infty} \frac{e^x}{n!} = \infty$

73. First we will find $\displaystyle\lim_{n \to \infty} \left(1 + \frac{i}{n}\right)^{nt}$, which is of the form 1^{∞}. $y = \left(1 + \frac{i}{n}\right)^{nt} \implies \ln y = nt \ln\left(1 + \frac{i}{n}\right)$, so

$$\lim_{n \to \infty} \ln y = \lim_{n \to \infty} nt \ln\left(1 + \frac{i}{n}\right) = t \lim_{n \to \infty} \frac{\ln(1 + i/n)}{1/n} \overset{H}{=} t \lim_{n \to \infty} \frac{(-i/n^2)}{(1 + i/n)(-1/n^2)} = t \lim_{n \to \infty} \frac{i}{1 + i/n} = ti \implies$$

$$\lim_{n \to \infty} y = e^{it}. \text{ Thus, as } n \to \infty, A = A_0\left(1 + \frac{i}{n}\right)^{nt} \to A_0 e^{it}.$$

75. We see that both numerator and denominator approach 0, so we can use l'Hospital's Rule:

$$\lim_{x \to a} \frac{\sqrt{2a^3 x - x^4} - a\sqrt[3]{aax}}{a - \sqrt[4]{ax^3}} \overset{H}{=} \lim_{x \to a} \frac{\frac{1}{2}(2a^3 x - x^4)^{-1/2}(2a^3 - 4x^3) - a\left(\frac{1}{3}\right)(aax)^{-2/3} a^2}{-\frac{1}{4}(ax^3)^{-3/4}(3ax^2)}$$

$$= \frac{\frac{1}{2}(2a^3 a - a^4)^{-1/2}(2a^3 - 4a^3) - \frac{1}{3}a^3(a^2 a)^{-2/3}}{-\frac{1}{4}(aa^3)^{-3/4}(3aa^2)}$$

$$= \frac{(a^4)^{-1/2}(-a^3) - \frac{1}{3}a^3(a^3)^{-2/3}}{-\frac{3}{4}a^3(a^4)^{-3/4}} = \frac{-a - \frac{1}{3}a}{-\frac{3}{4}} = \frac{4}{3}\left(\frac{4}{3}a\right) = \frac{16}{9}a$$

77. Since $\lim\limits_{h\to 0}[f(x+h)-f(x-h)]=f(x)-f(x)=0$ (f is differentiable and hence continuous) and

$\lim\limits_{h\to 0}2h=0$, we use l'Hospital's Rule:

$$\lim_{h\to 0}\frac{f(x+h)-f(x-h)}{2h}\overset{\text{H}}{=}\lim_{h\to 0}\frac{f'(x+h)(1)-f'(x-h)(-1)}{2}=\frac{f'(x)+f'(x)}{2}=\frac{2f'(x)}{2}=f'(x)$$

$\dfrac{f(x+h)-f(x-h)}{2h}$ is the slope of the secant line between $(x-h,f(x-h))$ and $(x+h,f(x+h))$. As $h\to 0$, this line gets closer to the tangent line and its slope approaches $f'(x)$.

79. (a) We show that $\lim\limits_{x\to 0}\dfrac{f(x)}{x^n}=0$ for every integer $n\geq 0$. Let $y=\dfrac{1}{x^2}$. Then

$$\lim_{x\to 0}\frac{f(x)}{x^{2n}}=\lim_{x\to 0}\frac{e^{-1/x^2}}{(x^2)^n}=\lim_{y\to\infty}\frac{y^n}{e^y}\overset{\text{H}}{=}\lim_{y\to\infty}\frac{ny^{n-1}}{e^y}\overset{\text{H}}{=}\cdots\overset{\text{H}}{=}\lim_{y\to\infty}\frac{n!}{e^y}=0\Rightarrow$$

$$\lim_{x\to 0}\frac{f(x)}{x^n}=\lim_{x\to 0}x^n\frac{f(x)}{x^{2n}}=\lim_{x\to 0}x^n\lim_{x\to 0}\frac{f(x)}{x^{2n}}=0.\text{ Thus, }f'(0)=\lim_{x\to 0}\frac{f(x)-f(0)}{x-0}=\lim_{x\to 0}\frac{f(x)}{x}=0.$$

(b) Using the Chain Rule and the Quotient Rule we see that $f^{(n)}(x)$ exists for $x\neq 0$. In fact, we prove by induction that for each $n\geq 0$, there is a polynomial p_n and a non-negative integer k_n with $f^{(n)}(x)=p_n(x)f(x)/x^{k_n}$ for $x\neq 0$. This is true for $n=0$; suppose it is true for the nth derivative. Then

$$f^{(n+1)}(x)=\left[x^{k_n}\left[p_n'(x)f(x)+p_n(x)f'(x)\right]-k_nx^{k_n-1}p_n(x)f(x)\right]x^{-2k_n}$$

$$=\left[x^{k_n}p_n'(x)+p_n(x)\left(2/x^3\right)-k_nx^{k_n-1}p_n(x)\right]f(x)x^{-2k_n}$$

$$=\left[x^{k_n+3}p_n'(x)+2p_n(x)-k_nx^{k_n+2}p_n(x)\right]f(x)x^{-(2k_n+3)}$$

which has the desired form.

Now we show by induction that $f^{(n)}(0)=0$ for all n. By part (a), $f'(0)=0$. Suppose that $f^{(n)}(0)=0$. Then

$$f^{(n+1)}(0)=\lim_{x\to 0}\frac{f^{(n)}(x)-f^{(n)}(0)}{x-0}=\lim_{x\to 0}\frac{f^{(n)}(x)}{x}=\lim_{x\to 0}\frac{p_n(x)f(x)/x^{k_n}}{x}=\lim_{x\to 0}\frac{p_n(x)f(x)}{x^{k_n+1}}$$

$$=\lim_{x\to 0}p_n(x)\lim_{x\to 0}\frac{f(x)}{x^{k_n+1}}=p_n(0)\cdot 0=0$$

4.5 Summary of Curve Sketching

1. $y = f(x) = x^3 + x = x(x^2 + 1)$ **A.** f is a polynomial, so $D = \mathbb{R}$.
B. x-intercept $= 0$, y-intercept $= f(0) = 0$ **C.** $f(-x) = -f(x)$, so f
is odd; the curve is symmetric about the origin. **D.** f is a polynomial, so
there is no asymptote. **E.** $f'(x) = 3x^2 + 1 > 0$, so f is increasing on
$(-\infty, \infty)$. **F.** There is no critical number and hence, no local maximum
or minimum value. **G.** $f''(x) = 6x > 0$ on $(0, \infty)$ and $f''(x) < 0$ on
$(-\infty, 0)$, so f is CU on $(0, \infty)$ and CD on $(-\infty, 0)$. Since the concavity
changes at $x = 0$, there is an inflection point at $(0, 0)$.

H.

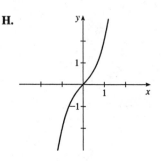

3. $y = f(x) = 2 - 15x + 9x^2 - x^3 = -(x - 2)(x^2 - 7x + 1)$ **A.** $D = \mathbb{R}$ **B.** y-intercept: $f(0) = 2$;
x-intercepts: $f(x) = 0 \Rightarrow x = 2$ or (by the quadratic formula)
$x = \frac{7 \pm \sqrt{45}}{2}$ ($\approx 0.15, 6.85$) **C.** No symmetry **D.** No asymptote
E. $f'(x) = -15 + 18x - 3x^2 = -3(x^2 - 6x + 5)$
$\qquad = -3(x - 1)(x - 5) > 0 \Leftrightarrow 1 < x < 5$
so f is increasing on $(1, 5)$ and decreasing on $(-\infty, 1)$ and $(5, \infty)$.
F. Local maximum $f(5) = 27$, local minimum $f(1) = -5$
G. $f''(x) = 18 - 6x = -6(x - 3) > 0 \Leftrightarrow x < 3$, so f is CU on
$(-\infty, 3)$ and CD on $(3, \infty)$. IP at $(3, 11)$

H.

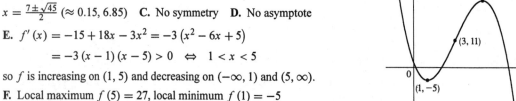

5. $y = f(x) = x^4 + 4x^3 = x^3(x + 4)$ **A.** $D = \mathbb{R}$ **B.** y-intercept:
$f(0) = 0$; x-intercepts: $f(x) = 0 \Leftrightarrow x = -4, 0$ **C.** No symmetry
D. No asymptote **E.** $f'(x) = 4x^3 + 12x^2 = 4x^2(x + 3) > 0 \Leftrightarrow$
$x > -3$, so f is increasing on $(-3, \infty)$ and decreasing on $(-\infty, -3)$.
F. Local minimum $f(-3) = -27$, no local maximum
G. $f''(x) = 12x^2 + 24x = 12x(x + 2) < 0 \Leftrightarrow -2 < x < 0$, so f is
CD on $(-2, 0)$ and CU on $(-\infty, -2)$ and $(0, \infty)$. IP at $(0, 0)$ and
$(-2, -16)$

H.

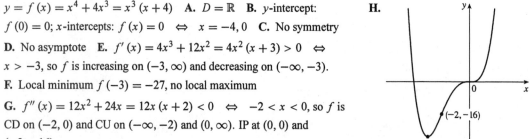

7. $y = f(x) = x/(x - 1)$ **A.** $D = \{x \mid x \neq 1\} = (-\infty, 1) \cup (1, \infty)$ **B.** x-intercept$= 0$, y-intercept$= f(0) = 0$
C. No symmetry **D.** $\displaystyle\lim_{x \to \pm\infty} \frac{x}{x - 1} = 1$, so $y = 1$ is a HA. $\displaystyle\lim_{x \to 1^-} \frac{x}{x - 1} = -\infty$,

$\displaystyle\lim_{x \to 1^+} \frac{x}{x - 1} = \infty$, so $x = 1$ is a VA.

E. $f'(x) = \dfrac{(x - 1) - x}{(x - 1)^2} = \dfrac{-1}{(x - 1)^2} < 0$ for $x \neq 1$, so f is decreasing

on $(-\infty, 1)$ and $(1, \infty)$. **F.** No extremum **G.** $f''(x) = \dfrac{2}{(x - 1)^3} > 0$

$\Leftrightarrow x > 1$, so f is CU on $(1, \infty)$ and CD on $(-\infty, 1)$. No IP

H.

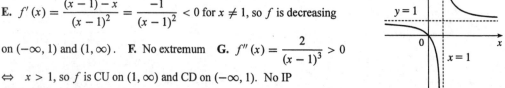

9. $y = f(x) = 1/(x^2 - 9)$ **A.** $D = \{x \mid x \neq \pm3\} = (-\infty, -3) \cup (-3, 3) \cup (3, \infty)$

B. y-intercept $= f(0) = -\frac{1}{9}$, no x-intercept   **C.** $f(-x) = f(x) \Rightarrow f$ is even; the curve is symmetric about

the y-axis.   **D.** $\displaystyle\lim_{x \to \pm\infty} \frac{1}{x^2 - 9} = 0$, so $y = 0$ is a HA. $\displaystyle\lim_{x \to 3^-} \frac{1}{x^2 - 9} = -\infty$, $\displaystyle\lim_{x \to 3^+} \frac{1}{x^2 - 9} = \infty$,

$\displaystyle\lim_{x \to -3^-} \frac{1}{x^2 - 9} = \infty$, $\displaystyle\lim_{x \to -3^+} \frac{1}{x^2 - 9} = -\infty$, so $x = 3$ and $x = -3$ are   **H.**

VA.   **E.** $f'(x) = -\dfrac{2x}{(x^2 - 9)^2} > 0 \Leftrightarrow x < 0 \ (x \neq -3)$ so f is

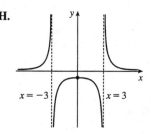

increasing on $(-\infty, -3)$ and $(-3, 0)$ and decreasing on $(0, 3)$ and $(3, \infty)$.

F. Local maximum $f(0) = -\frac{1}{9}$.

G. $y'' = \dfrac{-2(x^2 - 9)^2 + (2x) \, 2 \, (x^2 - 9) \, (2x)}{(x^2 - 9)^4} = \dfrac{6(x^2 + 3)}{(x^2 - 9)^3} > 0 \Leftrightarrow$

$x^2 > 9 \Leftrightarrow x > 3$ or $x < -3$, so f is CU on $(-\infty, -3)$ and $(3, \infty)$

and CD on $(-3, 3)$. No IP

11. $y = f(x) = x/(x^2 + 9)$ **A.** $D = \mathbb{R}$ **B.** y-intercept: $f(0) = 0$; x-intercept: $f(x) = 0 \Leftrightarrow x = 0$

C. $f(-x) = -f(x)$, so f is odd and the curve is symmetric about the origin.   **D.** $\displaystyle\lim_{x \to \pm\infty} [x/(x^2 + 9)] = 0$, so

$y = 0$ is a HA; no VA   **E.** $f'(x) = \dfrac{(x^2 + 9)(1) - x(2x)}{(x^2 + 9)^2} = \dfrac{9 - x^2}{(x^2 + 9)^2} = \dfrac{(3 + x)(3 - x)}{(x^2 + 9)^2} > 0 \Leftrightarrow$

$-3 < x < 3$, so f is increasing on $(-3, 3)$ and decreasing on $(-\infty, -3)$ and $(3, \infty)$.   **F.** Local minimum

$f(-3) = -\frac{1}{6}$, local maximum $f(3) = \frac{1}{6}$

G. $f''(x) = \dfrac{(x^2 + 9)^2 (-2x) - (9 - x^2) \cdot 2 (x^2 + 9) (2x)}{\left[(x^2 + 9)^2\right]^2} = \dfrac{(2x)(x^2 + 9)\left[-(x^2 + 9) - 2(9 - x^2)\right]}{(x^2 + 9)^4}$

$= \dfrac{2x(x^2 - 27)}{(x^2 + 9)^3} = 0 \Leftrightarrow x = 0, \pm\sqrt{27} = \pm3\sqrt{3}$

$f''(x) > 0 \Leftrightarrow -3\sqrt{3} < x < 0$ or $x > 3\sqrt{3}$, so f is CU on $\left(-3\sqrt{3}, 0\right)$ **H.**

and $\left(3\sqrt{3}, \infty\right)$, and CD on $\left(-\infty, -3\sqrt{3}\right)$ and $\left(0, 3\sqrt{3}\right)$. There are three

inflection points: $(0, 0)$ and $\left(\pm3\sqrt{3}, \pm\frac{1}{12}\sqrt{3}\right)$.

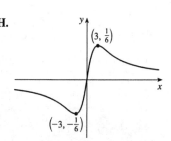

13. $y = f(x) = \dfrac{1}{(x-1)(x+2)} = \dfrac{1}{x^2+x-2}$ **A.** $D = \{x \mid x \neq -2, 1\} = (-\infty, -2) \cup (-2, 1) \cup (1, \infty)$

B. y-intercept: $f(0) = -\frac{1}{2}$; no x-intercept   **C.** No symmetry

D. $\displaystyle\lim_{x \to \pm\infty} \dfrac{1}{x^2+2x-2} = \lim_{x \to \pm\infty} \dfrac{1/x^2}{1 + 1/x - 2/x^2} = \dfrac{0}{1} = 0$, so $y = 0$ is a HA. $x = -2$ and $x = 1$ are

VA.   **E.** $f'(x) = \dfrac{(x^2+x-2)\cdot 0 - 1(2x+1)}{(x-1)^2(x+2)^2} = -\dfrac{2x+1}{(x-1)^2(x+2)^2} > 0 \iff x < -\frac{1}{2}\ (x \neq -2);$

$f'(x) < 0 \iff x > -\frac{1}{2}\ (x \neq 1)$. So f is increasing on $(-\infty, -2)$ and $\left(-2, -\frac{1}{2}\right)$, and f is decreasing on

$\left(-\frac{1}{2}, 1\right)$ and $(1, \infty)$.   **F.** $f\left(-\frac{1}{2}\right) = -\frac{4}{9}$ is a local maximum.

G. $f''(x) = \dfrac{(x^2+x-2)^2(-2) - [-(2x+1)](2)(x^2+x-2)(2x+1)}{\left[(x-1)^2(x+2)^2\right]^2}$

$= \dfrac{2(x^2+x-2)\left[-1(x^2+x-2) + (2x+1)^2\right]}{(x-1)^4(x+2)^4} = \dfrac{2(-x^2-x+2+4x^2+4x+1)}{(x-1)^3(x+2)^3}$

$= \dfrac{2(3x^2+3x+3)}{(x-1)^3(x+2)^3} = \dfrac{6(x^2+x+1)}{(x-1)^3(x+2)^3}$

The numerator is always positive, so the sign of f'' is determined by the   **H.**
denominator, which is negative only for $-2 < x < 1$. Thus, f is CD on
$(-2, 1)$ and CU on $(-\infty, -2)$ and $(1, \infty)$. No IP

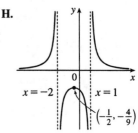

15. $y = f(x) = \dfrac{1+x^2}{1-x^2} = -1 + \dfrac{2}{1-x^2}$ **A.** $D = \{x \mid x \neq \pm 1\}$ **B.** No x-intercept,

y-intercept $= f(0) = 1$ **C.** $f(-x) = f(x)$, so f is even and the curve is symmetric about the y-axis.

D. $\displaystyle\lim_{x \to \pm\infty} \dfrac{1+x^2}{1-x^2} = \lim_{x \to \pm\infty} \dfrac{(1/x^2)+1}{(1/x^2)-1} = -1$, so $y = -1$ is a HA. $\displaystyle\lim_{x \to 1^-} \dfrac{1+x^2}{1-x^2} = \infty$, $\displaystyle\lim_{x \to 1^+} \dfrac{1+x^2}{1-x^2} = -\infty$,

$\displaystyle\lim_{x \to -1^-} \dfrac{1+x^2}{1-x^2} = -\infty$, $\displaystyle\lim_{x \to -1^+} \dfrac{1+x^2}{1-x^2} = \infty$. So $x = 1$ and $x = -1$ are VA.

E. $f'(x) = \dfrac{4x}{(1-x^2)^2} > 0 \iff x > 0\ (x \neq 1)$, so f increases on   **H.**

$(0, 1)$ and $(1, \infty)$, and decreases on $(-\infty, -1)$ and $(-1, 0)$.

F. $f(0) = 1$ is a local minimum.

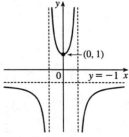

G. $y'' = \dfrac{4(1-x^2)^2 - 4x \cdot 2(1-x^2)(-2x)}{(1-x^2)^4} = \dfrac{4(1+3x^2)}{(1-x^2)^3} > 0 \iff$

$x^2 < 1 \iff -1 < x < 1$, so f is CU on $(-1, 1)$ and CD on $(-\infty, -1)$
and $(1, \infty)$. No IP

17. $y = f(x) = \dfrac{1}{x^3 - x} = \dfrac{1}{x(x-1)(x+1)}$ **A.** $D = \{x \mid x \neq 0, \pm 1\}$ **B.** No intercept **C.** $f(-x) = -f(x)$,

symmetric about $(0,0)$ **D.** $\displaystyle\lim_{x \to \pm\infty} \dfrac{1}{x^3 - x} = 0$, so $y = 0$ is a HA. $\displaystyle\lim_{x \to 0^-} \dfrac{1}{x^3 - x} = \infty$, $\displaystyle\lim_{x \to 0^+} \dfrac{1}{x^3 - x} = -\infty$,

$\displaystyle\lim_{x \to 1^-} \dfrac{1}{x^3 - x} = -\infty$, $\displaystyle\lim_{x \to 1^+} \dfrac{1}{x^3 - x} = \infty$, $\displaystyle\lim_{x \to -1^-} \dfrac{1}{x^3 - x} = -\infty$, $\displaystyle\lim_{x \to -1^+} \dfrac{1}{x^3 - x} = \infty$. So $x = 0$, $x = 1$, and

$x = -1$ are VA. **E.** $f'(x) = \dfrac{1 - 3x^2}{\left(x^3 - x\right)^2}$ $\Rightarrow$ $f'(x) > 0 \iff x^2 < \frac{1}{3} \iff -\frac{1}{\sqrt{3}} < x < \frac{1}{\sqrt{3}}$ $(x \neq 0)$,

so f is increasing on $\left(-\frac{1}{\sqrt{3}}, 0\right)$, $\left(0, \frac{1}{\sqrt{3}}\right)$ and decreasing on $(-\infty, -1)$, **H.**

$\left(-1, -\frac{1}{\sqrt{3}}\right)$, $\left(\frac{1}{\sqrt{3}}, 1\right)$, and $(1, \infty)$. **F.** Local minimum

$f\left(-\frac{1}{\sqrt{3}}\right) = \frac{3\sqrt{3}}{2}$, local maximum $f\left(\frac{1}{\sqrt{3}}\right) = -\frac{3\sqrt{3}}{2}$

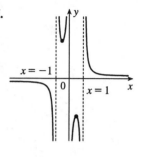

$x = -1$ $x = 1$

G. $f''(x) = \dfrac{2\left(6x^4 - 3x^2 + 1\right)}{\left(x^3 - x\right)^3}$. Since $6x^4 - 3x^2 + 1$ has negative

discriminant as a quadratic in x^2, it is positive, so $f''(x) > 0$ $\iff$

$x^3 - x > 0$ $\iff$ $x > 1$ or $-1 < x < 0$. f is CU on $(-1, 0)$ and $(1, \infty)$,

and CD on $(-\infty, -1)$ and $(0, 1)$. No IP

19. $y = f(x) = x\sqrt{5 - x}$ **A.** The domain is $\{x \mid 5 - x \geq 0\} = (-\infty, 5]$ **B.** y-intercept: $f(0) = 0$;

x-intercepts: $f(x) = 0 \iff x = 0, 5$ **C.** No symmetry **D.** No asymptote

E. $f'(x) = x \cdot \frac{1}{2}(5 - x)^{-1/2}(-1) + (5 - x)^{1/2} \cdot 1 = \frac{1}{2}(5 - x)^{-1/2}[-x + 2(5 - x)] = \dfrac{10 - 3x}{2\sqrt{5 - x}} > 0 \iff$

$x < \frac{10}{3}$, so f is increasing on $\left(-\infty, \frac{10}{3}\right)$ and decreasing on $\left(\frac{10}{3}, 5\right)$.

F. Local maximum $f\left(\frac{10}{3}\right) = \frac{10}{9}\sqrt{15} \approx 4.3$; no local minimum **H.**

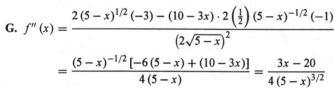

$\left(\frac{10}{3}, \frac{10\sqrt{15}}{9}\right)$

G. $f''(x) = \dfrac{2(5 - x)^{1/2}(-3) - (10 - 3x) \cdot 2\left(\frac{1}{2}\right)(5 - x)^{-1/2}(-1)}{\left(2\sqrt{5 - x}\right)^2}$

$= \dfrac{(5 - x)^{-1/2}[-6(5 - x) + (10 - 3x)]}{4(5 - x)} = \dfrac{3x - 20}{4(5 - x)^{3/2}}$

$f''(x) < 0$ for $x < 5$, so f is CD on $(-\infty, 5)$. No IP

21. $y = f(x) = \sqrt{x^2 + 1} - x$ **A.** $D = \mathbb{R}$ **B.** No x-intercept,

y-intercept $= 1$ **C.** No symmetry **D.** $\displaystyle\lim_{x \to -\infty}\left(\sqrt{x^2 + 1} - x\right) = \infty$ and

$\displaystyle\lim_{x \to \infty}\left(\sqrt{x^2 + 1} - x\right) = \lim_{x \to \infty}\left(\sqrt{x^2 + 1} - x\right)\dfrac{\sqrt{x^2 + 1} + x}{\sqrt{x^2 + 1} + x} = \lim_{x \to \infty}\dfrac{1}{\sqrt{x^2 + 1} + x} = 0$,

so $y = 0$ is a HA. **E.** $f'(x) = \dfrac{x}{\sqrt{x^2 + 1}} - 1 = \dfrac{x - \sqrt{x^2 + 1}}{\sqrt{x^2 + 1}}$ $\Rightarrow$ **H.**

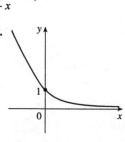

$f'(x) < 0$, so f is decreasing on $\mathbb{R}$. **F.** No extremum

G. $f''(x) = \dfrac{1}{\left(x^2 + 1\right)^{3/2}} > 0$, so f is CU on $\mathbb{R}$. No IP

23. $y = f(x) = \sqrt[4]{x^2 - 25}$ **A.** $D = \{x \mid x^2 \geq 25\} = (-\infty, -5] \cup [5, \infty)$ **B.** x-intercepts are ± 5, no y-intercept

C. $f(-x) = f(x)$, so the curve is symmetric about the y-axis. **D.** $\lim\limits_{x \to \pm\infty} \sqrt[4]{x^2 - 25} = \infty$, no asymptote

E. $f'(x) = \frac{1}{4}(x^2 - 25)^{-3/4}(2x) = \dfrac{x}{2(x^2 - 25)^{3/4}} > 0$ if $x > 5$, so f is increasing on $(5, \infty)$ and decreasing on

$(-\infty, -5)$. **F.** No local extremum

G. $y'' = \dfrac{2(x^2 - 25)^{3/4} - 3x^2(x^2 - 25)^{-1/4}}{4(x^2 - 25)^{3/2}}$

$= -\dfrac{x^2 + 50}{4(x^2 - 25)^{7/4}} < 0$

so f is CD on $(-\infty, -5)$ and $(5, \infty)$. No IP

H.

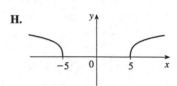

25. $y = f(x) = \dfrac{\sqrt{1 - x^2}}{x}$ **A.** $D = \{x \mid |x| \leq 1, x \neq 0\} = [-1, 0) \cup (0, 1]$ **B.** x-intercepts ± 1, no y-intercept

C. $f(-x) = -f(x)$, so the curve is symmetric about $(0, 0)$. **D.** $\lim\limits_{x \to 0^+} \dfrac{\sqrt{1 - x^2}}{x} = \infty$, $\lim\limits_{x \to 0^-} \dfrac{\sqrt{1 - x^2}}{x} = -\infty$,

so $x = 0$ is a VA. **E.** $f'(x) = \dfrac{(-x^2/\sqrt{1 - x^2}) - \sqrt{1 - x^2}}{x^2} = -\dfrac{1}{x^2\sqrt{1 - x^2}} < 0$, so f is decreasing on $(-1, 0)$

and $(0, 1)$. **F.** No extremum

G. $f''(x) = \dfrac{2 - 3x^2}{x^3(1 - x^2)^{3/2}} > 0 \iff -1 < x < -\sqrt{\frac{2}{3}}$ or

$0 < x < \sqrt{\frac{2}{3}}$, so f is CU on $\left(-1, -\sqrt{\frac{2}{3}}\right)$ and $\left(0, \sqrt{\frac{2}{3}}\right)$ and CD on

$\left(-\sqrt{\frac{2}{3}}, 0\right)$ and $\left(\sqrt{\frac{2}{3}}, 1\right)$. IP $\left(\pm\sqrt{\frac{2}{3}}, \pm\frac{1}{\sqrt{2}}\right)$.

H.

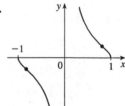

27. $y = f(x) = x + 3x^{2/3}$ **A.** $D = \mathbb{R}$ **B.** $y = x + 3x^{2/3} = x^{2/3}(x^{1/3} + 3) = 0$ if $x = 0$ or -27 (x-intercepts),

y-intercept $= f(0) = 0$ **C.** No symmetry **D.** $\lim\limits_{x \to \infty}(x + 3x^{2/3}) = \infty$,

$\lim\limits_{x \to -\infty}(x + 3x^{2/3}) = \lim\limits_{x \to -\infty} x^{2/3}(x^{1/3} + 3) = -\infty$, no asymptote

E. $f'(x) = 1 + 2x^{-1/3} = (x^{1/3} + 2)/x^{1/3} > 0 \iff x > 0$ or

$x < -8$, so f increases on $(-\infty, -8)$, $(0, \infty)$ and decreases on

$(-8, 0)$. **F.** Local maximum $f(-8) = 4$, local minimum $f(0) = 0$

G. $f''(x) = -\frac{2}{3}x^{-4/3} < 0$ ($x \neq 0$) so f is CD on $(-\infty, 0)$ and $(0, \infty)$.

No IP

H.

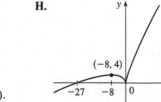

29. $y = f(x) = x + \sqrt{|x|}$ **A.** $D = \mathbb{R}$ **B.** x-intercepts $= 0, -1$, y-intercept 0 **C.** No symmetry

D. $\lim\limits_{x \to \infty} (x + \sqrt{|x|}) = \infty$, $\lim\limits_{x \to -\infty} (x + \sqrt{|x|}) = -\infty$. No asymptote **E.** For $x > 0$, $f(x) = x + \sqrt{x}$ $\Rightarrow$

$f'(x) = 1 + \dfrac{1}{2\sqrt{x}} > 0$, so f increases on $(0, \infty)$.

For $x < 0$, $f(x) = x + \sqrt{-x} \to f'(x) = 1 - \dfrac{1}{2\sqrt{-x}} > 0$ $\Leftrightarrow$ **H.**

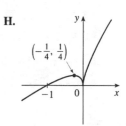

$2\sqrt{-x} > 1$ $\Leftrightarrow$ $-x > \frac{1}{4}$ $\Leftrightarrow$ $x < -\frac{1}{4}$, so f increases on $\left(-\infty, -\frac{1}{4}\right)$

and decreases on $\left(-\frac{1}{4}, 0\right)$. **F.** $f\left(-\frac{1}{4}\right) = \frac{1}{4}$ is a local maximum,

$f(0) = 0$ is a local minimum. **G.** For $x > 0$, $f''(x) = -\frac{1}{4}x^{-3/2}$ $\Rightarrow$

$f''(x) < 0$, so f is CD on $(0, \infty)$. For $x < 0$, $f''(x) = -\frac{1}{4}(-x)^{-3/2}$

$\Rightarrow$ $f''(x) < 0$, so f is CD on $(-\infty, 0)$. No IP

31. $y = f(x) = \cos x - \sin x$ **A.** $D = \mathbb{R}$ **B.** $y = 0$ $\Leftrightarrow$ $\cos x = \sin x$ $\Leftrightarrow$ $x = n\pi + \frac{\pi}{4}$, n an integer

(x-intercepts), y-intercept $= f(0) = 1$. **C.** Periodic with period 2π **D.** No asymptote

E. $f'(x) = -\sin x - \cos x = 0$ $\Leftrightarrow$ $\cos x = -\sin x$ $\Leftrightarrow$ $x = 2n\pi + \frac{3\pi}{4}$ or $2n\pi + \frac{7\pi}{4}$. $f'(x) > 0$ $\Leftrightarrow$

$\cos x < -\sin x$ $\Leftrightarrow$ $2n\pi + \frac{3\pi}{4} < x < 2n\pi + \frac{7\pi}{4}$, so f is increasing on $\left(2n\pi + \frac{3\pi}{4}, 2n\pi + \frac{7\pi}{4}\right)$ and decreasing

on $\left(2n\pi - \frac{\pi}{4}, 2n\pi + \frac{3\pi}{4}\right)$. **F.** Local maxima $f\left(2n\pi - \frac{\pi}{4}\right) = \sqrt{2}$, local **H.**

minima $f\left(2n\pi + \frac{3\pi}{4}\right) = -\sqrt{2}$. **G.** $f''(x) = -\cos x + \sin x > 0$ $\Leftrightarrow$

$\sin x > \cos x$ $\Leftrightarrow$ $x \in \left(2n\pi + \frac{\pi}{4}, 2n\pi + \frac{5\pi}{4}\right)$, so f is CU on these

intervals and CD on $\left(2n\pi - \frac{3\pi}{4}, 2n\pi + \frac{\pi}{4}\right)$. IP $\left(n\pi + \frac{\pi}{4}, 0\right)$

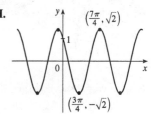

33. $y = f(x) = x \tan x$, $-\frac{\pi}{2} < x < \frac{\pi}{2}$ **A.** $D = \left(-\frac{\pi}{2}, \frac{\pi}{2}\right)$ **B.** Intercepts **H.**

are 0 **C.** $f(-x) = f(x)$, so the curve is symmetric about the y-axis.

D. $\lim\limits_{x \to \pi/2^-} x \tan x = \infty$ and $\lim\limits_{x \to -\pi/2^+} x \tan x = \infty$, so $x = \frac{\pi}{2}$ and

$x = -\frac{\pi}{2}$ are VA. **E.** $f'(x) = \tan x + x \sec^2 x > 0$ $\Leftrightarrow$ $0 < x < \frac{\pi}{2}$, so

f increases on $\left(0, \frac{\pi}{2}\right)$ and decreases on $\left(-\frac{\pi}{2}, 0\right)$. **F.** Absolute minimum

$f(0) = 0$. **G.** $y'' = 2\sec^2 x + 2x \tan x \sec^2 x > 0$ for $-\frac{\pi}{2} < x < \frac{\pi}{2}$, so

f is CU on $\left(-\frac{\pi}{2}, \frac{\pi}{2}\right)$. No IP

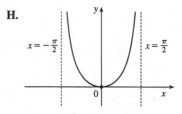

35. $y = f(x) = x/2 - \sin x,\ 0 < x < 3\pi$ **A.** $D = (0, 3\pi)$ **B.** No y-intercept. The x-intercept, approximately

1.9, can be found using Newton's Method.   **C.** No symmetry   **D.** No asymptote   **E.** $f'(x) = \frac{1}{2} - \cos x > 0$

$\Leftrightarrow\ \cos x < \frac{1}{2}\ \Leftrightarrow\ \frac{\pi}{3} < x < \frac{5\pi}{3}$ or $\frac{7\pi}{3} < x < 3\pi$, so f is increasing on $\left(\frac{\pi}{3}, \frac{5\pi}{3}\right)$ and $\left(\frac{7\pi}{3}, 3\pi\right)$ and decreasing

on $\left(0, \frac{\pi}{3}\right)$ and $\left(\frac{5\pi}{3}, \frac{7\pi}{3}\right)$.   **F.** $f\left(\frac{\pi}{3}\right) = \frac{\pi}{6} - \frac{\sqrt{3}}{2}$ is a local minimum,

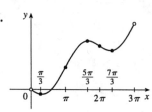

$f\left(\frac{5\pi}{3}\right) = \frac{5\pi}{6} + \frac{\sqrt{3}}{2}$ is a local maximum, $f\left(\frac{7\pi}{3}\right) = \frac{7\pi}{6} - \frac{\sqrt{3}}{2}$ is a local

minimum.   **G.** $f''(x) = \sin x > 0\ \Leftrightarrow\ 0 < x < \pi$ or $2\pi < x < 3\pi$,

so f is CU on $(0, \pi)$ and $(2\pi, 3\pi)$ and CD on $(\pi, 2\pi)$. IP $\left(\pi, \frac{\pi}{2}\right)$ and

$(2\pi, \pi)$.

37. $y = f(x) = 2\cos x + \sin^2 x$ **A.** $D = \mathbb{R}$ **B.** y-intercept $= f(0) = 2$ **C.** $f(-x) = f(x)$, so the

curve is symmetric about the y-axis.   Periodic with period 2π **D.** No asymptote

E. $f'(x) = -2\sin x + 2\sin x \cos x = 2\sin x (\cos x - 1) > 0\ \Leftrightarrow\ \sin x < 0\ \Leftrightarrow\ (2n-1)\pi < x < 2n\pi$, so

f is increasing on $((2n-1)\pi, 2n\pi)$ and decreasing on $(2n\pi, (2n+1)\pi)$.   **F.** $f(2n\pi) = 2$ is a local

maximum. $f((2n+1)\pi) = -2$ is a local minimum.

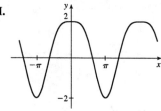

G. $f''(x) = -2\cos x + 2\cos 2x = 2(2\cos^2 x - \cos x - 1)$

$\qquad = 2(2\cos x + 1)(\cos x - 1) > 0$

$\Leftrightarrow\ \cos x < -\frac{1}{2}\ \Leftrightarrow\ x \in \left(2n\pi + \frac{2\pi}{3}, 2n\pi + \frac{4\pi}{3}\right)$, so f is CU on these

intervals and CD on $\left(2n\pi - \frac{2\pi}{3}, 2n\pi + \frac{2\pi}{3}\right)$. IP when $x = 2n\pi \pm \frac{2\pi}{3}$

39. $y = f(x) = \sin 2x - 2\sin x$ **A.** $D = \mathbb{R}$ **B.** y-intercept $= f(0) = 0$. $y = 0\ \Leftrightarrow$

$2\sin x = \sin 2x = 2\sin x \cos x\ \Leftrightarrow\ \sin x = 0$ or $\cos x = 1\ \Leftrightarrow\ x = n\pi$ (x-intercepts)

C. $f(-x) = -f(x)$, so the curve is symmetric about $(0, 0)$. *Note:* f is periodic

with period 2π, so we determine D–G for $-\pi \le x \le \pi$.   **D.** No asymptotes

E. $f'(x) = 2\cos 2x - 2\cos x = 2(2\cos^2 x - 1 - \cos x) = 2(2\cos x + 1)(\cos x - 1) > 0\ \Leftrightarrow\ \cos x < -\frac{1}{2}$

$\Leftrightarrow\ -\pi < x < -\frac{2\pi}{3}$ or $\frac{2\pi}{3} < x < \pi$, so f is increasing on $\left(-\pi, -\frac{2\pi}{3}\right)$ and $\left(\frac{2\pi}{3}, \pi\right)$ and decreasing on

$\left(-\frac{2\pi}{3}, \frac{2\pi}{3}\right)$.   **F.** $f\left(-\frac{2\pi}{3}\right) = \frac{3\sqrt{3}}{2}$ is a local maximum, $f\left(\frac{2\pi}{3}\right) = -\frac{3\sqrt{3}}{2}$ is a local minimum.

G. $f''(x) = -4\sin 2x + 2\sin x = 2\sin x (1 - 4\cos x) = 0$ when $x = 0$,   **H.**

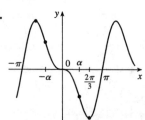

$\pm\pi$ or $\cos x = \frac{1}{4}$. If $\alpha = \cos^{-1}\frac{1}{4}$, then f is CU on $(-\alpha, 0)$ and (α, π)

and CD on $(-\pi, -\alpha)$ and $(0, \alpha)$. IP $(0, 0), (\pi, 0), \left(\alpha, -\frac{3\sqrt{15}}{8}\right)$,

$\left(-\alpha, \frac{3\sqrt{15}}{8}\right)$.

41. $y = 1/(1+e^{-x})$ **A.** $D = \mathbb{R}$ **B.** No x-intercept; y-intercept $= f(0) = \frac{1}{2}$.   **C.** No symmetry

D. $\lim\limits_{x \to \infty} 1/(1+e^{-x}) = \frac{1}{1+0} = 1$ and $\lim\limits_{x \to -\infty} 1/(1+e^{-x}) = 0$ (since $\lim\limits_{x \to -\infty} e^{-x} = \infty$,) so f has horizontal

asymptotes $y = 0$ and $y = 1$.   **E.** $f'(x) = -(1+e^{-x})^{-2}(-e^{-x}) = e^{-x}/(1+e^{-x})^2$. This is positive for all x,

so f is increasing on $\mathbb{R}$.   **F.** No extremum

G. $f''(x) = \dfrac{(1+e^{-x})^2(-e^{-x}) - e^{-x}(2)(1+e^{-x})(-e^{-x})}{(1+e^{-x})^4}$

$\qquad = \dfrac{e^{-x}(e^{-x}-1)}{(1+e^{-x})^3}$

H.

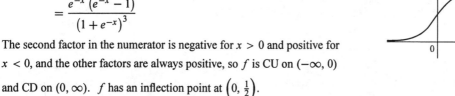

The second factor in the numerator is negative for $x > 0$ and positive for $x < 0$, and the other factors are always positive, so f is CU on $(-\infty, 0)$ and CD on $(0, \infty)$. f has an inflection point at $\left(0, \frac{1}{2}\right)$.

43. $y = f(x) = x \ln x$ **A.** $D = (0, \infty)$ **B.** x-intercept when $\ln x = 0 \iff x = 1$, no y-intercept

C. No symmetry   **D.** $\lim\limits_{x \to \infty} x \ln x = \infty$, $\lim\limits_{x \to 0^+} x \ln x = \lim\limits_{x \to 0^+} \dfrac{\ln x}{1/x} \overset{\text{H}}{=} \lim\limits_{x \to 0^+} \dfrac{1/x}{-1/x^2} = \lim\limits_{x \to 0^+} (-x) = 0$, no

asymptote.   **E.** $f'(x) = \ln x + 1 = 0$ when $\ln x = -1 \iff x = e^{-1}$.

$f'(x) > 0 \iff \ln x > -1 \iff x > e^{-1}$, so f is increasing on $(1/e, \infty)$ and decreasing on $(0, 1/e)$.   **F.** $f(1/e) = -1/e$ is an absolute and local minimum.   **G.** $f''(x) = 1/x > 0$, so f is CU on $(0, \infty)$. No IP

H.

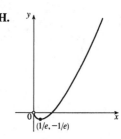

45. $y = f(x) = xe^{-x}$ **A.** $D = \mathbb{R}$ **B.** Intercepts are 0   **C.** No symmetry

D. $\lim\limits_{x \to \infty} xe^{-x} = \lim\limits_{x \to \infty} \dfrac{x}{e^x} \overset{\text{H}}{=} \lim\limits_{x \to \infty} \dfrac{1}{e^x} = 0$, so $y = 0$ is a HA. $\lim\limits_{x \to -\infty} xe^{-x} = -\infty$

E. $f'(x) = e^{-x} - xe^{-x} = e^{-x}(1-x) > 0 \iff x < 1$, so f is increasing on $(-\infty, 1)$ and decreasing on $(1, \infty)$.   **F.** Absolute maximum $f(1) = 1/e$.   **G.** $f''(x) = e^{-x}(x-2) > 0 \iff x > 2$, so f is CU on $(2, \infty)$ and CD on $(-\infty, 2)$. IP is $(2, 2/e^2)$.

H.

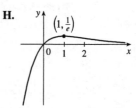

47. $y = f(x) = \ln(x^2 - x)$ **A.** $\{x \mid x^2 - x > 0\} = \{x \mid x < 0 \text{ or } x > 1\} = (-\infty, 0) \cup (1, \infty)$. **B.** x-intercepts

occur when $x^2 - x = 1$ $\Leftrightarrow$ $x^2 - x - 1 = 0$ $\Leftrightarrow$ $x = \frac{1}{2}\left(1 \pm \sqrt{5}\right)$. No y-intercept **C.** No symmetry

D. $\displaystyle\lim_{x \to \infty} \ln(x^2 - x) = \infty$, no HA. $\displaystyle\lim_{x \to 0^-} \ln(x^2 - x) = -\infty$, $\displaystyle\lim_{x \to 1^+} \ln(x^2 - x) = -\infty$, so $x = 0$

and $x = 1$ are VA. **E.** $f'(x) = \dfrac{2x - 1}{x^2 - x} > 0$ when $x > 1$ and $f'(x) < 0$ **H.**

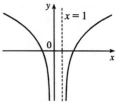

when $x < 0$, so f is increasing on $(1, \infty)$ and decreasing on $(-\infty, 0)$.

F. No extrema **G.** $f''(x) = \dfrac{2(x^2 - x) - (2x - 1)^2}{(x^2 - x)^2} = \dfrac{-2x^2 + 2x - 1}{(x^2 - x)^2}$

$\Rightarrow$ $f''(x) < 0$ for all x since $-2x^2 + 2x - 1$ has a negative discriminant.
So f is CD on $(-\infty, 0)$ and $(1, \infty)$. No IP.

49. $y = f(x) = xe^{-x^2}$ **A.** $D = \mathbb{R}$ **B.** Intercepts are 0 **C.** $f(-x) = -f(x)$, so the curve is symmetric about the

origin. **D.** $\displaystyle\lim_{x \to \pm\infty} xe^{-x^2} = \lim_{x \to \pm\infty} \dfrac{x}{e^{x^2}} \overset{H}{=} \lim_{x \to \pm\infty} \dfrac{1}{2xe^{x^2}} = 0$, so $y = 0$ is a HA.

E. $f'(x) = e^{-x^2} - 2x^2 e^{-x^2} = e^{-x^2}\left(1 - 2x^2\right) > 0$ $\Leftrightarrow$ $x^2 < \frac{1}{2}$ $\Leftrightarrow$ $|x| < \frac{1}{\sqrt{2}}$, so f is increasing on

$\left(-\frac{1}{\sqrt{2}}, \frac{1}{\sqrt{2}}\right)$ and decreasing on $\left(-\infty, -\frac{1}{\sqrt{2}}\right)$ and $\left(\frac{1}{\sqrt{2}}, \infty\right)$. **F.** $f\left(\frac{1}{\sqrt{2}}\right) = 1/\sqrt{2e}$ is a local maximum,

$f\left(-\frac{1}{\sqrt{2}}\right) = -1/\sqrt{2e}$ is a local minimum. **G.** $f''(x) = -2xe^{-x^2}\left(1 - 2x^2\right) - 4xe^{-x^2} = 2xe^{-x^2}\left(2x^2 - 3\right) > 0$

$\Leftrightarrow$ $x > \sqrt{\frac{3}{2}}$ or $-\sqrt{\frac{3}{2}} < x < 0$, so f is CU on $\left(\sqrt{\frac{3}{2}}, \infty\right)$ and **H.**

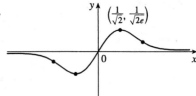

$\left(-\sqrt{\frac{3}{2}}, 0\right)$ and CD on $\left(-\infty, -\sqrt{\frac{3}{2}}\right)$ and $\left(0, \sqrt{\frac{3}{2}}\right)$. IP are

$(0, 0)$ and $\left(\pm\sqrt{\frac{3}{2}}, \pm\sqrt{\frac{3}{2}}e^{-3/2}\right)$.

51. $y = -\dfrac{W}{24EI}x^4 + \dfrac{WL}{12EI}x^3 - \dfrac{WL^2}{24EI}x^2 = -\dfrac{W}{24EI}x^2\left(x^2 - 2Lx + L^2\right)$

$= \dfrac{-W}{24EI}x^2(X - L)^2 = cx^2(x - L)^2$

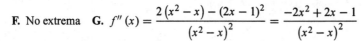

where $c = -\dfrac{W}{24EI}$ is a negative constant and $0 \leq x \leq L$. We sketch

$f(x) = cx^2(x - L)^2$ for $c = -1$. $f(0) = f(L) = 0$.

$f'(x) = cx^2[2(x - L)] + (x - L)^2(2cx) = 2cx(x - L)[x + (x - L)] = 2cx(x - L)(2x - L)$. So for

$0 < x < L$, $f'(x) > 0$ $\Leftrightarrow$ $x(x - L)(2x - L) < 0$ (since $c < 0$) $\Leftrightarrow$ $L/2 < x < L$ and $f'(x) < 0$ $\Leftrightarrow$

$0 < x < L/2$. So f is increasing on $(L/2, L)$ and decreasing on $(0, L/2)$, and there is a local and absolute

minimum at $(L/2, f(L/2)) = (L/2, cL^4/16)$.

$f'(x) = 2c[x(x - L)(2x - L)]$ $\Rightarrow$

$f''(x) = 2c[1(x - L)(2x - L) + x(1)(2x - L) + x(x - L)(2)] = 2c(6x^2 - 6Lx + L^2) = 0$ $\Leftrightarrow$

$x = \dfrac{6L \pm \sqrt{12L^2}}{12} = \frac{1}{2}L \pm \frac{\sqrt{3}}{6}L$, and these are the x-coordinates of the two inflection points.

53. $y = f(x) = x^3/(x^2 - 1)$ **A.** $D = \{x \mid x \neq \pm 1\} = (-\infty, -1) \cup (-1, 1) \cup (1, \infty)$ **B.** x-intercept $= 0$, y-intercept $= 0$ **C.** $f(-x) = -f(x)$ $\Rightarrow$ f is odd, so the curve is symmetric about the origin.

D. $\lim\limits_{x \to \infty} \dfrac{x^3}{x^2 - 1} = \infty$ but long division gives $\dfrac{x^3}{x^2 - 1} = x + \dfrac{x}{x^2 - 1}$ so $f(x) - x = \dfrac{x}{x^2 - 1} \to 0$ as $x \to \pm\infty$

$\Rightarrow$ $y = x$ is a slant asymptote. $\lim\limits_{x \to 1^-} \dfrac{x^3}{x^2 - 1} = -\infty$, $\lim\limits_{x \to 1^+} \dfrac{x^3}{x^2 - 1} = \infty$, $\lim\limits_{x \to -1^-} \dfrac{x^3}{x^2 - 1} = -\infty$,

$\lim\limits_{x \to -1^+} \dfrac{x^3}{x^2 - 1} = \infty$, so $x = 1$ and $x = -1$ are VA. **E.** $f'(x) = \dfrac{3x^2(x^2 - 1) - x^3(2x)}{(x^2 - 1)^2} = \dfrac{x^2(x^2 - 3)}{(x^2 - 1)^2}$ $\Rightarrow$

$f'(x) > 0$ $\Leftrightarrow$ $x^2 > 3$ $\Leftrightarrow$ $x > \sqrt{3}$ or $x < -\sqrt{3}$, so f is increasing on $\left(-\infty, -\sqrt{3}\right)$ and $\left(\sqrt{3}, \infty\right)$

and decreasing on $\left(-\sqrt{3}, -1\right)$, $(-1, 1)$, and $\left(1, \sqrt{3}\right)$. **H.**

F. $f\left(-\sqrt{3}\right) = -\frac{3\sqrt{3}}{2}$ is a local maximum and $f\left(\sqrt{3}\right) = \frac{3\sqrt{3}}{2}$ is a local

minimum. **G.** $y'' = \dfrac{2x(x^2 + 3)}{(x^2 - 1)^3} > 0$ $\Leftrightarrow$ $x > 1$ or $-1 < x < 0$, so f

is CU on $(-1, 0)$ and $(1, \infty)$ and CD on $(-\infty, -1)$ and $(0, 1)$. IP $(0, 0)$

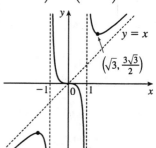

55. $y = f(x) = (x^2 + 4)/x = x + 4/x$ **A.** $D = \{x \mid x \neq 0\} = (-\infty, 0) \cup (0, \infty)$ **B.** No intercept
C. $f(-x) = -f(x)$ $\Rightarrow$ symmetry about the origin **D.** $\lim\limits_{x \to \infty} (x + 4/x) = \infty$ but $f(x) - x = 4/x \to 0$ as

$x \to \pm\infty$, so $y = x$ is a slant asymptote. $\lim\limits_{x \to 0^+} (x + 4/x) = \infty$ and **H.**

$\lim\limits_{x \to 0^-} (x + 4/x) = -\infty$, so $x = 0$ is a VA. **E.** $f'(x) = 1 - 4/x^2 > 0$

$\Leftrightarrow$ $x^2 > 4$ $\Leftrightarrow$ $x > 2$ or $x < -2$, so f is increasing on $(-\infty, -2)$
and $(2, \infty)$ and decreasing on $(-2, 0)$ and $(0, 2)$. **F.** $f(-2) = -4$ is a
local maximum and $f(2) = 4$ is a local minimum.

G. $f''(x) = 8/x^3 > 0$ $\Leftrightarrow$ $x > 0$ so f is CU on $(0, \infty)$ and CD on
$(-\infty, 0)$. No IP

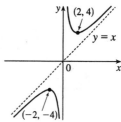

57. $y = \dfrac{1}{x - 1} - x$ **A.** $D = \{x \mid x \neq 1\}$ **B.** $y = 0$ $\Leftrightarrow$ $x = \dfrac{1}{x - 1}$ $\Leftrightarrow$ $x^2 - x - 1 = 0$ $\Rightarrow$ $x = \frac{1 \pm \sqrt{5}}{2}$

(x-intercepts), y-intercept $= f(0) = -1$ **C.** No symmetry **D.** $y - (-x) = \dfrac{1}{x - 1} \to 0$ as

$x \to \pm\infty$, so $y = -x$ is a slant asymptote. $\lim\limits_{x \to 1^+} \left(\dfrac{1}{x - 1} - x\right) = \infty$ and **H.**

$\lim\limits_{x \to 1^-} \left(\dfrac{1}{x - 1} - x\right) = -\infty$, so $x = 1$ is a VA.

E. $f'(x) = -1 - 1/(x - 1)^2 < 0$ for all $x \neq 1$, so f is decreasing on
$(-\infty, 1)$ and $(1, \infty)$. **F.** No local extremum

G. $f''(x) = \dfrac{2}{(x - 1)^3} > 0$ $\Leftrightarrow$ $x > 1$, so f is CU on $(1, \infty)$ and CD

on $(-\infty, 1)$. No IP

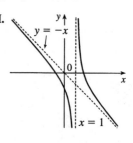

59. $\dfrac{x^2}{a^2} - \dfrac{y^2}{b^2} = 1 \Rightarrow y = \pm\dfrac{b}{a}\sqrt{x^2 - a^2}$. Now

$$\lim_{x\to\infty}\left[\dfrac{b}{a}\sqrt{x^2 - a^2} - \dfrac{b}{a}x\right] = \dfrac{b}{a}\cdot\lim_{x\to\infty}\left(\sqrt{x^2 - a^2} - x\right)\dfrac{\sqrt{x^2 - a^2} + x}{\sqrt{x^2 - a^2} + x} = \dfrac{b}{a}\cdot\lim_{x\to\infty}\dfrac{-a^2}{\sqrt{x^2 - a^2} + x} = 0,\ \text{which}$$

shows that $y = \dfrac{b}{a}x$ is a slant asymptote. Similarly

$$\lim_{x\to\infty}\left[-\dfrac{b}{a}\sqrt{x^2 - a^2} - \left(-\dfrac{b}{a}x\right)\right] = -\dfrac{b}{a}\cdot\lim_{x\to\infty}\dfrac{-a^2}{\sqrt{x^2 - a^2} + x} = 0,\ \text{so}\ y = -\dfrac{b}{a}x\ \text{is a slant asymptote.}$$

61. $\displaystyle\lim_{x\to\pm\infty}\left[f(x) - x^3\right] = \lim_{x\to\pm\infty}\dfrac{x^4 + 1}{x} - \dfrac{x^4}{x} = \lim_{x\to\pm\infty}\dfrac{1}{x} = 0$, so the graph of f is asymptotic to that of $y = x^3$.

A. $D = \{x \mid x \neq 0\}$ **B.** No intercept **C.** f is symmetric about the origin. **D.** $\displaystyle\lim_{x\to0^-}\left(x^3 + \dfrac{1}{x}\right) = -\infty$ and

$\displaystyle\lim_{x\to0^+}\left(x^3 + \dfrac{1}{x}\right) = \infty$, so $x = 0$ is a vertical asymptote, and as shown above, the graph of

f is asymptotic to that of $y = x^3$. **E.** $f'(x) = 3x^2 - 1/x^2 > 0 \Leftrightarrow$

$x^4 > \frac{1}{3} \Leftrightarrow |x| > \frac{1}{\sqrt[4]{3}}$, so f is increasing on $\left(-\infty, -\frac{1}{\sqrt[4]{3}}\right)$ and

$\left(\frac{1}{\sqrt[4]{3}}, \infty\right)$ and decreasing on $\left(-\frac{1}{\sqrt[4]{3}}, 0\right)$ and $\left(0, \frac{1}{\sqrt[4]{3}}\right)$. **F.** Local

maximum $f\left(-\frac{1}{\sqrt[4]{3}}\right) = -4 \cdot 3^{-5/4}$, local minimum $f\left(\frac{1}{\sqrt[4]{3}}\right) = 4 \cdot 3^{-5/4}$

G. $f''(x) = 6x + 2/x^3 > 0 \Leftrightarrow x > 0$, so f is CU on $(0, \infty)$ and CD on $(-\infty, 0)$. No IP

H.

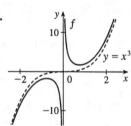

4.6 Graphing with Calculus *and* Calculators

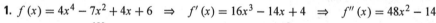

1. $f(x) = 4x^4 - 7x^2 + 4x + 6 \Rightarrow f'(x) = 16x^3 - 14x + 4 \Rightarrow f''(x) = 48x^2 - 14$

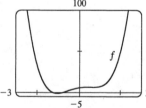

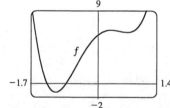

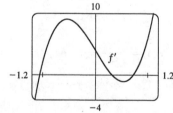

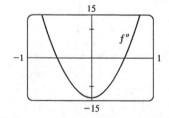

After finding suitable viewing rectangles (by ensuring that we have located all of the x-values where either $f' = 0$ or $f'' = 0$) we estimate from the graph of f' that f is increasing on $(-1.1, 0.3)$ and $(0.7, \infty)$ and decreasing on $(-\infty, -1.1)$ and $(0.3, 0.7)$, with a local maximum of $f(0.3) \approx 6.6$ and minima of $f(-1.1) \approx -1.0$ and $f(0.7) \approx 6.3$. We estimate from the graph of f'' that f is CU on $(-\infty, -0.5)$ and $(0.5, \infty)$ and CD on $(-0.5, 0.5)$, and that f has inflection points at about $(-0.5, 2.1)$ and $(0.5, 6.5)$.

3. $f(x) = \sqrt[3]{x^2 - 3x - 5}$ $\Rightarrow$ $f'(x) = \dfrac{1}{3}\dfrac{2x - 3}{(x^2 - 3x - 5)^{2/3}}$ $\Rightarrow$ $f''(x) = -\dfrac{2}{9}\dfrac{x^2 - 3x + 24}{(x^2 - 3x - 5)^{5/3}}$

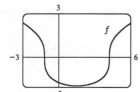 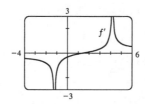 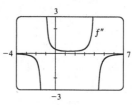

Note: With some CAS's, including Maple, it is necessary to define $f(x) = \dfrac{x^2 - 3x - 5}{|x^2 - 3x - 5|}|x^2 - 3x - 5|^{1/3}$, since the CAS does not compute real cube roots of negative numbers. We estimate from the graph of f' that f is increasing on $(1.5, 4.2)$ and $(4.2, \infty)$, and decreasing on $(-\infty, -1.2)$ and $(-1.2, 1.5)$. f has no maximum. Minimum: $f(1.5) \approx -1.9$. From the graph of f'', we estimate that f is CU on $(-1.2, 4.2)$ and CD on $(-\infty, -1.2)$ and $(4.2, \infty)$. IP $(-1.2, 0)$ and $(4.2, 0)$.

5. $f(x) = \dfrac{x}{x^3 - x^2 - 4x + 1}$ $\Rightarrow$ $f'(x) = \dfrac{-2x^3 + x^2 + 1}{(x^3 - x^2 - 4x + 1)^2}$ $\Rightarrow$

$f''(x) = \dfrac{2(3x^5 - 3x^4 + 5x^3 - 6x^2 + 3x + 4)}{(x^3 - x^2 - 4x + 1)^3}$

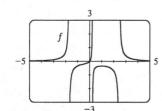 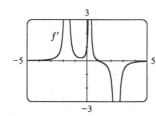 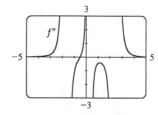

We estimate from the graph of f that $y = 0$ is a horizontal asymptote, and that there are vertical asymptotes at $x = -1.7$, $x = 0.24$, and $x = 2.46$. From the graph of f', we estimate that f is increasing on $(-\infty, -1.7)$, $(-1.7, 0.24)$, and $(0.24, 1)$, and that f is decreasing on $(1, 2.46)$ and $(2.46, \infty)$. There is a local maximum at $f(1) = -\frac{1}{3}$. From the graph of f'', we estimate that f is CU on $(-\infty, -1.7)$, $(-0.506, 0.24)$, and $(2.46, \infty)$, and that f is CD on $(-1.7, -0.506)$ and $(0.24, 2.46)$. There is an inflection point at $(-0.506, -0.192)$.

7. $f(x) = x^2 \sin x$ $\Rightarrow$ $f'(x) = 2x \sin x + x^2 \cos x$ $\Rightarrow$ $f''(x) = 2 \sin x + 4x \cos x - x^2 \sin x$

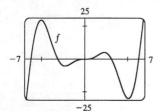 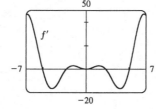 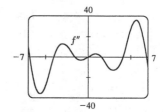

We estimate from the graph of f' that f is increasing on $(-7, -5.1)$, $(-2.3, 2.3)$, and $(5.1, 7)$ and decreasing on $(-5.1, -2.3)$, and $(2.3, 5.1)$. Local maxima: $f(-5.1) \approx 24.1$, $f(2.3) \approx 3.9$. Local minima: $f(-2.3) \approx -3.9$, $f(5.1) \approx -24.1$. From the graph of f'', we estimate that f is CU on $(-7, -6.8)$, $(-4.0, -1.5)$, $(0, 1.5)$, and $(4.0, 6.8)$, and CD on $(-6.8, -4.0)$, $(-1.5, 0)$, $(1.5, 4.0)$, and $(6.8, 7)$. f has IP at $(-6.8, -24.4)$, $(-4.0, 12.0)$, $(-1.5, -2.3)$, $(0, 0)$, $(1.5, 2.3)$, $(4.0, -12.0)$ and $(6.8, 24.4)$.

9. $f(x) = 8x^3 - 3x^2 - 10 \implies f'(x) = 24x^2 - 6x \implies f''(x) = 48x - 6$

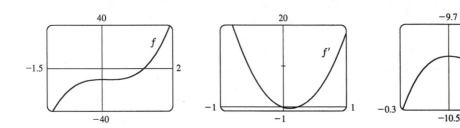

From the graphs, it appears that $f(x) = 8x^3 - 3x^2 - 10$ increases on $(-\infty, 0)$ and $(0.25, \infty)$ and decreases on $(0, 0.25)$; that f has a local maximum of $f(0) = -10.0$ and a local minimum of $f(0.25) \approx -10.1$; that f is CU on $(0.1, \infty)$ and CD on $(-\infty, 0.1)$; and that f has an IP at $(0.1, -10)$. $f(x) = 8x^3 - 3x^2 - 10 \implies f'(x) = 24x^2 - 6x = 6x(4x - 1)$, which is positive ($f$ is increasing) for $(-\infty, 0)$ and $\left(\frac{1}{4}, \infty\right)$, and negative ($f$ is decreasing) on $\left(0, \frac{1}{4}\right)$. By the FDT, f has a local maximum at $x = 0$: $f(0) = 8(0)^3 - 3(0)^2 - 10 = -10$; and f has a local minimum at $\frac{1}{4}$: $f\left(\frac{1}{4}\right) = \frac{1}{8} - \frac{3}{16} - 10 = -\frac{161}{16}$. $f'(x) = 24x^2 - 6x \implies$ $f''(x) = 48x - 6 = 6(8x - 1)$, which is positive ($f$ is CU) on $\left(\frac{1}{8}, \infty\right)$, and negative ($f$ is CD) on $\left(-\infty, \frac{1}{8}\right)$. f has an IP at $\left(\frac{1}{8}, f\left(\frac{1}{8}\right)\right) = \left(\frac{1}{8}, -\frac{321}{32}\right)$.

11.

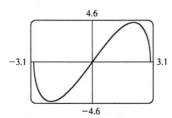

From the graph, it appears that f increases on $(-2.1, 2.1)$ and decreases on $(-3, -2.1)$ and $(2.1, 3)$; that f has a local maximum of $f(2.1) \approx 4.5$ and a local minimum of $f(-2.1) \approx -4.5$; that f is CU on $(-3.0, 0)$ and CD on $(0, 3.0)$, and that f has an IP at $(0, 0)$. $f(x) = x\sqrt{9 - x^2} \implies f'(x) = \dfrac{-x^2}{\sqrt{9 - x^2}} + \sqrt{9 - x^2} = \dfrac{9 - 2x^2}{\sqrt{9 - x^2}}$, which is positive ($f$ is increasing) on $\left(\frac{-3\sqrt{2}}{2}, \frac{3\sqrt{2}}{2}\right)$ and negative (f is decreasing) on $\left(-3, \frac{-3\sqrt{2}}{2}\right)$ and $\left(\frac{3\sqrt{2}}{2}, 3\right)$. By the FDT, f has a local maximum of $f\left(\frac{3\sqrt{2}}{2}\right) = \frac{3\sqrt{2}}{2}\sqrt{9 - \left(\frac{3\sqrt{2}}{2}\right)^2} = \frac{9}{2}$; and f has a local minimum of $f\left(\frac{-3\sqrt{2}}{2}\right) = -\frac{9}{2}$ (since f is an odd function.) $f'(x) = \dfrac{-x^2}{\sqrt{9 - x^2}} + \sqrt{9 - x^2} \implies$

$$f''(x) = \dfrac{\sqrt{9 - x^2}(-2x) + x^2\left(\frac{1}{2}\right)(9 - x^2)^{-1/2}(-2x)}{9 - x^2} - x(9 - x^2)^{-1/2} = \dfrac{-2x - x^3(9 - x^2)^{-1} - x}{\sqrt{9 - x^2}}$$

$$= \dfrac{-3x}{\sqrt{9 - x^2}} - \dfrac{x^3}{(9 - x^2)^{3/2}} = \dfrac{x(2x^2 - 27)}{(9 - x^2)^{3/2}}$$

is positive (f is CU) on $(-3, 0)$, and negative (f is CD) on $(0, 3)$. f has an IP at $(0, 0)$.

13.

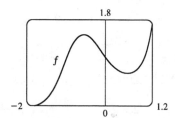

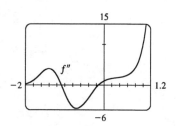

$f(x) = e^{x^3 - x} \to 0$ as $x \to -\infty$, and $f(x) \to \infty$ as $x \to \infty$. From the graph, it appears that f has a local

minimum of about $f(0.58) = 0.68$, and a local maximum of about $f(-0.58) = 1.47$. To find the exact values, we

calculate $f'(x) = (3x^2 - 1) e^{x^3 - x}$, which is 0 when $3x^2 - 1 = 0$ ⇔ $x = \pm \frac{1}{\sqrt{3}}$. The negative root corresponds

to the local maximum $f\left(-\frac{1}{\sqrt{3}}\right) = e^{(-1/\sqrt{3})^3 - (-1/\sqrt{3})} = e^{2\sqrt{3}/9}$, and the positive root corresponds to the local

minimum $f\left(\frac{1}{\sqrt{3}}\right) = e^{(1/\sqrt{3})^3 - (1/\sqrt{3})} = e^{-2\sqrt{3}/9}$. To estimate the inflection points, we calculate and graph

$f''(x) = \frac{d}{dx}\left[(3x^2 - 1) e^{x^3 - x}\right] = (3x^2 - 1) e^{x^3 - x} (3x^2 - 1) + e^{x^3 - x} (6x) = e^{x^3 - x} (9x^4 - 6x^2 + 6x + 1)$. From

the graph, it appears that $f''(x)$ changes sign (and thus f has inflection points) at $x \approx -0.15$ and $x \approx -1.09$.

From the graph of f, we see that these x-values correspond to inflection points at about $(-0.15, 1.15)$ and

$(-1.09, 0.82)$.

15. (a) $f(x) = x^2 \ln x$

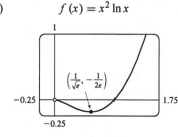

$\left(\frac{1}{\sqrt{e}}, -\frac{1}{2e}\right)$

(b) $\lim_{x \to \infty} x^2 \ln x = \infty$,

$$\lim_{x \to 0^+} x^2 \ln x = \lim_{x \to 0^+} \frac{\ln x}{1/x^2} \overset{H}{=} \lim_{x \to 0^+} \frac{1/x}{-2/x^3} = \lim_{x \to 0^+} \left(-\frac{x^2}{2}\right) = 0.$$

There is a hole at $(0, 0)$.

(c) It appears that there is an IP at about $(0.2, -0.06)$ and a local minimum at $(0.6, -0.18)$. $f(x) = x^2 \ln x$ ⇒
$f'(x) = 2x \ln x + x = x(2 \ln x + 1) > 0$ ⇔ $\ln x > -\frac{1}{2}$ ⇔ $x > e^{-1/2}$, so f is increasing on
$(1/\sqrt{e}, \infty)$, decreasing on $(0, 1/\sqrt{e})$. By the FDT, $f(1/\sqrt{e}) = -1/(2e)$ is a local minimum. This
point is approximately $(0.6065, -0.1839)$, which agrees with our estimate.
$f''(x) = x(2/x) + (2 \ln x + 1) = 2 \ln x + 3 > 0$ ⇔ $\ln x > -\frac{3}{2}$ ⇔ $x > e^{-3/2}$, so f is CU on
$(e^{-3/2}, \infty)$ and CD on $(0, e^{-3/2})$. IP is $(e^{-3/2}, -3/(2e^3)) \approx (0.2231, -0.0747)$.

17.

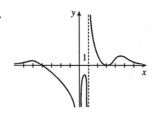

$f(x) = \dfrac{(x+4)(x-3)^2}{x^4(x-1)}$ has VA at $x = 0$ and at $x = 1$ since

$\lim\limits_{x \to 0} f(x) = -\infty$, $\lim\limits_{x \to 1^-} f(x) = -\infty$ and $\lim\limits_{x \to 1^+} f(x) = \infty$.

$f(x) = \dfrac{(1 + 4/x)(1 - 3/x)^2}{x(x-1)} \to 0^+$ as $x \to \pm\infty$, so f is asymptotic to

the x-axis. Since f is undefined at $x = 0$, it has no y-intercept.

$f(x) = 0 \;\Rightarrow\; (x+4)(x-3)^2 = 0 \;\Rightarrow\; x = -4$ or $x = 3$, so f has x-intercepts -4 and 3. Note, however, that the graph of f is only tangent to the x-axis and does not cross it at $x = 3$, since f is positive as $x \to 3^-$ and as $x \to 3^+$.

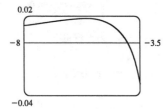
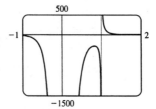
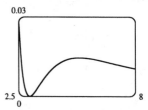

From these graphs, it appears that f has three maxima and one minimum. The maxima are approximately $f(-5.6) = 0.0182$, $f(0.82) = -281.5$ and $f(5.2) = 0.0145$ and we know (since the graph is tangent to the x-axis at $x = 3$) that the minimum is $f(3) = 0$.

19. $f(x) = \dfrac{x^2(x+1)^3}{(x-2)^2(x-4)^4} \;\Rightarrow\; f'(x) = -\dfrac{x(x+1)^2(x^3 + 18x^2 - 44x - 16)}{(x-2)^3(x-4)^5}$ (from CAS).

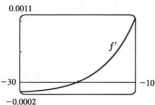
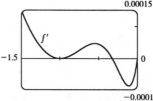
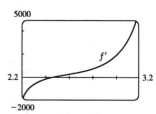

From the graphs of f', it seems that the critical points which indicate extrema occur at $x \approx -20$, -0.3, and 2.5, as estimated in Example 3. (There is another critical point at $x = -1$, but the sign of f' does not change there.) We differentiate again, obtaining $f''(x) = 2\dfrac{(x+1)(x^6 + 36x^5 + 6x^4 - 628x^3 + 684x^2 + 672x + 64)}{(x-2)^4(x-4)^6}$.

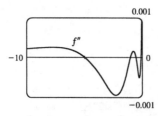

From the graphs of f'', it appears that f is CU on $(-\infty, -5.0)$, $(-1.0, -0.5)$, $(-0.1, 2.0)$, $(2.0, 4.0)$ and $(4.0, \infty)$ and CD on $(-5.0, -1.0)$ and $(-0.5, -0.1)$. We check back on the graphs of f to find the y-coordinates of the inflection points, and find that these points are approximately $(-5, -0.005)$, $(-1, 0)$, $(-0.5, 0.00001)$, and $(-0.1, 0.0000066)$.

21. $y = f(x) = \dfrac{\sin^2 x}{\sqrt{x^2 + 1}}$ with $0 \le x \le 3\pi$. From a CAS, $y' = \dfrac{\sin x \left[2\left(x^2 + 1\right)\cos x - x \sin x\right]}{\left(x^2 + 1\right)^{3/2}}$ and

$y'' = \dfrac{\left(4x^4 + 6x^2 + 5\right)\cos^2 x - 4x\left(x^2 + 1\right)\sin x \cos x - 2x^4 - 2x^2 - 3}{\left(x^2 + 1\right)^{5/2}}$.

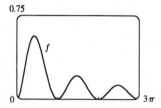

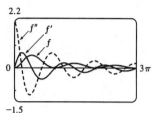

From the graph of f' and the formula for y', we determine that $y' = 0$ when $x = \pi, 2\pi, 3\pi$, or $x \approx 1.3, 4.6$, or 7.8. So f is increasing on $(0, 1.3)$, $(\pi, 4.6)$, and $(2\pi, 7.8)$. f is decreasing on $(1.3, \pi)$, $(4.6, 2\pi)$, and $(7.8, 3\pi)$. Local maxima: $f(1.3) \approx 0.6$, $f(4.6) \approx 0.21$, and $f(7.8) \approx 0.13$. Local minima: $f(\pi) = f(2\pi) = f(3\pi) = 0$. From the graph of f'', we see that $y'' = 0 \Leftrightarrow x \approx 0.6, 2.1, 3.8, 5.4, 7.0$, or 8.6. So f is CU on $(0, 0.6)$, $(2.1, 3.8)$, $(5.4, 7.0)$, and $(8.6, 3\pi)$. f is CD on $(0.6, 2.1)$, $(3.8, 5.4)$, and $(7.0, 8.6)$. There are IP at $(0.6, 0.25)$, $(2.1, 0.31)$, $(3.8, 0.10)$, $(5.4, 0.11)$, $(7.0, 0.061)$, and $(8.6, 0.065)$.

23. (a) $f(x) = x^{1/x}$

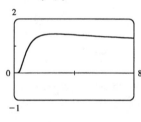

(b) Recall that $a^b = e^{b \ln a}$. $\lim\limits_{x \to 0^+} x^{1/x} = \lim\limits_{x \to 0^+} e^{(1/x)\ln x}$. As $x \to 0^+$,

$\dfrac{\ln x}{x} \to -\infty$, so $x^{1/x} \to 0$. This indicates that there is a hole at

$(0, 0)$. As $x \to \infty$, we have the indeterminate form ∞^0.

$\lim\limits_{x \to \infty} x^{1/x} = \lim\limits_{x \to \infty} e^{(1/x)\ln x}$, but $\lim\limits_{x \to \infty} \dfrac{\ln x}{x} \overset{H}{=} \lim\limits_{x \to \infty} \dfrac{1/x}{1} = 0$, so

$\lim\limits_{x \to \infty} x^{1/x} = e^0 = 1$. This indicates that $y = 1$ is a HA.

(c) Estimated maximum: $(2.72, 1.45)$. No estimated minimum. We use logarithmic differentiation to find any

critical numbers. $y = x^{1/x} \Rightarrow \ln y = \dfrac{1}{x}\ln x \Rightarrow \dfrac{y'}{y} = \dfrac{1}{x} \cdot \dfrac{1}{x} + (\ln x)\left(-\dfrac{1}{x^2}\right) \Rightarrow$

$y' = x^{1/x}\left(\dfrac{1 - \ln x}{x^2}\right) = 0 \Rightarrow \ln x = 1 \Rightarrow x = e$. For $0 < x < e$, $y' > 0$ and for $x > e$, $y' < 0$, so

$f(e) = e^{1/e}$ is a local maximum. This point is approximately $(2.7183, 1.4447)$, which agrees with our estimate.

(d)

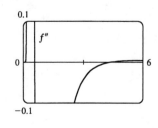

From the graph, we see that $f''(x) = 0$ at $x \approx 0.58$ and $x \approx 4.37$. Since f'' changes sign at these values, they are x-coordinates of inflection points.

25.

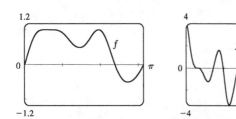 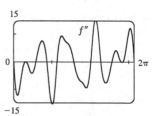

From the graph of $f(x) = \sin(x + \sin 3x)$ in the viewing rectangle $[0, \pi]$ by $[-1.2, 1.2]$, it looks like f has two maxima and two minima. If we calculate and graph $f'(x) = [\cos(x + \sin 3x)](1 + 3\cos 3x)$ on $[0, 2\pi]$, we see that the graph of f' appears to be almost tangent to the x-axis at about $x = 0.7$. The graph of $f'' = -[\sin(x + \sin 3x)](1 + 3\cos 3x)^2 + \cos(x + \sin 3x)(-9\sin 3x)$ is even more interesting near this x-value: it seems to just touch the x-axis.

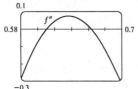 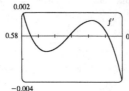 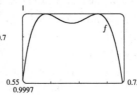 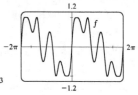

If we zoom in on this place on the graph of f'', we see that f'' actually does cross the axis twice near $x = 0.65$, indicating a change in concavity for a very short interval. If we look at the graph of f' on the same interval, we see that it changes sign three times near $x = 0.65$, indicating that what we had thought was a broad extremum at about $x = 0.7$ actually consists of three extrema (two maxima and a minimum). These maxima are roughly $f(0.59) = 1$ and $f(0.68) = 1$, and the minimum is roughly $f(0.64) = 0.99996$. There are also a maximum of about $f(1.96) = 1$ and minima of about $f(1.46) = 0.49$ and $f(2.73) = -0.51$. The points of inflection on $(0, \pi)$ are about $(0.61, 0.99998)$, $(0.66, 0.99998)$, $(1.17, 0.72)$, $(1.75, 0.77)$, and $(2.28, 0.34)$. On $(\pi, 2\pi)$, they are about $(4.01, -0.34)$, $(4.54, -0.77)$, $(5.11, -0.72)$, $(5.62, -0.99998)$, and $(5.67, -0.99998)$. There are also IP at $(0, 0)$ and $(\pi, 0)$. Note that the function is odd and periodic with period 2π, and it is also rotationally symmetric about all points of the form $((2n + 1)\pi, 0)$, n an integer.

27. $f(x) = x^4 + cx^2 = x^2(x^2 + c)$. Note that f is an even function. For $c \geq 0$, the only x-intercept is the point $(0, 0)$. We calculate $f'(x) = 4x^3 + 2cx = 4x\left(x^2 + \frac{1}{2}c\right)$ $\Rightarrow$ $f''(x) = 12x^2 + 2c$. If $c \geq 0$, $x = 0$ is the only critical point and there are no inflection points. As we can see from the examples, there is no change in the basic shape of the graph

for $c \geq 0$; it merely becomes steeper as c increases. For $c = 0$, the graph is the simple curve $y = x^4$. For $c < 0$, there are x-intercepts at 0 and at $\pm\sqrt{-c}$. Also, there is a maximum at $(0, 0)$, and there are minima at $\left(\pm\sqrt{-\frac{1}{2}c}, -\frac{1}{4}c^2\right)$. As $c \to -\infty$, the x-coordinates of these minima get larger in absolute value, and the minimum points move downward. There are inflection points at $\left(\pm\sqrt{-\frac{1}{6}c}, -\frac{5}{36}c^2\right)$, which also move away from the origin as $c \to -\infty$.

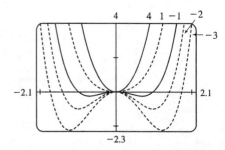

29.

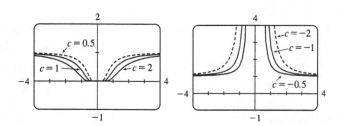

$c = 0$ is a transitional value — we get the graph of $y = 1$. For $c > 0$, we see that there is a HA at $y = 1$, and that the graph spreads out as c increases. At first glance there appears to be a minimum at $(0, 0)$, but $f(0)$ is undefined, so there is no minimum or maximum. For $c < 0$, we still have the HA at $y = 1$, but the range is $(1, \infty)$ rather than $(0, 1)$. We also have a VA at $x = 0$. $f(x) = e^{-c/x^2}$ $\Rightarrow$ $f'(x) = e^{-c/x^2} \left(-2c/x^3\right)$ $\Rightarrow$

$f''(x) = \dfrac{2c\left(2c - 3x^2\right)}{x^6 e^{c/x^2}}$. $f'(x) \neq 0$ and $f'(x)$ exists for all $x \neq 0$ (and 0 is not in the domain of f), so there are

no maxima or minima. $f''(x) = 0$ $\Rightarrow$ $x = \pm\sqrt{2c/3}$, so if $c > 0$, the inflection points spread out as c increases, and if $c < 0$, there are no IP. For $c > 0$, there are IP at $\left(\pm\sqrt{2c/3}, e^{-3/2}\right)$. Note that the y-coordinate of the IP is constant.

31. Note that $c = 0$ is a transitional value at which the graph consists of the x-axis. Also, we can see that if we

substitute $-c$ for c, the function $f(x) = \dfrac{cx}{1 + c^2 x^2}$ will be reflected in the x-axis, so we investigate only positive

values of c (except $c = -1$, as a demonstration of this reflective property). Also, f is an odd

function. $\lim\limits_{x \to \pm\infty} f(x) = 0$, so $y = 0$ is a horizontal asymptote for all c. We calculate

$f'(x) = \dfrac{c\left(1 + c^2 x^2\right) - cx\left(2c^2 x\right)}{\left(1 + c^2 x^2\right)^2} = -\dfrac{c\left(c^2 x^2 - 1\right)}{\left(1 + c^2 x^2\right)^2}$. $f'(x) = 0$ $\Leftrightarrow$ $c^2 x^2 - 1 = 0$ $\Leftrightarrow$ $x = \pm 1/c$. So there

is an absolute maximum of $f(1/c) = \frac{1}{2}$ and an absolute minimum of $f(-1/c) = -\frac{1}{2}$. These extrema have the

same value regardless of c, but the maximum points move closer to the y-axis as c increases.

$f''(x) = \dfrac{\left(-2c^3 x\right)\left(1 + c^2 x^2\right)^2 - \left(-c^3 x^2 + c\right)\left[2\left(1 + c^2 x^2\right)\left(2c^2 x\right)\right]}{\left(1 + c^2 x^2\right)^4}$

$= \dfrac{\left(-2c^3 x\right)\left(1 + c^2 x^2\right) + \left(c^3 x^2 - c\right)\left(4c^2 x\right)}{\left(1 + c^2 x^2\right)^3} = \dfrac{2c^3 x\left(c^2 x^2 - 3\right)}{\left(1 + c^2 x^2\right)^3}$

$f''(x) = 0$ $\Leftrightarrow$ $x = 0$ or $\pm\sqrt{3}/c$, so there are inflection points at

$(0, 0)$ and at $\left(\pm\sqrt{3}/c, \pm\sqrt{3}/4\right)$.

Again, the y-coordinate of the inflection points does not depend on c, but as c increases, both inflection points approach the y-axis.

33. $f(x) = cx + \sin x \quad \Rightarrow \quad f'(x) = c + \cos x \quad \Rightarrow \quad f''(x) = -\sin x$

$f(-x) = -f(x)$, so f is an odd function and its graph is symmetric with respect to the origin.

$f(x) = 0 \quad \Leftrightarrow \quad \sin x = -cx$, so 0 is always an x-intercept.

$f'(x) = 0 \quad \Leftrightarrow \quad \cos x = -c$, so there is no critical number when $|c| > 1$. If $|c| \le 1$, then there are infinitely many critical numbers. If x_1 is the unique solution of $\cos x = -c$ in the interval $[0, \pi]$, then the critical numbers are $2n\pi \pm x_1$, where n ranges over the integers. (Special cases: When $c = 1$, $x_1 = 0$; when $c = 0$, $x = \frac{\pi}{2}$; and when $c = -1$, $x_1 = \pi$.)

$f''(x) < 0 \quad \Leftrightarrow \quad \sin x > 0$, so f is CD on intervals of the form $(2n\pi, (2n+1)\pi)$. f is CU on intervals of the form $((2n-1)\pi, 2n\pi)$. The inflection points of f are the points $(2n\pi, 2n\pi c)$, where n is an integer.

If $c \ge 1$, then $f'(x) \ge 0$ for all x, so f is increasing and has no extremum. If $c \le -1$, then $f'(x) \le 0$ for all x, so f is decreasing and has no extremum. If $|c| < 1$, then $f'(x) > 0 \quad \Leftrightarrow \quad \cos x > -c \quad \Leftrightarrow \quad x$ is in an interval of the form $(2n\pi - x_1, 2n\pi + x_1)$ for some integer n. These are the intervals on which f is increasing. Similarly, we find that f is decreasing on the intervals of the form $(2n\pi + x_1, 2(n+1)\pi - x_1)$. Thus, f has local maxima at the points $2n\pi + x_1$, where f has the values $c(2n\pi + x_1) + \sin x_1 = c(2n\pi + x_1) + \sqrt{1 - c^2}$, and f has local minima at the points $2n\pi - x_1$, where we have $f(2n\pi - x_1) = c(2n\pi - x_1) - \sin x_1 = c(2n\pi - x_1) - \sqrt{1 - c^2}$.

The transitional values of c are -1 and 1. The inflection points move vertically, but not horizontally, when c changes. When $|c| \ge 1$, there is no extremum. For $|c| < 1$, the maxima are spaced 2π apart horizontally, as are the minima. The horizontal spacing between maxima and adjacent minima is regular (and equals π) when $c = 0$, but the horizontal space between a local maximum and the nearest local minimum shrinks as $|c|$ approaches 1.

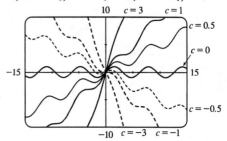

35. If $c < 0$, then $\lim\limits_{x \to -\infty} f(x) = \lim\limits_{x \to -\infty} \dfrac{x}{e^{cx}} \overset{\text{H}}{=} \lim\limits_{x \to -\infty} \dfrac{1}{ce^{cx}} = 0$, and $\lim\limits_{x \to \infty} f(x) = \infty$.

If $c > 0$, then $\lim\limits_{x \to -\infty} f(x) = -\infty$, and $\lim\limits_{x \to \infty} f(x) \overset{\text{H}}{=} \lim\limits_{x \to \infty} \dfrac{1}{ce^{cx}} = 0$.

If $c = 0$, then $f(x) = x$, so $\lim\limits_{x \to \pm\infty} f(x) = \pm\infty$ respectively.

So we see that $c = 0$ is a transitional value. We now exclude the case $c = 0$, since we know how the function behaves in that case. To find the maxima and minima of f, we differentiate: $f(x) = xe^{-cx} \quad \Rightarrow$ $f'(x) = x\left(-ce^{-cx}\right) + e^{-cx} = (1 - cx)e^{-cx}$. This is 0 when $1 - cx = 0 \quad \Leftrightarrow \quad x = 1/c$. If $c < 0$ then this represents a minimum of $f(1/c) = 1/(ce)$, since $f'(x)$ changes from negative to positive at $x = 1/c$; and if $c > 0$, it represents a maximum. As $|c|$ increases, the maximum or minimum gets closer to the origin. To find the inflection points, we differentiate again: $f'(x) = e^{-cx}(1 - cx) \quad \Rightarrow$ $f''(x) = e^{-cx}(-c) + (1 - cx)\left(-ce^{-cx}\right) = (cx - 2)ce^{-cx}$. This changes sign when $cx - 2 = 0 \quad \Leftrightarrow \quad x = 2/c$. So as $|c|$ increases, the points of inflection get closer to the origin.

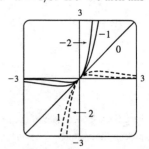

37. (a) $f(x) = cx^4 - 2x^2 + 1$. For $c = 0$, $f(x) = -2x^2 + 1$, a parabola whose vertex, $(0, 1)$, is the absolute maximum. For $c > 0$, $f(x) = cx^4 - 2x^2 + 1$ opens upward with two minimum points. As $c \to 0$, the minimum points spread apart and move downward; they are below the x-axis for $0 < c < 1$ and above for $c > 1$. For $c < 0$, the graph opens downward, and has an absolute maximum at $x = 0$ and no local minimum.

(b) $f'(x) = 4cx^3 - 4x = 4cx(x^2 - 1/c)$ $(c \neq 0)$. If $c \leq 0$, 0 is the only

critical number. $f''(x) = 12cx^2 - 4$, so $f''(0) = -4$ and there is a

local maximum at $(0, f(0)) = (0, 1)$, which lies on $y = 1 - x^2$. If

$c > 0$, the critical numbers are 0 and $\pm 1/\sqrt{c}$. As before, there is a

local maximum at $(0, f(0)) = (0, 1)$, which lies on $y = 1 - x^2$.

$f''(\pm 1/\sqrt{c}) = 12 - 4 = 8 > 0$, so there is a local minimum at

$x = \pm 1/\sqrt{c}$. Here $f(\pm 1/\sqrt{c}) = c(1/c^2) - 2/c + 1 = -1/c + 1$.

But $(\pm 1/\sqrt{c}, -1/c + 1)$ lies on $y = 1 - x^2$ since

$1 - (\pm 1/\sqrt{c})^2 = 1 - 1/c$.

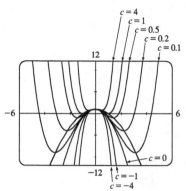

4.7 Optimization Problems

1. (a)

First Number	Second Number	Product
1	22	22
2	21	42
3	20	60
4	19	76
5	18	90
6	17	102
7	16	112
8	15	120
9	14	126
10	13	130
11	12	132

We needn't consider pairs where the first number is larger than the second, since we can just interchange the numbers in such cases. The answer appears to be 11 and 12, but we have considered only integers in the table.

(b) Call the two numbers x and y. Then $x + y = 23$, so $y = 23 - x$. Call the product P. Then $P = xy = x(23 - x) = 23x - x^2$, so we wish to maximize the function $P(x) = 23x - x^2$. Since $P'(x) = 23 - 2x$, we see that $P'(x) = 0$ ⟺ $x = \frac{23}{2} = 11.5$. Thus, the maximum value of P is $P(11.5) = (11.5)^2 = 132.25$ and it occurs when $x = y = 11.5$.

Or: Note that $P''(x) = -2 < 0$ for all x, so P is everywhere concave downward and the local maximum at $x = 11.5$ must be an absolute maximum.

3. The two numbers are x and $\dfrac{100}{x}$, where $x > 0$. Minimize $f(x) = x + \dfrac{100}{x}$. $f'(x) = 1 - \dfrac{100}{x^2} = \dfrac{x^2 - 100}{x^2}$.
The critical number is $x = 10$. Since $f'(x) < 0$ for $0 < x < 10$ and $f'(x) > 0$ for $x > 10$, there is an absolute minimum at $x = 10$. The numbers are 10 and 10.

5. If the rectangle has dimensions x and y, then its perimeter is $2x + 2y = 100$ m, so $y = 50 - x$. Thus, the area is
$A = xy = x(50 - x)$. We wish to maximize the function $A(x) = x(50 - x) = 50x - x^2$, where $0 < x < 50$.
Since $A'(x) = 50 - 2x = -2(x - 25)$, $A'(x) > 0$ for $0 < x < 25$ and $A'(x) < 0$ for $25 < x < 50$. Thus, A has
an absolute maximum at $x = 25$, and $A(25) = 25^2 = 625$ m^2. The dimensions of the rectangle that maximize its
area are $x = y = 25$ m. (The rectangle is a square.)

7. (a)

The areas of the three figures are 12,500, 12,500, and 9000 ft^2. There appears to be a maximum area of at least
12,500 ft^2.

(b) Let x denote the length of each of two sides and three dividers. Let y
denote the length of the other two sides.

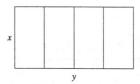

(c) Area $A = $ length $\times$ width $= y \cdot x$

(d) Length of fencing $= 750 \quad \Rightarrow \quad 5x + 2y = 750$

(e) $5x + 2y = 750 \quad \Rightarrow \quad y = 375 - \frac{5}{2}x \quad \Rightarrow \quad A(x) = \left(375 - \frac{5}{2}x\right)x = 375x - \frac{5}{2}x^2$

(f) $A'(x) = 375 - 5x = 0 \quad \Rightarrow \quad x = 75$. Since $A''(x) = -5 < 0$ there is an absolute maximum when $x = 75$.
Then $y = \frac{375}{2} = 187.5$. The largest area is $75\left(\frac{375}{2}\right) = 14,062.5$ ft^2. These values of x and y are between the
values in the first and second figures in part (a). Our original estimate was low.

9.

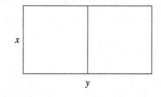

$xy = 1.5 \times 10^6$, so $y = 1.5 \times 10^6/x$. Minimize the amount of fencing,
which is $3x + 2y = 3x + 2(1.5 \times 10^6/x) = 3x + 3 \times 10^6/x = F(x)$.
$F'(x) = 3 - 3 \times 10^6/x^2 = 3(x^2 - 10^6)/x^2$. The critical number is
$x = 10^3$ and $F'(x) < 0$ for $0 < x < 10^3$ and $F'(x) > 0$ if $x > 10^3$, so the
absolute minimum occurs when $x = 10^3$ and $y = 1.5 \times 10^3$.

The field should be 1000 feet by 1500 feet with the middle fence parallel to the short side of the field.

11. Let b be the base of the box and h the height. The surface area is $1200 = b^2 + 4hb \quad \Rightarrow \quad h = (1200 - b^2)/(4b)$.
The volume is $V = b^2h = b^2(1200 - b^2)/4b = 300b - b^3/4 \quad \Rightarrow \quad V'(b) = 300 - \frac{3}{4}b^2$. $V'(b) = 0 \quad \Rightarrow$
$b = \sqrt{400} = 20$. Since $V'(b) > 0$ for $0 < b < 20$ and $V'(b) < 0$ for $b > 20$, there is an absolute maximum when
$b = 20$ by the First Derivative Test for Absolute Maximum or Minimum Values. If $b = 20$, then
$h = (1200 - 20^2)/(4 \cdot 20) = 10$, so the largest possible volume is $b^2h = (20)^2(10) = 4000$ cm^3.

13.

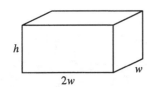

$10 = (2w)(w)h = 2w^2 h$, so $h = 5/w^2$. The cost is

$$C(w) = 10\left(2w^2\right) + 6\left[2\left(2wh\right) + 2hw\right] + 6\left(2w^2\right)$$

$$= 32w^2 + 36wh = 32w^2 + 180/w$$

$$C'(w) = 64w - 180/w^2 = 4\left(16w^3 - 45\right)/w^2 \quad \Rightarrow \quad w = \sqrt[3]{\tfrac{45}{16}} \text{ is the}$$

critical number. $C'(w) < 0$ for $0 < w < \sqrt[3]{\tfrac{45}{16}}$ and $C'(w) > 0$ for $w > \sqrt[3]{\tfrac{45}{16}}$. The minimum cost is

$$C\left(\sqrt[3]{\tfrac{45}{16}}\right) = 32\,(2.8125)^{2/3} + 180 \Big/ \sqrt{2.8125} \approx \$191.28.$$

---

15. The distance from a point (x, y) on the line to the origin is $\sqrt{(x-0)^2 + (y-0)^2} = \sqrt{x^2 + y^2}$.

However, it is easier to work with the *square* of the distance, that is,

$$D(x) = \left(\sqrt{x^2 + y^2}\right)^2 = x^2 + y^2 = x^2 + (4x+7)^2 = 17x^2 + 56x + 49. \text{ Because the distance is positive, its}$$

minimum value will occur at the same point as the minimum value of D.

$D'(x) = 34x + 56$, so $D'(x) = 0 \Leftrightarrow x = -\tfrac{28}{17}$. $D''(x) = 34 > 0$, so D is concave upward for all x. Thus, D

has an absolute minimum at $x = -\tfrac{28}{17}$. The point closest to the origin is $\left(-\tfrac{28}{17}, \tfrac{7}{17}\right)$.

---

17. By symmetry, the points are (x, y) and $(x, -y)$, where $y > 0$. The square of the distance is

$D(x) = (x-2)^2 + y^2 = (x-2)^2 + \left(4 + x^2\right) = 2x^2 - 4x + 8$. So $D'(x) = 4x - 4 = 0 \quad \Rightarrow \quad x = 1$ and

$y = \pm\sqrt{4+1} = \pm\sqrt{5}$. The points are $\left(1, \pm\sqrt{5}\right)$.

19.

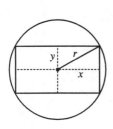

Area of rectangle is $(2x)(2y) = 4xy$. Also $r^2 = x^2 + y^2$ so

$y = \sqrt{r^2 - x^2}$, so the area is $A(x) = 4x\sqrt{r^2 - x^2}$. Now

$$A'(x) = 4\left(\sqrt{r^2 - x^2} - \frac{x^2}{\sqrt{r^2 - x^2}}\right) = 4\frac{r^2 - 2x^2}{\sqrt{r^2 - x^2}}. \text{ The critical number}$$

is $x = \tfrac{1}{\sqrt{2}}r$. Clearly this gives a maximum.

$y = \sqrt{r^2 - \left(\tfrac{1}{\sqrt{2}}r\right)^2} = \sqrt{\tfrac{1}{2}r^2} = \tfrac{1}{\sqrt{2}}r = x$, which tells us that the rectangle

is a square. The dimensions are $2x = \sqrt{2}r$ and $2y = \sqrt{2}r$.

21.

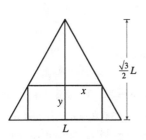

$\dfrac{\tfrac{\sqrt{3}}{2}L - y}{x} = \dfrac{\tfrac{\sqrt{3}}{2}L}{L/2} = \sqrt{3}$ (similar triangles) $\sqrt{3}x = \tfrac{\sqrt{3}}{2}L - y \quad \Rightarrow$

$y = \tfrac{\sqrt{3}}{2}(L - 2x)$. The area of the inscribed rectangle is

$A(x) = (2x)y = \sqrt{3}x(L - 2x)$ where $0 \le x \le L/2$. Now

$0 = A'(x) = \sqrt{3}L - 4\sqrt{3}x \quad \Rightarrow \quad x = \sqrt{3}L\Big/\left(4\sqrt{3}\right) = L/4$. Since

$A(0) = A(L/2) = 0$, the maximum occurs when $x = L/4$, and

$y = \tfrac{\sqrt{3}}{2}L - \tfrac{\sqrt{3}}{4}L = \tfrac{\sqrt{3}}{4}L$, so the dimensions are $L/2$ and $\tfrac{\sqrt{3}}{4}L$.

23.

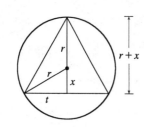

The area of the triangle is

$$A(x) = \tfrac{1}{2}(2t)(r+x) = t(r+x) = \sqrt{r^2 - x^2}\,(r+x). \text{ Then}$$

$$0 = A'(x) = r\frac{-2x}{2\sqrt{r^2 - x^2}} + \sqrt{r^2 - x^2} + x\frac{-2x}{2\sqrt{r^2 - x^2}}$$

$$= -\frac{x^2 + rx}{\sqrt{r^2 - x^2}} + \sqrt{r^2 - x^2} \quad \Rightarrow$$

$$\frac{x^2 + rx}{\sqrt{r^2 - x^2}} = \sqrt{r^2 - x^2} \quad \Rightarrow \quad x^2 + rx = r^2 - x^2 \quad \Rightarrow \quad 0 = 2x^2 + rx - r^2 = (2x - r)(x + r) \quad \Rightarrow \quad x = \tfrac{1}{2}r \text{ or}$$

$x = -r$. Now $A(r) = 0 = A(-r) \quad \Rightarrow \quad$ the maximum occurs where $x = \tfrac{1}{2}r$, so the triangle has height

$r + \tfrac{1}{2}r = \tfrac{3}{2}r$ and base $2\sqrt{r^2 - \left(\tfrac{1}{2}r\right)^2} = 2\sqrt{\tfrac{3}{4}r^2} = \sqrt{3}\,r.$

25.

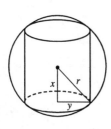

The cylinder has volume $V = \pi y^2 (2x)$. Also $x^2 + y^2 = r^2 \quad \Rightarrow$

$y^2 = r^2 - x^2$, so $V(x) = \pi (r^2 - x^2)(2x) = 2\pi (r^2 x - x^3)$, where

$0 \le x \le r$. $V'(x) = 2\pi (r^2 - 3x^2) = 0 \quad \Rightarrow \quad x = r/\sqrt{3}$. Now

$V(0) = V(r) = 0$, so there is a maximum when $x = r \left/ \sqrt{3}\right.$ and

$$V\left(r\left/\sqrt{3}\right.\right) = \pi (r^2 - r^2/3)\left(2r\left/\sqrt{3}\right.\right) = 4\pi r^3 \left/\left(3\sqrt{3}\right)\right..$$

27.

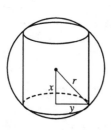

The cylinder has surface area $2\pi y^2 + 2\pi y (2x)$. Now $x^2 + y^2 = r^2 \quad \Rightarrow$

$y = \sqrt{r^2 - x^2}$, so the surface area is

$$S(x) = 2\pi (r^2 - x^2) + 4\pi x \sqrt{r^2 - x^2}, 0 \le x \le r.$$

$$S'(x) = -4\pi x + 4\pi \sqrt{r^2 - x^2} - 4\pi x^2 \!\left/\sqrt{r^2 - x^2}\right.$$

$$= \frac{4\pi \left(r^2 - 2x^2 - x\sqrt{r^2 - x^2}\right)}{\sqrt{r^2 - x^2}} = 0 \quad \Rightarrow$$

$x\sqrt{r^2 - x^2} = r^2 - 2x^2$ (★) $\Rightarrow \quad x^2(r^2 - x^2) = r^4 - 4r^2 x^2 + 4x^4 \quad \Rightarrow \quad 5x^4 - 5r^2 x^2 + r^4 = 0.$ By the

quadratic formula, $x^2 = \frac{5 \pm \sqrt{5}}{10} r^2$, but we reject the root with the + sign since it doesn't satisfy (★). So

$x = \sqrt{\frac{5 - \sqrt{5}}{10}} r$. Since $S(0) = S(r) = 0$, the maximum occurs at the critical number and $x^2 = \frac{5 - \sqrt{5}}{10} r^2 \quad \Rightarrow$

$y^2 = \frac{5 + \sqrt{5}}{10} r^2 \quad \Rightarrow \quad$ the surface area is $2\pi \left(\frac{5 + \sqrt{5}}{10}\right) r^2 + 4\pi \sqrt{\frac{5 - \sqrt{5}}{10}} \sqrt{\frac{5 + \sqrt{5}}{10}} r^2 = \pi r^2 \left(1 + \sqrt{5}\right).$

29.

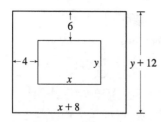

$xy = 384 \quad \Rightarrow \quad y = 384/x$. Total area is

$A(x) = (8 + x)(12 + 384/x) = 12(40 + x + 256/x)$, so

$A'(x) = 12(1 - 256/x^2) = 0 \quad \Rightarrow \quad x = 16$. There is an absolute

minimum when $x = 16$ since $A'(x) < 0$ for $0 < x < 16$ and $A'(x) > 0$

for $x > 16$. When $x = 16$, $y = 384/16 = 24$, so the dimensions are

24 cm and 36 cm.

31.

Let x be the length of the wire used for the square. The total area is

$$A(x) = \left(\frac{x}{4}\right)^2 + \frac{1}{2}\left(\frac{10-x}{3}\right)\frac{\sqrt{3}}{2}\left(\frac{10-x}{3}\right)$$

$$= \frac{1}{16}x^2 + \frac{\sqrt{3}}{36}(10-x)^2 , 0 \le x \le 10$$

$A'(x) = \frac{1}{8}x - \frac{\sqrt{3}}{18}(10-x) = 0 \iff \frac{9}{72}x + \frac{4\sqrt{3}}{72}x - \frac{40\sqrt{3}}{72} = 0 \iff x = \frac{40\sqrt{3}}{9+4\sqrt{3}}$. Now

$A(0) = \left(\frac{\sqrt{3}}{36}\right)100 \approx 4.81$, $A(10) = \frac{100}{16} = 6.25$ and $A\left(\frac{40\sqrt{3}}{9+4\sqrt{3}}\right) \approx 2.72$, so

(a) The maximum area occurs when $x = 10$ m, and all the wire is used for the square.

(b) The minimum area occurs when $x = \frac{40\sqrt{3}}{9+4\sqrt{3}} \approx 4.35$ m.

33.

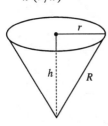

The volume is $V = \pi r^2 h$ and the surface area is

$$S(r) = \pi r^2 + 2\pi rh = \pi r^2 + 2\pi r\left(\frac{V}{\pi r^2}\right) = \pi r^2 + \frac{2V}{r}.$$

$$S'(r) = 2\pi r - \frac{2V}{r^2} = 0 \implies 2\pi r^3 = 2V \implies r = \sqrt[3]{\frac{V}{\pi}} \text{ cm}.$$

This gives an absolute minimum since $S'(r) < 0$ for $0 < r < \sqrt[3]{\frac{V}{\pi}}$ and $S'(r) > 0$ for $r > \sqrt[3]{\frac{V}{\pi}}$. When $r = \sqrt[3]{\frac{V}{\pi}}$,

$$h = \frac{V}{\pi r^2} = \frac{V}{\pi (V/\pi)^{2/3}} = \sqrt[3]{\frac{V}{\pi}} \text{ cm}.$$

35.

$h^2 + r^2 = R^2 \implies V = \frac{\pi}{3}r^2 h = \frac{\pi}{3}(R^2 - h^2)h = \frac{\pi}{3}(R^2 h - h^3)$.

$V'(h) = \frac{\pi}{3}(R^2 - 3h^2) = 0$ when $h = \frac{1}{\sqrt{3}}R$. This gives an absolute

maximum, since $V'(h) > 0$ for $0 < h < \frac{1}{\sqrt{3}}R$ and $V'(h) < 0$ for

$h > \frac{1}{\sqrt{3}}R$. Maximum volume is

$$V\left(\frac{1}{\sqrt{3}}R\right) = \frac{\pi}{3}\left(\frac{1}{\sqrt{3}}R^3 - \frac{1}{3\sqrt{3}}R^3\right) = \frac{2}{9\sqrt{3}}\pi R^3.$$

37. $S = 6sh - \frac{3}{2}s^2 \cot\theta + 3s^2\frac{\sqrt{3}}{2}\csc\theta$

(a) $\frac{dS}{d\theta} = \frac{3}{2}s^2 \csc^2\theta - 3s^2\frac{\sqrt{3}}{2}\csc\theta\cot\theta$ or $\frac{3}{2}s^2\csc\theta\left(\csc\theta - \sqrt{3}\cot\theta\right)$.

(b) $\frac{dS}{d\theta} = 0$ when $\csc\theta - \sqrt{3}\cot\theta = 0 \implies \frac{1}{\sin\theta} - \sqrt{3}\frac{\cos\theta}{\sin\theta} = 0 \implies \cos\theta = \frac{1}{\sqrt{3}}$. The First Derivative Test

shows that the minimum surface area occurs when $\theta = \cos^{-1}\frac{1}{\sqrt{3}} \approx 55°$.

(c)

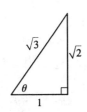

If $\cos\theta = \frac{1}{\sqrt{3}}$, then $\cot\theta = \frac{1}{\sqrt{2}}$ and $\csc\theta = \frac{\sqrt{3}}{\sqrt{2}}$, so the surface area

is $S = 6sh - \frac{3}{2}s^2\frac{1}{\sqrt{2}} + 3s^2\frac{\sqrt{3}}{2}\frac{\sqrt{3}}{\sqrt{2}} = 6sh - \frac{3}{2\sqrt{2}}s^2 + \frac{9}{2\sqrt{2}}s^2 = 6s\left(h + \frac{1}{2\sqrt{2}}s\right)$.

39. Here $T(x) = \dfrac{\sqrt{x^2+25}}{6} + \dfrac{5-x}{8}, 0 \le x \le 5 \Rightarrow T'(x) = \dfrac{x}{6\sqrt{x^2+25}} - \dfrac{1}{8} = 0 \Leftrightarrow 8x = 6\sqrt{x^2+25} \Leftrightarrow$

$16x^2 = 9(x^2+25) \Leftrightarrow x = \dfrac{15}{\sqrt{7}}$. But $\dfrac{15}{\sqrt{7}} > 5$, so T has no critical number. Since $T(0) \approx 1.46$ and

$T(5) \approx 1.18$, he should row directly to B.

41.

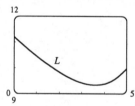

The total illumination is $I(x) = \dfrac{3k}{x^2} + \dfrac{k}{(10-x)^2}, 0 < x < 10$. Then

$I'(x) = \dfrac{-6k}{x^3} + \dfrac{2k}{(10-x)^3} = 0 \Rightarrow 6k(10-x)^3 = 2kx^3 \Rightarrow$

$\sqrt[3]{3}(10-x) = x \Rightarrow x = \dfrac{10\sqrt[3]{3}}{1+\sqrt[3]{3}} \approx 5.9$ ft. This gives a minimum

since $I''(x) > 0$ for $0 < x < 10$.

43.

Here $s^2 = h^2 + b^2/4$, so $h^2 = s^2 - b^2/4$. The area is $A = \frac{1}{2}b\sqrt{s^2 - b^2/4}$. Let the

perimeter be p, so $2s + b = p$ or $s = (p-b)/2 \Rightarrow$

$A(b) = \frac{1}{2}b\sqrt{(p-b)^2/4 - b^2/4} = b\sqrt{p^2 - 2pb}/4$. Now

$A'(b) = \dfrac{\sqrt{p^2-2pb}}{4} - \dfrac{bp/4}{\sqrt{p^2-2pb}} = \dfrac{-3pb+p^2}{4\sqrt{p^2-2pb}}$. Therefore, $A'(b) = 0$

$\Rightarrow -3pb + p^2 = 0 \Rightarrow b = p/3$. Since $A'(b) > 0$ for $b < p/3$ and $A'(b) < 0$ for $b > p/3$, there is an absolute

maximum when $b = p/3$. But then $2s + p/3 = p$ so $s = p/3 \Rightarrow s = b \Rightarrow$ the triangle is equilateral.

45. $L(x) = |AP| + |BP| + |CP| = x + \sqrt{(5-x)^2 + 2^2} + \sqrt{(5-x)^2 + 3^2}$

$= x + \sqrt{x^2 - 10x + 29} + \sqrt{x^2 - 10x + 34} \Rightarrow$

$L'(x) = 1 + \dfrac{x-5}{\sqrt{x^2-10x+29}} + \dfrac{x-5}{\sqrt{x^2-10x+34}}$

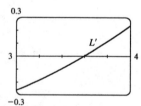

From the graphs of L and L', it seems that the minimum value of L is about $L(3.59) = 9.35$ m.

47.

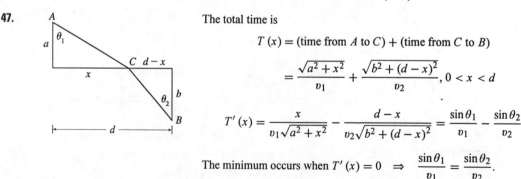

The total time is

$T(x) = (\text{time from } A \text{ to } C) + (\text{time from } C \text{ to } B)$

$= \dfrac{\sqrt{a^2+x^2}}{v_1} + \dfrac{\sqrt{b^2+(d-x)^2}}{v_2}, 0 < x < d$

$T'(x) = \dfrac{x}{v_1\sqrt{a^2+x^2}} - \dfrac{d-x}{v_2\sqrt{b^2+(d-x)^2}} = \dfrac{\sin\theta_1}{v_1} - \dfrac{\sin\theta_2}{v_2}$

The minimum occurs when $T'(x) = 0 \Rightarrow \dfrac{\sin\theta_1}{v_1} = \dfrac{\sin\theta_2}{v_2}$.

49.

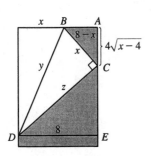

$y^2 = x^2 + z^2$, but triangles CDE and BCA are similar, so

$z/8 = x / (4\sqrt{x-4})$ $\Rightarrow$ $z = 2x / \sqrt{x-4}$. Thus, we minimize

$f(x) = y^2 = x^2 + 4x^2/(x-4) = x^3/(x-4)$, $4 < x \le 8$.

$f'(x) = \dfrac{3x^2(x-4) - x^3}{(x-4)^2} = \dfrac{2x^2(x-6)}{(x-4)^2} = 0$ when $x = 6$. $f'(x) < 0$

when $x < 6$, $f'(x) > 0$ when $x > 6$, so the minimum occurs when

$x = 6$ in.

51.

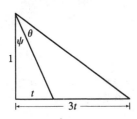

It suffices to maximize $\tan\theta$. Now

$\dfrac{3t}{1} = \tan(\psi + \theta) = \dfrac{\tan\psi + \tan\theta}{1 - \tan\psi\tan\theta} = \dfrac{t + \tan\theta}{1 - t\tan\theta}$. So

$3t(1 - t\tan\theta) = t + \tan\theta$ $\Rightarrow$ $2t = (1 + 3t^2)\tan\theta$ $\Rightarrow$

$\tan\theta = \dfrac{2t}{1 + 3t^2}$. Let $f(t) = \tan\theta = \dfrac{2t}{1 + 3t^2}$ $\Rightarrow$

$f'(t) = \dfrac{2(1 + 3t^2) - 2t(6t)}{(1 + 3t^2)^2} = \dfrac{2(1 - 3t^2)}{(1 + 3t^2)^2} = 0$ $\Leftrightarrow$ $1 - 3t^2 = 0$ $\Leftrightarrow$ $t = \frac{1}{\sqrt{3}}$ since $t \ge 0$. Now $f'(t) > 0$

for $0 \le t < \frac{1}{\sqrt{3}}$ and $f'(t) < 0$ for $t > \frac{1}{\sqrt{3}}$, so f has an absolute maximum when $t = \frac{1}{\sqrt{3}}$ and

$\tan\theta = \dfrac{2(1/\sqrt{3})}{1 + 3(1/\sqrt{3})^2} = \frac{1}{\sqrt{3}}$ $\Rightarrow$ $\theta = \frac{\pi}{6}$. Substituting for t and θ in $3t = \tan(\psi + \theta)$ gives us $\sqrt{3} = \tan\left(\psi + \frac{\pi}{6}\right)$

$\Rightarrow$ $\psi = \frac{\pi}{6}$.

53.

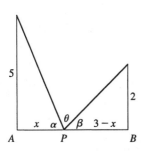

$\tan\alpha = \dfrac{5}{x}$, $\tan\beta = \dfrac{2}{3-x}$ $\Rightarrow$ $\theta = \pi - \tan^{-1}\left(\dfrac{5}{x}\right) - \tan^{-1}\left(\dfrac{2}{3-x}\right)$

$\Rightarrow$ $\dfrac{d\theta}{dx} = -\dfrac{1}{1 + \left(\dfrac{5}{x}\right)^2}\left(-\dfrac{5}{x^2}\right) - \dfrac{1}{1 + \left(\dfrac{2}{3-x}\right)^2}\left[\dfrac{2}{(3-x)^2}\right] \dfrac{d\theta}{dx} = 0$

$\Rightarrow$ $\dfrac{5}{x^2 + 25} = \dfrac{2}{x^2 - 6x + 13}$ $\Rightarrow$ $2x^2 + 50 = 5x^2 - 30x + 65$ $\Rightarrow$

$x^2 - 10x + 5 = 0$ $\Rightarrow$ $x = 5 \pm 2\sqrt{5}$. We reject the root with the $+$ sign,

since it is larger than 3. $d\theta/dx > 0$ for $x < 5 - 2\sqrt{5}$ and $d\theta/dx < 0$ for

$x > 5 - 2\sqrt{5}$, so θ is maximized when $|AP| = x = 5 - 2\sqrt{5} \approx 0.53$.

55.

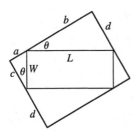

$a = W\sin\theta$, $c = W\cos\theta$, $b = L\cos\theta$, $d = L\sin\theta$, so the area of the

circumscribed rectangle is

$A(\theta) = (a + b)(c + d) = (W\sin\theta + L\cos\theta)(W\cos\theta + L\sin\theta)$

$= LW\sin^2\theta + LW\cos^2\theta + \left(L^2 + W^2\right)\sin\theta\cos\theta$

$= LW + \frac{1}{2}\left(L^2 + W^2\right)\sin 2\theta$, $0 \le \theta \le \frac{\pi}{2}$

This expression shows, without calculus, that the maximum value of $A(\theta)$ occurs when $\sin 2\theta = 1$ $\Leftrightarrow$ $2\theta = \frac{\pi}{2}$

$\Rightarrow$ $x = \frac{\pi}{4}$. So the maximum area is $A\left(\frac{\pi}{4}\right) = LW + \frac{1}{2}\left(L^2 + W^2\right) = \frac{1}{2}(L + W)^2$.

57. (a)

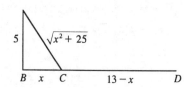

If $k =$ energy/km over land, then energy/km over water $= 1.4k$. So the total energy is

$$E = 1.4k\sqrt{25 + x^2} + k(13 - x),\ 0 \le x \le 13,\ \text{and so}\ \frac{dE}{dx} = \frac{1.4kx}{(25 + x^2)^{1/2}} - k.\ \text{Set}\ \frac{dE}{dx} = 0:$$

$1.4kx = k(25 + x^2)^{1/2} \Rightarrow 1.96x^2 = x^2 + 25 \Rightarrow 0.96x^2 = 25 \Rightarrow x = \frac{5}{\sqrt{0.96}} \approx 5.1$. Testing against the value of E at the endpoints: $E(0) = 1.4k(5) + 13k = 20k$, $E(5.1) \approx 17.9k$, $E(13) \approx 19.5k$. Thus, to minimize energy, the bird should fly to a point about 5.1 km from B.

(b) If W/L is large, the bird would fly to a point C that is closer to B than to D to minimize the energy used flying over water. If W/L is small, the bird would fly to a point C that is closer to D than to B to minimize the distance of the flight. $E = W\sqrt{25 + x^2} + L(13 - x) \Rightarrow \frac{dE}{dx} = \frac{Wx}{\sqrt{25 + x^2}} - L = 0$ when

$\frac{W}{L} = \frac{\sqrt{25 + x^2}}{x}$. By the same sort of argument as in part (a), this ratio will give the minimal expenditure of energy if the bird heads for the point x km from B.

(c) For flight direct to D, $x = 13$, so from part (b), $W/L = \frac{\sqrt{25 + 13^2}}{13} \approx 1.07$. There is no value of W/L for which the bird should fly directly to B. But note that $\lim_{x \to 0^+} (W/L) = \infty$, so if the point at which E is a minimum is close to B, then W/L is large.

(d) Assuming that the birds instinctively choose the path that minimizes the energy expenditure, we can use the equation for $dE/dx = 0$ from part (a) with $1.4k = c$, $x = 4$, and $k = 1$: $c(4) = 1 \cdot (25 + 4^2)^{1/2} \Rightarrow c = \sqrt{41}/4 \approx 1.6$.

4.8 Applications to Economics

1. (a) $C(0)$ represents the fixed costs of production, such as rent, utilities, machinery etc., which are incurred even when nothing is produced.

(b) The inflection point is the point at which $C''(x)$ changes from negative to positive, that is, the marginal cost $C'(x)$ changes from decreasing to increasing. So the marginal cost is minimized.

(c) The marginal cost function is $C'(x)$. We graph it as in Example 1 in Section 2.9.

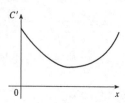

3. $c(x) = 21.4 - 0.002x$ and $c(x) = C(x)/x \Rightarrow C(x) = 21.4x - 0.002x^2$. $C'(x) = 21.4 - 0.004x$ and $C'(1000) = 17.4$. This is an estimate of the cost of producing the 1001st unit.

5. (a) The cost function is $C(x) = 40,000 + 300x + x^2$, so the cost at a production level of 1000 is
$C(1000) = \$1,340,000$. The average cost function is $c(x) = \dfrac{C(x)}{x} = \dfrac{40,000}{x} + 300 + x$ and
$c(1000) = \$1340/\text{unit}$. The marginal cost function is $C'(x) = 300 + 2x$ and $C'(1000) = \$2300/\text{unit}$.

(b) We must have $C'(x) = c(x)$ $\Leftrightarrow$ $300 + 2x = \dfrac{40,000}{x} + 300 + x$ $\Leftrightarrow$ $x = \dfrac{40,000}{x}$ $\Rightarrow$ $x^2 = 40,000$

$\Rightarrow$ $x = \sqrt{40,000} = 200$. This gives a minimum value of the average cost function $c(x)$ since
$c''(x) = \dfrac{80,000}{x^3} > 0$.

(c) The minimum average cost is $c(200) = \$700/\text{unit}$.

7. (a) $C(x) = 45 + \dfrac{x}{2} + \dfrac{x^2}{560}$, $C(1000) = \$2330.71$. $c(x) = \dfrac{45}{x} + \dfrac{1}{2} + \dfrac{x}{560}$, $c(1000) = \$2.33/\text{unit}$.

$C'(x) = \dfrac{1}{2} + \dfrac{x}{280}$, $C'(1000) = \$4.07/\text{unit}$.

(b) We must have $C'(x) = c(x)$ $\Rightarrow$ $\dfrac{1}{2} + \dfrac{x}{280} = \dfrac{45}{x} + \dfrac{1}{2} + \dfrac{x}{560}$ $\Rightarrow$ $\dfrac{45}{x} = \dfrac{x}{560}$ $\Rightarrow$ $x^2 = (45)(560)$

$\Rightarrow$ $x = \sqrt{25,200} \approx 159$. This is a minimum since $c''(x) = 90/x^2 > 0$.

(c) The minimum average cost is $c(159) = \$1.07/\text{unit}$.

9. (a) $C(x) = 2\sqrt{x} + \dfrac{x^2}{8000}$, $C(1000) = \$188.25$. $c(x) = \dfrac{2}{\sqrt{x}} + \dfrac{x}{8000}$, $c(1000) = \$0.19/\text{unit}$.

$C'(x) = \dfrac{1}{\sqrt{x}} + \dfrac{x}{4000}$, $C'(1000) = \$0.28/\text{unit}$.

(b) We must have $C'(x) = c(x)$ $\Rightarrow$ $\dfrac{1}{\sqrt{x}} + \dfrac{x}{4000} = \dfrac{2}{\sqrt{x}} + \dfrac{x}{8000}$ $\Rightarrow$ $\dfrac{x}{8000} = \dfrac{1}{\sqrt{x}}$ $\Rightarrow$ $x^{3/2} = 8000$ $\Rightarrow$

$x = (8000)^{2/3} = 400$. This is a minimum since $c''(x) = \frac{3}{2}x^{-5/2} > 0$.

(c) The minimum average cost is $c(400) = \$0.15/\text{unit}$.

11. (a) $C(x) = 3700 + 5x - 0.04x^2 + 0.0003x^3$ $\Rightarrow$ $C'(x) = 5 - 0.08x + 0.0009x^2$ (marginal cost).
$c(x) = \dfrac{C(x)}{x} = \dfrac{3700}{x} + 5 - 0.04x + 0.0003x^2$ (average cost).

(b)

(c) $c'(x) = -\dfrac{3700}{x^2} - 0.04 + 0.0006x = 0$ $\Rightarrow$

$x_1 = 208.51$. $c(x_1) \approx \$27.45/\text{unit}$.

(d) $C''(x) = -0.08 + 0.0018x = 0$ $\Rightarrow$

$x_1 = \frac{800}{18} = 44.4\overline{4}$. $C'(x_1) = \$3.22/\text{unit}$.

$C''(x) = 0.0018 > 0$ for all x, so this is the
minimum marginal cost.

The graphs intersect at $(208.51, 27.45)$, so the
production level that minimizes average cost
is about 209 units.

13. $C(x) = 680 + 4x + 0.01x^2$, $p(x) = 12$ $\Rightarrow$ $R(x) = xp(x) = 12x$. If the profit is maximum, then
$R'(x) = C'(x)$ $\Rightarrow$ $12 = 4 + 0.02x$ $\Rightarrow$ $0.02x = 8$ $\Rightarrow$ $x = 400$. The profit is maximized if $P''(x) < 0$,
but since $P''(x) = R''(x) - C''(x)$, we can just check the condition $R''(x) < C''(x)$. Now
$R''(x) = 0 < 0.02 = C''(x)$, so $x = 400$ gives a maximum.

15. $C(x) = 1450 + 36x - x^2 + 0.001x^3$, $p(x) = 60 - 0.01x$. Then $R(x) = xp(x) = 60x - 0.01x^2$. If the profit is maximum, then $R'(x) = C'(x)$ $\Leftrightarrow$ $60 - 0.02x = 36 - 2x + 0.003x^2$ $\Rightarrow$ $0.003x^2 - 1.98x - 24 = 0$. By the quadratic formula, $x = \dfrac{1.98 \pm \sqrt{(-1.98)^2 + 4(0.003)(24)}}{2(0.003)} = \dfrac{1.98 \pm \sqrt{4.2084}}{0.006}$. Since $x > 0$, $x \approx (1.98 + 2.05)/0.006 \approx 672$. Now $R''(x) = -0.02$ and $C''(x) = -2 + 0.006x$ $\Rightarrow$ $C''(672) = 2.032$ $\Rightarrow$ $R''(672) < C''(672)$ $\Rightarrow$ there is a maximum at $x = 672$.

17. $C(x) = 0.001x^3 - 0.3x^2 + 6x + 900$. The marginal cost is $C'(x) = 0.003x^2 - 0.6x + 6$. $C'(x)$ is increasing when $C''(x) > 0$ $\Leftrightarrow$ $0.006x - 0.6 > 0$ $\Leftrightarrow$ $x > 0.6/0.006 = 100$. So $C'(x)$ starts to increase when $x = 100$.

19. (a) $C(x) = 1200 + 12x - 0.1x^2 + 0.0005x^3$. $R(x) = xp(x) = 29x - 0.00021x^2$.

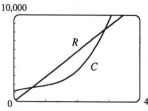

10,000

0   400

Since the profit is maximized when $R'(x) = C'(x)$, we examine the curves R and C in the figure, looking for x-values at which the slopes of the tangent lines are equal. It appears that $x = 200$ is a good estimate.

(b) $R'(x) = C'(x)$ $\Rightarrow$ $29 - 0.00042x = 12 - 0.2x + 0.0015x^2$ $\Rightarrow$ $0.0015x^2 - 0.19958x - 17 = 0$ $\Rightarrow$ $x \approx 192.06$ (for $x > 0$). As in Exercise 13, $R''(x) < C''(x)$ $\Rightarrow$ $-0.00042 < -0.2 + 0.003x$ $\Leftrightarrow$ $0.003x > 0.19958$ $\Leftrightarrow$ $x > 66.5$. Our value of 192 is in this range, so we have a maximum profit when we produce 192 yards of fabric.

21. (a) We are given that the demand function p is linear and $p(27{,}000) = 10$, $p(33{,}000) = 8$, so the slope is $\dfrac{10 - 8}{27{,}000 - 33{,}000} = -\dfrac{1}{3000}$ and an equation of the line is $y - 10 = \left(-\dfrac{1}{3000}\right)(x - 27{,}000)$ $\Rightarrow$ $y = p(x) = -\dfrac{1}{3000}x + 19 = 19 - x/3000$.

(b) The revenue is $R(x) = xp(x) = 19x - x^2/3000$ $\Rightarrow$ $R'(x) = 19 - x/1500 = 0$ when $x = 28{,}500$. Since $R''(x) = -1/1500 < 0$, the maximum revenue occurs when $x = 28{,}500$ $\Rightarrow$ the price is $p(28{,}500) = \$9.50$.

23. (a) As in Example 3, we see that the demand function p is linear. We are given that $p(1000) = 450$ and deduce that $p(1100) = 440$, since a \$10 reduction in price increases sales by 100 per week. The slope for p is $\dfrac{440 - 450}{1100 - 1000} = -\dfrac{1}{10}$, so an equation is $p - 450 = -\dfrac{1}{10}(x - 1000)$ or $p(x) = -\dfrac{1}{10}x + 550$.

(b) $R(x) = xp(x) = 500x - x^2/10$. $R'(x) = 550 - x/5 = 0$ when $x = 5(550) = 2750$. $p(2750) = 275$, so the rebate should be $450 - 275 = \$175$.

(c) $C(x) = 68{,}000 + 150x$ $\Rightarrow$
$P(x) = R(x) - C(x) = 550x - x^2/10 - 68{,}000 - 150x = 400x - x^2/10 - 68{,}000$, $P'(x) = 400 - x/5 = 0$ when $x = 2000$. $p(2000) = 350$. Therefore, the rebate to maximize profits should be $450 - 350 = \$100$.

4.9 Newton's Method

1.

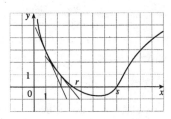

The tangent line at $x = 1$ intersects the x-axis at $x \approx 2.3$, so $x_2 \approx 2.3$. The tangent line at $x = 2.3$ intersects the x-axis at $x \approx 3$, so $x_3 \approx 3.0$.

3. Since $x_1 = 3$ and $y = 5x - 4$ is tangent to $y = f(x)$ at $x = 3$, we simply need to find where the tangent line intersects the x-axis. $y = 0 \Rightarrow 5x_2 - 4 = 0 \Rightarrow x_2 = \frac{4}{5}$.

5. $f(x) = x^3 + x + 1 \Rightarrow f'(x) = 3x^2 + 1$, so $x_{n+1} = x_n - \dfrac{x_n^3 + x_n + 1}{3x_n^2 + 1}$. $x_1 = -1 \Rightarrow$

$x_2 = -1 - \dfrac{-1 - 1 + 1}{3 \cdot 1 + 1} = -0.75 \Rightarrow x_3 = -0.75 - \dfrac{(-0.75)^3 - 0.75 + 1}{3(-0.75)^2 + 1} \approx -0.6860$. Here is a quick and

easy method for finding the iterations on a programmable calculator. (The screens shown are from the TI-82, but the method is similar on other calculators.) Assign $x^3 + x + 1$ to Y_1 and $3x^2 + 1$ to Y_2. Now store -1 in X and then enter $X - Y_1/Y_2 \to X$ to get -0.75. By successively pressing the ENTER key, you get the approximations $x_1, x_2, x_3, \ldots$.

```
Y1■X^3+X+1
Y2■3X^2+1
Y3=■
Y4=
Y5=
Y6=
Y7=
Y8=
```

```
-1→X
              -1
X-Y1/Y2→X
            -.75
   -.6860465116
   -.6823395826
   -.6823278039
```

7. $f(x) = x^4 - 20 \Rightarrow f'(x) = 4x^3$, so $x_{n+1} = x_n - \dfrac{x_n^4 - 20}{4x_n^3}$. $x_1 = 2 \Rightarrow x_2 = 2 - \dfrac{2^4 - 20}{4(2)^3} = 2.1250 \Rightarrow$

$x_3 = 2.125 - \dfrac{(2.125)^4 - 20}{4(2.125)^3} = 2.1148$.

9. To approximate $x = \sqrt[3]{30}$ (so that $x^3 = 30$), we can take $f(x) = x^3 - 30$. So $f'(x) = 3x^2$, and thus,

$x_{n+1} = x_n - \dfrac{x_n^3 - 30}{3x_n^2}$. Since $\sqrt[3]{27} = 3$ and 27 is close to 30, we'll use $x_1 = 3$. We need to find approximations

until they agree to eight decimal places. $x_1 = 3 \Rightarrow x_2 \approx 3.11111111$, $x_3 \approx 3.10723734$,

$x_4 \approx 3.10723251 \approx x_5$. So $\sqrt[3]{30} \approx 3.10723251$, to eight decimal places.

11. $f(x) = 2x^3 - 6x^2 + 3x + 1 \Rightarrow f'(x) = 6x^2 - 12x + 3 \Rightarrow x_{n+1} = x_n - \dfrac{2x_n^3 - 6x_n^2 + 3x_n + 1}{6x_n^2 - 12x_n + 3}$. We need

to find approximations until they agree to six decimal places. $x_1 = 2.5 \Rightarrow x_2 \approx 2.285714$, $x_3 \approx 2.228824$,

$x_4 \approx 2.224765$, $x_5 \approx 2.224745 \approx x_6$. So the root is 2.224745, to six decimal places.

13.

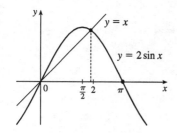

From the graph it appears that there is a root near 2, so we take $x_1 = 2$.
Write the equation as $f(x) = 2\sin x - x = 0$. Then $f'(x) = 2\cos x - 1$,
so $x_{n+1} = x_n - \dfrac{2\sin x_n - x_n}{2\cos x_n - 1}$ $\Rightarrow$ $x_1 = 2$, $x_2 \approx 1.900996$,
$x_3 \approx 1.895512$, $x_4 \approx 1.895494 \approx x_5$. So the root is 1.895494, to six
decimal places.

15.

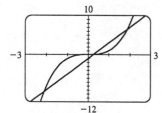

From the graph, we see that $y = x^3$ and $y = 4x - 1$ intersect three times.
Good first approximations are $x = -2$, $x = 0$, and $x = 2$.
$f(x) = x^3 - 4x + 1$ $\Rightarrow$ $f'(x) = 3x^2 - 4$, so
$$x_{n+1} = x_n - \frac{x_n^3 - 4x_n + 1}{3x_n^2 - 4}.$$

$x_1 = -2$	$x_1 = 0$	$x_1 = 2$
$x_2 = -2.125$	$x_2 = 0.25$	$x_2 = 1.875$
$x_3 \approx -2.114975$	$x_3 \approx 0.254098$	$x_3 \approx 1.860979$
$x_4 \approx -2.114908 \approx x_5$	$x_4 \approx 0.254102 \approx x_5$	$x_4 \approx 1.860806 \approx x_5$

To six decimal places, the roots are -2.114908, 0.254102, and 1.860806.

17.

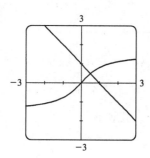

From the graph, there appears to be a point of intersection near $x = 0.5$.
Now solving $\tan^{-1} x = 1 - x$ is the same as solving
$f(x) = \tan^{-1} x + x - 1 = 0$. $f(x) = \tan^{-1} x + x - 1$ $\Rightarrow$
$f'(x) = \dfrac{1}{1 + x^2} + 1$, so $x_{n+1} = x_n - \dfrac{\tan^{-1} x_n + x_n - 1}{1/(1 + x_n^2) + 1}$.
$x_1 = 0.5$ $\Rightarrow$ $x_2 \approx 0.520196$, $x_3 \approx 0.520269 \approx x_4$. So to six decimal
places, the root is 0.520269.

19.

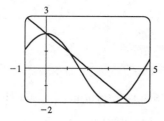

From the graph and by inspection, $x = 0$ is a root. Also, $y = 2\cos x$ and
$y = 2 - x$ intersect at $x \approx 1$ and at $x \approx 3.5$. $f(x) = 2\cos x + x - 2$ $\Rightarrow$
$f'(x) = -2\sin x + 1$, so $x_{n+1} = x_n - \dfrac{2\cos x_n + x_n - 2}{-2\sin x_n + 1}$.

$x_1 = 1$	$x_1 = 3.5$
$x_2 \approx 1.118026$	$x_2 \approx 3.719159$
$x_3 \approx 1.109188$	$x_3 \approx 3.698331$
$x_4 \approx 1.109144 \approx x_5$	$x_4 \approx 3.698154 \approx x_5$

To six decimal places, the roots are 0, 1.109144, and 3.698154.

21.

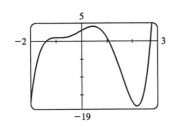

$f(x) = x^5 - x^4 - 5x^3 - x^2 + 4x + 3 \implies$
$f'(x) = 5x^4 - 4x^3 - 15x^2 - 2x + 4 \implies$
$x_{n+1} = x_n - \dfrac{x_n^5 - x_n^4 - 5x_n^3 - x_n^2 + 4x_n + 3}{5x_n^4 - 4x_n^3 - 15x_n^2 - 2x_n + 4}$. From the graph of f, there

appear to be roots near -1.4, 1.1, and 2.7.

$x_1 = -1.4$	$x_1 = 1.1$	$x_1 = 2.7$
$x_2 \approx -1.39210970$	$x_2 = 1.07780402$	$x_2 \approx 2.72046250$
$x_3 \approx -1.39194698$	$x_3 \approx 1.07739442$	$x_3 \approx 2.71987870$
$x_4 \approx -1.39194691 \approx x_5$	$x_4 \approx 1.07739428 \approx x_5$	$x_4 \approx 2.71987822 \approx x_5$

To eight decimal places, the roots are -1.39194691, 1.07739428, and 2.71987822.

23.

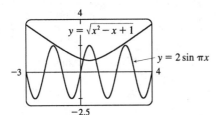

From the graph, we see that there are roots of this equation

near 0.2 and 0.8. $f(x) = \sqrt{x^2 - x + 1} - 2\sin \pi x \implies$

$f'(x) = \dfrac{2x - 1}{2\sqrt{x^2 - x + 1}} - 2\pi \cos \pi x$, so

$x_{n+1} = x_n - \dfrac{\sqrt{x_n^2 - x_n + 1} - 2\sin \pi x_n}{\dfrac{2x_n - 1}{2\sqrt{x_n^2 - x_n + 1}} - 2\pi \cos \pi x_n}$.

Taking $x_1 = 0.2$, we get $x_2 \approx 0.15212015$, $x_3 \approx 0.15438067$, $x_4 \approx 0.15438500 \approx x_5$. Taking $x_1 = 0.8$, we get
$x_2 \approx 0.84787985$, $x_3 \approx 0.84561933$, $x_4 \approx 0.84561500 \approx x_5$. So, to eight decimal places, the roots of the equation
are 0.15438500 and 0.84561500.

25.

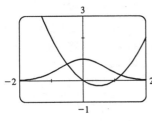

From the graph, we see that $y = e^{-x^2}$ and $y = x^2 - x$ intersect twice.
Good first approximations are $x = -0.5$ and $x = 1.1$.
$f(x) = e^{-x^2} - x^2 + x \implies f'(x) = -2xe^{-x^2} - 2x + 1$, so
$x_{n+1} = x_n - \dfrac{e^{-x_n^2} - x_n^2 + x_n}{-2x_ne^{-x_n^2} - 2x_n + 1}$.

$x_1 = -0.5$	$x_1 = 1.1$
$x_2 \approx -0.51036446$	$x_2 \approx 1.20139754$
$x_3 \approx -0.51031156 \approx x_4$	$x_3 \approx 1.19844118$
	$x_4 \approx 1.19843871 \approx x_5$

To eight decimal places, the roots are -0.51031156 and 1.19843871.

27. (a) $f(x) = x^2 - a \implies f'(x) = 2x$, so Newton's method gives

$x_{n+1} = x_n - \dfrac{x_n^2 - a}{2x_n} = x_n - \dfrac{1}{2}x_n + \dfrac{a}{2x_n} = \dfrac{1}{2}x_n + \dfrac{a}{2x_n} = \dfrac{1}{2}\left(x_n + \dfrac{a}{x_n}\right)$.

(b) Using (a) with $a = 1000$ and $x_1 = \sqrt{900} = 30$, we get $x_2 \approx 31.666667$, $x_3 \approx 31.622807$, and
$x_4 \approx 31.622777 \approx x_5$. So $\sqrt{1000} \approx 31.622777$.

29. $f(x) = x^3 - 3x + 6 \implies f'(x) = 3x^2 - 3$. If $x_1 = 1$, then $f'(x_1) = 0$ and the tangent line used for
approximating x_2 is horizontal. Attempting to find x_2 results in trying to divide by zero.

31. For $f(x) = x^{1/3}$, $f'(x) = \frac{1}{3}x^{-2/3}$ and

$$x_{n+1} = x_n - \frac{f(x_n)}{f'(x_n)} = x_n - \frac{x_n^{1/3}}{\frac{1}{3}x_n^{-2/3}} = x_n - 3x_n = -2x_n. \text{ Therefore,}$$

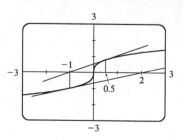

each successive approximation becomes twice as large as the previous one in absolute value, so the sequence of approximations fails to converge to the root, which is 0. In the figure, we have $x_1 = 0.5$, $x_2 = -2(0.5) = -1$, and $x_3 = -2(-1) = 2$.

33. (a) $f(x) = 3x^4 - 28x^3 + 6x^2 + 24x \Rightarrow f'(x) = 12x^3 - 84x^2 + 12x + 24 \Rightarrow$

$f''(x) = 36x^2 - 168x + 12$. Now to solve $f'(x) = 0$, try $x_1 = \frac{1}{2} \Rightarrow x_2 = x_1 - \frac{f'(x_1)}{f''(x_1)} = \frac{2}{3} \Rightarrow$

$x_3 \approx 0.6455 \Rightarrow x_4 \approx 0.6452 \Rightarrow x_5 \approx 0.6452$. Now try $x_1 = 6 \Rightarrow x_2 = 7.12 \Rightarrow x_3 \approx 6.8353$

$\Rightarrow x_4 \approx 6.8102 \Rightarrow x_5 \approx 6.8100$. Finally try $x_1 = -0.5 \Rightarrow x_2 \approx -0.4571 \Rightarrow x_3 \approx -0.4552 \Rightarrow$

$x_4 \approx -0.4552$. Therefore, $x = -0.455$, 6.810 and 0.645 are all critical numbers correct to three decimal places.

(b) $f(-1) = 13$, $f(7) = -1939$, $f(6.810) \approx -1949.07$, $f(-0.455) \approx -6.912$, $f(0.645) \approx 10.982$.

Therefore, $f(6.810) \approx -1949.07$ is the absolute minimum correct to two decimal places.

35.

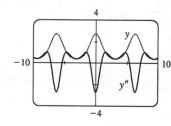

From the figure, we see that $y = f(x) = e^{\cos x}$ is periodic with period 2π. To find the x-coordinates of the IP, we only need to approximate the zeros of y'' on $[0, \pi]$. $f'(x) = -e^{\cos x}\sin x \Rightarrow$

$f''(x) = e^{\cos x}(\sin^2 x - \cos x)$. Since $e^{\cos x} \neq 0$, we will use Newton's method with $g(x) = \sin^2 x - \cos x$, $g'(x) = 2\sin x \cos x + \sin x$, and $x_1 = 1$. $x_2 \approx 0.904173$, $x_3 \approx 0.904557 \approx x_4$. Thus, $(0.904557, 1.855277)$ is the IP.

37.

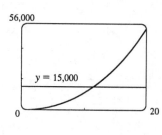

The volume of the silo, in terms of its radius, is

$$V(r) = \pi r^2 (30) + \frac{1}{2}\left(\frac{4}{3}\pi r^3\right) = 30\pi r^2 + \frac{2}{3}\pi r^3.$$

From a graph of V, we see that $V(r) = 15{,}000$ at $r \approx 11$ ft. Now we use Newton's method to solve the equation $V(r) - 15{,}000 = 0$.

$\frac{dV}{dr} = 60\pi r + 2\pi r^2$, so $r_{n+1} = r_n - \dfrac{30\pi r_n^2 + \frac{2}{3}\pi r_n^3 - 15{,}000}{60\pi r_n + 2\pi r_n^2}$. Taking

$r_1 = 11$, we get $r_2 = 11.2853$, $r_3 = 11.2807 \approx r_4$. So in order for the silo to hold $15{,}000$ ft^3 of grain, its radius must be about 11.2807 ft.

39. In this case, $A = 18,000$, $R = 375$, and $n = 5(12) = 60$. So the formula becomes $18,000 = \dfrac{375}{x}\left[1 - (1+x)^{-60}\right]$

$\Leftrightarrow \quad 48x = 1 - (1+x)^{-60} \quad \Leftrightarrow \quad 48x(1+x)^{60} - (1+x)^{60} + 1 = 0$.

Let the LHS be called $f(x)$, so that

$$f'(x) = 48x(60)(1+x)^{59} + 48(1+x)^{60} - 60(1+x)^{59}$$

$$= 12(1+x)^{59}[4x(60) + 4(1+x) - 5] = 12(1+x)^{59}(244x - 1)$$

$x_{n+1} = x_n - \dfrac{48x_n(1+x_n)^{60} - (1+x_n)^{60} + 1}{12(1+x_n)^{59}(244x_n - 1)}$. An interest rate of 1%/month seems like a reasonable estimate

for $x = i$. So let $x_1 = 1\% = 0.01$, and we get $x_2 = 0.0082202$, $x_3 \approx 0.0076802$, $x_4 \approx 0.0076291$,
$x_5 \approx 0.0076286 \approx x_6$. Thus, the dealer is charging a monthly interest rate of 0.76286% (or 9.55%/year, compounded monthly).

4.10 Antiderivatives

1. $f(x) = 6x^2 - 8x + 3 \quad \Rightarrow \quad F(x) = 6\dfrac{x^{2+1}}{2+1} - 8\dfrac{x^{1+1}}{1+1} + 3x + C = 2x^3 - 4x^2 + 3x + C$

Check: $F'(x) = 2 \cdot 3x^2 - 4 \cdot 2x + 3 + 0 = 6x^2 - 8x + 3 = f(x)$

3. $f(x) = 1 - x^3 + 5x^5 - 3x^7 \quad \Rightarrow \quad F(x) = x - \dfrac{x^{3+1}}{3+1} + 5\dfrac{x^{5+1}}{5+1} - 3\dfrac{x^{7+1}}{7+1} + C = x - \tfrac{1}{4}x^4 + \tfrac{5}{6}x^6 - \tfrac{3}{8}x^8 + C$

5. $f(x) = 5x^{1/4} - 7x^{3/4} \quad \Rightarrow \quad F(x) = 5\dfrac{x^{1/4+1}}{\frac{1}{4}+1} - 7\dfrac{x^{3/4+1}}{\frac{3}{4}+1} + C = 5\dfrac{x^{5/4}}{5/4} - 7\dfrac{x^{7/4}}{7/4} + C = 4x^{5/4} - 4x^{7/4} + C$

7. $f(x) = \sqrt{x} + \sqrt[3]{x} = x^{1/2} + x^{1/3} \quad \Rightarrow \quad F(x) = \tfrac{1}{3/2}x^{3/2} + \tfrac{1}{4/3}x^{4/3} + C = \tfrac{2}{3}x^{3/2} + \tfrac{3}{4}x^{4/3} + C$

9. $f(x) = \dfrac{10}{x^9} = 10x^{-9}$ has domain $(-\infty, 0) \cup (0, \infty)$, so $F(x) = \begin{cases} \dfrac{10x^{-8}}{-8} + C_1 = -\dfrac{5}{4x^8} + C_1 & \text{if } x < 0 \\[2mm] -\dfrac{5}{4x^8} + C_2 & \text{if } x > 0 \end{cases}$

See Example 1(c) for a similar exercise.

11. $g(t) = \dfrac{t^3 + 2t^2}{\sqrt{t}} = t^{5/2} + 2t^{3/2} \quad \Rightarrow \quad G(t) = \dfrac{t^{7/2}}{7/2} + \dfrac{2t^{5/2}}{5/2} + C = \tfrac{2}{7}t^{7/2} + \tfrac{4}{5}t^{5/2} + C$

Note that g has domain $(0, \infty)$.

13. $f(t) = 3\cos t - 4\sin t \quad \Rightarrow \quad F(t) = 3(\sin t) - 4(-\cos t) + C = 3\sin t + 4\cos t + C$

15. $f(x) = 2x + \dfrac{5}{\sqrt{1-x^2}} \quad \Rightarrow \quad F(x) = x^2 + 5\sin^{-1}x + C$

17. $f(x) = 5x^4 - 2x^5 \quad \Rightarrow \quad F(x) = x^5 - \tfrac{1}{3}x^6 + C.\ F(0) = 4 \quad \Rightarrow$

$C = 4$, so $F(x) = x^5 - \tfrac{1}{3}x^6 + 4$. The graph confirms our answer since
$f(x) = 0$ when F has a local maximum, f is positive when F is
increasing, and f is negative when F is decreasing.

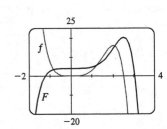

19. $f''(x) = 6x + 12x^2 \quad \Rightarrow \quad f'(x) = 3x^2 + 4x^3 + C \quad \Rightarrow \quad f(x) = x^3 + x^4 + Cx + D$

21. $f''(x) = 1 + x^{4/5}$ ⟹ $f'(x) = x + \frac{5}{9}x^{9/5} + C$ ⟹
$f(x) = \frac{1}{2}x^2 + \frac{5}{9} \cdot \frac{5}{14}x^{14/5} + Cx + D = \frac{1}{2}x^2 + \frac{25}{126}x^{14/5} + Cx + D$

23. $f'''(t) = e^t$ ⟹ $f''(t) = e^t + C$ ⟹ $f'(t) = e^t + Ct + D$ ⟹ $f(t) = e^t + \frac{1}{2}Ct^2 + Dt + E$

25. $f'(x) = 1 - 6x$ ⟹ $f(x) = x - 3x^2 + C$. $f(0) = C$ and $f(0) = 8$ ⟹ $C = 8$, so $f(x) = x - 3x^2 + 8$.

27. $f'(x) = 3\sqrt{x} - 1/\sqrt{x} = 3x^{1/2} - x^{-1/2}$ ⟹ $f(x) = 3\left(\frac{1}{3/2}\right)x^{3/2} - \frac{1}{1/2}x^{1/2} + C$ ⟹
$2 = f(1) = 2 - 2 + C = C$ ⟹ $f(x) = 2x^{3/2} - 2x^{1/2} + 2$

29. $f'(x) = 3\cos x + 5\sin x$ ⟹ $f(x) = 3\sin x - 5\cos x + C$ ⟹ $4 = f(0) = -5 + C$ ⟹ $C = 9$ ⟹
$f(x) = 3\sin x - 5\cos x + 9$

31. $f'(x) = 2/x$ ⟹ $f(x) = 2\ln|x| + C = 2\ln(-x) + C$ (since $x < 0$). Now
$f(-1) = 2\ln 1 + C = 2(0) + C = 7$ ⟹ $C = 7$. Therefore, $f(x) = 2\ln(-x) + 7, x < 0$.

33. $f''(x) = x$ ⟹ $f'(x) = \frac{1}{2}x^2 + C$ ⟹ $2 = f'(0) = C$ ⟹ $f'(x) = \frac{1}{2}x^2 + 2$ ⟹
$f(x) = \frac{1}{6}x^3 + 2x + D$ ⟹ $-3 = f(0) = D$ ⟹ $f(x) = \frac{1}{6}x^3 + 2x - 3$

35. $f''(x) = x^2 + 3\cos x$ ⟹ $f'(x) = \frac{1}{3}x^3 + 3\sin x + C$ ⟹ $3 = f'(0) = C$ ⟹ $f'(x) = \frac{1}{3}x^3 + 3\sin x + 3$
⟹ $f(x) = \frac{1}{12}x^4 - 3\cos x + 3x + D$ ⟹ $2 = f(0) = -3 + D$ ⟹ $D = 5$ ⟹
$f(x) = \frac{1}{12}x^4 - 3\cos x + 3x + 5$

37. $f''(x) = 6x + 6$ ⟹ $f'(x) = 3x^2 + 6x + C$ ⟹ $f(x) = x^3 + 3x^2 + Cx + D$. $4 = f(0) = D$ and
$3 = f(1) = 1 + 3 + C + D = 4 + C + 4$ ⟹ $C = -5$, so $f(x) = x^3 + 3x^2 - 5x + 4$.

39. $f''(x) = x^{-3}$ ⟹ $f'(x) = -\frac{1}{2}x^{-2} + C$ ⟹ $f(x) = \frac{1}{2}x^{-1} + Cx + D$ ⟹ $0 = f(1) = \frac{1}{2} + C + D$ and
$0 = f(2) = \frac{1}{4} + 2C + D$. Solving these equations, we get $C = \frac{1}{4}, D = -\frac{3}{4}$, so $f(x) = 1/(2x) + \frac{1}{4}x - \frac{3}{4}$.

41. $f''(x) = x^{-2}, x > 0$ ⟹ $f'(x) = -1/x + C$ ⟹ $f(x) = -\ln|x| + Cx + D = -\ln x + Cx + D$ (since
$x > 0$). $0 = f(1) = C + D$ and $0 = f(2) = -\ln 2 + 2C + D = -\ln 2 + 2C - C$ (since $D = -C$) $= -\ln 2 + C$
⟹ $C = \ln 2$ and $D = -\ln 2$. So $f(x) = -\ln x + (\ln 2)x - \ln 2$.

43. Given $f'(x) = 2x + 1$, we have $f(x) = x^2 + x + C$. Since f passes through $(1, 6), 6 = f(1) = 1^2 + 1 + C$ ⟹
$C = 4$. Therefore, $f(x) = x^2 + x + 4$ and $f(2) = 2^2 + 2 + 4 = 10$.

45. b is the antiderivative of f. For small x, f is negative, so the graph of its antiderivative must be decreasing. But
both a and c are increasing for small x, so only b can be f's antiderivative. Also, f is positive where b is
increasing, which supports our conclusion.

47. The graph of F will have a minimum at 0 and a maximum at 2, since $f = F'$ goes from negative to positive at
$x = 0$, and from positive to negative at $x = 2$.

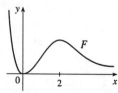

49.

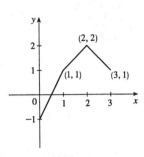

$$f'(x) = \begin{cases} 2 & \text{if } 0 \le x < 1 \\ 1 & \text{if } 1 < x < 2 \\ -1 & \text{if } 2 < x \le 3 \end{cases} \Rightarrow f(x) = \begin{cases} 2x + C & \text{if } 0 \le x < 1 \\ x + D & \text{if } 1 < x < 2 \\ -x + E & \text{if } 2 < x \le 3 \end{cases}$$

$f(0) = -1 \Rightarrow 2(0) + C = -1 \Rightarrow C = -1$. Starting at the point $(0, -1)$ and moving to the right on a line with slope 2 gets us to the point $(1, 1)$. The slope for $1 < x < 2$ is 1, so we get to the point $(2, 2)$. The line connecting $(1, 1)$ to $(2, 2)$ is $y = x$, so $D = 0$. The slope for $2 < x \le 3$ is -1, so we get to $(3, 1)$. $f(3) = 1 \Rightarrow -3 + E = 1 \Rightarrow E = 4$.

Thus, $f(x) = \begin{cases} 2x - 1 & \text{if } 0 \le x < 1 \\ x & \text{if } 1 < x < 2 \\ -x + 4 & \text{if } 2 < x \le 3 \end{cases}$

Note that f is continuous, but $f'(x)$ does not exist at $x = 1$ or at $x = 2$.

51.

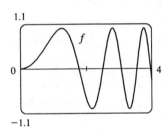

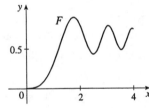

53.

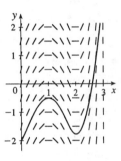

55.

x	$f(x)$
0	1
0.5	0.959
1.0	0.841
1.5	0.665
2.0	0.455
2.5	0.239
3.0	0.047

x	$f(x)$
3.5	−0.100
4.0	−0.189
4.5	−0.217
5.0	−0.192
5.5	−0.128
6.0	−0.047

We compute slopes (values of f) as in the table and draw a direction field as in Example 6. Then we use the direction field to graph F starting at $(0, 0)$.

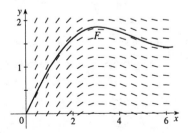

57.

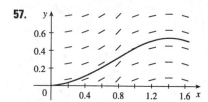

Remember that the values of f are the slopes of F at any x. For example, at $x = 1.4$, the slope of F is $f(1.4) = 0$.

59. $v(t) = s'(t) = \sin t - \cos t \Rightarrow s(t) = -\cos t - \sin t + C.$ $s(0) = -1 + C$ and $s(0) = 0 \Rightarrow C = 1$, so
$s(t) = -\cos t - \sin t + 1.$

61. $a(t) = v'(t) = t - 2 \Rightarrow v(t) = \frac{1}{2}t^2 - 2t + C.$ $v(0) = C$ and $v(0) = 3 \Rightarrow C = 3$, so $v(t) = \frac{1}{2}t^2 - 2t + 3$
and $s(t) = \frac{1}{6}t^3 - t^2 + 3t + D.$ $s(0) = D$ and $s(0) = 1 \Rightarrow D = 1$, and $s(t) = \frac{1}{6}t^3 - t^2 + 3t + 1.$

63. $a(t) = v'(t) = 10 \sin t + 3 \cos t \Rightarrow v(t) = -10 \cos t + 3 \sin t + C \Rightarrow$
$s(t) = -10 \sin t - 3 \cos t + Ct + D.$ $s(0) = -3 + D = 0$ and $s(2\pi) = -3 + 2\pi C + D = 12 \Rightarrow D = 3$ and
$C = \frac{6}{\pi}.$ Thus, $s(t) = -10 \sin t - 3 \cos t + \frac{6}{\pi}t + 3.$

65. (a) We first observe that since the stone is dropped 450 m above the ground, $v(0) = 0$ and $s(0) = 450.$
$v'(t) = a(t) = -9.8 \Rightarrow v(t) = -9.8t + C$, but $C = v(0) = 0$, so $v(t) = -9.8t \Rightarrow$
$s(t) = -4.9t^2 + D \Rightarrow D = s(0) = 450 \Rightarrow s(t) = 450 - 4.9t^2.$

(b) It reaches the ground when $0 = s(t) = 450 - 4.9t^2 \Rightarrow t^2 = 450/4.9 \Rightarrow t_1 = \sqrt{450/4.9} \approx 9.58$ s.

(c) $v(t_1) = -9.8\sqrt{450/4.9} \approx -93.9$ m/s

(d) This is just reworking parts (a) and (b) with $v(0) = -5.$ $v(t) = -9.8t + C \Rightarrow -5 = 0 + C \Rightarrow$
$v(t) = -9.8t - 5.$ $s(t) = -4.9t^2 - 5t + D \Rightarrow 450 = s(0) = D \Rightarrow s(t) = -4.9t^2 - 5t + 450.$
$s(t) = 0 \Rightarrow t = \left(5 \pm \sqrt{8845}\right)\big/(-9.8) \Rightarrow t_1 \approx 9.09$ s.

67. By Exercise 66, $s(t) = -4.9t^2 + v_0t + s_0$ and $v(t) = s'(t) = -9.8t + v_0.$ So
$[v(t)]^2 = (-9.8t + v_0)^2 = (9.8)^2 t^2 - 19.6v_0t + v_0^2 = v_0^2 - 19.6\left(v_0t - 4.9t^2\right).$ But $-4.9t^2 + v_0t$ is just $s(t)$
without the s_0 term, that is, $s(t) - s_0.$ Thus, $[v(t)]^2 = v_0^2 - 19.6[s(t) - s_0].$

69. Marginal cost $= 1.92 - 0.002x = C'(x) \Rightarrow C(x) = 1.92x - 0.001x^2 + K.$ But
$C(1) = 1.92 - 0.001 + K = 562 \Rightarrow K = 560.081.$ Therefore, $C(x) = 1.92x - 0.001x^2 + 560.081 \Rightarrow$
$C(100) = 742.081$, so the cost of producing 100 items is \$742.08.

71. Taking the upward direction to be positive we have that for $0 \leq t \leq 10$ (using the subscript 1 to refer to
$0 \leq t \leq 10$), $a_1(t) = -(9 - 0.9t) = v_1'(t) \Rightarrow v_1(t) = -9t + 0.45t^2 + v_0$, but $v_1(0) = v_0 = -10 \Rightarrow$
$v_1(t) = -9t + 0.45t^2 - 10 = s_1'(t) \Rightarrow s_1(t) = -\frac{9}{2}t^2 + 0.15t^3 - 10t + s_0.$ But $s_1(0) = 500 = s_0 \Rightarrow$
$s_1(t) = -\frac{9}{2}t^2 + 0.15t^3 - 10t + 500.$ $s_1(10) = 100$, so it takes more than 10 seconds for the raindrop to fall. Now
for $t > 10$, $a(t) = 0 = v'(t) \Rightarrow v(t) = \text{constant} = v_1(10) = -9(10) + 0.45(10)^2 - 10 = -55 \Rightarrow$
$v(t) = -55.$ At 55 ft/s, it will take $100/55 \approx 1.8$ s to fall the last 100 ft. Hence, the total time is 11.8 s.

73. $a(t) = k$, the initial velocity is 30 mi/h $= 30 \cdot \frac{5280}{3600} = 44$ ft/s, and the final velocity is
50 mi/h$= 50 \cdot \frac{5280}{3600} = \frac{220}{3}$ ft/s. So $v(t) = kt + C$ and $v(0) = 44 \Rightarrow C = 44.$ Thus, $v(t) = kt + 44 \Rightarrow$
$\frac{220}{3} = v(5) = 5k + 44 \Rightarrow k = \frac{88}{15} \approx 5.87$ ft/s^2.

75. Using Exercise 66 with $a = -32$, $v_0 = 0$, and $s_0 = h$ (the height of the cliff), we know that the height at time t is
$s(t) = -16t^2 + h.$ $v(t) = s'(t) = -32t \Rightarrow -32t = -120 \Rightarrow t = 3.75$, so
$0 = s(3.75) = -16(3.75)^2 + h \Rightarrow h = 16(3.75)^2 = 225$ ft.

77. (a) First note that 90 mi/h $= 90 \times \frac{5280}{3600}$ ft/s $= 132$ ft/s. Then $a\,(t) = 4$ ft/s^2 $\Rightarrow$ $v\,(t) = 4t + C$, but $v\,(0) = 0$
$\Rightarrow$ $C = 0$. Now $4t = 132$ when $t = \frac{132}{4} = 33$ s, so it takes 33 s to reach 132 ft/s. Therefore, taking
$s\,(0) = 0$, we have $s\,(t) = 2t^2$, $0 \le t \le 33$. So $s\,(33) = 2178$ ft. 15 minutes $= 15\,(60) = 900$ s, so for
$33 \le t \le 933$ we have $v\,(t) = 132$ ft/s $\Rightarrow$ $s\,(933) = 132\,(900) + 2178 = 120{,}978$ ft $= 22.9125$ mi.

(b) As in part (a), the train accelerates for 33 s and travels 2178 ft while doing so. Similarly, it decelerates for 33 s
and travels 2178 ft at the end of its trip. During the remaining $900 - 66 = 834$ s it travels at 132 ft/s, so
the distance traveled is $132 \cdot 834 = 110{,}088$ ft. Thus, the total distance is
$2178 + 110{,}088 + 2178 = 114{,}444$ ft $= 21.675$ mi.

(c) 45 mi $= 45\,(5280) = 237{,}600$ ft. Subtract $2\,(2178)$ to take care of the speeding up and slowing down, and we
have 233,244 ft at 132 ft/s for a trip of $233{,}244/132 = 1767$ s at 90 mi/h. The total time is
$1767 + 2\,(33) = 1833$ s or 30.55 min.

(d) $37.5\,(60) = 2250$ s. $2250 - 2\,(33) = 2184$ s at maximum speed. $2184\,(132) + 2\,(2178) = 292{,}644$ total feet or
$292{,}644/5280 = 55.425$ mi.

4 Review

CONCEPT CHECK

1. A function f has an **absolute maximum** at $x = c$ if $f\,(c)$ is the largest function value on the entire domain of f,
whereas f has a **local maximum** at c if $f\,(c)$ is the largest function value when x is near c. See Figure 4 in
Section 4.1.

2. (a) See Theorem 4.1.3.

(b) See the Closed Interval Method before Example 8 in Section 4.1.

3. (a) See Theorem 4.1.4.

(b) See Definition 4.1.6.

4. (a) See Rolle's Theorem at the beginning of Section 4.2.

(b) See the Mean Value Theorem in Section 4.2. Geometrical interpretation — there is some point P on the graph
of a function f [on the interval (a, b)] where the tangent line is parallel to the secant line that connects
$(a, f\,(a))$ and $(b, f\,(b))$.

5. (a) See the I/D Test before Example 1 in Section 4.3.

(b) See the Concavity Test just before Example 4 in Section 4.3.

6. (a) See the First Derivative Test after Example 1 in Section 4.3.

(b) See the Second Derivative Test before Example 6 in Section 4.3.

(c) See the note before Example 7 in Section 4.3.

7. (a) See page 306.

(b) Write fg as $\dfrac{f}{1/g}$ or $\dfrac{g}{1/f}$.

(c) Convert the difference into a quotient using a common denominator, rationalizing, factoring, or some other
method.

(d) Convert the power to a product by taking the natural logarithm of both sides of $y = f^g$ or by writing f^g as
$e^{g \ln f}$.

8. Without calculus you could get misleading graphs that fail to show the most interesting features of a function. See the discussions on pages 315 and 323.

9. (a) See Figure 3 in Section 4.9.

(b) $x_2 = x_1 - \dfrac{f(x_1)}{f'(x_1)}$

(c) $x_{n+1} = x_n - \dfrac{f(x_n)}{f'(x_n)}$

(d) Newton's method is likely to fail or to work very slowly when $f'(x_1)$ is close to 0.

10. (a) See the definition at the beginning of Section 4.10.

(b) If F_1 and F_2 are both antiderivatives of f on an interval I, then they differ by a constant.

TRUE-FALSE QUIZ

1. False. For example, take $f(x) = x^3$, then $f'(x) = 3x^2$ and $f'(0) = 0$, but $f(0) = 0$ is not a maximum or minimum; $(0, 0)$ is an inflection point.

3. False. For example, $f(x) = x$ is continuous on $(0, 1)$ but attains neither a maximum nor a minimum value on $(0, 1)$. Don't confuse this with f being continuous on the *closed* interval $[a, b]$, which would make the statement true.

5. True by the ID Test.

7. False. $f'(x) = g'(x) \;\Rightarrow\; f(x) = g(x) + C$. For example, $f(x) = x + 2$, $g(x) = x + 1 \;\Rightarrow\;$ $f'(x) = g'(x) = 1$, but $f(x) \neq g(x)$.

9. True. The graph of one such function is sketched.

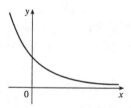

11. True. Let $x_1 < x_2$ where $x_1, x_2 \in I$. Then $f(x_1) < f(x_2)$ and $g(x_1) < g(x_2)$ (since f and g are increasing on I), so $(f + g)(x_1) = f(x_1) + g(x_1) < f(x_2) + g(x_2) = (f + g)(x_2)$.

13. False. Take $f(x) = x$ and $g(x) = x - 1$. Then both f and g are increasing on $(0, 1)$. But $f(x)g(x) = x(x - 1)$ is not increasing on $(0, 1)$.

15. True. Let $x_1, x_2 \in I$ and $x_1 < x_2$. Then $f(x_1) < f(x_2)$ (f is increasing) $\;\Rightarrow\; \dfrac{1}{f(x_1)} > \dfrac{1}{f(x_2)}$ (f is positive) $\;\Rightarrow\; g(x_1) > g(x_2) \;\Rightarrow\; g(x) = 1/f(x)$ is decreasing on I.

17. True. If $f'(x)$ exists and is nonzero for all x, then $f'(x)$ is either positive everywhere or negative everywhere. Hence, f is either strictly increasing everywhere or strictly decreasing everywhere, so $f(0)$ cannot equal $f(1)$.

EXERCISES

1. $f(x) = 10 + 27x - x^3$, $0 \le x \le 4$. $f'(x) = 27 - 3x^2 = -3(x^2 - 9) = -3(x+3)(x-3) = 0$ only when $x = 3$ (since -3 is not in the domain). $f'(x) > 0$ for $x < 3$ and $f'(x) < 0$ for $x > 3$, so $f(3) = 64$ is a local and absolute maximum value. Checking the endpoints, we find $f(0) = 10$ and $f(4) = 54$. Thus, $f(0) = 10$ is the absolute minimum value.

3. $f(x) = \dfrac{x}{x^2 + x + 1}$, $-2 \le x \le 0$. $f'(x) = \dfrac{(x^2 + x + 1)(1) - x(2x + 1)}{(x^2 + x + 1)^2} = \dfrac{1 - x^2}{(x^2 + x + 1)^2} = 0 \iff x = -1$

(since 1 is not in the domain). $f'(x) < 0$ for $-2 < x < -1$ and $f'(x) > 0$ for $-1 < x < 0$, so $f(-1) = -1$ is a local and absolute minimum value. $f(-2) = -\frac{2}{3}$ and $f(0) = 0$, $f(0) = 0$ is an absolute maximum value.

5. $f(x) = x - \sqrt{2}\sin x$, $0 \le x \le \pi$. $f'(x) = 1 - \sqrt{2}\cos x = 0 \Rightarrow \cos x = \dfrac{1}{\sqrt{2}} \Rightarrow x = \dfrac{\pi}{4}$.

$f''\left(\frac{\pi}{4}\right) = \sqrt{2}\sin\frac{\pi}{4} = 1 > 0$, so $f\left(\frac{\pi}{4}\right) = \frac{\pi}{4} - 1$ is a local minimum. Also $f(0) = 0$ and $f(\pi) = \pi$, so the absolute minimum is $f\left(\frac{\pi}{4}\right) = \frac{\pi}{4} - 1$, the absolute maximum is $f(\pi) = \pi$.

7. $\lim\limits_{x \to \pi} \dfrac{\sin x}{x^2 - \pi^2} \overset{H}{=} \lim\limits_{x \to \pi} \dfrac{\cos x}{2x} = -\dfrac{1}{2\pi}$

9. $\lim\limits_{x \to \infty} \dfrac{\ln(\ln x)}{\ln x} \overset{H}{=} \lim\limits_{x \to \infty} \dfrac{1/(x \ln x)}{1/x} = \lim\limits_{x \to \infty} \dfrac{1}{\ln x} = 0$

11. $\lim\limits_{x \to 0} \dfrac{\ln(1 - x) + x + \frac{1}{2}x^2}{x^3} \overset{H}{=} \lim\limits_{x \to 0} \dfrac{-\dfrac{1}{1 - x} + 1 + x}{3x^2} \overset{H}{=} \lim\limits_{x \to 0} \dfrac{-\dfrac{1}{(1 - x)^2} + 1}{6x} \overset{H}{=} \lim\limits_{x \to 0} \dfrac{-\dfrac{2}{(1 - x)^3}}{6} = -\dfrac{2}{6} = -\dfrac{1}{3}$

13. $\lim\limits_{x \to 0}\left(\csc^2 x - x^{-2}\right) = \lim\limits_{x \to 0}\left[\dfrac{1}{\sin^2 x} - \dfrac{1}{x^2}\right] = \lim\limits_{x \to 0} \dfrac{x^2 - \sin^2 x}{x^2 \sin^2 x} \overset{H}{=} \lim\limits_{x \to 0} \dfrac{2x - \sin 2x}{2x \sin^2 x + x^2 \sin 2x}$

$\overset{H}{=} \lim\limits_{x \to 0} \dfrac{2 - 2\cos 2x}{2\sin^2 x + 4x \sin 2x + 2x^2 \cos 2x} \overset{H}{=} \lim\limits_{x \to 0} \dfrac{4\sin 2x}{6\sin 2x + 12x \cos 2x - 4x^2 \sin 2x}$

$\overset{H}{=} \lim\limits_{x \to 0} \dfrac{8\cos 2x}{24\cos 2x - 32x \sin 2x - 8x^2 \cos 2x} = \dfrac{8}{24} = \dfrac{1}{3}$

15.

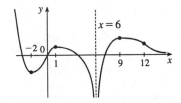

17.

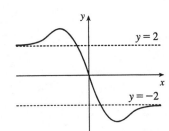

19. $y = f(x) = 2 - 2x - x^3$ **A.** $D = \mathbb{R}$ **B.** y-intercept: $f(0) = 2$. The x-intercept (approximately 0.770917) can be found using Newton's Method. **C.** No symmetry **D.** No asymptote

E. $f'(x) = -2 - 3x^2 = -1(3x^2 + 2) < 0$, so f is decreasing on $\mathbb{R}$.

F. No local extremum **G.** $f''(x) = -6x < 0$ on $(0, \infty)$ and $f''(x) > 0$ on $(-\infty, 0)$, so f is CD on $(0, \infty)$ and CU on $(-\infty, 0)$. There is an IP at $(0, 2)$.

H.

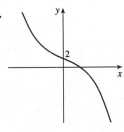

21. $y = f(x) = x^4 - 3x^3 + 3x^2 - x = x(x-1)^3$ **A.** $D = \mathbb{R}$ **B.** y-intercept: $f(0) = 0$; x-intercepts: $f(x) = 0$
$\Leftrightarrow$ $x = 0$ or $x = 1$ **C.** No symmetry **D.** f is a polynomial function and hence, it has no asymptote.
E. $f'(x) = 4x^3 - 9x^2 + 6x - 1$. Since the sum of the coefficients is 0, 1 is a root of f', so
$f'(x) = (x-1)(4x^2 - 5x + 1) = (x-1)^2(4x-1)$. $f'(x) < 0 \Rightarrow x < \frac{1}{4}$, so f is decreasing on
$\left(-\infty, \frac{1}{4}\right)$ and f is increasing on $\left(\frac{1}{4}, \infty\right)$. **F.** $f'(x)$ does not change **H.**

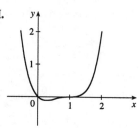

sign at $x = 1$, so there is not a local extremum there. $f\left(\frac{1}{4}\right) = -\frac{27}{256}$ is a

local minimum. **G.** $f''(x) = 12x^2 - 18x + 6 = 6(2x-1)(x-1)$.
$f''(x) = 0 \Leftrightarrow x = \frac{1}{2}$ or 1. $f''(x) < 0 \Leftrightarrow \frac{1}{2} < x < 1 \Rightarrow f$ is

CD on $\left(\frac{1}{2}, 1\right)$ and CU on $\left(-\infty, \frac{1}{2}\right)$ and $(1, \infty)$. There are inflection

points at $\left(\frac{1}{2}, -\frac{1}{16}\right)$ and $(1, 0)$.

23. $y = f(x) = \dfrac{1}{x(x-3)^2}$ **A.** $D = \{x \mid x \neq 0, 3\} = (-\infty, 0) \cup (0, 3) \cup (3, \infty)$ **B.** No intercepts. **C.** No

symmetry. **D.** $\displaystyle\lim_{x \to \pm\infty} \dfrac{1}{x(x-3)^2} = 0$, so $y = 0$ is a HA. $\displaystyle\lim_{x \to 0^+} \dfrac{1}{x(x-3)^2} = \infty$, $\displaystyle\lim_{x \to 0^-} \dfrac{1}{x(x-3)^2} = -\infty$,

$\displaystyle\lim_{x \to 3} \dfrac{1}{x(x-3)^2} = \infty$, so $x = 0$ and $x = 3$ are VA. **E.** $f'(x) = -\dfrac{(x-3)^2 + 2x(x-3)}{x^2(x-3)^4} = \dfrac{3(1-x)}{x^2(x-3)^3} \Rightarrow$

$f'(x) > 0 \Leftrightarrow 1 < x < 3$, so f is increasing

on $(1, 3)$ and decreasing on $(-\infty, 0)$, $(0, 1)$, and $(3, \infty)$. **F.** $f(1) = \frac{1}{4}$ **H.**

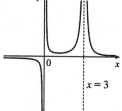

is a local minimum. **G.** $f''(x) = \dfrac{6(2x^2 - 4x + 3)}{x^3(x-3)^4}$. Note that

$2x^2 - 4x + 3 > 0$ for all x since it has negative discriminant. So
$f''(x) > 0 \Leftrightarrow x > 0 \Rightarrow f$ is CU on $(0, 3)$ and $(3, \infty)$ and CD on
$(-\infty, 0)$. No IP

25. $y = f(x) = \dfrac{x^2}{x+8} = x - 8 + \dfrac{64}{x+8}$ **A.** $D = \{x \mid x \neq -8\}$ **B.** Intercepts are 0 **C.** No symmetry

D. $\displaystyle\lim_{x \to \infty} \dfrac{x^2}{x+8} = \infty$, but $f(x) - (x-8) = \dfrac{64}{x+8} \Rightarrow 0$ as $x \to \infty$, so $y = x - 8$ is a slant

asymptote. $\displaystyle\lim_{x \to -8^+} \dfrac{x^2}{x+8} = \infty$ and $\displaystyle\lim_{x \to -8^-} \dfrac{x^2}{x+8} = -\infty$, so $x = -8$ is a VA.

E. $f'(x) = 1 - \dfrac{64}{(x+8)^2} = \dfrac{x(x+16)}{(x+8)^2} > 0$

$\Leftrightarrow x > 0$ or $x < -16$, so f is increasing on $(-\infty, -16)$ and $(0, \infty)$ and **H.**

decreasing on $(-16, -8)$ and $(-8, 0)$. **F.** $f(-16) = -32$ is a local

maximum, $f(0) = 0$ is a local

minimum. **G.** $f''(x) = 128/(x+8)^3 > 0 \Leftrightarrow x > -8$, so f is CU

on $(-8, \infty)$ and CD on $(-\infty, -8)$. No IP

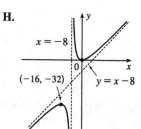

27. $y = f(x) = x\sqrt{2+x}$ **A.** $D = [-2, \infty)$ **B.** y-intercept: $f(0) = 0$; x-intercepts: -2 and 0 **C.** No symmetry **D.** No asymptote **E.** $f'(x) = \dfrac{x}{2\sqrt{2+x}} + \sqrt{2+x} = \dfrac{1}{2\sqrt{2+x}}[x + 2(2+x)] = \dfrac{3x+4}{2\sqrt{2+x}} = 0$

when $x = -\frac{4}{3}$, so f is decreasing on $\left(-2, -\frac{4}{3}\right)$ and increasing on $\left(-\frac{4}{3}, \infty\right)$. **F.** Local minimum

$f\left(-\frac{4}{3}\right) = -\frac{4}{3}\sqrt{\frac{2}{3}} = -\frac{4\sqrt{6}}{9} \approx -1.09$, no local maximum

H.

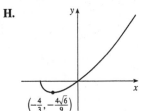

G. $f''(x) = \dfrac{2\sqrt{2+x} \cdot 3 - (3x+4)\dfrac{1}{\sqrt{2+x}}}{4(2+x)}$

$= \dfrac{6(2+x) - (3x+4)}{4(2+x)^{3/2}} = \dfrac{3x+8}{4(2+x)^{3/2}}$

$\left(-\frac{4}{3}, -\frac{4\sqrt{6}}{9}\right)$

$f''(x) > 0$ for $x > -2$, so f is CU everywhere. No IP

29. $y = f(x) = \sin^2 x - 2\cos x$ **A.** $D = \mathbb{R}$ **B.** y-intercept: $f(0) = -2$ **C.** $f(-x) = f(x)$, so f is symmetric with respect to the y-axis. f has period 2π. **D.** No asymptote
E. $y' = 2\sin x \cos x + 2\sin x = 2\sin x (\cos x + 1)$. $y' = 0 \Leftrightarrow \sin x = 0$ or $\cos x = -1 \Leftrightarrow x = n\pi$ or $x = (2n+1)\pi$. $y' > 0$ when $\sin x > 0$, since $\cos x + 1 \geq 0$ for all x. Therefore, $y' > 0$ (and so f is increasing) on $(2n\pi, (2n+1)\pi)$; $y' < 0$ (and so f is decreasing) on $((2n-1)\pi, 2n\pi)$. **F.** Local maxima are $f((2n+1)\pi) = 2$; local minima are $f(2n\pi) = -2$. **G.** $y' = \sin 2x + 2\sin x \Rightarrow$

$$y'' = 2\cos 2x + 2\cos x = 2\left(2\cos^2 x - 1\right) + 2\cos x = 4\cos^2 x + 2\cos x - 2$$

$$= 2\left(2\cos^2 x + \cos x - 1\right) = 2(2\cos x - 1)(\cos x + 1)$$

$y'' = 0 \Leftrightarrow \cos x = \frac{1}{2}$ or $-1 \Leftrightarrow x = 2n\pi \pm \frac{\pi}{3}$ or $x = (2n+1)\pi$. $y'' > 0$ (and so f is CU) on $\left(2n\pi - \frac{\pi}{3}, 2n\pi + \frac{\pi}{3}\right)$; $y'' \leq 0$ (and so f is CD) on $\left(2n\pi + \frac{\pi}{3}, 2n\pi + \frac{5\pi}{3}\right)$. There are inflection points at $\left(2n\pi \pm \frac{\pi}{3}, -\frac{1}{4}\right)$.

H.

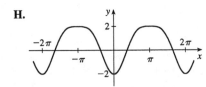

31. $y = f(x) = \sin^{-1}(1/x)$ **A.** $D = \{x \mid -1 \leq 1/x \leq 1\} = (-\infty, -1] \cup [1, \infty)$. **B.** No intercept
C. $f(-x) = -f(x)$, symmetric about the origin **D.** $\lim\limits_{x \to \pm\infty} \sin^{-1}(1/x) = \sin^{-1}(0) = 0$, so $y = 0$ is a HA.

E. $f'(x) = \dfrac{1}{\sqrt{1-(1/x)^2}}\left(-\dfrac{1}{x^2}\right) = \dfrac{-1}{\sqrt{x^4 - x^2}} < 0$, so f is decreasing on $(-\infty, -1)$ and $(1, \infty)$.

F. No local extremum, but $f(1) = \frac{\pi}{2}$ is the absolute maximum and
$f(-1) = -\frac{\pi}{2}$ is the absolute minimum.

H.

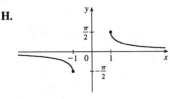

G. $f''(x) = \dfrac{4x^3 - 2x}{2(x^4 - x^2)^{3/2}} = \dfrac{x(2x^2 - 1)}{(x^4 - x^2)^{3/2}} > 0$ for $x > 1$ and

$f''(x) < 0$ for $x < -1$, so f is CU on $(1, \infty)$ and CD on $(-\infty, -1)$. No IP

33. $y = f(x) = e^x + e^{-3x}$ **A.** $D = \mathbb{R}$ **B.** y-intercept 2; no x-intercept   **C.** No symmetry

D. $\lim\limits_{x \to \pm\infty} (e^x + e^{-3x}) = \infty$, no asymptote   **E.** $y = f(x) = e^x + e^{-3x}$ $\Rightarrow$

$f'(x) = e^x - 3e^{-3x} = e^{-3x}(e^{4x} - 3) > 0 \iff e^{4x} > 3 \iff$

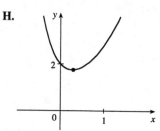

H.

$4x > \ln 3 \iff x > \frac{1}{4}\ln 3$, so f is increasing on $\left(\frac{1}{4}\ln 3, \infty\right)$ and

decreasing on $\left(-\infty, \frac{1}{4}\ln 3\right)$.   **F.** Absolute minimum

$f\left(\frac{1}{4}\ln 3\right) = 3^{1/4} + 3^{-3/4} \approx 1.75$.   **G.** $f''(x) = e^x + 9e^{-3x} > 0$, so f

is CU on $(-\infty, \infty)$. No IP.

35. $f(x) = \dfrac{x^2 - 1}{x^3}$ $\Rightarrow$ $f'(x) = \dfrac{x^3(2x) - (x^2 - 1)3x^2}{x^6} = \dfrac{3 - x^2}{x^4}$ $\Rightarrow$

$f''(x) = \dfrac{x^4(-2x) - (3 - x^2)4x^3}{x^8} = \dfrac{2x^2 - 12}{x^5}$

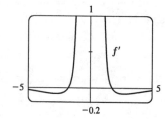

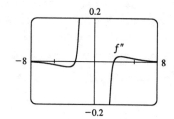

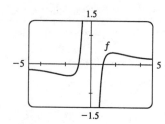

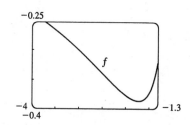

Estimates: From the graphs of f' and f'', it appears that f is increasing on $(-1.73, 0)$ and $(0, 1.73)$ and decreasing on $(-\infty, -1.73)$ and $(1.73, \infty)$; f has a local maximum of about $f(1.73) = 0.38$ and a local minimum of about $f(-1.7) = -0.38$; f is CU on $(-2.45, 0)$ and $(2.45, \infty)$, and CD on $(-\infty, -2.45)$ and $(0, 2.45)$; and f has inflection points at about $(-2.45, -0.34)$ and $(2.45, 0.34)$.

Exact: Now $f'(x) = \dfrac{3 - x^2}{x^4}$ is positive for $0 < x^2 < 3$, that is, f is increasing on $\left(-\sqrt{3}, 0\right)$ and $\left(0, \sqrt{3}\right)$; and

$f'(x)$ is negative (and so f is decreasing) on $\left(-\infty, -\sqrt{3}\right)$ and $\left(\sqrt{3}, \infty\right)$. $f'(x) = 0$ when $x = \pm\sqrt{3}$. f' goes

from positive to negative at $x = \sqrt{3}$, so f has a local maximum of $f(\sqrt{3}) = \dfrac{(\sqrt{3})^2 - 1}{(\sqrt{3})^3} = \dfrac{2\sqrt{3}}{9}$; and since f is odd,

we know that maxima on the interval $(0, \infty)$ correspond to minima on $(-\infty, 0)$, so f has a local minimum of

$f(-\sqrt{3}) = -\dfrac{2\sqrt{3}}{9}$. Also, $f''(x) = \dfrac{2x^2 - 12}{x^5}$ is positive (so f is CU) on $\left(-\sqrt{6}, 0\right)$ and $\left(\sqrt{6}, \infty\right)$, and negative

(so f is CD) on $\left(-\infty, -\sqrt{6}\right)$ and $\left(0, \sqrt{6}\right)$. There are IP at $\left(\sqrt{6}, \dfrac{5\sqrt{6}}{36}\right)$ and $\left(-\sqrt{6}, -\dfrac{5\sqrt{6}}{36}\right)$.

37. $f(x) = 3x^6 - 5x^5 + x^4 - 5x^3 - 2x^2 + 2$, $f'(x) = 18x^5 - 25x^4 + 4x^3 - 15x^2 - 4x$,

$f''(x) = 90x^4 - 100x^3 + 12x^2 - 30x - 4$

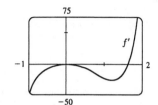

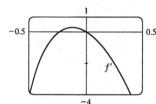

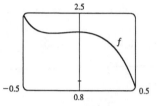

From the graphs of f' and f'', it appears that f is increasing on $(-0.23, 0)$ and $(1.62, \infty)$ and decreasing on $(-\infty, -0.23)$ and $(0, 1.62)$; f has a local maximum of about $f(0) = 2$ and local minima

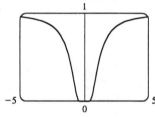

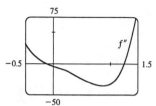

of about $f(-0.23) = 1.96$ and $f(1.62) = -19.2$; f is CU on $(-\infty, -0.12)$ and $(1.24, \infty)$ and CD on $(-0.12, 1.24)$; and f has inflection points at about $(-0.12, 1.98)$ and $(1.2, -12.1)$.

39.

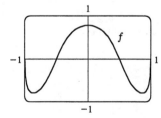

From the graph, we estimate the points of inflection to be about $(\pm 0.8, 0.2)$. $f(x) = e^{-1/x^2} \Rightarrow f'(x) = 2x^{-3}e^{-1/x^2} \Rightarrow f''(x) = 2\left[x^{-3}\left(2x^{-3}\right)e^{-1/x^2} + e^{-1/x^2}\left(-3x^{-4}\right)\right] = 2x^{-6}e^{-1/x^2}\left(2 - 3x^2\right)$. This is 0 when $2 - 3x^2 = 0 \Leftrightarrow x = \pm\sqrt{\frac{2}{3}}$, so the inflection points are $\left(\pm\sqrt{\frac{2}{3}}, e^{-3/2}\right)$.

41. $f(x) = \arctan(\cos(3\arcsin x))$. We use a CAS to compute f' and f'', and to graph f, f', and f'':

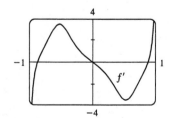

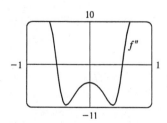

From the graph of f', it appears that the only maximum occurs at $x = 0$ and there are minima at $x = \pm 0.87$. From the graph of f'', it appears that there are inflection points at $x = \pm 0.52$.

43.

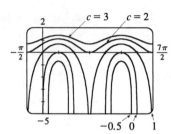

The family of functions $f(x) = \ln(\sin x + C)$ all have the same period and all have maximum values at $x = \frac{\pi}{2} + 2\pi n$. Since the domain of ln is $(0, \infty)$, f has a graph only if $\sin x + C > 0$ somewhere. Since $-1 \le \sin x \le 1$, this happens if $C > -1$, that is, f has no graph if $C \le -1$. Similarly, if $C > 1$, then $\sin x + C > 0$ and f is continuous on $(-\infty, \infty)$. As C increases, the graph of f is shifted vertically upward and flattens out. If $-1 < C \le 1$, f is defined where $\sin x + C > 0$ $\Leftrightarrow$ $\sin x > -C$ $\Leftrightarrow$ $\sin^{-1}(-C) < x < \pi - \sin^{-1}(-C)$. Since the period is 2π, the domain of f is $\left(2n\pi + \sin^{-1}(-C), (2n+1)\pi - \sin^{-1}(-C)\right)$, n an integer.

45. $f(x) = x^{101} + x^{51} + x - 1 = 0$. Since f is continuous and $f(0) = -1$ and $f(1) = 2$, the equation has at least one root in $(0, 1)$, by the Intermediate Value Theorem. Suppose the equation has two roots, a and b, with $a < b$. Then $f(a) = 0 = f(b)$, so by the Mean Value Theorem, $f'(x) = \dfrac{f(b) - f(a)}{b - a} = \dfrac{0}{b - a} = 0$, so $f'(x)$ has a root in (a, b). But this is impossible since $f'(x) = 101x^{100} + 51x^{50} + 1 \ge 1$ for all x.

47. Since f is continuous on $[32, 33]$ and differentiable on $(32, 33)$, then by the Mean Value Theorem there exists a number c in $(32, 33)$ such that $f'(c) = \frac{1}{5}c^{-4/5} = \dfrac{\sqrt[5]{33} - \sqrt[5]{32}}{33 - 32} = \sqrt[5]{33} - 2$, but $\frac{1}{5}c^{-4/5} > 0$ $\Rightarrow$ $\sqrt[5]{33} - 2 > 0$ $\Rightarrow$ $\sqrt[5]{33} > 2$. Also f' is decreasing, so that $f'(c) < f'(32) = \frac{1}{5}(32)^{-4/5} = 0.0125$ $\Rightarrow$ $0.0125 > f'(c) = \sqrt[5]{33} - 2$ $\Rightarrow$ $\sqrt[5]{33} < 2.0125$. Therefore, $2 < \sqrt[5]{33} < 2.0125$.

49. (a) $g(x) = f(x^2)$ $\Rightarrow$ $g'(x) = 2xf'(x^2)$ by the Chain Rule. Since $f'(x) > 0$ for all $x \ne 0$, we must have $f'(x^2) > 0$ for $x \ne 0$, so $g'(x) = 0$ $\Leftrightarrow$ $x = 0$. Now $g'(x)$ changes sign (from negative to positive) at $x = 0$, since one of its factors, $f'(x^2)$, is positive for all x, and its other factor, $2x$, changes from negative to positive at this point, so by the First Derivative Test, f has a local and absolute minimum at $x = 0$.

(b) $g'(x) = 2xf'(x^2)$ $\Rightarrow$ $g''(x) = 2[xf''(x^2)(2x) + f'(x^2)] = 4x^2 f''(x^2) + 2f'(x^2)$ by the Product Rule and the Chain Rule. But $x^2 > 0$ for all $x \ne 0$, $f''(x^2) > 0$ (since f is CU for $x > 0$), and $f'(x^2) > 0$ for all $x \ne 0$, so since all of its factors are positive, $g''(x) > 0$ for $x \ne 0$. Whether $g''(0)$ is positive or 0 doesn't matter (since the sign of g'' does not change there); g is concave upward on $\mathbb{R}$.

51. If $B = 0$, the line is vertical and the distance from $x = -\dfrac{C}{A}$ to (x_1, y_1) is $\left| x_1 + \dfrac{C}{A} \right| = \dfrac{|Ax_1 + By_1 + C|}{\sqrt{A^2 + B^2}}$, so

assume $B \neq 0$. The square of the distance from (x_1, y_1) to the line is $f(x) = (x - x_1)^2 + (y - y_1)^2$ where

$Ax + By + C = 0$, so we minimize $f(x) = (x - x_1)^2 + \left(-\dfrac{A}{B}x - \dfrac{C}{B} - y_1 \right)^2 \Rightarrow$

$f'(x) = 2(x - x_1) + 2\left(-\dfrac{A}{B}x - \dfrac{C}{B} - y_1 \right)\left(-\dfrac{A}{B} \right)$. $f'(x) = 0 \Rightarrow x = \dfrac{B^2 x_1 - ABy_1 - AC}{A^2 + B^2}$ and this gives a

minimum since $f''(x) = 2\left(1 + \dfrac{A^2}{B^2} \right) > 0$. Substituting this value of x into $f(x)$ and simplifying gives

$f(x) = \dfrac{(Ax_1 + By_1 + C)^2}{A^2 + B^2}$, so the minimum distance is $\sqrt{f(x)} = \dfrac{|Ax_1 + By_1 + C|}{\sqrt{A^2 + B^2}}$.

53.

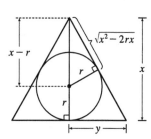

By similar triangles, $\dfrac{y}{x} = \dfrac{r}{\sqrt{x^2 - 2rx}}$, so the area of the triangle

is $A(x) = \tfrac{1}{2}(2y)x = xy = \dfrac{rx^2}{\sqrt{x^2 - 2rx}} \Rightarrow$

$A'(x) = \dfrac{2rx\sqrt{x^2 - 2rx} - rx^2(x - r)\Big/\sqrt{x^2 - 2rx}}{x^2 - 2rx} = \dfrac{rx^2(x - 3r)}{\left(x^2 - 2rx \right)^{3/2}}$

$= 0$ when $x = 3r$.

$A'(x) < 0$ when $2r < x < 3r$, $A'(x) > 0$ when $x > 3r$. So

$x = 3r$ gives a minimum and

$A(3r) = r(9r^2)\Big/\left(\sqrt{3}r \right) = 3\sqrt{3}r^2$.

55.

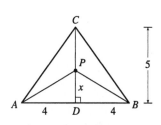

We minimize

$L(x) = |PA| + |PB| + |PC| = 2\sqrt{x^2 + 16} + (5 - x)$,

$0 \le x \le 5$. $L'(x) = 2x\Big/\sqrt{x^2 + 16} - 1 = 0 \Leftrightarrow$

$2x = \sqrt{x^2 + 16} \Leftrightarrow 4x^2 = x^2 + 16 \Leftrightarrow x = \dfrac{4}{\sqrt{3}}$.

$L(0) = 13$, $L\left(\dfrac{4}{\sqrt{3}} \right) \approx 11.9$, $L(5) \approx 12.8$, so the minimum

occurs when $x = \dfrac{4}{\sqrt{3}} \approx 2.3$.

57. $v = K\sqrt{\dfrac{L}{C} + \dfrac{C}{L}} \Rightarrow \dfrac{dv}{dL} = \dfrac{K}{2\sqrt{(L/C) + (C/L)}}\left(\dfrac{1}{C} - \dfrac{C}{L^2} \right) = 0 \Leftrightarrow \dfrac{1}{C} = \dfrac{C}{L^2} \Leftrightarrow L^2 = C^2 \Leftrightarrow$

$L = C$. This gives the minimum velocity since $v' < 0$ for $0 < L < C$ and $v' > 0$ for $L > C$.

59. Let $x =$ selling price of ticket. Then $12 - x$ is the amount the ticket price has been lowered, so the number
of tickets sold is $11{,}000 + 1000(12 - x) = 23{,}000 - 1000x$. The revenue is
$R(x) = x(23{,}000 - 1000x) = 23{,}000x - 1000x^2$, so $R'(x) = 23{,}000 - 2000x = 0$ when $x = 11.5$. Since
$R''(x) = -2000 < 0$, the maximum revenue occurs when the ticket prices are \$11.50.

61. $f(x) = x^4 + x - 1 \Rightarrow f'(x) = 4x^3 + 1 \Rightarrow x_{n+1} = x_n - \dfrac{x_n^4 + x_n - 1}{4x_n^3 + 1}$. If $x_1 = 0.5$ then $x_2 \approx 0.791667$,

$x_3 \approx 0.729862$, $x_4 \approx 0.724528$, $x_5 \approx 0.724492$ and $x_6 \approx 0.724492$, so, to six decimal places, the root is 0.724492.

63. $f(x) = x^6 + 2x^2 - 8x + 3 \implies f'(x) = 6x^5 + 4x - 8$. We want to find the minimum of f, so we examine the graph of f' looking for values at which f' changes from negative to positive.

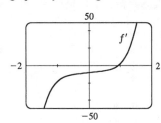

From the graph, we see that this occurs at $x \approx 1$. So we will use Newton's method with $g(x) = f'(x) = 6x^5 + 4x - 8$, $g'(x) = 30x^4 + 4$, and $x_1 = 1$. $x_{n+1} = x_n - \dfrac{6x_n^5 + 4x_n - 8}{30x_n^4 + 4}$ gives us $x_2 \approx 0.941176$,

$x_3 \approx 0.934068$, $x_4 \approx 0.933975 \approx x_5$. Thus, $f(x_5) \approx -2.063421$ is the absolute minimum value of f.

65. $f'(x) = \sqrt{x^5} - 4/\sqrt[5]{x} = x^{5/2} - 4x^{-1/5} \implies f(x) = \frac{2}{7}x^{7/2} - 4\left(\frac{5}{4}x^{4/5}\right) + C = \frac{2}{7}x^{7/2} - 5x^{4/5} + C$

67. $f(x) = e^x - 1/x \implies F(x) = \begin{cases} e^x - \ln|x| + C_1 \text{ if } x < 0 \\ e^x - \ln|x| + C_2 \text{ if } x > 0 \end{cases}$

69. $f'(x) = (1+x)/\sqrt{x} = x^{-1/2} + x^{1/2} \implies f(x) = 2x^{1/2} + \frac{2}{3}x^{3/2} + C \implies 0 = f(1) = 2 + \frac{2}{3} + C \implies C = -\frac{8}{3} \implies f(x) = 2x^{1/2} + \frac{2}{3}x^{3/2} - \frac{8}{3}$

71. $f''(x) = x^3 + x \implies f'(x) = \frac{1}{4}x^4 + \frac{1}{2}x^2 + C \implies 1 = f'(0) = C \implies f'(x) = \frac{1}{4}x^4 + \frac{1}{2}x^2 + 1 \implies f(x) = \frac{1}{20}x^5 + \frac{1}{6}x^3 + x + D \implies -1 = f(0) = D \implies f(x) = \frac{1}{20}x^5 + \frac{1}{6}x^3 + x - 1$

73. (a)

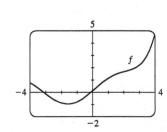 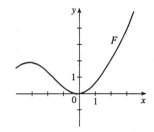

Since f is 0 just to the left of the y-axis, we must have a minimum of F at the same place since we are increasing through $(0, 0)$ on F. There must be a local maximum to the left of $x = -3$, since f changes from positive to negative there.

(b) $f(x) = 0.1e^x + \sin x \implies F(x) = 0.1e^x - \cos x + C$.
$F(0) = 0 \implies 0.1 - 1 + C = 0 \implies C = 0.9$, so
$F(x) = 0.1e^x - \cos x + 0.9$.

(c)

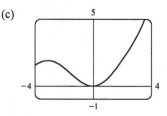

75. Choosing the positive direction to be upward, we have $a(t) = -9.8 \implies v(t) = -9.8t + v_0$, but $v(0) = 0 = v_0$ $\implies v(t) = -9.8t = s'(t) \implies s(t) = -4.9t^2 + s_0$, but $s(0) = s_0 = 500 \implies s(t) = -4.9t^2 + 500$. When $s = 0$, $-4.9t^2 + 500 = 0 \implies t_1 = \sqrt{\frac{500}{4.9}} \approx 10.1 \implies v(t_1) = -9.8\sqrt{\frac{500}{4.9}} \approx -98.995$ m/s. Therefore, the canister will not burst.

77. (a)

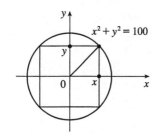

The cross-sectional area is $A = 2x \cdot 2y = 4xy = 4x\sqrt{100 - x^2}$, $0 \le x \le 10$, so

$$\frac{dA}{dx} = 4x \left(\tfrac{1}{2}\right) (100 - x^2)^{-1/2} (-2x) + (100 - x^2)^{1/2} \cdot 4$$

$$= \frac{-4x^2}{(100 - x^2)^{1/2}} + 4 (100 - x^2)^{1/2}$$

$$= 0 \text{ when } -x^2 + (100 - x^2) = 0 \quad \Rightarrow$$

$x^2 = 50 \quad \Rightarrow \quad x = \sqrt{50} \approx 7.07 \quad \Rightarrow \quad y = \sqrt{100 - \left(\sqrt{50}\right)^2} = \sqrt{50}$. Since $A(0) = A(10) = 0$, the

rectangle of maximum area is a square.

(b)

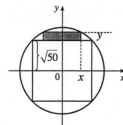

The cross-sectional area of each rectangular plank (shaded in the figure) is

$$A = 2x \left(y - \sqrt{50}\right) = 2x \left[\sqrt{100 - x^2} - \sqrt{50}\right], 0 \le x \le \sqrt{50}, \text{ so}$$

$$\frac{dA}{dx} = 2 \left(\sqrt{100 - x^2} - \sqrt{50}\right) + 2x \left(\tfrac{1}{2}\right) (100 - x^2)^{-1/2} (-2x)$$

$$= 2 (100 - x^2)^{1/2} - 2\sqrt{50} - \frac{2x^2}{(100 - x^2)^{1/2}}$$

Set $\dfrac{dA}{dx} = 0$: $(100 - x^2) - \sqrt{50} (100 - x^2)^{1/2} - x^2 = 0 \quad \Rightarrow \quad 100 - 2x^2 = \sqrt{50} (100 - x^2)^{1/2} \quad \Rightarrow$

$10{,}000 - 400x^2 + 4x^4 = 50 (100 - x^2) \quad \Rightarrow \quad 2500 - 175x^2 + 2x^4 = 0 \quad \Rightarrow$

$x^2 = \dfrac{175 \pm \sqrt{10{,}625}}{4} \approx 69.52$ or $17.98 \quad \Rightarrow \quad x \approx 8.34$ or 4.24.

But $8.34 > \sqrt{50}$, so $x_1 \approx 4.24 \quad \Rightarrow \quad y - \sqrt{50} = \sqrt{100 - x_1^2} - \sqrt{50} \approx 1.99$. Each plank should have

dimensions about $8\frac{1}{2}$ inches by 2 inches.

(c) From the figure in part (a), the width is $2x$ and the depth is $2y$, so the strength is

$S = k (2x) (2y)^2 = 8kxy^2 = 8kx (100 - x^2) = 800kx - 8kx^3, 0 \le x \le 10$. $dS/dx = 800k - 24kx^2 = 0$

when $24kx^2 = 800k \quad \Rightarrow \quad x^2 = \frac{100}{3} \quad \Rightarrow \quad x = \frac{10}{\sqrt{3}} \quad \Rightarrow \quad y = \sqrt{\frac{200}{3}} = \frac{10\sqrt{2}}{\sqrt{3}} = \sqrt{2}x$ and

$S(0) = S(10) = 0$, so the maximum strength occurs when $x = \frac{10}{\sqrt{3}}$. The dimensions should be

$\frac{20}{\sqrt{3}} \approx 11.55$ inches by $\frac{20\sqrt{2}}{\sqrt{3}} \approx 16.33$ inches.

79. (a) $I = \dfrac{k \cos\theta}{d^2} = \dfrac{k\,(h/d)}{d^2} = k\dfrac{h}{d^3} = k\dfrac{h}{\left(\sqrt{40^2 + h^2}\right)^3} = k\dfrac{h}{(1600 + h^2)^{3/2}}$ ⇒

$$\frac{dI}{dh} = k\frac{(1600 + h^2)^{3/2} - h\frac{3}{2}(1600 + h^2)^{1/2} \cdot 2h}{(1600 + h^2)^3} = \frac{k(1600 + h^2)^{1/2}(1600 + h^2 - 3h^2)}{(1600 + h^2)^3}$$

$$= \frac{k(1600 - 2h^2)}{(1600 + h^2)^{5/2}}$$

Set $dI/dh = 0$: $1600 - 2h^2 = 0 \Rightarrow h^2 = 800 \Rightarrow h = \sqrt{800} = 20\sqrt{2}$. By the First Derivative Test, I has a local maximum at $h = 20\sqrt{2} \approx 28$ ft.

(b)

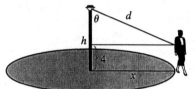

$$\frac{dx}{dt} = 4 \text{ ft/s}$$

$$I = \frac{k\cos\theta}{d^2} = \frac{k\,[(h-4)/d]}{d^2} = \frac{k\,(h-4)}{d^3} = \frac{k\,(h-4)}{[(h-4)^2 + x^2]^{3/2}} = k\,(h-4)\,[(h-4)^2 + x^2]^{-3/2}$$

$$\frac{dI}{dt} = \frac{dI}{dx} \cdot \frac{dx}{dt} = k\,(h-4)\left(-\tfrac{3}{2}\right)[(h-4)^2 + x^2]^{-5/2} \cdot 2x \cdot \frac{dx}{dt}$$

$$= k\,(h-4)\,(-3x)\,[(h-4)^2 + x^2]^{-5/2} \cdot 4 = \frac{-12xk\,(h-4)}{[(h-4)^2 + x^2]^{5/2}}$$

$$\left.\frac{dI}{dt}\right|_{x=40} = -\frac{480k\,(h-4)}{[(h-4)^2 + 1600]^{5/2}}$$

Problems Plus

1. Let $y = f(x) = e^{-x^2}$. The area of the rectangle under the curve from $-x$ to x is $A(x) = 2xe^{-x^2}$ where $x \geq 0$.

We maximize $A(x)$: $A'(x) = 2e^{-x^2} - 4x^2 e^{-x^2} = 2e^{-x^2}(1 - 2x^2) = 0 \Rightarrow x = \frac{1}{\sqrt{2}}$. This gives a maximum

since $A'(x) > 0$ for $0 \leq x < \frac{1}{\sqrt{2}}$ and $A'(x) < 0$ for $x > \frac{1}{\sqrt{2}}$. We next determine the points of inflection of $f(x)$.

Notice that $f'(x) = -2xe^{-x^2} = -A(x)$. So $f''(x) = -A'(x)$ and hence, $f''(x) < 0$ for $-\frac{1}{\sqrt{2}} < x < \frac{1}{\sqrt{2}}$ and

$f''(x) > 0$ for $x < -\frac{1}{\sqrt{2}}$ and $x > \frac{1}{\sqrt{2}}$. So $f(x)$ changes concavity at $x = \pm\frac{1}{\sqrt{2}}$, and the two vertices of the

rectangle of largest area are at the inflection points.

3. First, we recognize some symmetry in the inequality: $\dfrac{e^{x+y}}{xy} \geq e^2 \Leftrightarrow \dfrac{e^x}{x} \cdot \dfrac{e^y}{y} \geq e \cdot e$. This suggests that we

need to show that $\dfrac{e^x}{x} \geq e$ for $x > 0$. If we can do this, then the inequality $\dfrac{e^y}{y} \geq e$ is true, and the given inequality

follows.

$f(x) = \dfrac{e^x}{x} \Rightarrow f'(x) = \dfrac{xe^x - e^x}{x^2} = \dfrac{e^x(x-1)}{x^2} = 0 \Rightarrow x = 1$. By the First Derivative Test, we have a

minimum of $f(1) = e$, so $e^x/x \geq e$ for all x.

5. First we show that $x(1-x) \leq \frac{1}{4}$ for all x. Let $f(x) = x(1-x) = x - x^2$. Then $f'(x) = 1 - 2x$. This is 0

when $x = \frac{1}{2}$ and $f'(x) > 0$ for $x < \frac{1}{2}$, $f'(x) < 0$ for $x > \frac{1}{2}$, so the absolute maximum of f is $f\left(\frac{1}{2}\right) = \frac{1}{4}$. Thus,

$x(1-x) \leq \frac{1}{4}$ for all x.

Now suppose that the given assertion is false, that is, $a(1-b) > \frac{1}{4}$ and $b(1-a) > \frac{1}{4}$. Multiply these

inequalities: $a(1-b)b(1-a) > \frac{1}{16} \Rightarrow [a(1-a)][b(1-b)] > \frac{1}{16}$. But we know that $a(1-a) \leq \frac{1}{4}$ and

$b(1-b) \leq \frac{1}{4} \Rightarrow [a(1-a)][b(1-b)] \leq \frac{1}{16}$. Thus, we have a contradiction, so the given assertion is proved.

7. Differentiating $x^2 + xy + y^2 = 12$ implicitly with respect to x gives $2x + y + x\dfrac{dy}{dx} + 2y\dfrac{dy}{dx} = 0$, so

$\dfrac{dy}{dx} = -\dfrac{2x+y}{x+2y}$. At a highest or lowest point, $\dfrac{dy}{dx} = 0 \Leftrightarrow y = -2x$. Substituting this into the original

equation gives $x^2 + x(-2x) + (-2x)^2 = 12$, so $3x^2 = 12$ and $x = \pm 2$. If $x = 2$, then $y = -2x = -4$, and if

$x = -2$ then $y = 4$. Thus, the highest and lowest points are $(-2, 4)$ and $(2, -4)$.

9. $f(x) = \dfrac{1}{1+|x|} + \dfrac{1}{1+|x-2|}$

$$= \begin{cases} \dfrac{1}{1-x} + \dfrac{1}{1-(x-2)} & \text{if } x < 0 \\[2mm] \dfrac{1}{1+x} + \dfrac{1}{1-(x-2)} & \text{if } 0 \le x < 2 \\[2mm] \dfrac{1}{1+x} + \dfrac{1}{1+(x-2)} & \text{if } x \ge 2 \end{cases} \Rightarrow f'(x) = \begin{cases} \dfrac{1}{(1-x)^2} + \dfrac{1}{(3-.x)^2} & \text{if } x < 0 \\[2mm] \dfrac{-1}{(1+x)^2} + \dfrac{1}{(3-x)^2} & \text{if } 0 < x < 2 \\[2mm] \dfrac{-1}{(1+x)^2} - \dfrac{1}{(x-1)^2} & \text{if } x > 2 \end{cases}$$

We see that $f'(x) > 0$ for $x < 0$ and $f'(x) < 0$ for $x > 2$. For $0 < x < 2$, we have

$f'(x) = \dfrac{1}{(3-x)^2} - \dfrac{1}{(x+1)^2} = \dfrac{(x^2+2x+1)-(x^2-6x+9)}{(3-x)^2(x+1)^2} = \dfrac{8(x-1)}{(3-x)^2(x+1)^2}$, so $f'(x) < 0$ for

$0 < x < 1$, $f'(1) = 0$ and $f'(x) > 0$ for $1 < x < 2$. We have shown that $f'(x) > 0$ for $x < 0$; $f'(x) < 0$ for

$0 < x < 1$; $f'(x) > 0$ for $1 < x < 2$; and $f'(x) < 0$ for $x > 2$. Therefore, by the First Derivative Test, the local

maxima of f are at $x = 0$ and $x = 2$, where f takes the value $\frac{4}{3}$. Therefore, $\frac{4}{3}$ is the absolute maximum value

of f.

11. We first show that $\dfrac{x}{1+x^2} < \tan^{-1} x$ for $x > 0$. Let $f(x) = \tan^{-1} x - \dfrac{x}{1+x^2}$. Then

$f'(x) = \dfrac{1}{1+x^2} - \dfrac{1(1+x^2) - x(2x)}{(1+x^2)^2} = \dfrac{(1+x^2)-(1-x^2)}{(1+x^2)^2} = \dfrac{2x^2}{(1+x^2)^2} > 0$ for $x > 0$. So $f(x)$ is

increasing on $(0, \infty)$. Hence, $0 < x \Rightarrow 0 = f(0) < f(x) = \tan^{-1} x - \dfrac{x}{1+x^2}$. So $\dfrac{x}{1+x^2} < \tan^{-1} x$ for

$0 < x$. We next show that $\tan^{-1} x < x$ for $x > 0$. Let $h(x) = x - \tan^{-1} x$. Then

$h'(x) = 1 - \dfrac{1}{1+x^2} = \dfrac{x^2}{1+x^2} > 0$. Hence, $h(x)$ is increasing on $(0, \infty)$. So for $0 < x$,

$0 = h(0) < h(x) = x - \tan^{-1} x$. Hence, $\tan^{-1} x < x$ for $x > 0$, and we conclude that $\dfrac{x}{1+x^2} < \tan^{-1} x < x$ for

$x > 0$.

13. $A = (x_1, x_1^2)$ and $B = (x_2, x_2^2)$, where x_1 and x_2 are the solutions of the quadratic equation $x^2 = mx + b$. Let

$P = (x, x^2)$ and set $A_1 = (x_1, 0)$, $B_1 = (x_2, 0)$, and $P_1 = (x, 0)$. Let $f(x)$ denote the area of triangle PAB.

Then $f(x)$ can be expressed in terms of the areas of three trapezoids as follows:

$$f(x) = \text{area}(A_1ABB_1) - \text{area}(A_1APP_1) - \text{area}(B_1BPP_1)$$

$$= \tfrac{1}{2}\left(x_1^2 + x_2^2\right)(x_2 - x_1) - \tfrac{1}{2}\left(x_1^2 + x^2\right)(x - x_1) - \tfrac{1}{2}\left(x^2 + x_2^2\right)(x_2 - x)$$

After expanding and canceling terms, we get

$f(x) = \tfrac{1}{2}\left(x_2 x_1^2 - x_1 x_2^2 - x x_1^2 + x_1 x^2 - x_2 x^2 + x x_2^2\right) = \tfrac{1}{2}\left[x_1^2(x_2 - x) + x_2^2(x - x_1) + x^2(x_1 - x_2)\right]$

$f'(x) = \tfrac{1}{2}\left[-x_1^2 + x_2^2 + 2x(x_1 - x_2)\right]$. $f''(x) = \tfrac{1}{2}[2(x_1 - x_2)] = x_1 - x_2 < 0$ since $x_2 > x_1$.

$f'(x) = 0 \Rightarrow 2x(x_1 - x_2) = x_1^2 - x_2^2 \Rightarrow x_P = \tfrac{1}{2}(x_1 + x_2)$.

$$f(x_P) = \frac{1}{2}\left(x_1^2\left[\frac{1}{2}(x_2 - x_1)\right] + x_2^2\left[\frac{1}{2}(x_2 - x_1)\right] + \frac{1}{4}(x_1 + x_2)^2(x_1 - x_2)\right)$$

$$= \frac{1}{2}\left[\frac{1}{2}(x_2 - x_1)\left(x_1^2 + x_2^2\right) - \frac{1}{4}(x_2 - x_1)(x_1 + x_2)^2\right]$$

$$= \frac{1}{8}(x_2 - x_1)\left[2\left(x_1^2 + x_2^2\right) - \left(x_1^2 + 2x_1x_2 + x_2^2\right)\right]$$

$$= \frac{1}{8}(x_2 - x_1)\left(x_1^2 - 2x_1x_2 + x_2^2\right) = \frac{1}{8}(x_2 - x_1)(x_1 - x_2)^2 = \frac{1}{8}(x_2 - x_1)(x_2 - x_1)^2$$

$$= \frac{1}{8}(x_2 - x_1)^3$$

To put this in terms of m and b, we solve the system $y = x_1^2$ and $y = mx_1 + b$, giving us $x_1^2 - mx_1 - b = 0 \Rightarrow$

$x_1 = \frac{1}{2}\left(m - \sqrt{m^2 + 4b}\right)$. Similarly, $x_2 = \frac{1}{2}\left(m + \sqrt{m^2 + 4b}\right)$. The area is then $\frac{1}{8}(x_2 - x_1)^3 = \frac{1}{8}\left(\sqrt{m^2 + 4b}\right)^3$,

and is attained at the point $P\left(x_P, x_P^2\right) = P\left(\frac{1}{2}m, \frac{1}{4}m^2\right)$.

Note: Another way to get an expression for $f(x)$ is to use the formula for an area of a triangle in terms of the

coordinates of the vertices: $f(x) = \frac{1}{2}\left[(x_2x_1^2 - x_1x_2^2) + (x_1x^2 - xx_1^2) + (xx_2^2 - x_2x^2)\right]$.

15. $f(x) = (a^2 + a - 6)\cos 2x + (a - 2)x + \cos 1 \Rightarrow f'(x) = -(a^2 + a - 6)\sin 2x \,(2) + (a - 2)$.

The derivative exists for all x, so the only possible critical points will occur where $f'(x) = 0 \Leftrightarrow$

$2(a - 2)(a + 3)\sin 2x = a - 2 \Leftrightarrow$ either $a = 2$ or $2(a + 3)\sin 2x = 1$, with the latter implying that

$\sin 2x = \dfrac{1}{2(a + 3)}$. Since the range of $\sin 2x$ is $[-1, 1]$, this equation has no solution whenever either

$\dfrac{1}{2(a + 3)} > 1$ or $\dfrac{1}{2(a + 3)} > 1$. Solving these inequalities, we get $-\frac{7}{2} < a < -\frac{5}{2}$.

17. (a) Let $y = |AD|$, $x = |AB|$, and $1/x = |AC|$, so that $|AB| \cdot |AC| = 1$. We

compute the area $\mathcal{A}$ of $\triangle ABC$ in two ways. First,

$\mathcal{A} = \frac{1}{2}|AB||AC|\sin\frac{2\pi}{3} = \frac{1}{2} \cdot 1 \cdot \frac{\sqrt{3}}{2} = \frac{\sqrt{3}}{4}$. Second,

$\mathcal{A} = (\text{area of } \triangle ABD) + (\text{area of } \triangle ACD)$

$= \frac{1}{2}|AB||AD|\sin\frac{\pi}{3} + \frac{1}{2}|AD||AC|\sin\frac{\pi}{3}$

$= \frac{1}{2}xy\frac{\sqrt{3}}{2} + \frac{1}{2}y(1/x)\frac{\sqrt{3}}{2} = \frac{\sqrt{3}}{4}y(x + 1/x)$

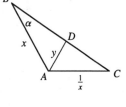

Equating the two expressions for the area, we get $\frac{\sqrt{3}}{4}y\left(x + \dfrac{1}{x}\right) = \frac{\sqrt{3}}{4} \Leftrightarrow y = \dfrac{1}{x + 1/x} = \dfrac{x}{x^2 + 1}, x > 0$.

Another Method: Use the Law of Sines on the triangles ABD and ABC. In $\triangle ABD$, we have

$\angle A + \angle B + \angle D = 180° \Leftrightarrow 60° + \alpha + \angle D = 180° \Leftrightarrow \angle D = 120° - \alpha$. Thus,

$\dfrac{x}{y} = \dfrac{\sin(120° - \alpha)}{\sin\alpha} = \dfrac{\sin 120° \cos\alpha - \cos 120° \sin\alpha}{\sin\alpha} = \dfrac{\frac{\sqrt{3}}{2}\cos\alpha + \frac{1}{2}\sin\alpha}{\sin\alpha} \Rightarrow \dfrac{x}{y} = \frac{\sqrt{3}}{2}\cot\alpha + \frac{1}{2}$, and

by a similar argument with $\triangle ABC$, $\frac{\sqrt{3}}{2}\cot\alpha = x^2 + \frac{1}{2}$. Eliminating $\cot\alpha$ gives $\dfrac{x}{y} = \left(x^2 + \frac{1}{2}\right) + \frac{1}{2} \Rightarrow$

$y = \dfrac{x}{x^2 + 1}, x > 0$.

(b) We differentiate our expression for y with respect to x to find the maximum:

$$\frac{dy}{dx} = \frac{(x^2 + 1) - x\,(2x)}{(x^2 + 1)^2} = \frac{1 - x^2}{(x^2 + 1)^2} = 0 \text{ when } x = 1.\ \text{This indicates a maximum by the First Derivative}$$

Test, since $y'\,(x) > 0$ for $0 < x < 1$ and $y'\,(x) < 0$ for $x > 1$, so the maximum value of y is $y\,(1) = \frac{1}{2}$.

19. (a) $A = \frac{1}{2}bh$ with $\sin\theta = h/c$, so $A = \frac{1}{2}bc\sin\theta$. But A is a

constant, so differentiating this equation with respect to t, we

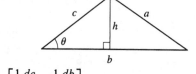

get $\dfrac{dA}{dt} = 0 = \dfrac{1}{2}\left[bc\cos\theta\dfrac{d\theta}{dt} + b\dfrac{dc}{dt}\sin\theta + \dfrac{db}{dt}c\sin\theta\right]$

$\Rightarrow\ bc\cos\theta\dfrac{d\theta}{dt} = -\sin\theta\left[b\dfrac{dc}{dt} + c\dfrac{db}{dt}\right] \Rightarrow \dfrac{d\theta}{dt} = -\tan\theta\left[\dfrac{1}{c}\dfrac{dc}{dt} + \dfrac{1}{b}\dfrac{db}{dt}\right].$

(b) We use the Law of Cosines to get the length of side a in terms of those of b and c, and then

we differentiate implicitly with respect to t: $a^2 = b^2 + c^2 - 2bc\cos\theta \Rightarrow$

$2a\dfrac{da}{dt} = 2b\dfrac{db}{dt} + 2c\dfrac{dc}{dt} - 2\left[bc\,(-\sin\theta)\dfrac{d\theta}{dt} + b\dfrac{dc}{dt}\cos\theta + \dfrac{db}{dt}c\cos\theta\right] \Rightarrow$

$\dfrac{da}{dt} = \dfrac{1}{a}\left(b\dfrac{db}{dt} + c\dfrac{dc}{dt} + bc\sin\theta\dfrac{d\theta}{dt} - b\dfrac{dc}{dt}\cos\theta - c\dfrac{db}{dt}\cos\theta\right).$ Now we substitute our value of a from the

Law of Cosines and the value of $d\theta/dt$ from part (a), and simplify (primes signify differentiation by t):

$$\frac{da}{dt} = \frac{bb' + cc' + bc\sin\theta\left[-\tan\theta\,(c'/c + b'/b)\right] - (bc' + cb')\,(\cos\theta)}{\sqrt{b^2 + c^2 - 2bc\cos\theta}}$$

$$= \frac{bb' + cc' - [\sin^2\theta\,(bc' + cb') + \cos^2\theta\,(bc' + cb')]/\cos\theta}{\sqrt{b^2 + c^2 - 2bc\cos\theta}} = \frac{bb' + cc' - (bc' + cb')\sec\theta}{\sqrt{b^2 + c^2 - 2bc\cos\theta}}$$

21. Suppose that the curve $y = a^x$ intersects the line $y = x$. Then $a^{x_0} = x_0$ for some $x_0 > 0$, and hence $a = x_0^{1/x_0}$.

We find the maximum value of $g\,(x) = x^{1/x}, > 0$, because if a is larger than the maximum

value of this function, then the curve $y = a^x$ does not intersect the line $y = x$.

$g'\,(x) = e^{(1/x)\ln x}\left(-\dfrac{1}{x^2}\ln x + \dfrac{1}{x}\cdot\dfrac{1}{x}\right) = x^{1/x}\left(\dfrac{1}{x^2}\right)(1 - \ln x).$ This is 0 only where $x = e$, and for $0 < x < e$,

$f'\,(x) > 0$, while for $x > e$, $f'\,(x) < 0$, so g has an absolute maximum of $g\,(e) = e^{1/e}$. So if $y = a^x$ intersects

$y = x$, we must have $0 < a \le e^{1/e}$. Conversely, suppose that $0 < a \le e^{1/e}$. Then $a^e \le e$, so the graph of $y = a^x$

lies below or touches the graph of $y = x$ at $x = e$. Also $a^0 = 1 > 0$, so the graph of $y = a^x$ lies above that of

$y = x$ at $x = 0$. Therefore, by the Intermediate Value Theorem, the graphs of $y = a^x$ and $y = x$ must intersect

somewhere between $x = 0$ and $x = e$.

23. Note that $f(0) = 0$, so for $x \neq 0$, $\left| \dfrac{f(x) - f(0)}{x - 0} \right| = \left| \dfrac{f(x)}{x} \right| = \dfrac{|f(x)|}{|x|} \leq \dfrac{|\sin x|}{|x|} = \dfrac{\sin x}{x}$. Therefore,

$|f'(0)| = \left| \lim\limits_{x \to 0} \dfrac{f(x) - f(0)}{x - 0} \right| = \lim\limits_{x \to 0} \left| \dfrac{f(x) - f(0)}{x - 0} \right| \leq \lim\limits_{x \to 0} \dfrac{\sin x}{x} = 1$. But

$f'(x) = a_1 \cos x + 2a_2 \cos 2x + \cdots + na_n \cos nx$, so $|f'(0)| = |a_1 + 2a_2 + \cdots + na_n| \leq 1$.

Another Solution: We are given that $\left| \sum_{k=1}^{n} a_k \sin kx \right| \leq |\sin x|$. So for x close to 0, and $x \neq 0$, we have

$\left| \sum\limits_{k=1}^{n} a_k \dfrac{\sin kx}{\sin x} \right| \leq 1 \quad \Rightarrow \quad \lim\limits_{x \to 0} \left| \sum\limits_{k=1}^{n} a_k \dfrac{\sin kx}{\sin x} \right| \leq 1 \quad \Rightarrow \quad \left| \sum\limits_{k=1}^{n} a_k \lim\limits_{x \to 0} \dfrac{\sin kx}{\sin x} \right| \leq 1$. But by l'Hospital's Rule,

$\lim\limits_{x \to 0} \dfrac{\sin kx}{\sin x} = \lim\limits_{x \to 0} \dfrac{k \cos kx}{\cos x} = k$, so $\left| \sum\limits_{k=1}^{n} ka_k \right| \leq 1$.

5 Integrals

5.1 Areas and Distances

1. (a) Since f is *increasing*, we can obtain a *lower* estimate by using *left* endpoints.

$$L_5 = \sum_{i=1}^{5} f(x_{i-1})\, \Delta x \quad [\Delta x = \tfrac{10-0}{5} = 2]$$

$$= f(x_0) \cdot 2 + f(x_1) \cdot 2 + f(x_2) \cdot 2 + f(x_3) \cdot 2 + f(x_4) \cdot 2$$

$$= 2[f(0) + f(2) + f(4) + f(6) + f(8)]$$

$$\approx 2(1 + 3 + 4.3 + 5.4 + 6.3) = 2(20) = 40$$

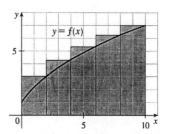

Since f is *increasing*, we can obtain an *upper* estimate by using *right* endpoints.

$$R_5 = \sum_{i=1}^{5} f(x_i)\, \Delta x$$

$$= 2[f(x_1) + f(x_2) + f(x_3) + f(x_4) + f(x_5)]$$

$$= 2[f(2) + f(4) + f(6) + f(8) + f(10)]$$

$$\approx 2(3 + 4.3 + 5.4 + 6.3 + 7) = 2(26) = 52$$

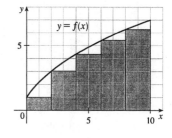

(b) $L_{10} = \sum_{i=1}^{10} f(x_{i-1})\, \Delta x \quad [\Delta x = \tfrac{10-0}{10} = 1]$

$$= 1[f(x_0) + f(x_1) + \cdots + f(x_9)]$$

$$= f(0) + f(1) + \cdots + f(9)$$

$$\approx 1 + 2.1 + 3 + 3.7 + 4.3 + 4.9 + 5.4 + 5.8 + 6.3 + 6.7$$

$$= 43.2$$

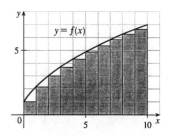

$$R_{10} = \sum_{i=1}^{10} f(x_i)\, \Delta x = f(1) + f(2) + \cdots + f(10)$$

$$= L_{10} + 1 \cdot f(10) - 1 \cdot f(0) \quad \begin{bmatrix} \text{add rightmost rectangle,} \\ \text{subtract leftmost} \end{bmatrix}$$

$$= 43.2 + 7 - 1 = 49.2$$

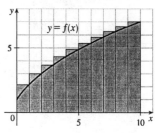

3. (a) $R_4 = \sum_{i=1}^{4} f(x_i)\,\Delta x \quad [\Delta x = \frac{5-1}{4} = 1]$

$\quad = f(x_1) \cdot 1 + f(x_2) \cdot 1 + f(x_3) \cdot 1 + f(x_4) \cdot 1$

$\quad = f(2) + f(3) + f(4) + f(5)$

$\quad = \frac{1}{2} + \frac{1}{3} + \frac{1}{4} + \frac{1}{5} = \frac{77}{60} = 1.28\overline{3}$

Since f is decreasing on $[1, 5]$, R_4 is an underestimate.

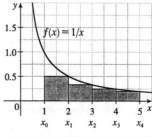

(b) $L_4 = \sum_{i=1}^{4} f(x_{i-1})\,\Delta x$

$\quad = f(1) + f(2) + f(3) + f(4)$

$\quad = 1 + \frac{1}{2} + \frac{1}{3} + \frac{1}{4} = \frac{25}{12} = 2.08\overline{3}$

L_4 is an overestimate. Alternatively, we could just add the leftmost rectangle and subtract the rightmost; that is,

$L_4 = R_4 + f(1) \cdot 1 - f(5) \cdot 1.$

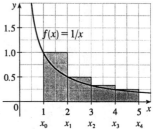

5. (a) $f(x) = x^3 + 2$ and $\Delta x = \dfrac{2-(-1)}{3} = 1 \;\Rightarrow$

$R_3 = 1 \cdot f(0) + 1 \cdot f(1) + 1 \cdot f(2) = 1 \cdot 2 + 1 \cdot 3 + 1 \cdot 10 = 15.$

$\Delta x = \dfrac{2-(-1)}{6} = 0.5 \;\Rightarrow$

$R_6 = 0.5\,[f(-0.5) + f(0) + f(0.5) + f(1) + f(1.5) + f(2)]$

$\quad = 0.5\,(1.875 + 2 + 2.125 + 3 + 5.375 + 10)$

$\quad = 0.5\,(24.375) = 12.1875$

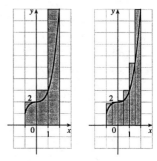

(b) $L_3 = 1 \cdot f(-1) + 1 \cdot f(0) + 1 \cdot f(1) = 1 \cdot 1 + 1 \cdot 2 + 1 \cdot 3 = 6.$

$L_6 = 0.5\,[f(-1) + f(-0.5) + f(0) + f(0.5) + f(1) + f(1.5)]$

$\quad = 0.5\,(1 + 1.875 + 2 + 2.125 + 3 + 5.375)$

$\quad = 0.5\,(15.375) = 7.6875$

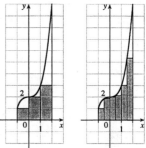

(c) $M_3 = 1 \cdot f(-0.5) + 1 \cdot f(0.5) + 1 \cdot f(1.5)$

$\quad = 1 \cdot 1.875 + 1 \cdot 2.125 + 1 \cdot 5.375 = 9.375.$

$M_6 = 0.5\,[f(-0.75) + f(-0.25) + f(0.25)$

$\qquad\qquad + f(0.75) + f(1.25) + f(1.75)]$

$\quad = 0.5\,(1.578125 + 1.984375 + 2.015625$

$\qquad\qquad + 2.421875 + 3.953125 + 7.359375)$

$\quad = 0.5\,(19.3125) = 9.65625$

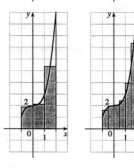

(d) M_6 appears to be the best estimate.

7. Here is one possible algorithm (ordered sequence of operations) for calculating the sums:

1 Let SUM = 0, X_MIN = 0, X_MAX = π, N = 10 (or 30 or 50, depending on which sum we are calculating),
DELTA_X = (X_MAX − X_MIN) /N, and RIGHT_ENDPOINT = X_MIN + DELTA_X.

2 Repeat steps 2a, 2b in sequence until RIGHT_ENDPOINT > X_MAX.

2a Add sin (RIGHT_ENDPOINT) to SUM.

2b Add DELTA_X to RIGHT_ENDPOINT.

At the end of this procedure, (DELTA_X) · (SUM) is equal to the answer we are looking for. We find that
$R_{10} = \frac{\pi}{10} \sum_{i=1}^{10} \sin \left(\frac{i\pi}{10}\right) \approx 1.9835$, $R_{30} = \frac{\pi}{30} \sum_{i=1}^{30} \sin \left(\frac{i\pi}{30}\right) \approx 1.9982$, and $R_{50} = \frac{\pi}{50} \sum_{i=1}^{50} \sin \left(\frac{i\pi}{50}\right) \approx 1.9993$.
It appears that the exact area is 2.

9. In Maple, we have to perform a number of steps before getting a numerical answer. After
loading the student package [command: `with(student);`] we use the command
`left_sum:=leftsum(x^(1/2),x=1..4,10 [or 30, or 50]);` which gives us the expression in
summation notation. To get a numerical approximation to the sum, we use `evalf(left_sum);`. Mathematica
does not have a special command for these sums, so we must type them in manually. For example, the first left
sum is given by `(3/10)*Sum[Sqrt[1+3(i-1)/10],{i,1,10}]`, and we use the `N` command on the
resulting output to get a numerical approximation.

In Derive, we use the `LEFT_RIEMANN` command to get the left sums, but must define the right sums ourselves.
(We can define a new function using `LEFT_RIEMANN` with k ranging from 1 to n instead of from 0 to $n − 1$.)

(a) With $f(x) = \sqrt{x}$, $1 \le x \le 4$, the left sums are of the form $L_n = \frac{3}{n} \sum_{i=1}^{n} \sqrt{1 + \frac{3(i-1)}{10}}$. Specifically,

$L_{10} \approx 4.5148$, $L_{30} \approx 4.6165$, and $L_{50} \approx 4.6366$. The right sums are of the form $R_n = \frac{3}{n} \sum_{i=1}^{n} \sqrt{1 + \frac{3i}{n}}$.

Specifically, $R_{10} \approx 4.8148$, $R_{30} \approx 4.7165$, and $R_{50} \approx 4.6966$.

(b) In Maple, we use the `leftbox` and `rightbox` commands (with the same arguments as `leftsum` and
`rightsum` above) to generate the graphs.

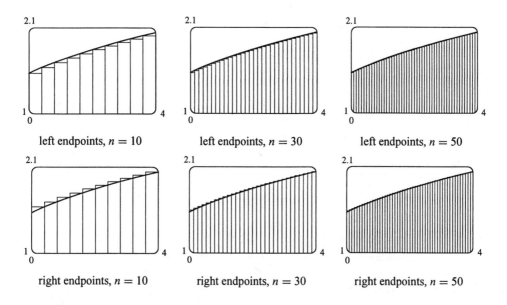

| left endpoints, $n = 10$ | left endpoints, $n = 30$ | left endpoints, $n = 50$ |

| right endpoints, $n = 10$ | right endpoints, $n = 30$ | right endpoints, $n = 50$ |

(c) We know that since $\sqrt{x}$ is an increasing function on $(1, 4)$, all of the left sums are smaller than the actual area, and all of the right sums are larger than the actual area. Since the left sum with $n = 50$ is about $4.637 > 4.6$ and the right sum with $n = 50$ is about $4.697 < 4.7$, we conclude that $4.6 < L_{50} <$ actual area $< R_{50} < 4.7$, so the actual area is between 4.6 and 4.7.

11. Since v is an increasing function, L_6 will give us a lower estimate and R_6 will give us an upper estimate.

$L_6 = (0 \text{ ft/s}) (0.5 \text{ s}) + (6.2) (0.5) + (10.8) (0.5) + (14.9) (0.5) + (18.1) (0.5) + (19.4) (0.5)$

$\quad = 0.5 (69.4) = 34.7 \text{ ft}$

$R_6 = 0.5 (6.2 + 10.8 + 14.9 + 18.1 + 19.4 + 20.2) = 0.5 (89.6) = 44.8 \text{ ft}$

13. For a decreasing function, using left endpoints gives us an overestimate and using right endpoints results in an underestimate. We will use M_6 to get an estimate. $\Delta t = 1$, so

$$M_6 = 1 \left[v (0.5) + v (1.5) + v (2.5) + v (3.5) + v (4.5) + v (5.5) \right]$$
$$\approx 55 + 40 + 28 + 18 + 10 + 4 = 155 \text{ ft}$$

For a very rough check on the above calculation, we can draw a line from $(0, 70)$ to $(6, 0)$ and calculate the area of the triangle: $\frac{1}{2} (70) (6) = 210$. This is clearly an overestimate, so our midpoint estimate of 155 is reasonable.

15. $f (x) = \sqrt[3]{x}, 0 \le x \le 8 \;\Rightarrow\; \Delta x = \dfrac{8 - 0}{n} = \dfrac{8}{n}, x_i = 0 + i \, \Delta x = \dfrac{8i}{n} \;\Rightarrow$

$R_n = f (x_1) \Delta x + f (x_2) \Delta x + \cdots + f (x_n) \Delta x = \displaystyle\sum_{i=1}^{n} f (x_i) \Delta x = \sum_{i=1}^{n} \sqrt[3]{\dfrac{8i}{n}} \cdot \dfrac{8}{n}$. Thus,

$A = \displaystyle\lim_{n\to\infty} R_n = \lim_{n\to\infty} \dfrac{8}{n} \sum_{i=1}^{n} \sqrt[3]{\dfrac{8i}{n}}$.

17. $f (x) = x + \ln x, 2 \le x \le 6 \;\Rightarrow\; \Delta x = \dfrac{6 - 2}{n} = \dfrac{4}{n}, x_i = 2 + i \, \Delta x = 2 + \dfrac{4i}{n} \;\Rightarrow$

$R_n = f (x_1) \Delta x + f (x_2) \Delta x + \cdots + f (x_n) \Delta x = \displaystyle\sum_{i=1}^{n} f (x_i) \Delta x = \sum_{i=1}^{n} \left[2 + \dfrac{4i}{n} + \ln \left(2 + \dfrac{4i}{n} \right) \right] \cdot \dfrac{4}{n}$. Thus,

$A = \displaystyle\lim_{n\to\infty} R_n = \lim_{n\to\infty} \dfrac{4}{n} \sum_{i=1}^{n} \left[2 + \dfrac{4i}{n} + \ln \left(2 + \dfrac{4i}{n} \right) \right]$.

19. $\displaystyle\lim_{n\to\infty} \sum_{i=1}^{n} \dfrac{\pi}{4n} \tan \dfrac{\pi i}{4n}$ can be interpreted as the area of the region lying under the graph of $y = \tan x$ on the interval

$\left[0, \frac{\pi}{4} \right]$, since for $y = \tan x$ on $\left[0, \frac{\pi}{4} \right]$ with partition points $x_i = \left(\dfrac{\pi}{4} \right) \dfrac{i}{n}$, $\Delta x = \dfrac{\pi}{4n}$ and $x_i^* = x_i$, the expression for

the area is $A = \displaystyle\lim_{n\to\infty} \sum_{i=1}^{n} f (x_i^*) \Delta x = \lim_{n\to\infty} \sum_{i=1}^{n} \tan \left(\dfrac{\pi i}{4n} \right) \dfrac{\pi}{4n}$. Note that this answer is not unique, since the

expression for the area is the same for the function $y = \tan (k\pi + x)$ on the interval $\left[k\pi, k\pi + \frac{\pi}{4} \right]$ where k is any integer.

21. (a) $\Delta x = \dfrac{2 - 0}{n} = \dfrac{2}{n}$ and $x_i = 0 + i \, \Delta x = \dfrac{2i}{n}$.

$A = \displaystyle\lim_{n\to\infty} R_n = \lim_{n\to\infty} \sum_{i=1}^{n} f (x_i) \Delta x = \lim_{n\to\infty} \sum_{i=1}^{n} \left(\dfrac{2i}{n} \right)^5 \cdot \dfrac{2}{n} = \lim_{n\to\infty} \dfrac{64}{n^6} \sum_{i=1}^{n} i^5$.

(b) $\displaystyle\sum_{i=1}^{n} i^5 = \dfrac{n^2 (n + 1)^2 (2n^2 + 2n - 1)}{12}$

(c) $\displaystyle\lim_{n\to\infty}\frac{64}{n^6}\cdot\frac{n^2\,(n+1)^2\,(2n^2+2n-1)}{12}=\frac{64}{12}\lim_{n\to\infty}\frac{(n^2+2n+1)\,(2n^2+2n-1)}{n^2\cdot n^2}$

$$=\frac{16}{3}\lim_{n\to\infty}\left(1+\frac{2}{n}+\frac{1}{n^2}\right)\left(2+\frac{2}{n}-\frac{1}{n^2}\right)=\tfrac{16}{3}\cdot1\cdot2=\tfrac{32}{3}$$

23. $\Delta x=\dfrac{b-0}{n}=\dfrac{b}{n}$ and $x_i=0+i\,\Delta x=\dfrac{bi}{n}$.

$$A=\lim_{n\to\infty}R_n=\lim_{n\to\infty}\sum_{i=1}^{n}f\,(x_i)\,\Delta x=\lim_{n\to\infty}\sum_{i=1}^{n}\cos\left(\frac{bi}{n}\right)\cdot\frac{b}{n}=\lim_{n\to\infty}\left[\frac{b\sin\left(b\left(\dfrac{1}{2n}+1\right)\right)}{2n\sin\left(\dfrac{b}{2n}\right)}-\frac{b}{2n}\right]=\sin b$$

If $b=\frac{\pi}{2}$, then $A=\sin\frac{\pi}{2}=1$.

5.2 The Definite Integral

1. $R_4=\displaystyle\sum_{i=1}^{4}f\,(x_i)\,\Delta x\qquad[x_i^*=x_i$ is a right endpoint and $\Delta x=0.5]$

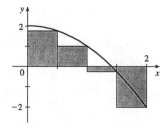

$=0.5\,[f\,(0.5)+f\,(1)+f\,(1.5)+f\,(2)]\qquad[f\,(x)=2-x^2]$

$=0.5\,[1.75+1+(-0.25)+(-2)]$

$=0.5\,(0.5)=0.25$

The Riemann sum represents the area of the two rectangles above the x-axis minus the area of the two rectangles below the x-axis.

3. $M_5=\displaystyle\sum_{i=1}^{5}f\,(\bar{x}_i)\,\Delta x\qquad[x_i^*=\bar{x}_i=\frac{1}{2}\,(x_{i-1}+x_i)$ is a midpoint and $\Delta x=1]$

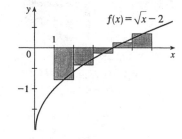

$=1\,[f\,(1.5)+f\,(2.5)+f\,(3.5)$

$\qquad+f\,(4.5)+f\,(5.5)]\qquad[f\,(x)=\sqrt{x}-2]$

≈-0.856759

The Riemann sum represents the area of the two rectangles above the x-axis minus the area of the three rectangles below the x-axis.

5. (a) Using the right endpoints to approximate $\int_0^8 f\,(x)\,dx$, we have

$$\sum_{i=1}^{4}f\,(x_i)\,\Delta x=2\,[f\,(2)+f\,(4)+f\,(6)+f\,(8)]\approx2\,(1+2-2+1)=4.$$

(b) Using the left endpoints to approximate $\int_0^8 f\,(x)\,dx$, we have

$$\sum_{i=1}^{4}f\,(x_{i-1})\,\Delta x=2\,[f\,(0)+f\,(2)+f\,(4)+f\,(6)]\approx2\,(2+1+2-2)=6.$$

(c) Using the midpoint of each subinterval to approximate $\int_0^8 f\,(x)\,dx$, we have

$$\sum_{i=1}^{4}f\,(\bar{x}_i)\,\Delta x=2\,[f\,(1)+f\,(3)+f\,(5)+f\,(7)]\approx2\,(3+2+1-1)=10.$$

7. Since f is increasing, $L_5 \le \int_0^{25} f(x)\,dx \le R_5$.

$$\text{Lower estimate} = L_5 = \sum_{i=1}^{5} f(x_{i-1})\,\Delta x = 5[f(0) + f(5) + f(10) + f(15) + f(20)]$$

$$= 5(-42 - 37 - 25 - 6 + 15) = 5(-95) = -475$$

$$\text{Upper estimate} = R_5 = \sum_{i=1}^{5} f(x_i)\,\Delta x = 5[f(5) + f(10) + f(15) + f(20) + f(25)]$$

$$= 5(-37 - 25 - 6 + 15 + 36) = 5(-17) = -85$$

9. $\Delta x = (10 - 0)/5 = 2$, so the endpoints are 0, 2, 4, 6, 8, and 10, and the midpoints are 1, 3, 5, 7, and 9. The Midpoint Rule gives

$$\int_0^{10} \sin\sqrt{x}\,dx \approx \sum_{i=1}^{5} f(\overline{x}_i)\,\Delta x = 2\left(\sin\sqrt{1} + \sin\sqrt{3} + \sin\sqrt{5} + \sin\sqrt{7} + \sin\sqrt{9}\right) \approx 6.4643.$$

11. $\Delta x = (2 - 1)/10 = 0.1$, so the endpoints are $1.0, 1.1, \ldots, 2.0$ and the midpoints are $1.05, 1.15, \ldots, 1.95$. The Midpoint Rule gives

$$\int_1^2 \sqrt{1 + x^2}\,dx \approx \sum_{i=1}^{10} f(\overline{x}_i)\,\Delta x = 0.1\left[\sqrt{1 + (1.05)^2} + \sqrt{1 + (1.15)^2} + \cdots + \sqrt{1 + (1.95)^2}\right] \approx 1.8100.$$

13. In Maple, we use the command with(student); to load the sum and box commands, then m:=middlesum(sqrt(1+x^2),x=1..2,10); which gives us the sum in summation notation, then M:=evalf(m); which gives $M_{10} \approx 1.81001414$, confirming the result of Exercise 11. The command middlebox(sqrt(1+x^2),x=1..2,10); generates the graph. Repeating for $n = 20$ and $n = 30$ gives $M_{20} \approx 1.81007263$ and $M_{30} \approx 1.81008347$.

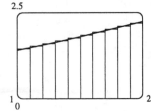
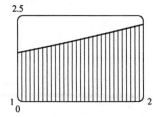
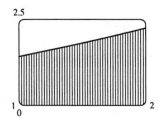

15. On $[0, \pi]$, $\displaystyle\lim_{n\to\infty}\sum_{i=1}^{n} x_i \sin x_i\,\Delta x = \int_0^\pi (x\sin x)\,dx$.

17. On $[0, 1]$, $\displaystyle\lim_{n\to\infty}\sum_{i=1}^{n}\left[2(x_i^*)^2 - 5x_i^*\right]\Delta x = \int_0^1 (2x^2 - 5x)\,dx$.

19. Note that $\Delta x = \dfrac{5 - (-1)}{n} = \dfrac{6}{n}$ and $x_i = -1 + i\,\Delta x = -1 + \dfrac{6i}{n}$.

$$\int_{-1}^{5} (1 + 3x)\,dx = \lim_{n\to\infty} \sum_{i=1}^{n} f(x_i)\,\Delta x = \lim_{n\to\infty} \sum_{i=1}^{n} \left[1 + 3\left(-1 + \frac{6i}{n}\right)\right]\frac{6}{n}$$

$$= \lim_{n\to\infty} \frac{6}{n} \sum_{i=1}^{n} \left[-2 + \frac{18i}{n}\right] = \lim_{n\to\infty} \frac{6}{n} \left[\sum_{i=1}^{n}(-2) + \sum_{i=1}^{n}\frac{18i}{n}\right]$$

$$= \lim_{n\to\infty} \frac{6}{n} \left[-2n + \frac{18}{n} \sum_{i=1}^{n} i\right] = \lim_{n\to\infty} \frac{6}{n} \left[-2n + \frac{18}{n} \cdot \frac{n(n+1)}{2}\right]$$

$$= \lim_{n\to\infty} \left[-12 + \frac{108}{n^2} \cdot \frac{n(n+1)}{2}\right] = \lim_{n\to\infty} \left[-12 + 54\frac{n+1}{n}\right]$$

$$= \lim_{n\to\infty} \left[-12 + 54\left(1 + \frac{1}{n}\right)\right] = -12 + 54 \cdot 1 = 42$$

21. $\displaystyle\int_{0}^{2} \left(2 - x^2\right)dx = \lim_{n\to\infty} \sum_{i=1}^{n} f(x_i)\,\Delta x \quad [\Delta x = 2/n \text{ and } x_i = 2i/n] \quad = \lim_{n\to\infty} \sum_{i=1}^{n} \left(2 - \frac{4i^2}{n^2}\right)\left(\frac{2}{n}\right)$

$$= \lim_{n\to\infty} \frac{2}{n}\left(2n - \frac{8}{n^3}\sum_{i=1}^{n} i^2\right) = \lim_{n\to\infty}\left[4 - \frac{8}{n^3} \cdot \frac{n(n+1)(2n+1)}{6}\right]$$

$$= \lim_{n\to\infty}\left(4 - \frac{4}{3} \cdot \frac{n+1}{n} \cdot \frac{2n+1}{n}\right) = \lim_{n\to\infty}\left[4 - \frac{4}{3}\left(1 + \frac{1}{n}\right)\left(2 + \frac{1}{n}\right)\right] = 4 - \frac{4}{3}\cdot 2 = \frac{4}{3}$$

23. Note that $x_i = 1 + i\,\Delta x = 1 + i\,(1/n) = 1 + i/n$.

$$\int_{1}^{2} x^3\,dx = \lim_{n\to\infty} \sum_{i=1}^{n} f(x_i)\,\Delta x = \lim_{n\to\infty} \sum_{i=1}^{n}\left(1 + \frac{i}{n}\right)^3\left(\frac{1}{n}\right) = \lim_{n\to\infty}\frac{1}{n}\sum_{i=1}^{n}\left(\frac{n+i}{n}\right)^3$$

$$= \lim_{n\to\infty}\frac{1}{n^4}\sum_{i=1}^{n}\left(n^3 + 3n^2 i + 3ni^2 + i^3\right) = \lim_{n\to\infty}\frac{1}{n^4}\left[n\cdot n^3 + 3n^2\sum_{i=1}^{n} i + 3n\sum_{i=1}^{n} i^2 + \sum_{i=1}^{n} i^3\right]$$

$$= \lim_{n\to\infty}\left[1 + \frac{3}{n^2} \cdot \frac{n(n+1)}{2} + \frac{3}{n^3} \cdot \frac{n(n+1)(2n+1)}{6} + \frac{1}{n^4} \cdot \frac{n^2(n+1)^2}{4}\right]$$

$$= \lim_{n\to\infty}\left[1 + \frac{3}{2} \cdot \frac{n+1}{n} + \frac{1}{2} \cdot \frac{n+1}{n} \cdot \frac{2n+1}{n} + \frac{1}{4} \cdot \frac{(n+1)^2}{n^2}\right]$$

$$= \lim_{n\to\infty}\left[1 + \frac{3}{2}\left(1 + \frac{1}{n}\right) + \frac{1}{2}\left(1 + \frac{1}{n}\right)\left(2 + \frac{1}{n}\right) + \frac{1}{4}\left(1 + \frac{1}{n}\right)^2\right] = 1 + \frac{3}{2} + \frac{1}{2}\cdot 2 + \frac{1}{4} = 3.75$$

25. $\displaystyle\int_{a}^{b} x\,dx = \lim_{n\to\infty}\frac{b-a}{n}\sum_{i=1}^{n}\left[a + \frac{b-a}{n}i\right] = \lim_{n\to\infty}\left[\frac{a(b-a)}{n}\sum_{i=1}^{n} 1 + \frac{(b-a)^2}{n^2}\sum_{i=1}^{n} i\right]$

$$= \lim_{n\to\infty}\left[\frac{a(b-a)}{n}n + \frac{(b-a)^2}{n^2} \cdot \frac{n(n+1)}{2}\right] = a(b-a) + \lim_{n\to\infty}\frac{(b-a)^2}{2}\left(1 + \frac{1}{n}\right)$$

$$= a(b-a) + \tfrac{1}{2}(b-a)^2 = (b-a)\left(a + \tfrac{1}{2}b - \tfrac{1}{2}a\right) = (b-a)\tfrac{1}{2}(b+a) = \tfrac{1}{2}(b^2 - a^2)$$

27. $\Delta x = (\pi - 0)/n = \pi/n$ and $x_i^* = x_i = \pi i/n$.

$$\int_0^\pi \sin 5x\, dx = \lim_{n\to\infty} \sum_{i=1}^n (\sin 5x_i)\left(\frac{\pi}{n}\right) = \pi \lim_{n\to\infty} \frac{1}{n} \sum_{i=1}^n \sin\left(\frac{5\pi i}{n}\right) = \pi \lim_{n\to\infty} \frac{1}{n} \cot\left(\frac{5\pi}{2n}\right) = \pi \left(\frac{2}{5\pi}\right) = \frac{2}{5}$$

29. (a) Think of $\int_0^2 f(x)\, dx$ as the area of a trapezoid with bases 1 and 3 and height 2.

$\int_0^2 f(x)\, dx = \frac{1}{2}(1+3)2 = 4$.

(b) $\int_0^5 f(x)\, dx = \int_0^2 f(x)\, dx + \int_2^3 f(x)\, dx + \int_3^5 f(x)\, dx$

                 trapezoid   rectangle   triangle

           $= \frac{1}{2}(1+3)2 + \quad 3\cdot 1 \quad + \quad \frac{1}{2}\cdot 2\cdot 3 \quad = 10$

(c) $\int_5^7 f(x)\, dx$ is the negative of the area of the triangle with base 2 and height 3. $\int_5^7 f(x)\, dx = -\frac{1}{2}\cdot 2\cdot 3 = -3$.

(d) As in parts (a) and (c), $\int_7^9 f(x)\, dx = -\frac{1}{2}(3+2)2 = -5$. Now

$\int_0^9 f(x)\, dx = \int_0^5 f(x)\, dx + \int_5^7 f(x)\, dx + \int_7^9 f(x)\, dx = 10 - 3 - 5 = 2$.

31. $\int_1^3 (1+2x)\, dx$ can be interpreted as the area under the graph of $f(x) = 1+2x$ between $x = 1$ and $x = 3$. This is equal to the area of the rectangle plus the area of the triangle, so $\int_1^3 (1+2x)\, dx = A = 2\cdot 3 + \frac{1}{2}\cdot 2\cdot 4 = 10$.

Or: Use the formula for the area of a trapezoid: $a = \frac{1}{2}(2)(3+7) = 10$.

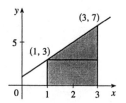

33. $\int_{-3}^0 \left(1+\sqrt{9-x^2}\right) dx$ can be interpreted as the area under the graph of

$f(x) = 1+\sqrt{9-x^2}$ between $x = -3$ and $x = 0$. This is equal to one-quarter the area of the circle with radius 3, plus the area of the rectangle, so

$\int_{-3}^0 \left(1+\sqrt{9-x^2}\right) dx = \frac{1}{4}\pi\cdot 3^2 + 1\cdot 3 = 3 + \frac{9}{4}\pi$.

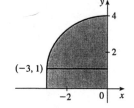

35. $\int_{-2}^2 (1-|x|)\, dx$ can be interpreted as the area of the middle triangle minus the

areas of the outside ones, so $\int_{-2}^2 (1-|x|)\, dx = \frac{1}{2}\cdot 2\cdot 1 - 2\cdot \frac{1}{2}\cdot 1\cdot 1 = 0$.

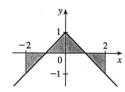

37. $\int_9^4 \sqrt{t}\, dt = -\int_4^9 \sqrt{t}\, dt$ (because we reversed the limits of integration) $= -\frac{38}{3}$

39. $\int_0^1 (5-6x^2)\, dx = \int_0^1 5\, dx - 6\int_0^1 x^2\, dx = 5(1-0) - 6\left(\frac{1}{3}\right) = 5 - 2 = 3$

41. $\int_1^3 e^{x+2}\, dx = e^2 \int_1^3 e^x\, dx = e^2(e^3 - e) = e^5 - e^3$

43. $\int_1^3 f(x)\, dx + \int_3^6 f(x)\, dx + \int_6^{12} f(x)\, dx = \int_1^6 f(x)\, dx + \int_6^{12} f(x)\, dx = \int_1^{12} f(x)\, dx$

45. $\int_2^5 f(x)\, dx + \int_5^8 f(x)\, dx = \int_2^8 f(x)\, dx \implies \int_2^5 f(x)\, dx + 2.5 = 1.7 \implies \int_2^5 f(x)\, dx = -0.8$

47. $0 \le \sin x < 1$ on $\left[0, \frac{\pi}{4}\right]$, so $\sin^3 x \le \sin^2 x$ on $\left[0, \frac{\pi}{4}\right]$. Hence, $\int_0^{\pi/4} \sin^3 x\, dx \le \int_0^{\pi/4} \sin^2 x\, dx$ (Property 7).

49. If $-1 \le x \le 1$, then $0 \le x^2 \le 1$ and $1 \le 1+x^2 \le 2$, so $1 \le \sqrt{1+x^2} \le \sqrt{2}$ and

$1[1-(-1)] \le \int_{-1}^1 \sqrt{1+x^2}\, dx \le \sqrt{2}[1-(-1)]$ [Property 8]; that is, $2 \le \int_{-1}^1 \sqrt{1+x^2}\, dx \le 2\sqrt{2}$.

51. If $1 \le x \le 2$, then $\frac{1}{2} \le \frac{1}{x} \le 1$, so $\frac{1}{2}(2-1) \le \int_1^2 \frac{1}{x}\,dx \le 1(2-1)$ or $\frac{1}{2} \le \int_1^2 \frac{1}{x}\,dx \le 1$.

53. If $f(x) = x^2 + 2x$, $-3 \le x \le 0$, then $f'(x) = 2x + 2 = 0$ when $x = -1$, and $f(-1) = -1$. At the endpoints, $f(-3) = 3$, $f(0) = 0$. Thus, the absolute minimum is $m = -1$ and the absolute maximum is $M = 3$. Thus, $-1[0-(-3)] \le \int_{-3}^0 (x^2+2x)\,dx \le 3[0-(-3)]$ or $-3 \le \int_{-3}^0 (x^2+2x)\,dx \le 9$.

55. The only critical number of $f(x) = xe^{-x}$ on $[0,2]$ is $x = 1$. Since $f(0) = 0$, $f(1) = e^{-1} \approx 0.368$, and $f(2) = 2e^{-2} \approx 0.271$, we know that the absolute minimum value of f on $[0,2]$ is 0, and the absolute maximum is e^{-1}. By Property 8, $0 \le xe^{-x} \le e^{-1}$ for $0 \le x \le 2$ $\Rightarrow$ $0(2-0) \le \int_0^2 xe^{-x}\,dx \le e^{-1}(2-0)$ $\Rightarrow$ $0 \le \int_0^2 xe^{-x}\,dx \le 2/e$.

57. $\sqrt{x^4+1} \ge \sqrt{x^4} = x^2$, so $\int_1^3 \sqrt{x^4+1}\,dx \ge \int_1^3 x^2\,dx = \frac{1}{3}(3^3 - 1^3) = \frac{26}{3}$.

59. Using a regular partition and right endpoints as in the proof of Property 2, we calculate
$$\int_a^b cf(x)\,dx = \lim_{n\to\infty} \sum_{i=1}^n cf(x_i)\,\Delta x_i = \lim_{n\to\infty} c\sum_{i=1}^n f(x_i)\,\Delta x_i = c\lim_{n\to\infty}\sum_{i=1}^n f(x_i)\,\Delta x_i = c\int_a^b f(x)\,dx.$$

61. Since $-|f(x)| \le f(x) \le |f(x)|$, it follows from Property 7 that
$$-\int_a^b |f(x)|\,dx \le \int_a^b f(x)\,dx \le \int_a^b |f(x)|\,dx \quad\Rightarrow\quad \left|\int_a^b f(x)\,dx\right| \le \int_a^b |f(x)|\,dx$$

Note that the definite integral is a real number, and so the following property applies: $-a \le b \le a$ $\Rightarrow$ $|b| \le a$ for all real numbers b and nonnegative numbers a.

63. $\lim_{n\to\infty} \sum_{i=1}^n \frac{i^4}{n^5} = \lim_{n\to\infty} \frac{1}{n}\sum_{i=1}^n \left(\frac{i}{n}\right)^4 = \int_0^1 x^4\,dx$

65. Choose $x_i = 1 + \frac{i}{n}$ and $x_i^* = \sqrt{x_{i-1}x_i} = \sqrt{\left(1+\frac{i-1}{n}\right)\left(1+\frac{i}{n}\right)}$. Then
$$\int_1^2 x^{-2}\,dx = \lim_{n\to\infty}\frac{1}{n}\sum_{i=1}^n \frac{1}{\left(1+\frac{i-1}{n}\right)\left(1+\frac{i}{n}\right)} = \lim_{n\to\infty} n\sum_{i=1}^n \frac{1}{(n+i-1)(n+i)}$$
$$= \lim_{n\to\infty} n\sum_{i=1}^n \left(\frac{1}{n+i-1} - \frac{1}{n+i}\right) \quad\text{(by the hint)}$$
$$= \lim_{n\to\infty} n\left(\sum_{i=0}^{n-1}\frac{1}{n+i} - \sum_{i=1}^n \frac{1}{n+i}\right) = \lim_{n\to\infty} n\left(\frac{1}{n} - \frac{1}{2n}\right) = \lim_{n\to\infty}\left(1 - \frac{1}{2}\right) = \frac{1}{2}$$

The Fundamental Theorem of Calculus

1. (a) $g(0) = \int_0^0 f(t)\,dt = 0$, $g(1) = \int_0^1 f(t)\,dt = 1 \cdot 2 = 2$,

$g(2) = \int_0^2 f(t)\,dt = \int_0^1 f(t)\,dt + \int_1^2 f(t)\,dt = g(1) + \int_1^2 f(t)\,dt = 2 + 1 \cdot 2 + \frac{1}{2} \cdot 1 \cdot 2 = 5$,

$g(3) = \int_0^3 f(t)\,dt = g(2) + \int_2^3 f(t)\,dt = 5 + \frac{1}{2} \cdot 1 \cdot 4 = 7$,   (d)

$g(6) = g(3) + \int_3^6 f(t)\,dt = 7 + \left[-\left(\frac{1}{2} \cdot 2 \cdot 2 + 1 \cdot 2\right) \right] = 7 - 4 = 3$

(b) g is increasing on $(0, 3)$ because as x increases from 0 to 3, we keep
adding more area.

(c) g has a maximum value when we start subtracting area, that is, at $x = 3$.

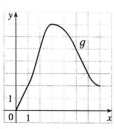

3.

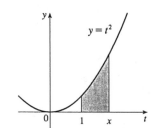

(a) By FTC1 with $f(t) = t^2$ and $a = 1$, $g(x) = \int_1^x t^2\,dt \Rightarrow$

$\qquad g'(x) = f(x) = x^2$.

(b) Using FTC2, $g(x) = \int_1^x t^2\,dt = \left[\frac{1}{3}t^3 \right]_1^x = \frac{1}{3}x^3 - \frac{1}{3} \Rightarrow$

$\qquad g'(x) = x^2$.

5. $f(t) = \sqrt{1 + 2t}$ and $g(x) = \int_0^x \sqrt{1 + 2t}\,dt$, so by FTC1, $g'(x) = f(x) = \sqrt{1 + 2x}$.

7. $f(t) = t^2 \sin t$ and $g(y) = \int_2^y t^2 \sin t\,dt$, so by FTC1, $g'(y) = f(y) = y^2 \sin y$.

9. $F(x) = \int_x^2 \cos(t^2)\,dt = -\int_2^x \cos(t^2)\,dt \Rightarrow F'(x) = -\cos(x^2)$

11. Let $u = \dfrac{1}{x}$. Then $\dfrac{du}{dx} = -\dfrac{1}{x^2}$. Also, $\dfrac{dh}{dx} = \dfrac{dh}{du}\dfrac{du}{dx}$, so

$h'(x) = \dfrac{d}{dx} \displaystyle\int_2^{1/x} \arctan t\,dt = \dfrac{d}{du} \displaystyle\int_2^u \arctan t\,dt \cdot \dfrac{du}{dx} = \arctan u\,\dfrac{du}{dx} = -\dfrac{\arctan(1/x)}{x^2}$.

13. Let $u = \sqrt{x}$. Then $\dfrac{du}{dx} = \dfrac{1}{2\sqrt{x}}$. Also, $\dfrac{dy}{dx} = \dfrac{dy}{du}\dfrac{du}{dx}$, so

$y' = \dfrac{d}{dx} \displaystyle\int_3^{\sqrt{x}} \dfrac{\cos t}{t}\,dt = \dfrac{d}{du} \displaystyle\int_3^u \dfrac{\cos t}{t}\,dt \cdot \dfrac{du}{dx} = \dfrac{\cos u}{u} \cdot \dfrac{1}{2\sqrt{x}} = \dfrac{\cos \sqrt{x}}{\sqrt{x}} \cdot \dfrac{1}{2\sqrt{x}} = \dfrac{\cos \sqrt{x}}{2x}$.

15. Let $w = 1 - 3x$. Then $\dfrac{dw}{dx} = -3$. Also, $\dfrac{dy}{dx} = \dfrac{dy}{dw}\dfrac{dw}{dx}$, so

$$y' = \dfrac{d}{dx} \int_{1-3x}^1 \dfrac{u^3}{1 + u^2}\,du = \dfrac{d}{dw} \int_w^1 \dfrac{u^3}{1 + u^2}\,du \cdot \dfrac{dw}{dx}$$

$$= -\dfrac{d}{dw} \int_1^w \dfrac{u^3}{1 + u^2}\,du \cdot \dfrac{dw}{dx} = -\dfrac{w^3}{1 + w^2}(-3) = \dfrac{3(1 - 3x)^3}{1 + (1 - 3x)^2}$$

17. $\displaystyle\int_{-1}^3 x^5\,dx = \left[\dfrac{x^6}{6} \right]_{-1}^3 = \dfrac{3^6}{6} - \dfrac{(-1)^6}{6} = \dfrac{729 - 1}{6} = \dfrac{364}{3}$

19. $\int_2^8 (4x + 3)\,dx = \left[\frac{4}{2}x^2 + 3x \right]_2^8 = [(2 \cdot 8^2 + 3 \cdot 8) - (2 \cdot 2^2 + 3 \cdot 2)] = 152 - 14 = 138$

21. $\int_0^4 \sqrt{x}\, dx = \int_0^4 x^{1/2}\, dx = \left[\dfrac{x^{3/2}}{3/2}\right]_0^4 = \left[\dfrac{2x^{3/2}}{3}\right]_0^4 = \dfrac{2(4)^{3/2}}{3} - 0 = \dfrac{16}{3}$

23. $\int_1^2 \dfrac{3}{t^4}\, dt = 3\int_1^2 t^{-4}\, dt = 3\left[\dfrac{t^{-3}}{-3}\right]_1^2 = \dfrac{3}{-3}\left[\dfrac{1}{t^3}\right]_1^2 = -1\left(\dfrac{1}{8} - 1\right) = \dfrac{7}{8}$

25. $\int_3^3 \sqrt{x^5 + 2}\, dx = 0$ since the lower and upper limits are equal.

27. $\int_{-4}^2 \dfrac{2}{x^6}\, dx$ does not exist since $f(x) = \dfrac{2}{x^6}$ has an infinite discontinuity at 0.

29. $\int_{\pi/4}^{\pi/3} \sin t\, dt = [-\cos t]_{\pi/4}^{\pi/3} = -\cos\dfrac{\pi}{3} + \cos\dfrac{\pi}{4} = -\dfrac{1}{2} + \dfrac{1}{\sqrt{2}} = \dfrac{\sqrt{2}-1}{2}$

31. $\int_{\pi/2}^{\pi} \sec x \tan x\, dx$ does not exist since $\sec x \tan x$ has an infinite discontinuity at $\dfrac{\pi}{2}$.

33. $\int_1^9 \dfrac{1}{2x}\, dx = \dfrac{1}{2}\int_1^9 \dfrac{1}{x}\, dx = \dfrac{1}{2}\left[\ln|x|\right]_1^9 = \dfrac{1}{2}(\ln 9 - \ln 1) = \dfrac{1}{2}\ln 9 - 0 = \ln 9^{1/2} = \ln 3$

35. $\int_8^9 2^t\, dt = \left[\dfrac{1}{\ln 2} 2^t\right]_8^9 = \dfrac{1}{\ln 2}\left(2^9 - 2^8\right) = \dfrac{2^8}{\ln 2}$

37. $\int_1^{\sqrt{3}} \dfrac{6}{1+x^2}\, dx = 6\left[\tan^{-1} x\right]_1^{\sqrt{3}} = 6\tan^{-1}\sqrt{3} - 6\tan^{-1} 1 = 6\cdot\dfrac{\pi}{3} - 6\cdot\dfrac{\pi}{4} = \dfrac{\pi}{2}$

39. $\int_0^2 f(x)\, dx = \int_0^1 x^4 dx + \int_1^2 x^5 dx = \left[\dfrac{1}{5}x^5\right]_0^1 + \left[\dfrac{1}{6}x^6\right]_1^2 = \left(\dfrac{1}{5} - 0\right) + \left(\dfrac{64}{6} - \dfrac{1}{6}\right) = 10.7$

41. From the graph, it appears that the area is about 60. The actual area is

$\int_0^{27} x^{1/3} dx = \left[\dfrac{3}{4}x^{4/3}\right]_0^{27} = \dfrac{3}{4}\cdot 81 - 0 = \dfrac{243}{4} = 60.75$. This is $\dfrac{3}{4}$ of the

area of the viewing rectangle.

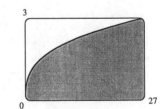

43. It appears that the area under the graph is about $\dfrac{2}{3}$ of the area of the

viewing rectangle, or about $\dfrac{2}{3}\pi \approx 2.1$. The actual area is

$\int_0^{\pi} \sin x\, dx = [-\cos x]_0^{\pi} = -\cos\pi + \cos 0 = -(-1) + 1 = 2.$

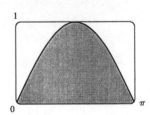

45. $\int_{-1}^2 x^3\, dx = \left[\dfrac{1}{4}x^4\right]_{-1}^2 = 4 - \dfrac{1}{4} = \dfrac{15}{4} = 3.75$

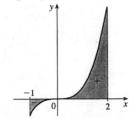

47. $g(x) = \int_{2x}^{3x} \frac{u^2-1}{u^2+1}\,du = \int_{2x}^{0} \frac{u^2-1}{u^2+1}\,du + \int_{0}^{3x} \frac{u^2-1}{u^2+1}\,du = -\int_{0}^{2x} \frac{u^2-1}{u^2+1}\,du + \int_{0}^{3x} \frac{u^2-1}{u^2+1}\,du \;\Rightarrow$

$g'(x) = -\dfrac{(2x)^2-1}{(2x)^2+1}\cdot \dfrac{d}{dx}(2x) + \dfrac{(3x)^2-1}{(3x)^2+1}\cdot \dfrac{d}{dx}(3x) = -2\cdot\dfrac{4x^2-1}{4x^2+1} + 3\cdot\dfrac{9x^2-1}{9x^2+1}$

49. $y = \int_{\sqrt{x}}^{x^3} \sqrt{t}\,\sin t\,dt = \int_{\sqrt{x}}^{1} \sqrt{t}\,\sin t\,dt + \int_{1}^{x^3} \sqrt{t}\,\sin t\,dt = -\int_{1}^{\sqrt{x}} \sqrt{t}\,\sin t\,dt + \int_{1}^{x^3} \sqrt{t}\,\sin t\,dt \;\Rightarrow$

$y' = -\sqrt[4]{x}\,(\sin\sqrt{x})\cdot\dfrac{d}{dx}(\sqrt{x}) + x^{3/2}\sin(x^3)\cdot\dfrac{d}{dx}(x^3) = -\dfrac{\sqrt[4]{x}\,\sin\sqrt{x}}{2\sqrt{x}} + x^{3/2}\sin(x^3)(3x^2)$

$= 3x^{7/2}\sin(x^3) - \dfrac{\sin\sqrt{x}}{2\sqrt[4]{x}}$

51. $F(x) = \int_{1}^{x} f(t)\,dt \;\Rightarrow\; F'(x) = f(x) = \int_{1}^{x^2} \dfrac{\sqrt{1+u^4}}{u}\,du \;\Rightarrow$

$F''(x) = f'(x) = \dfrac{\sqrt{1+(x^2)^4}}{x^2}\cdot\dfrac{d}{dx}(x^2) = \dfrac{2\sqrt{1+x^8}}{x}$. So $F''(2) = \sqrt{1+2^8} = \sqrt{257}$.

53. (a) The Fresnel Function $S(x) = \int_{0}^{x} \sin(\tfrac{\pi}{2}t^2)\,dt$ has local maximum values where $0 = S'(x) = \sin(\tfrac{\pi}{2}x^2)$ and S' changes from positive to negative. For $x > 0$, this happens when $\tfrac{\pi}{2}x^2 = (2n-1)\pi$ [odd multiples of π] $\Leftrightarrow$ $x = \sqrt{2(2n-1)}$, n any positive integer. For $x < 0$, S' changes from positive to negative where $\tfrac{\pi}{2}x^2 = 2n\pi$ [even multiples of π] $\Leftrightarrow$ $x = -2\sqrt{n}$, since if $x < 0$, then as x increases, x^2 decreases. S' does not change sign at $x = 0$.

(b) S is concave upward on those intervals where $S''(x) > 0$. Differentiating our expression for $S'(x)$, we get $S''(x) = \cos(\tfrac{\pi}{2}x^2)(2\tfrac{\pi}{2}x) = \pi x\cos(\tfrac{\pi}{2}x^2)$. For $x > 0$, $S''(x) > 0$ where $\cos(\tfrac{\pi}{2}x^2) > 0$ $\Leftrightarrow$ $0 < \tfrac{\pi}{2}x^2 < \tfrac{\pi}{2}$ or $\left(2n-\tfrac{1}{2}\right)\pi < \tfrac{\pi}{2}x^2 < \left(2n+\tfrac{1}{2}\right)\pi$, n any integer $\Leftrightarrow$ $0 < x < 1$ or $\sqrt{4n-1} < x < \sqrt{4n+1}$, n any positive integer. For $x < 0$, as x increases, x^2 decreases, so the intervals of upward concavity for $x < 0$ are $(-\sqrt{4n-1}, -\sqrt{4n-3})$, n any positive integer. To summarize: S is concave upward on the intervals $(0, 1)$, $(-\sqrt{3}, -1)$, $(\sqrt{3}, \sqrt{5})$, $(-\sqrt{7}, -\sqrt{5})$, $(\sqrt{7}, 3)$, ….

(c) In Maple, we use `plot({int(sin(Pi*t^2/2),t=0..x),0.2},x=0..2);`. Note that Maple recognizes the Fresnel function, calling it `FresnelS(x)`. In Mathematica, we use `Plot[{Integrate[Sin[Pi*t^2/2],{t,0,x}],0.2},{x,0,2}]`. In Derive, we load the utility file FRESNEL and plot FRESNEL_SIN(x). From the graphs, we see that $\int_{0}^{x} \sin(\tfrac{\pi}{2}t^2)\,dt = 0.2$ at $x \approx 0.74$.

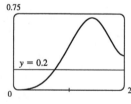

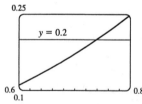

55. (a) By FTC1, $g'(x) = f(x)$. So $g'(x) = f(x) = 0$ at $x = 1, 3, 5, 7$, and 9. g has local maxima at $x = 1$ and 5 (since $f = g'$ changes from positive to negative there) and local minima at $x = 3$ and 7. There is no local maximum or minimum at $x = 9$, since f is not defined for $x > 9$.

(b) We can see from the graph that $\left|\int_{0}^{1} f\,dt\right| < \left|\int_{1}^{3} f\,dt\right| < \left|\int_{3}^{5} f\,dt\right| < \left|\int_{5}^{7} f\,dt\right| < \left|\int_{7}^{9} f\,dt\right|$. So $g(1) = \left|\int_{0}^{1} f\,dt\right|$, $g(5) = \int_{0}^{5} f\,dt = g(1) - \left|\int_{1}^{3} f\,dt\right| + \left|\int_{3}^{5} f\,dt\right|$, and

$g(9) = \int_0^9 f \, dt = g(5) - \left|\int_5^7 f \, dt\right| + \left|\int_7^9 f \, dt\right|$. Thus, $g(1) < g(5) < g(9)$, and so the absolute maximum of $g(x)$ occurs at $x = 9$.

(c) g is concave downward on those intervals where $g'' < 0$. But $g'(x) = f(x)$, so $g''(x) = f'(x)$, which is negative on (approximately) $\left(\frac{1}{2}, 2\right)$, $(4, 6)$ and $(8, 9)$. So g is concave downward on these intervals.

(d)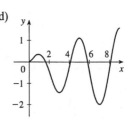

57. $\lim\limits_{n\to\infty} \sum\limits_{i=1}^{n} \dfrac{i^3}{n^4} = \lim\limits_{n\to\infty} \dfrac{1-0}{n} \sum\limits_{i=1}^{n} \left(\dfrac{i}{n}\right)^3 = \int_0^1 x^3 \, dx = \left[\dfrac{x^4}{4}\right]_0^1 = \dfrac{1}{4}$

59. Suppose $h < 0$. Since f is continuous on $[x + h, x]$, the Extreme Value Theorem says that there are numbers u and v in $[x + h, x]$ such that $f(u) = m$ and $f(v) = M$, where m and M are the absolute minimum and maximum values of f on $[x + h, x]$. By Property 8 of integrals, $m(-h) \le \int_{x+h}^{x} f(t) \, dt \le M(-h)$; that is, $f(u)(-h) \le -\int_x^{x+h} f(t) \, dt \le f(v)(-h)$. Since $-h > 0$, we can divide this inequality by $-h$:

$f(u) \le \dfrac{1}{h} \int_x^{x+h} f(t) \, dt \le f(v)$. By Equation 2, $\dfrac{g(x+h) - g(x)}{h} = \dfrac{1}{h} \int_x^{x+h} f(t) \, dt$ for $h \ne 0$, and hence

$f(u) \le \dfrac{g(x+h) - g(x)}{h} \le f(v)$, which is Equation 3 in the case where $h < 0$.

61. (a) Let $f(x) = \sqrt{x}$ $\Rightarrow$ $f'(x) = 1/(2\sqrt{x}) > 0$ for $x > 0$ $\Rightarrow$ f is increasing on $(0, \infty)$. If $x \ge 0$, then $x^3 \ge 0$, so $1 + x^3 \ge 1$ and since f is increasing, this means that $f(1 + x^3) \ge f(1)$ $\Rightarrow$ $\sqrt{1 + x^3} \ge 1$ for $x \ge 0$. Next let $g(t) = t^2 - t$ $\Rightarrow$ $g'(t) = 2t - 1$ $\Rightarrow$ $g'(t) > 0$ when $t \ge 1$. Thus, g is increasing on $(1, \infty)$. And since $g(1) = 0$, $g(t) \ge 0$ when $t \ge 1$. Now let $t = \sqrt{1 + x^3}$, where $x \ge 0$. $\sqrt{1 + x^3} \ge 1$ (from above) $\Rightarrow$ $t \ge 1$ $\Rightarrow$ $g(t) \ge 0$ $\Rightarrow$ $(1 + x^3) - \sqrt{1 + x^3} \ge 0$ for $x \ge 0$. Therefore, $1 \le \sqrt{1 + x^3} \le 1 + x^3$ for $x \ge 0$.

(b) From part (a) and Property 7: $\int_0^1 1 \, dx \le \int_0^1 \sqrt{1 + x^3} \, dx \le \int_0^1 (1 + x^3) \, dx$ $\Leftrightarrow$

$[x]_0^1 \le \int_0^1 \sqrt{1 + x^3} \, dx \le \left[x + \frac{1}{4}x^4\right]_0^1$ $\Leftrightarrow$ $1 \le \int_0^1 \sqrt{1 + x^3} \, dx \le 1 + \frac{1}{4} = 1.25$.

63. Using FTC1, we differentiate both sides of $6 + \int_a^x \dfrac{f(t)}{t^2} \, dt = 2\sqrt{x}$ to get $\dfrac{f(x)}{x^2} = 2\dfrac{1}{2\sqrt{x}}$ $\Rightarrow$ $f(x) = x^{3/2}$.

To find a, we substitute $x = a$ in the original equation to obtain $6 + \int_a^a \dfrac{f(t)}{t^2} \, dt = 2\sqrt{a}$ $\Rightarrow$ $6 + 0 = 2\sqrt{a}$ $\Rightarrow$ $a = 9$.

65. (a) Let $F(t) = \int_0^t f(s) \, ds$. Then, by FTC1, $F'(t) = f(t) =$ rate of depreciation, so $F(t)$ represents the loss in value over the interval $[0, t]$.

(b) $C(t) = \dfrac{A + F(t)}{t}$ represents the average expenditure per unit of t during the interval $[0, t]$, assuming that there has been only one overhaul during that time period. The company wants to minimize average expenditure.

(c) $C(t) = \dfrac{1}{t}\left[A + \int_0^t f(s) \, ds\right]$. Using FTC1, we have $C'(t) = -\dfrac{1}{t^2}\left[A + \int_0^t f(s) \, ds\right] + \dfrac{1}{t}f(t)$. $C'(t) = 0$ $\Rightarrow$ $t f(t) = A + \int_0^t f(s) \, ds$ $\Rightarrow$ $f(t) = \dfrac{1}{t}\left[A + \int_0^t f(s) \, ds\right] = C(t)$.

5.4 Indefinite Integrals and the Total Change Theorem

1. $\dfrac{d}{dx}\left[\sqrt{x^2+1}+C\right]=\frac{1}{2}\left(x^2+1\right)^{-1/2}\cdot 2x=\dfrac{x}{\sqrt{x^2+1}}$

3. $\dfrac{d}{dx}\left[\dfrac{x}{a^2\sqrt{a^2-x^2}}+C\right]=\dfrac{1}{a^2}\dfrac{\sqrt{a^2-x^2}-x\,(-x)\Big/\sqrt{a^2-x^2}}{a^2-x^2}=\dfrac{1}{a^2}\dfrac{\left(a^2-x^2\right)+x^2}{\left(a^2-x^2\right)^{3/2}}=\dfrac{1}{\sqrt{\left(a^2-x^2\right)^3}}$

5. $\int x^{-3/4}\,dx=\dfrac{x^{-3/4+1}}{-3/4+1}+C=\dfrac{x^{1/4}}{1/4}+C=4x^{1/4}+C$

7. $\int\left(x^3+6x+1\right)dx=\dfrac{x^4}{4}+6\dfrac{x^2}{2}+x+C=\frac{1}{4}x^4+3x^2+x+C$

9. $\int\left(1-t\right)\left(2+t^2\right)dt=\int\left(2-2t+t^2-t^3\right)dt=2t-2\dfrac{t^2}{2}+\dfrac{t^3}{3}-\dfrac{t^4}{4}+C=2t-t^2+\frac{1}{3}t^3-\frac{1}{4}t^4+C$

11. $\int\left(2-\sqrt{x}\right)^2dx=\int\left(4-4\sqrt{x}+x\right)dx=4x-4\dfrac{x^{3/2}}{3/2}+\dfrac{x^2}{2}+C=4x-\frac{8}{3}x^{3/2}+\frac{1}{2}x^2+C$

13. $\displaystyle\int\dfrac{\sin x}{1-\sin^2 x}\,dx=\int\dfrac{\sin x}{\cos^2 x}\,dx=\int\dfrac{1}{\cos x}\cdot\dfrac{\sin x}{\cos x}\,dx=\int\sec x\tan x\,dx=\sec x+C$

15. $\int x\sqrt{x}\,dx=\int x^{3/2}\,dx=\frac{2}{5}x^{5/2}+C$

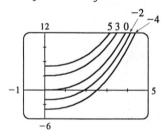

17. $\int_0^1\left(1-2x-3x^2\right)dx=\left[x-2\cdot\frac{1}{2}x^2-3\cdot\frac{1}{3}x^3\right]_0^1=\left[x-x^2-x^3\right]_0^1=(1-1-1)-0=-1$

19. $\int_{-1}^0\left(2x-e^x\right)dx=\left[x^2-e^x\right]_{-1}^0=(0-1)-\left(1-e^{-1}\right)=-2+1/e$

21. $\displaystyle\int_1^3\left(\dfrac{1}{t^2}-\dfrac{1}{t^4}\right)dt=\int_1^3\left(t^{-2}-t^{-4}\right)dt=\left[\dfrac{t^{-1}}{-1}-\dfrac{t^{-3}}{-3}\right]_1^3=\left[\dfrac{1}{3t^3}-\dfrac{1}{t}\right]_1^3=\left(\dfrac{1}{81}-\dfrac{1}{3}\right)-\left(\dfrac{1}{3}-1\right)=\dfrac{28}{81}$

23. $\displaystyle\int_1^2\dfrac{x^2+1}{\sqrt{x}}\,dx=\int_1^2\left(x^{3/2}+x^{-1/2}\right)dx=\left[\dfrac{x^{5/2}}{5/2}+\dfrac{x^{1/2}}{1/2}\right]_1^2=\left[\frac{2}{5}x^{5/2}+2x^{1/2}\right]_1^2$

$=\left(\frac{2}{5}4\sqrt{2}+2\sqrt{2}\right)-\left(\frac{2}{5}+2\right)=\dfrac{18\sqrt{2}-12}{5}=\frac{6}{5}\left(3\sqrt{2}-2\right)$

25. $\int_0^1 u\left(\sqrt{u}+\sqrt[3]{u}\right)du=\int_0^1\left(u^{3/2}+u^{4/3}\right)du=\left[\dfrac{u^{5/2}}{5/2}+\dfrac{u^{7/3}}{7/3}\right]_0^1=\left[\frac{2}{5}u^{5/2}+\frac{3}{7}u^{7/3}\right]_0^1=\frac{2}{5}+\frac{3}{7}=\frac{29}{35}$

27. $\int_1^4\sqrt{5/x}\,dx=\sqrt{5}\int_1^4 x^{-1/2}\,dx=\sqrt{5}\left[2\sqrt{x}\right]_1^4=\sqrt{5}\left(2\cdot 2-2\cdot 1\right)=2\sqrt{5}$

29. $\int_{-2}^{3} |x^2 - 1|\, dx = \int_{-2}^{-1} (x^2 - 1)\, dx + \int_{-1}^{1} (1 - x^2)\, dx + \int_{1}^{3} (x^2 - 1)\, dx$

$$= \left[\frac{x^3}{3} - x\right]_{-2}^{-1} + \left[x - \frac{x^3}{3}\right]_{-1}^{1} + \left[\frac{x^3}{3} - x\right]_{1}^{3}$$

$$= \left(-\tfrac{1}{3} + 1\right) - \left(-\tfrac{8}{3} + 2\right) + \left(1 - \tfrac{1}{3}\right) - \left(-1 + \tfrac{1}{3}\right) + (9 - 3) - \left(\tfrac{1}{3} - 1\right) = \tfrac{28}{3}$$

31. $\int_{1}^{4} \left(\sqrt{t} - \dfrac{2}{\sqrt{t}}\right) dt = \int_{1}^{4} \left(t^{1/2} - 2t^{-1/2}\right) dt = \left[\dfrac{t^{3/2}}{3/2} - 2\dfrac{t^{1/2}}{1/2}\right]_{1}^{4} = \left[\tfrac{2}{3} t^{3/2} - 4t^{1/2}\right]_{1}^{4}$

$$= \left(\tfrac{2}{3} \cdot 8 - 4 \cdot 2\right) - \left(\tfrac{2}{3} - 4\right) = \tfrac{2}{3}$$

33. $\int_{-1}^{0} (x + 1)^3\, dx = \int_{-1}^{0} (x^3 + 3x^2 + 3x + 1)\, dx = \left[\dfrac{x^4}{4} + 3\dfrac{x^3}{3} + 3\dfrac{x^2}{2} + x\right]_{-1}^{0} = 0 - \left[\tfrac{1}{4} - 1 + \tfrac{3}{2} - 1\right]$

$$= 2 - \tfrac{7}{4} = \tfrac{1}{4}$$

35. $\int_{\pi/6}^{\pi/3} \csc^2 \theta\, d\theta = [-\cot\theta]_{\pi/6}^{\pi/3} = -\cot\frac{\pi}{3} + \cot\frac{\pi}{6} = -\tfrac{1}{3}\sqrt{3} + \sqrt{3} = \tfrac{2}{3}\sqrt{3}$

37. $\int_{0}^{\pi/4} \dfrac{1 + \cos^2\theta}{\cos^2\theta}\, d\theta = \int_{0}^{\pi/4} \left(\dfrac{1}{\cos^2\theta} + \dfrac{\cos^2\theta}{\cos^2\theta}\right) d\theta = \int_{0}^{\pi/4} \left(\sec^2\theta + 1\right) d\theta = [\tan\theta + \theta]_{0}^{\pi/4}$

$$= \left(1 + \tfrac{\pi}{4}\right) - (0 + 0) = 1 + \tfrac{\pi}{4}$$

39. $\int_{1}^{e} \dfrac{x^2 + x + 1}{x}\, dx = \int_{1}^{e} \left(x + 1 + \dfrac{1}{x}\right) dx = \left[\dfrac{1}{2}x^2 + x + \ln x\right]_{1}^{e}$

$$= \left(\tfrac{1}{2}e^2 + e + \ln e\right) - \left(\tfrac{1}{2} + 1 + \ln 1\right) = \tfrac{1}{2}e^2 + e - \tfrac{1}{2}$$

41. $\int_{-1}^{2} (x - 2\,|x|)\, dx = \int_{-1}^{0} 3x\, dx + \int_{0}^{2} (-x)\, dx = 3\left[\tfrac{1}{2}x^2\right]_{-1}^{0} - \left[\tfrac{1}{2}x^2\right]_{0}^{2} = \left(3 \cdot 0 - 3 \cdot \tfrac{1}{2}\right) - (2 - 0) = -\tfrac{7}{2} = -3.5$

43. The graph shows that $y = x + x^2 - x^4$ has x-intercepts at $x = 0$ and at $x \approx 1.32$. So the area of the region below the curve and above the x-axis is about

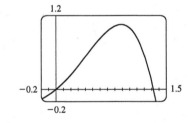

$$\int_{0}^{1.32} (x + x^2 - x^4)\, dx = \left[\tfrac{1}{2}x^2 + \tfrac{1}{3}x^3 - \tfrac{1}{5}x^5\right]_{0}^{1.32}$$

$$= \left[\tfrac{1}{2}(1.32)^2 + \tfrac{1}{3}(1.32)^3 - \tfrac{1}{5}(1.32)^5\right] - 0$$

$$\approx 0.84$$

45. $A = \int_{0}^{2} (2y - y^2)\, dy = \left[y^2 - \tfrac{1}{3}y^3\right]_{0}^{2} = \left(4 - \tfrac{8}{3}\right) - 0 = \tfrac{4}{3}$

47. If $w'(t)$ is the rate of change of weight in pounds per year, then $w(t)$ represents the weight in pounds of the child at age t. We know from the Total Change Theorem that $\int_{5}^{10} w'(t)\, dt = w(10) - w(5)$, so the integral represents the increase in the child's weight between the ages of 5 and 10.

49. Since $r(t)$ is the rate at which oil leaks, we can write $r(t) = -V'(t)$, where $V(t)$ is the volume of oil at time t. [Note that the minus sign is needed because V is decreasing, so $V'(t)$ is negative, but $r(t)$ is positive.] Thus, by the Total Change Theorem, $\int_{0}^{120} r(t)\, dt = -\int_{0}^{120} V'(t)\, dt = -[V(120) - V(0)] = V(0) - V(120)$, which is the number of gallons of oil that leaked from the tank in the first two hours.

51. By the Total Change Theorem, $\int_{1000}^{5000} R'(x)\, dx = R(5000) - R(1000)$, so it represents the increase in revenue when production is increased from 1000 units to 5000 units.

53. (a) displacement $= \int_0^3 (3t - 5)\, dt = \left[\frac{3}{2}t^2 - 5t\right]_0^3 = \frac{27}{2} - 15 = -\frac{3}{2}$ m

(b) distance traveled $= \int_0^3 |3t - 5|\, dt = \int_0^{5/3} (5 - 3t)\, dt + \int_{5/3}^3 (3t - 5)\, dt$

$$= \left[5t - \frac{3}{2}t^2\right]_0^{5/3} + \left[\frac{3}{2}t^2 - 5t\right]_{5/3}^3 = \frac{25}{3} - \frac{3}{2} \cdot \frac{25}{9} + \frac{27}{2} - 15 - \left(\frac{3}{2} \cdot \frac{25}{9} - \frac{25}{3}\right) = \frac{41}{6}$$ m

55. (a) $v'(t) = a(t) = t + 4 \;\Rightarrow\; v(t) = \frac{1}{2}t^2 + 4t + C \;\Rightarrow\; v(0) = C = 5 \;\Rightarrow\; v(t) = \frac{1}{2}t^2 + 4t + 5$ m/s

(b) distance traveled $= \int_0^{10} |v(t)|\, dt = \int_0^{10} \left|\frac{1}{2}t^2 + 4t + 5\right|\, dt = \int_0^{10} \left(\frac{1}{2}t^2 + 4t + 5\right) dt$

$$= \left[\frac{1}{6}t^3 + 2t^2 + 5t\right]_0^{10} = \frac{500}{3} + 200 + 50 = 416\frac{2}{3}$$ m

57. Since $m'(x) = \rho(x)$, $m = \int_0^4 \rho(x)\, dx = \int_0^4 (9 + 2\sqrt{x})\, dx = \left[9x + \frac{4}{3}x^{3/2}\right]_0^4 = 36 + \frac{32}{3} - 0 = \frac{140}{3} = 46\frac{2}{3}$ kg.

59. Let s be the position of the car. We know from Equation 2 that $s(100) - s(0) = \int_0^{100} v(t)\, dt$. We use the Midpoint Rule for $0 \le t \le 100$ with $n = 5$. Note that the length of each of the five time intervals is 20 seconds $= \frac{1}{180}$ hour. So the distance traveled is

$$\int_0^{100} v(t)\, dt \approx \frac{1}{180}\left[v(10) + v(30) + v(50) + v(70) + v(90)\right] = \frac{1}{180}(38 + 58 + 51 + 53 + 47)$$

$$= \frac{247}{180} \approx 1.4 \text{ miles}$$

61. $\int_{2000}^{4000} C'(x)\, dx = \int_{2000}^{4000} (3 - 0.01x + 0.000006x^2)\, dx = \left[3x - 0.005x^2 + 0.000002x^3\right]_{2000}^{4000}$

$$= 60{,}000 - 2{,}000 = \$58{,}000$$

63. (a) We can find the area between the Lorenz curve and the line $y = x$ by subtracting the area under $y = L(x)$ from the area under $y = x$. Thus,

$$\text{coefficient of inequality} = \frac{\text{area between Lorenz curve and straight line}}{\text{area under straight line}} = \frac{\int_0^1 [x - L(x)]\, dx}{\int_0^1 x\, dx}$$

$$= \frac{\int_0^1 [x - L(x)]\, dx}{[x^2/2]_0^1} = \frac{\int_0^1 [x - L(x)]\, dx}{1/2} = 2\int_0^1 [x - L(x)]\, dx$$

(b) $L(x) = \frac{5}{12}x^2 + \frac{7}{12}x \;\Rightarrow\; L\left(\frac{1}{2}\right) = \frac{5}{48} + \frac{7}{24} = \frac{19}{48} = 0.39583$, so the bottom 50% of the households receive about 40% of the income.

$$\text{coefficient of inequality} = 2\int_0^1 [x - L(x)]\, dx = 2\int_0^1 \left(x - \frac{5}{12}x^2 - \frac{7}{12}x\right) dx = 2\int_0^1 \frac{5}{12}(x - x^2)\, dx$$

$$= \frac{5}{6}\left[\frac{1}{2}x^2 - \frac{1}{3}x^3\right]_0^1 = \frac{5}{36}$$

5.5 The Substitution Rule

1. Let $u = 3x$. Then $du = 3\, dx$, so $\int \cos 3x\, dx = \int \cos u\left(\frac{1}{3}\, du\right) = \frac{1}{3}\sin u + C = \frac{1}{3}\sin 3x + C$.

3. Let $u = x^3 + 1$. Then $du = 3x^2\, dx$, so

$$\int x^2 \sqrt{x^3 + 1}\, dx = \int \sqrt{u}\left(\frac{1}{3}\, du\right) = \frac{1}{3}\frac{u^{3/2}}{3/2} + C = \frac{1}{3} \cdot \frac{2}{3}u^{3/2} + C = \frac{2}{9}(x^3 + 1)^{3/2} + C$$

5. Let $u = 1 + 2x$. Then $du = 2\,dx$, so

$$\int \frac{4}{(1+2x)^3}\,dx = 4\int u^{-3}\left(\tfrac{1}{2}\,du\right) = 2\frac{u^{-2}}{-2} + C = -\frac{1}{u^2} + C = -\frac{1}{(1+2x)^2} + C$$

7. Let $u = x^2 + 3$. Then $du = 2x\,dx$, so $\int 2x\left(x^2+3\right)^4 dx = \int u^4\,du = \tfrac{1}{5}u^5 + C = \tfrac{1}{5}\left(x^2+3\right)^5 + C$.

9. Let $u = x - 1$. Then $du = dx$, so $\int \sqrt{x-1}\,dx = \int u^{1/2}\,du = \tfrac{2}{3}u^{3/2} + C = \tfrac{2}{3}(x-1)^{3/2} + C$.

11. Let $u = 5 - 3x$. Then $du = -3\,dx$, so $\displaystyle\int \frac{dx}{5-3x} = -\frac{1}{3}\int \frac{1}{u}\,du = -\tfrac{1}{3}\ln|u| + C = -\tfrac{1}{3}\ln|5-3x| + C$.

13. Let $u = 1 + x + 2x^2$. Then $du = (1 + 4x)\,dx$, so

$$\int \frac{1+4x}{\sqrt{1+x+2x^2}}\,dx = \int \frac{du}{\sqrt{u}} = \int u^{-1/2}\,du = \frac{u^{1/2}}{1/2} + C = 2\sqrt{1+x+2x^2} + C$$

15. Let $u = t + 1$. Then $du = dt$, so $\displaystyle\int \frac{2}{(t+1)^6}\,dt = 2\int u^{-6}\,du = -\tfrac{2}{5}u^{-5} + C = -\frac{2}{5(t+1)^5} + C$.

17. Let $u = 1 - 2y$. Then $du = -2\,dy$, so

$$\int (1-2y)^{1.3}\,dy = \int u^{1.3}\left(-\tfrac{1}{2}\,du\right) = -\frac{1}{2}\left(\frac{u^{2.3}}{2.3}\right) + C = -\frac{(1-2y)^{2.3}}{4.6} + C$$

19. Let $u = 2\theta$. Then $du = 2\,d\theta$, so $\int \cos 2\theta\,d\theta = \int \cos u\left(\tfrac{1}{2}\,du\right) = \tfrac{1}{2}\sin u + C = \tfrac{1}{2}\sin 2\theta + C$.

21. Let $u = \ln x$. Then $du = \dfrac{dx}{x}$, so $\displaystyle\int \frac{(\ln x)^2}{x}\,dx = \int u^2\,du = \tfrac{1}{3}u^3 + C = \tfrac{1}{3}(\ln x)^3 + C$.

23. Let $u = t^2$. Then $du = 2t\,dt$, so $\int t\sin\left(t^2\right)dt = \int \sin u\left(\tfrac{1}{2}\,du\right) = -\tfrac{1}{2}\cos u + C = -\tfrac{1}{2}\cos\left(t^2\right) + C$.

25. Let $u = 1 + \sec x$. Then $du = \sec x\tan x\,dx$, so

$$\int \sec x\tan x\sqrt{1+\sec x}\,dx = \int u^{1/2}\,du = \tfrac{2}{3}u^{3/2} + C = \tfrac{2}{3}(1+\sec x)^{3/2} + C$$

27. Let $u = 1 + e^x$. Then $du = e^x\,dx$, so $\int e^x\sqrt{1+e^x}\,dx = \int \sqrt{u}\,du = \tfrac{2}{3}u^{3/2} + C = \tfrac{2}{3}(1+e^x)^{3/2} + C$.

Or: Let $u = \sqrt{1+e^x}$. Then $u^2 = 1 + e^x$ and $2u\,du = e^x\,dx$, so

$\int e^x\sqrt{1+e^x}\,dx = \int u\cdot 2u\,du = \tfrac{2}{3}u^3 + C = \tfrac{2}{3}(1+e^x)^{3/2} + C$.

29. Let $u = \cos x$. Then $du = -\sin x\,dx$, so $\int \cos^4 x\sin x\,dx = \int u^4(-du) = -\tfrac{1}{5}u^5 + C = -\tfrac{1}{5}\cos^5 x + C$.

31. Let $u = \ln x$. Then $du = \dfrac{dx}{x}$, so $\displaystyle\int \frac{dx}{x\ln x} = \int \frac{du}{u} = \ln|u| + C = \ln|\ln x| + C$.

33. Let $u = \cot x$. Then $du = -\csc^2 x\,dx$, so $\int \sqrt{\cot x}\,\csc^2 x\,dx = \int \sqrt{u}\,(-du) = -\dfrac{u^{3/2}}{3/2} + C = -\tfrac{2}{3}(\cot x)^{3/2} + C$.

35. Let $u = x^2 + 2x$. Then $du = 2(x+1)\,dx$, so $\displaystyle\int \frac{x+1}{x^2+2x}\,dx = \int \frac{\tfrac{1}{2}\,du}{u} = \frac{1}{2}\ln|u| + C = \frac{1}{2}\ln\left|x^2+2x\right| + C$.

37. Let $u = \sec x$. Then $du = \sec x\tan x\,dx$, so

$$\int \sec^3 x\tan x\,dx = \int \sec^2 x\,(\sec x\tan x)\,dx = \int u^2\,du = \tfrac{1}{3}u^3 + C = \tfrac{1}{3}\sec^3 x + C$$

39. Let $u = b + cx^{a+1}$. Then $du = (a+1)cx^a\,dx$, so

$$\int x^a\sqrt{b+cx^{a+1}}\,dx = \int u^{1/2}\frac{1}{(a+1)c}\,du = \frac{1}{(a+1)c}\left(\tfrac{2}{3}u^{3/2}\right) + C = \frac{2}{3c(a+1)}\left(b+cx^{a+1}\right)^{3/2} + C$$

41. $\int \dfrac{1+x}{1+x^2}\,dx = \int \dfrac{1}{1+x^2}\,dx + \int \dfrac{x}{1+x^2}\,dx = \tan^{-1}x + \dfrac{1}{2}\int \dfrac{2x\,dx}{1+x^2} = \tan^{-1}x + \dfrac{1}{2}\ln\left(1+x^2\right) + C$

(In the last step, we evaluate $\int du/u$ where $u = 1+x^2$.)

43. Let $u = x+2$. Then $du = dx$, so

$$\int \dfrac{x}{\sqrt[4]{x+2}}\,dx = \int \dfrac{u-2}{\sqrt[4]{u}}\,du = \int \left(u^{3/4} - 2u^{-1/4}\right)du = \tfrac{4}{7}u^{7/4} - 2\cdot\tfrac{4}{3}u^{3/4} + C$$

$$= \tfrac{4}{7}(x+2)^{7/4} - \tfrac{8}{3}(x+2)^{3/4} + C$$

In Exercises 45 and 47, let $f(x)$ denote the integrand and $F(x)$ its antiderivative (with $C = 0$).

45. $f(x) = \dfrac{3x-1}{(3x^2-2x+1)^4}$. $u = 3x^2 - 2x + 1 \;\Rightarrow\; du = 2(3x-1)\,dx$,

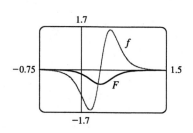

so

$$\int \dfrac{3x-1}{(3x^2-2x+1)^4}\,dx = \int \dfrac{1}{u^4}\left(\tfrac{1}{2}\,du\right) = \dfrac{1}{2}\int u^{-4}\,du$$

$$= -\tfrac{1}{6}u^{-3} + C = -\dfrac{1}{6(3x^2-2x+1)^3} + C$$

Notice that at $x = \tfrac{1}{3}$, f changes from negative to positive, and F has a local minimum.

47. $f(x) = \sin^3 x \cos x$. $u = \sin x \;\Rightarrow\; du = \cos x\,dx$, so

$$\int \sin^3 x \cos x\,dx = \int u^3\,du = \tfrac{1}{4}u^4 + C = \tfrac{1}{4}\sin^4 x + C$$

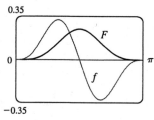

Note that at $x = \tfrac{\pi}{2}$, f changes from positive to negative and F has a local maximum. Also, both f and F are periodic with period π, so at $x = 0$ and at $x = \pi$, f changes from negative to positive and F has local minima.

49. Let $u = x - 1$, so $du = dx$. When $x = 0$, $u = -1$; when $x = 2$, $u = 1$. Therefore, $\int_0^2 (x-1)^{25}\,dx = \int_{-1}^1 u^{25}\,du = 0$ by Theorem 6(b), since $f(u) = u^{25}$ is an odd function.

51. Let $u = 1 + 2x^3$, so $du = 6x^2\,dx$. When $x = 0$, $u = 1$; when $x = 1$, $u = 3$. Therefore, $\int_0^1 x^2\left(1+2x^3\right)^5\,dx = \int_1^3 u^5\left(\tfrac{1}{6}\,du\right) = \tfrac{1}{6}\left[\tfrac{1}{6}u^6\right]_1^3 = \tfrac{1}{36}\left(3^6 - 1^6\right) = \tfrac{1}{36}(729 - 1) = \tfrac{728}{36} = \tfrac{182}{9}$

53. Let $u = \pi t$, so $du = \pi\,dt$. When $t = 0$, $u = 0$; when $t = 1$, $u = \pi$. Therefore, $\int_0^1 \cos \pi t\,dt = \int_0^\pi \cos u\left(\tfrac{1}{\pi}\,du\right) = \tfrac{1}{\pi}[\sin u]_0^\pi = \tfrac{1}{\pi}(0 - 0) = 0$

55. Let $u = 1 + \dfrac{1}{x}$, so $du = -\dfrac{dx}{x^2}$. When $x = 1$, $u = 2$; when $x = 4$, $u = \tfrac{5}{4}$. Therefore,

$$\int_1^4 \dfrac{1}{x^2}\sqrt{1+\dfrac{1}{x}}\,dx = \int_2^{5/4} u^{1/2}\,(-du) = \int_{5/4}^2 u^{1/2}\,du$$

$$= \left[\tfrac{2}{3}u^{3/2}\right]_{5/4}^2 = \tfrac{2}{3}\left(2\sqrt{2} - \tfrac{5\sqrt{5}}{8}\right) = \tfrac{4\sqrt{2}}{3} - \tfrac{5\sqrt{5}}{12}$$

57. Let $u = 2x + 3$, so $du = 2\,dx$. When $x = 0$, $u = 3$; when $x = 3$, $u = 9$. Therefore,

$$\int_0^3 \frac{dx}{2x+3} = \int_3^9 \frac{\frac{1}{2}\,du}{u} = \left[\tfrac{1}{2}\ln u\right]_3^9 = \tfrac{1}{2}\,(\ln 9 - \ln 3) = \tfrac{1}{2}\ln\tfrac{9}{3} = \tfrac{1}{2}\ln 3 \quad (\text{or } \ln\sqrt{3})$$

59. Let $u = \cos\theta$, so $du = -\sin\theta\,d\theta$. When $\theta = 0$, $u = 1$; when $\theta = \frac{\pi}{3}$, $u = \frac{1}{2}$. Therefore,

$$\int_0^{\pi/3} \frac{\sin\theta}{\cos^2\theta}\,d\theta = \int_1^{1/2} \frac{-du}{u^2} = \int_{1/2}^1 u^{-2}\,du = \left[-\frac{1}{u}\right]_{1/2}^1 = -1 + 2 = 1$$

61. Let $u = 1 + 2x$, so $du = 2\,dx$. When $x = 0$, $u = 1$; when $x = 13$, $u = 27$. Therefore,

$$\int_0^{13} \frac{dx}{\sqrt[3]{(1+2x)^2}} = \int_1^{27} u^{-2/3}\left(\tfrac{1}{2}\,du\right) = \left[\tfrac{1}{2}\cdot 3u^{1/3}\right]_1^{27} = \tfrac{3}{2}\,(3 - 1) = 3$$

63. Let $u = x - 1$, so $du = dx$. When $x = 1$, $u = 0$; when $x = 2$, $u = 1$. Therefore,

$$\int_1^2 x\sqrt{x-1}\,dx = \int_0^1 (u + 1)\sqrt{u}\,du = \int_0^1 (u^{3/2} + u^{1/2})\,du = \left[\tfrac{2}{5}u^{5/2} + \tfrac{2}{3}u^{3/2}\right]_0^1 = \tfrac{2}{5} + \tfrac{2}{3} = \tfrac{16}{15}$$

65. Let $u = \ln x$, so $du = \dfrac{dx}{x}$. When $x = e$, $u = 1$; when $x = e^4$; $u = 4$. Therefore,

$$\int_e^{e^4} \frac{dx}{x\sqrt{\ln x}} = \int_1^4 u^{-1/2}\,du = 2\left[u^{1/2}\right]_1^4 = 2\,(2 - 1) = 2$$

67. $\displaystyle\int_0^4 \frac{dx}{(x-2)^3}$ does not exist since $\dfrac{1}{(x-2)^3}$ has an infinite discontinuity at $x = 2$.

69. Let $u = x^2 + a^2$, so $du = 2x\,dx$. When $x = 0$, $u = a^2$; when $x = a$, $u = 2a^2$. Therefore,

$$\int_0^a x\sqrt{x^2+a^2}\,dx = \int_{a^2}^{2a^2} u^{1/2}\left(\tfrac{1}{2}\,du\right) = \tfrac{1}{2}\left[\tfrac{2}{3}u^{3/2}\right]_{a^2}^{2a^2} = \left[\tfrac{1}{3}u^{3/2}\right]_{a^2}^{2a^2} = \tfrac{1}{3}\left(2\sqrt{2} - 1\right)a^3$$

71. From the graph, it appears that the area under the curve is about

$1 + \left(\text{a little more than } \tfrac{1}{2}\cdot 1\cdot 0.7\right)$, or about 1.4. The exact area is given

by $A = \int_0^1 \sqrt{2x+1}\,dx$. Let $u = 2x + 1$, so $du = 2\,dx$, the limits change

to $2\cdot 0 + 1 = 1$ and $2\cdot 1 + 1 = 3$, and

$A = \int_1^3 \sqrt{u}\left(\tfrac{1}{2}\,du\right) = \left[\tfrac{1}{3}u^{3/2}\right]_1^3 = \tfrac{1}{3}\left(3\sqrt{3} - 1\right) = \sqrt{3} - \tfrac{1}{3} \approx 1.399.$

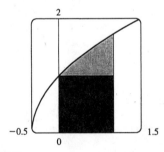

73. We split the integral: $\int_{-2}^2 (x + 3)\sqrt{4-x^2}\,dx = \int_{-2}^2 x\sqrt{4-x^2}\,dx + \int_{-2}^2 3\sqrt{4-x^2}\,dx$. The first integral is 0 by Theorem 6(b), since $f(x) = x\sqrt{4-x^2}$ is an odd function and we are integrating from $x = -2$ to $x = 2$. The second integral we interpret as three times the area of a semicircle with radius 2, so the original integral is equal to $0 + 3\cdot\tfrac{1}{2}\left(\pi\cdot 2^2\right) = 6\pi$.

75. The volume of inhaled air in the lungs at time t is

$$V(t) = \int_0^t f(u)\,du = \int_0^t \frac{1}{2}\sin\left(\tfrac{2}{5}\pi u\right)\,du = \int_0^{2\pi t/5} \frac{1}{2}\sin v\left(\tfrac{5}{2\pi}\,dv\right) \quad [\text{substitute } v = \tfrac{2\pi}{5}u,\ dv = \tfrac{2\pi}{5}\,du]$$

$$= \tfrac{5}{4\pi}\left[-\cos v\right]_0^{2\pi t/5} = \tfrac{5}{4\pi}\left[-\cos\left(\tfrac{2}{5}\pi t\right) + 1\right] = \tfrac{5}{4\pi}\left[1 - \cos\left(\tfrac{2}{5}\pi t\right)\right] \text{ liters}$$

77. Let $u = 2x$. Then $du = 2\,dx$, so $\int_0^2 f(2x)\,dx = \int_0^4 f(u)\left(\tfrac{1}{2}\,du\right) = \tfrac{1}{2}\int_0^4 f(u)\,du = \tfrac{1}{2}\,(10) = 5$.

79. (a) Let $u = -x$. Then $du = -dx$. When $x = a$, $u = -a$; when $x = b$, $u = -b$. So

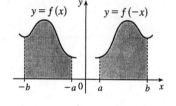

$$\int_a^b f(-x)\,dx = \int_{-a}^{-b} f(u)\,(-du) = \int_{-b}^{-a} f(u)\,du = \int_{-b}^{-a} f(x)\,dx$$

From the diagram, we see that the equality follows from the fact that we are reflecting the graph of f, and the limits of integration, about the y-axis.

(b) Let $u = x + c$. Then $du = dx$. When $x = a$, $u = a + c$; when $x = b$, $u = b + c$. So

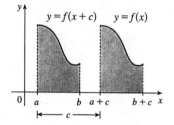

$$\int_a^b f(x+c)\,dx = \int_{a+c}^{b+c} f(u)\,du = \int_{a+c}^{b+c} f(x)\,dx$$

From the diagram, we see that the equality follows from the fact that we are translating the graph of f, and the limits of integration, by a distance c.

81. Let $u = 1 - x$. Then $du = -dx$, so

$$\int_0^1 x^a (1-x)^b\,dx = -\int_1^0 (1-u)^a u^b\,du = \int_0^1 u^b (1-u)^a\,du = \int_0^1 x^b (1-x)^a\,dx.$$

83. $\dfrac{x\sin x}{1 + \cos^2 x} = x \cdot \dfrac{\sin x}{2 - \sin^2 x} = xf(\sin x)$, where $f(t) = \dfrac{t}{2 - t^2}$. By Exercise 82,

$$\int_0^\pi \frac{x\sin x}{1 + \cos^2 x}\,dx = \int_0^\pi xf(\sin x)\,dx = \frac{\pi}{2}\int_0^\pi f(\sin x)\,dx = \frac{\pi}{2}\int_0^\pi \frac{\sin x}{1 + \cos^2 x}\,dx$$

Let $u = \cos x$. Then $du = -\sin x\,dx$. When $x = \pi$, $u = -1$ and when $x = 0$, $u = 1$. So

$$\frac{\pi}{2}\int_0^\pi \frac{\sin x}{1 + \cos^2 x}\,dx = -\frac{\pi}{2}\int_1^{-1} \frac{du}{1 + u^2} = \frac{\pi}{2}\int_{-1}^1 \frac{du}{1 + u^2} = \frac{\pi}{2}\Big[\tan^{-1} u\Big]_{-1}^1$$

$$= \frac{\pi}{2}\Big[\tan^{-1} 1 - \tan^{-1}(-1)\Big] = \frac{\pi}{2}\Big[\frac{\pi}{4} - \Big(-\frac{\pi}{4}\Big)\Big] = \frac{\pi^2}{4}$$

5.6 The Logarithm Defined as an Integral

1. (a)

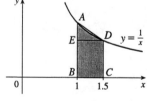

We interpret $\ln 1.5$ as the area under the curve $y = 1/x$ from $x = 1$ to $x = 1.5$. The area of the rectangle $BCDE$ is $\frac{1}{2} \cdot \frac{2}{3} = \frac{1}{3}$. The area of the trapezoid $ABCD$ is $\frac{1}{2} \cdot \frac{1}{2}\Big(1 + \frac{2}{3}\Big) = \frac{5}{12}$. Thus, by comparing areas, we observe that $\frac{1}{3} < \ln 1.5 < \frac{5}{12}$.

(b) With $f(t) = 1/t$, $n = 10$, and $\Delta x = 0.05$, we have

$$\ln 1.5 = \int_1^{1.5} (1/t)\,dt \approx (0.05)\,[f(1.025) + f(1.075) + \cdots + f(1.475)]$$

$$= (0.05)\Big[\tfrac{1}{1.025} + \tfrac{1}{1.075} + \cdots + \tfrac{1}{1.475}\Big] \approx 0.4054$$

3. (a)

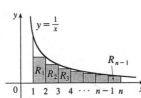

The area of R_i is $\dfrac{1}{i+1}$ and so $\dfrac{1}{2}+\dfrac{1}{3}+\cdots+\dfrac{1}{n} < \displaystyle\int_1^n \dfrac{1}{t}\,dt = \ln n.$

(b)

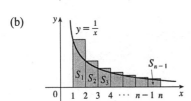

The area of S_i is $\dfrac{1}{i}$ and so $1+\dfrac{1}{2}+\cdots+\dfrac{1}{n-1} > \displaystyle\int_1^n \dfrac{1}{t}\,dt = \ln n.$

5. If $f(x) = \ln(x^r)$, then $f'(x) = (1/x^r)(rx^{r-1}) = r/x$. But if $g(x) = r\ln x$, then $g'(x) = r/x$. So f and g must differ by a constant: $\ln(x^r) = r\ln x + C$. Put $x = 1$: $\ln(1^r) = r\ln 1 + C \;\Rightarrow\; C = 0$, so $\ln(x^r) = r\ln x$.

7. Using the third law of logarithms and Equation 10, we have $\ln e^{rx} = rx = r\ln e^x = \ln(e^x)^r$. Since $\ln$ is a one-to-one function, it follows that $e^{rx} = (e^x)^r$.

9. Using Definition 13, the first law of logarithms, and the first law of exponents for e^x, we have
$$(ab)^x = e^{x\ln(ab)} = e^{x(\ln a + \ln b)} = e^{x\ln a + x\ln b} = e^{x\ln a}e^{x\ln b} = a^x b^x.$$

 Review

CONCEPT CHECK

1. (a) $\sum_{i=1}^{n} f(x_i^*)\,\Delta x$ is an expression for a Riemann sum of a function f. x_i^* is a point in the ith subinterval $[x_{i-1}, x_i]$ and Δx is the length of the subintervals.

(b) See Figure 1 in Section 5.2.

(c) In Section 5.2, see Figure 3 and the paragraph to the right of it.

2. (a) See Definition 5.2.2.

(b) See Figure 2 in Section 5.2.

(c) In Section 5.2, see Figure 4 and the paragraph to the right of it.

3. See page 397.

4. (a) See the Total Change Theorem on page 404.

(b) $\int_{t_1}^{t_2} r(t)\,dt$ represents the change in the amount of water in the reservoir between time t_1 and time t_2.

5. (a) $\int_{60}^{120} v(t)\,dt$ represents the change in position of the particle from $t = 60$ to $t = 120$ seconds.

(b) $\int_{60}^{120} |v(t)|\,dt$ represents the total distance traveled by the particle from $t = 60$ to 120 seconds.

(c) $\int_{60}^{120} a(t)\,dt$ represents the change in the velocity of the particle from $t = 60$ to $t = 120$ seconds.

6. (a) $\int f(x)\,dx$ is the family of functions $\{F \mid F' = f\}$. Any two such functions differ by a constant.

(b) The connection is given by the Evaluation Theorem: $\int_a^b f(x)\,dx = \left[\int f(x)\,dx\right]_a^b$ if f is continuous..

7. The precise version of this statement is given by the Fundamental Theorem of Calculus. See the statement of this theorem and the paragraph that follows it on pages 397 and 398.

8. See the Substitution Rule (5.5.4). This says that it is permissible to operate with the dx after an integral sign as if it were a differential.

TRUE-FALSE QUIZ

1. True by Property 2 of the Integral in Section 5.2.

3. True by Property 3 of the Integral in Section 5.2.

5. False. For example, let $f(x) = x^2$. Then $\int_0^1 \sqrt{x^2}\,dx = \int_0^1 x\,dx = \frac{1}{2}$, but $\sqrt{\int_0^1 x^2\,dx} = \sqrt{\frac{1}{3}} = \frac{1}{\sqrt{3}}$.

7. True by Comparison Property 7 of the Integral in Section 5.2.

9. True. The integrand is an odd function that is continuous on $[-1, 1]$, so the result follows from Equation 5.5.7(b).

11. False. The function $f(x) = 1/x^4$ is not bounded on the interval $[-2, 1]$. It has an infinite discontinuity at $x = 0$, so it is not integrable on the interval. (If the integral were to exist, a positive value would be expected, by Comparison Property 6 of Integrals.)

13. False. For example, the function $y = |x|$ is continuous on $\mathbb{R}$, but has no derivative at $x = 0$.

EXERCISES

1. (a)

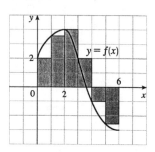

$$L_6 = \sum_{i=1}^{6} f(x_{i-1})\,\Delta x \quad [\Delta x = \tfrac{6-0}{6} = 1]$$

$$= f(x_0) \cdot 1 + f(x_1) \cdot 1 + f(x_2) \cdot 1$$
$$+ f(x_3) \cdot 1 + f(x_4) \cdot 1 + f(x_5) \cdot 1$$
$$\approx 2 + 3.5 + 4 + 2 + (-1) + (-2.5) = 8$$

The Riemann sum represents the sum of the areas of the first four rectangles and the negatives of the areas of the last two rectangles.

(b)

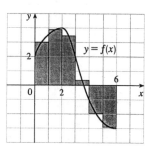

$$M_6 = \sum_{i=1}^{6} f(\overline{x}_i)\,\Delta x \quad [\Delta x = \tfrac{6-0}{6} = 1]$$

$$= f(\overline{x}_1) \cdot 1 + f(\overline{x}_2) \cdot 1 + f(\overline{x}_3) \cdot 1$$
$$+ f(\overline{x}_4) \cdot 1 + f(\overline{x}_5) \cdot 1 + f(\overline{x}_6) \cdot 1$$
$$= f(0.5) + f(1.5) + f(2.5) + f(3.5) + f(4.5) + f(5.5)$$
$$= 3 + 3.9 + 3.3 + 0.2 + (-2) + (-2.8) = 5.6$$

The Riemann sum represents the sum of the areas of the first four rectangles and the negatives of the areas of the last two rectangles.

3. $\int_0^1 \left(x + \sqrt{1 - x^2}\right) dx = \int_0^1 x\,dx + \int_0^1 \sqrt{1 - x^2}\,dx = I_1 + I_2$.

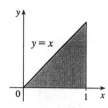
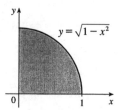

I_1 can be interpreted as the area of the triangle shown in the figure and I_2 can be interpreted as the area of the quarter-circle. Area $= \frac{1}{2}(1)(1) + \frac{1}{4}(\pi)(1)^2 = \frac{1}{2} + \frac{\pi}{4}$.

5. $\int_0^6 f(x)\,dx = \int_0^4 f(x)\,dx + \int_4^6 f(x)\,dx \quad \Rightarrow \quad 10 = 7 + \int_4^6 f(x)\,dx \quad \Rightarrow \quad \int_4^6 f(x)\,dx = 10 - 7 = 3$

7. First note that either a or b must be the graph of $\int_0^x f(t)\, dt$, since $\int_0^0 f(t)\, dt = 0$, and $c(0) \neq 0$. Now notice that $b > 0$ when c is increasing, and that $c > 0$ when a is increasing. It follows that c is the graph of $f(x)$, b is the graph of $f'(x)$, and a is the graph of $\int_0^x f(t)\, dt$.

9. $\int_1^2 (8x^3 + 3x^2)\, dx = \left[\frac{8}{4}x^4 + \frac{3}{3}x^3\right]_1^2 = (2 \cdot 2^4 + 2^3) - (2 + 1) = 37$

11. $\int_0^1 (1 - x^9)\, dx = \left[x - \frac{1}{10}x^{10}\right]_0^1 = 1 - \frac{1}{10} = \frac{9}{10}$

13. $\int_1^8 \sqrt[3]{x}\, (x - 1)\, dx = \int_1^8 (x^{4/3} - x^{1/3})\, dx = \left[\frac{3}{7}x^{7/3} - \frac{3}{4}x^{4/3}\right]_1^8 = \left(\frac{3}{7} \cdot 128 - \frac{3}{4} \cdot 16\right) - \left(\frac{3}{7} - \frac{3}{4}\right) = \frac{1209}{28}$

15. Let $u = 1 + 2x^3$. Then $du = 6x^2\, dx$, so
$$\int_0^2 x^2 (1 + 2x^3)^3\, dx = \int_1^{17} u^3 \left(\frac{1}{6}\, du\right) = \left[\frac{1}{24}u^4\right]_1^{17} = \frac{1}{24}(17^4 - 1) = 3480.$$

17. Let $u = 2x + 3$. Then $du = 2\, dx$, so $\displaystyle\int_3^{11} \frac{dx}{\sqrt{2x + 3}} = \int_9^{25} u^{-1/2} \left(\frac{1}{2}\, du\right) = \left[u^{1/2}\right]_9^{25} = 5 - 3 = 2.$

19. $\displaystyle\int_{-2}^{-1} \frac{dx}{(2x + 3)^4}$ does not exist since the integrand has an infinite discontinuity at $x = -\frac{3}{2}$.

21. $\int_0^1 e^{\pi t}\, dt = \left[\frac{1}{\pi}e^{\pi t}\right]_0^1 = \frac{1}{\pi}(e^\pi - 1)$

23. $\displaystyle\int_2^4 \frac{1 + x - x^2}{x^2}\, dx = \int_2^4 \left(x^{-2} + \frac{1}{x} - 1\right) dx = \left[-\frac{1}{x} + \ln x - x\right]_2^4$
$$= \left(-\frac{1}{4} + \ln 4 - 4\right) - \left(-\frac{1}{2} + \ln 2 - 2\right) = \ln 2 - \frac{7}{4}$$

25. Let $u = 2 + x^5$. Then $du = 5x^4\, dx$, so
$$\int \frac{x^4\, dx}{(2 + x^5)^6} = \int u^{-6} \left(\frac{1}{5}\, du\right) = \frac{1}{5}\left(\frac{u^{-5}}{-5}\right) + C = -\frac{1}{25u^5} + C = -\frac{1}{25(2 + x^5)^5} + C.$$

27. Let $u = \pi x$. Then $du = \pi\, dx$, so $\displaystyle\int \sin \pi x\, dx = \int \frac{\sin u\, du}{\pi} = \frac{-\cos u}{\pi} + C = -\frac{\cos \pi x}{\pi} + C.$

29. Let $u = \frac{1}{t}$. Then $du = -\frac{1}{t^2}\, dt$, so $\displaystyle\int \frac{\cos(1/t)}{t^2}\, dt = \int \cos u\, (-du) = -\sin u + C = -\sin\left(\frac{1}{t}\right) + C.$

31. $u = \sqrt{x} \ \Rightarrow \ du = \dfrac{dx}{2\sqrt{x}}$, so $\displaystyle\int \frac{e^{\sqrt{x}}}{\sqrt{x}}\, dx = 2 \int e^u\, du = 2e^u + C = 2e^{\sqrt{x}} + C.$

33. Let $u = \ln(\cos x)$. Then $du = \dfrac{-\sin x}{\cos x}\, dx = -\tan x\, dx \ \Rightarrow$
$\int \tan x \ln(\cos x)\, dx = -\int u\, du = -\frac{1}{2}u^2 + C = -\frac{1}{2}[\ln(\cos x)]^2 + C.$

35. Let $u = 1 + x^4$. Then $du = 4x^3\, dx \ \Rightarrow \ \displaystyle\int \frac{x^3}{1 + x^4}\, dx = \frac{1}{4}\int \frac{1}{u}\, du = \frac{1}{4}\ln|u| + C = \frac{1}{4}\ln\left(1 + x^4\right) + C.$

37. Let $u = 1 + \sec\theta$, so $du = \sec\theta \tan\theta\, d\theta \ \Rightarrow$
$$\int \frac{\sec\theta \tan\theta}{1 + \sec\theta}\, d\theta = \int \frac{1}{u}\, du = \ln|u| + C = \ln|1 + \sec\theta| + C$$

39. $\int_0^{2\pi} |\sin x|\, dx = \int_0^\pi \sin x\, dx - \int_\pi^{2\pi} \sin x\, dx = 2\int_0^\pi \sin x\, dx = -2\,[\cos x]_0^\pi = -2\,[(-1) - 1] = 4$

In Exercise 41, let $f(x)$ denote the integrand and $F(x)$ its antiderivative (with $C = 0$).

41. $u = 1 + \sin x \implies du = \cos x\, dx$, so

$$\int \frac{\cos x\, dx}{\sqrt{1 + \sin x}} = \int u^{-1/2}\, du = 2u^{1/2} + C = 2\sqrt{1 + \sin x} + C.$$

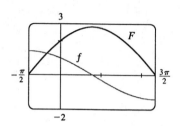

43. From the graph, it appears that the area under the curve $y = x\sqrt{x}$
between $x = 0$ and $x = 4$ is somewhat less than half the area of an
8×4 rectangle, so perhaps about 13 or 14. To find the exact value, we
evaluate

$$\int_0^4 x\sqrt{x}\, dx = \int_0^4 x^{3/2}\, dx = \left[\tfrac{2}{5} x^{5/2}\right]_0^4 = \tfrac{2}{5}(4)^{5/2} = \tfrac{64}{5} = 12.8.$$

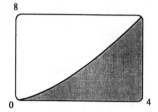

45. By FTC1, $F(x) = \int_1^x \sqrt{1 + t^4}\, dt \implies F'(x) = \sqrt{1 + x^4}.$

47. $g(x) = \displaystyle\int_0^{x^3} \frac{t\, dt}{\sqrt{1 + t^3}}$. Let $y = g(u)$ and $u = x^3$.

Then $g'(x) = \dfrac{dy}{dx} = \dfrac{dy}{du}\dfrac{du}{dx} = \dfrac{u}{\sqrt{1 + u^3}} 3x^2 = \dfrac{x^3}{\sqrt{1 + x^9}} 3x^2 = \dfrac{3x^5}{\sqrt{1 + x^9}}.$

49. $y = \displaystyle\int_{\sqrt{x}}^x \frac{e^t}{t}\, dt = \int_{\sqrt{x}}^1 \frac{e^t}{t}\, dt + \int_1^x \frac{e^t}{t}\, dt = -\int_1^{\sqrt{x}} \frac{e^t}{t}\, dt + \int_1^x \frac{e^t}{t}\, dt \implies$

$\dfrac{dy}{dx} = -\dfrac{d}{dx}\left(\displaystyle\int_1^{\sqrt{x}} \frac{e^t}{t}\, dt\right) + \dfrac{d}{dx}\left(\displaystyle\int_1^x \frac{e^t}{t}\, dt\right).$ Let $u = \sqrt{x}$. Then

$$\dfrac{d}{dx}\int_1^{\sqrt{x}} \frac{e^t}{t}\, dt = \dfrac{d}{dx}\int_1^u \frac{e^t}{t}\, dt = \dfrac{d}{du}\left(\int_1^u \frac{e^t}{t}\, dt\right)\dfrac{du}{dx} = \dfrac{e^u}{u} \cdot \dfrac{1}{2\sqrt{x}} = \dfrac{e^{\sqrt{x}}}{\sqrt{x}} \cdot \dfrac{1}{2\sqrt{x}} = \dfrac{e^{\sqrt{x}}}{2x}$$

so $\dfrac{dy}{dx} = -\dfrac{e^{\sqrt{x}}}{2x} + \dfrac{e^x}{x}.$

51. If $1 \le x \le 3$, then $2 \le \sqrt{x^2 + 3} \le 2\sqrt{3}$, so $2(3 - 1) \le \int_1^3 \sqrt{x^2 + 3}\, dx \le 2\sqrt{3}(3 - 1)$; that is,
$4 \le \int_1^3 \sqrt{x^2 + 3}\, dx \le 4\sqrt{3}.$

53. $0 \le x \le 1 \implies 0 \le \cos x \le 1 \implies x^2 \cos x \le x^2 \implies \int_0^1 x^2 \cos x\, dx \le \int_0^1 x^2\, dx = \tfrac{1}{3}\left[x^3\right]_0^1 = \tfrac{1}{3}$ [Property
7].

55. $\cos x \le 1 \implies e^x \cos x \le e^x \implies \int_0^1 e^x \cos x\, dx \le \int_0^1 e^x\, dx = [e^x]_0^1 = e - 1$

57. Let $f(x) = \sqrt{1 + x^3}$ on $[0, 1]$. The Midpoint Rule with $n = 5$ gives

$$\int_0^1 \sqrt{1 + x^3}\, dx \approx \tfrac{1}{5}[f(0.1) + f(0.3) + f(0.5) + f(0.7) + f(0.9)]$$

$$= \tfrac{1}{5}\left[\sqrt{1 + (0.1)^3} + \sqrt{1 + (0.3)^3} + \cdots + \sqrt{1 + (0.9)^3}\right] \approx 1.110$$

59. Total percentage increase $= \int_{1991}^{1997} r(t)\, dt \approx M_3 = \frac{1997 - 1991}{3} \left[r(1992) + r(1994) + r(1996) \right]$

$$= 2\,(7.4 + 4.8 + 3.0) = 30.4$$

61. We use the Midpoint Rule with $n = 6$ and $\Delta t = \frac{24 - 0}{6} = 4$.

Increase in bee population $= \int_0^{24} r(t)\, dt \approx M_6 = 4\,[f(2) + f(6) + f(10) + f(14) + f(18) + f(22)]$

$$\approx 4\,[60.15 + 976.81 + 7023.11 + 8550.88 + 1374.96 + 152.74]$$

$$= 4\,(18{,}138.65) = 72{,}554.6$$

For the answer in the back of the text, we used the estimates 50, 1000, 7000, 8550, 1350, and 150 to get 72,400.

63. By the Fundamental Theorem of Calculus, we know that $F(x) = \int_a^x t^2 \sin(t^2)\, dt$ is an antiderivative of $f(x) = x^2 \sin(x^2)$. This integral cannot be expressed in any simpler form. Since $\int_a^a f\, dt = 0$ for any a, we can take $a = 1$, and then $F(1) = 0$, as required. So $F(x) = \int_1^x t^2 \sin(t^2)\, dt$ is the desired function.

65. $1 = \int_0^1 \sinh cx\, dx = \frac{1}{c} \left[\cosh cx \right]_0^1 = \frac{1}{c}\,(\cosh c - 1) \quad \Rightarrow \quad c = \cosh c - 1 \quad \Rightarrow \quad \cosh c = c + 1 \quad \Rightarrow$
$c \approx 1.6161$

67. We differentiate both sides of the given equation using FTC1, and get $f(x) = e^{2x} + 2xe^{2x} + e^{-x} f(x) \quad \Rightarrow$

$$f(x)\,(1 - e^{-x}) = e^{2x} + 2xe^{2x} \quad \Rightarrow \quad f(x) = \frac{e^{2x}\,(1 + 2x)}{1 - e^{-x}}.$$

69. Let $u = f(x)$ and $du = f'(x)\, dx$. So $2 \int_a^b f(x) f'(x)\, dx = 2 \int_{f(a)}^{f(b)} u\, du = \left[u^2 \right]_{f(a)}^{f(b)} = [f(b)]^2 - [f(a)]^2$.

71. Let $u = 1 - x$. Then $du = -dx$, so $\int_0^1 f(1 - x)\, dx = \int_1^0 f(u)\,(-du) = \int_0^1 f(u)\, du = \int_0^1 f(x)\, dx$.

Problems Plus

1. Differentiating both sides of the equation $x \sin \pi x = \int_0^{x^2} f(t)\, dt$ (using FTC1 and the Chain Rule for the right side) gives $\sin \pi x + \pi x \cos \pi x = 2x f(x^2)$. Letting $x = 2$ so that $f(x^2) = f(4)$, we obtain $\sin 2\pi + 2\pi \cos 2\pi = 4f(4)$, so $f(4) = \frac{1}{4}(0 + 2\pi \cdot 1) = \frac{\pi}{2}$.

3. For $1 \le x \le 2$, we have $x^4 \le 2^4 = 16$, so $1 + x^4 \le 17$ and $\dfrac{1}{1+x^4} \ge \dfrac{1}{17}$. Thus,

$$\int_1^2 \frac{1}{1+x^4}\, dx \ge \int_1^2 \frac{1}{17}\, dx = \frac{1}{17}. \text{ Also } 1 + x^4 > x^4 \text{ for } 1 \le x \le 2, \text{ so } \frac{1}{1+x^4} < \frac{1}{x^4} \text{ and}$$

$$\int_1^2 \frac{1}{1+x^4}\, dx < \int_1^2 x^{-4}\, dx = \left[\frac{x^{-3}}{-3}\right]_1^2 = -\frac{1}{24} + \frac{1}{3} = \frac{7}{24}. \text{ Thus, we have the estimate}$$

$$\frac{1}{17} \le \int_1^2 \frac{1}{1+x^4}\, dx \le \frac{7}{24}.$$

5. $f(x) = \int_0^{g(x)} \dfrac{1}{\sqrt{1+t^3}}\, dt$, where $g(x) = \int_0^{\cos x} \left[1 + \sin(t^2)\right] dt$. Using FTC1 and the Chain Rule (twice) we

have $f'(x) = \dfrac{1}{\sqrt{1+[g(x)]^3}} g'(x) = \dfrac{1}{\sqrt{1+[g(x)]^3}} \left[1 + \sin(\cos^2 x)\right](-\sin x)$. Now

$g\left(\frac{\pi}{2}\right) = \int_0^0 \left[1 + \sin(t^2)\right] dt = 0$, so $f'\left(\frac{\pi}{2}\right) = \frac{1}{\sqrt{1+0}}(1 + \sin 0)(-1) = 1 \cdot 1 \cdot (-1) = -1$.

7. By l'Hospital's Rule and the Fundamental Theorem, using the notation $\exp(y) = e^y$,

$$\lim_{x \to 0} \frac{\int_0^x (1 - \tan 2t)^{1/t}\, dt}{x} \overset{H}{=} \lim_{x \to 0} \frac{(1 - \tan 2x)^{1/x}}{1} = \exp\left(\lim_{x \to 0} \frac{\ln(1 - \tan 2x)}{x}\right)$$

$$\overset{H}{=} \exp\left(\lim_{x \to 0} \frac{-2 \sec^2 2x}{1 - \tan 2x}\right) = \exp\left(\frac{-2 \cdot 1^2}{1 - 0}\right) = e^{-2}$$

9. $f(x) = 2 + x - x^2 = (-x + 2)(x + 1) = 0 \iff x = 2$ or $x = -1$. $f(x) \ge 0$ for $x \in [-1, 2]$ and $f(x) < 0$ everywhere else. The integral $\int_a^b (2 + x - x^2)\, dx$ has a maximum on the interval where the integrand is positive, which is $[-1, 2]$. So $a = -1$, $b = 2$. (Any larger interval gives a smaller integral since $f(x) < 0$ outside $[-1, 2]$. Any smaller interval also gives a smaller integral since $f(x) \ge 0$ in $[-1, 2]$.)

11. (a) We can split the integral $\int_0^n [\![x]\!]\, dx$ into the sum $\displaystyle\sum_{i=1}^n \left[\int_{i-1}^i [\![x]\!]\, dx\right]$. But on each of the intervals $[i - 1, i)$ of integration, $[\![x]\!]$ is a constant function, namely $i - 1$. So the ith integral in the sum is equal to

$(i-1)[i - (i-1)] = (i-1)$. So the original integral is equal to $\displaystyle\sum_{i=1}^n (i-1) = \sum_{i=1}^{n-1} i = \frac{(n-1)n}{2}$.

(b) We can write $\int_a^b [\![x]\!]\, dx = \int_0^b [\![x]\!]\, dx - \int_0^a [\![x]\!]\, dx$.

Now $\int_0^b [\![x]\!]\, dx = \int_0^{[\![b]\!]} [\![x]\!]\, dx + \int_{[\![b]\!]}^b [\![x]\!]\, dx$. The first of these integrals is equal to $\frac{1}{2}([\![b]\!] - 1)[\![b]\!]$, by part (a), and since $[\![x]\!] = [\![b]\!]$ on $[[\![b]\!], b]$, the second integral is just $[\![b]\!](b - [\![b]\!])$. So

$\int_0^b [\![x]\!]\, dx = \frac{1}{2}([\![b]\!] - 1)[\![b]\!] + [\![b]\!](b - [\![b]\!]) = \frac{1}{2}[\![b]\!](2b - [\![b]\!] - 1)$ and similarly

$\int_0^a [\![x]\!]\, dx = \frac{1}{2}[\![a]\!](2a - [\![a]\!] - 1)$. Therefore $\int_a^b [\![x]\!]\, dx = \frac{1}{2}[\![b]\!](2b - [\![b]\!] - 1) - \frac{1}{2}[\![a]\!](2a - [\![a]\!] - 1)$.

13. Differentiating the equation $\int_0^x f(t)\,dt = [f(x)]^2$ using FTC1 gives $f(x) = 2f(x)f'(x)$ ⇒

$f(x)[2f'(x) - 1] = 0$, so $f(x) = 0$ or $f'(x) = \frac{1}{2}$. $f'(x) = \frac{1}{2}$ ⇒ $f(x) = \frac{1}{2}x + C$. To find C we substitute

into the original equation to get $\int_0^x \left(\frac{1}{2}t + C\right) dt = \left(\frac{1}{2}x + C\right)^2$ ⟺ $\frac{1}{4}x^2 + Cx = \frac{1}{4}x^2 + Cx + C^2$. It follows

that $C = 0$, so $f(x) = \frac{1}{2}x$. Therefore, $f(x) = 0$ or $f(x) = \frac{1}{2}x$.

15. Note that $\frac{d}{dx}\left(\int_0^x \left[\int_0^u f(t)\,dt\right] du\right) = \int_0^x f(t)\,dt$ by FTC1, while

$$\frac{d}{dx}\left[\int_0^x f(u)(x-u)\,du\right] = \frac{d}{dx}\left[x\int_0^x f(u)\,du\right] - \frac{d}{dx}\left[\int_0^x f(u)u\,du\right] = \int_0^x f(u)\,du + xf(x) - f(x)x$$

$$= \int_0^x f(u)\,du$$

Hence, $\int_0^x f(u)(x-u)\,du = \int_0^x \left[\int_0^u f(t)\,dt\right] du + C$. Setting $x = 0$ gives $C = 0$.

17. $\lim\limits_{n\to\infty} \left(\dfrac{1}{\sqrt{n}\sqrt{n+1}} + \dfrac{1}{\sqrt{n}\sqrt{n+2}} + \cdots + \dfrac{1}{\sqrt{n}\sqrt{n+n}}\right) = \lim\limits_{n\to\infty} \dfrac{1}{n}\left(\sqrt{\dfrac{n}{n+1}} + \sqrt{\dfrac{n}{n+2}} + \cdots + \sqrt{\dfrac{n}{n+n}}\right)$

$= \lim\limits_{n\to\infty} \dfrac{1}{n}\left(\dfrac{1}{\sqrt{1+1/n}} + \dfrac{1}{\sqrt{1+2/n}} + \cdots + \dfrac{1}{\sqrt{1+1}}\right) = \lim\limits_{n\to\infty} \dfrac{1}{n}\sum\limits_{i=1}^{n} f\left(\dfrac{i}{n}\right)$ $\left(\text{where } f(x) = \dfrac{1}{\sqrt{1+x}}\right)$

$= \int_0^1 \dfrac{1}{\sqrt{1+x}}\,dx = \left[2\sqrt{1+x}\right]_0^1 = 2\left(\sqrt{2} - 1\right)$

19. The shaded region has area $\int_0^1 f(x)\,dx = \frac{1}{3}$. The integral $\int_0^1 f^{-1}(y)\,dy$

gives the area of the unshaded region, which we know to be $1 - \frac{1}{3} = \frac{2}{3}$. So

$\int_0^1 f^{-1}(y)\,dy = \frac{2}{3}$.

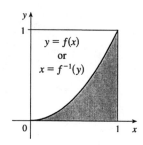

6 Applications of Integration

6.1 Areas between Curves

1. $A = \int_{-1}^{1} [(x^2 + 3) - x] \, dx = 2 \int_0^1 (x^2 + 3) \, dx$ [by Theorem 5.5.7(a)] $= 2 \left[\frac{1}{3}x^3 + 3x \right]_0^1 = 2 \left(\frac{1}{3} + 3 \right) = \frac{20}{3}$

3. $A = \int_{-1}^{1} [(1 - y^4) - (y^3 - y)] \, dy = 2 \int_0^1 (1 - y^4) \, dx$ [by Theorem 5.5.7(a)]

$= 2 \left[-\frac{1}{5}y^5 + y \right]_0^1 = 2 \left(-\frac{1}{5} + 1 \right) = \frac{8}{5}$

5. $A = \int_{-1}^{2} [(9 - x^2) - (x + 1)] \, dx$

$= \int_{-1}^{2} (8 - x - x^2) \, dx$

$= \left[8x - \frac{x^2}{2} - \frac{x^3}{3} \right]_{-1}^{2}$

$= \left(16 - 2 - \frac{8}{3} \right) - \left(-8 - \frac{1}{2} + \frac{1}{3} \right)$

$= 22 - 3 + \frac{1}{2} = \frac{39}{2}$

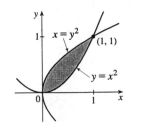
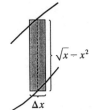

7. $A = \int_0^1 (x - x^2) \, dx$

$= \left[\frac{1}{2}x^2 - \frac{1}{3}x^3 \right]_0^1$

$= \frac{1}{2} - \frac{1}{3}$

$= \frac{1}{6}$

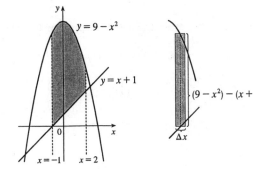

9. $A = \int_1^2 \left(\frac{1}{x} - \frac{1}{x^2} \right) dx = \left[\ln x + \frac{1}{x} \right]_1^2$

$= \left(\ln 2 + \frac{1}{2} \right) - (\ln 1 + 1)$

$= \ln 2 - \frac{1}{2} \approx 0.19$

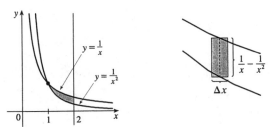

11. $A = \int_0^1 (\sqrt{x} - x^2) \, dx$

$= \left[\frac{2}{3}x^{3/2} - \frac{1}{3}x^3 \right]_0^1$

$= \frac{2}{3} - \frac{1}{3}$

$= \frac{1}{3}$

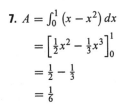
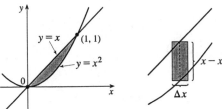

13. $A = \int_{-1}^{1} \left[(x^2 + 3) - 4x^2\right] dx$

$\qquad = 2\int_0^1 (3 - 3x^2) \, dx$

$\qquad = 2\left[3x - x^3\right]_0^1 = 2(3 - 1) = 4$

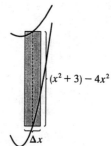

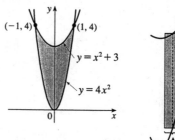

15. $x + 1 = (x - 1)^2 \quad \Rightarrow \quad x + 1 = x^2 - 2x + 1 \quad \Rightarrow \quad 0 = x^2 - 3x \quad \Rightarrow \quad 0 = x(x - 3) \quad \Rightarrow \quad x = 0 \text{ or } 3.$

$$A = \int_{-1}^{2} \left|(x+1) - (x-1)^2\right| dx = \int_{-1}^{0} \left[(x-1)^2 - (x+1)\right] dx + \int_{0}^{2} \left[(x+1) - (x-1)^2\right] dx$$

$$= \int_{-1}^{0} (x^2 - 3x) \, dx + \int_{0}^{2} (3x - x^2) \, dx = \left[\tfrac{1}{3}x^3 - \tfrac{3}{2}x^2\right]_{-1}^{0} + \left[\tfrac{3}{2}x^2 - \tfrac{1}{3}x^3\right]_{0}^{2}$$

$$= 0 - \left(-\tfrac{1}{3} - \tfrac{3}{2}\right) + \left(6 - \tfrac{8}{3}\right) - 0 = \tfrac{31}{6}$$

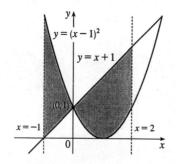

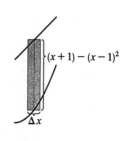

17. $A = \int_{-1}^{3} \left[(2y + 3) - y^2\right] dy$

$\qquad = \left[y^2 + 3y - \tfrac{1}{3}y^3\right]_{-1}^{3}$

$\qquad = (9 + 9 - 9) - \left(1 - 3 + \tfrac{1}{3}\right)$

$\qquad = \tfrac{32}{3}$

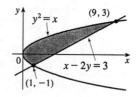

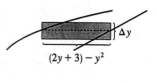

19. $A = \int_{-1}^{1} \left[(1 - y^2) - (y^2 - 1)\right] dy$

$\qquad = \int_{-1}^{1} 2(1 - y^2) \, dy$

$\qquad = 4\int_0^1 (1 - y^2) \, dy$

$\qquad = 4\left[y - \tfrac{1}{3}y^3\right]_0^1 = 4\left(1 - \tfrac{1}{3}\right) = \tfrac{8}{3}$

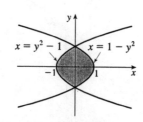

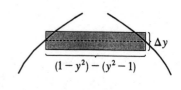

21. Notice that $\cos x = \sin 2x = 2\sin x \cos x \;\Leftrightarrow\; 2\sin x = 1$ or $\cos x = 0 \;\Leftrightarrow\;$
$x = \frac{\pi}{6}$ or $\frac{\pi}{2}$.

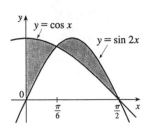

$$A = \int_0^{\pi/6}(\cos x - \sin 2x)\,dx + \int_{\pi/6}^{\pi/2}(\sin 2x - \cos x)\,dx$$

$$= \left[\sin x + \tfrac{1}{2}\cos 2x\right]_0^{\pi/6} + \left[-\tfrac{1}{2}\cos 2x - \sin x\right]_{\pi/6}^{\pi/2}$$

$$= \tfrac{1}{2} + \tfrac{1}{2}\cdot\tfrac{1}{2} - \left(0 + \tfrac{1}{2}\cdot 1\right) + \left(\tfrac{1}{2} - 1\right) - \left(-\tfrac{1}{2}\cdot\tfrac{1}{2} - \tfrac{1}{2}\right) = \tfrac{1}{2}$$

23. From the graph, we see that the curves intersect at $x = 0$, $x = \frac{\pi}{2}$, and $x = \pi$. By symmetry,

$$A = \int_0^{\pi}\left|\cos x - \left(1 - \frac{2x}{\pi}\right)\right|dx = 2\int_0^{\pi/2}\left[\cos x - \left(1 - \frac{2x}{\pi}\right)\right]dx = 2\int_0^{\pi/2}\left(\cos x - 1 + \frac{2x}{\pi}\right)dx$$

$$= 2\left[\sin x - x + \tfrac{1}{\pi}x^2\right]_0^{\pi/2} = 2\left[\left(1 - \tfrac{\pi}{2} + \tfrac{1}{\pi}\cdot\tfrac{\pi^2}{4}\right) - 0\right] = 2\left(1 - \tfrac{\pi}{2} + \tfrac{\pi}{4}\right) = 2 - \tfrac{\pi}{2}$$

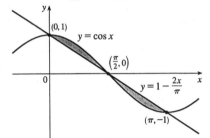

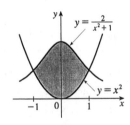

25. $A = \int_{-1}^{1}\left(\dfrac{2}{x^2+1} - x^2\right)dx$

$$= 2\int_0^1\left(\frac{2}{x^2+1} - x^2\right)dx$$

$$= 2\left[2\tan^{-1}x - \tfrac{1}{3}x^3\right]_0^1 = 2\left(2\cdot\tfrac{\pi}{4} - \tfrac{1}{3}\right)$$

$$= \pi - \tfrac{2}{3} \approx 2.47$$

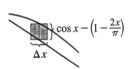

27. An equation of the line through $(0,0)$ and $(2,1)$ is $y = \tfrac{1}{2}x$; through $(0,0)$
and $(-1,6)$ is $y = -6x$; through $(2,1)$ and $(-1,6)$ is $y = -\tfrac{5}{3}x + \tfrac{13}{3}$.

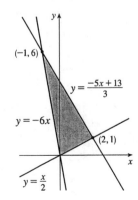

$$A = \int_{-1}^{0}\left[\left(-\tfrac{5}{3}x + \tfrac{13}{3}\right) - (-6x)\right]dx + \int_0^2\left[\left(-\tfrac{5}{3}x + \tfrac{13}{3}\right) - \tfrac{1}{2}x\right]dx$$

$$= \int_{-1}^{0}\left(\tfrac{13}{3}x + \tfrac{13}{3}\right)dx + \int_0^2\left(-\tfrac{13}{6}x + \tfrac{13}{3}\right)dx$$

$$= \tfrac{13}{3}\int_{-1}^{0}(x+1)\,dx + \tfrac{13}{3}\int_0^2\left(-\tfrac{1}{2}x + 1\right)dx$$

$$= \tfrac{13}{3}\left[\tfrac{1}{2}x^2 + x\right]_{-1}^{0} + \tfrac{13}{3}\left[-\tfrac{1}{4}x^2 + x\right]_0^2$$

$$= \tfrac{13}{3}\left[0 - \left(\tfrac{1}{2} - 1\right)\right] + \tfrac{13}{3}\left[(-1+2) - 0\right] = \tfrac{13}{3}\cdot\tfrac{1}{2} + \tfrac{13}{3}\cdot 1 = \tfrac{13}{2}$$

29. $A = \int_{-1}^{1} |x^3 - x| \, dx = 2 \int_{0}^{1} (x - x^3) \, dx$ [by symmetry]

$= 2 \left[\frac{1}{2}x^2 - \frac{1}{4}x^4 \right]_0^1 = 2 \left(\frac{1}{2} - \frac{1}{4} \right) = \frac{1}{2}$

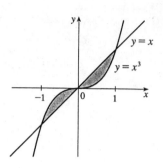

31. Let $f(x) = \sqrt{1 + x^3} - (1 - x)$, $\Delta x = \frac{2-0}{4} = \frac{1}{2}$.

$$A = \int_0^2 \left[\sqrt{1+x^3} - (1-x) \right] dx \approx \frac{1}{2} \left[f\left(\frac{1}{4}\right) + f\left(\frac{3}{4}\right) + f\left(\frac{5}{4}\right) + f\left(\frac{7}{4}\right) \right]$$

$$= \frac{1}{2} \left[\left(\frac{\sqrt{65}}{8} - \frac{3}{4} \right) + \left(\frac{\sqrt{91}}{8} - \frac{1}{4} \right) + \left(\frac{3\sqrt{21}}{8} + \frac{1}{4} \right) + \left(\frac{\sqrt{407}}{8} + \frac{3}{4} \right) \right]$$

$$= \frac{1}{16} \left(\sqrt{65} + \sqrt{91} + 3\sqrt{21} + \sqrt{407} \right) \approx 3.22$$

33.

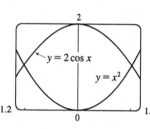

From the graph, we see that the curves intersect at $x \approx \pm 1.02$, with $2 \cos x > x^2$ on $(-1.02, 1.02)$. So the area of the region bounded by the curves is

$$A \approx \int_{-1.02}^{1.02} (2\cos x - x^2) \, dx = 2 \int_0^{1.02} (2\cos x - x^2) \, dx$$

$$= 2 \left[2 \sin x - \frac{1}{3} x^3 \right]_0^{1.02} \approx 2.70$$

35. A typical graphing calculator solution is as follows. Assign X^2 to

Y1 ($y_1 = x^2$) and Exp(-X^2) to Y2 ($y_2 = e^{-x^2}$). Graph the functions and find (and store) the x-coordinates of the points of intersection. In this case, we have some symmetry, so we need to find only one point of intersection. Store $x \approx 0.75308916$ in memory location B. Now use the appropriate integration command to approximate the area:

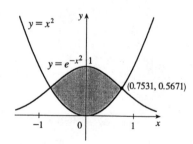

$A = 2 \int_0^B (y_2 - y_1) \, dx = 2*\text{Int}(Y2-Y1,X,0,B) \approx 0.979263.$

37.

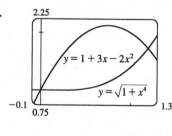

From the graph, we see that the curves intersect at $x = 0$ and at $x \approx 1.19$, with $1 + 3x - 2x^2 > \sqrt{1 + x^4}$ on $(0, 1.19)$. So, using the Midpoint Rule with $f(x) = 1 + 3x - 2x^2 - \sqrt{1 + x^4}$ on $[0, 1.19]$ with $n = 4$, we calculate the approximate area between the curves:

$$A \approx \int_0^{1.19} \left(1 + 3x - 2x^2 - \sqrt{1 + x^4} \right) dx$$

$$\approx \frac{1.19}{4} \left[f\left(\frac{1.19}{8}\right) + f\left(\frac{3 \cdot 1.19}{8}\right) + f\left(\frac{5 \cdot 1.19}{8}\right) + f\left(\frac{7 \cdot 1.19}{8}\right) \right] \approx 0.83$$

39. 1 second $= \frac{1}{3600}$ hour, so 10 s $= \frac{1}{360}$ h. With the given data, we can take $n = 5$ to use the Midpoint Rule.
$\Delta t = \frac{1/360-0}{5} = \frac{1}{1800}$, so

$$\text{distance}_{\text{Kelly}} - \text{distance}_{\text{Chris}} = \int_0^{1/360} v_K \, dt - \int_0^{1/360} v_C \, dt = \int_0^{1/360} (v_K - v_C) \, dt$$

$$\approx M_5 = \frac{1}{1800} \left[(v_K - v_C)(1) + (v_K - v_C)(3) + (v_K - v_C)(5) \right.$$
$$\left. + (v_K - v_C)(7) + (v_K - v_C)(9) \right]$$

$$= \frac{1}{1800} \left[(22 - 20) + (52 - 46) + (71 - 62) + (86 - 75) + (98 - 86) \right]$$

$$= \frac{1}{1800} (2 + 6 + 9 + 11 + 12) = \frac{1}{1800} (40) = \frac{1}{45} \text{ mile, or } 117\frac{1}{3} \text{ feet}$$

41. We know that the area under curve A between $t = 0$ and $t = x$ is $\int_0^x v_A(t) \, dt = s_A(x)$, where $v_A(t)$ is the velocity of car A and s_A is its displacement. Similarly, the area under curve B between $t = 0$ and $t = x$ is $\int_0^x v_B(t) \, dt = s_B(x)$.

(a) After one minute, the area under curve A is greater than the area under curve B. So A is ahead after one minute.

(b) Its numerical value is $s_A(1) - s_B(1)$, which is the distance by which A is ahead of B after 1 minute.

(c) After two minutes, car B is traveling faster than car A and has gained some ground, but the area under curve A from $t = 0$ to $t = 2$ is still greater than the corresponding area for curve B, so car A is still ahead.

(d) From the graph, it appears that the area between curves A and B for $0 \le t \le 1$ (when car A is going faster), which corresponds to the distance by which car A is ahead, seems to be about 3 squares. Therefore, the cars will be side by side at the time x where the area between the curves for $1 \le t \le x$ (when car B is going faster) is the same as the area for $0 \le t \le 1$. From the graph, it appears that this time is $x \approx 2.2$. So the cars are side by side when $t \approx 2.2$ minutes.

43.

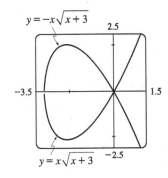

$y = -x\sqrt{x+3}$

$y = x\sqrt{x+3}$

To graph this function, we must first express it as a combination of explicit functions of y; namely, $y = \pm x\sqrt{x+3}$. We can see from the graph that the loop extends from $x = -3$ to $x = 0$, and that by symmetry, the area we seek is just twice the area under the top half of the curve on this interval, the equation of the top half being $y = -x\sqrt{x+3}$. So the area is
$A = 2 \int_{-3}^0 \left(-x\sqrt{x+3} \right) dx$. We substitute $u = x + 3$, so $du = dx$ and the limits change to 0 and 3, and we get

$$A = -2 \int_0^3 \left[(u - 3)\sqrt{u} \right] du = -2 \int_0^3 \left(u^{3/2} - 3u^{1/2} \right) du$$

$$= -2 \left[\frac{2}{5} u^{5/2} - 2u^{3/2} \right]_0^3 = -2 \left[\frac{2}{5} \left(3^2\sqrt{3} \right) - 2 \left(3\sqrt{3} \right) \right] = \frac{24}{5}\sqrt{3}$$

45. By the symmetry of the problem, we consider only the first quadrant where $y = x^2 \Rightarrow x = \sqrt{y}$. We are looking for a number b such that $\int_0^4 x \, dy = 2 \int_0^b x \, dy \Rightarrow \int_0^4 \sqrt{y} \, dy = 2 \int_0^b \sqrt{y} \, dy \Rightarrow \frac{2}{3} \left[y^{3/2} \right]_0^4 = \frac{4}{3} \left[y^{3/2} \right]_0^b$
$\Rightarrow \frac{2}{3} (8 - 0) = \frac{4}{3} \left(b^{3/2} - 0 \right) \Rightarrow b^{3/2} = 4 \Rightarrow b = 4^{2/3} \approx 2.52$.

47. We first assume that $c > 0$, since c can be replaced by $-c$ in both equations without changing the graphs, and if $c = 0$ the curves do not enclose a region. We see from the graph that the enclosed area lies between $x = -c$ and $x = c$, and by symmetry, it is equal to twice the area under the top half of the graph (whose equation is $y = c^2 - x^2$). The enclosed area is

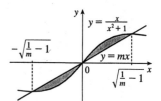

$$2 \int_{-c}^{c} \left(c^2 - x^2\right) dx = 4 \int_{0}^{c} \left(c^2 - x^2\right) dx = 4\left[c^2 x - \tfrac{1}{3}x^3\right]_{0}^{c}$$

$$= 4\left(c^3 - \tfrac{1}{3}c^3\right) = \tfrac{8}{3}c^3$$

which is equal to 576 when $c = \sqrt[3]{216} = 6$. Note that $c = -6$ is another solution, since the graphs are the same.

49. The curve and the line will determine a region when they intersect at two or more points. So we solve the equation $x / \left(x^2 + 1\right) = mx \;\Rightarrow$

$x = 0$ or $mx^2 + m - 1 = 0 \;\Rightarrow\; x = 0$ or $x^2 = \dfrac{1 - m}{m} \;\Rightarrow$

$x = 0$ or $x = \pm\sqrt{\dfrac{1}{m} - 1}$. Note that if $m = 1$, this has only

the solution $x = 0$, and no region is determined. But if $1/m - 1 > 0 \;\Leftrightarrow\; 1/m > 1 \;\Leftrightarrow\; 0 < m < 1$, then there are two solutions. [Another way of seeing this is to observe that the slope of the tangent to $y = x / \left(x^2 + 1\right)$ at the origin is $y' = 1$ and therefore we must have $0 < m < 1$.] Note that we cannot just integrate between the positive and negative roots, since the curve and the line cross at the origin. Since mx and $x / \left(x^2 + 1\right)$ are both odd functions, the total area is twice the area between the curves on the interval $\left[0, \sqrt{1/m - 1}\right]$. So the total area enclosed is

$$2 \int_{0}^{\sqrt{1/m-1}} \left[\frac{x}{x^2 + 1} - mx\right] dx = 2\left[\tfrac{1}{2} \ln \left(x^2 + 1\right) - \tfrac{1}{2}mx^2\right]_{0}^{\sqrt{1/m-1}}$$

$$= \left[\ln \left(1/m - 1 + 1\right) - m \left(1/m - 1\right)\right] - (\ln 1 - 0)$$

$$= \ln \left(1/m\right) + m - 1 = m - \ln m - 1$$

6.2 Volumes

1. A cross-section is circular with radius x^2, so its area is $A\left(x\right) = \pi \left(x^2\right)^2$.

$$V = \int_{0}^{1} A\left(x\right) dx = \int_{0}^{1} \pi \left(x^2\right)^2 dx = \pi \int_{0}^{1} x^4 dx = \pi \left[\tfrac{1}{5}x^5\right]_{0}^{1} = \tfrac{\pi}{5}$$

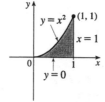

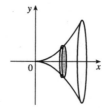

3. A cross-section is circular with radius $1/x$, so its area is $A(x) = \pi (1/x)^2$.

$$V = \int_1^2 A(x)\,dx = \int_1^2 \pi \left(\frac{1}{x}\right)^2 dx = \pi \int_1^2 \frac{1}{x^2}\,dx = \pi \left[-\frac{1}{x}\right]_1^2 = \pi \left[-\frac{1}{2} - (-1)\right] = \frac{\pi}{2}$$

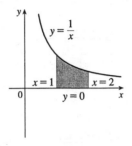
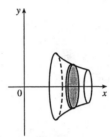

5. A cross-section is circular with radius $\sqrt{y}$, so its area is $A(y) = \pi \left(\sqrt{y}\right)^2$.

$$V = \int_0^4 A(y)\,dy = \int_0^4 \pi \left(\sqrt{y}\right)^2 dy = \pi \int_0^4 y\,dy = \pi \left[\frac{1}{2}y^2\right]_0^4 = 8\pi$$

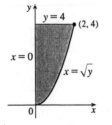

7. A cross-section is an annulus with inner radius x^2 and outer radius $\sqrt{x}$, so its area is

$$A(x) = \pi \left(\sqrt{x}\right)^2 - \pi \left(x^2\right)^2 = \pi \left(x - x^4\right).$$

$$V = \int_0^1 A(x)\,dx = \pi \int_0^1 \left(x - x^4\right) dx = \pi \left[\frac{1}{2}x^2 - \frac{1}{5}x^5\right]_0^1 = \pi \left(\frac{1}{2} - \frac{1}{5}\right) = \frac{3\pi}{10}$$

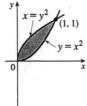
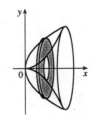

9. A cross-section is an annulus with inner radius y^2 and outer radius $2y$, so its area is

$A(y) = \pi (2y)^2 - \pi (y^2)^2 = \pi (4y^2 - y^4)$.

$$V = \int_0^2 A(y)\, dy = \pi \int_0^2 (4y^2 - y^4)\, dy = \pi \left[\tfrac{4}{3}y^3 - \tfrac{1}{5}y^5 \right]_0^2 = \pi \left(\tfrac{32}{3} - \tfrac{32}{5} \right) = \tfrac{64\pi}{15}$$

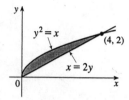

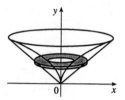

11. A cross-section is an annulus with inner radius $1 - \sqrt{x}$ and outer radius $1 - x$, so its area is

$A(x) = \pi (1-x)^2 - \pi (1 - \sqrt{x})^2 = \pi \left[(1 - 2x + x^2) - (1 - 2\sqrt{x} + x) \right] = \pi (-3x + x^2 + 2\sqrt{x})$.

$$V = \int_0^1 A(x)\, dx = \pi \int_0^1 (-3x + x^2 + 2\sqrt{x})\, dx = \pi \left[-\tfrac{3}{2}x^2 + \tfrac{1}{3}x^3 + \tfrac{4}{3}x^{3/2} \right]_0^1 = \pi \left(-\tfrac{3}{2} + \tfrac{5}{3} \right) = \tfrac{\pi}{6}$$

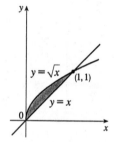

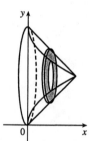

13. A cross-section is an annulus with inner radius $2 - 1$ and outer radius $2 - x^4$, so its area is

$A(x) = \pi (2 - x^4)^2 - \pi (2-1)^2 = \pi (3 - 4x^4 + x^8)$.

$$V = \int_{-1}^1 A(x)\, dx = 2 \int_0^1 A(x)\, dx = 2\pi \int_0^1 (3 - 4x^4 + x^8)\, dx = 2\pi \left[3x - \tfrac{4}{5}x^5 + \tfrac{1}{9}x^9 \right]_0^1$$

$$= 2\pi \left(3 - \tfrac{4}{5} + \tfrac{1}{9} \right) = \tfrac{208}{45}\pi$$

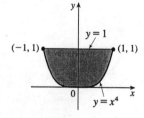

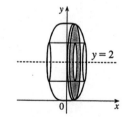

15. $V = \int_{-1}^{1} \pi \left(1 - y^2\right)^2 dy = 2 \int_{0}^{1} \pi \left(1 - y^2\right)^2 dy = 2\pi \int_{0}^{1} \left(1 - 2y^2 + y^4\right) dy$

$\quad = 2\pi \left[y - \frac{2}{3}y^3 + \frac{1}{5}y^5\right]_0^1 = 2\pi \cdot \frac{8}{15} = \frac{16}{15}\pi$

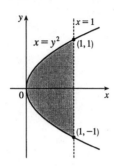 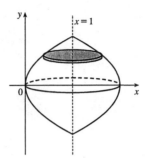

17. $V = \int_{0}^{1} \pi \left\{ \left[\sqrt{y} - (-1)\right]^2 - \left[y^2 - (-1)\right]^2 \right\} dy = \int_{0}^{1} \pi \left[\left(\sqrt{y} + 1\right)^2 - \left(y^2 + 1\right)^2\right] dy$

$\quad = \pi \int_{0}^{1} \left(y + 2\sqrt{y} + 1 - y^4 - 2y^2 - 1\right) dy = \pi \int_{0}^{1} \left(y + 2\sqrt{y} - y^4 - 2y^2\right) dy$

$\quad = \pi \left[\frac{1}{2}y^2 + \frac{4}{3}y^{3/2} - \frac{1}{5}y^5 - \frac{2}{3}y^3\right]_0^1 = \pi \left(\frac{1}{2} + \frac{4}{3} - \frac{1}{5} - \frac{2}{3}\right) = \frac{29}{30}\pi$

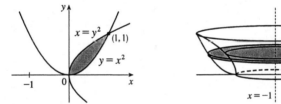 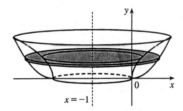

19. $V = \pi \int_{0}^{8} \left(\frac{1}{4}x\right)^2 dx = \frac{\pi}{16}\left[\frac{1}{3}x^3\right]_0^8 = \frac{32}{3}\pi$

21. $V = \pi \int_{0}^{2} (8 - 4y)^2 dy = \pi \left[64y - 32y^2 + \frac{16}{3}y^3\right]_0^2 = \pi \left(128 - 128 + \frac{128}{3}\right) = \frac{128}{3}\pi$

23. $V = \pi \int_{0}^{8} \left[\left(\sqrt[3]{x}\right)^2 - \left(\frac{1}{4}x\right)^2\right] = \pi \int_{0}^{8} \left(x^{2/3} - \frac{1}{16}x^2\right) dx = \pi \left[\frac{3}{5}x^{5/3} - \frac{1}{48}x^3\right]_0^8 = \pi \left(\frac{96}{5} - \frac{32}{3}\right) = \frac{128}{15}\pi$

25. $V = \pi \int_{0}^{8} \left[\left(2 - \frac{1}{4}x\right)^2 - \left(2 - \sqrt[3]{x}\right)^2\right] dx = \pi \int_{0}^{8} \left(-x + \frac{1}{16}x^2 + 4x^{1/3} - x^{2/3}\right) dx$

$\quad = \pi \left[-\frac{1}{2}x^2 + \frac{1}{48}x^3 + 3x^{4/3} - \frac{3}{5}x^{5/3}\right]_0^8 = \pi \left(-32 + \frac{32}{3} + 48 - \frac{96}{5}\right) = \frac{112}{15}\pi$

27. $V = \pi \int_{0}^{8} \left(2^2 - x^{2/3}\right) dx = \pi \left[4x - \frac{3}{5}x^{5/3}\right]_0^8 = \pi \left(32 - \frac{96}{5}\right) = \frac{64}{5}\pi$

29. $V = \pi \int_{0}^{8} \left(2 - \sqrt[3]{x}\right)^2 dx = \pi \int_{0}^{8} \left(4 - 4x^{1/3} + x^{2/3}\right) dx = \pi \left[4x - 3x^{4/3} + \frac{3}{5}x^{5/3}\right]_0^8 = \pi \left(32 - 48 + \frac{96}{5}\right) = \frac{16}{5}\pi$

31. $V = \pi \int_{1}^{e} \left[1^2 - (\ln x)^2\right] dx$

33. $V = \pi \int_0^\pi \left[(1-0)^2 - (1-\sin x)^2\right] dx = \pi \int_0^\pi \left[1^2 - (1-\sin x)^2\right] dx$

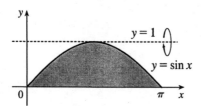

35. $V = \pi \int_{-\sqrt{8}}^{\sqrt{8}} \left\{ [3-(-2)]^2 - \left[\sqrt{y^2+1}-(-2)\right]^2 \right\} dy = \pi \int_{-2\sqrt{2}}^{2\sqrt{2}} \left[5^2 - \left(\sqrt{1+y^2}+2\right)^2\right] dy$

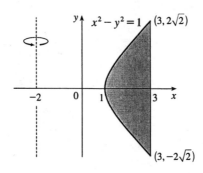

37.

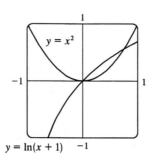

$y = x^2$ and $y = \ln(x+1)$ intersect at $x = 0$ and at $x \approx 0.747$.

$$V \approx \pi \int_0^{0.747} \left[[\ln(x+1)]^2 - \left(x^2\right)^2 \right] dx \approx 0.132$$

39. $\pi \int_0^{\pi/2} \cos^2 x \, dx$ describes the volume of the solid obtained by rotating the region $\mathcal{R} = \{(x,y) \mid 0 \leq x \leq \frac{\pi}{2}, 0 \leq y \leq \cos x\}$ of the xy-plane about the x-axis.

41. $\pi \int_0^1 \left(y^4 - y^8\right) dy = \pi \int_0^1 \left[\left(y^2\right)^2 - \left(y^4\right)^2\right] dy$ describes the volume of the solid obtained by rotating the region $\mathcal{R} = \{(x,y) \mid 0 \leq y \leq 1, y^4 \leq x \leq y^2\}$ of the xy-plane about the y-axis.

43. $V = \int_0^{15} A(x) \, dx \approx M_5 = 3 \left[A(1.5) + A(4.5) + A(7.5) + A(10.5) + A(13.5)\right]$
$$= 3(18 + 79 + 106 + 128 + 39) = 3 \cdot 370 = 1110 \text{ cm}^3$$

45. $V = \pi \int_0^h \left(-\frac{r}{h}y + r\right)^2 dy = \pi \int_0^h \left[\frac{r^2}{h^2}y^2 - \frac{2r^2}{h}y + r^2\right] dy$

$= \pi \left[\frac{r^2}{3h^2}y^3 - \frac{r^2}{h}y^2 + r^2y\right]_0^h = \pi \left(\frac{1}{3}r^2h - r^2h + r^2h\right) = \frac{1}{3}\pi r^2 h$

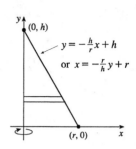

$y = -\frac{h}{r}x + h$

or $x = -\frac{r}{h}y + r$

$(0, h)$

$(r, 0)$

47. $x^2 + y^2 = r^2 \implies x^2 = r^2 - y^2 \implies$

$V = \pi \int_{r-h}^r \left(r^2 - y^2\right) dy = \pi \left[r^2 y - \frac{y^3}{3}\right]_{r-h}^r$

$= \pi \left(\left[r^3 - \frac{r^3}{3}\right] - \left[r^2(r-h) - \frac{(r-h)^3}{3}\right]\right)$

$= \pi \left(\frac{2}{3}r^3 - \frac{1}{3}(r-h)\left[3r^2 - (r-h)^2\right]\right)$

$= \frac{1}{3}\pi \left(2r^3 - (r-h)\left[3r^2 - (r^2 - 2rh + h^2)\right]\right)$

$= \frac{1}{3}\pi h^2 (3r - h)$, or, equivalently, $\pi h^2 \left(r - \frac{h}{3}\right)$

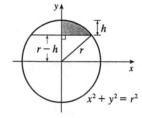

$r-h$ r

$x^2 + y^2 = r^2$

49. For a cross-section at height y, we see from similar triangles that $\dfrac{a/2}{b/2} = \dfrac{h-y}{h}$, so $a = b\left(1 - \dfrac{y}{h}\right)$. Similarly, for

cross-sections having $2b$ as their base and β replacing a, $\beta = 2b\left(1 - \dfrac{y}{h}\right)$. So

$V = \int_0^h A(y)\, dy = \int_0^h \left[b\left(1 - \frac{y}{h}\right)\right]\left[2b\left(1 - \frac{y}{h}\right)\right] dy = \int_0^h 2b^2 \left(1 - \frac{y}{h}\right)^2 dy$

$= 2b^2 \int_0^h \left(1 - \frac{2y}{h} + \frac{y^2}{h^2}\right) dy = 2b^2 \left[y - \frac{y^2}{h} + \frac{y^3}{3h^2}\right]_0^h = 2b^2 \left[h - h + \frac{1}{3}h\right]$

$= \frac{2}{3}b^2 h$ ($= \frac{1}{3}Bh$ where B is the area of the base, as with any pyramid.)

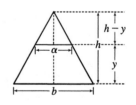

$h-y$

α

h

y

b

51. A cross-section at height z is a triangle similar to the base, so its area is

$A(z) = \frac{1}{2} \cdot 3\left(\frac{5-z}{5}\right) \cdot 4\left(\frac{5-z}{5}\right) = 6\left(1 - \frac{z}{5}\right)^2$, so

$V = \int_0^5 A(z)\, dz = 6 \int_0^5 (1 - z/5)^2\, dz = 6\left[(-5)\frac{1}{3}\left(1 - \frac{1}{5}z\right)^3\right]_0^5$

$= -10(-1) = 10 \text{ cm}^3$

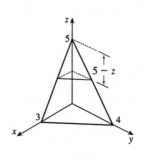

5

$5 - z$

3

4

x

y

z

53. If l is a leg of the isosceles right triangle and $2y$ is the hypotenuse, then

$$l^2 + l^2 = (2y)^2 \quad \Rightarrow \quad l = \sqrt{2}y.$$

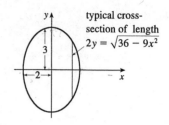

typical cross-section of length $2y = \sqrt{36 - 9x^2}$

$$V = \int_{-2}^{2} A(x)\, dx = 2\int_{0}^{2} A(x)\, dx = 2\int_{0}^{2} \tfrac{1}{2}\left(\sqrt{2}y\right)^2 dx = 2\int_{0}^{2} y^2\, dx$$

$$= \tfrac{1}{2}\int_{0}^{2}(36 - 9x^2)\, dx = \tfrac{9}{2}\int_{0}^{2}(4 - x^2)\, dx = \tfrac{9}{2}\left[4x - \tfrac{1}{3}x^3\right]_0^2$$

$$= \tfrac{9}{2}\left(8 - \tfrac{8}{3}\right) = 24$$

55. The square has area $A(y) = \left(2\sqrt{y}\right)^2 = 4y$, so $V = \int_0^1 A(y)\, dy = \int_0^1 4y\, dy = [2y^2]_0^1 = 2.$

57. A typical cross-section perpendicular to the y-axis in the base has length
$\ell(y) = 3 - \tfrac{3}{2}y$. This length is the leg of an isosceles right triangle, so

$$A(y) = \tfrac{1}{2}[\ell(y)]^2 \quad (\tfrac{1}{2}bh \text{ with base} = \text{height})$$

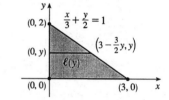

$$= \tfrac{1}{2}\left[3\left(1 - \tfrac{1}{2}y\right)\right]^2 = \tfrac{9}{2}\left(1 - \tfrac{1}{2}y\right)^2 = \tfrac{9}{2}\left(1 - y + \tfrac{1}{4}y^2\right)$$

Thus,

$$V = \int_0^2 A(y)\, dy = \tfrac{9}{2}\int_0^2\left(1 - y + \tfrac{1}{4}y^2\right) dy = \tfrac{9}{2}\left[y - \tfrac{1}{2}y^2 + \tfrac{1}{12}y^3\right]_0^2 = \tfrac{9}{2}\left[\left(2 - 2 + \tfrac{2}{3}\right) - 0\right] = \tfrac{9}{2}\cdot\tfrac{2}{3} = 3$$

59. (a) The torus is obtained by rotating the circle $(x - R)^2 + y^2 = r^2$ about the
y-axis. Solving for y, we see that the right half of the circle is given by

$$x = R + \sqrt{r^2 - y^2} = f(y) \text{ and the left half by}$$

$$x = R - \sqrt{r^2 - y^2} = g(y). \text{ So}$$

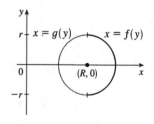

$$V = \pi \int_{-r}^{r} \left([f(y)]^2 - [g(y)]^2\right) dy = 2\pi \int_0^r 4R\sqrt{r^2 - y^2}\, dy$$

$$= 8\pi R \int_0^r \sqrt{r^2 - y^2}\, dy$$

(b) Observe that the integral represents a quarter of the area of a circle with radius r, so
$8\pi R \int_0^r \sqrt{r^2 - y^2}\, dy = 8\pi R \cdot \tfrac{1}{4}\left(\pi r^2\right) = 2\pi^2 r^2 R.$

61. (a) Volume$(S_1) = \int_0^h A(z)\, dz =$ Volume(S_2) since the cross-sectional area $A(z)$ at height z is the same for both solids.

(b) By Cavalieri's Principle, the volume of the cylinder in the figure is the same as that of a right circular cylinder with radius r and height h, that is, $\pi r^2 h$.

63. The volume is obtained by rotating the area common to two circles of radius r, as shown. The volume of the right half is

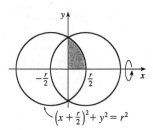

$$V_{\text{right}} = \pi \int_0^{r/2} y^2 \, dx = \pi \int_0^{r/2} \left[r^2 - \left(\tfrac{1}{2}r + x \right)^2 \right] dx$$

$$= \pi \left[r^2 x - \tfrac{1}{3} \left(\tfrac{1}{2}r + x \right)^3 \right]_0^{r/2} = \pi \left[\left(\tfrac{1}{2}r^3 - \tfrac{1}{3}r^3 \right) - \left(0 - \tfrac{1}{24}r^3 \right) \right] = \tfrac{5}{24}\pi r^3$$

$$\left(x + \tfrac{r}{2} \right)^2 + y^2 = r^2$$

So by symmetry, the total volume is twice this, or $\tfrac{5}{12}\pi r^3$.

Another Solution: We observe that the volume is the twice the volume of a cap of a sphere, so we can use the formula from Exercise 47 with $h = \tfrac{1}{2}r$: $V = 2 \cdot \tfrac{1}{3}\pi r h^2 (3r - h) = \tfrac{2}{3}\pi \left(\tfrac{1}{2}r \right)^2 \left(3r - \tfrac{1}{2}r \right) = \tfrac{5}{12}\pi r^3$.

65. Take the x-axis to be the axis of the cylindrical hole of radius r.
A quarter of the cross-section through y, perpendicular to the y-axis, is the rectangle shown. Using Pythagoras twice, we see that the dimensions of this rectangle are $x = \sqrt{R^2 - y^2}$ and $z = \sqrt{r^2 - y^2}$, so $\tfrac{1}{4}A(y) = xz = \sqrt{r^2 - y^2}\sqrt{R^2 - y^2}$, and

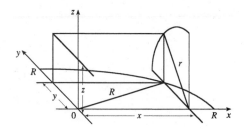

$$V = \int_{-r}^{r} A(y) \, dy = \int_{-r}^{r} 4\sqrt{r^2 - y^2}\sqrt{R^2 - y^2} \, dy$$

$$= 8 \int_0^r \sqrt{r^2 - y^2}\sqrt{R^2 - y^2} \, dy$$

67. (a) The radius of the barrel is the same at each end by symmetry, since the function $y = R - cx^2$ is even. Since the barrel is obtained by rotating the function y about the x-axis, this radius is equal to the value of y at $x = \tfrac{1}{2}h$, which is $R - c \left(\tfrac{1}{2}h \right)^2 = R - d = r$.

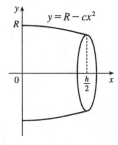

(b) The barrel is symmetric about the y-axis, so its volume is twice the volume of that part of the barrel for $x > 0$. Also, the barrel is a volume of rotation, so

$$V = 2 \int_0^{h/2} \pi \left(y^2 \right) dx = 2\pi \int_0^{h/2} \left(R - cx^2 \right)^2 dx = 2\pi \left[R^2 x - \tfrac{2}{3}Rcx^3 + \tfrac{1}{5}c^2 x^5 \right]_0^{h/2}$$

$$= 2\pi \left(\tfrac{1}{2}R^2 h - \tfrac{1}{12}Rch^3 + \tfrac{1}{160}c^2 h^5 \right)$$

Trying to make this look more like the expression we want, we rewrite it as $V = \tfrac{1}{3}\pi h \left[2R^2 + \left(R^2 - \tfrac{1}{2}Rch^2 + \tfrac{3}{80}c^2 h^4 \right) \right]$. But

$$R^2 - \tfrac{1}{2}Rch^2 + \tfrac{3}{80}c^2 h^4 = \left(R - \tfrac{1}{4}ch^2 \right)^2 - \tfrac{1}{40}c^2 h^4 = (R - d)^2 - \tfrac{2}{5}\left(\tfrac{1}{4}ch^2 \right)^2 = r^2 - \tfrac{2}{5}d^2.$$

Substituting this back into V, we see that $V = \tfrac{1}{3}\pi h \left(2R^2 + r^2 - \tfrac{2}{5}d^2 \right)$, as required.

69. We are given that the rate of change of the volume of water is $\dfrac{dV}{dt} = -kA(x)$, where k is some positive constant and $A(x)$ is the area of the surface when the water has depth x. Now we are concerned with the rate of change of the depth of the water with respect to time, that is, $\dfrac{dx}{dt}$. But by the Chain Rule, $\dfrac{dV}{dt} = \dfrac{dV}{dx}\dfrac{dx}{dt}$, so the first equation can be written $\dfrac{dV}{dx}\dfrac{dx}{dt} = -kA(x)$ (★). Also, we know that the total volume of water up to a depth x is $V(x) = \int_0^x A(s)\,ds$, where $A(s)$ is the area of a cross-section of the water at a depth s. Differentiating this equation with respect to x, we get $\dfrac{dV}{dx} = A(x)$. Substituting this into equation ★, we get $A(x)\dfrac{dx}{dt} = -kA(x)$

$\Rightarrow \dfrac{dx}{dt} = -k$, a constant.

6.3 Volumes by Cylindrical Shells

1.

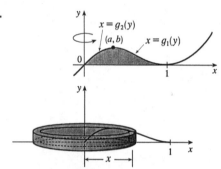

If we were to use the "washer" method, we would first have to locate the local maximum point (a, b) of $y = x(x-1)^2$ using the methods of Chapter 4. Then we would have to solve the equation $y = x(x-1)^2$ for x in terms of y to obtain the functions $x = g_1(y)$ and $x = g_2(y)$ shown in the first figure. This step would be difficult because it involves the cubic formula. Finally we would find the volume using

$$V = \pi \int_0^b \{[g_1(y)]^2 - [g_2(y)]^2\}\,dy.$$

Using shells, we find that a typical approximating shell has radius x, so its circumference is $2\pi x$. Its height is y, that is, $x(x-1)^2$. So the total volume is

$$V = \int_0^1 2\pi x\left[x(x-1)^2\right]dx = 2\pi \int_0^1 (x^4 - 2x^3 + x^2)\,dx = 2\pi\left[\frac{x^5}{5} - 2\frac{x^4}{4} + \frac{x^3}{3}\right]_0^1 = \frac{\pi}{15}$$

3. $V = \displaystyle\int_1^2 2\pi x \cdot \frac{1}{x}\,dx = 2\pi \int_1^2 1\,dx$

$= 2\pi\,[x]_1^2 = 2\pi(2-1) = 2\pi$

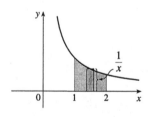

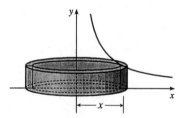

5. $V = \int_0^1 2\pi x e^{-x^2}\,dx.$ Let $u = x^2.$ Thus, $du = 2x\,dx,$ so $V = \pi \int_0^1 e^{-u}\,du = \pi\left[-e^{-u}\right]_0^1 = \pi(1 - 1/e).$

7. $V = \int_0^4 2\pi x\left(\sqrt{x} - \frac{1}{2}x\right)dx$

$= 2\pi \int_0^4 x^{3/2}\,dx - \pi \int_0^4 x^2\,dx$

$= 2\pi\left[\frac{2}{5}x^{5/2}\right]_0^4 - \pi\left[\frac{1}{3}x^3\right]_0^4$

$= \frac{4}{5}\pi(32) - \frac{64}{3}\pi = \frac{64}{15}\pi$

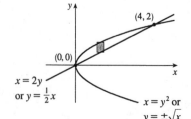

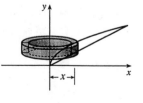

9. $V = \int_1^2 2\pi y \left(1 + y^2\right) dy = 2\pi \int_1^2 \left(y + y^3\right) dy = 2\pi \left[\frac{1}{2}y^2 + \frac{1}{4}y^4\right]_1^2$

$= 2\pi \left[(2 + 4) - \left(\frac{1}{2} + \frac{1}{4}\right)\right] = 2\pi \left(\frac{21}{4}\right) = \frac{21\pi}{2}$

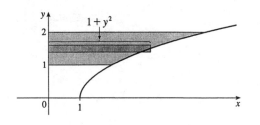
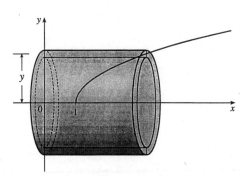

11. $V = \int_0^9 2\pi y \cdot 2\sqrt{y}\, dy$

$= 4\pi \int_0^9 y^{3/2}\, dy$

$= 4\pi \left[\frac{2}{5}y^{5/2}\right]_0^9$

$= \frac{8}{5}\pi (243 - 0)$

$= \frac{1944}{5}\pi$

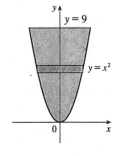
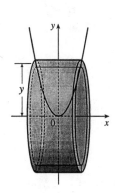

13. $V = \int_0^1 2\pi y \left[(2 - y) - y^2\right] dy$

$= 2\pi \left[y^2 - \frac{1}{3}y^3 - \frac{1}{4}y^4\right]_0^1$

$= 2\pi \left(1 - \frac{1}{3} - \frac{1}{4}\right) = \frac{5}{6}\pi$

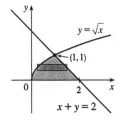
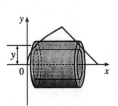

15. $V = \int_1^2 2\pi (x - 1) x^2\, dx = 2\pi \left[\frac{1}{4}x^4 - \frac{1}{3}x^3\right]_1^2$

$= 2\pi \left[\left(4 - \frac{8}{3}\right) - \left(\frac{1}{4} - \frac{1}{3}\right)\right] = \frac{17}{6}\pi$

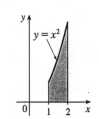

17. $V = \int_1^2 2\pi \, (4-x) \, x^2 \, dx = 2\pi \left[\frac{4}{3}x^3 - \frac{1}{4}x^4 \right]_1^2$

$= 2\pi \left[\left(\frac{32}{3} - 4 \right) - \left(\frac{4}{3} - \frac{1}{4} \right) \right] = \frac{67}{6}\pi$

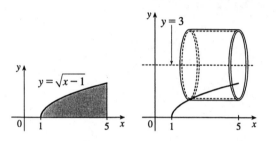

19. $V = \int_0^2 2\pi \, (3-y) \, (5-x) \, dy$

$= \int_0^2 2\pi \, (3-y) \, (5 - y^2 - 1) \, dy$

$= \int_0^2 2\pi \, (12 - 4y - 3y^2 + y^3) \, dy$

$= 2\pi \left[12y - 2y^2 - y^3 + \frac{1}{4}y^4 \right]_0^2$

$= 2\pi \, (24 - 8 - 8 + 4) = 24\pi$

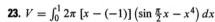

21. $V = \int_1^2 2\pi x \ln x \, dx$

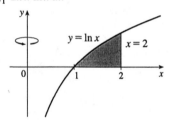

23. $V = \int_0^1 2\pi \, [x - (-1)] \left(\sin \frac{\pi}{2}x - x^4 \right) dx$

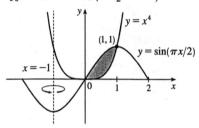

25. $V = \int_0^\pi 2\pi \, (4-y) \, \sqrt{\sin y} \, dy$

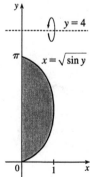

27. $\Delta x = \dfrac{\pi/4 - 0}{4} = \dfrac{\pi}{16}$.

$V = \int_0^{\pi/4} 2\pi x \tan x \, dx \approx 2\pi \cdot \frac{\pi}{16} \left(\frac{\pi}{32} \tan \frac{\pi}{32} + \frac{3\pi}{32} \tan \frac{3\pi}{32} + \frac{5\pi}{32} \tan \frac{5\pi}{32} + \frac{7\pi}{32} \tan \frac{7\pi}{32} \right) \approx 1.142$

29. The solid is obtained by rotating the region bounded by the curve $y = \cos x$ and the line $y = 0$, from $x = 0$ to $x = \frac{\pi}{2}$, about the y-axis.

31. The solid is obtained by rotating the region in the first quadrant bounded by the curves $y = x^2$ and $y = x^6$ about the y-axis.

33.

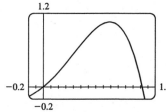

From the graph, it appears that the curves intersect at $x = 0$ and at $x \approx 1.32$, with $x + x^2 - x^4 > 0$ on $(0, 1.32)$. So the volume of the solid obtained by rotating the region about the y-axis is

$$V \approx 2\pi \int_0^{1.32} x \left(x + x^2 - x^4\right) dx = 2\pi \left[\tfrac{1}{3}x^3 + \tfrac{1}{4}x^4 - \tfrac{1}{6}x^6\right]_0^{1.32} \approx 4.05.$$

35. Use disks:

$$V = \int_{-2}^{1} \pi \left(x^2 + x - 2\right)^2 dx = \pi \int_{-2}^{1} \left(x^4 + 2x^3 - 3x^2 - 4x + 4\right) dx$$

$$= \pi \left[\tfrac{1}{5}x^5 + \tfrac{1}{2}x^4 - x^3 - 2x^2 + 4x\right]_{-2}^{1} = \pi \left[\left(\tfrac{1}{5} + \tfrac{1}{2} - 1 - 2 + 4\right) - \left(-\tfrac{32}{5} + 8 + 8 - 8 - 8\right)\right]$$

$$= \pi \left(\tfrac{33}{5} + \tfrac{3}{2}\right) = \tfrac{81}{10}\pi$$

37. Use shells:

$$V = \int_1^4 2\pi \left[x - (-1)\right]\left[5 - (x + 4/x)\right] dx = 2\pi \int_1^4 (x + 1)(5 - x - 4/x) dx$$

$$= 2\pi \int_1^4 \left(5x - x^2 - 4 + 5 - x - 4/x\right) dx$$

$$= 2\pi \int_1^4 \left(-x^2 + 4x + 1 - 4/x\right) dx = 2\pi \left[-\tfrac{1}{3}x^3 + 2x^2 + x - 4\ln x\right]_1^4$$

$$= 2\pi \left[\left(-\tfrac{64}{3} + 32 + 4 - 4\ln 4\right) - \left(-\tfrac{1}{3} + 2 + 1 - 0\right)\right]$$

$$= 2\pi \left(12 - 4\ln 4\right) = 8\pi \left(3 - \ln 4\right)$$

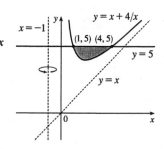

39. Use disks: $V = \pi \displaystyle\int_0^2 \left[\sqrt{1 - (y - 1)^2}\right]^2 dy = \pi \int_0^2 \left(2y - y^2\right) dy = \pi \left[y^2 - \tfrac{1}{3}y^3\right]_0^2 = \pi \left(4 - \dfrac{8}{3}\right) = \dfrac{4}{3}\pi$

41. $V = 2\displaystyle\int_0^r 2\pi x\sqrt{r^2 - x^2}\, dx = -2\pi \int_0^r \left(r^2 - x^2\right)^{1/2}(-2x)\, dx = \left[-2\pi \cdot \tfrac{2}{3}\left(r^2 - x^2\right)^{3/2}\right]_0^r = -\tfrac{4}{3}\pi \left(0 - r^3\right) =$
$\tfrac{4}{3}\pi r^3$

43. $V = 2\pi \displaystyle\int_0^r x\left(-\dfrac{h}{r}x + h\right) dx = 2\pi h \int_0^r \left(-\dfrac{x^2}{r} + x\right) dx = 2\pi h \left[-\dfrac{x^3}{3r} + \dfrac{x^2}{2}\right]_0^r = 2\pi h \dfrac{r^2}{6} = \dfrac{\pi r^2 h}{3}$

6.4 Work

1. By Equation 2, $W = Fd = (900)(8) = 7200$ J.

3. By Equation 4,

$$W = \int_a^b f(x)\,dx = \int_0^9 \frac{10}{(1+x)^2}\,dx = 10 \int_1^{10} \frac{1}{u^2}\,du \quad (u = 1+x,\ du = dx)$$

$$= 10\left[-\frac{1}{u}\right]_1^{10} = 10\left(-\frac{1}{10} + 1\right) = 9 \text{ ft-lb}$$

5. $10 = f(x) = kx = \frac{1}{3}k$ (4 inches $= \frac{1}{3}$ foot), so $k = 30$ lb/ft and $f(x) = 30x$. Now 6 inches $= \frac{1}{2}$ foot, so $W = \int_0^{1/2} 30x\,dx = \left[15x^2\right]_0^{1/2} = \frac{15}{4}$ ft-lb.

7. If $\int_0^{0.12} kx\,dx = 2$ J, then $2 = \left[\frac{1}{2}kx^2\right]_0^{0.12} = \frac{1}{2}k\,(0.0144) = 0.0072k$ and $k = \frac{2}{0.0072} = \frac{2500}{9} \approx 277.78$. Thus, the work needed to stretch the spring from 35 cm to 40 cm is

$$\int_{0.05}^{0.10} \frac{2500}{9}x\,dx = \left[\frac{1250}{9}x^2\right]_{1/20}^{1/10} = \frac{1250}{9}\left(\frac{1}{100} - \frac{1}{400}\right) = \frac{25}{24} \approx 1.04 \text{ J}.$$

9. $f(x) = kx$, so $30 = \frac{2500}{9}x$ and $x = \frac{270}{2500}$ m $= 10.8$ cm

In Exercises 11–15, n is the number of subintervals of length Δx, and x_i^* is a sample point in the ith subinterval $[x_{i-1}, x_i]$.

11. First notice that the exact height of the building does not matter (as long as it is more than 50 ft). The portion of the rope from x ft to $(x + \Delta x)$ ft below the top of the building weighs $\frac{1}{2}\Delta x$ lb and must be lifted x_i^* ft, so its contribution to the total work is $\frac{1}{2}x_i^*\Delta x$ ft-lb. The total work is

$$W = \lim_{n \to \infty} \sum_{i=1}^n \frac{1}{2}x_i^*\Delta x = \int_0^{50} \frac{1}{2}x\,dx = \left[\frac{1}{4}x^2\right]_0^{50} = \frac{2500}{4} = 625 \text{ ft-lb}$$

13. The work needed to lift the cable is $\lim_{n \to \infty} \sum_{i=1}^n 2x_i^*\Delta x = \int_0^{500} 2x\,dx = \left[x^2\right]_0^{500} = 250{,}000$ ft-lb. The work needed to lift the coal is 800 lb $\cdot 500$ ft $= 400{,}000$ ft-lb. Thus, the total work required is $250{,}000 + 400{,}000 = 650{,}000$ ft-lb.

15. A "slice" of water Δx m thick and lying at a depth of x_i^* m (where $0 \le x_i^* \le \frac{1}{2}$) has volume $2\Delta x$ m^3, a mass of $2000\Delta x$ kg, weighs about $(9.8)(2000\Delta x) = 19{,}600\Delta x$ N, and thus requires about $19{,}600x_i^*\Delta x$ J of work for its removal. So

$$W = \lim_{n \to \infty} \sum_{i=1}^n 19{,}600x_i^*\Delta x = \int_0^{1/2} 19{,}600x\,dx = \left[9800x^2\right]_0^{1/2} = 2450 \text{ J}$$

17. A rectangular "slice" of water Δx m thick and lying x ft above the bottom has width x ft and volume $8x\,\Delta x$ m^3. It weighs about $(9.8 \times 10^3)(8x\,\Delta x)$ N, and must be lifted $(5 - x)$ m by the pump, so the work needed is about $(9.8 \times 10^3)(5 - x)(8x\,\Delta x)$ J. The total work required is

$$W \approx \int_0^3 (9.8 \times 10^3)(5 - x)8x\,dx = (9.8 \times 10^3)\int_0^3 (40x - 8x^2)\,dx = (9.8 \times 10^3)\left[20x^2 - \frac{8}{3}x^3\right]_0^3$$

$$= \left(9.8 \times 10^3\right)(180 - 72) = \left(9.8 \times 10^3\right)(108) = 1058.4 \times 10^3 \approx 1.06 \times 10^6 \text{ J}$$

19. Measure depth x downward from the flat top of the tank, so that $0 \le x \le 2$ ft. Then
$$\Delta W = (62.5)\left(2\sqrt{4-x^2}\right)(8\Delta x)(x+1) \text{ ft-lb, so}$$

$$W \approx (62.5)(16)\int_0^2 (x+1)\sqrt{4-x^2}\,dx = 1000\left(\int_0^2 x\sqrt{4-x^2}\,dx + \int_0^2 \sqrt{4-x^2}\,dx\right)$$

$$= 1000\left[\int_0^4 u^{1/2}\left(\tfrac{1}{2}\,du\right) + \tfrac{1}{4}\pi\,(2^2)\right] \quad (\text{Put } u = 4-x^2, \text{ so } du = -2x\,dx)$$

$$= 1000\left(\left[\tfrac{1}{2}\cdot\tfrac{2}{3}u^{3/2}\right]_0^4 + \pi\right) = 1000\left(\tfrac{8}{3}+\pi\right) \approx 5.8 \times 10^3 \text{ ft-lb}$$

Note: The second integral represents the area of a quarter-circle of radius 2.

21. If only 4.7×10^5 J of work is done, then only the water above a certain level (call it h) will be pumped out. So we use the same formula as in Exercise 17, except that the work is fixed, and we are trying to find the lower limit of integration: $4.7 \times 10^5 \approx \int_h^3 (9.8 \times 10^3)(5-x)8x\,dx = (9.8 \times 10^3)\left[20x^2 - \tfrac{8}{3}x^3\right]_h^3 \iff$

$\tfrac{4.7}{9.8} \times 10^2 \approx 48 = \left(20\cdot 3^2 - \tfrac{8}{3}\cdot 3^3\right) - \left(20h^2 - \tfrac{8}{3}h^3\right) \iff$

$2h^3 - 15h^2 + 45 = 0$. To find the solution of this equation, we plot

$2h^3 - 15h^2 + 45$ between $h = 0$ and $h = 3$. We see that the equation is satisfied for $h \approx 2.0$. So the depth of water remaining in the tank is about 2.0 m.

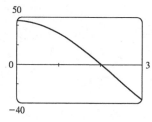

23. $V = \pi r^2 x$, so V is a function of x and P can also be regarded as a function of x. If $V_1 = \pi r^2 x_1$ and $V_2 = \pi r^2 x_2$, then

$$W = \int_{x_1}^{x_2} F(x)\,dx = \int_{x_1}^{x_2} \pi r^2 P\,(V(x))\,dx = \int_{x_1}^{x_2} P\,(V(x))\,dV(x) \quad [\text{Put } V(x) = \pi r^2 x, \text{ so } dV(x) = \pi r^2\,dx]$$

$$= \int_{V_1}^{V_2} P(V)\,dV \text{ by the Substitution Rule.}$$

25. $W = \displaystyle\int_a^b F(r)\,dr = \int_a^b G\frac{m_1 m_2}{r^2}\,dr = Gm_1 m_2\left[\frac{-1}{r}\right]_a^b = Gm_1 m_2\left(\frac{1}{a} - \frac{1}{b}\right)$

6.5 Average Value of a Function

1. $f_{\text{ave}} = \dfrac{1}{b-a}\displaystyle\int_a^b f(x)\,dx = \dfrac{1}{1-(-1)}\int_{-1}^{1} x^2\,dx = \tfrac{1}{2}\cdot 2\int_0^1 x^2\,dx = \left[\tfrac{1}{3}x^3\right]_0^1 = \tfrac{1}{3}$

3. $g_{\text{ave}} = \dfrac{1}{\pi/2-0}\displaystyle\int_0^{\pi/2} \cos x\,dx = \tfrac{2}{\pi}[\sin x]_0^{\pi/2} = \tfrac{2}{\pi}(1-0) = \tfrac{2}{\pi}$

5. $f_{\text{ave}} = \dfrac{1}{5-0}\displaystyle\int_0^5 te^{-t^2}\,dt = \tfrac{1}{5}\int_0^{-25}\left(-\tfrac{1}{2}\right)e^u\,du \quad [u = -t^2,\, du = -2t\,dt]$

$= -\tfrac{1}{10}[e^u]_0^{-25} = -\tfrac{1}{10}(e^{-25} - 1) = \tfrac{1}{10}(1 - e^{-25})$

7. $h_{\text{ave}} = \dfrac{1}{\pi-0}\displaystyle\int_0^\pi \cos^4 x \sin x\,dx = \tfrac{1}{\pi}\int_1^{-1} u^4\,(-du) \quad [u = \cos x,\, du = -\sin x\,dx] = \tfrac{1}{\pi}\int_{-1}^{1} u^4\,du$

$= \tfrac{1}{\pi}\cdot 2\displaystyle\int_0^1 u^4\,du = \tfrac{2}{\pi}\left[\tfrac{1}{5}u^5\right]_0^1 = \tfrac{2}{5\pi}$

9. (a) $f_{ave} = \frac{1}{2-0} \int_0^2 (4 - x^2) \, dx$

$= \frac{1}{2} \left[4x - \frac{1}{3}x^3 \right]_0^2$

$= \frac{1}{2} \left(8 - \frac{8}{3} \right) = \frac{8}{3}$

(b) $f_{ave} = f(c) \iff \frac{8}{3} = 4 - c^2 \iff c^2 = \frac{4}{3} \iff c = \frac{2}{\sqrt{3}} \approx 1.15$

(c)
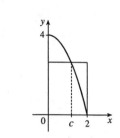

11. (a) $f_{ave} = \frac{1}{2-0} \int_0^2 (x^3 - x + 1) \, dx$

$= \frac{1}{2} \left[\frac{1}{4}x^4 - \frac{1}{2}x^2 + x \right]_0^2$

$= \frac{1}{2} (4 - 2 + 2) = 2$

(b) From the graph, $f(x) = 2$ at $x \approx 1.32$.

(c)
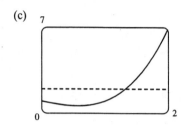

13. Since f is continuous on $[1, 3]$, by the Mean Value Theorem for Integrals there exists a number c in $[1, 3]$ such that $\int_1^3 f(x) \, dx = f(c) (3 - 1) \Rightarrow 8 = 2f(c)$; that is, there is a number c such that $f(c) = \frac{8}{2} = 4$.

15. $T_{ave} = \frac{1}{12} \int_0^{12} \left[50 + 14 \sin \frac{1}{12} \pi t \right] dt = \frac{1}{12} \left[50t - 14 \cdot \frac{12}{\pi} \cos \frac{1}{12} \pi t \right]_0^{12}$

$= \frac{1}{12} \left[50 \cdot 12 + 14 \cdot \frac{12}{\pi} + 14 \cdot \frac{12}{\pi} \right] = \left(50 + \frac{28}{\pi} \right) {}^{\circ}F \approx 59^{\circ} F$

17. $\rho_{ave} = \frac{1}{8} \int_0^8 \frac{12}{\sqrt{x+1}} \, dx = \frac{3}{2} \int_0^8 (x+1)^{-1/2} \, dx = \left[3\sqrt{x+1} \right]_0^8 = 9 - 3 = 6 \text{ kg/m}$

19. $V_{ave} = \frac{1}{5} \int_0^5 V(t) \, dt = \frac{1}{5} \int_0^5 \frac{5}{4\pi} \left[1 - \cos \left(\frac{2}{5}\pi t \right) \right] dt = \frac{1}{4\pi} \int_0^5 \left[1 - \cos \left(\frac{2}{5}\pi t \right) \right] dt$

$= \frac{1}{4\pi} \left[t - \frac{5}{2\pi} \sin \left(\frac{2}{5}\pi t \right) \right]_0^5 = \frac{1}{4\pi} [(5 - 0) - 0] = \frac{5}{4\pi} \approx 0.4 \text{ L}$

21. Let $F(x) = \int_a^x f(t) \, dt$ for x in $[a, b]$. Then F is continuous on $[a, b]$ and differentiable on (a, b), so by the Mean Value Theorem there is a number c in (a, b) such that $F(b) - F(a) = F'(c)(b - a)$. But $F'(x) = f(x)$ by the Fundamental Theorem of Calculus. Therefore, $\int_a^b f(t) \, dt - 0 = f(c)(b - a)$.

6 Review

CONCEPT CHECK

1. (a) See Section 6.1, Figure 2 and Equations 6.1.1 and 6.1.2.

(b) Instead of using "top minus bottom" and integrating from left to right, we use "right minus left" and integrate from bottom to top. See Figures 11 and 12 in Section 6.1.

2. The numerical value of the area represents the number of meters by which Sue is ahead of Kathy after 1 minute.

3. (a) See the discussion in Section 6.2, near Figures 2 and 3, ending in the Definition of Volume.

(b) See the discussion between Examples 5 and 6 in Section 6.2. If the cross-section is a disk, find the radius in terms of x or y and use $A = \pi \, (\text{radius})^2$. If the cross-section is a washer, find the inner radius r_{in} and outer radius r_{out} and use $A = \pi \left(r_{out}^2 \right) - \pi \left(r_{in}^2 \right)$.

4. (a) $V = 2\pi r h \, \Delta r =$ (circumference) (height) (thickness)

(b) For a typical shell, find the circumference and height in terms of x or y and calculate
$V = \int_a^b$ (circumference) (height) $(dx$ or $dy)$, where a and b are the limits on x or y.

(c) Sometimes slicing produces washers or disks whose radii are difficult (or impossible) to find explicitly. On other occasions, the cylindrical shell method leads to an easier integral than slicing does.

5. $\int_0^6 f(x) \, dx$ represents the amount of work done. Its units are newton-meters, or joules.

6. (a) See the boxed equation preceding Example 1 in Section 6.5.

(b) The Mean Value Theorem for Integrals says that there is a number c at which the value of f is exactly equal to the average value of the function, that is, $f(c) = f_{ave}$. For a geometric interpretation of the Mean Value Theorem for Integrals , see Figure 2 in Section 6.5 and the discussion which accompanies it. :

EXERCISES

---

1. $0 = x^2 - x - 6 = (x-3)(x+2) \Leftrightarrow x = 3$ or -2. So

$A = \int_{-2}^3 \left[0 - (x^2 - x - 6)\right] dx = \int_{-2}^3 (-x^2 + x + 6) \, dx$

$= \left[-\frac{1}{3}x^3 + \frac{1}{2}x^2 + 6x\right]_{-2}^3$

$= \left(-9 + \frac{9}{2} + 18\right) - \left(\frac{8}{3} + 2 - 12\right)$

$= \frac{125}{6}$

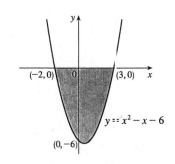

3. $A = \int_0^1 \left[(e^x - 1) - (x^2 - x)\right] dx$

$= \int_0^1 (e^x - 1 - x^2 + x) \, dx = \left[e^x - x - \frac{1}{3}x^3 + \frac{1}{2}x^2\right]_0^1$

$= \left(e - 1 - \frac{1}{3} + \frac{1}{2}\right) - (1 - 0 - 0 + 0) = e - \frac{11}{6}$

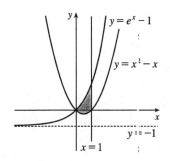

5. $A = \int_0^\pi |\sin x - (-\cos x)| \, dx = \int_0^{3\pi/4} (\sin x + \cos x) \, dx - \int_{3\pi/4}^\pi (\sin x + \cos x) \, dx$

$= [\sin x - \cos x]_0^{3\pi/4} - [-\cos x + \sin x]_{3\pi/4}^\pi$

$= \left(\frac{1}{\sqrt{2}} + \frac{1}{\sqrt{2}}\right) - (0 - 1) - (1 + 0) + \left(\frac{1}{\sqrt{2}} + \frac{1}{\sqrt{2}}\right) = \sqrt{2} + 1 - 1 + \sqrt{2} = 2\sqrt{2}$

7. $V = \int_1^3 \pi \left(\sqrt{x-1}\right)^2 dx = \pi \int_1^3 (x-1) \, dx = \pi \left[\frac{1}{2}x^2 - x\right]_1^3 = \pi \left[\left(\frac{9}{2} - 3\right) - \left(\frac{1}{2} - 1\right)\right] = 2\pi$ ι

9. $V = \int_1^3 2\pi y \left(-y^2 + 4y - 3\right) dy = 2\pi \int_1^3 \left(-y^3 + 4y^2 - 3y\right) dy = 2\pi \left[-\frac{1}{4}y^4 + \frac{4}{3}y^3 - \frac{3}{2}y^2\right]_1^3$

$= 2\pi \left[\left(-\frac{81}{4} + 36 - \frac{27}{2}\right) - \left(-\frac{1}{4} + \frac{4}{3} - \frac{3}{2}\right)\right] = \frac{16\pi}{3}$

11. $V = \int_a^{a+h} 2\pi x \cdot 2\sqrt{x^2 - a^2}\, dx = 2\pi \int_0^{2ah+h^2} u^{1/2}\, du$ (Put $u = x^2 - a^2$, so $du = 2x\, dx$)

$= 2\pi \left[\frac{2}{3}u^{3/2}\right]_0^{2ah+h^2} = \frac{4}{3}\pi \left(2ah + h^2\right)^{3/2}$

13. $V = \int_0^1 \pi \left[\left(1 - x^3\right)^2 - \left(1 - x^2\right)^2\right] dx$

15. (a) $V = \int_0^1 \pi \left[(x)^2 - \left(x^2\right)^2\right] dx = \int_0^1 \pi \left(x^2 - x^4\right) dx = \pi \left[\frac{1}{3}x^3 - \frac{1}{5}x^5\right]_0^1 = \pi \left[\frac{1}{3} - \frac{1}{5}\right] = \frac{2\pi}{15}$

(b) $V = \int_0^1 \pi \left[\left(\sqrt{y}\right)^2 - y^2\right] dy = \int_0^1 \pi \left(y - y^2\right) dy = \pi \left[\frac{1}{2}y^2 - \frac{1}{3}y^3\right]_0^1 = \pi \left[\frac{1}{2} - \frac{1}{3}\right] = \frac{\pi}{6}$

(c) $V = \int_0^1 \pi \left[\left(2 - x^2\right)^2 - (2 - x)^2\right] dx = \int_0^1 \pi \left(x^4 - 5x^2 + 4x\right) dx = \pi \left[\frac{1}{5}x^5 - \frac{5}{3}x^3 + 2x^2\right]_0^1$

$= \pi \left[\frac{1}{5} - \frac{5}{3} + 2\right] = \frac{8\pi}{15}$

17. (a) Using the Midpoint Rule on $[0, 1]$ with $f(x) = \tan\left(x^2\right)$ and $n = 4$, we estimate

$A = \int_0^1 \tan\left(x^2\right) dx \approx \frac{1}{4}\left[\tan\left(\left(\frac{1}{8}\right)^2\right) + \tan\left(\left(\frac{3}{8}\right)^2\right) + \tan\left(\left(\frac{5}{8}\right)^2\right) + \tan\left(\left(\frac{7}{8}\right)^2\right)\right] \approx \frac{1}{4}(1.53) \approx 0.38$

(b) Using the Midpoint Rule on $[0, 1]$ with $f(x) = \pi \tan^2\left(x^2\right)$ (for disks) and $n = 4$, we estimate

$V = \int_0^1 f(x)\, dx \approx \frac{1}{4}\pi \left[\tan^2\left(\left(\frac{1}{8}\right)^2\right) + \tan^2\left(\left(\frac{3}{8}\right)^2\right) + \tan^2\left(\left(\frac{5}{8}\right)^2\right) + \tan^2\left(\left(\frac{7}{8}\right)^2\right)\right] \approx \frac{\pi}{4}(1.114) \approx 0.87$

19. The solid is obtained by rotating the region $\mathcal{R} = \left\{(x, y) \mid 0 \le x \le \frac{\pi}{2}, 0 \le y \le \cos x\right\}$ about the y-axis.

21. The solid is obtained by rotating the region under the curve $y = \sin x$, above $y = 0$, from $x = 0$ to $x = \pi$, about the x-axis.

23. Take the base to be the disk $x^2 + y^2 \le 9$. Then $V = \int_{-3}^3 A(x)\, dx$, where $A(x_0)$ is the area of the isosceles right triangle whose hypotenuse lies along the line $x = x_0$ in the xy-plane. $A(x) = \frac{1}{2}\left(\sqrt{2}\sqrt{9 - x^2}\right)^2 = 9 - x^2$, so

$V = 2\int_0^3 A(x)\, dx = 2\int_0^3 \left(9 - x^2\right) dx = 2\left[9x - \frac{1}{3}x^3\right]_0^3 = 2(27 - 9) = 36.$

25. Equilateral triangles with sides measuring $\frac{1}{4}x$ meters have height $\frac{1}{4}x \sin 60° = \frac{\sqrt{3}}{8}x$. Therefore,

$A(x) = \frac{1}{2} \cdot \frac{1}{4}x \cdot \frac{\sqrt{3}}{8}x = \frac{\sqrt{3}}{64}x^2$. $V = \int_0^{20} A(x)\, dx = \frac{\sqrt{3}}{64}\int_0^{20} x^2\, dx = \frac{\sqrt{3}}{64}\left[\frac{1}{3}x^3\right]_0^{20} = \frac{8000\sqrt{3}}{64 \cdot 3} = \frac{125\sqrt{3}}{3}$ m^3

27. $f(x) = kx \Rightarrow 30\,\text{N} = k(15 - 12)\,\text{cm} \Rightarrow k = 10\,\text{N/cm} = 1000\,\text{N/m}$. $20\,\text{cm} - 12\,\text{cm} = 0.08\,\text{m} \Rightarrow$

$W = \int_0^{0.08} kx\, dx = 1000 \int_0^{0.08} x\, dx = 500\left[x^2\right]_0^{0.08} = 500\,(0.08)^2 = 3.2\,\text{N-m} = 3.2\,\text{J}.$

29. (a) The parabola has equation $y = ax^2$ with vertex at the origin and passing

through $(4, 4)$. $4 = a \cdot 4^2 \;\Rightarrow\; a = \frac{1}{4} \;\Rightarrow\; y = \frac{1}{4}x^2 \;\Rightarrow\; x^2 = 4y$

$\Rightarrow\; x = 2\sqrt{y}$. Each circular disk has radius $2\sqrt{y}$ and is moved $4 - y$ ft.

$$W = \int_0^4 \pi \left(2\sqrt{y}\right)^2 62.5 \,(4 - y) \, dy = 250\pi \int_0^4 y \,(4 - y) \, dy$$

$$= 250\pi \left[2y^2 - \tfrac{1}{3}y^3\right]_0^4 = 250\pi \left(32 - \tfrac{64}{3}\right) = \tfrac{8000\pi}{3} \approx 8377.6 \text{ ft-lb}$$

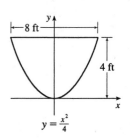

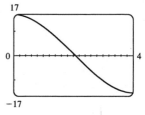

(b) In part (a) we knew the final water level (0) but not the amount of work
done. Here we use the same equation, except with the work fixed, and the
lower limit of integration (that is, the final water level — call it h)

unknown: $W = 4000 \;\Leftrightarrow\; 250\pi \left[2y^2 - \tfrac{1}{3}y^3\right]_h^4 = 4000 \;\Leftrightarrow$

$\tfrac{16}{\pi} = \left[\left(32 - \tfrac{64}{3}\right) - \left(2h^2 - \tfrac{1}{3}h^3\right)\right] \;\Leftrightarrow\; h^3 - 6h^2 + 32 - \tfrac{48}{\pi} = 0.$

We plot the graph of the function $f\,(h) = h^3 - 6h^2 + 32 - \tfrac{48}{\pi}$ on the interval $[0, 4]$ to see where it is 0. From
the graph, $f\,(h) = 0$ for $h \approx 2.06$. So the depth of water remaining is about 2.06 ft.

31. $\lim\limits_{h \to 0} f_{\text{ave}} = \lim\limits_{h \to 0} \dfrac{1}{h} \int_x^{x+h} f\,(t) \, dt = \lim\limits_{h \to 0} \dfrac{F\,(x+h) - F\,(x)}{h}$, where $F\,(x) = \int_a^x f\,(t) \, dt$. But we recognize this

limit as being $F'\,(x)$ by the definition of a derivative. Therefore, $\lim\limits_{h \to 0} f_{\text{ave}} = F'\,(x) = f\,(x)$ by FTC1.

Problems Plus

1. The area under the graph of f from 0 to t is equal to $\int_0^t f(x)\,dx$, so the requirement is that $\int_0^t f(x)\,dx = t^3$ for all t. We differentiate this equation with respect to t (with the help of FTC1) to get $f(t) = 3t^2$. This function is positive and continuous, as required.

3. The volume generated from $x = 0$ to $x = b$ is $\int_0^b \pi\,[f(x)]^2\,dx$. Hence, we are given that $b^2 = \int_0^b \pi\,[f(x)]^2\,dx$ for all $b > 0$. Differentiating both sides of this equation using the Fundamental Theorem of Calculus gives $2b = \pi\,[f(b)]^2 \;\Rightarrow\; f(b) = \sqrt{2b/\pi}$, since f is positive. Therefore, $f(x) = \sqrt{2x/\pi}$.

5. (a) $V = \pi h^2 (r - h/3) = \tfrac{1}{3}\pi h^2 (3r - h)$. See the solution to Exercise 6.2.47.

(b) The smaller segment has height $h = 1 - x$ and so by part (a) its volume is
$V = \tfrac{1}{3}\pi\,(1 - x)^2\,[3\,(1) - (1 - x)] = \tfrac{1}{3}\pi\,(x - 1)^2\,(x + 2)$. This volume must be $\tfrac{1}{3}$ of the total volume of the sphere, which is $\tfrac{4}{3}\pi\,(1)^3$. So $\tfrac{1}{3}\pi\,(x - 1)^2\,(x + 2) = \tfrac{1}{3}\left(\tfrac{4}{3}\pi\right) \;\Rightarrow\; (x^2 - 2x + 1)\,(x + 2) = \tfrac{4}{3} \;\Rightarrow$
$x^3 - 3x + 2 = \tfrac{4}{3} \;\Rightarrow\; 3x^3 - 9x + 2 = 0$. Using Newton's method with $f(x) = 3x^3 - 9x + 2$,
$f'(x) = 9x^2 - 9$, we get $x_{n+1} = x_n - \dfrac{3x_n^3 - 9x_n + 2}{9x_n^2 - 9}$. Taking $x_1 = 0$, we get $x_2 \approx 0.2222$, and $x_3 \approx 0.2261 \approx x_4$, so, correct to four decimal places, $x \approx 0.2261$.

(c) With $r = 0.5$ and $s = 0.75$, the equation $x^3 - 3rx^2 + 4r^3 s = 0$ becomes $x^3 - 3\,(0.5)\,x^2 + 4\,(0.5)^3\,(0.75) = 0$
$\Rightarrow\; x^3 - \tfrac{3}{2}x^2 + 4\left(\tfrac{1}{8}\right)\tfrac{3}{4} = 0 \;\Rightarrow\; 8x^3 - 12x^2 + 3 = 0$. We use Newton's method with
$f(x) = 8x^3 - 12x^2 + 3$, $f'(x) = 24x^2 - 24x$, so $x_{n+1} = x_n - \dfrac{8x_n^3 - 12x_n^2 + 3}{24x_n^2 - 24x_n}$. Take $x_1 = 0.5$. Then
$x_2 \approx 0.6667$, and $x_3 \approx 0.6736 \approx x_4$. So to four decimal places the depth is 0.6736 m.

(d) (i) From part (a) with $r = 5$ in., the volume of water in the bowl is
$V = \tfrac{1}{3}\pi h^2\,(3r - h) = \tfrac{1}{3}\pi h^2\,(15 - h) = 5\pi h^2 - \tfrac{1}{3}\pi h^3$. We are given that $\dfrac{dV}{dt} = 0.2$ m^3/s and we want to find $\dfrac{dh}{dt}$ when $h = 3$. Now $\dfrac{dV}{dt} = 10\pi h\dfrac{dh}{dt} - \pi h^2\dfrac{dh}{dt}$, so $\dfrac{dh}{dt} = \dfrac{0.2}{\pi\,(10h - h^2)}$. When $h = 3$, we have
$\dfrac{dh}{dt} = \dfrac{0.2}{\pi\,(10 \cdot 3 - 3^2)} = \dfrac{1}{105\pi} \approx 0.003$ in/s.

(ii) From part (a), the volume of water required to fill the bowl from the instant that the water is 4 in. deep is
$V = \tfrac{1}{2} \cdot \tfrac{4}{3}\pi\,(5)^3 - \tfrac{1}{3}\pi\,(4)^2\,(15 - 4) = \tfrac{2}{3} \cdot 125\pi - \tfrac{16}{3} \cdot 11\pi = \tfrac{74}{3}\pi$. To find the time required to fill the bowl we divide this volume by the rate: Time $= \dfrac{74\pi/3}{0.2} = \dfrac{370\pi}{3} \approx 387$ s ≈ 6.5 min

7. We are given that the rate of change of the volume of water is $\dfrac{dV}{dt} = -kA(x)$, where k is some positive constant and $A(x)$ is the area of the surface when the water has depth x. Now we are concerned with the rate of change of the depth of the water with respect to time, that is, $\dfrac{dx}{dt}$. But by the Chain Rule, $\dfrac{dV}{dt} = \dfrac{dV}{dx}\dfrac{dx}{dt}$, so the first equation can be written $\dfrac{dV}{dx}\dfrac{dx}{dt} = -kA(x)$ ($\bigstar$). Also, we know that the total volume of water up to a depth x is $V(x) = \int_0^x A(s)\,ds$, where $A(s)$ is the area of a cross-section of the water at a depth s. Differentiating this equation with respect to x, we get $dV/dx = A(x)$. Substituting this into equation $\bigstar$, we get $A(x)\,(dx/dt) = -kA(x) \;\Rightarrow\; dx/dt = -k$, a constant.

9. We must find expressions for the areas A and B, and then set them equal and see what this says about the curve C. If $P = (a, 2a^2)$, then area A is just $\int_0^a (2x^2 - x^2)\,dx = \int_0^a x^2\,dx = \frac{1}{3}a^3$. To find area B, we use y as the variable of integration. So we find the equation of the middle curve as a function of y: $y = 2x^2 \Leftrightarrow x = \sqrt{y/2}$, since we are concerned with the first quadrant only. We can express area B as

$$\int_0^{2a^2} \left[\sqrt{y/2} - C\,(y)\right] dy = \left[\frac{4}{3}(y/2)^{3/2}\right]_0^{2a^2} - \int_0^{2a^2} C\,(y)\,dy = \frac{4}{3}a^3 - \int_0^{2a^2} C\,(y)\,dy,\ \text{where } C\,(y) \text{ is the function}$$

with graph C. Setting $A = B$, we get $\frac{1}{3}a^3 = \frac{4}{3}a^3 - \int_0^{2a^2} C\,(y)\,dy \Leftrightarrow \int_0^{2a^2} C\,(y)\,dy = a^3$. Now we differentiate this equation with respect to a using the Chain Rule and the Fundamental Theorem:

$C\,(2a^2)\,(4a) = 3a^2 \Rightarrow C\,(y) = \frac{3}{4}\sqrt{y/2}$, where $y = 2a^2$. Now we can solve for y: $x = \frac{3}{4}\sqrt{y/2} \Rightarrow x^2 = \frac{9}{16}(y/2) \Rightarrow y = \frac{32}{9}x^2$.

11. (a) Stacking disks along the y-axis gives us $V = \int_0^h \pi\,[f\,(y)]^2\,dy$.

(b) Using the Chain Rule, $\dfrac{dV}{dt} = \dfrac{dV}{dh} \cdot \dfrac{dh}{dt} = \pi\,[f\,(h)]^2\,\dfrac{dh}{dt}$.

(c) $kA\sqrt{h} = \pi\,[f\,(h)]^2\,\dfrac{dh}{dt}$. Set $\dfrac{dh}{dt} = C$: $\pi\,[f\,(h)]^2\,C = kA\sqrt{h} \Rightarrow [f\,(h)]^2 = \dfrac{kA}{\pi C}\sqrt{h} \Rightarrow$

$f\,(h) = \sqrt{\dfrac{kA}{\pi C}}\,h^{1/4}$, that is, $f\,(y) = \sqrt{\dfrac{kA}{\pi C}}\,y^{1/4}$. The advantage of having $\dfrac{dh}{dt} = C$ is that the markings on the container are evenly spaced.

13. We assume that P lies in the region of positive x. Since $y = x^3$ is an odd function, this assumption will not affect the result of the calculation. Let $P = (a, a^3)$. The slope of the tangent to the curve $y = x^3$ at P is $3a^2$, and so the equation of the tangent is $y - a^3 = 3a^2\,(x - a) \Leftrightarrow y = 3a^2x - 2a^3$. We solve this simultaneously with $y = x^3$ to find the other point of intersection:

$x^3 = 3a^2x - 2a^3 \Leftrightarrow (x - a)^2\,(x + 2a) = 0$. So $Q = (-2a, -8a^3)$ is the other point of intersection. The equation of the tangent at Q is

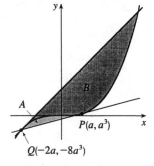

$y - (-8a^3) = 12a^2\,[x - (-2a)] \Leftrightarrow y = 12a^2x + 16a^3$.

By symmetry, this tangent will intersect the curve again at $x = -2\,(-2a) = 4a$. The curve lies above the first tangent, and below the second, so we are looking for a relationship between $A = \int_{-2a}^{a} [x^3 - (3a^2x - 2a^3)]\,dx$ and

$B = \int_{-2a}^{4a} [(12a^2x + 16a^3) - x^3]\,dx$. We calculate $A = \left[\frac{1}{4}x^4 - \frac{3}{2}a^2x^2 + 2a^3x\right]_{-2a}^{a} = \frac{3}{4}a^4 - (-6a^4) = \frac{27}{4}a^4$,

and $B = \left[6a^2x^2 + 16a^3x - \frac{1}{4}x^4\right]_{-2a}^{4a} = 96a^4 - (-12a^4) = 108a^4$. We see that $B = 16A = 2^4 A$. This is because our calculation of area B was essentially the same as that of area A, with a replaced by $-2a$, so if we replace a with $-2a$ in our expression for A, we get $\frac{27}{4}(-2a)^4 = 108a^4 = B$.

Techniques of Integration

7.1 Integration by Parts

1. Let $u = \ln x$, $dv = x\,dx$ $\Rightarrow$ $du = dx/x$, $v = \frac{1}{2}x^2$. Then by Equation 2,

$$\int x \ln x\,dx = \frac{1}{2}x^2 \ln x - \int \frac{1}{2}x^2\,(dx/x) = \frac{1}{2}x^2 \ln x - \frac{1}{2}\int x\,dx = \frac{1}{2}x^2 \ln x - \frac{1}{2} \cdot \frac{1}{2}x^2 + C$$
$$= \frac{1}{2}x^2 \ln x - \frac{1}{4}x^2 + C$$

3. Let $u = x$, $dv = e^{2x}\,dx$ $\Rightarrow$ $du = dx$, $v = \frac{1}{2}e^{2x}$. Then by Equation 2,
$$\int x e^{2x}\,dx = \frac{1}{2}x e^{2x} - \int \frac{1}{2}e^{2x}\,dx = \frac{1}{2}x e^{2x} - \frac{1}{4}e^{2x} + C.$$

5. Let $u = x$, $dv = \sin 4x\,dx$ $\Rightarrow$ $du = dx$, $v = -\frac{1}{4}\cos 4x$. Then
$$\int x \sin 4x\,dx = -\frac{1}{4}x \cos 4x - \int \left(-\frac{1}{4}\cos 4x\right) dx = -\frac{1}{4}x \cos 4x + \frac{1}{16}\sin 4x + C.$$

7. Let $u = x^2$, $dv = \cos 3x\,dx$ $\Rightarrow$ $du = 2x\,dx$, $v = \frac{1}{3}\sin 3x$.
Then $I = \int x^2 \cos 3x\,dx = \frac{1}{3}x^2 \sin 3x - \frac{2}{3}\int x \sin 3x\,dx$ by Equation 2. Next let
$U = x$, $dV = \sin 3x\,dx$ $\Rightarrow$ $dU = dx$, $V = -\frac{1}{3}\cos 3x$ to get
$\int x \sin 3x\,dx = -\frac{1}{3}x \cos 3x + \frac{1}{3}\int \cos 3x\,dx = -\frac{1}{3}x \cos 3x + \frac{1}{9}\sin 3x + C_1$. Substituting for $\int x \sin 3x\,dx$, we get
$I = \frac{1}{3}x^2 \sin 3x - \frac{2}{3}\left(-\frac{1}{3}x \cos 3x + \frac{1}{9}\sin 3x + C_1\right) = \frac{1}{3}x^2 \sin 3x + \frac{2}{9}x \cos 3x - \frac{2}{27}\sin 3x + C$, where
$C = -\frac{2}{3}C_1$.

9. Let $u = (\ln x)^2$, $dv = dx$ $\Rightarrow$ $du = 2\ln x \cdot \frac{1}{x}\,dx$, $v = x$.
Then $I = \int (\ln x)^2\,dx = x(\ln x)^2 - 2\int \ln x\,dx$. Next let $U = \ln x$, $dV = dx$ $\Rightarrow$ $dU = 1/x\,dx$, $V = x$ to get
$\int \ln x\,dx = x \ln x - \int x \cdot (1/x)\,dx = x \ln x - x + C_1$. Thus, $I = x(\ln x)^2 - 2x \ln x + 2x + C$, where
$C = -2C_1$.

11. First let $u = \sin 3\theta$, $dv = e^{2\theta}\,d\theta$ $\Rightarrow$ $du = 3\cos 3\theta\,d\theta$, $v = \frac{1}{2}e^{2\theta}$. Then
$I = \int e^{2\theta}\sin 3\theta\,d\theta = \frac{1}{2}e^{2\theta}\sin 3\theta - \frac{3}{2}\int e^{2\theta}\cos 3\theta\,d\theta$. Next let $U = \cos 3\theta$,
$dV = e^{2\theta}\,d\theta$ $\Rightarrow$ $dU = -3\sin 3\theta\,d\theta$, $V = \frac{1}{2}e^{2\theta}$ to get
$\int e^{2\theta}\cos 3\theta\,d\theta = \frac{1}{2}e^{2\theta}\cos 3\theta + \frac{3}{2}\int e^{2\theta}\sin 3\theta\,d\theta$. Substituting in the previous formula gives
$I = \frac{1}{2}e^{2\theta}\sin 3\theta - \frac{3}{4}e^{2\theta}\cos 3\theta - \frac{9}{4}\int e^{2\theta}\sin 3\theta\,d\theta = \frac{1}{2}e^{2\theta}\sin 3\theta - \frac{3}{4}e^{2\theta}\cos 3\theta - \frac{9}{4}I$ $\Rightarrow$
$\frac{13}{4}I = \frac{1}{2}e^{2\theta}\sin 3\theta - \frac{3}{4}e^{2\theta}\cos 3\theta + C_1$. Hence, $I = \frac{1}{13}e^{2\theta}(2\sin 3\theta - 3\cos 3\theta) + C$, where $C = \frac{4}{13}C_1$.

13. Let $u = y$, $dv = \sinh y\,dy$ $\Rightarrow$ $du = dy$, $v = \cosh y$. Then
$\int y \sinh y\,dy = y \cosh y - \int \cosh y\,dy = y \cosh y - \sinh y + C.$

15. Let $u = t$, $dv = e^{-t}\,dt$ $\Rightarrow$ $du = dt$, $v = -e^{-t}$. By Formula 6,
$\int_0^1 t e^{-t}\,dt = \left[-t e^{-t}\right]_0^1 + \int_0^1 e^{-t}\,dt = -1/e + \left[-e^{-t}\right]_0^1 = -1/e - 1/e + 1 = 1 - 2/e.$

17. Let $u = \ln x$, $dv = x^{-2}\,dx$ $\Rightarrow$ $du = \dfrac{1}{x}\,dx$, $v = -x^{-1}$. By (6),

$$\int_1^2 \frac{\ln x}{x^2}\,dx = \left[-\frac{\ln x}{x}\right]_1^2 + \int_1^2 x^{-2}\,dx = -\tfrac{1}{2}\ln 2 + \ln 1 + \left[-\frac{1}{x}\right]_1^2 = -\tfrac{1}{2}\ln 2 + 0 - \tfrac{1}{2} + 1 = \tfrac{1}{2} - \tfrac{1}{2}\ln 2.$$

19. $I = \int_1^4 \ln\sqrt{x}\,dx = \tfrac{1}{2}\int_1^4 \ln x\,dx = \tfrac{1}{2}[x\ln x - x]_1^4$ as in Example 2. So
$I = \tfrac{1}{2}[(4\ln 4 - 4) - (0 - 1)] = 2\ln 4 - \tfrac{3}{2}$.

21. Let $u = \cos^{-1}x$, $dv = dx$ $\Rightarrow$ $du = -\dfrac{dx}{\sqrt{1-x^2}}$, $v = x$. Then

$$I = \int_0^{1/2} \cos^{-1}x\,dx = \left[x\cos^{-1}x\right]_0^{1/2} + \int_0^{1/2} \frac{x\,dx}{\sqrt{1-x^2}} = \tfrac{1}{2}\cdot\tfrac{\pi}{3} + \int_1^{3/4} t^{-1/2}\left[-\frac{1}{2}\,dt\right], \text{ where } t = 1 - x^2 \Rightarrow$$

$dt = -2x\,dx$. Thus, $I = \tfrac{\pi}{6} + \tfrac{1}{2}\int_{3/4}^1 t^{-1/2}\,dt = [\sqrt{t}]_{3/4}^1 = \tfrac{\pi}{6} + 1 - \tfrac{\sqrt{3}}{2} = \tfrac{1}{6}\left(\pi + 6 - 3\sqrt{3}\right)$.

23. Let $u = \ln(\sin x)$, $dv = \cos x\,dx$ $\Rightarrow$ $du = \dfrac{\cos x}{\sin x}\,dx$, $v = \sin x$. Then

$I = \int \cos x\ln(\sin x)\,dx = \sin x\ln(\sin x) - \int \cos x\,dx = \sin x\ln(\sin x) - \sin x + C$.

Another Method: Substitute $t = \sin x$, so $dt = \cos x\,dx$. Then $I = \int \ln t\,dt = t\ln t - t + C$ (see Example 2) and
so $I = \sin x(\ln\sin x - 1) + C$.

25. Let $w = \ln x$ $\Leftrightarrow$ $dw = dx/x$. Then $x = e^w$ and $dx = e^w\,dw$, so

$$\int \cos(\ln x)\,dx = \int e^w\cos w\,dw = \tfrac{1}{2}e^w(\sin w + \cos w) + C \text{ (by the method of Example 4)}$$

$$= \tfrac{1}{2}x[\sin(\ln x) + \cos(\ln x)] + C$$

27. Let $u = (\ln x)^2$, $dv = x^4\,dx$ $\Rightarrow$ $du = 2\dfrac{\ln x}{x}\,dx$, $v = \dfrac{x^5}{5}$. By (6),

$$\int_1^2 x^4(\ln x)^2\,dx = \left[\frac{x^5}{5}(\ln x)^2\right]_1^2 - 2\int_1^2 \frac{x^4}{5}\ln x\,dx = \tfrac{32}{5}(\ln 2)^2 - 0 - 2\int_1^2 \frac{x^4}{5}\ln x\,dx.$$

Let $U = \ln x$, $dV = \dfrac{x^4}{5}\,dx$ $\Rightarrow$ $dU = \dfrac{1}{x}\,dx$, $V = \dfrac{x^5}{25}$. So

$$2\int_1^2 \frac{x^4}{5}\ln x\,dx = \left[\frac{x^5}{25}\ln x\right]_1^2 - \int_1^2 \frac{x^4}{25}\,dx = \tfrac{32}{25}\ln 2 - 0 - \left[\frac{x^5}{125}\right]_1^2 = \tfrac{32}{25}\ln 2 - \left(\tfrac{32}{125} - \tfrac{1}{125}\right). \text{ So}$$

$\int_1^2 x^4(\ln x)^2\,dx = \tfrac{32}{5}(\ln 2)^2 - 2\left(\tfrac{32}{25}\ln 2 - \tfrac{31}{125}\right) = \tfrac{32}{5}(\ln 2)^2 - \tfrac{64}{25}\ln 2 + \tfrac{62}{125}$.

29. Let $w = \sqrt{x}$, so that $x = w^2$ and $dx = 2w\,dw$. Thus, $\int \sin\sqrt{x}\,dx = \int 2w\sin w\,dw$.
Now use parts with $u = 2w$, $dv = \sin w\,dw$, $du = 2\,dw$, $v = -\cos w$ to get
$\int 2w\sin w\,dw = -2w\cos w + \int 2\cos w\,dw = -2w\cos w + 2\sin w + C = -2\sqrt{x}\cos\sqrt{x} + 2\sin\sqrt{x} + C$.

31. Let $x = \theta^2$, so $dx = 2\theta\,d\theta$. So $\int_{\sqrt{\pi/2}}^{\sqrt{\pi}} \theta^3\cos(\theta^2)\,d\theta = \tfrac{1}{2}\int_{\pi/2}^{\pi} x\cos x\,dx$. Let $u = x$, $dv = \cos x\,dx$ $\Rightarrow$
$du = dx$, $v = \sin x$. So

$$\tfrac{1}{2}\int_{\pi/2}^{\pi} x\cos x\,dx = \tfrac{1}{2}\left([x\sin x]_{\pi/2}^{\pi} - \int_{\pi/2}^{\pi} \sin x\,dx\right) = \tfrac{1}{2}[x\sin x + \cos x]_{\pi/2}^{\pi}$$

$$= \tfrac{1}{2}(\pi\sin\pi + \cos\pi) - \tfrac{1}{2}\left(\tfrac{\pi}{2}\sin\tfrac{\pi}{2} + \cos\tfrac{\pi}{2}\right) = \tfrac{1}{2}(\pi\cdot 0 - 1) - \tfrac{1}{2}\left(\tfrac{\pi}{2}\cdot 1 + 0\right) = -\tfrac{1}{2} - \tfrac{\pi}{4}$$

In Exercises 33 and 35, let $f(x)$ denote the integrand and $F(x)$ its antiderivative (with $C = 0$).

33. Let $u = x$, $dv = \cos \pi x \, dx$ $\Rightarrow$ $du = dx$, $v = (\sin \pi x)/x$. Then

$$\int x \cos \pi x \, dx = x \cdot \frac{\sin \pi x}{\pi} - \int \frac{\sin \pi x}{\pi} \, dx = \frac{x \sin \pi x}{\pi} + \frac{\cos \pi x}{\pi^2} + C.$$

We see from the graph that this is reasonable, since F has extreme values where f is 0.

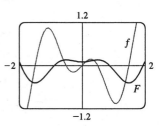

35. Let $u = 2x + 3$, $dv = e^x \, dx$ $\Rightarrow$ $du = 2 \, dx$, $v = e^x$. Then
$\int (2x + 3) e^x \, dx = (2x + 3) e^x - 2 \int e^x \, dx = (2x + 3) e^x - 2e^x + C = (2x + 1) e^x + C$. We see from the graph that this is reasonable, since F has a minimum where f changes from negative to positive.

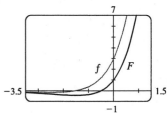

37. (a) Take $n = 2$ in Example 6 to get $\int \sin^2 x \, dx = -\frac{1}{2} \cos x \sin x + \frac{1}{2} \int 1 \, dx = \frac{x}{2} - \frac{\sin 2x}{4} + C$.

(b) $\int \sin^4 x \, dx = -\frac{1}{4} \cos x \sin^3 x + \frac{3}{4} \int \sin^2 x \, dx = -\frac{1}{4} \cos x \sin^3 x + \frac{3}{8} x - \frac{3}{16} \sin 2x + C$.

39. (a) $\int_0^{\pi/2} \sin^n x \, dx = \left[-\frac{\cos x \sin^{n-1} x}{n} \right]_0^{\pi/2} + \frac{n-1}{n} \int_0^{\pi/2} \sin^{n-2} x \, dx = \frac{n-1}{n} \int_0^{\pi/2} \sin^{n-2} x \, dx$

(b) $\int_0^{\pi/2} \sin^3 x \, dx = \frac{2}{3} \int_0^{\pi/2} \sin x \, dx = \left[-\frac{2}{3} \cos x \right]_0^{\pi/2} = \frac{2}{3}$; $\int_0^{\pi/2} \sin^5 x \, dx = \frac{4}{5} \int_0^{\pi/2} \sin^3 x \, dx = \frac{4}{5} \cdot \frac{2}{3} = \frac{8}{15}$

(c) The formula holds for $n = 1$ (that is, $2n + 1 = 3$) by (b). Assume it holds for some $k \geq 1$. Then $\int_0^{\pi/2} \sin^{2k+1} x \, dx = \frac{2 \cdot 4 \cdot 6 \cdots (2k)}{3 \cdot 5 \cdot 7 \cdots (2k+1)}$. By Example 6,

$\int_0^{\pi/2} \sin^{2k+3} x \, dx = \frac{2k+2}{2k+3} \int_0^{\pi/2} \sin^{2k+1} x \, dx = \frac{2 \cdot 4 \cdot 6 \cdots [2(k+1)]}{2 \cdot 4 \cdot 6 \cdots [2(k+1)+1]}$ as desired. By induction, the formula holds for all $n \geq 1$.

41. Let $u = (\ln x)^n$, $dv = dx$ $\Rightarrow$ $du = n (\ln x)^{n-1} (dx/x)$, $v = x$. By Equation 2, $\int (\ln x)^n \, dx = x (\ln x)^n - n \int (\ln x)^{n-1} \, dx$.

43. Let $u = (x^2 + a^2)^n$, $dv = dx$ $\Rightarrow$ $du = n (x^2 + a^2)^{n-1} 2x \, dx$, $v = x$. Then

$$\int (x^2 + a^2)^n \, dx = x (x^2 + a^2)^n - 2n \int x^2 (x^2 + a^2)^{n-1} \, dx$$

$$= x (x^2 + a^2)^n - 2n \left[\int (x^2 + a^2)^n \, dx - a^2 \int (x^2 + a^2)^{n-1} \, dx \right] \quad \text{[since } x^2 = (x^2 + a^2) - a^2]$$

$\Rightarrow$ $(2n + 1) \int (x^2 + a^2)^n \, dx = x (x^2 + a^2)^n + 2na^2 \int (x^2 + a^2)^{n-1} \, dx$, and

$$\int (x^2 + a^2)^n \, dx = \frac{x (x^2 + a^2)^n}{2n + 1} + \frac{2na^2}{2n + 1} \int (x^2 + a^2)^{n-1} \, dx \quad \text{(provided } 2n + 1 \neq 0).$$

45. Take $n = 3$ in Exercise 41 to get
$\int (\ln x)^3 \, dx = x (\ln x)^3 - 3 \int (\ln x)^2 \, dx = x (\ln x)^3 - 3x (\ln x)^2 + 6x \ln x - 6x + C$ (by Exercise 9).
Or: Instead of using Exercise 9, apply Exercise 41 again with $n = 2$.

47. Let $u = \sin^{-1} x$, $dv = dx$ $\Rightarrow$ $du = \dfrac{dx}{\sqrt{1-x^2}}$, $v = x$. Then

$$\text{area} = \int_0^{1/2} \sin^{-1} x \, dx = \left[x \sin^{-1} x \right]_0^{1/2} - \int_0^{1/2} \frac{x}{\sqrt{1-x^2}} \, dx$$

$$= \tfrac{1}{2} \left(\tfrac{\pi}{6} \right) + \left[\sqrt{1-x^2} \right]_0^{1/2} = \tfrac{\pi}{12} + \tfrac{\sqrt{3}}{2} - 1 = \tfrac{1}{12} \left(\pi + 6\sqrt{3} - 12 \right)$$

49.

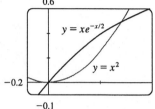
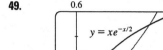

From the graph, we see that the curves intersect at approximately $x = 0$ and $x = 0.70$, with $xe^{-x/2} > x^2$ on $(0, 0.70)$. So the area bounded by the curves is approximately $A = \int_0^{0.70} \left(xe^{-x/2} - x^2 \right) dx$. We separate this into two integrals, and evaluate the first one by parts with $u = x$,

$$dv = e^{-x/2} \, dx \quad \Rightarrow \quad du = dx, v = -2e^{-x/2}:$$

$$A = \left[-2xe^{-x/2} \right]_0^{0.70} - \int_0^{0.70} \left(-2e^{-x/2} \right) dx - \left[\tfrac{1}{3} x^3 \right]_0^{0.70}$$

$$= \left[-2 \left(0.70 \right) e^{-0.35} - 0 \right] - \left[4e^{-x/2} \right]_0^{0.70} - \tfrac{1}{3} \left[0.70^3 - 0 \right] \approx 0.080$$

51. Volume $= \int_{2\pi}^{3\pi} 2\pi x \sin x \, dx$. Let $u = x$, $dv = \sin x \, dx$ $\Rightarrow$ $du = dx, v = -\cos x$ $\Rightarrow$
$V = 2\pi \left[-x \cos x + \sin x \right]_{2\pi}^{3\pi} = 2\pi \left[(3\pi + 0) - (-2\pi + 0) \right] = 2\pi \left(5\pi \right) = 10\pi^2$.

53. Volume $= \int_{-1}^{0} 2\pi \left(1 - x \right) e^{-x} \, dx$. Let $u = 1 - x$, $dv = e^{-x} \, dx$ $\Rightarrow$ $du = -dx, v = -e^{-x}$ $\Rightarrow$
$V = 2\pi \left[xe^{-x} \right]_{-1}^{0} = 2\pi \left(0 + e \right) = 2\pi e$

55. Let $u = x$, $dv = \cos 2x \, dx$ $\Rightarrow$ $du = dx, v = \tfrac{1}{2} \sin 2x \, dx$. Then
$\int_0^{\pi/2} x \cos 2x \, dx = \left[\tfrac{1}{2} x \sin 2x \right]_0^{\pi/2} - \tfrac{1}{2} \int_0^{\pi/2} \sin 2x \, dx = 0 + \left[\tfrac{1}{4} \cos 2x \right]_0^{\pi/2} = \tfrac{1}{4} \left(-1 - 1 \right) = -\tfrac{1}{2}$. Hence, the
average value of f is $\dfrac{-1/2}{\pi/2 - 0} = -\dfrac{1}{\pi}$.

57. Since $v(t) > 0$ for all t, the desired distance $s(t) = \int_0^t v(w) \, dw = \int_0^t w^2 e^{-w} \, dw$. Let $u = w^2$, $dv = e^{-w} \, dw$
$\Rightarrow$ $du = 2w \, dw, v = -e^{-w}$. Then $s(t) = \left[-w^2 e^{-w} \right]_0^t + 2 \int_0^t we^{-w} \, dw$. Now let $U = w, dV = e^{-w} \, dw$ $\Rightarrow$
$dU = dw, V = -e^{-w}$. Then

$$s(t) = -t^2 e^{-t} + 2 \left(\left[-we^{-w} \right]_0^t + \int_0^t e^{-w} \, dw \right) = -t^2 e^{-t} - 2te^{-t} - 2e^{-t} + 2$$

$$= 2 - e^{-t} \left(t^2 + 2t + 2 \right) \text{ meters}$$

59. Take $g(x) = x$ and $g'(x) = 1$ in Equation 1.

61. By Exercise 60, $\int_1^e \ln x \, dx = e \ln e - 1 \ln 1 - \int_{\ln 1}^{\ln e} e^y \, dy = e - \int_0^1 e^y \, dy = e - \left[e^y \right]_0^1 = e - (e - 1) = 1$.

63. Using the formula for volumes of rotation and the figure, we see that

Volume $= \int_0^d \pi b^2 \, dy - \int_0^c \pi a^2 \, dy - \int_c^d \pi \, [g\,(y)]^2 \, dy = \pi b^2 d - \pi a^2 c - \int_c^d \pi \, [g\,(y)]^2 \, dy$. Let $y = f\,(x)$, which

gives $dy = f'\,(x)\, dx$ and $g\,(y) = x$, so that $V = \pi b^2 d - \pi a^2 c - \pi \int_a^b x^2 f'\,(x)\, dx$. Now integrate

by parts with $u = x^2$, and $dv = f'\,(x)\, dx \;\Rightarrow\; du = 2x\, dx, v = f\,(x)$, and

$\int_a^b x^2 f'\,(x)\, dx = \left[x^2 f\,(x)\right]_a^b - \int_a^b 2x \, f\,(x)\, dx = b^2 f\,(b) - a^2 f\,(a) - \int_a^b 2x \, f\,(x)\, dx$, but $f\,(a) = c$ and

$f\,(b) = d \;\Rightarrow\; V = \pi b^2 d - \pi a^2 c - \pi \left[b^2 d - a^2 c - \int_a^b 2xf\,(x)\, dx\right] = \int_a^b 2\pi x f\,(x)\, dx$.

7.2 Trigonometric Integrals

1. $\int \sin^3 x \cos^2 x \, dx = \int \sin^2 x \cos^2 x \sin x \, dx = \int \left(1 - \cos^2 x\right) \cos^2 x \sin x \, dx \overset{c}{=} \int \left(1 - u^2\right) u^2 \left(-du\right)$

$= \int \left(u^2 - 1\right) u^2 \, du = \int \left(u^4 - u^2\right) du = \frac{1}{5} u^5 - \frac{1}{3} u^3 + C = \frac{1}{5} \cos^5 x - \frac{1}{3} \cos^3 x + C$

3. $\int_{\pi/2}^{3\pi/4} \sin^5 x \cos^3 x \, dx = \int_{\pi/2}^{3\pi/4} \sin^5 x \cos^2 x \cos x \, dx = \int_{\pi/2}^{3\pi/4} \sin^5 x \left(1 - \sin^2 x\right) \cos x \, dx \overset{s}{=} \int_1^{\sqrt{2}/2} u^5 \left(1 - u^2\right) du$

$= \int_1^{\sqrt{2}/2} \left(u^5 - u^7\right) du = \left[\frac{1}{6} u^6 - \frac{1}{8} u^8\right]_1^{\sqrt{2}/2} = \left(\frac{1/8}{6} - \frac{1/16}{8}\right) - \left(\frac{1}{6} - \frac{1}{8}\right) = -\frac{11}{384}$

5. $\int \cos^5 x \sin^4 x \, dx = \int \cos^4 x \sin^4 x \cos x \, dx = \int \left(1 - \sin^2 x\right)^2 \sin^4 x \cos x \, dx \overset{s}{=} \int \left(1 - u^2\right)^2 u^4 \, du$

$= \int \left(1 - 2u^2 + u^4\right) u^4 \, du = \int \left(u^4 - 2u^6 + u^8\right) du = \frac{1}{5} u^5 - \frac{2}{7} u^7 + \frac{1}{9} u^9 + C$

$= \frac{1}{5} \sin^5 x - \frac{2}{7} \sin^7 x + \frac{1}{9} \sin^9 x + C$

7. $\int_0^{\pi/2} \sin^2 3x \, dx = \int_0^{\pi/2} \frac{1}{2} \left(1 - \cos 6x\right) dx = \left[\frac{1}{2} x - \frac{1}{12} \sin 6x\right]_0^{\pi/2} = \frac{\pi}{4}$

9. $\int \cos^4 t \, dt = \int \left[\frac{1}{2} \left(1 + \cos 2t\right)\right]^2 dt = \frac{1}{4} \int \left(1 + 2 \cos 2t + \cos^2 2t\right) dt$

$= \frac{1}{4} t + \frac{1}{4} \sin 2t + \frac{1}{4} \int \frac{1}{2} \left(1 + \cos 4t\right) dt = \frac{1}{4} \left[t + \sin 2t + \frac{1}{2} t + \frac{1}{8} \sin 4t\right] + C$

$= \frac{3}{8} t + \frac{1}{4} \sin 2t + \frac{1}{32} \sin 4t + C$

11. $\int \left(1 - \sin 2x\right)^2 dx = \int \left(1 - 2 \sin 2x + \sin^2 2x\right) dx = \int \left[1 - 2 \sin 2x + \frac{1}{2} \left(1 - \cos 4x\right)\right] dx$

$= \int \left[\frac{3}{2} - 2 \sin 2x - \frac{1}{2} \cos 4x\right] dx = \frac{3}{2} x + \cos 2x - \frac{1}{8} \sin 4x + C$

13. $\int_0^{\pi/4} \sin^4 x \cos^2 x \, dx = \int_0^{\pi/4} \sin^2 x \left(\sin x \cos x\right)^2 dx = \int_0^{\pi/4} \frac{1}{2} \left(1 - \cos 2x\right) \left(\frac{1}{2} \sin 2x\right)^2 dx$

$= \frac{1}{8} \int_0^{\pi/4} \left(1 - \cos 2x\right) \sin^2 2x \, dx = \frac{1}{8} \int_0^{\pi/4} \sin^2 2x \, dx - \frac{1}{8} \int_0^{\pi/4} \sin^2 2x \cos 2x \, dx$

$= \frac{1}{16} \int_0^{\pi/4} \left(1 - \cos 4x\right) dx - \frac{1}{16} \left[\frac{1}{3} \sin^3 2x\right]_0^{\pi/4} = \frac{1}{16} \left[x - \frac{1}{4} \sin 4x - \frac{1}{3} \sin^3 2x\right]_0^{\pi/4}$

$= \frac{1}{16} \left(\frac{\pi}{4} - 0 - \frac{1}{3}\right) = \frac{1}{192} \left(3\pi - 4\right)$

15. $\int \sin^3 x \sqrt{\cos x} \, dx = \int \left(1 - \cos^2 x\right) \sqrt{\cos x} \sin x \, dx \overset{c}{=} \int \left(1 - u^2\right) u^{1/2} \left(-du\right) = \int \left(u^{5/2} - u^{1/2}\right) du$

$= \frac{2}{7} u^{7/2} - \frac{2}{3} u^{3/2} + C = \frac{2}{7} \left(\cos x\right)^{7/2} - \frac{2}{3} \left(\cos x\right)^{3/2} + C = \left[\frac{2}{7} \cos^3 x - \frac{2}{3} \cos x\right] \sqrt{\cos x} + C$

17. $\int \cos^2 x \tan^3 x \, dx = \int \frac{\sin^3 x}{\cos x} dx \overset{c}{=} \int \frac{\left(1 - u^2\right) \left(-du\right)}{u} = \int \left[\frac{-1}{u} + u\right] du$

$= -\ln|u| + \frac{1}{2} u^2 + C = \frac{1}{2} \cos^2 x - \ln|\cos x| + C$

19. $\int \dfrac{1 - \sin x}{\cos x} \, dx = \int (\sec x - \tan x) \, dx = \ln|\sec x + \tan x| - \ln|\sec x| + C \quad \left[\begin{array}{l}\text{by (1) and the boxed}\\ \text{formula above it}\end{array}\right]$

$$= \ln|(\sec x + \tan x)\cos x| + C = \ln|1 + \sin x| + C$$

$$= \ln(1 + \sin x) + C \text{ since } 1 + \sin x \geq 0$$

Or: $\int \dfrac{1 - \sin x}{\cos x} \, dx = \int \dfrac{1 - \sin x}{\cos x} \cdot \dfrac{1 + \sin x}{1 + \sin x} \, dx = \int \dfrac{(1 - \sin^2 x)\, dx}{\cos x \,(1 + \sin x)} = \int \dfrac{\cos x \, dx}{1 + \sin x}$

$$= \int \dfrac{dw}{w} \quad (\text{where } w = 1 + \sin x, \, dw = \cos x \, dx)$$

$$= \ln|w| + C = \ln|1 + \sin x| + C = \ln(1 + \sin x) + C$$

21. $\int \tan^2 x \, dx = \int (\sec^2 x - 1) \, dx = \tan x - x + C$

23. $\int \sec^4 x \, dx = \int (\tan^2 x + 1) \sec^2 x \, dx = \int \tan^2 x \sec^2 x \, dx + \int \sec^2 x \, dx = \frac{1}{3} \tan^3 x + \tan x + C$

25. Let $u = \tan t \Rightarrow du = \sec^2 t \, dt$. Then $\int_0^{\pi/4} \tan^4 t \sec^2 t \, dt = \int_0^1 u^4 \, du = \left[\frac{1}{5}u^5\right]_0^1 = \frac{1}{5}$.

27. $\int \tan^3 x \sec x \, dx = \int \tan^2 x \sec x \tan x \, dx = \int (\sec^2 x - 1) \sec x \tan x \, dx$

$$= \int (u^2 - 1) \, du \quad [u = \sec x, \, du = \sec x \tan x \, dx] = \frac{1}{3}u^3 - u + C = \frac{1}{3}\sec^3 x - \sec x + C$$

29. Let $u = \sec x \Rightarrow du = \sec x \tan x \, dx$. Then

$$\int_0^{\pi/3} \tan^5 x \sec x \, dx = \int_0^{\pi/3} (\sec^2 x - 1)^2 \sec x \tan x \, dx = \int_1^2 (u^2 - 1)^2 \, du = \int_1^2 (u^4 - 2u^2 + 1) \, du$$

$$= \left[\frac{1}{5}u^5 - \frac{2}{3}u^3 + u\right]_1^2 = \left(\frac{32}{5} - \frac{16}{3} + 2\right) - \left(\frac{1}{5} - \frac{2}{3} + 1\right) = \frac{38}{15}$$

31. $\int \tan^5 x \, dx = \int (\sec^2 x - 1)^2 \tan x \, dx = \int \sec^4 x \tan x \, dx - 2 \int \sec^2 x \tan x \, dx + \int \tan x \, dx$

$$= \int \sec^3 x \sec x \tan x \, dx - 2 \int \tan x \sec^2 x \, dx + \int \tan x \, dx$$

$$= \frac{1}{4} \sec^4 x - \tan^2 x + \ln|\sec x| + C \quad [\text{or } \frac{1}{4} \sec^4 x - \sec^2 x + \ln|\sec x| + C]$$

33. Let $u = \tan x \Rightarrow du = \sec^2 x \, dx$. Then

$$\int \dfrac{\sec^2 x}{\cot x} \, dx = \int \tan x \sec^2 x \, dx = \int u \, du = \frac{1}{2}u^2 + C = \frac{1}{2} \tan^2 x + C.$$

35. $\int_{\pi/6}^{\pi/2} \cot^2 x \, dx = \int_{\pi/6}^{\pi/2} (\csc^2 x - 1) \, dx = [-\cot x - x]_{\pi/6}^{\pi/2} = \left(0 - \frac{\pi}{2}\right) - \left(-\sqrt{3} - \frac{\pi}{6}\right) = \sqrt{3} - \frac{\pi}{3}$

37. $\int \cot^2 w \csc^4 w \, dw = \int \cot^2 w \csc^2 w \csc^2 w \, dw = \int \cot^2 w \, (1 + \cot^2 w) \csc^2 w \, dw$

$$= \int u^2 \, (1 + u^2) \, (-du) \quad [u = \cot w, \, du = -\csc^2 w \, dw] = -\int (u^2 + u^4) \, du$$

$$= -\frac{1}{3}u^3 - \frac{1}{5}u^5 + C = -\frac{1}{3} \cot^3 w - \frac{1}{5} \cot^5 w + C$$

39. $I = \displaystyle\int \csc x \, dx = \int \dfrac{\csc x \, (\csc x - \cot x)}{\csc x - \cot x} \, dx = \int \dfrac{-\csc x \cot x + \csc^2 x}{\csc x - \cot x} \, dx.$ Let $u = \csc x - \cot x \Rightarrow$

$du = (-\csc x \cot x + \csc^2 x) \, dx$. Then $I = \int du/u = \ln|u| = \ln|\csc x - \cot x| + C.$

41. $\int \sin 5x \sin 2x \, dx = \int \frac{1}{2} [\cos(5x - 2x) - \cos(5x + 2x)] \, dx = \frac{1}{2} \int (\cos 3x - \cos 7x) \, dx$

$$= \frac{1}{6} \sin 3x - \frac{1}{14} \sin 7x + C$$

43. $\int \cos 7\theta \cos 5\theta \, d\theta = \int \frac{1}{2} [\cos(7\theta - 5\theta) + \cos(7\theta + 5\theta)] \, d\theta = \frac{1}{2} \int (\cos 2\theta + \cos 12\theta) \, d\theta$

$$= \frac{1}{2} \left(\frac{1}{2} \sin 2\theta + \frac{1}{12} \sin 12\theta\right) + C = \frac{1}{4} \sin 2\theta + \frac{1}{24} \sin 12\theta + C$$

45. $\int \dfrac{1 - \tan^2 x}{\sec^2 x} \, dx = \int (\cos^2 x - \sin^2 x) \, dx = \int \cos 2x \, dx = \dfrac{1}{2} \sin 2x + C$

47. Let $u = \cos x \;\Rightarrow\; du = -\sin x\,dx$. Then

$$\int \sin^5 x\,dx = \int \left(1 - \cos^2 x\right)^2 \sin x\,dx = \int \left(1 - u^2\right)^2 (-du)$$

$$= \int \left(-1 + 2u^2 - u^4\right) du = -\tfrac{1}{5}u^5 + \tfrac{2}{3}u^3 - u + C$$

$$= -\tfrac{1}{5}\cos^5 x + \tfrac{2}{3}\cos^3 x - \cos x + C$$

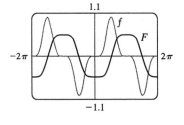

Notice that F is increasing when $f(x) > 0$, so the graphs serve as a check on our work.

49. $\int \sin 3x \sin 6x\,dx = \int \tfrac{1}{2}\left[\cos(3x - 6x) - \cos(3x + 6x)\right] dx$

$$= \tfrac{1}{2}\int (\cos 3x - \cos 9x)\,dx$$

$$= \tfrac{1}{6}\sin 3x - \tfrac{1}{18}\sin 9x + C$$

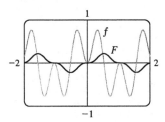

Notice that $f(x) = 0$ whenever F has a horizontal tangent.

51. $f_{\text{ave}} = \tfrac{1}{2\pi} \int_{-\pi}^{\pi} \sin^2 x \cos^3 x\,dx = \tfrac{1}{2\pi} \int_{-\pi}^{\pi} \sin^2 x \left(1 - \sin^2 x\right) \cos x\,dx = \tfrac{1}{2\pi} \int_0^0 u^2 \left(1 - u^2\right) du$ (where $u = \sin x$) $= 0$

53. For $0 < x < \tfrac{\pi}{2}$, we have $0 < \sin x < 1$, so $\sin^3 x < \sin x$. Hence the area is

$$\int_0^{\pi/2} \left(\sin x - \sin^3 x\right) dx = \int_0^{\pi/2} \sin x \left(1 - \sin^2 x\right) dx = \int_0^{\pi/2} \cos^2 x \sin x\,dx. \text{ Now let } u = \cos x \;\Rightarrow\;$$

$du = -\sin x\,dx$. Then area $= \int_1^0 u^2\,(-du) = \int_0^1 u^2\,du = \left[\tfrac{1}{3}u^3\right]_0^1 = \tfrac{1}{3}$.

55.

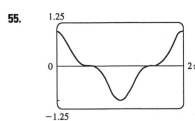

It seems from the graph that $\int_0^{2\pi} \cos^3 x\,dx = 0$, since the area below the x-axis and above the graph looks about equal to the area above the axis and below the graph. By Example 1, the integral is

$$\left[\sin x - \tfrac{1}{3}\sin^3 x\right]_0^{2\pi} = 0. \text{ Note that due to symmetry, the integral of any}$$

odd power of $\sin x$ or $\cos x$ between limits which differ by $2n\pi$ (n any integer) is 0.

57. $V = \int_{\pi/2}^{\pi} \pi \sin^2 x\,dx = \pi \int_{\pi/2}^{\pi} \tfrac{1}{2}(1 - \cos 2x)\,dx = \pi \left[\tfrac{1}{2}x - \tfrac{1}{4}\sin 2x\right]_{\pi/2}^{\pi} = \pi \left(\tfrac{\pi}{2} - 0 - \tfrac{\pi}{4} + 0\right) = \tfrac{\pi^2}{4}$

59. Volume $= \pi \int_0^{\pi/2} \left[(1 + \cos x)^2 - 1^2\right] dx = \pi \int_0^{\pi/2} \left(2\cos x + \cos^2 x\right) dx$

$$= \pi \left[2\sin x + \tfrac{1}{2}x + \tfrac{1}{4}\sin 2x\right]_0^{\pi/2} = \pi \left(2 + \tfrac{\pi}{4}\right) = 2\pi + \tfrac{\pi^2}{4}$$

61. $s = f(t) = \int_0^t \sin \omega u \cos^2 \omega u\,du$. Let $y = \cos \omega u \;\Rightarrow\; dy = -\omega \sin \omega u\,du$. Then

$$s = -\tfrac{1}{\omega} \int_1^{\cos \omega t} y^2\,dy = -\tfrac{1}{\omega} \left[\tfrac{1}{3}y^3\right]_1^{\cos \omega t} = \tfrac{1}{3\omega}\left(1 - \cos^3 \omega t\right).$$

63. Just note that the integrand is odd $[f(-x) = -f(x)]$.

Or: If $m \ne n$, calculate

$$\int_{-\pi}^{\pi} \sin mx \cos nx\,dx = \int_{-\pi}^{\pi} \tfrac{1}{2}\left[\sin(m - n)x + \sin(m + n)x\right] dx$$

$$= \tfrac{1}{2}\left[-\frac{\cos(m - n)x}{m - n} - \frac{\cos(m + n)x}{m + n}\right]_{-\pi}^{\pi} = 0$$

If $m = n$, then the first term in each set of brackets is zero.

65. $\int_{-\pi}^{\pi} \cos mx \cos nx \, dx = \int_{-\pi}^{\pi} \frac{1}{2} \left[\cos (m-n)x + \cos (m+n)x \right] dx.$ If $m \neq n$,

this is equal to $\dfrac{1}{2} \left[\dfrac{\sin (m-n)x}{m-n} + \dfrac{\sin (m+n)x}{m+n} \right]_{-\pi}^{\pi} = 0.$ If $m = n$, we get

$\int_{-\pi}^{\pi} \frac{1}{2} \left[1 + \cos (m+n)x \right] dx = \left[\frac{1}{2}x \right]_{-\pi}^{\pi} + \left[\dfrac{\sin (m+n)x}{2(m+n)} \right]_{-\pi}^{\pi} = \pi + 0 = \pi.$

7.3 Trigonometric Substitution

1. Let $x = 3 \sec \theta$, where $0 \le \theta < \frac{\pi}{2}$ or $\pi \le \theta < \frac{3\pi}{2}$. Then

$dx = 3 \sec \theta \tan \theta \, d\theta$ and

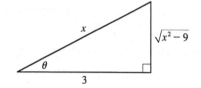

$\sqrt{x^2 - 9} = \sqrt{9 \sec^2 \theta - 9} = \sqrt{9 (\sec^2 \theta - 1)} = \sqrt{9 \tan^2 \theta}$

$\qquad = 3 \, |\tan \theta| = 3 \tan \theta$ for the relevant values of θ.

$\displaystyle\int \dfrac{1}{x^2 \sqrt{x^2 - 9}} \, dx = \int \dfrac{1}{9 \sec^2 \theta \cdot 3 \tan \theta} 3 \sec \theta \tan \theta \, d\theta = \frac{1}{9} \int \cos \theta \, d\theta = \frac{1}{9} \sin \theta + C = \dfrac{1}{9} \dfrac{\sqrt{x^2 - 9}}{x} + C$

Note that $-\sec (\theta + \pi) = \sec \theta$, so the figure is sufficient for the case $\pi \le \theta < \frac{3\pi}{2}$.

3. Let $x = 3 \tan \theta$, where $-\frac{\pi}{2} < \theta < \frac{\pi}{2}$. Then $dx = 3 \sec^2 \theta \, d\theta$ and

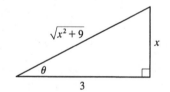

$\sqrt{x^2 + 9} = \sqrt{9 \tan^2 \theta + 9} = \sqrt{9 (\tan^2 \theta + 1)} = \sqrt{9 \sec^2 \theta}$

$\qquad = 3 \, |\sec \theta| = 3 \sec \theta$ for the relevant values of θ.

$\displaystyle\int \dfrac{x^3}{\sqrt{x^2 + 9}} \, dx = \int \dfrac{3^3 \tan^3 \theta}{3 \sec \theta} 3 \sec^2 \theta \, d\theta = 3^3 \int \tan^3 \theta \sec \theta \, d\theta = 3^3 \int \tan^2 \theta \tan \theta \sec \theta \, d\theta$

$\qquad = 3^3 \int (\sec^2 \theta - 1) \tan \theta \sec \theta \, d\theta = 3^3 \int (u^2 - 1) \, du \quad [u = \sec \theta, \, du = \sec \theta \tan \theta \, d\theta]$

$\qquad = 3^3 \left(\frac{1}{3} u^3 - u \right) + C = 3^3 \left(\frac{1}{3} \sec^3 \theta - \sec \theta \right) + C = 3^3 \left[\dfrac{1}{3} \dfrac{(x^2 + 9)^{3/2}}{3^3} - \dfrac{\sqrt{x^2 + 9}}{3} \right] + C$

$\qquad = \frac{1}{3} \left(x^2 + 9 \right)^{3/2} - 9\sqrt{x^2 + 9} + C$ or $\frac{1}{3} \left(x^2 - 18 \right) \sqrt{x^2 + 9} + C$

5. Let $t = \sec \theta$, so $dt = \sec \theta \tan \theta \, d\theta$, $t = \sqrt{2} \, \Rightarrow \, \theta = \frac{\pi}{4}$, and $t = 2 \, \Rightarrow \, \theta = \frac{\pi}{3}$. Then

$\displaystyle\int_{\sqrt{2}}^{2} \dfrac{1}{t^3 \sqrt{t^2 - 1}} \, dt = \int_{\pi/4}^{\pi/3} \dfrac{1}{\sec^3 \theta \tan \theta} \sec \theta \tan \theta \, d\theta = \int_{\pi/4}^{\pi/3} \cos^2 \theta \, d\theta = \int_{\pi/4}^{\pi/3} \frac{1}{2} (1 + \cos 2\theta) \, d\theta$

$\qquad = \frac{1}{2} \left[\theta + \frac{1}{2} \sin 2\theta \right]_{\pi/4}^{\pi/3} = \frac{1}{2} \left[\left(\frac{\pi}{3} + \frac{1}{2} \frac{\sqrt{3}}{2} \right) - \left(\frac{\pi}{4} + \frac{1}{2} \cdot 1 \right) \right]$

$\qquad = \frac{1}{2} \left(\frac{\pi}{12} + \frac{\sqrt{3}}{4} - \frac{1}{2} \right) = \frac{\pi}{24} + \frac{\sqrt{3}}{8} - \frac{1}{4}$

7. Let $x = 5\sin\theta$, so $dx = 5\cos\theta\,d\theta$. Then

$$\int \frac{1}{x^2\sqrt{25-x^2}}\,dx = \int \frac{1}{5^2\sin^2\theta \cdot 5\cos\theta} 5\cos\theta\,d\theta$$

$$= \tfrac{1}{25}\int \csc^2\theta\,d\theta = -\tfrac{1}{25}\cot\theta + C$$

$$= -\frac{1}{25}\frac{\sqrt{25-x^2}}{x} + C$$

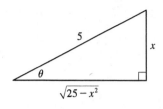

9. Let $x = \sqrt{3}\tan\theta$, where $-\frac{\pi}{2} < \theta < \frac{\pi}{2}$. Then

$$\int \frac{dx}{x\sqrt{x^2+3}} = \int \frac{\sqrt{3}\sec^2\theta\,d\theta}{\sqrt{3}\tan\theta\sqrt{3}\sec\theta}$$

$$= \tfrac{1}{\sqrt{3}}\int \csc\theta\,d\theta = \tfrac{1}{\sqrt{3}}\ln|\csc\theta - \cot\theta| + C$$

$$= \frac{1}{\sqrt{3}}\ln\left|\frac{\sqrt{x^2+3}-\sqrt{3}}{x}\right| + C$$

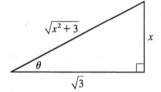

11. Let $2x = \sin\theta$, where $-\frac{\pi}{2} \le \theta \le \frac{\pi}{2}$. Then $x = \frac{1}{2}\sin\theta$,

$dx = \frac{1}{2}\cos\theta\,d\theta$, and $\sqrt{1-4x^2} = \sqrt{1-(2x)^2} = \cos\theta$.

$$\int \sqrt{1-4x^2}\,dx = \int \cos\theta \left(\tfrac{1}{2}\cos\theta\right) d\theta = \tfrac{1}{4}\int (1+\cos 2\theta)\,d\theta$$

$$= \tfrac{1}{4}\left(\theta + \tfrac{1}{2}\sin 2\theta\right) + C = \tfrac{1}{4}(\theta + \sin\theta\cos\theta) + C$$

$$= \tfrac{1}{4}\left[\sin^{-1}(2x) + 2x\sqrt{1-4x^2}\right] + C$$

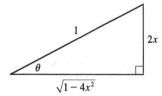

13. $9x^2 - 4 = (3x)^2 - 4$, so let $3x = 2\sec\theta$, where $0 \le \theta < \frac{\pi}{2}$ or

$\pi \le \theta < \frac{3\pi}{2}$. Then

$dx = \frac{2}{3}\sec\theta\tan\theta\,d\theta$ and $\sqrt{9x^2-4} = 2\tan\theta$.

$$\int \frac{\sqrt{9x^2-4}}{x}\,dx = \int \frac{2\tan\theta}{\frac{2}{3}\sec\theta} \cdot \frac{2}{3}\sec\theta\tan\theta\,d\theta$$

$$= 2\int \tan^2\theta\,d\theta = 2\int \left(\sec^2\theta - 1\right) d\theta = 2(\tan\theta - \theta) + C$$

$$= \sqrt{9x^2-4} - 2\sec^{-1}\left(\frac{3x}{2}\right) + C$$

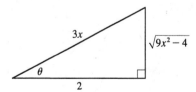

15. Let $x = a \sin\theta$, where $-\frac{\pi}{2} \leq \theta \leq \frac{\pi}{2}$. Then $dx = a\cos\theta\, d\theta$ and

$$\int \frac{x^2\, dx}{(a^2 - x^2)^{3/2}} = \int \frac{a^2 \sin^2\theta a\cos\theta\, d\theta}{a^3 \cos^3\theta} = \int \tan^2\theta\, d\theta$$

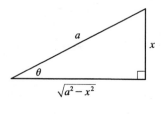

$$= \int \left(\sec^2\theta - 1\right) d\theta = \tan\theta - \theta + C$$

$$= \frac{x}{\sqrt{a^2 - x^2}} - \sin^{-1}\frac{x}{a} + C$$

17. Let $u = x^2 - 7$, so $du = 2x\, dx$. Then $\displaystyle\int \frac{x}{\sqrt{x^2 - 7}}\, dx = \frac{1}{2}\int \frac{1}{\sqrt{u}}\, du = \frac{1}{2}\cdot 2\sqrt{u} + C = \sqrt{x^2 - 7} + C.$

19. Let $x = 3\tan\theta$, where $-\frac{\pi}{2} < \theta < \frac{\pi}{2}$. Then $dx = 3\sec^2\theta\, d\theta$ and $\sqrt{9 + x^2} = 3\sec\theta$.

$$\int_0^3 \frac{dx}{\sqrt{9 + x^2}} = \int_0^{\pi/4} \frac{3\sec^2\theta\, d\theta}{3\sec\theta} = \int_0^{\pi/4} \sec\theta\, d\theta = [\ln|\sec\theta + \tan\theta|]_0^{\pi/4}$$

$$= \ln\left(\sqrt{2} + 1\right) - \ln 1 = \ln\left(\sqrt{2} + 1\right)$$

21. Let $u = 4 - 9x^2 \ \Rightarrow\ du = -18x\, dx$. Then $x^2 = \frac{1}{9}(4 - u)$ and

$$\int_0^{2/3} x^3\sqrt{4 - 9x^2}\, dx = \int_4^0 \frac{1}{9}(4 - u)u^{1/2}\left(-\frac{1}{18}\right) du = \frac{1}{162}\int_0^4 \left(4u^{1/2} - u^{3/2}\right) du$$

$$= \frac{1}{162}\left[\frac{8}{3}u^{3/2} - \frac{2}{5}u^{5/2}\right]_0^4 = \frac{1}{162}\left[\frac{64}{3} - \frac{64}{5}\right] = \frac{64}{1215}$$

Or: Let $3x = 2\sin\theta$, where $-\frac{\pi}{2} \leq \theta \leq \frac{\pi}{2}$.

23. $2x - x^2 = -\left(x^2 - 2x + 1\right) + 1 = 1 - (x - 1)^2$. Let $u = x - 1$. Then $du = dx$ and

$$\int \sqrt{2x - x^2}\, dx = \int \sqrt{1 - u^2}\, du = \int \cos^2\theta\, d\theta \quad (\text{where } u = \sin\theta,\ -\frac{\pi}{2} \leq \theta \leq \frac{\pi}{2})$$

$$= \frac{1}{2}\int (1 + \cos 2\theta)\, d\theta = \frac{1}{2}\left(\theta + \frac{1}{2}\sin 2\theta\right) + C = \frac{1}{2}\left(\sin^{-1} u + u\sqrt{1 - u^2}\right) + C$$

$$= \frac{1}{2}\left[\sin^{-1}(x - 1) + (x - 1)\sqrt{2x - x^2}\right] + C$$

25. $9x^2 + 6x - 8 = (3x + 1)^2 - 9$, so let $u = 3x + 1$, $du = 3dx$. Then $\displaystyle\int \frac{dx}{\sqrt{9x^2 + 6x - 8}} = \int \frac{\frac{1}{3}du}{\sqrt{u^2 - 9}}$. Now let $u = 3\sec\theta$, where $0 \leq \theta < \frac{\pi}{2}$ or $\pi \leq \theta < \frac{3\pi}{2}$. Then $du = 3\sec\theta\tan\theta\, d\theta$ and $\sqrt{u^2 - 9} = 3\tan\theta$, so

$$\int \frac{\frac{1}{3}du}{\sqrt{u^2 - 9}} = \int \frac{\sec\theta\tan\theta\, d\theta}{3\tan\theta} = \frac{1}{3}\int \sec\theta\, d\theta = \frac{1}{3}\ln|\sec\theta + \tan\theta| + C_1 = \frac{1}{3}\ln\left|\frac{u + \sqrt{u^2 - 9}}{3}\right| + C_1$$

$$= \frac{1}{3}\ln\left|u + \sqrt{u^2 - 9}\right| + C = \frac{1}{3}\ln\left|3x + 1 + \sqrt{9x^2 + 6x - 8}\right| + C$$

27. $x^2 + 2x + 2 = (x + 1)^2 + 1$. Let $u = x + 1$, $du = dx$. Then

$$\int \frac{dx}{(x^2 + 2x + 2)^2} = \int \frac{du}{(u^2 + 1)^2} = \int \frac{\sec^2 \theta d\theta}{\sec^4 \theta} \quad \begin{pmatrix} \text{where } u = \tan \theta, du = \sec^2 \theta \, d\theta, \\ \text{and } u^2 + 1 = \sec^2 \theta \end{pmatrix}$$

$$= \int \cos^2 \theta \, d\theta = \tfrac{1}{2} \int (1 + \cos 2\theta) \, d\theta = \tfrac{1}{2} (\theta + \sin \theta \cos \theta) + C$$

$$= \frac{1}{2} \left[\tan^{-1} u + \frac{u}{1 + u^2} \right] + C = \frac{1}{2} \left[\tan^{-1} (x + 1) + \frac{x + 1}{x^2 + 2x + 2} \right] + C$$

29. Let $u = e^t \implies du = e^t dt$. Then

$$\int e^t \sqrt{9 - e^{2t}} \, dt = \int \sqrt{9 - u^2} \, du = \int (3 \cos \theta) \, 3 \cos \theta \, d\theta \quad (\text{where } u = 3 \sin \theta, \; -\tfrac{\pi}{2} \le \theta \le \tfrac{\pi}{2})$$

$$= 9 \int \cos^2 \theta \, d\theta = \tfrac{9}{2} \int (1 + \cos 2\theta) \, d\theta = \tfrac{9}{2} (\theta + \sin \theta \cos \theta) + C$$

$$= \frac{9}{2} \left[\sin^{-1} \left(\frac{u}{3} \right) + \frac{u}{3} \cdot \frac{\sqrt{9 - u^2}}{3} \right] + C = \tfrac{9}{2} \sin^{-1} \left(\tfrac{1}{3} e^t \right) + \tfrac{1}{2} e^t \sqrt{9 - e^{2t}} + C$$

31. (a) Let $x = a \tan \theta$, where $-\tfrac{\pi}{2} < \theta < \tfrac{\pi}{2}$. Then $\sqrt{x^2 + a^2} = a \sec \theta$ and

$$\int \frac{dx}{\sqrt{x^2 + a^2}} = \int \frac{a \sec^2 \theta \, d\theta}{a \sec \theta} = \int \sec \theta \, d\theta = \ln |\sec \theta + \tan \theta| + C_1 = \ln \left| \frac{\sqrt{x^2 + a^2}}{a} + \frac{x}{a} \right| + C_1$$

$$= \ln \left(x + \sqrt{x^2 + a^2} \right) + C \quad \text{where } C = C_1 - \ln |a|$$

(b) Let $x = a \sinh t$, so that $dx = a \cosh t \, dt$ and $\sqrt{x^2 + a^2} = a \cosh t$. Then

$$\int \frac{dx}{\sqrt{x^2 + a^2}} = \int \frac{a \cosh t \, dt}{a \cosh t} = t + C = \sinh^{-1} \frac{x}{a} + C.$$

33. $f(x) = \left(4 - x^2 \right)^{3/2}$ on $[0, 2] \implies f_{\text{ave}} = \frac{1}{2 - 0} \int_0^2 \left(4 - x^2 \right)^{3/2} dx$. Let $x = 2 \sin \theta \implies dx = 2 \cos \theta \, d\theta$ and $\left(4 - x^2 \right)^{3/2} = (2 \cos \theta)^3$. So

$$f_{\text{ave}} = \tfrac{1}{2} \int_0^{\pi/2} (2 \cos \theta)^3 \, 2 \cos \theta \, d\theta = 8 \int_0^{\pi/2} \cos^4 \theta \, d\theta$$

$$= 8 \left[\tfrac{3}{8} \theta + \tfrac{1}{4} \sin 2\theta + \tfrac{1}{32} \sin 4\theta \right]_0^{\pi/2} \quad \text{[by Exercise 7.2.9]} \quad = 8 \left[\left(\tfrac{3\pi}{16} + 0 + 0 \right) - (0 + 0 + 0) \right] = \tfrac{3\pi}{2}$$

35. Area of $\triangle POQ = \tfrac{1}{2} (r \cos \theta)(r \sin \theta) = \tfrac{1}{2} r^2 \sin \theta \cos \theta$. Area of region $PQR = \int_{r \cos \theta}^r \sqrt{r^2 - x^2} \, dx$.
Let $x = r \cos u \implies dx = -r \sin u \, du$ for $0 \le u \le \tfrac{\pi}{2}$. Then we obtain

$$\int \sqrt{r^2 - x^2} \, dx = \int r \sin u \, (-r \sin u) \, du = -r^2 \int \sin^2 u \, du = -\tfrac{1}{2} r^2 (u - \sin u \cos u) + C$$

$$= -\tfrac{1}{2} r^2 \cos^{-1} (x/r) + \tfrac{1}{2} x \sqrt{r^2 - x^2} + C$$

so

$$\text{area of region } PQR = \tfrac{1}{2} \left[-r^2 \cos^{-1} (x/r) + x \sqrt{r^2 - x^2} \right]_{r \cos \theta}^r = \tfrac{1}{2} \left[0 - \left(-r^2 \theta + r \cos \theta \, r \sin \theta \right) \right]$$

$$= \tfrac{1}{2} r^2 \theta - \tfrac{1}{2} r^2 \sin \theta \cos \theta$$

and thus, (area of sector POR) = (area of $\triangle POQ$) + (area of region PQR) = $\tfrac{1}{2} r^2 \theta$.

37. From the graph, it appears that the curve $y = x^2\sqrt{4-x^2}$ and the line

$y = 2 - x$ intersect at about $x = 0.81$ and $x = 2$, with $x^2\sqrt{4-x^2} > 2 - x$ on

$(0.81, 2)$. So the area bounded by the curve and the line is

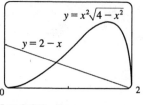

$y = x^2\sqrt{4-x^2}$

$y = 2 - x$

$A \approx \int_{0.81}^{2}\left[x^2\sqrt{4-x^2} - (2-x)\right]dx = \int_{0.81}^{2} x^2\sqrt{4-x^2}\,dx - \left[2x - \tfrac{1}{2}x^2\right]_{0.81}^{2}$.

To evaluate the integral, we put $x = 2\sin\theta$, where $-\tfrac{\pi}{2} \le \theta \le \tfrac{\pi}{2}$. Then

$dx = 2\cos\theta\,d\theta, x = 2 \;\Rightarrow\; \theta = \sin^{-1}1 = \tfrac{\pi}{2}$, and $x = 0.81 \;\Rightarrow\; \theta = \sin^{-1}0.405 \approx 0.417$. So

$\int_{0.81}^{2} x^2\sqrt{4-x^2}\,dx \approx \int_{0.417}^{\pi/2} 4\sin^2\theta\,(2\cos\theta)(2\cos\theta\,d\theta) = 4\int_{0.417}^{\pi/2}\sin^2 2\theta\,d\theta = 4\int_{0.417}^{\pi/2}\tfrac{1}{2}(1-\cos 4\theta)\,d\theta$

$= 2\left[\theta - \tfrac{1}{4}\sin 4\theta\right]_{0.417}^{\pi/2} = 2\left[\left(\tfrac{\pi}{2} - 0\right) - \left(0.417 - \tfrac{1}{4}(0.995)\right)\right] \approx 2.81$

Thus, $A \approx 2.81 - \left[\left(2\cdot 2 - \tfrac{1}{2}\cdot 2^2\right) - \left(2\cdot 0.81 - \tfrac{1}{2}\cdot 0.81^2\right)\right] \approx 2.10$.

39. Let the equation of the large circle be $x^2 + y^2 = R^2$. Then the equation of the small circle is $x^2 + (y-b)^2 = r^2$, where $b = \sqrt{R^2 - r^2}$ is the distance between the centers of the circles. The desired area is

$A = \int_{-r}^{r}\left[\left(b + \sqrt{r^2 - x^2}\right) - \sqrt{R^2 - x^2}\right]dx = 2\int_{0}^{r}\left(b + \sqrt{r^2 - x^2} - \sqrt{R^2 - x^2}\right)dx$

$= 2\int_{0}^{r} b\,dx + 2\int_{0}^{r}\sqrt{r^2 - x^2}\,dx - 2\int_{0}^{r}\sqrt{R^2 - x^2}\,dx$

The first integral is just $2br = 2r\sqrt{R^2 - r^2}$. To evaluate the other two integrals, note that

$\int\sqrt{a^2 - x^2}\,dx = \int a^2\cos^2\theta\,d\theta \;\; (x = a\sin\theta, dx = a\cos\theta\,d\theta) \;\; = \tfrac{1}{2}\left(\tfrac{1}{2}a^2\right)\int(1 + \cos 2\theta)$

$= \tfrac{1}{2}a^2\left(\theta + \tfrac{1}{2}\sin 2\theta\right) + C = \tfrac{1}{2}a^2\left(\theta + \sin\theta\cos\theta\right) + C$

$= \dfrac{a^2}{2}\arcsin\left(\dfrac{x}{a}\right) + \dfrac{a^2}{2}\left(\dfrac{x}{a}\right)\dfrac{\sqrt{a^2 - x^2}}{a} + C = \dfrac{a^2}{2}\arcsin\left(\dfrac{x}{a}\right) + \dfrac{x}{2}\sqrt{a^2 - x^2} + C$

so the desired area is

$A = 2r\sqrt{R^2 - r^2} + \left[r^2\arcsin(x/r) + x\sqrt{r^2 - x^2}\right]_{0}^{r} - \left[R^2\arcsin(x/R) + x\sqrt{R^2 - x^2}\right]_{0}^{r}$

$= 2r\sqrt{R^2 - r^2} + r^2\left(\tfrac{\pi}{2}\right) - \left[R^2\arcsin(r/R) + r\sqrt{R^2 - r^2}\right] = r\sqrt{R^2 - r^2} + \tfrac{\pi}{2}r^2 - R^2\arcsin(r/R)$

41. We use cylindrical shells and assume that $R > r$. $x^2 = r^2 - (y-R)^2 \;\Rightarrow\; x = \pm\sqrt{r^2 - (y-R)^2}$, so

$g(y) = 2\sqrt{r^2 - (y-R)^2}$ and

$V = \int_{R-r}^{R+r} 2\pi y \cdot 2\sqrt{r^2 - (y-R)^2}\,dy = \int_{-r}^{r} 4\pi\,(u+R)\sqrt{r^2 - u^2}\,du$ (where $u = y - R$)

$= 4\pi\int_{-r}^{r} u\sqrt{r^2 - u^2}\,du + 4\pi R\int_{-r}^{r}\sqrt{r^2 - u^2}\,du$ $\left(\begin{array}{l}\text{where } u = r\sin\theta, du = r\cos\theta\,d\theta \\ \text{in the second integral}\end{array}\right)$

$= 4\pi\left[-\tfrac{1}{3}\left(r^2 - u^2\right)^{3/2}\right]_{-r}^{r} + 4\pi R\int_{-\pi/2}^{\pi/2} r^2\cos^2\theta\,d\theta = -\tfrac{4\pi}{3}(0 - 0) + 4\pi Rr^2\int_{-\pi/2}^{\pi/2}\cos^2\theta\,d\theta$

$= 2\pi Rr^2\int_{-\pi/2}^{\pi/2}(1 + \cos 2\theta)\,d\theta = 2\pi Rr^2\left[\theta + \tfrac{1}{2}\sin 2\theta\right]_{-\pi/2}^{\pi/2} = 2\pi^2 Rr^2$

Another Method: Use washers instead of shells, so $V = 8\pi R\int_{0}^{r}\sqrt{r^2 - y^2}\,dy$ as in Exercise 6.2.59(a), but evaluate the integral using $y = r\sin\theta$.

7.4 Integration of Rational Functions by Partial Fractions

1. $\dfrac{3}{(2x+3)\,(x-1)} = \dfrac{A}{2x+3} + \dfrac{B}{x-1}$

3. $\dfrac{x^2+9x-12}{(3x-1)\,(x+6)^2} = \dfrac{A}{3x-1} + \dfrac{B}{x+6} + \dfrac{C}{(x+6)^2}$

5. $\dfrac{1}{x^4-x^3} = \dfrac{1}{x^3\,(x-1)} = \dfrac{A}{x} + \dfrac{B}{x^2} + \dfrac{C}{x^3} + \dfrac{D}{x-1}$

7. $\dfrac{x^2+1}{x^2-1} = 1 + \dfrac{2}{(x-1)\,(x+1)} = 1 + \dfrac{A}{x-1} + \dfrac{B}{x+1}$

9. $\dfrac{t^4+t^2+1}{(t^2+1)\,(t^2+4)^2} = \dfrac{At+B}{t^2+1} + \dfrac{Ct+D}{t^2+4} + \dfrac{Et+F}{(t^2+4)^2}$

11. $\dfrac{x^4}{(x^2+9)^3} = \dfrac{Ax+B}{x^2+9} + \dfrac{Cx+D}{(x^2+9)^2} + \dfrac{Ex+F}{(x^2+9)^3}$

13. $\displaystyle\int \frac{x^2}{x+1}\,dx = \int\left(x-1+\frac{1}{x+1}\right)dx = \tfrac{1}{2}x^2 - x + \ln|x+1| + C$

15. $\dfrac{x-9}{(x+5)\,(x-2)} = \dfrac{A}{x+5} + \dfrac{B}{x-2}$. Multiply both sides by $(x+5)\,(x-2)$ to get $x-9 = A\,(x-2) + B\,(x+5)$.
Substituting 2 for x gives $-7 = 7B \iff B = -1$. Substituting -5 for x gives $-14 = -7A \iff A = 2$. Thus,

$$\int \frac{x-9}{(x+5)\,(x-2)}\,dx = \int\left(\frac{2}{x+5} + \frac{-1}{x-2}\right)dx = 2\ln|x+5| - \ln|x-2| + C$$

17. $\dfrac{x^2+1}{x^2-x} = 1 + \dfrac{x+1}{x\,(x-1)} = 1 - \dfrac{1}{x} + \dfrac{2}{x-1}$, so

$$\int \frac{x^2+1}{x^2-x}\,dx = x - \ln|x| + 2\ln|x-1| + C = x + \ln\frac{(x-1)^2}{|x|} + C$$

19. $\dfrac{2x+3}{(x+1)^2} = \dfrac{A}{x+1} + \dfrac{B}{(x+1)^2} \Rightarrow 2x+3 = A\,(x+1) + B$. Take $x = -1$ to get $B = 1$, and equate coefficients of x to get $A = 2$. Now

$$\int_0^1 \frac{2x+3}{(x+1)^2}\,dx = \int_0^1\left[\frac{2}{x+1} + \frac{1}{(x+1)^2}\right]dx = \left[2\ln(x+1) - \frac{1}{x+1}\right]_0^1$$

$$= 2\ln 2 - \tfrac{1}{2} - (2\ln 1 - 1) = 2\ln 2 + \tfrac{1}{2}$$

21. $\dfrac{4y^2-7y-12}{y\,(y+2)\,(y-3)} = \dfrac{A}{y} + \dfrac{B}{y+2} + \dfrac{C}{y-3} \Rightarrow 4y^2 - 7y - 12 = A\,(y+2)\,(y-3) + By\,(y-3) + Cy\,(y+2)$.
Setting $y = 0$ gives $-12 = -6A$, so $A = 2$. Setting $y = -2$ gives $18 = 10B$, so $B = \tfrac{9}{5}$. Setting $y = 3$ gives $3 = 15C$, so $C = \tfrac{1}{5}$. Now

$$\int_1^2 \frac{4y^2-7y-12}{y\,(y+2)\,(y-3)}\,dy = \int_1^2\left(\frac{2}{y} + \frac{9/5}{y+2} + \frac{1/5}{y-3}\right)dy = \left[2\ln|y| + \tfrac{9}{5}\ln|y+2| + \tfrac{1}{5}\ln|y-3|\right]_1^2$$

$$= 2\ln 2 + \tfrac{9}{5}\ln 4 + \tfrac{1}{5}\ln 1 - 2\ln 1 - \tfrac{9}{5}\ln 3 - \tfrac{1}{5}\ln 2$$

$$= 2\ln 2 + \tfrac{18}{5}\ln 2 - \tfrac{1}{5}\ln 2 - \tfrac{9}{5}\ln 3 = \tfrac{27}{5}\ln 2 - \tfrac{9}{5}\ln 3 = \tfrac{9}{5}(3\ln 2 - \ln 3) = \tfrac{9}{5}\ln\tfrac{8}{3}$$

23. $\dfrac{1}{(x+5)^2(x-1)} = \dfrac{A}{x+5} + \dfrac{B}{(x+5)^2} + \dfrac{C}{x-1} \Rightarrow 1 = A(x+5)(x-1) + B(x-1) + C(x+5)^2$. Setting

$x = -5$ gives $1 = -6B$, so $B = -\frac{1}{6}$. Setting $x = 1$ gives $1 = 36C$, so $C = \frac{1}{36}$. Setting $x = -2$ gives

$1 = A(3)(-3) + B(-3) + C(3^2) = -9A - 3B + 9C = -9A + \frac{1}{2} + \frac{1}{4} = -9A + \frac{3}{4}$, so $9A = -\frac{1}{4}$ and $A = -\frac{1}{36}$.

Now

$$\int \frac{1}{(x+5)^2(x-1)}\,dx = \int \left[\frac{-1/36}{x+5} - \frac{1/6}{(x+5)^2} + \frac{1/36}{x-1} \right] dx$$

$$= -\tfrac{1}{36}\ln|x+5| + \frac{1}{6(x+5)} + \tfrac{1}{36}\ln|x-1| + C$$

25. $\dfrac{5x^2 + 3x - 2}{x^3 + 2x^2} = \dfrac{5x^2 + 3x - 2}{x^2(x+2)} = \dfrac{A}{x} + \dfrac{B}{x^2} + \dfrac{C}{x+2}$. Multiply by $x^2(x+2)$ to get

$5x^2 + 3x - 2 = Ax(x+2) + B(x+2) + Cx^2$. Set $x = -2$ to get $C = 3$, and take $x = 0$ to get

$B = -1$. Equating the coefficients of x^2 gives $5 = A + C \Rightarrow A = 2$. So

$$\int \frac{5x^2 + 3x - 2}{x^3 + 2x^2}\,dx = \int \left(\frac{2}{x} - \frac{1}{x^2} + \frac{3}{x+2} \right) dx = 2\ln|x| + \frac{1}{x} + 3\ln|x+2| + C.$$

27. $\dfrac{x^2}{(x+1)^3} = \dfrac{A}{x+1} + \dfrac{B}{(x+1)^2} + \dfrac{C}{(x+1)^3}$. Multiply by $(x+1)^3$ to get $x^2 = A(x+1)^2 + B(x+1) + C$.

Setting $x = -1$ gives $C = 1$. Equating the coefficients of x^2 gives $A = 1$, and setting $x = 0$ gives $B = -2$. Now

$$\int \frac{x^2\,dx}{(x+1)^3} = \int \left[\frac{1}{x+1} - \frac{2}{(x+1)^2} + \frac{1}{(x+1)^3} \right] dx = \ln|x+1| + \frac{2}{x+1} - \frac{1}{2(x+1)^2} + C.$$

29. $\dfrac{x^3}{x^2+1} = \dfrac{(x^3+x)-x}{x^2+1} = x - \dfrac{x}{x^2+1}$, so

$$\int_0^1 \frac{x^3}{x^2+1}\,dx = \int_0^1 x\,dx - \int_0^1 \frac{x\,dx}{x^2+1} = \left[\tfrac{1}{2}x^2\right]_0^1 - \tfrac{1}{2}\int_1^2 \frac{1}{u}\,du \quad \text{(where } u = x^2 + 1,\, du = 2x\,dx)$$

$$= \tfrac{1}{2} - \left[\tfrac{1}{2}\ln u\right]_1^2 = \tfrac{1}{2} - \tfrac{1}{2}\ln 2 = \tfrac{1}{2}(1 - \ln 2)$$

31. $\dfrac{3x^2 - 4x + 5}{(x-1)(x^2+1)} = \dfrac{A}{x-1} + \dfrac{Bx+C}{x^2+1} \Rightarrow 3x^2 - 4x + 5 = A(x^2+1) + (Bx+C)(x-1)$. Take $x = 1$ to get

$4 = 2A$ or $A = 2$. Now $(Bx + C)(x-1) = 3x^2 - 4x + 5 - 2(x^2+1) = x^2 - 4x + 3$. Equating coefficients of

x^2 and then comparing the constant terms, we get $B = 1$ and $C = -3$. Hence,

$$\int \frac{3x^2 - 4x + 5}{(x-1)(x^2+1)}\,dx = \int \left[\frac{2}{x-1} + \frac{x-3}{x^2+1} \right] dx = 2\ln|x-1| + \int \frac{x\,dx}{x^2+1} - 3\int \frac{dx}{x^2+1}$$

$$= 2\ln|x-1| + \tfrac{1}{2}\ln\left(x^2+1\right) - 3\tan^{-1}x + C$$

$$= \ln(x-1)^2 + \ln\sqrt{x^2+1} - 3\tan^{-1}x + C$$

33. $\dfrac{2t^3 - t^2 + 3t - 1}{(t^2 + 1)(t^2 + 2)} = \dfrac{At + B}{t^2 + 1} + \dfrac{Ct + D}{t^2 + 2}$ $\Rightarrow$

$2t^3 - t^2 + 3t - 1 = (At + B)(t^2 + 2) + (Ct + D)(t^2 + 1) = (A + C)t^3 + (B + D)t^2 + (2A + C)t + (2B + D)$

$\Rightarrow$ $A + C = 2$, $B + D = -1$, $2A + C = 3$, and $2B + D = -1$ $\Rightarrow$ $A = 1$, $C = 1$, $B = 0$, and $D = -1$. Now

$$\int \frac{2t^3 - t^2 + 3t - 1}{(t^2 + 1)(t^2 + 2)}\, dt = \int \left(\frac{t}{t^2 + 1} + \frac{t - 1}{t^2 + 2}\right) dt = \frac{1}{2}\int \frac{2t\, dt}{t^2 + 1} + \frac{1}{2}\int \frac{2t\, dt}{t^2 + 2} - \int \frac{dt}{t^2 + 2}$$

$$= \tfrac{1}{2}\ln\left(t^2 + 1\right) + \tfrac{1}{2}\ln\left(t^2 + 2\right) - \tfrac{1}{\sqrt{2}}\tan^{-1}\left(\tfrac{1}{\sqrt{2}}t\right) + C$$

$$\text{or } \tfrac{1}{2}\ln\left(\left(t^2 + 1\right)\left(t^2 + 2\right)\right) - \tfrac{\sqrt{2}}{2}\tan^{-1}\left(\tfrac{1}{\sqrt{2}}t\right) + C$$

35. $\dfrac{1}{x^3 - 1} = \dfrac{1}{(x - 1)(x^2 + x + 1)} = \dfrac{A}{x - 1} + \dfrac{Bx + C}{x^2 + x + 1}$ $\Rightarrow$ $1 = A(x^2 + x + 1) + (Bx + C)(x - 1)$. Take

$x = 1$ to get $A = \frac{1}{3}$. Equating coefficients of x^2 and then comparing the constant terms, we get $0 = \frac{1}{3} + B$,

$1 = \frac{1}{3} - C$, so $B = -\frac{1}{3}$, $C = -\frac{2}{3}$ $\Rightarrow$

$$\int \frac{dx}{x^3 - 1} = \int \frac{\frac{1}{3}}{x - 1}\, dx + \int \frac{-\frac{1}{3}x - \frac{2}{3}}{x^2 + x + 1}\, dx = \tfrac{1}{3}\ln|x - 1| - \frac{1}{3}\int \frac{x + 2}{x^2 + x + 1}\, dx$$

$$= \tfrac{1}{3}\ln|x - 1| - \frac{1}{3}\int \frac{x + 1/2}{x^2 + x + 1}\, dx - \frac{1}{3}\int \frac{(3/2)\, dx}{(x + 1/2)^2 + 3/4}$$

$$= \tfrac{1}{3}\ln|x - 1| - \tfrac{1}{6}\ln\left(x^2 + x + 1\right) - \tfrac{1}{2}\left(\tfrac{2}{\sqrt{3}}\right)\tan^{-1}\left(\frac{x + \frac{1}{2}}{\sqrt{3}/2}\right) + K$$

$$= \tfrac{1}{3}\ln|x - 1| - \tfrac{1}{6}\ln\left(x^2 + x + 1\right) - \tfrac{1}{\sqrt{3}}\tan^{-1}\left(\tfrac{1}{\sqrt{3}}(2x + 1)\right) + K$$

37. Let $u = x^3 + 3x^2 + 4$. Then $du = 3\left(x^2 + 2x\right) dx$ $\Rightarrow$

$$\int_2^5 \frac{x^2 + 2x}{x^3 + 3x^2 + 4}\, dx = \frac{1}{3}\int_{24}^{204} \frac{du}{u} = \tfrac{1}{3}[\ln u]_{24}^{204} = \tfrac{1}{3}(\ln 204 - \ln 24) = \tfrac{1}{3}\ln \tfrac{204}{24} = \tfrac{1}{3}\ln \tfrac{17}{2}$$

39. $\dfrac{1}{x^4 - x^2} = \dfrac{1}{x^2(x - 1)(x + 1)} = \dfrac{A}{x} + \dfrac{B}{x^2} + \dfrac{C}{x - 1} + \dfrac{D}{x + 1}$. Multiply by $x^2(x - 1)(x + 1)$ to get

$1 = Ax(x - 1)(x + 1) + B(x - 1)(x + 1) + Cx^2(x + 1) + Dx^2(x - 1)$. Setting $x = 1$ gives $C = \frac{1}{2}$, taking

$x = -1$ gives $D = -\frac{1}{2}$. Equating the coefficients of x^3 gives $0 = A + C + D = A$. Finally, setting $x = 0$ yields

$B = -1$. Now $\displaystyle\int \frac{dx}{x^4 - x^2} = \int \left[\frac{-1}{x^2} + \frac{1/2}{x - 1} - \frac{1/2}{x + 1}\right] dx = \frac{1}{x} + \tfrac{1}{2}\ln\left|\frac{x - 1}{x + 1}\right| + C$.

41. $\displaystyle\int \frac{x-3}{(x^2+2x+4)^2}\,dx = \int \frac{x-3}{\left[(x+1)^2+3\right]^2}\,dx = \int \frac{u-4}{(u^2+3)^2}\,du$ (with $u=x+1$)

$\displaystyle = \int \frac{u\,du}{(u^2+3)^2} - 4\int \frac{du}{(u^2+3)^2} = \frac{1}{2}\int \frac{dv}{v^2} - 4\int \frac{\sqrt{3}\,\sec^2\theta\,d\theta}{9\sec^4\theta}$ $\begin{bmatrix} v=u^2+3 \text{ in the first integral;} \\ u=\sqrt{3}\tan\theta \text{ in the second} \end{bmatrix}$

$\displaystyle = \frac{-1}{(2v)} - \frac{4\sqrt{3}}{9}\int \cos^2\theta\,d\theta = \frac{-1}{2(u^2+3)} - \frac{2\sqrt{3}}{9}(\theta + \sin\theta\cos\theta) + C$

$\displaystyle = \frac{-1}{2(x^2+2x+4)} - \frac{2\sqrt{3}}{9}\left[\tan^{-1}\left(\frac{x+1}{\sqrt{3}}\right) + \frac{\sqrt{3}(x+1)}{x^2+2x+4}\right] + C$

$\displaystyle = \frac{-1}{2(x^2+2x+4)} - \frac{2\sqrt{3}}{9}\tan^{-1}\left(\frac{x+1}{\sqrt{3}}\right) - \frac{2(x+1)}{3(x^2+2x+4)} + C$

43. Let $u=\sqrt{x+1}$. Then $x=u^2-1$, $dx=2u\,du$ $\Rightarrow$

$\displaystyle \int \frac{dx}{x\sqrt{x+1}} = \int \frac{2u\,du}{(u^2-1)\,u} = 2\int \frac{du}{u^2-1} = \ln\left|\frac{u-1}{u+1}\right| + C = \ln\left|\frac{\sqrt{x+1}-1}{\sqrt{x+1}+1}\right| + C.$

45. Let $u=\sqrt{x}$, so $u^2=x$ and $dx=2u\,du$. Thus,

$\displaystyle \int_9^{16} \frac{\sqrt{x}}{x-4}\,dx = \int_3^4 \frac{u}{u^2-4}2u\,du = 2\int_3^4 \frac{u^2}{u^2-4}\,du = 2\int_3^4 \left(1 + \frac{4}{u^2-4}\right)du = 2 + 8\int_3^4 \frac{du}{(u+2)(u-2)}$

$\displaystyle = 2 + 8\int_3^4 \left(\frac{-1/4}{u+2} + \frac{1/4}{u-2}\right) \text{ (by partial fractions)} = 2 + 8\left[-\tfrac{1}{4}\ln|u+2| + \tfrac{1}{4}\ln|u-2|\right]_3^4$

$\displaystyle = 2 + [2\ln|u-2| - 2\ln|u+2|]_3^4 = 2 + 2\left[\ln\left|\frac{u-2}{u+2}\right|\right]_3^4 = 2 + 2\left(\ln\tfrac{2}{6} - \ln\tfrac{1}{5}\right)$

$\displaystyle = 2 + 2\ln\tfrac{5}{3} \text{ or } 2 + \ln\tfrac{25}{9}$

47. Let $u=\sqrt[3]{x^2+1}$. Then $x^2=u^3-1$, $2x\,dx=3u^2\,du$ $\Rightarrow$

$\displaystyle \int \frac{x^3\,dx}{\sqrt[3]{x^2+1}} = \int \frac{(u^3-1)\tfrac{3}{2}u^2\,du}{u} = \tfrac{3}{2}\int (u^4-u)\,du = \tfrac{3}{10}u^5 - \tfrac{3}{4}u^2 + C$

$\displaystyle = \tfrac{3}{10}\left(x^2+1\right)^{5/3} - \tfrac{3}{4}\left(x^2+1\right)^{2/3} + C$

49. If we were to substitute $u=\sqrt{x}$, then the square root would disappear but a cube root would remain. On the other hand, the substitution $u=\sqrt[3]{x}$ would eliminate the cube root but leave a square root. We can eliminate both roots by means of the substitution $u=\sqrt[6]{x}$. (Note that 6 is the least common multiple of 2 and 3.)

Let $u=\sqrt[6]{x}$. Then $x=u^6$, so $dx=6u^5\,du$ and $\sqrt{x}=u^3$, $\sqrt[3]{x}=u^2$. Thus,

$\displaystyle \int \frac{dx}{\sqrt{x}-\sqrt[3]{x}} = \int \frac{6u^5\,du}{u^3-u^2} = 6\int \frac{u^5}{u^2(u-1)}\,du = 6\int \frac{u^3}{u-1}\,du$

$\displaystyle = 6\int \left(u^2+u+1+\frac{1}{u-1}\right)du \text{ (by long division)}$

$\displaystyle = 6\left(\tfrac{1}{3}u^3 + \tfrac{1}{2}u^2 + u + \ln|u-1|\right) + C = 2\sqrt{x} + 3\sqrt[3]{x} + 6\sqrt[6]{x} + 6\ln\left|\sqrt[6]{x}-1\right| + C$

51. Let $u = e^x$. Then $x = \ln u$, $dx = \dfrac{du}{u}$ $\Rightarrow$

$$\int \frac{e^{2x}\, dx}{e^{2x} + 3e^x + 2} = \int \frac{u^2\,(du/u)}{u^2 + 3u + 2} = \int \frac{u\, du}{(u+1)(u+2)} = \int \left[\frac{-1}{u+1} + \frac{2}{u+2}\right] du$$

$$= 2\ln|u+2| - \ln|u+1| + C = \ln\left[(e^x + 2)^2 / (e^x + 1)\right] + C$$

53. From the graph, we see that the integral will be negative, and we guess that the area is about the same as that of a rectangle with width 2 and height 0.3, so we estimate the integral to be $-(2 \cdot 0.3) = -0.6$. Now

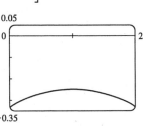

$$\frac{1}{x^2 - 2x - 3} = \frac{1}{(x-3)(x+1)} = \frac{A}{x-3} + \frac{B}{x+1} \quad \Leftrightarrow$$

$1 = (A+B)x + A - 3B$, so $A = -B$ and $A - 3B = 1$ $\Leftrightarrow$ $A = \frac{1}{4}$ and $B = -\frac{1}{4}$, so the integral becomes

$$\int_0^2 \frac{dx}{x^2 - 2x - 3} = \frac{1}{4}\int_0^2 \frac{dx}{x-3} - \frac{1}{4}\int_0^2 \frac{dx}{x+1} = \frac{1}{4}\left[\ln|x-3| - \ln|x+1|\right]_0^2$$

$$= \frac{1}{4}\left[\ln\left|\frac{x-3}{x+1}\right|\right]_0^2 = \frac{1}{4}\left(\ln\frac{1}{3} - \ln 3\right) = -\frac{1}{2}\ln 3 \approx -0.55$$

55. $\displaystyle \int \frac{dx}{x^2 - 2x} = \int \frac{dx}{(x-1)^2 - 1} = \int \frac{du}{u^2 - 1}$ (put $u = x - 1$)

$$= \frac{1}{2}\ln\left|\frac{u-1}{u+1}\right| + C \text{ (by Equation 6)} = \frac{1}{2}\ln\left|\frac{x-2}{x}\right| + C$$

57. (a) If $t = \tan\left(\dfrac{x}{2}\right)$, then $\dfrac{x}{2} = \tan^{-1} t$. The figure gives $\cos\left(\dfrac{x}{2}\right) = \dfrac{1}{\sqrt{1+t^2}}$ and

$\sin\left(\dfrac{x}{2}\right) = \dfrac{t}{\sqrt{1+t^2}}$.

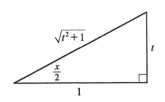

(b) $\cos x = \cos\left(2 \cdot \dfrac{x}{2}\right) = 2\cos^2\left(\dfrac{x}{2}\right) - 1$

$$= 2\left(\frac{1}{\sqrt{1+t^2}}\right)^2 - 1 = \frac{2}{1+t^2} - 1 = \frac{1-t^2}{1+t^2}$$

$\sin x = \sin\left(2 \cdot \dfrac{x}{2}\right) = 2\sin\left(\dfrac{x}{2}\right)\cos\left(\dfrac{x}{2}\right)$

$$= 2\frac{t}{\sqrt{1+t^2}}\frac{1}{\sqrt{1+t^2}} = \frac{2t}{1+t^2}$$

(c) $\dfrac{x}{2} = \arctan t$ $\Rightarrow$ $x = 2\arctan t$ $\Rightarrow$ $dx = \dfrac{2}{1+t^2}\, dt$

59. Let $t = \tan(x/2)$. Then, using the expressions in Exercise 57, we have

$$\int \frac{1}{3\sin x - 4\cos x}\, dx = \int \frac{1}{3\left(\dfrac{2t}{1+t^2}\right) - 4\left(\dfrac{1-t^2}{1+t^2}\right)} \frac{2\, dt}{1+t^2} = 2\int \frac{dt}{3(2t) - 4(1-t^2)} = \int \frac{dt}{2t^2 + 3t - 2}$$

$$= \int \frac{dt}{(2t-1)(t+2)} = \int \left[\frac{2}{5}\frac{1}{2t-1} - \frac{1}{5}\frac{1}{t+2}\right] dt \text{ (using partial fractions)}$$

$$= \frac{1}{5}\left[\ln|2t-1| - \ln|t+2|\right] + C = \frac{1}{5}\ln\left|\frac{2t-1}{t+2}\right| + C = \frac{1}{5}\ln\left|\frac{2\tan(x/2) - 1}{\tan(x/2) + 2}\right| + C$$

61. Let $t = \tan\left(\frac{x}{2}\right)$. Then, by Exercise 57,

$$\int \frac{dx}{2\sin x + \sin 2x} = \frac{1}{2}\int \frac{dx}{\sin x + \sin x \cos x} = \frac{1}{2}\int \frac{2\,dt/(1+t^2)}{2t/(1+t^2) + 2t(1-t^2)/(1+t^2)^2}$$

$$= \frac{1}{2}\int \frac{(1+t^2)\,dt}{t(1+t^2)+t(1-t^2)} = \frac{1}{4}\int \frac{(1+t^2)\,dt}{t} = \frac{1}{4}\int\left(\frac{1}{t}+t\right)dt$$

$$= \frac{1}{4}\ln|t| + \frac{1}{8}t^2 + C = \frac{1}{4}\ln\left|\tan\left(\frac{1}{2}x\right)\right| + \frac{1}{8}\tan^2\left(\frac{1}{2}x\right) + C$$

63. $\dfrac{x+1}{x-1} = 1 + \dfrac{2}{x-1} > 0$ for $2 \le x \le 3$, so

$$\text{area} = \int_2^3 \left[1 + \frac{2}{x-1}\right]dx = [x + 2\ln|x-1|]_2^3 = (3+2\ln 2)-(2+2\ln 1) = 1 + 2\ln 2.$$

65. $\dfrac{P+S}{P[(r-1)P-S]} = \dfrac{A}{P} + \dfrac{B}{(r-1)P-S}$ ⇒ $P+S = A[(r-1)P-S] + BP = [(r-1)A+B]P - AS$

⇒ $(r-1)A+B = 1, -A = 1$ ⇒ $A = -1, B = r$. Now

$$t = \int \frac{P+S}{P[(r-1)P-S]}\,dP = \int\left[\frac{-1}{P} + \frac{r}{(r-1)P-S}\right]dP = -\int\frac{dP}{P} + \frac{r}{r-1}\int\frac{r-1}{(r-1)P-S}\,dP$$

so $t = -\ln P + \dfrac{r}{r-1}\ln|(r-1)P-S| + C$. Here $r = 0.10$ and $S = 900$, so

$$t = -\ln P + \frac{0.1}{-0.9}\ln|-0.9P-900| + C = -\ln P - \frac{1}{9}\ln(|-1||0.9P+900|) = -\ln P - \frac{1}{9}\ln(0.9P+900) + C$$

When $t = 0$, $P = 10{,}000$, so $0 = -\ln 10{,}000 - \frac{1}{9}\ln(9900) + C$. Thus, $C = \ln 10{,}000 + \frac{1}{9}\ln 9900$ [≈ 10.2326], so our equation becomes

$$t = \ln 10{,}000 - \ln P + \frac{1}{9}\ln 9900 - \frac{1}{9}\ln(0.9P+900) = \ln\frac{10{,}000}{P} + \frac{1}{9}\ln\frac{9900}{0.9P+900}$$

$$= \ln\frac{10{,}000}{P} + \frac{1}{9}\ln\frac{1100}{0.1P+100} = \ln\frac{10{,}000}{P} + \frac{1}{9}\ln\frac{11{,}000}{P+1000}$$

67. (a) In Maple, we define $f(x)$, and then use `convert(f,parfrac,x);` to obtain

$$f(x) = \frac{24{,}110/4879}{5x+2} - \frac{668/323}{2x+1} - \frac{9438/80{,}155}{3x-7} + \frac{(22{,}098x+48{,}935)/260{,}015}{x^2+x+5}.$$

In Mathematica, we use the command `Apart`, and in Derive, we use `Expand`.

(b) $\int f(x)\,dx = \frac{24{,}110}{4879}\cdot\frac15\ln|5x+2| - \frac{668}{323}\cdot\frac12\ln|2x+1| - \frac{9438}{80{,}155}\cdot\frac13\ln|3x-7|$

$$+\frac{1}{260{,}015}\int\frac{22{,}098\left(x+\frac12\right)+37{,}886}{\left(x+\frac12\right)^2+\frac{19}{4}}\,dx + C$$

$$= \frac{24{,}110}{4879}\cdot\frac15\ln|5x+2| - \frac{668}{323}\cdot\frac12\ln|2x+1| - \frac{9438}{80{,}155}\cdot\frac13\ln|3x-7|$$

$$+\frac{1}{260{,}015}\left[22{,}098\cdot\frac12\ln\left(x^2+x+5\right)+37{,}886\cdot\sqrt{\frac{4}{19}}\tan^{-1}\left(\frac{1}{\sqrt{19/4}}\left(x+\frac12\right)\right)\right]+C$$

$$= \frac{4822}{4879}\ln|5x+2| - \frac{334}{323}\ln|2x+1| - \frac{3146}{80{,}155}\ln|3x-7| + \frac{11{,}049}{260{,}015}\ln\left(x^2+x+5\right)$$

$$+\frac{75{,}772}{260{,}015\sqrt{19}}\tan^{-1}\left[\frac{1}{\sqrt{19}}(2x+1)\right]+C.$$

Using a CAS, we get

$$\frac{4822\ln(5x+2)}{4879} - \frac{334\ln(2x+1)}{323} - \frac{3146\ln(3x-7)}{80{,}155}$$

$$+\frac{11{,}049\ln\left(x^2+x+5\right)}{260{,}015} + \frac{3988\sqrt{19}}{260{,}115}\tan^{-1}\left[\frac{\sqrt{19}}{19}(2x+1)\right]$$

The main difference in this answer is that the absolute value signs and the constant of integration have been omitted. Also, the fractions have been reduced and the denominators rationalized.

69. There are only finitely many values of x where $Q(x)=0$ (assuming that Q is not the zero polynomial). At all other values of x, $F(x)/Q(x)=G(x)/Q(x)$, so $F(x)=G(x)$. In other words, the values of F and G agree at all except perhaps finitely many values of x. By continuity of F and G, the polynomials F and G must agree at those values of x too.

More explicitly: if a is a value of x such that $Q(a)=0$, then $Q(x)\neq0$ for all x sufficiently close to a. Thus,

$$F(a) = \lim_{x\to a}F(x) \text{ (by continuity of } F) = \lim_{x\to a}G(x) \text{ [whenever } Q(x)\neq0]$$

$$= G(a) \text{ (by continuity of } G).$$

7.5 Strategy for Integration

1. Let $u=\sin x$. Then $\displaystyle\int\frac{\cos x\,dx}{1+\sin^2 x} = \int\frac{du}{1+u^2} = \tan^{-1}u+C = \tan^{-1}(\sin x)+C$.

3. Let $u=\arctan y$. Then $du=\dfrac{dy}{1+y^2}$ $\Rightarrow$ $\displaystyle\int_{-1}^{1}\frac{e^{\arctan y}}{1+y^2}\,dy = \int_{-\pi/4}^{\pi/4}e^u\,du = \left[e^u\right]_{-\pi/4}^{\pi/4} = e^{\pi/4}-e^{-\pi/4}$.

5. $\int\sin^2 x\cos^3 x\,dx = \int\sin^2 x\left(1-\sin^2 x\right)\cos x\,dx = \int u^2\left(1-u^2\right)du$ (put $u=\sin x$)

$$= \int\left(u^2-u^4\right)du = \frac13u^3-\frac15u^5+C = \frac13\sin^3 x-\frac15\sin^5 x+C$$

7. Let $u = \sqrt{9-x^2}$. Then $u^2 = 9 - x^2$, $u\,du = -x\,dx \Rightarrow$

$$\int \frac{\sqrt{9-x^2}}{x}\,dx = \int \frac{\sqrt{9-x^2}}{x^2}x\,dx = \int \frac{u}{9-u^2}(-u)\,du = \int \left[1 - \frac{9}{9-u^2}\right]du$$

$$= u + 9\int \frac{du}{u^2-9} = u + \frac{9}{2\cdot 3}\ln\left|\frac{u-3}{u+3}\right| + C = \sqrt{9-x^2} + \frac{3}{2}\ln\left|\frac{\sqrt{9-x^2}-3}{\sqrt{9-x^2}+3}\right| + C$$

$$= \sqrt{9-x^2} + \frac{3}{2}\ln\frac{(\sqrt{9-x^2}-3)^2}{x^2} + C = \sqrt{9-x^2} + 3\ln\left|\frac{3-\sqrt{9-x^2}}{x}\right| + C$$

Or: Put $x = 3\sin\theta$.

9. Let $u = 1 - x^2 \Rightarrow du = -2x\,dx$. Then

$$\int_0^{1/2} \frac{x}{\sqrt{1-x^2}}\,dx = -\frac{1}{2}\int_1^{3/4} \frac{1}{\sqrt{u}}\,du = \frac{1}{2}\int_{3/4}^1 u^{-1/2}\,du = \frac{1}{2}\left[2u^{1/2}\right]_{3/4}^1 = \left[\sqrt{u}\right]_{3/4}^1 = 1 - \frac{\sqrt{3}}{2}$$

11. $\displaystyle\int_0^2 \frac{2t}{(t-3)^2}\,dt = \int_{-3}^{-1} \frac{2(u+3)}{u^2}\,du$ $[u = t-3,\ du = dt]$ $= \int_{-3}^{-1}\left(\frac{2}{u} + \frac{6}{u^2}\right)du = \left[2\ln|u| - \frac{6}{u}\right]_{-3}^{-1}$

$$= (2\ln 1 + 6) - (2\ln 3 + 2) = 4 - 2\ln 3 \text{ or } 4 - \ln 9$$

13. $\displaystyle\int \frac{x-1}{x^2-4x+5}\,dx = \int \frac{(x-2)+1}{(x-2)^2+1}\,dx = \int \left(\frac{u}{u^2+1} + \frac{1}{u^2+1}\right)du$ $[u = x-2,\ du = dx]$

$$= \frac{1}{2}\ln(u^2+1) + \tan^{-1}u + C = \frac{1}{2}\ln(x^2-4x+5) + \tan^{-1}(x-2) + C$$

15. Let $u = e^x$. Then $\int e^{x+e^x}\,dx = \int e^{e^x}e^x\,dx = \int e^u\,du = e^u + C = e^{e^x} + C$.

17. Use integration by parts: $u = \ln(1+x^2)$, $dv = dx \Rightarrow du = \dfrac{2x}{1+x^2}\,dx$, $v = x$, so

$$\int \ln\left(1+x^2\right)dx = x\ln\left(1+x^2\right) - \int x\cdot\frac{2x\,dx}{1+x^2} = x\ln\left(1+x^2\right) - 2\int\left[1 - \frac{1}{1+x^2}\right]dx$$

$$= x\ln\left(1+x^2\right) - 2x + 2\tan^{-1}x + C$$

19. Integrate by parts three times, first with $u = t^3$, $dv = e^{-2t}\,dt$:

$$\int t^3e^{-2t}\,dt = -\frac{1}{2}t^3e^{-2t} + \frac{1}{2}\int 3t^2e^{-2t}\,dt = -\frac{1}{2}t^3e^{-2t} - \frac{3}{4}t^2e^{-2t} + \frac{1}{2}\int 3te^{-2t}\,dt$$

$$= -e^{-2t}\left[\tfrac{1}{2}t^3 + \tfrac{3}{4}t^2\right] - \frac{3}{4}te^{-2t} + \frac{3}{4}\int e^{-2t}\,dt = -e^{-2t}\left[\tfrac{1}{2}t^3 + \tfrac{3}{4}t^2 + \tfrac{3}{4}t + \tfrac{3}{8}\right] + C$$

$$= -\frac{1}{8}e^{-2t}\left(4t^3 + 6t^2 + 6t + 3\right) + C$$

21. Let $u = 1 + \sqrt{x}$. Then $x = (u-1)^2$, $dx = 2(u-1)\,du \Rightarrow$

$$\int_0^1 \left(1+\sqrt{x}\right)^8 dx = \int_1^2 u^8\cdot 2(u-1)\,du = 2\int_1^2 \left(u^9 - u^8\right)du = \left[\tfrac{1}{5}u^{10} - 2\cdot\tfrac{1}{9}u^9\right]_1^2$$

$$= \frac{1024}{5} - \frac{1024}{9} - \frac{1}{5} + \frac{2}{9} = \frac{4097}{45}$$

23. $\displaystyle\frac{3x^2-2}{x^2-2x-8} = 3 + \frac{6x+22}{(x-4)(x+2)} = 3 + \frac{A}{x-4} + \frac{B}{x+2} \Rightarrow 6x + 22 = A(x+2) + B(x-4)$. Setting

$x = 4$ gives $46 = 6A$, so $A = \frac{23}{3}$. Setting $x = -2$ gives $10 = -6B$, so $B = -\frac{5}{3}$. Now

$$\int \frac{3x^2-2}{x^2-2x-8}\,dx = \int \left(3 + \frac{23/3}{x-4} - \frac{5/3}{x+2}\right)dx = 3x + \frac{23}{3}\ln|x-4| - \frac{5}{3}\ln|x+2| + C.$$

25. Let $u = \ln(\sin x)$. Then $du = \cot x\, dx$ $\Rightarrow$ $\int \cot x \ln(\sin x)\, dx = \int u\, du = \frac{1}{2}u^2 + C = \frac{1}{2}[\ln(\sin x)]^2 + C$.

27. $\int_{-3}^{3} |x^3 + x^2 - 2x|\, dx = \int_{-3}^{3} |(x+2)x(x-1)|\, dx$

$\qquad = -\int_{-3}^{-2}(x^3 + x^2 - 2x)\, dx + \int_{-2}^{0}(x^3 + x^2 - 2x)\, dx - \int_{0}^{1}(x^3 + x^2 - 2x)\, dx + \int_{1}^{3}(x^3 + x^2 - 2x)\, dx$

Let $f(x) = \frac{1}{4}x^4 + \frac{1}{3}x^3 - x^2$. Then $f'(x) = x^3 + x^2 - 2x$, so

$\qquad \int_{-3}^{3} |x^3 + x^2 - 2x|\, dx = -f(-2) + f(-3) + f(0) - f(-2) - f(1) + f(0) + f(3) - f(1)$

$\qquad = f(-3) - 2f(-2) + 2f(0) - 2f(1) + f(3) = \frac{9}{4} - 2\left(-\frac{8}{3}\right) + 2 \cdot 0 - 2\left(-\frac{5}{12}\right) + \frac{81}{4} = \frac{86}{3}$

29. As in Example 5,

$$\int \sqrt{\frac{1+x}{1-x}}\, dx = \int \frac{\sqrt{1+x}}{\sqrt{1-x}} \cdot \frac{\sqrt{1+x}}{\sqrt{1+x}}\, dx = \int \frac{1+x}{\sqrt{1-x^2}}\, dx = \int \frac{dx}{\sqrt{1-x^2}} + \int \frac{x\, dx}{\sqrt{1-x^2}}$$

$$= \sin^{-1} x - \sqrt{1-x^2} + C$$

Another Method: Substitute $u = \sqrt{(1+x)/(1-x)}$.

31. $\displaystyle\int_0^5 \frac{3w-1}{w+2}\, dw = \int_0^5 \left(3 - \frac{7}{w+2}\right) dw = [3w - 7\ln|w+2|]_0^5 = 15 - 7\ln 7 + 7\ln 2 = 15 - 7\ln\frac{7}{2}$

33. $I = \int e^{2x} \sin 3x\, dx = -\frac{1}{3}e^{2x}\cos 3x + \frac{2}{3}\int e^{2x}\cos 3x\, dx$ $\quad\begin{bmatrix} u = e^{2x},\ dv = \sin 3x\, dx, \\ du = 2e^{2x},\ v = -\frac{1}{3}\cos 3x \end{bmatrix}$

$\qquad = -\frac{1}{3}e^{2x}\cos 3x + \frac{2}{3}\left(\frac{1}{3}e^{2x}\sin 3x - \frac{2}{3}\int e^{2x}\sin 3x\, dx\right)$ $\quad\begin{bmatrix} U = e^{2x},\ dV = \cos 3x\, dx, \\ dU = 2e^{2x},\ V = \frac{1}{3}\sin 3x \end{bmatrix}$

So $I = -\frac{1}{3}e^{2x}\cos 3x + \frac{2}{9}e^{2x}\sin 3x - \frac{4}{9}I$ $\Rightarrow$ $\frac{13}{9}I = \frac{1}{9}e^{2x}(2\sin 3x - 3\cos 3x) + C_1$ and $I = \frac{1}{13}e^{2x}(2\sin 3x - 3\cos 3x) + C$, where $C = \frac{9}{13}C_1$.

35. Because $f(x) = x^8 \sin x$ is the product of an even function and an odd function, it is odd. Therefore, $\int_{-1}^{1} x^8 \sin x\, dx = 0$ [by (5.5.7)(b)].

37. $\int_0^{\pi/4} \cos^2\theta \tan^2\theta\, d\theta = \int_0^{\pi/4} \sin^2\theta\, d\theta = \int_0^{\pi/4} \frac{1}{2}(1 - \cos 2\theta)\, d\theta = \left[\frac{1}{2}\theta - \frac{1}{4}\sin 2\theta\right]_0^{\pi/4}$

$\qquad = \left(\frac{\pi}{8} - \frac{1}{4}\right) - (0 - 0) = \frac{\pi}{8} - \frac{1}{4}$

39. Let $u = 1 - x^2$. Then $du = -2x\, dx$ $\Rightarrow$

$$\int \frac{x\, dx}{1 - x^2 + \sqrt{1-x^2}} = -\frac{1}{2}\int \frac{du}{u + \sqrt{u}} = -\int \frac{v\, dv}{v^2 + v} \quad (v = \sqrt{u},\ u = v^2,\ du = 2v\, dv)$$

$$= -\int \frac{dv}{v+1} = -\ln|v+1| + C = -\ln\left(\sqrt{1-x^2} + 1\right) + C$$

41. Let $u = \theta$, $dv = \tan^2\theta\, d\theta = (\sec^2\theta - 1)\, d\theta$ $\Rightarrow$ $du = d\theta$ and $v = \tan\theta - \theta$. So

$$\int \theta \tan^2\theta\, d\theta = \theta(\tan\theta - \theta) - \int(\tan\theta - \theta)\, d\theta = \theta\tan\theta - \theta^2 - \ln|\sec\theta| + \frac{1}{2}\theta^2 + C$$

$$= \theta\tan\theta - \frac{1}{2}\theta^2 - \ln|\sec\theta| + C$$

43. Let $t = x^3$. Then $dt = 3x^2\, dx$ $\Rightarrow$ $I = \int x^5 e^{-x^3}\, dx = \frac{1}{3}\int t e^{-t}\, dt$. Now integrate by parts with $u = t$, $dv = e^{-t}\, dt$: $I = -\frac{1}{3}te^{-t} + \frac{1}{3}\int e^{-t}\, dt = -\frac{1}{3}te^{-t} - \frac{1}{3}e^{-t} + C = -\frac{1}{3}e^{-x^3}(x^3 + 1) + C$.

45. $\int \dfrac{x+a}{x^2+a^2}\,dx = \dfrac{1}{2}\int \dfrac{2x\,dx}{x^2+a^2} + a\int \dfrac{dx}{x^2+a^2} = \dfrac{1}{2}\ln\left(x^2+a^2\right) + a\cdot\dfrac{1}{a}\tan^{-1}\left(\dfrac{x}{a}\right) + C$

$\qquad = \ln\sqrt{x^2+a^2} + \tan^{-1}(x/a) + C$

47. $\int \sin^2 \pi x \cos^4 \pi x\,dx = \int \dfrac{1}{2}(1-\cos 2\pi x)\left[\dfrac{1}{2}(1+\cos 2\pi x)\right]^2 dx$

$\qquad = \dfrac{1}{8}\int (1-\cos 2\pi x)\left(1+2\cos 2\pi x + \cos^2 2\pi x\right)dx$

$\qquad = \dfrac{1}{8}\int \left(1 + 2\cos 2\pi x + \cos^2 2\pi x - \cos 2\pi x - 2\cos^2 2\pi x - \cos^3 2\pi x\right)dx$

$\qquad = \dfrac{1}{8}\int \left(1 + \cos 2\pi x - \cos^2 2\pi x - \cos^3 2\pi x\right)dx$

$\qquad = \dfrac{1}{8}\int \left[1 + \cos 2\pi x - \dfrac{1}{2}(1+\cos 4\pi x)\right]dx - \dfrac{1}{8}\int \cos^3 2\pi x\,dx$

$\qquad = \dfrac{1}{8}\int \left(\dfrac{1}{2} + \cos 2\pi x - \dfrac{1}{2}\cos 4\pi x\right)dx - \dfrac{1}{8}\int (1-\sin^2 2\pi x)\cos 2\pi x\,dx$

$\qquad = \dfrac{1}{8}\left(\dfrac{1}{2}x + \dfrac{1}{2\pi}\sin 2\pi x - \dfrac{1}{8\pi}\sin 4\pi x\right) - \dfrac{1}{8}\int (1-u^2)\dfrac{1}{2\pi}\,du \quad [u = \sin 2\pi x,\ du = 2\pi\cos 2\pi x\,dx]$

$\qquad = \dfrac{1}{16}x + \dfrac{1}{16\pi}\sin 2\pi x - \dfrac{1}{64\pi}\sin 4\pi x - \dfrac{1}{16\pi}\left(u - \dfrac{1}{3}u^3\right) + C$

$\qquad = \dfrac{1}{16}x + \dfrac{1}{16\pi}\sin 2\pi x - \dfrac{1}{64\pi}\sin 4\pi x - \dfrac{1}{16\pi}\sin 2\pi x + \dfrac{1}{48\pi}\sin^3 2\pi x + C$

$\qquad = \dfrac{1}{16}x - \dfrac{1}{64\pi}\sin 4\pi x + \dfrac{1}{48\pi}\sin^3 2\pi x + C$

49. Let $u = \sqrt{4x+1} \;\Rightarrow\; u^2 = 4x+1 \;\Rightarrow\; 2u\,du = 4\,dx \;\Rightarrow\; dx = \dfrac{1}{2}u\,du.$ So

$$\int \dfrac{1}{x\sqrt{4x+1}}\,dx = \int \dfrac{\frac{1}{2}u\,du}{\frac{1}{4}(u^2-1)u} = 2\int \dfrac{du}{u^2-1} = 2\left(\dfrac{1}{2}\right)\ln\left|\dfrac{u-1}{u+1}\right| + C \quad \text{(by Formula 19)}$$

$$= \ln\left|\dfrac{\sqrt{4x+1}-1}{\sqrt{4x+1}+1}\right| + C$$

51. Let $2x = \tan\theta \;\Rightarrow\; x = \dfrac{1}{2}\tan\theta,\ dx = \dfrac{1}{2}\sec^2\theta\,d\theta,\ \sqrt{4x^2+1} = \sec\theta,$ so

$$\int \dfrac{dx}{x\sqrt{4x^2+1}} = \int \dfrac{\frac{1}{2}\sec^2\theta\,d\theta}{\frac{1}{2}\tan\theta\sec\theta} = \int \dfrac{\sec\theta}{\tan\theta}\,d\theta = \int \csc\theta\,d\theta$$

$$= -\ln|\csc\theta + \cot\theta| + C \quad [\text{or } \ln|\csc\theta - \cot\theta| + C]$$

$$= -\ln\left|\dfrac{\sqrt{4x^2+1}}{2x} + \dfrac{1}{2x}\right| + C \quad \left[\text{or } \ln\left|\dfrac{\sqrt{4x^2+1}}{2x} - \dfrac{1}{2x}\right| + C\right]$$

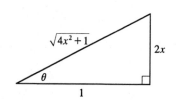

53. $\int x^2 \sinh(mx)\,dx = \dfrac{1}{m}x^2\cosh(mx) - \dfrac{2}{m}\int x\cosh(mx)\,dx \quad \begin{bmatrix} u = x^2,\ dv = \sinh(mx)\,dx, \\ du = 2x\,dx,\ v = \dfrac{1}{m}\cosh(mx) \end{bmatrix}$

$\qquad = \dfrac{1}{m}x^2\cosh(mx) - \dfrac{2}{m}\left(\dfrac{1}{m}x\sinh(mx) - \dfrac{1}{m}\int \sinh(mx)\,dx\right) \quad \begin{bmatrix} U = x,\ dV = \cosh(mx)\,dx, \\ dU = dx,\ V = \dfrac{1}{m}\sinh(mx) \end{bmatrix}$

$\qquad = \dfrac{1}{m}x^2\cosh(mx) - \dfrac{2}{m^2}x\sinh(mx) + \dfrac{2}{m^3}\cosh(mx) + C$

55. Let $u = \sqrt{x+1}.$ Then $x = u^2 - 1 \;\Rightarrow\;$

$$\int \dfrac{dx}{x+4+4\sqrt{x+1}} = \int \dfrac{2u\,du}{u^2+3+4u} = \int \left[\dfrac{-1}{u+1} + \dfrac{3}{u+3}\right]du$$

$$= 3\ln|u+3| - \ln|u+1| + C = 3\ln\left(\sqrt{x+1}+3\right) - \ln\left(\sqrt{x+1}+1\right) + C$$

57. Let $u = \sqrt[3]{x+c}$. Then $x = u^3 - c$ $\Rightarrow$

$$\int x\sqrt[3]{x+c}\,dx = \int (u^3 - c)\,u \cdot 3u^2\,du = 3\int (u^6 - cu^3)\,du = \tfrac{3}{7}u^7 - \tfrac{3}{4}cu^4 + C$$
$$= \tfrac{3}{7}(x+c)^{7/3} - \tfrac{3}{4}c\,(x+c)^{4/3} + C$$

59. Let $u = e^x$. Then $x = \ln u$, $dx = du/u$ $\Rightarrow$

$$\int \frac{dx}{e^{3x} - e^x} = \int \frac{du/u}{u^3 - u} = \int \frac{du}{(u-1)\,u^2\,(u+1)} = \int \left[\frac{1/2}{u-1} - \frac{1}{u^2} - \frac{1/2}{u+1} \right] du$$
$$= \frac{1}{u} + \frac{1}{2}\ln\left|\frac{u-1}{u+1}\right| + C = e^{-x} + \frac{1}{2}\ln\left|\frac{e^x - 1}{e^x + 1}\right| + C$$

61. Let $u = x^5$. Then $du = 5x^4\,dx$ $\Rightarrow$

$$\int \frac{x^4\,dx}{x^{10} + 16} = \int \frac{\tfrac{1}{5}\,du}{u^2 + 16} = \tfrac{1}{5}\cdot\tfrac{1}{4}\tan^{-1}\left(\tfrac{1}{4}u\right) + C = \tfrac{1}{20}\tan^{-1}\left(\tfrac{1}{4}x^5\right) + C.$$

63. Let $u = \csc 2x$. Then $du = -2\cot 2x\,\csc 2x\,dx$ $\Rightarrow$

$$\int \cot^3 2x\,\csc^3 2x\,dx = \int \csc^2 2x\,(\csc^2 2x - 1)\cot 2x\,\csc 2x\,dx = \int u^2\,(u^2 - 1)\left(-\tfrac{1}{2}\,du\right)$$
$$= -\frac{1}{2}\int \left(u^4 - u^2\right)du = -\frac{1}{2}\left[\tfrac{1}{5}u^5 - \tfrac{1}{3}u^3\right] + C = \tfrac{1}{6}\csc^3 2x - \tfrac{1}{10}\csc^5 2x + C$$

65. $\displaystyle\int \frac{dx}{\sqrt{x+1} + \sqrt{x}} = \int \left(\sqrt{x+1} - \sqrt{x}\right)dx = \frac{2}{3}\left[(x+1)^{3/2} - x^{3/2}\right] + C$

67. Let $u = \sqrt{t}$. Then $du = dt/\left(2\sqrt{t}\right)$ $\Rightarrow$

$$\int \frac{\arctan\sqrt{t}}{\sqrt{t}}\,dt = \int \tan^{-1} u\,2\,du = 2u\tan^{-1} u - \int \frac{2u\,du}{1 + u^2}\ \ \text{(by parts)}$$
$$= 2u\tan^{-1} u - \ln\left(1 + u^2\right) + C = 2\sqrt{t}\tan^{-1}\sqrt{t} - \ln(1 + t) + C$$

69. Let $u = e^x$. Then $x = \ln u$, $dx = du/u$ $\Rightarrow$

$$\int \frac{e^{2x}}{1 + e^x}\,dx = \int \frac{u^2}{1 + u}\frac{du}{u} = \int \frac{u}{1 + u}\,du = \int \left(1 - \frac{1}{1 + u}\right)du$$
$$= u - \ln|1 + u| + C = e^x - \ln(1 + e^x) + C$$

71. $\displaystyle\frac{x}{x^4 + 4x^2 + 3} = \frac{x}{(x^2 + 3)\,(x^2 + 1)} = \frac{Ax + B}{x^2 + 3} + \frac{Cx + D}{x^2 + 1}$ $\Rightarrow$

$$x = (Ax + B)\left(x^2 + 1\right) + (Cx + D)\left(x^2 + 3\right) = \left(Ax^3 + Bx^2 + Ax + B\right) + \left(Cx^3 + Dx^2 + 3Cx + 3D\right)$$
$$= (A + C)x^3 + (B + D)x^2 + (A + 3C)x + (B + 3D) \Rightarrow$$

$A + C = 0$, $B + D = 0$, $A + 3C = 1$, $B + 3D = 0$ $\Rightarrow$ $A = -\tfrac{1}{2}$, $C = \tfrac{1}{2}$, $B = 0$, $D = 0$. Thus,

$$\int \frac{x}{x^4 + 4x^2 + 3}\,dx = \int \left(\frac{-\tfrac{1}{2}x}{x^2 + 3} + \frac{\tfrac{1}{2}x}{x^2 + 1}\right)dx$$
$$= -\tfrac{1}{4}\ln\left(x^2 + 3\right) + \tfrac{1}{4}\ln\left(x^2 + 1\right) + C \ \text{or} \ \frac{1}{4}\ln\left(\frac{x^2 + 1}{x^2 + 3}\right) + C$$

73. $\dfrac{1}{(x-2)(x^2+4)} = \dfrac{A}{x-2} + \dfrac{Bx+C}{x^2+4} \Rightarrow$

$1 = A(x^2+4) + (Bx+C)(x-2) = (A+B)x^2 + (C-2B)x + (4A-2C)$. So $0 = A+B = C-2B$,

$1 = 4A - 2C$. Setting $x=2$ gives $A = \frac{1}{8} \Rightarrow B = -\frac{1}{8}$ and $C = -\frac{1}{4}$. So

$$\int \frac{1}{(x-2)(x^2+4)}\,dx = \int \left(\frac{\frac{1}{8}}{x-2} + \frac{-\frac{1}{8}x - \frac{1}{4}}{x^2+4} \right) dx = \frac{1}{8}\int \frac{dx}{x-2} - \frac{1}{16}\int \frac{2x\,dx}{x^2+4} - \frac{1}{4}\int \frac{dx}{x^2+4}$$

$$= \tfrac{1}{8}\ln|x-2| - \tfrac{1}{16}\ln\left(x^2+4\right) - \tfrac{1}{8}\tan^{-1}(x/2) + C$$

75. $\int \sin x \sin 2x \sin 3x\,dx = \int \sin x \cdot \frac{1}{2}\left[\cos(2x-3x) - \cos(2x+3x)\right] dx = \frac{1}{2}\int (\sin x \cos x - \sin x \cos 5x)\,dx$

$$= \tfrac{1}{4}\int \sin 2x\,dx - \tfrac{1}{2}\int \tfrac{1}{2}\left[\sin(x+5x) + \sin(x-5x)\right] dx$$

$$= -\tfrac{1}{8}\cos 2x - \tfrac{1}{4}\int (\sin 6x - \sin 4x)\,dx = -\tfrac{1}{8}\cos 2x + \tfrac{1}{24}\cos 6x - \tfrac{1}{16}\cos 4x + C$$

7.6 Integration Using Tables and Computer Algebra Systems

Keep in mind that there are several ways to approach many of these exercises, and different methods can lead to different forms of the answer.

1. We could make the substitution $u = \sqrt{2}x$ to obtain the radical $\sqrt{7-u^2}$ and then use Formula 33 with $a = \sqrt{7}$. Alternatively, we will factor $\sqrt{2}$ out of the radical and use $a = \sqrt{\frac{7}{2}}$.

$$\int \frac{\sqrt{7-2x^2}}{x^2}\,dx = \sqrt{2}\int \frac{\sqrt{\frac{7}{2}-x^2}}{x^2}\,dx \overset{33}{=} \sqrt{2}\left[-\frac{1}{x}\sqrt{\tfrac{7}{2}-x^2} - \sin^{-1}\frac{x}{\sqrt{\frac{7}{2}}} \right] + C$$

$$= -\frac{1}{x}\sqrt{7-2x^2} - \sqrt{2}\sin^{-1}\left(\sqrt{\tfrac{2}{7}}x\right) + C$$

3. Let $u = \pi x \Rightarrow du = \pi\,dx$, so

$$\int \sec^3(\pi x)\,dx = \frac{1}{\pi}\int \sec^3 u\,du \overset{71}{=} \frac{1}{\pi}\left(\tfrac{1}{2}\sec u \tan u + \tfrac{1}{2}\ln|\sec u + \tan u| \right) + C$$

$$= \frac{1}{2\pi}\sec \pi x \tan \pi x + \frac{1}{2\pi}\ln|\sec \pi x + \tan \pi x| + C$$

5. $\displaystyle\int_0^1 2x \cos^{-1} x\,dx \overset{91}{=} 2\left[\frac{2x^2-1}{4}\cos^{-1}x - \frac{x\sqrt{1-x^2}}{4} \right]_0^1 = 2\left[\left(\tfrac{1}{4}\cdot 0 - 0\right) - \left(-\tfrac{1}{4}\cdot\tfrac{\pi}{2} - 0\right) \right] = 2\left(\tfrac{\pi}{8}\right) = \tfrac{\pi}{4}$

7. By Formula 99 with $a = -3$ and $b = 4$,

$$\int e^{-3x}\cos 4x\,dx = \frac{e^{-3x}}{(-3)^2 + 4^2}(-3\cos 4x + 4\sin 4x) + C = \frac{e^{-3x}}{25}(-3\cos 4x + 4\sin 4x) + C.$$

9. Let $u = x^2$. Then $du = 2x\,dx$, so

$$\int x \sin^{-1}(x^2)\,dx = \tfrac{1}{2}\int \sin^{-1} u\,du \overset{87}{=} \tfrac{1}{2}\left(u\sin^{-1} u + \sqrt{1-u^2} \right) + C$$

$$= \tfrac{1}{2}\left(x^2 \sin^{-1}(x^2) + \sqrt{1-x^4} \right) + C$$

11. $\int_{-1}^{0} t^2 e^{-t}\, dt \overset{97}{=} \left[\frac{1}{-1}t^2 e^{-t}\right]_{-1}^{0} - \frac{2}{-1}\int_{-1}^{0} t e^{-t}\, dt = e + 2\int_{-1}^{0} t e^{-t}\, dt \overset{96}{=} e + 2\left[\frac{1}{(-1)^2}(-t-1)e^{-t}\right]_{-1}^{0}$

$$= e + 2\left[-e^0 + 0\right] = e - 2$$

13. Let $u = 3x$. Then $du = 3\,dx$, so

$$\int \frac{\sqrt{9x^2-1}}{x^2}\, dx = \int \frac{\sqrt{u^2-1}}{u^2/9}\frac{du}{3} = 3\int \frac{\sqrt{u^2-1}}{u^2}\, du \overset{42}{=} -\frac{3\sqrt{u^2-1}}{u} + 3\ln\left|u + \sqrt{u^2-1}\right| + C$$

$$= -\frac{\sqrt{9x^2-1}}{x} + 3\ln\left|3x + \sqrt{9x^2-1}\right| + C$$

15. Let $u = e^x$. Then $du = e^x\,dx$, so

$$\int e^x \operatorname{sech}(e^x)\, dx = \int \operatorname{sech} u\, du \overset{107}{=} \tan^{-1}|\sinh u| + C = \tan^{-1}[\sinh(e^x)] + C$$

17. $\int_{-2}^{1} \sqrt{5-4x-x^2}\, dx = \int_{-2}^{1}\sqrt{5-(x^2+4x)}\, dx = \int_{-2}^{1}\sqrt{5+4-(x^2+4x+4)}\, dx$

$$= \int_{-2}^{1}\sqrt{9-(x+2)^2}\, dx = \int_{0}^{3}\sqrt{3^2-u^2}\, du \quad [u=x+2, du=dx]$$

$$\overset{30}{=} \left[\frac{u}{2}\sqrt{9-u^2} + \frac{9}{2}\sin^{-1}\frac{u}{3}\right]_{0}^{3} = \left[\left(0 + \frac{9}{2}\cdot\frac{\pi}{2}\right) - (0+0)\right] = \frac{9\pi}{4}$$

19. Let $u = \sin x$. Then $du = \cos x\,dx$, so

$$\int \sin^2 x\cos x \ln(\sin x)\, dx = \int u^2 \ln u\, du \overset{101}{=} \frac{1}{9}u^3(3\ln u - 1) + C = \frac{1}{9}\sin^3 x\,[3\ln(\sin x) - 1] + C.$$

21. $\displaystyle\int \sqrt{2+3\cos x}\,\tan x\, dx = -\int \frac{\sqrt{2+3\cos x}}{\cos x}(-\sin x\, dx) = -\int \frac{\sqrt{2+3u}}{u}\, du$ $(u = \cos x,\ du = -\sin x\, dx)$

$$\overset{58}{=} -2\sqrt{2+3u} - 2\int \frac{du}{u\sqrt{2+3u}} \overset{57}{=} -2\sqrt{2+3u} - 2\cdot\frac{1}{\sqrt{2}}\ln\left|\frac{\sqrt{2+3u}-\sqrt{2}}{\sqrt{2+3u}+\sqrt{2}}\right| + C$$

$$= -2\sqrt{2+3\cos x} - \sqrt{2}\ln\left|\frac{\sqrt{2+3\cos x}-\sqrt{2}}{\sqrt{2+3\cos x}+\sqrt{2}}\right| + C$$

23. $\int \sec^5 x\, dx \overset{77}{=} \frac{1}{4}\tan x\sec^3 x + \frac{3}{4}\int \sec^3 x\, dx \overset{77}{=} \frac{1}{4}\tan x\sec^3 x + \frac{3}{4}\left(\frac{1}{2}\tan x\sec x + \frac{1}{2}\int \sec x\, dx\right)$

$$\overset{14}{=} \frac{1}{4}\tan x\sec^3 x + \frac{3}{8}\tan x\sec x + \frac{3}{8}\ln|\sec x + \tan x| + C$$

25. $\int_{0}^{\pi/2}\cos^5 x\, dx \overset{74}{=} \frac{1}{5}\left[\cos^4 x\sin x\right]_{0}^{\pi/2} + \frac{4}{5}\int_{0}^{\pi/2}\cos^3 x\, dx \overset{68}{=} 0 + \frac{4}{5}\left[\frac{1}{3}(2+\cos^2 x)\sin x\right]_{0}^{\pi/2} = \frac{4}{15}(2-0) = \frac{8}{15}$

27. Let $u = e^x$ $\Rightarrow$ $\ln u = x$ $\Rightarrow$ $dx = \dfrac{du}{u}$. Then

$$\int \sqrt{e^{2x}-1}\, dx = \int \frac{\sqrt{u^2-1}}{u}\, du \overset{41}{=} \sqrt{u^2-1} - \cos^{-1}(1/u) + C = \sqrt{e^{2x}-1} - \cos^{-1}(e^{-x}) + C.$$

29. Let $u = x^5$, $du = 5x^4\,dx$. Then

$$\int \frac{x^4\, dx}{\sqrt{x^{10}-2}} = \frac{1}{5}\int \frac{du}{\sqrt{u^2-2}} \overset{43}{=} \frac{1}{5}\ln\left|u + \sqrt{u^2-2}\right| + C = \frac{1}{5}\ln\left|x^5 + \sqrt{x^{10}-2}\right| + C.$$

31. Volume $= \displaystyle\int_{0}^{1}\frac{2\pi x}{(1+5x)^2}\, dx \overset{51}{=} 2\pi\left[\frac{1}{25(1+5x)} + \frac{1}{25}\ln|1+5x|\right]_{0}^{1} = \frac{2\pi}{25}\left(\frac{1}{6} + \ln 6 - 1 - \ln 1\right)$

$$= \frac{2\pi}{25}\left(\ln 6 - \frac{5}{6}\right)$$

33. (a) $\dfrac{d}{du}\left[\dfrac{1}{b^3}\left(a+bu-\dfrac{a^2}{a+bu}-2a\ln|a+bu|\right)+C\right]=\dfrac{1}{b^3}\left[b+\dfrac{ba^2}{(a+bu)^2}-\dfrac{2ab}{(a+bu)}\right]$

$$=\dfrac{1}{b^3}\left[\dfrac{b\,(a+bu)^2+ba^2-(a+bu)\,2ab}{(a+bu)^2}\right]=\dfrac{1}{b^3}\left[\dfrac{b^3u^2}{(a+bu)^2}\right]=\dfrac{u^2}{(a+bu)^2}$$

(b) Let $t=a+bu \Rightarrow dt=b\,du$.

$$\int\dfrac{u^2\,du}{(a+bu)^2}=\dfrac{1}{b^3}\int\dfrac{(t-a)^2}{t^2}\,dt=\dfrac{1}{b^3}\int\left(1-\dfrac{2a}{t}+\dfrac{a^2}{t^2}\right)dt=\dfrac{1}{b^3}\left(t-2a\ln|t|-\dfrac{a^2}{t}\right)+C$$

$$=\dfrac{1}{b^3}\left(a+bu-\dfrac{a^2}{a+bu}-2a\ln|a+bu|\right)+C$$

35. Maple, Mathematica and Derive all give $\int x^2\sqrt{5-x^2}\,dx=-\frac{1}{4}x\,(5-x^2)^{3/2}+\frac{5}{8}x\sqrt{5-x^2}+\frac{25}{8}\sin^{-1}\left(\frac{1}{\sqrt{5}}x\right)$.

Using Formula 31, we get $\int x^2\sqrt{5-x^2}\,dx=\frac{1}{8}x\,(2x^2-5)\sqrt{5-x^2}+\frac{1}{8}\,(5^2)\sin^{-1}\left(\frac{1}{\sqrt{5}}x\right)+C$. But

$-\frac{1}{4}x\,(5-x^2)^{3/2}+\frac{5}{8}x\sqrt{5-x^2}=\frac{1}{8}x\sqrt{5-x^2}\,[5-2\,(5-x^2)]=\frac{1}{8}x\,(2x^2-5)\sqrt{5-x^2}$, and the $\sin^{-1}$ terms are the same in each expression, so the answers are equivalent.

37. Maple and Derive both give $\int\sin^3 x\cos^2 x\,dx=-\frac{1}{5}\sin^2 x\cos^3 x-\frac{2}{15}\cos^3 x$ (although Derive factors the expression), and Mathematica gives $\int\sin^3 x\cos^2 x\,dx=-\frac{1}{8}\cos x-\frac{1}{48}\cos 3x+\frac{1}{80}\cos 5x$. We can use a CAS to show that both of these expressions are equal to $-\frac{1}{3}\cos^3 x+\frac{1}{5}\cos^5 x$. Using Formula 86, we write

$$\int\sin^3 x\cos^2 x\,dx=-\frac{1}{5}\sin^2 x\cos^3 x+\frac{2}{5}\int\sin x\cos^2 x\,dx=-\frac{1}{5}\sin^2 x\cos^3 x+\frac{2}{5}\left(-\frac{1}{3}\cos^3 x\right)+C$$

$$=-\frac{1}{5}\sin^2 x\cos^3 x-\frac{2}{15}\cos^3 x+C$$

39. Maple gives $\int x\sqrt{1+2x}\,dx=\frac{1}{10}\,(1+2x)^{5/2}-\frac{1}{6}\,(1+2x)^{3/2}$, Mathematica gives $\sqrt{1+2x}\left(\frac{2}{5}x^2+\frac{1}{15}x-\frac{1}{15}\right)$, and Derive gives $\frac{1}{15}\,(1+2x)^{3/2}\,(3x-1)$. The first two expressions can be simplified to Derive's result. If we use Formula 54, we get

$$\int x\sqrt{1+2x}\,dx=\frac{2}{15(2)^2}\,(3\cdot 2x-2\cdot 1)\,(1+2x)^{3/2}+C=\frac{1}{30}\,(6x-2)\,(1+2x)^{3/2}+C$$

$$=\frac{1}{15}\,(3x-1)\,(1+2x)^{3/2}$$

41. Maple gives $\int\tan^5 x\,dx=\frac{1}{4}\tan^4 x-\frac{1}{2}\tan^2 x+\frac{1}{2}\ln\left(1+\tan^2 x\right)$, Mathematica gives $\int\tan^5 x\,dx=\frac{1}{4}\,[-1-2\cos(2x)]\sec^4 x-\ln(\cos x)$, and Derive gives $\int\tan^5 x\,dx=\frac{1}{4}\tan^4 x-\frac{1}{2}\tan^2 x-\ln(\cos x)$. These expressions are equivalent, and none includes absolute value bars or a constant of integration. Note that Mathematica's and Derive's expressions suggest that the integral is undefined where $\cos x<0$, which is not the case.

Using Formula 75, $\int\tan^5 x\,dx=\frac{1}{5-1}\tan^{5-1}x-\int\tan^{5-2}x\,dx=\frac{1}{4}\tan^4 x-\int\tan^3 x\,dx$. Using Formula 69, $\int\tan^3 x\,dx=\frac{1}{2}\tan^2 x+\ln|\cos x|+C$, so $\int\tan^5 x\,dx=\frac{1}{4}\tan^4 x-\frac{1}{2}\tan^2 x-\ln|\cos x|+C$.

43. Derive gives $I=\int 2^x\sqrt{4^x-1}\,dx=\dfrac{2^{x-1}\sqrt{2^{2x}-1}}{\ln 2}-\dfrac{\ln\left(\sqrt{2^{2x}-1}+2^x\right)}{2\ln 2}$ immediately. Neither Maple nor Mathematica is able to evaluate I in its given form. However, if we instead write I as $\int 2^x\sqrt{(2^x)^2-1}\,dx$, both systems give the same answer as Derive (after minor simplification). Our trick works because the CAS now recognizes 2^x as a promising substitution.

45. Maple gives the antiderivative

$$F(x) = \int \frac{x^2 - 1}{x^4 + x^2 + 1}\, dx = -\tfrac{1}{2}\ln\left(x^2 + x + 1\right) + \tfrac{1}{2}\ln\left(x^2 - x + 1\right).$$

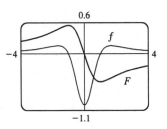

We can see that at 0, this antiderivative is 0. From the graphs, it appears that F has a maximum at $x = -1$ and a minimum at $x = 1$ [since $F'(x) = f(x)$ changes sign at these x-values], and that F has inflection points at $x \approx -1.7$, $x = 0$ and $x \approx 1.7$ [since $f(x)$ has extrema at these x-values].

47. Since f is everywhere positive, we know that its antiderivative F is increasing. Maple gives

$$\int \sin^4 x \cos^6 x\, dx = -\tfrac{1}{10}\sin^3 x \cos^7 x - \tfrac{3}{80}\sin x \cos^7 x + \tfrac{1}{160}\cos^5 x \sin x$$
$$+ \tfrac{1}{128}\cos^3 x \sin x + \tfrac{3}{256}\cos x \sin x + \tfrac{3}{256}x$$

and this is 0 at $x = 0$.

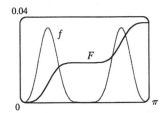

Approximate Integration

1. (a) $L_2 = \sum\limits_{i=1}^{2} f(x_{i-1})\,\Delta x = f(x_0)\cdot 2 + f(x_1)\cdot 2 = 2\,[f(0) + f(2)] = 2\,(0.5 + 2.5) = 6$

$R_2 = \sum\limits_{i=1}^{2} f(x_i)\,\Delta x = f(x_1)\cdot 2 + f(x_2)\cdot 2 = 2\,[f(2) + f(4)] = 2\,(2.5 + 3.5) = 12$

$M_2 = \sum\limits_{i=1}^{2} f(\overline{x}_i)\,\Delta x = f(\overline{x}_1)\cdot 2 + f(\overline{x}_2)\cdot 2 = 2\,[f(1) + f(3)] \approx 2\,(1.7 + 3.2) = 9.8$

(b)

L_2 is an underestimate, since the area under the small rectangles is less than the area under the curve, and R_2 is an overestimate, since the area under the large rectangles is greater than the area under the curve. It appears that M_2 is an overestimate, though it is fairly close to I. See the solution to Exercise 43 for a proof of the fact that if f is concave down on (a, b), then the Midpoint Rule is an overestimate of $\int_a^b f(x)\,dx$.

(c) $T_2 = \left(\tfrac{1}{2}\Delta x\right)[f(x_0) + 2f(x_1) + f(x_2)] = \tfrac{1}{2}\,[f(0) + 2f(2) + f(4)] = 0.5 + 2\,(2.5) + 3.5 = 9$.

This approximation is an underestimate, since the graph is concave down. See the solution to Exercise 43 for a general proof of this conclusion.

(d) For any n, we will have $L_n < T_n < I < M_n < R_n$.

3. $f(x) = \cos(x^2)$, $\Delta x = \frac{1-0}{4} = \frac{1}{4}$

(a) $T_4 = \frac{1}{4 \cdot 2}\left[f(0) + 2f\left(\frac{1}{4}\right) + 2f\left(\frac{2}{4}\right) + 2f\left(\frac{3}{4}\right) + f(1)\right] \approx 0.895759$

(b) $M_4 = \frac{1}{4}\left[f\left(\frac{1}{8}\right) + f\left(\frac{3}{8}\right) + f\left(\frac{5}{8}\right) + f\left(\frac{7}{8}\right)\right] \approx 0.908907$

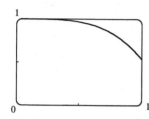

The graph shows that f is concave down on $[0, 1]$. So T_4 is an underestimate and M_4 is an overestimate. We can conclude that $0.895759 < \int_0^1 \cos(x^2)\,dx < 0.908907$.

5. $f(x) = x^2 \sin x$, $\Delta x = \frac{b-a}{n} = \frac{\pi - 0}{8} = \frac{\pi}{8}$

(a) $M_8 = \frac{\pi}{8}\left[f\left(\frac{\pi}{16}\right) + f\left(\frac{3\pi}{16}\right) + f\left(\frac{5\pi}{16}\right) + \cdots + f\left(\frac{15\pi}{16}\right)\right] \approx 5.932957$

(b) $S_8 = \frac{\pi}{8 \cdot 3}\left[f(0) + 4f\left(\frac{\pi}{8}\right) + 2f\left(\frac{2\pi}{8}\right) + 4f\left(\frac{3\pi}{8}\right) + 2f\left(\frac{4\pi}{8}\right) + 4f\left(\frac{5\pi}{8}\right) + 2f\left(\frac{6\pi}{8}\right) + 4f\left(\frac{7\pi}{8}\right) + f(\pi)\right]$
≈ 5.869247

Actual: $\int_0^\pi x^2 \sin x\,dx \overset{84}{=} \left[-x^2 \cos x\right]_0^\pi + 2\int_0^\pi x \cos x\,dx \overset{83}{=} \left[-\pi^2(-1) - 0\right] + 2\left[\cos x + x \sin x\right]_0^\pi$
$= \pi^2 + 2\left[(-1 + 0) - (1 + 0)\right] = \pi^2 - 4 \approx 5.869604$

Errors: $E_M = $ actual $- M_8 = \int_0^\pi x^2 \sin x\,dx - M_8 \approx -0.063353$
$E_S = $ actual $- S_8 = \int_0^\pi x^2 \sin x\,dx - S_8 \approx 0.000357$

7. $f(x) = \sqrt{1 + x^3}$, $\Delta x = \frac{1 - (-1)}{8} = \frac{1}{4}$

(a) $T_8 = \frac{0.25}{2}\left[f(-1) + 2f\left(-\frac{3}{4}\right) + 2f\left(-\frac{1}{2}\right) + \cdots + 2f\left(\frac{1}{2}\right) + 2f\left(\frac{3}{4}\right) + f(1)\right] \approx 1.913972$

(b) $S_8 = \frac{0.25}{3}\left[f(-1) + 4f\left(-\frac{3}{4}\right) + 2f\left(-\frac{1}{2}\right) + 4f\left(-\frac{1}{4}\right) + 2f(0) + 4f\left(\frac{1}{4}\right) + 2f\left(\frac{1}{2}\right) + 4f\left(\frac{3}{4}\right) + f(1)\right]$
≈ 1.934766

9. $f(x) = \frac{\sin x}{x}$, $\Delta x = \frac{\pi - \pi/2}{6} = \frac{\pi}{12}$

(a) $T_6 = \frac{\pi}{24}\left[f\left(\frac{\pi}{2}\right) + 2f\left(\frac{7\pi}{12}\right) + 2f\left(\frac{2\pi}{3}\right) + 2f\left(\frac{3\pi}{4}\right) + 2f\left(\frac{5\pi}{6}\right) + 2f\left(\frac{11\pi}{12}\right) + f(\pi)\right] \approx 0.481672$

(b) $S_6 = \frac{\pi}{36}\left[f\left(\frac{\pi}{2}\right) + 4f\left(\frac{7\pi}{12}\right) + 2f\left(\frac{2\pi}{3}\right) + 4f\left(\frac{3\pi}{4}\right) + 2f\left(\frac{5\pi}{6}\right) + 4f\left(\frac{11\pi}{12}\right) + f(\pi)\right] \approx 0.481172$

11. $f(x) = e^{-x^2}$, $\Delta x = \frac{1 - 0}{10} = \frac{1}{10}$

(a) $T_{10} = \frac{1}{10 \cdot 2}\left[f(0) + 2f(0.1) + 2f(0.2) + \cdots + 2f(0.8) + 2f(0.9) + f(1)\right] \approx 0.746211$

(b) $M_{10} = \frac{1}{10}\left[f(0.05) + f(0.15) + f(0.25) + \cdots + f(0.75) + f(0.85) + f(0.95)\right] \approx 0.747131$

(c) $S_{10} = \frac{1}{10 \cdot 3}\left[f(0) + 4f(0.1) + 2f(0.2) + 4f(0.3) + 2f(0.4) + 4f(0.5)\right.$
$\left. + 2f(0.6) + 4f(0.7) + 2f(0.8) + 4f(0.9) + f(1)\right] \approx 0.746825$

13. $f(t) = \sin\left(e^{t/2}\right)$, $\Delta t = \dfrac{1/2 - 0}{8} = \dfrac{1}{16}$

(a) $T_8 = \frac{1}{16 \cdot 2}\left[f(0) + 2f\left(\frac{1}{16}\right) + 2f\left(\frac{2}{16}\right) + \cdots + 2f\left(\frac{7}{16}\right) + f\left(\frac{1}{2}\right)\right] \approx 0.451948$

(b) $M_8 = \frac{1}{16}\left[f\left(\frac{1}{32}\right) + f\left(\frac{3}{32}\right) + f\left(\frac{5}{32}\right) + \cdots + f\left(\frac{13}{32}\right) + f\left(\frac{15}{32}\right)\right] \approx 0.451991$

(c) $S_8 = \frac{1}{16 \cdot 3}\left[f(0) + 4f\left(\frac{1}{16}\right) + 2f\left(\frac{2}{16}\right) + \cdots + 4f\left(\frac{7}{16}\right) + f\left(\frac{1}{2}\right)\right] \approx 0.451976$

15. $f(x) = e^{1/x}$, $\Delta x = \dfrac{2 - 1}{4} = \dfrac{1}{4}$

(a) $T_4 = \frac{1}{4 \cdot 2}\left[f(1) + 2f(1.25) + 2f(1.5) + 2f(1.75) + f(2)\right] \approx 2.031893$

(b) $M_4 = \frac{1}{4}\left[f(1.125) + f(1.375) + f(1.625) + f(1.875)\right] \approx 2.014207$

(c) $S_4 = \frac{1}{4 \cdot 3}\left[f(1) + 4f(1.25) + 2f(1.5) + 4f(1.75) + f(2)\right] \approx 2.020651$

17. $f(x) = x^5 e^x$, $\Delta x = \dfrac{1 - 0}{10} = \dfrac{1}{10}$

(a) $T_{10} = \frac{1}{10 \cdot 2}\left[f(0) + 2f(0.1) + 2f(0.2) + \cdots + 2f(0.9) + f(1)\right] \approx 0.409140$

(b) $M_{10} = \frac{1}{10}\left[f(0.05) + f(0.15) + f(0.25) + \cdots + f(0.95)\right] \approx 0.388849$

(c) $S_{10} = \frac{1}{10 \cdot 3}\left[f(0) + 4f(0.1) + 2f(0.2) + 4f(0.3) + 2f(0.4) + 4f(0.5)\right.$
$\left. + 2f(0.6) + 4f(0.7) + 2f(0.8) + 4f(0.9) + f(1)\right] \approx 0.395802$

19. $f(y) = \dfrac{1}{1 + y^5}$, $\Delta y = \dfrac{3 - 0}{6} = \dfrac{1}{2}$

(a) $T_6 = \frac{1}{2 \cdot 2}\left[f(0) + 2f\left(\frac{1}{2}\right) + 2f\left(\frac{2}{2}\right) + 2f\left(\frac{3}{2}\right) + 2f\left(\frac{4}{2}\right) + 2f\left(\frac{5}{2}\right) + f(3)\right] \approx 1.064275$

(b) $M_6 = \frac{1}{2}\left[f\left(\frac{1}{4}\right) + f\left(\frac{3}{4}\right) + f\left(\frac{5}{4}\right) + f\left(\frac{7}{4}\right) + f\left(\frac{9}{4}\right) + f\left(\frac{11}{4}\right)\right] \approx 1.067416$

(c) $S_6 = \frac{1}{2 \cdot 3}\left[f(0) + 4f\left(\frac{1}{2}\right) + 2f\left(\frac{2}{2}\right) + 4f\left(\frac{3}{2}\right) + 2f\left(\frac{4}{2}\right) + 4f\left(\frac{5}{2}\right) + f(3)\right] \approx 1.074915$

21. $f(x) = e^{-x^2}$, $\Delta x = \dfrac{2 - 0}{10} = \dfrac{1}{5}$

(a) $T_{10} = \frac{1}{5 \cdot 2}\{f(0) + 2[f(0.2) + f(0.4) + \cdots + f(1.8)] + f(2)\} \approx 0.881839$

$M_{10} = \frac{1}{5}\left[f(0.1) + f(0.3) + f(0.5) + \cdots + f(1.7) + f(1.9)\right] \approx 0.882202$

(b) $f(x) = e^{-x^2}$, $f'(x) = -2xe^{-x^2}$, $f''(x) = \left(4x^2 - 2\right)e^{-x^2}$, $f'''(x) = 4x\left(3 - 2x^2\right)e^{-x^2}$. $f'''(x) = 0$ $\Leftrightarrow$

$x = 0$ or $x = \pm\sqrt{\frac{3}{2}}$. So to find the maximum value of $\left|f''(x)\right|$ on $[0, 2]$, we need only consider its values at

$x = 0$, $x = 2$, and $x = \sqrt{\frac{3}{2}}$. $|f''(0)| = 2$, $|f''(2)| \approx 0.2564$ and $\left|f''\left(\sqrt{\frac{3}{2}}\right)\right| \approx 0.8925$. Thus, taking $K = 2$,

$a = 0$, $b = 2$, and $n = 10$ in Theorem 3, we get $|E_T| \le 2 \cdot 2^3 / \left(12 \cdot 10^2\right) = \frac{1}{75} = 0.01\overline{3}$, and

$|E_M| \le |E_T|/2 \le 0.00\overline{6}$.

(c) Take $K = 2$ [as in part (b)] in Theorem 3. $|E_T| \le \dfrac{K(b-a)^3}{12n^2} \le 10^{-5}$ $\Leftrightarrow$ $\dfrac{2(2-0)^3}{12n^2} \le 10^{-5}$ $\Leftrightarrow$

$\frac{3}{4}n^2 \ge 10^5$ $\Leftrightarrow$ $n \ge 365.1\ldots$ $\Leftrightarrow$ $n \ge 366$. Take $n = 366$ for T_n. For E_M, again take $K = 2$ in

Theorem 3 to get $|E_M| \le 10^{-5}$ $\Leftrightarrow$ $\frac{3}{2}n^2 \ge 10^5$ $\Leftrightarrow$ $n \ge 258.2$ $\Rightarrow$ $n \ge 259$. Take $n = 259$ for M_n.

23. (a) $T_{10} = \frac{1}{10 \cdot 2} \{f(0) + 2[f(0.1) + f(0.2) + \cdots + f(0.9)] + f(1)\} \approx 1.71971349$

$S_{10} = \frac{1}{10 \cdot 3} [f(0) + 4f(0.1) + 2f(0.2) + 4f(0.3) + \cdots + 4f(0.9) + f(1)] \approx 1.71828278$

Since $I = \int_0^1 e^x \, dx = [e^x]_0^1 = e - 1 \approx 1.71828183$, $E_T = I - T_{10} \approx -0.00143166$ and

$E_S = I - S_{10} \approx -0.00000095$.

(b) $f(x) = e^x \Rightarrow f''(x) = e^x \le e$ for $0 \le x \le 1$. Taking $K = e$, $a = 0$, $b = 1$, and $n = 10$ in Theorem 3, we get $|E_T| \le e(1)^3/(12 \cdot 10^2) \approx 0.002265 > 0.00143166$ [actual $|E_T|$ from (a)]. $f^{(4)}(x) = e^x < e$ for $0 \le x \le 1$. Using Theorem 4, we have $|E_S| \le e(1)^5/(180 \cdot 10^4) \approx 0.0000015 > 0.00000095$ [actual $|E_S|$ from (a)]. We see that the actual errors are about two-thirds the size of the error estimates.

(c) From part (b), we take $K = e$ to get $|E_T| \le \dfrac{K(b-a)^3}{12n^2} \le 0.00001 \Rightarrow n^2 \ge \dfrac{e(1^3)}{12(0.00001)} \Rightarrow$

$n \ge 150.5$. Take $n = 151$ for T_n. Now $|E_M| \le \dfrac{K(b-a)^3}{24n^2} \le 0.00001 \Rightarrow n \ge 106.4$. Take $n = 107$ for

M_n. Finally, $|E_S| \le \dfrac{K(b-a)^5}{180n^4} \le 0.00001 \Rightarrow n^4 \ge \dfrac{e(1^5)}{180(0.00001)} \Rightarrow n \ge 6.23$. Take $n = 8$ for S_n

(since n has to be even for Simpson's Rule).

25. (a) Using the CAS, we differentiate $f(x) = e^{\cos x}$ twice, and find that $f''(x) = e^{\cos x}(\sin^2 x - \cos x)$. From the graph, we see that the maximum value of $|f''(x)|$ occurs at the endpoints of the interval $[0, 2\pi]$. Since $f''(0) = -e$, we can use $K = e$ or $K = 2.8$.

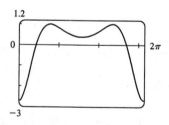

(b) A CAS gives $M_{10} \approx 7.954926518$. (In Maple, use `student[middlesum]`.)

(c) Using Theorem 3 for the Midpoint Rule, with $K = e$, we get $|E_M| \le \dfrac{e(2\pi - 0)^3}{24 \cdot 10^2} \approx 0.280945995$. With

$K = 2.8$, we get $|E_M| \le \dfrac{2.8(2\pi - 0)^3}{24 \cdot 10^2} = 0.289391916$.

(d) A CAS gives $I \approx 7.954926521$.

(e) The actual error is only about 3×10^{-9}, much less than the estimate in part (c).

(f) We use the CAS to differentiate twice more, and then graph
$f^{(4)}(x) = e^{\cos x}\left(\sin^4 x - 6\sin^2 x \cos x + 3 - 7\sin^2 x + \cos x\right)$. From the graph, we see that the maximum
value of $\left|f^{(4)}(x)\right|$ occurs at the endpoints of the interval $[0, 2\pi]$. Since $f^{(4)}(0) = 4e$, we can use $K = 4e$ or
$K = 10.9$.

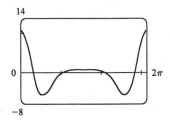

(g) A CAS gives $S_{10} \approx 7.953789422$. (In Maple, use `student[simpson]`.)

(h) Using Theorem 4 with $K = 4e$, we get $|E_S| \leq \dfrac{4e\,(2\pi - 0)^5}{180 \cdot 10^4} \approx 0.059153618$. With $K = 10.9$, we get
$|E_S| \leq \dfrac{10.9\,(2\pi - 0)^5}{180 \cdot 10^4} = 0.059299814.$

(i) The actual error is about $7.954926521 - 7.953789422 \approx 0.00114$. This is quite a bit smaller than the estimate
in part (h), though the difference is not nearly as great as it was in the case of the Midpoint Rule.

(j) To ensure that $|E_S| \leq 0.0001$, we use Theorem 4: $|E_S| \leq \dfrac{4e\,(2\pi)^5}{180 \cdot n^4} \leq 0.0001 \quad\Rightarrow\quad \dfrac{4e\,(2\pi)^5}{180 \cdot 0.0001} \leq n^4 \quad\Rightarrow$
$n^4 \geq 5{,}915{,}362 \quad\Leftrightarrow\quad n \geq 49.3$. So we must take $n \geq 50$ to ensure that $|I - S_n| \leq 0.0001$. ($K = 10.9$ leads to
the same value of n.)

27. $I = \int_0^1 x^3 dx = \left[\frac{1}{4}x^4\right]_0^1 = 0.25.$ $f(x) = x^3.$

$n = 4$: $L_4 = \frac{1}{4}\left[0^3 + \left(\frac{1}{4}\right)^3 + \left(\frac{1}{2}\right)^3 + \left(\frac{3}{4}\right)^3\right] = 0.140625$

$R_4 = \frac{1}{4}\left[\left(\frac{1}{4}\right)^3 + \left(\frac{1}{2}\right)^3 + \left(\frac{3}{4}\right)^3 + 1^3\right] = 0.390625$

$T_4 = \frac{1}{4\cdot 2}\left[0^3 + 2\left(\frac{1}{4}\right)^3 + 2\left(\frac{1}{2}\right)^3 + 2\left(\frac{3}{4}\right)^3 + 1^3\right] = 0.265625,$

$M_4 = \frac{1}{4}\left[\left(\frac{1}{8}\right)^3 + \left(\frac{3}{8}\right)^3 + \left(\frac{5}{8}\right)^3 + \left(\frac{7}{8}\right)^3\right] = 0.2421875,$

$E_L = I - L_4 = \frac{1}{4} - 0.140625 = 0.109375,$ $E_R = \frac{1}{4} - 0.390625 = -0.140625,$

$E_T = \frac{1}{4} - 0.265625 = -0.015625,$ $E_M = \frac{1}{4} - 0.2421875 = 0.0078125$

$n = 8$: $L_8 = \frac{1}{8}\left[f(0) + f\left(\frac{1}{8}\right) + f\left(\frac{2}{8}\right) + \cdots + f\left(\frac{7}{8}\right)\right] \approx 0.191406$

$R_8 = \frac{1}{8}\left[f\left(\frac{1}{8}\right) + f\left(\frac{2}{8}\right) + \cdots + f\left(\frac{7}{8}\right) + f(1)\right] \approx 0.316406$

$T_8 = \frac{1}{8\cdot 2}\left\{f(0) + 2\left[f\left(\frac{1}{8}\right) + f\left(\frac{2}{8}\right) + \cdots + f\left(\frac{7}{8}\right)\right] + f(1)\right\} \approx 0.253906$

$M_8 = \frac{1}{8}\left[f\left(\frac{1}{16}\right) + f\left(\frac{3}{16}\right) + \cdots + f\left(\frac{13}{16}\right) + f\left(\frac{15}{16}\right)\right] = 0.248047$

$E_L \approx \frac{1}{4} - 0.191406 \approx 0.058594,$ $E_R \approx \frac{1}{4} - 0.316406 \approx -0.066406,$

$E_T \approx \frac{1}{4} - 0.253906 \approx -0.003906,$ $E_M \approx \frac{1}{4} - 0.248047 \approx 0.001953.$

$n = 16$: $L_{16} = \frac{1}{16}\left[f(0) + f\left(\frac{1}{16}\right) + f\left(\frac{2}{16}\right) + \cdots + f\left(\frac{15}{16}\right)\right] \approx 0.219727$

$R_{16} = \frac{1}{16}\left[f\left(\frac{1}{16}\right) + f\left(\frac{2}{16}\right) + \cdots + f\left(\frac{15}{16}\right) + f(1)\right] \approx 0.282227$

$T_{16} = \frac{1}{16\cdot 2}\left\{f(0) + 2\left[f\left(\frac{1}{16}\right) + f\left(\frac{2}{16}\right) + \cdots + f\left(\frac{15}{16}\right)\right] + f(1)\right\} \approx 0.250977$

$M_{16} = \frac{1}{16}\left[f\left(\frac{1}{32}\right) + f\left(\frac{3}{32}\right) + \cdots + f\left(\frac{31}{32}\right)\right] \approx 0.249512$

$E_L \approx \frac{1}{4} - 0.219727 \approx 0.030273,$ $E_R \approx \frac{1}{4} - 0.282227 \approx -0.032227,$

$E_T \approx \frac{1}{4} - 0.250977 \approx -0.000977,$ $E_M \approx \frac{1}{4} - 0.249512 \approx 0.000488.$

n	L_n	R_n	T_n	M_n
4	0.140625	0.390625	0.265625	0.242188
8	0.191406	0.316406	0.253906	0.248047
16	0.219727	0.282227	0.250977	0.249512

n	E_L	E_R	E_T	E_M
4	0.109375	−0.140625	−0.015625	0.007813
8	0.058594	−0.066406	−0.003906	0.001953
16	0.030273	−0.032227	−0.000977	0.000488

Observations:

1. E_L and E_R are always opposite in sign, as are E_T and E_M.

2. As n is doubled, E_L and E_R are decreased by about a factor of 2, and E_T and E_M are decreased by a factor of about 4.

3. The Midpoint approximation is about twice as accurate as the Trapezoidal approximation.

4. All the approximations become more accurate as the value of n increases.

5. The Midpoint and Trapezoidal approximations are much more accurate than the endpoint approximations.

29. $\int_1^4 \sqrt{x}\, dx = \left[\frac{2}{3}x^{3/2}\right]_1^4 = \frac{2}{3}(8-1) = \frac{14}{3} \approx 4.666667$

$n = 6$: $\quad \Delta x = (4-1)/6 = \frac{1}{2}$

$$T_6 = \frac{1}{2\cdot 2}\left[\sqrt{1} + 2\sqrt{1.5} + 2\sqrt{2} + 2\sqrt{2.5} + 2\sqrt{3} + 2\sqrt{3.5} + \sqrt{4}\right] \approx 4.661488$$

$$M_6 = \frac{1}{2}\left[\sqrt{1.25} + \sqrt{1.75} + \sqrt{2.25} + \sqrt{2.75} + \sqrt{3.25} + \sqrt{3.75}\right] \approx 4.669245$$

$$S_6 = \frac{1}{2\cdot 3}\left[\sqrt{1} + 4\sqrt{1.5} + 2\sqrt{2} + 4\sqrt{2.5} + 2\sqrt{3} + 4\sqrt{3.5} + \sqrt{4}\right] \approx 4.666563$$

$E_T \approx \frac{14}{3} - 4.661488 \approx 0.005178$, $\quad E_M \approx \frac{14}{3} - 4.669245 \approx -0.002578$,

$E_S \approx \frac{14}{3} - 4.666563 \approx 0.000104$.

$n = 12$: $\quad \Delta x = (4-1)/12 = \frac{1}{4}$

$$T_{12} = \frac{1}{4\cdot 2}\left(f(1) + 2[f(1.25) + f(1.5) + \cdots + f(3.5) + f(3.75)] + f(4)\right) \approx 4.665367$$

$$M_{12} = \frac{1}{4}[f(1.125) + f(1.375) + f(1.625) + \cdots + f(3.875)] \approx 4.667316$$

$$S_{12} = \frac{1}{4\cdot 3}[f(1) + 4f(1.25) + 2f(1.5) + 4f(1.75) + \cdots + 4f(3.75) + f(4)] \approx 4.666659$$

$E_T \approx \frac{14}{3} - 4.665367 \approx 0.001300$, $\quad E_M \approx \frac{14}{3} - 4.667316 \approx -0.000649$,

$E_S \approx \frac{14}{3} - 4.666659 \approx 0.000007$.

Note: These errors were computed more precisely and then rounded to six places. That is, they were not computed by comparing the rounded values of T_n, M_n, and S_n with the rounded value of the actual definite integral.

n	T_n	M_n	S_n
6	4.661488	4.669245	4.666563
12	4.665367	4.667316	4.666659

n	E_T	E_M	E_S
6	0.005178	−0.002578	0.000104
12	0.001300	−0.000649	0.000007

Observations:

1. E_T and E_M are opposite in sign and decrease by a factor of about 4 as n is doubled.

2. The Simpson's approximation is much more accurate than the Midpoint and Trapezoidal approximations, and seems to decrease by a factor of about 16 as n is doubled.

31. $\Delta x = (4-0)/4 = 1$

(a) $T_4 = \frac{1}{2}[f(0) + 2f(1) + 2f(2) + 2f(3) + f(4)] \approx \frac{1}{2}[0 + 2(3) + 2(5) + 2(3) + 1] = 11.5$

(b) $M_4 = 1 \cdot [f(0.5) + f(1.5) + f(2.5) + f(3.5)] \approx 1 + 4.5 + 4.5 + 2 = 12$

(c) $S_4 = \frac{1}{3}[f(0) + 4f(1) + 2f(2) + 4f(3) + f(4)] \approx \frac{1}{3}[0 + 4(3) + 2(5) + 4(3) + 1] = 11.\overline{6}$

33. (a) $\int_1^{3.2} y\, dx \approx \frac{0.2}{2}[4.9 + 2(5.4) + 2(5.8) + 2(6.2) + 2(6.7) + 2(7.0)$

$$+ 2(7.3) + 2(7.5) + 2(8.0) + 2(8.2) + 2(8.3) + 8.3] = 15.4$$

(b) $-1 \le f''(x) \le 3 \quad \Rightarrow \quad |f''(x)| \le 3$, so use $K = 3$, $a = 1$, $b = 3.2$, and $n = 11$ in Theorem 3. So

$$|E_T| \le \frac{3(3.2-1)^3}{12(11)^2} = 0.022.$$

35. By the Total Change Theorem, the increase in velocity is equal to $\int_0^6 a(t)\, dt$. We use Simpson's Rule with $n = 6$ and $\Delta t = 1$ to estimate this integral:

$$\int_0^6 a(t)\, dt \approx S_6 = \frac{1}{1\cdot 3}[a(0) + 4a(1) + 2a(2) + 4a(3) + 2a(4) + 4a(5) + a(6)]$$

$$\approx \frac{1}{3}[0 + 4(0.5) + 2(4.1) + 4(9.8) + 2(12.9) + 4(9.5) + 0] = \frac{1}{3}(113.2) = 37.7\overline{3} \text{ ft/s}$$

37. By the Total Change Theorem, the total percentage increase is equal to $\int_{1987}^{1997} r\,(t)\,dt$. We use Simpson's Rule with $n = 10$ and $\Delta t = 1$ to estimate this integral:

$\int_{1987}^{1997} r\,(t)\,dt \approx S_{10} = \frac{1}{3}\,[r\,(1987) + 4r\,(1988) + 2r\,(1989) + \cdots + 4r\,(1996) + r\,(1997)]$

$\approx \frac{1}{3}\,[4.0 + 4\,(4.1) + 2\,(5.7) + 4\,(5.8) + 2\,(3.6) + 4\,(1.4) + 2\,(2.1) + 4\,(2.3) + 2\,(2.8) + 4\,(3.2) + 2.6]$

$= \frac{1}{3}\,(102.2) = 34.0\overline{6}\% \approx 34.1\%$

39. Volume $= \pi \int_0^2 \left(\sqrt[3]{1+x^3}\right)^2\,dx = \pi \int_0^2 \left(1+x^3\right)^{2/3}\,dx$. $V \approx \pi \cdot S_{10}$ where $f\,(x) = \left(1+x^3\right)^{2/3}$ and $\Delta x = (2-0)\,/10 = \frac{1}{5}$. Therefore,

$V \approx \pi \cdot S_{10} = \pi\,\dfrac{1}{5 \cdot 3}\,[f\,(0) + 4f\,(0.2) + 2f\,(0.4) + 4f\,(0.6) + 2f\,(0.8) + 4f\,(1)$

$+ 2f\,(1.2) + 4f\,(1.4) + 2f\,(1.6) + 4f\,(1.8) + f\,(2)] \approx 12.325078$

41. $I\,(\theta) = \dfrac{N^2 \sin^2 k}{k^2}$, where $k = \dfrac{\pi N d \sin \theta}{\lambda}$, $N = 10{,}000$, $d = 10^{-4}$, and $\lambda = 632.8 \times 10^{-9}$. So

$I\,(\theta) = \dfrac{\left(10^4\right)^2 \sin^2 k}{k^2}$, where $k = \dfrac{\pi\,\left(10^4\right)\,\left(10^{-4}\right)\sin\theta}{632.8 \times 10^{-9}}$. Now $n = 10$ and $\Delta\theta = \dfrac{10^{-6} - \left(-10^{-6}\right)}{10} = 2 \times 10^{-7}$, so

$M_{10} = 2 \times 10^{-7}\,[I\,(-0.0000009) + I\,(-0.0000007) + \cdots + I\,(0.0000009)] \approx 59.4$.

43. Since the Trapezoidal and Midpoint approximations on the interval $[a, b]$ are the sums of the Trapezoidal and Midpoint approximations on the subintervals $[x_{i-1}, x_i]$, $i = 1, 2, \ldots, n$, we can focus our attention on one such interval. The condition $f''\,(x) < 0$ for $a \le x \le b$ means that the graph of f is concave down as in Figure 5. In that figure, T_n is the area of the trapezoid $AQRD$, $\int_a^b f\,(x)\,dx$ is the area of the region $AQPRD$, and M_n is the area of the trapezoid $ABCD$, so $T_n < \int_a^b f\,(x)\,dx < M_n$. In general, the condition $f'' < 0$ implies that the graph of f on $[a, b]$ lies above the chord joining the points $(a, f\,(a))$ and $(b, f\,(b))$. Thus, $\int_a^b f\,(x)\,dx > T_n$. Since M_n is the area under a tangent to the graph, and since $f'' < 0$ implies that the tangent lies above the graph, we also have $M_n > \int_a^b f\,(x)\,dx$. Thus, $T_n < \int_a^b f\,(x)\,dx < M_n$.

45. $T_n = \frac{1}{2}\Delta x\,[f\,(x_0) + 2f\,(x_1) + \cdots + 2f\,(x_{n-1}) + f\,(x_n)]$ and

$M_n = \Delta x\,[f\,(\overline{x}_1) + f\,(\overline{x}_2) + \cdots + f\,(\overline{x}_{n-1}) + f\,(\overline{x}_n)]$, where $\overline{x}_i = \frac{1}{2}\,(x_{i-1} + x_i)$. Now

$T_{2n} = \frac{1}{2}\left(\frac{1}{2}\Delta x\right)[f\,(x_0) + 2f\,(\overline{x}_1) + 2f\,(x_1) + 2f\,(\overline{x}_2) + 2f\,(x_2) + \cdots$

$+ 2f\,(\overline{x}_{n-1}) + 2f\,(x_{n-1}) + 2f\,(\overline{x}_n) + f\,(x_n)]$

so

$\frac{1}{2}\,(T_n + M_n) = \frac{1}{2}T_n + \frac{1}{2}M_n$

$= \frac{1}{4}\Delta x\,[f\,(x_0) + 2f\,(x_1) + \cdots + 2f\,(x_{n-1}) + f\,(x_n)]$

$+ \frac{1}{4}\Delta x\,[2f\,(\overline{x}_1) + 2f\,(\overline{x}_2) + \cdots + 2f\,(\overline{x}_{n-1}) + 2f\,(\overline{x}_n)]$

$= T_{2n}$

7.8 Improper Integrals

1. (a) Since $\int_1^\infty x^4 e^{-x^4}\,dx$ has an infinite interval of integration, this is an improper integral of Type I.

(b) Since $y = \sec x$ has an infinite discontinuity at $x = \frac{\pi}{2}$, $\int_0^{\pi/2} \sec x\,dx$ is a Type II improper integral.

(c) Since $y = \dfrac{x}{(x-2)(x-3)}$ has an infinite discontinuity at $x = 2$, $\displaystyle\int_0^2 \dfrac{x}{x^2 - 5x + 6}\,dx$ is a Type II improper integral.

(d) Since $\displaystyle\int_{-\infty}^0 \dfrac{1}{x^2 + 25}\,dx$ has an infinite interval of integration, it is an improper integral of Type I.

3. The area under the graph of $y = 1/x^3 = x^{-3}$ between $x = 1$ and $x = t$ is

$A(t) = \int_1^t x^{-3}\,dx = \left[-\frac{1}{2}x^{-2}\right]_1^t = \frac{1}{2} - 1/(2t^2)$. So the area for $1 \le x \le 10$ is $A(10) = 0.5 - 0.005 = 0.495$,

the area for $1 \le x \le 100$ is $A(100) = 0.5 - 0.00005 = 0.49995$, and the area for $1 \le x \le 1000$ is

$A(1000) = 0.5 - 0.0000005 = 0.4999995$. The total area under the curve for $x \ge 1$ is

$\displaystyle\lim_{t\to\infty} A(t) = \lim_{t\to\infty}\left[\frac{1}{2} - 1/(2t^2)\right] = \frac{1}{2}$.

5. $\displaystyle\int_1^\infty \frac{1}{(3x+1)^2}\,dx = \lim_{t\to\infty}\int_1^t \frac{1}{(3x+1)^2}\,dx = \lim_{t\to\infty}\int_4^{3t+1} \frac{1}{u^2}\cdot\frac{1}{3}\,du$ $[u = 3x+1, du = 3\,dx]$

$\displaystyle = \frac{1}{3}\lim_{t\to\infty}\left[-\frac{1}{u}\right]_4^{3t+1} = \frac{1}{3}\lim_{t\to\infty}\left[-\frac{1}{3t+1} + \frac{1}{4}\right] = \frac{1}{3}\left(0 + \frac{1}{4}\right) = \frac{1}{12}$. Convergent

7. $\displaystyle\int_{-\infty}^{-1} \frac{1}{\sqrt{2-w}}\,dw = \lim_{t\to-\infty}\int_t^{-1} \frac{1}{\sqrt{2-w}}\,dw = \lim_{t\to-\infty}\left[-2\sqrt{2-w}\right]_t^{-1}$ $[u = 2 - w, du = -dw]$

$\displaystyle = \lim_{t\to-\infty}\left[-2\sqrt{3} + 2\sqrt{2-t}\right] = \infty$. Divergent

9. $\displaystyle\int_0^\infty e^{-x}\,dx = \lim_{t\to\infty}\int_0^t e^{-x}\,dx = \lim_{t\to\infty}\left[-e^{-x}\right]_0^t = \lim_{t\to\infty}(-e^{-t} + 1) = 1$

11. $\displaystyle\int_{-\infty}^\infty x^3\,dx = \int_{-\infty}^0 x^3\,dx + \int_0^\infty x^3\,dx$, but $\displaystyle\int_{-\infty}^0 x^3\,dx = \lim_{t\to-\infty}\left[\frac{1}{4}x^4\right]_t^0 = \lim_{t\to-\infty}\left(-\frac{1}{4}t^4\right) = -\infty$. Divergent

13. $\displaystyle\int_{-\infty}^\infty xe^{-x^2}\,dx = \int_{-\infty}^0 xe^{-x^2}\,dx + \int_0^\infty xe^{-x^2}\,dx$. $\displaystyle\int_{-\infty}^0 xe^{-x^2}\,dx = \lim_{t\to-\infty}\left(-\frac{1}{2}\right)\left[e^{-x^2}\right]_t^0$

$\displaystyle = \lim_{t\to-\infty}\left(-\frac{1}{2}\right)\left(1 - e^{-t^2}\right) = -\frac{1}{2}$, and

$\displaystyle\int_0^\infty xe^{-x^2}\,dx = \lim_{t\to\infty}\left(-\frac{1}{2}\right)\left[e^{-x^2}\right]_0^t = \lim_{t\to\infty}\left(-\frac{1}{2}\right)\left(e^{-t^2} - 1\right) = \frac{1}{2}$. Therefore, $\displaystyle\int_{-\infty}^\infty xe^{-x^2}\,dx = -\frac{1}{2} + \frac{1}{2} = 0$.

15. $\displaystyle\int_0^\infty \frac{dx}{(x+2)(x+3)} = \lim_{t\to\infty}\int_0^t \left[\frac{1}{x+2} - \frac{1}{x+3}\right]dx = \lim_{t\to\infty}\left[\ln\left(\frac{x+2}{x+3}\right)\right]_0^t = \lim_{t\to\infty}\left[\ln\left(\frac{t+2}{t+3}\right) - \ln\frac{2}{3}\right]$

$\displaystyle = \ln 1 - \ln\frac{2}{3} = -\ln\frac{2}{3}$

17. $\displaystyle\int_0^\infty \cos x\,dx = \lim_{t\to\infty}[\sin x]_0^t = \lim_{t\to\infty}\sin t$, which does not exist. Divergent

19. $\displaystyle\int_{-\infty}^1 xe^{2x}\,dx = \lim_{t\to-\infty}\int_t^1 xe^{2x}\,dx = \lim_{t\to-\infty}\left[\frac{1}{2}xe^{2x} - \frac{1}{4}e^{2x}\right]_t^1$ (by parts)

$\displaystyle = \lim_{t\to-\infty}\left[\frac{1}{2}e^2 - \frac{1}{4}e^2 - \frac{1}{2}te^{2t} + \frac{1}{4}e^{2t}\right] = \frac{1}{4}e^2 - 0 + 0 = \frac{1}{4}e^2$

since $\displaystyle\lim_{t\to-\infty} te^{2t} = \lim_{t\to-\infty}\frac{t}{e^{-2t}} \overset{H}{=} \lim_{t\to-\infty}\frac{1}{-2e^{-2t}} = \lim_{t\to-\infty}-\frac{1}{2}e^{2t} = 0$.

21. $\displaystyle\int_1^\infty \frac{\ln x}{x}\, dx = \lim_{t\to\infty}\left[\frac{(\ln x)^2}{2}\right]_1^t = \lim_{t\to\infty}\frac{(\ln t)^2}{2} = \infty.$ Divergent

23. $\displaystyle\int_{-\infty}^\infty \frac{x\, dx}{1+x^2} = \int_{-\infty}^0 \frac{x\, dx}{1+x^2} + \int_0^\infty \frac{x\, dx}{1+x^2}$ and

$\displaystyle\int_{-\infty}^0 \frac{x\, dx}{1+x^2} = \lim_{t\to -\infty}\left[\tfrac{1}{2}\ln\left(1+x^2\right)\right]_t^0 = \lim_{t\to -\infty}\left[0 - \tfrac{1}{2}\ln\left(1+t^2\right)\right] = -\infty.$ Divergent

25. Integrate by parts with $u = \ln x,\ dv = dx/x^2\ \Rightarrow\ du = dx/x,\ v = -1/x.$

$$\int_1^\infty \frac{\ln x}{x^2}\, dx = \lim_{t\to\infty}\int_1^t \frac{\ln x}{x^2}\, dx = \lim_{t\to\infty}\left[-\frac{\ln x}{x} - \frac{1}{x}\right]_1^t = \lim_{t\to\infty}\left(-\frac{\ln t}{t} - \frac{1}{t} + 0 + 1\right)$$
$$= -0 - 0 + 0 + 1 = 1$$

since $\displaystyle\lim_{t\to\infty}\frac{\ln t}{t} \overset{\text{H}}{=} \lim_{t\to\infty}\frac{1/t}{1} = 0.$

27. $\displaystyle\int_0^3 \frac{dx}{\sqrt{x}} = \lim_{t\to 0^+}\int_t^3 \frac{dx}{\sqrt{x}} = \lim_{t\to 0^+}\left[2\sqrt{x}\right]_t^3 = \lim_{t\to 0^+}\left(2\sqrt{3} - 2\sqrt{t}\right) = 2\sqrt{3}$

29. $\displaystyle\int_{-1}^0 \frac{dx}{x^2} = \lim_{t\to 0^-}\int_{-1}^t \frac{dx}{x^2} = \lim_{t\to 0^-}\left[\frac{-1}{x}\right]_{-1}^t = \lim_{t\to 0^-}\left[-\frac{1}{t} + \frac{1}{-1}\right] = \infty.$ Divergent

31. $\displaystyle\int_0^{\pi/4} \csc^2 t\, dt = \lim_{s\to 0^+}\int_s^{\pi/4}\csc^2 t\, dt = \lim_{s\to 0^+}\left[-\cot t\right]_s^{\pi/4} = \lim_{s\to 0^+}\left[-\cot\tfrac{\pi}{4} + \cot s\right] = \infty.$ Divergent

33. $\displaystyle\int_{-2}^3 \frac{dx}{x^4} = \int_{-2}^0 \frac{dx}{x^4} + \int_0^3 \frac{dx}{x^4},$ but $\displaystyle\int_{-2}^0 \frac{dx}{x^4} = \lim_{t\to 0^-}\left[-\frac{x^{-3}}{3}\right]_{-2}^t = \lim_{t\to 0^-}\left[-\frac{1}{3t^3} - \frac{1}{24}\right] = \infty.$ Divergent

35. $\displaystyle\int_0^\pi \sec x\, dx = \int_0^{\pi/2}\sec x\, dx + \int_{\pi/2}^\pi \sec x\, dx.\ \int_0^{\pi/2}\sec x\, dx = \lim_{t\to \pi/2^-}\int_0^t \sec x\, dx$

$$= \lim_{t\to \pi/2^-}\left[\ln|\sec x + \tan x|\right]_0^t = \lim_{t\to \pi/2^-}\ln|\sec t + \tan t| = \infty.\ \text{Divergent}$$

37. $\displaystyle\int_{-2}^2 \frac{dx}{x^2-1} = \int_{-2}^{-1}\frac{dx}{x^2-1} + \int_{-1}^0 \frac{dx}{x^2-1} + \int_0^1 \frac{dx}{x^2-1} + \int_1^2 \frac{dx}{x^2-1},$

and $\displaystyle\int \frac{dx}{x^2-1} = \int\frac{dx}{(x-1)(x+1)} = \frac{1}{2}\ln\left|\frac{x-1}{x+1}\right| + C,$ so

$\displaystyle\int_0^1 \frac{dx}{x^2-1} = \lim_{t\to 1^-}\left[\frac{1}{2}\ln\left|\frac{x-1}{x+1}\right|\right]_0^t = \lim_{t\to 1^-}\frac{1}{2}\ln\left|\frac{t-1}{t+1}\right| = -\infty.$ Divergent

39. $I = \int_0^2 z^2 \ln z\, dz = \lim_{t\to 0^+}\int_t^2 z^2 \ln z\, dz \overset{101}{=} \lim_{t\to 0^+}\left[\frac{z^3}{3^2}(3\ln z - 1)\right]_t^2$

$\displaystyle = \lim_{t\to 0^+}\left[\tfrac{8}{9}(3\ln 2 - 1) - \tfrac{1}{9}t^3(3\ln t - 1)\right] = \tfrac{8}{3}\ln 2 - \tfrac{8}{9} - \tfrac{1}{9}\lim_{t\to 0^+}\left[t^3(3\ln t - 1)\right] = \tfrac{8}{3}\ln 2 - \tfrac{8}{9} - \tfrac{1}{9}L$

Now $\displaystyle L = \lim_{t\to 0^+}\left[t^3(3\ln t - 1)\right] = \lim_{t\to 0^+}\frac{3\ln t - 1}{t^{-3}} \overset{\text{H}}{=} \lim_{t\to 0^+}\frac{3/t}{-3/t^4} = \lim_{t\to 0^+}(-t^3) = 0.$ Thus, $L = 0$ and

$I = \tfrac{8}{3}\ln 2 - \tfrac{8}{9}.$

41.

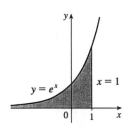

Area $= \int_{-\infty}^{1} e^x \, dx = \lim_{t \to -\infty} [e^x]_t^1$

$= e - \lim_{t \to -\infty} e^t = e$

43.

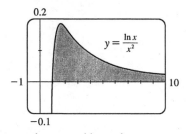

We integrate by parts with $u = \ln x$,

$dv = (1/x^2) \, dx, \, du = dx/x, \, v = -1/x$:

$$\int_1^{\infty} \frac{\ln x}{x^2} \, dx = \lim_{t \to \infty} \left(\left[-\frac{1}{x} \ln x \right]_1^t + \int_1^t \frac{1}{x^2} \, dx \right)$$

$$= \lim_{t \to \infty} \left(-\frac{\ln t}{t} - \frac{1}{t} + 1 \right) = 1$$

45.

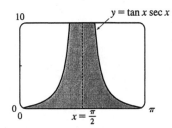

$\int_0^{\pi} \tan x \sec x \, dx = \int_0^{\pi/2} \tan x \sec x \, dx + \int_{\pi/2}^{\pi} \tan x \sec x \, dx$.

$\int_0^{\pi/2} \tan x \sec x \, dx = \lim_{t \to (\pi/2)^-} \int_0^t \tan x \sec x \, dx$

$\qquad = \lim_{t \to (\pi/2)^-} [\sec x]_0^t = \lim_{t \to (\pi/2)^-} (\sec t - 1)$

$\qquad = \infty$. Divergent

47. (a)

t	$\int_1^t g(x) \, dx$
2	0.447453
5	0.577101
10	0.621306
100	0.668479
1000	0.672957
10,000	0.673407

$g(x) = \frac{\sin^2 x}{x^2}$. It appears that the integral is

convergent.

(c)

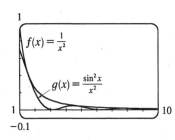

Since $\int_1^{\infty} f(x) \, dx$ is finite and the area under $g(x)$ is less than the area under $f(x)$ on any interval $[1, t]$, $\int_1^{\infty} g(x) \, dx$ must be finite; that is, the integral is convergent.

(b) $-1 \le \sin x \le 1 \Rightarrow 0 \le \sin^2 x \le 1 \Rightarrow 0 \le \frac{\sin^2 x}{x^2} \le \frac{1}{x^2}$. Since $\int_1^{\infty} \frac{1}{x^2} \, dx$ is convergent (Equation 2

with $p = 2 > 1$), $\int_1^{\infty} \frac{\sin^2 x}{x^2} \, dx$ is convergent by the Comparison Theorem.

49. For $x \ge 1$, $\frac{\cos^2 x}{1 + x^2} \le \frac{1}{1 + x^2} < \frac{1}{x^2}$. $\int_1^{\infty} \frac{1}{x^2} \, dx$ is convergent by Equation 2 with $p = 2 > 1$, so $\int_1^{\infty} \frac{\cos^2 x}{1 + x^2} \, dx$ is

convergent by the Comparison Theorem.

51. For $x \geq 1$, $x + e^{2x} > e^{2x} > 0$ $\Rightarrow$ $\dfrac{1}{x + e^{2x}} \leq \dfrac{1}{e^{2x}} = e^{-2x}$ on $[1, \infty)$.

$\int_1^\infty e^{-2x}\, dx = \lim\limits_{t \to \infty} \left[-\tfrac{1}{2} e^{-2x} \right]_1^t = \lim\limits_{t \to \infty} \left[-\tfrac{1}{2} e^{-2t} + \tfrac{1}{2} e^{-2} \right] = \tfrac{1}{2} e^{-2}$. Therefore, $\int_1^\infty e^{-2x}\, dx$ is convergent, and by

the Comparison Theorem, $\displaystyle\int_1^\infty \frac{dx}{x + e^{2x}}$ is also convergent.

53. $\dfrac{1}{x \sin x} \geq \dfrac{1}{x}$ on $\left(0, \tfrac{\pi}{2}\right]$ since $0 \leq \sin x \leq 1$. $\displaystyle\int_0^{\pi/2} \frac{dx}{x} = \lim\limits_{t \to 0^+} \int_t^{\pi/2} \frac{dx}{x} = \lim\limits_{t \to 0^+} \left[\ln x \right]_t^{\pi/2}$.

But $\ln t \to -\infty$ as $t \to 0^+$, so $\displaystyle\int_0^{\pi/2} \frac{dx}{x}$ is divergent, and by the Comparison Theorem, $\displaystyle\int_0^{\pi/2} \frac{dx}{x \sin x}$ is also

divergent.

55. $\displaystyle\int_0^\infty \frac{dx}{\sqrt{x}\,(1+x)} = \int_0^1 \frac{dx}{\sqrt{x}\,(1+x)} + \int_1^\infty \frac{dx}{\sqrt{x}\,(1+x)} = \lim\limits_{t \to 0^+} \int_t^1 \frac{dx}{\sqrt{x}\,(1+x)} + \lim\limits_{t \to \infty} \int_1^t \frac{dx}{\sqrt{x}\,(1+x)}$

$\displaystyle = \lim\limits_{t \to 0^+} \int_{\sqrt{t}}^1 \frac{2\, du}{1 + u^2} + \lim\limits_{t \to \infty} \int_1^{\sqrt{t}} \frac{2\, du}{1 + u^2} \quad (u = \sqrt{x}, \; x = u^2)$

$\displaystyle = \lim\limits_{t \to 0^+} \left[2 \tan^{-1} u \right]_{\sqrt{t}}^1 + \lim\limits_{t \to \infty} \left[2 \tan^{-1} u \right]_1^{\sqrt{t}}$

$\displaystyle = \lim\limits_{t \to 0^+} \left[2 \left(\tfrac{\pi}{4} \right) - 2 \tan^{-1} \sqrt{t} \right] + \lim\limits_{t \to \infty} \left[2 \tan^{-1} \sqrt{t} - 2 \left(\tfrac{\pi}{4} \right) \right] = \tfrac{\pi}{2} - 0 + 2 \left(\tfrac{\pi}{2} \right) - \tfrac{\pi}{2} = \pi$

57. If $p = 1$, then $\displaystyle\int_0^1 \frac{dx}{x^p} = \lim\limits_{t \to 0^+} \left[\ln x \right]_t^1 = \infty$. Divergent. If $p \neq 1$, then

$$\int_0^1 \frac{dx}{x^p} = \lim\limits_{t \to 0^+} \int_t^1 \frac{dx}{x^p} \quad \text{(note that the integral is not improper if } p < 0)$$

$$= \lim\limits_{t \to 0^+} \left[\frac{x^{-p+1}}{-p+1} \right]_t^1 = \lim\limits_{t \to 0^+} \frac{1}{1-p} \left[1 - \frac{1}{t^{p-1}} \right]$$

If $p > 1$, then $p - 1 > 0$, so $\dfrac{1}{t^{p-1}} \to \infty$ as $t \to 0^+$, and the integral diverges. Finally, if $p < 1$, then

$\displaystyle\int_0^1 \frac{dx}{x^p} = \frac{1}{1-p} \left[\lim\limits_{t \to 0^+} \left(1 - t^{1-p} \right) \right] = \frac{1}{1-p}$. Thus, the integral converges if and only if $p < 1$, and in that case

its value is $\dfrac{1}{1-p}$.

59. First suppose $p = -1$. Then

$\displaystyle\int_0^1 x^p \ln x\, dx = \int_0^1 \frac{\ln x}{x}\, dx = \lim\limits_{t \to 0^+} \int_t^1 \frac{\ln x}{x}\, dx = \lim\limits_{t \to 0^+} \left[\tfrac{1}{2} (\ln x)^2 \right]_t^1 = -\tfrac{1}{2} \lim\limits_{t \to 0^+} (\ln t)^2 = -\infty$, so

the integral diverges. Now suppose $p \neq -1$. Then integration by parts gives

$\displaystyle\int x^p \ln x\, dx = \frac{x^{p+1}}{p+1} \ln x - \int \frac{x^p}{p+1}\, dx = \frac{x^{p+1}}{p+1} \ln x - \frac{x^{p+1}}{(p+1)^2} + C$. If $p < -1$, then $p + 1 < 0$, so

$\displaystyle\int_0^1 x^p \ln x\, dx = \lim\limits_{t \to 0^+} \left[\frac{x^{p+1}}{p+1} \ln x - \frac{x^{p+1}}{(p+1)^2} \right]_t^1 = \frac{-1}{(p+1)^2} - \left(\frac{1}{p+1} \right) \lim\limits_{t \to 0^+} \left[t^{p+1} \left(\ln t - \frac{1}{p+1} \right) \right] = \infty$.

If $p > -1$, then $p + 1 > 0$ and

$$\int_0^1 x^p \ln x \, dx = \frac{-1}{(p+1)^2} - \left(\frac{1}{p+1}\right) \lim_{t \to 0^+} \frac{\ln t - 1/(p+1)}{t^{-(p+1)}} \stackrel{\text{H}}{=} \frac{-1}{(p+1)^2} - \left(\frac{1}{p+1}\right) \lim_{t \to 0^+} \frac{1/t}{-(p+1)t^{-(p+2)}}$$

$$= \frac{-1}{(p+1)^2} + \frac{1}{(p+1)^2} \lim_{t \to 0^+} t^{p+1} = \frac{-1}{(p+1)^2}$$

Thus, the integral converges to $-\dfrac{1}{(p+1)^2}$ if $p > -1$ and diverges otherwise.

61. (a) $\int_{-\infty}^{\infty} x \, dx = \int_{-\infty}^0 x \, dx + \int_0^{\infty} x \, dx$, and $\int_0^{\infty} x \, dx = \lim_{t \to \infty} \int_0^t x \, dx = \lim_{t \to \infty} \left[\frac{1}{2}t^2 - \frac{1}{2}(0^2)\right] = \infty$, so the integral

is divergent.

(b) $\int_{-t}^t x \, dx = \left[\frac{1}{2}x^2\right]_{-t}^t = \frac{1}{2}t^2 - \frac{1}{2}t^2 = 0$, so $\lim_{t \to \infty} \int_{-t}^t x \, dx = 0$. Therefore, $\int_{-\infty}^{\infty} x \, dx \neq \lim_{t \to \infty} \int_{-t}^t x \, dx$.

63. Volume $= \int_1^{\infty} \pi \left(\frac{1}{x}\right)^2 dx = \pi \lim_{t \to \infty} \int_1^t \frac{dx}{x^2} = \pi \lim_{t \to \infty} \left[-\frac{1}{x}\right]_1^t = \pi \lim_{t \to \infty} \left(1 - \frac{1}{t}\right) = \pi < \infty.$

65. Work $= \int_R^{\infty} F \, dr = \lim_{t \to \infty} \int_R^t \frac{GmM}{r^2} dr = \lim_{t \to \infty} GmM\left(\frac{1}{R} - \frac{1}{t}\right) = \frac{GmM}{R}$. The initial kinetic energy provides

the work, so $\frac{1}{2}mv_0^2 = \frac{GmM}{R} \quad \Rightarrow \quad v_0 = \sqrt{\frac{2GM}{R}}.$

67. (a)

(b) $r(t) = F'(t)$ is the rate at which the fraction $F(t)$ of burnt-out bulbs increases as t increases. This could be interpreted as a fractional burnout rate.

(c) $\int_0^{\infty} r(t) \, dt = \lim_{x \to \infty} F(x) = 1$, since all of the bulbs will eventually burn out.

69. $I = \int_a^{\infty} \frac{1}{x^2 + 1} dx = \lim_{t \to \infty} \int_a^t \frac{1}{x^2 + 1} dx = \lim_{t \to \infty} \left[\tan^{-1} x\right]_a^t = \lim_{t \to \infty} \left(\tan^{-1} t - \tan^{-1} a\right) = \frac{\pi}{2} - \tan^{-1} a.$

$I < 0.001 \quad \Rightarrow \quad \frac{\pi}{2} - \tan^{-1} a < 0.01 \quad \Rightarrow \quad \tan^{-1} a > \frac{\pi}{2} - 0.001 \quad \Rightarrow \quad a > \tan\left(\frac{\pi}{2} - 0.001\right) \approx 1000.$

71. (a) $F(s) = \int_0^{\infty} f(t) e^{-st} \, dt = \int_0^{\infty} e^{-st} \, dt = \lim_{n \to \infty} \left[-\frac{e^{-st}}{s}\right]_0^n = \lim_{n \to \infty} \left(\frac{e^{-sn}}{-s} + \frac{1}{s}\right)$. This converges to $\frac{1}{s}$ only

if $s > 0$. Therefore $F(s) = \frac{1}{s}$ with domain $\{s \mid s > 0\}$.

(b) $F(s) = \int_0^{\infty} f(t) e^{-st} \, dt = \int_0^{\infty} e^t e^{-st} \, dt = \lim_{n \to \infty} \int_0^n e^{t(1-s)} \, dt = \lim_{n \to \infty} \left[\frac{1}{1-s} e^{t(1-s)}\right]_0^n$

$= \lim_{n \to \infty} \left(\frac{e^{(1-s)n}}{1-s} - \frac{1}{1-s}\right)$

This converges only if $1 - s < 0 \quad \Rightarrow \quad s > 1$, in which case $F(s) = \frac{1}{s-1}$ with domain $\{s \mid s > 1\}$.

(c) $F(s) = \int_0^{\infty} f(t) e^{-st} \, dt = \lim_{n \to \infty} \int_0^n t e^{-st} \, dt$. Use integration by parts: let $u = t$, $dv = e^{-st} \, dt \quad \Rightarrow$

$du = dt$, $v = -\dfrac{e^{-st}}{s}$. Then $F(s) = \lim_{n \to \infty} \left[-\frac{t}{s}e^{-st} - \frac{1}{s^2}e^{-st}\right]_0^n = \lim_{n \to \infty} \left(\frac{-n}{se^{sn}} - \frac{1}{s^2 e^{sn}} + 0 + \frac{1}{s^2}\right) = \frac{1}{s^2}$

only if $s > 0$. Therefore, $F(s) = \dfrac{1}{s^2}$ and the domain of F is $\{s \mid s > 0\}$.

73. $G(s) = \int_0^\infty f'(t)\,e^{-st}\,dt$. Integrate by parts with $u = e^{-st}$, $dv = f'(t)\,dt$ $\Rightarrow$ $du = -se^{-st}$, $v = f(t)$:

$$G(s) = \lim_{n\to\infty}\left[f(t)\,e^{-st}\right]_0^n + s\int_0^\infty f(t)\,e^{-st}\,dt = \lim_{n\to\infty} f(n)\,e^{-sn} - f(0) + sF(s)$$

But $0 \le f(t) \le Me^{at}$ $\Rightarrow$ $0 \le f(t)\,e^{-st} \le Me^{at}e^{-st}$ and $\lim_{t\to\infty} Me^{t(a-s)} = 0$ for $s > a$. So by the Squeeze Theorem, $\lim_{t\to\infty} f(t)\,e^{-st} = 0$ for $s > a$ $\Rightarrow$ $G(s) = 0 - f(0) + sF(s) = sF(s) - f(0)$ for $s > a$.

75. We use integration by parts: let $u = x$, $dv = xe^{-x^2}\,dx$, $du = dx$, $v = -\tfrac{1}{2}e^{-x^2}$. So

$$\int_0^\infty x^2 e^{-x^2}\,dx = \lim_{t\to\infty}\left[-\tfrac{1}{2}xe^{-x^2}\right]_0^t + \tfrac{1}{2}\int_0^\infty e^{-x^2}\,dx = \lim_{t\to\infty} -t\Big/\left(2e^{t^2}\right) + \tfrac{1}{2}\int_0^\infty e^{-x^2}\,dx = \tfrac{1}{2}\int_0^\infty e^{-x^2}\,dx$$

(The limit is 0 by l'Hospital's Rule.)

77. For the first part of the integral, let $x = 2\tan\theta$ $\Rightarrow$ $dx = 2\sec^2\theta\,d\theta$.

$$\int \frac{1}{\sqrt{x^2+4}}\,dx = \int \sec\theta = \ln|\sec\theta + \tan\theta|. \text{ But } \tan\theta = \tfrac{1}{2}x, \text{ and}$$

$\sec\theta = \sqrt{1+\tan^2\theta} = \sqrt{1+\tfrac{1}{4}x^2} = \tfrac{1}{2}\sqrt{x^2+4}$. So

$$\int_0^\infty \left(\frac{1}{\sqrt{x^2+4}} - \frac{C}{x+2}\right)dx = \lim_{t\to\infty}\left[\ln\left|\frac{\sqrt{x^2+4}}{2} + \frac{x}{2}\right| - C\ln|x+2|\right]_0^t$$

$$= \lim_{t\to\infty}\ln\left(\frac{\sqrt{t^2+4}+t}{2(t+2)^C}\right) - (\ln 1 - C\ln 2)$$

$$= \ln\left(\lim_{t\to\infty}\frac{t+\sqrt{t^2+4}}{(t+2)^C}\right) + \ln 2^{C-1}$$

By l'Hospital's Rule, $\displaystyle\lim_{t\to\infty}\frac{t+\sqrt{t^2+4}}{(t+2)^C} = \lim_{t\to\infty}\frac{1+t/\sqrt{t^2+4}}{C(t+2)^{C-1}} = \frac{2}{C\displaystyle\lim_{t\to\infty}(t+2)^{C-1}}$.

If $C < 1$, we get ∞ and the interval diverges. If $C = 1$, we get 2, so the original integral converges to $\ln 2 + \ln 2^0 = \ln 2$. If $C > 1$, we get 0, so the original integral diverges to $-\infty$.

7 Review

CONCEPT CHECK

1. See Formula 7.1.1 or 7.1.2. We try to choose $u = f(x)$ to be a function that becomes simpler when differentiated (or at least not more complicated) as long as $dv = g'(x)\,dx$ can be readily integrated to give v.

2. See the Strategy for Evaluating $\int \sin^m x \cos^n x\,dx$ on page 480.

3. If $\sqrt{a^2 - x^2}$ occurs, try $x = a\sin\theta$; if $\sqrt{a^2 + x^2}$ occurs, try $x = a\tan\theta$, and if $\sqrt{x^2 - a^2}$ occurs, try $x = a\sec\theta$. See the Table of Trigonometric Substitutions on page 484.

4. See Equation 2 and Expressions 7, 9, and 11 in Section 7.4.

5. See the Midpoint Rule, the Trapezoidal Rule, and Simpson's Rule, as well as their associated error bounds, all in Section 7.7. We would expect the best estimate to be given by Simpson's Rule.

6. See Definitions 1(a), (b), and (c) in Section 7.8.

7. See Definitions 3(b), (a), and (c) in Section 7.8.

8. See the Comparison Theorem on page 530.

TRUE-FALSE QUIZ

1. False. Since the numerator has a higher degree than the denominator,

$$\frac{x\,(x^2+4)}{x^2-4} = x + \frac{8x}{x^2-4} = x + \frac{A}{x+2} + \frac{B}{x-2}.$$

3. False. It can be put in the form $\dfrac{A}{x} + \dfrac{B}{x^2} + \dfrac{C}{x-4}.$

5. False. This is an improper integral, since the denominator vanishes at $x = 1$.

$$\int_0^4 \frac{x}{x^2-1}\,dx = \int_0^1 \frac{x}{x^2-1}\,dx + \int_1^4 \frac{x}{x^2-1}\,dx \text{ and}$$

$$\int_0^1 \frac{x}{x^2-1} = \lim_{t\to 1^-} \int_0^t \frac{x}{x^2-1}\,dx = \lim_{t\to 1^-} \left[\tfrac{1}{2}\ln\left|x^2-1\right|\right]_0^t = \lim_{t\to 1^-} \tfrac{1}{2}\ln\left|t^2-1\right| = \infty$$

So the integral diverges.

7. False. See Exercise 61 in Section 7.8.

9. False. Examples include the functions $f\,(x) = e^{x^2}$, $g\,(x) = \sin\,(x^2)$, and $h\,(x) = \dfrac{\sin x}{x}$.

EXERCISES

1. $\displaystyle\int \frac{6x+1}{3x+2}\,dx = \int \left(2 - \frac{3}{3x+2}\right) dx = 2x - 3 \cdot \tfrac{1}{3}\ln|3x+2| + C = 2x - \ln|3x+2| + C$

3. $\int \cot^2 x\,dx = \int (\csc^2 x - 1)\,dx = -\cot x - x + C$

5. Use integration by parts with $u = \ln x$, $dv = x^4\,dx \;\Rightarrow\; du = dx/x,\, v = x^5/5$:

$$\int x^4 \ln x\,dx = \tfrac{1}{5}x^5 \ln x - \tfrac{1}{5}\int x^4\,dx = \tfrac{1}{5}x^5 \ln x - \tfrac{1}{25}x^5 + C = \tfrac{1}{25}x^5\,(5\ln x - 1) + C$$

7. Let $u = \sec x$. Then $du = \sec x \tan x\,dx$, so

$$\int \tan^7 x \sec^3 x\,dx = \int \tan^6 x \sec^2 x \sec x \tan x\,dx = \int (u^2-1)^3 u^2\,du = \int (u^8 - 3u^6 + 3u^4 - u^2)\,du$$

$$= \tfrac{1}{9}u^9 - \tfrac{3}{7}u^7 + \tfrac{3}{5}u^5 - \tfrac{1}{3}u^3 + C = \tfrac{1}{9}\sec^9 x - \tfrac{3}{7}\sec^7 x + \tfrac{3}{5}\sec^5 x - \tfrac{1}{3}\sec^3 x + C$$

9. Let $u = x^2$. Then $du = 2x\,dx$, so $\int x \sin\,(x^2)\,dx = \tfrac{1}{2}\int \sin u\,du = -\tfrac{1}{2}\cos u + C = -\tfrac{1}{2}\cos\,(x^2) + C.$

11. $\displaystyle\int \frac{dx}{x^3+x} = \int \left(\frac{1}{x} - \frac{x}{x^2+1}\right) dx = \ln|x| - \tfrac{1}{2}\ln\,(x^2+1) + C$

13. $\int \sin^2 \theta \cos^5 \theta\,d\theta = \int \sin^2 \theta\,(\cos^2 \theta)^2 \cos \theta\,d\theta = \int \sin^2 \theta\,(1 - \sin^2 \theta)^2 \cos \theta\,d\theta$

$$= \int u^2\,(1-u^2)^2\,du \;\; [u = \sin\theta,\, du = \cos\theta\,d\theta] \;\; = \int u^2\,(1 - 2u^2 + u^4)\,du$$

$$= \int (u^2 - 2u^4 + u^6)\,du = \tfrac{1}{3}u^3 - \tfrac{2}{5}u^5 + \tfrac{1}{7}u^7 + C = \tfrac{1}{3}\sin^3 \theta - \tfrac{2}{5}\sin^5 \theta + \tfrac{1}{7}\sin^7 \theta + C$$

15. Integrate by parts with $u = x$, $dv = \sec x \tan x\,dx \;\Rightarrow\; du = dx, v = \sec x$:

$$\int x \sec x \tan x\,dx = x \sec x - \int \sec x\,dx = x \sec x - \ln|\sec x + \tan x| + C$$

17. Let $u = \arctan x$. Then $du = \dfrac{dx}{1+x^2}$, so $\displaystyle\int \frac{(\arctan x)^5}{1+x^2}\,dx = \int u^5\,du = \tfrac{1}{6}u^6 + C = \tfrac{1}{6}\,(\arctan x)^6 + C.$

19.

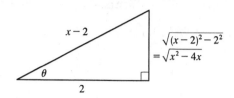

$$\int \frac{dx}{\sqrt{x^2 - 4x}} = \int \frac{dx}{\sqrt{(x^2 - 4x + 4) - 4}} = \int \frac{dx}{\sqrt{(x - 2)^2 - 2^2}}$$

$$= \int \frac{2 \sec \theta \tan \theta \, d\theta}{2 \tan \theta} \quad \begin{bmatrix} x - 2 = 2 \sec \theta, \\ dx = 2 \sec \theta \tan \theta \, d\theta \end{bmatrix} = \int \sec \theta \, d\theta = \ln |\sec \theta + \tan \theta| + C_1$$

$$= \ln \left| \frac{x - 2}{2} + \frac{\sqrt{x^2 - 4x}}{2} \right| + C_1 = \ln \left| x - 2 + \sqrt{x^2 - 4x} \right| + C, \text{ where } C = C_1 - \ln 2$$

21. Let $u = \cot 4x$. Then $du = -4 \csc^2 4x \, dx \quad \Rightarrow$

$$\int \csc^4 4x \, dx = \int \left(\cot^2 4x + 1 \right) \csc^2 4x \, dx = \int \left(u^2 + 1 \right) \left(-\tfrac{1}{4} \, du \right)$$

$$= -\tfrac{1}{4} \left(\tfrac{1}{3} u^3 + u \right) + C = -\tfrac{1}{12} \left(\cot^3 4x + 3 \cot 4x \right) + C$$

23. Let $u = \ln x$. Then $\displaystyle \int \frac{\ln (\ln x)}{x} \, dx = \int \ln u \, du$. Now use parts with $w = \ln u$, $dv = du \quad \Rightarrow \quad dw = du/u$,

$v = u \quad \Rightarrow \quad \int \ln u \, du = u \ln u - u + C = (\ln x) [\ln (\ln x) - 1] + C$.

25. $\displaystyle \frac{3x^3 - x^2 + 6x - 4}{(x^2 + 1)(x^2 + 2)} = \frac{Ax + B}{x^2 + 1} + \frac{Cx + D}{x^2 + 2} \quad \Rightarrow \quad 3x^3 - x^2 + 6x - 4 = (Ax + B)(x^2 + 2) + (Cx + D)(x^2 + 1)$.

Equating the coefficients gives $A + C = 3$, $B + D = -1$, $2A + C = 6$, and $2B + D = -4 \quad \Rightarrow \quad A = 3, C = 0$,

$B = -3$, and $D = 2$. Now

$$\int \frac{3x^3 - x^2 + 6x - 4}{(x^2 + 1)(x^2 + 2)} \, dx = 3 \int \frac{x - 1}{x^2 + 1} \, dx + 2 \int \frac{dx}{x^2 + 2}$$

$$= \tfrac{3}{2} \ln \left(x^2 + 1 \right) - 3 \tan^{-1} x + \sqrt{2} \tan^{-1} \left(\tfrac{1}{\sqrt{2}} x \right) + C$$

27. Let $u = e^{2r} \quad \Rightarrow \quad du = 2 e^{2r} \, dr$.

$$\int \frac{e^{2r}}{e^{4r} - 1} \, dr = \frac{1}{2} \int \frac{du}{u^2 - 1} = \frac{1}{2} \int \left[\frac{1}{2(u - 1)} - \frac{1}{2(u + 1)} \right] du$$

$$= \frac{1}{4} \ln \left| \frac{u - 1}{u + 1} \right| + C = \frac{1}{4} \ln \left| \frac{e^{2r} - 1}{e^{2r} + 1} \right| + C$$

29. Let $x = 2 \sin \theta \quad \Rightarrow \quad \left(4 - x^2 \right)^{3/2} = (2 \cos \theta)^3$, $dx = 2 \cos \theta \, d\theta$, so

$$\int \frac{x^2}{\left(4 - x^2 \right)^{3/2}} \, dx = \int \frac{4 \sin^2 \theta}{8 \cos^3 \theta} \, 2 \cos \theta \, d\theta = \int \tan^2 \theta \, d\theta = \int \left(\sec^2 \theta - 1 \right) d\theta$$

$$= \tan \theta - \theta + C = \frac{x}{\sqrt{4 - x^2}} - \sin^{-1} \left(\frac{x}{2} \right) + C$$

31. $\int (\cos x + \sin x)^2 \cos 2x \, dx = \int (\cos^2 x + 2\sin x \cos x + \sin^2 x) \cos 2x \, dx$

$\qquad = \int (1 + \sin 2x) \cos 2x \, dx = \int \cos 2x \, dx + \frac{1}{2} \int \sin 4x \, dx = \frac{1}{2} \sin 2x - \frac{1}{8} \cos 4x + C$

Or: $\int (\cos x + \sin x)^2 \cos 2x \, dx = \int (\cos x + \sin x)^2 (\cos^2 x - \sin^2 x) \, dx$

$\qquad = \int (\cos x + \sin x)^3 (\cos x - \sin x) \, dx = \frac{1}{4} (\cos x + \sin x)^4 + C_1$

33. $\displaystyle\int_1^\infty \frac{1}{(2x+1)^3} \, dx = \lim_{t\to\infty} \int_1^t \frac{1}{(2x+1)^3} \, dx = \lim_{t\to\infty} \int_1^t \frac{1}{2} (2x+1)^{-3} \, 2\, dx$

$\qquad = \lim_{t\to\infty} \left[-\frac{1}{4(2x+1)^2} \right]_1^t = -\frac{1}{4} \lim_{t\to\infty} \left[\frac{1}{(2t+1)^2} - \frac{1}{9} \right] = -\frac{1}{4} \left(0 - \frac{1}{9} \right) = \frac{1}{36}$

35. $\displaystyle\int_0^4 \frac{\ln x}{\sqrt{x}} \, dx = \lim_{t\to 0^+} \int_t^4 \frac{\ln x}{\sqrt{x}} \, dx \overset{*}{=} \lim_{t\to 0^+} \left[2\sqrt{x} \ln x - 4\sqrt{x} \right]_t^4$

$\qquad = \lim_{t\to 0^+} \left[(2 \cdot 2 \ln 4 - 4 \cdot 2) - (2\sqrt{t} \ln t - 4\sqrt{t}) \right] \overset{**}{=} (4 \ln 4 - 8) - (0 - 0) = 4 \ln 4 - 8$

(*) Let $u = \ln x$, $dv = \dfrac{1}{\sqrt{x}} \, dx \quad \Rightarrow \quad du = \dfrac{1}{x} \, dx$, $v = 2\sqrt{x}$. Then

$$\int \frac{\ln x}{\sqrt{x}} \, dx = 2\sqrt{x} \ln x - 2 \int \frac{dx}{\sqrt{x}} = 2\sqrt{x} \ln x - 4\sqrt{x} + C$$

(**) $\displaystyle\lim_{t\to 0^+} (2\sqrt{t} \ln t) = \lim_{t\to 0^+} \frac{2 \ln t}{t^{-1/2}} \overset{H}{=} \lim_{t\to 0^+} \frac{2/t}{-\frac{1}{2} t^{-3/2}} = \lim_{t\to 0^+} (-4\sqrt{t}) = 0$

37. $\displaystyle\int_0^1 \frac{t^2 - 1}{t^2 + 1} \, dt = \int_0^1 \left(1 - \frac{2}{t^2 + 1} \right) dt = \left[t - 2 \tan^{-1} t \right]_0^1 = (1 - 2 \cdot \frac{\pi}{4}) - 0 = 1 - \frac{\pi}{2}$

39. $\displaystyle\int_0^{\pi/2} \cos^3 x \sin 2x \, dx = \int_0^{\pi/2} 2 \cos^4 x \sin x \, dx = \left[-\frac{2}{5} \cos^5 x \right]_0^{\pi/2} = \frac{2}{5}$

41. $\displaystyle\int_0^3 \frac{dx}{x^2 - x - 2} = \int_0^3 \frac{dx}{(x+1)(x-2)} = \int_0^2 \frac{dx}{(x+1)(x-2)} + \int_2^3 \frac{dx}{(x+1)(x-2)}$, and

$\qquad \displaystyle\int_2^3 \frac{dx}{x^2 - x - 2} = \lim_{t\to 2^+} \int_t^3 \left[\frac{-1/3}{x+1} + \frac{1/3}{x-2} \right] dx = \lim_{t\to 2^+} \left[\frac{1}{3} \ln \left| \frac{x-2}{x+1} \right| \right]_t^3$

$\qquad = \displaystyle\lim_{t\to 2^+} \left[\frac{1}{3} \ln \frac{1}{4} - \frac{1}{3} \ln \left| \frac{t-2}{t+1} \right| \right] = \infty$

so $\displaystyle\int_0^3 \frac{dx}{x^2 - x - 2}$ diverges.

43. Let $u = \sqrt{x} + 2$. Then $x = (u-2)^2$, $dx = 2(u-2) \, du$, so

$$\int_1^4 \frac{\sqrt{x} \, dx}{\sqrt{x} + 2} = \int_3^4 \frac{2(u-2)^2 \, du}{u} = \int_3^4 \left[2u - 8 + \frac{8}{u} \right] du = \left[u^2 - 8u + 8 \ln u \right]_3^4$$

$$= (16 - 32 + 8 \ln 4) - (9 - 24 + 8 \ln 3) = -1 + 8 \ln 4 - 8 \ln 3 = 8 \ln \tfrac{4}{3} - 1$$

45. Let $x = \sec \theta$. Then

$$\int_1^2 \frac{\sqrt{x^2 - 1}}{x} \, dx = \int_0^{\pi/3} \frac{\tan \theta}{\sec \theta} \sec \theta \tan \theta \, d\theta = \int_0^{\pi/3} \tan^2 \theta \, d\theta = \int_0^{\pi/3} \left(\sec^2 \theta - 1 \right) d\theta$$

$$= [\tan \theta - \theta]_0^{\pi/3} = \sqrt{3} - \tfrac{\pi}{3}$$

47. $\int_0^{\pi/4} \tan^2\theta \sec^2\theta \, d\theta = \int_0^1 u^2 \, du \; [u = \tan\theta, \, du = \sec^2\theta \, d\theta] \; = \left[\frac{1}{3}u^3\right]_0^1 = \frac{1}{3} - 0 = \frac{1}{3}$

49. Let $u = 2x + 1$. Then

$$\int_{-\infty}^{\infty} \frac{dx}{4x^2 + 4x + 5} = \int_{-\infty}^{\infty} \frac{\frac{1}{2}\,du}{u^2 + 4} = \frac{1}{2}\int_{-\infty}^{0} \frac{du}{u^2 + 4} + \frac{1}{2}\int_0^{\infty} \frac{du}{u^2 + 4}$$

$$= \frac{1}{2}\lim_{t\to-\infty}\left[\frac{1}{2}\tan^{-1}\left(\frac{1}{2}u\right)\right]_t^0 + \frac{1}{2}\lim_{t\to\infty}\left[\frac{1}{2}\tan^{-1}\left(\frac{1}{2}u\right)\right]_0^t$$

$$= \frac{1}{4}\left[0 - \left(-\frac{\pi}{2}\right)\right] + \frac{1}{4}\left[\frac{\pi}{2} - 0\right] = \frac{\pi}{4}$$

51. We first make the substitution $t = x + 1$, so $\ln\left(x^2 + 2x + 2\right) = \ln\left[(x + 1)^2 + 1\right] = \ln\left(t^2 + 1\right)$. Then we use parts with $u = \ln\left(t^2 + 1\right)$, $dv = dt$:

$$\int \ln\left(t^2 + 1\right) dt = t\ln\left(t^2 + 1\right) - \int \frac{t\,(2t)\,dt}{t^2 + 1} = t\ln\left(t^2 + 1\right) - 2\int \frac{t^2\,dt}{t^2 + 1}$$

$$= t\ln\left(t^2 + 1\right) - 2\int \left(1 - \frac{1}{t^2 + 1}\right) dt = t\ln\left(t^2 + 1\right) - 2t + 2\arctan t + C$$

$$= (x + 1)\ln\left(x^2 + 2x + 2\right) - 2x + 2\arctan(x + 1) + K, \text{ where } K = C - 2$$

[Alternatively, we could have integrated by parts immediately with $u = \ln\left(x^2 + 2x + 2\right)$.] Notice from the graph that $f = 0$ where F has a horizontal tangent. Also, F is always increasing, and $f \geq 0$.

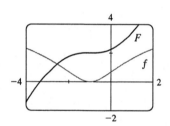

53. From the graph, it seems that $\int_0^{2\pi} \cos^2 x \sin^3 x \, dx = 0$. To evaluate the integral, we write the integral as $I = \int_0^{2\pi} \cos^2 x \left(1 - \cos^2 x\right)\sin x \, dx$ and let $u = \cos x \;\Rightarrow\; du = -\sin x \, dx$. Thus,

$$I = \int_1^1 u^2 \left(1 - u^2\right)(-du) = 0$$

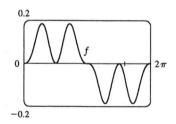

55. $u = e^x \;\Rightarrow\; du = e^x \, dx$, so

$$\int e^x \sqrt{1 - e^{2x}} \, dx = \int \sqrt{1 - u^2} \, du \overset{30}{=} \frac{1}{2}u\sqrt{1 - u^2} + \frac{1}{2}\sin^{-1} u + C = \frac{1}{2}\left[e^x\sqrt{1 - e^{2x}} + \sin^{-1}(e^x)\right] + C$$

57. $u = x + \frac{1}{2} \;\Rightarrow\; du = dx$, so

$$\int \sqrt{x^2 + x + 1}\, dx = \int \sqrt{\left(x + \frac{1}{2}\right)^2 + \frac{3}{4}}\, dx = \int \sqrt{u^2 + \left(\frac{\sqrt{3}}{2}\right)^2}\, du$$

$$\overset{21}{=} \frac{1}{2}u\sqrt{u^2 + \frac{3}{4}} + \frac{3}{8}\ln\left|u + \sqrt{u^2 + \frac{3}{4}}\right| + C$$

$$= \frac{2x + 1}{4}\sqrt{x^2 + x + 1} + \frac{3}{8}\ln\left|x + \frac{1}{2} + \sqrt{x^2 + x + 1}\right| + C$$

59. (a) $\dfrac{d}{du}\left[-\dfrac{1}{u}\sqrt{a^2-u^2}-\sin^{-1}\left(\dfrac{u}{a}\right)+C\right]=\dfrac{1}{u^2}\sqrt{a^2-u^2}+\dfrac{1}{\sqrt{a^2-u^2}}-\dfrac{1}{\sqrt{1-u^2/a^2}}\cdot\dfrac{1}{a}$

$$=\left(a^2-u^2\right)^{-1/2}\left[\dfrac{1}{u^2}\left(a^2-u^2\right)+1-1\right]=\dfrac{\sqrt{a^2-u^2}}{u^2}$$

(b) Let $u=a\sin\theta \Rightarrow du=a\cos\theta\,d\theta$, $a^2-u^2=a^2\left(1-\sin^2\theta\right)=a^2\cos^2\theta$.

$$\int\dfrac{\sqrt{a^2-u^2}}{u^2}\,du=\int\dfrac{a^2\cos^2\theta}{a^2\sin^2\theta}\,d\theta=\int\dfrac{1-\sin^2\theta}{\sin^2\theta}\,d\theta=\int\left(\csc^2\theta-1\right)d\theta=-\cot\theta-\theta+C$$

$$=-\dfrac{\sqrt{a^2-u^2}}{u}-\sin^{-1}\left(\dfrac{u}{a}\right)+C$$

61. For $n\geq 0$, $\int_0^\infty x^n\,dx=\lim\limits_{t\to\infty}\left[x^{n+1}/(n+1)\right]_0^t=\infty$. For $n<0$, $\int_0^\infty x^n\,dx=\int_0^1 x^n\,dx+\int_1^\infty x^n\,dx$. Both integrals are improper. By (7.8.2), the second integral diverges if $-1\leq n<0$. By Exercise 7.8.57, the first integral diverges if $n\leq -1$. Thus, $\int_0^\infty x^n\,dx$ is divergent for all values of n.

63. $f(x)=\sqrt{1+x^4}$, $\Delta x=\dfrac{b-a}{n}=\dfrac{1-0}{10}=\dfrac{1}{10}$

(a) $T_{10}=\dfrac{1}{10\cdot 2}\{f(0)+2[f(0.1)+f(0.2)+\cdots+f(0.9)]+f(1)\}\approx 1.090608$

(b) $M_{10}=\dfrac{1}{10}\left[f\left(\dfrac{1}{20}\right)+f\left(\dfrac{3}{20}\right)+f\left(\dfrac{5}{20}\right)+\cdots+f\left(\dfrac{19}{20}\right)\right]\approx 1.088840$

(c) $S_{10}=\dfrac{1}{10\cdot 3}[f(0)+4f(0.1)+2f(0.2)+\cdots+4f(0.9)+f(1)]\approx 1.089429$

f is concave upward, so the Trapezoidal Rule gives us an overestimate, the Midpoint Rule gives an underestimate, and we cannot tell whether Simpson's Rule gives us an overestimate or an underestimate.

65. $f(x)=\left(1+x^4\right)^{1/2}$, $f'(x)=\dfrac{1}{2}\left(1+x^4\right)^{-1/2}\left(4x^3\right)=2x^3\left(1+x^4\right)^{-1/2}$, $f''(x)=\left(2x^6+6x^2\right)\left(1+x^4\right)^{-3/2}$. A graph of f'' on $[0,1]$ shows that it has its maximum at $x=1$, so $\left|f''(x)\right|\leq f''(1)=\sqrt{8}$ on $[0,1]$. By taking $K=\sqrt{8}$, we find that the error in Exercise 63(a) is bounded by $\dfrac{K(b-a)^3}{12n^2}=\dfrac{\sqrt{8}}{1200}\approx 0.0024$, and in (b) by about $\dfrac{1}{2}(0.0024)=0.0012$.

Note: Another way to estimate K is to let $x=1$ in the factor $2x^6+6x^2$ (maximizing the numerator) and let $x=0$ in the factor $\left(1+x^4\right)^{-3/2}$ (minimizing the denominator). Doing so gives us $K=8$ and errors of 0.0067 and $0.003\overline{3}$.

Using $K=8$ for the Trapezoidal Rule, we have $|E_T|\leq\dfrac{K(b-a)^3}{12n^2}\leq 0.00001 \Leftrightarrow \dfrac{8(1-0)^3}{12n^2}\leq\dfrac{1}{100,000} \Leftrightarrow$

$n^2\geq\dfrac{800,000}{12} \Leftrightarrow n\gtrsim 258.2$, so we should take $n=259$.

For the Midpoint Rule, $|E_M|\leq\dfrac{K(b-a)^3}{24n^2}\leq 0.00001 \Leftrightarrow n^2\geq\dfrac{800,000}{24} \Leftrightarrow n\gtrsim 182.6$, so we should take $n=183$.

67. $\Delta t=\left(\dfrac{10}{60}-0\right)\Big/10=\dfrac{1}{60}$.

Distance traveled $=\int_0^{10} v\,dt\approx S_{10}=\dfrac{1}{60\cdot 3}[40+4(42)+2(45)+4(49)+2(52)$

$+4(54)+2(56)+4(57)+2(57)+4(55)+56]$

$=\dfrac{1}{180}(1544)=8.5\overline{7}$ mi

69. (a) $f(x) = \sin(\sin x)$. A CAS gives

$$f^{(4)}(x) = \sin(\sin x)\left[\cos^4 x + 7\cos^2 x - 3\right]$$

$$+ \cos(\sin x)\left[6\cos^2 x \sin x + \sin x\right]$$

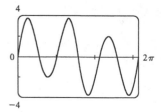

From the graph, we see that $\left|f^{(4)}(x)\right| < 3.8$ for $x \in [0, \pi]$.

(b) We use Simpson's Rule with $f(x) = \sin(\sin x)$ and $\Delta x = \frac{\pi}{10}$:

$$\int_0^\pi f(x)\,dx \approx \frac{\pi}{10 \cdot 3}\left[f(0) + 4f\left(\frac{\pi}{10}\right) + 2f\left(\frac{2\pi}{10}\right) + \cdots + 4f\left(\frac{9\pi}{10}\right) + f(\pi)\right] \approx 1.786721$$

From part (a), we know that $f^{(4)}(x) < 3.8$ on $[0, \pi]$, so we use Theorem 7.7.4 with $K = 3.8$, and estimate the error as $|E_S| \le \frac{3.8(\pi - 0)^5}{180(10)^4} \approx 0.000646$.

(c) If we want the error to be less than 0.00001, we must have $|E_S| \le \frac{3.8\pi^5}{180n^4} \le 0.00001$, so

$$n^4 \ge \frac{3.8\pi^5}{180(0.00001)} \approx 646{,}041.6 \quad\Rightarrow\quad n \ge 28.35.$$ Since n must be even for Simpson's Rule, we must have $n \ge 30$ to ensure the desired accuracy.

71. $\frac{x^3}{x^5 + 2} \le \frac{x^3}{x^5} = \frac{1}{x^2}$ for x in $[1, \infty)$. $\int_1^\infty \frac{1}{x^2}\,dx$ is convergent by (7.8.2) with $p = 2 > 1$. Therefore,

$\int_1^\infty \frac{x^3}{x^5 + 2}\,dx$ is convergent by the Comparison Theorem.

73. For x in $\left[0, \frac{\pi}{2}\right]$, $0 \le \cos^2 x \le \cos x$. For x in $\left[\frac{\pi}{2}, \pi\right]$, $\cos x \le 0 \le \cos^2 x$. Thus,

$$\text{area} = \int_0^{\pi/2}\left(\cos x - \cos^2 x\right)dx + \int_{\pi/2}^\pi\left(\cos^2 x - \cos x\right)dx$$

$$= \left[\sin x - \tfrac{1}{2}x - \tfrac{1}{4}\sin 2x\right]_0^{\pi/2} + \left[\tfrac{1}{2}x + \tfrac{1}{4}\sin 2x - \sin x\right]_{\pi/2}^\pi$$

$$= \left[\left(1 - \tfrac{\pi}{4}\right) - 0\right] + \left[\tfrac{\pi}{2} - \left(\tfrac{\pi}{4} - 1\right)\right] = 2$$

75. Using the formula for disks, the volume is

$$V = \int_0^{\pi/2} \pi\,[f(x)]^2\,dx = \pi\int_0^{\pi/2}\left(\cos^2 x\right)^2 dx = \pi\int_0^{\pi/2}\left[\tfrac{1}{2}(1 + \cos 2x)\right]^2 dx$$

$$= \tfrac{\pi}{4}\int_0^{\pi/2}\left(1 + \cos^2 2x + 2\cos 2x\right)dx = \tfrac{\pi}{4}\int_0^{\pi/2}\left[1 + \tfrac{1}{2}(1 + \cos 4x) + 2\cos 2x\right]dx$$

$$= \tfrac{\pi}{4}\left[\tfrac{3}{2}x + \tfrac{1}{2}\left(\tfrac{1}{4}\sin 4x\right) + 2\left(\tfrac{1}{2}\sin 2x\right)\right]_0^{\pi/2} = \tfrac{\pi}{4}\left[\left(\tfrac{3\pi}{4} + \tfrac{1}{8}\cdot 0 + 0\right) - 0\right] = \tfrac{3\pi^2}{16}$$

77. By the Fundamental Theorem of Calculus,

$$\int_0^\infty f'(x)\,dx = \lim_{t\to\infty}\int_0^t f'(x)\,dx = \lim_{t\to\infty}[f(t) - f(0)] = \lim_{t\to\infty} f(t) - f(0) = 0 - f(0) = -f(0)$$

79. Let $u = 1/x \Rightarrow x = 1/u \Rightarrow dx = -\left(1/u^2\right)du$.

$$\int_0^\infty \frac{\ln x}{1 + x^2}\,dx = \int_\infty^0 \frac{\ln(1/u)}{1 + 1/u^2}\left(-\frac{du}{u^2}\right) = \int_\infty^0 \frac{-\ln u}{u^2 + 1}(-du) = \int_\infty^0 \frac{\ln u}{1 + u^2}\,du = -\int_0^\infty \frac{\ln u}{1 + u^2}\,du$$

Therefore $\int_0^\infty \frac{\ln x}{1 + x^2}\,dx = -\int_0^\infty \frac{\ln x}{1 + x^2}\,dx = 0$.

Problems Plus

1.

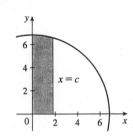

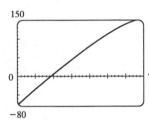

By symmetry, the problem can be reduced to finding the line $x = c$ such that the shaded area is one-third of the area of the quarter-circle. The equation of the circle is $y = \sqrt{49 - x^2}$, so we require that

$\int_0^c \sqrt{49 - x^2}\,dx = \frac{1}{3} \cdot \frac{1}{4}\pi\,(7)^2$ $\Leftrightarrow$

$\left[\frac{1}{2}x\sqrt{49 - x^2} + \frac{49}{2}\sin^{-1}(x/7)\right]_0^c = \frac{49}{12}\pi$ (by Formula 30) $\Leftrightarrow$

$\frac{1}{2}c\sqrt{49 - c^2} + \frac{49}{2}\sin^{-1}(c/7) = \frac{49}{12}\pi$.

This equation would be difficult to solve exactly, so we plot the left-hand side as a function of c, and find that the equation holds for $c \approx 1.85$. So the cuts should be made at distances of about 1.85 inches from the center of the pizza.

3. The given integral represents the difference of the shaded areas, which appears to be 0. It can be calculated by integrating with respect to either x or y, so we find x in terms of y for each curve: $y = \sqrt[3]{1 - x^7}$ $\Rightarrow$ $x = \sqrt[7]{1 - y^3}$ and $y = \sqrt[7]{1 - x^3}$ $\Rightarrow$ $x = \sqrt[3]{1 - y^7}$, so

$$\int_0^1 \left(\sqrt[3]{1 - y^7} - \sqrt[7]{1 - y^3}\right) dy = \int_0^1 \left(\sqrt[7]{1 - x^3} - \sqrt[3]{1 - x^7}\right) dx$$

But this equation is of the form $z = -z$. So

$$\int_0^1 \left(\sqrt[3]{1 - x^7} - \sqrt[7]{1 - x^3}\right) dx = 0$$

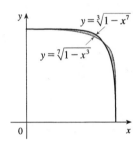

5. Recall that $\cos A \cos B = \frac{1}{2}[\cos(A + B) + \cos(A - B)]$. So

$$f(x) = \int_0^\pi \cos t \cos(x - t)\,dt = \frac{1}{2}\int_0^\pi [\cos(t + x - t) + \cos(t - x + t)]\,dt$$

$$= \frac{1}{2}\int_0^\pi [\cos x + \cos(2t - x)]\,dt = \frac{1}{2}\left[t\cos x + \frac{1}{2}\sin(2t - x)\right]_0^\pi$$

$$= \frac{\pi}{2}\cos x + \frac{1}{4}\sin(2\pi - x) - \frac{1}{4}\sin(-x) = \frac{\pi}{2}\cos x + \frac{1}{4}\sin(-x) - \frac{1}{4}\sin(-x)$$

$$= \frac{\pi}{2}\cos x$$

The minimum of $\cos x$ on this domain is -1, so the minimum value of $f(x)$ is $f(\pi) = -\frac{\pi}{2}$.

7. In accordance with the hint, we let $I_k = \int_0^1 \left(1 - x^2\right)^k dx$, and we find an expression for I_{k+1} in terms of I_k. We integrate I_{k+1} by parts with $u = \left(1 - x^2\right)^{k+1}$ $\Rightarrow$ $du = (k+1)\left(1-x^2\right)^k(-2x)$, $dv = dx$ $\Rightarrow$ $v = x$, and then split the remaining integral into identifiable quantities:

$$I_{k+1} = x\left(1-x^2\right)^{k+1}\Big|_0^1 + 2\,(k+1)\int_0^1 x^2\left(1-x^2\right)^k dx = (2k+2)\int_0^1\left(1-x^2\right)^k\left[1-\left(1-x^2\right)\right]dx$$

$$= (2k+2)\left(I_k - I_{k+1}\right)$$

So $I_{k+1}\left[1 + (2k+2)\right] = (2k+2)\,I_k$ $\Rightarrow$ $I_{k+1} = \frac{2k+2}{2k+3}I_k$. Now to complete the proof, we use induction:

$I_0 = 1 = \dfrac{2^0\,(0!)^2}{1!}$, so the formula holds for $n = 0$. Now suppose it holds for $n = k$. Then

$$I_{k+1} = \frac{2k+2}{2k+3}I_k = \frac{2k+2}{2k+3}\left[\frac{2^{2k}\,(k!)^2}{(2k+1)!}\right] = \frac{2\,(k+1)\,2^{2k}\,(k!)^2}{(2k+3)\,(2k+1)!} = \frac{2\,(k+1)}{2k+2}\cdot\frac{2\,(k+1)\,2^{2k}\,(k!)^2}{(2k+3)\,(2k+1)!}$$

$$= \frac{[2\,(k+1)]^2\,2^{2k}\,(k!)^2}{(2k+3)\,(2k+2)\,(2k+1)!} = \frac{2^{2(k+1)}\,[(k+1)!]^2}{[2\,(k+1)+1]!}$$

So by induction, the formula holds for all integers $n \geq 0$.

9. $0 < a < b$. Now

$$\int_0^1 [bx + a\,(1-x)]^t\,dx = \int_a^b \frac{u^t}{(b-a)}\,du \ \ [\text{put }u = bx + a\,(1-x)] = \left[\frac{u^{t+1}}{(t+1)\,(b-a)}\right]_a^b = \frac{b^{t+1} - a^{t+1}}{(t+1)\,(b-a)}.$$

Now let $y = \lim\limits_{t\to 0}\left[\dfrac{b^{t+1} - a^{t+1}}{(t+1)\,(b-a)}\right]^{1/t}$. Then $\ln y = \lim\limits_{t\to 0}\left[\dfrac{1}{t}\ln\dfrac{b^{t+1} - a^{t+1}}{(t+1)\,(b-a)}\right]$. This limit is of the form $0/0$, so we can apply l'Hospital's Rule to get

$$\ln y = \lim_{t\to 0}\left[\frac{b^{t+1}\ln b - a^{t+1}\ln a}{b^{t+1} - a^{t+1}} - \frac{1}{t+1}\right] = \frac{b\ln b - a\ln a}{b-a} - 1 = \frac{b\ln b}{b-a} - \frac{a\ln a}{b-a} - \ln e = \ln\frac{b^{b/(b-a)}}{e a^{a/(b-a)}}.$$

Therefore, $y = e^{-1}\left(\dfrac{b^b}{a^a}\right)^{1/(b-a)}$.

11. We integrate by parts with $u = \dfrac{1}{\ln(1+x+t)}$, $dv = \sin t\,dt$, so $du = \dfrac{-1}{(1+x+t)\,[\ln(1+x+t)]^2}$ and $v = -\cos t$. The integral becomes

$$I = \int_0^\infty \frac{\sin t\,dt}{\ln(1+x+t)} = \lim_{b\to\infty}\left(\left[\frac{-\cos t}{\ln(1+x+t)}\right]_0^b - \int_0^b \frac{\cos t\,dt}{(1+x+t)\,[\ln(1+x+t)]^2}\right)$$

$$= \lim_{b\to\infty}\frac{-\cos b}{\ln(1+x+b)} + \frac{1}{\ln(1+x)} + \int_0^\infty \frac{-\cos t\,dt}{(1+x+t)\,[\ln(1+x+t)]^2} = \frac{1}{\ln(1+x)} + J$$

where $J = \int_0^\infty \dfrac{-\cos t\,dt}{(1+x+t)\,[\ln(1+x+t)]^2}$. Now $-1 \leq -\cos t \leq 1$ for all t; in fact, the inequality is strict

except at isolated points. So $-\displaystyle\int_0^\infty \frac{dt}{(1+x+t)\,[\ln(1+x+t)]^2} < J < \int_0^\infty \frac{dt}{(1+x+t)\,[\ln(1+x+t)]^2}$ $\Leftrightarrow$

$-\dfrac{1}{\ln(1+x)} < J < \dfrac{1}{\ln(1+x)}$ $\Leftrightarrow$ $0 < I < \dfrac{2}{\ln(1+x)}$.

Further Applications of Integration

8.1 Arc Length

1. $y = 2 - 3x \implies L = \int_{-2}^{1} \sqrt{1 + (dy/dx)^2}\, dx = \int_{-2}^{1} \sqrt{1 + (-3)^2}\, dx = \sqrt{10}[1 - (-2)] = 3\sqrt{10}$.

The arc length can be calculated using the distance formula, since the curve is a line segment, so

$$L = [\text{distance from } (-2, 8) \text{ to } (1, -1)] = \sqrt{[1 - (-2)]^2 + [(-1) - 8]^2} = \sqrt{90} = 3\sqrt{10}$$

3. $y^2 = (x - 1)^3, y = (x - 1)^{3/2} \implies \dfrac{dy}{dx} = \frac{3}{2}(x - 1)^{1/2} \implies 1 + \left(\dfrac{dy}{dx}\right)^2 = 1 + \frac{9}{4}(x - 1)$. So

$$L = \int_1^2 \sqrt{1 + \tfrac{9}{4}(x - 1)}\, dx = \int_1^2 \sqrt{\tfrac{9}{4}x - \tfrac{5}{4}}\, dx = \left[\tfrac{4}{9} \cdot \tfrac{2}{3}\left(\tfrac{9}{4}x - \tfrac{5}{4}\right)^{3/2}\right]_1^2 = \tfrac{13\sqrt{13} - 8}{27}$$

5.

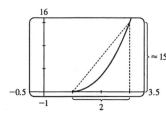

From the figure, the length of the curve is slightly larger than the hypotenuse of the triangle formed by the points $(1, 0)$, $(3, 0)$, and $(3, f(3)) \approx (3, 15)$, where $y = f(x) = \frac{2}{3}(x^2 - 1)^{3/2}$. This length is about $\sqrt{15^2 + 2^2} \approx 15$, so we might estimate the length to be 15.5.

$y = \frac{2}{3}(x^2 - 1)^{3/2} \implies y' = (x^2 - 1)^{1/2}(2x) \implies$

$1 + (y')^2 = 1 + 4x^2(x^2 - 1) = 4x^4 - 4x^2 + 1 = (2x^2 - 1)^2$, so, using the fact that $2x^2 - 1 > 0$ for $1 \le x \le 3$,

$$L = \int_1^3 \sqrt{(2x^2 - 1)^2}\, dx = \int_1^3 |2x^2 - 1|\, dx = \int_1^3 (2x^2 - 1)\, dx = \left[\tfrac{2}{3}x^3 - x\right]_1^3 = (18 - 3) - \left(\tfrac{2}{3} - 1\right) = \tfrac{46}{3} = 15.\overline{3}$$

7. $y = \frac{1}{3}(x^2 + 2)^{3/2} \implies dy/dx = \frac{1}{2}(x^2 + 2)^{1/2}(2x) = x\sqrt{x^2 + 2} \implies$

$1 + (dy/dx)^2 = 1 + x^2(x^2 + 2) = (x^2 + 1)^2$. So $L = \int_0^1 (x^2 + 1)\, dx = \left[\tfrac{1}{3}x^3 + x\right]_0^1 = \tfrac{4}{3}$.

9. $y = \dfrac{x^4}{4} + \dfrac{1}{8x^2} \implies \dfrac{dy}{dx} = x^3 - \dfrac{1}{4x^3} \implies 1 + \left(\dfrac{dy}{dx}\right)^2 = 1 + x^6 - \dfrac{1}{2} + \dfrac{1}{16x^6} = x^6 + \dfrac{1}{2} + \dfrac{1}{16x^6}$. So

$$L = \int_1^3 \left(x^3 + \tfrac{1}{4}x^{-3}\right) dx = \left[\tfrac{1}{4}x^4 - \tfrac{1}{8}x^{-2}\right]_1^3 = \left(\tfrac{81}{4} - \tfrac{1}{72}\right) - \left(\tfrac{1}{4} - \tfrac{1}{8}\right) = \tfrac{181}{9}$$

11. $y = \ln(\sec x) \implies \dfrac{dy}{dx} = \dfrac{\sec x \tan x}{\sec x} = \tan x \implies 1 + \left(\dfrac{dy}{dx}\right)^2 = 1 + \tan^2 x = \sec^2 x$, so

$$L = \int_0^{\pi/4} \sqrt{\sec^2 x}\, dx = \int_0^{\pi/4} \sec x\, dx = [\ln(\sec x + \tan x)]_0^{\pi/4} = \ln\left(\sqrt{2} + 1\right) - \ln(1 + 0) = \ln\left(\sqrt{2} + 1\right)$$

13. $y = \ln(1 - x^2) \Rightarrow \dfrac{dy}{dx} = \dfrac{-2x}{1 - x^2} \Rightarrow 1 + \left(\dfrac{dy}{dx}\right)^2 = 1 + \dfrac{4x^2}{(1 - x^2)^2} = \dfrac{(1 + x^2)^2}{(1 - x^2)^2}$. So

$$L = \int_0^{1/2} \frac{1 + x^2}{1 - x^2}\, dx = \int_0^{1/2} \left[-1 + \frac{2}{(1 - x)(1 + x)}\right] dx = \int_0^{1/2} \left[-1 + \frac{1}{1 + x} + \frac{1}{1 - x}\right] dx$$

$$= [-x + \ln(1 + x) - \ln(1 - x)]_0^{1/2} = -\tfrac{1}{2} + \ln\tfrac{3}{2} - \ln\tfrac{1}{2} - 0 = \ln 3 - \tfrac{1}{2}$$

15. $y = \cosh x \Rightarrow y' = \sinh x \Rightarrow 1 + (y')^2 = 1 + \sinh^2 x = \cosh^2 x$. So

$$L = \int_0^1 \cosh x\, dx = [\sinh x]_0^1 = \sinh 1 = \tfrac{1}{2}(e - 1/e)$$

17. $y = e^x \Rightarrow y' = e^x \Rightarrow 1 + (y')^2 = 1 + e^{2x}$. So

$$L = \int_0^1 \sqrt{1 + e^{2x}}\, dx = \int_1^e \sqrt{1 + u^2}\,\frac{du}{u} \quad [u = e^x, \text{ so } x = \ln u,\, dx = du/u]$$

$$= \int_1^e \frac{\sqrt{1 + u^2}}{u^2}\, u\, du = \int_{\sqrt{2}}^{\sqrt{1+e^2}} \frac{v}{v^2 - 1}\, v\, dv \quad \left[v = \sqrt{1 + u^2}, \text{ so } v^2 = 1 + u^2,\, v\, dv = u\, du\right]$$

$$= \int_{\sqrt{2}}^{\sqrt{1+e^2}} \left(1 + \frac{1/2}{v - 1} - \frac{1/2}{v + 1}\right) dv = \left[v + \frac{1}{2}\ln\frac{v - 1}{v + 1}\right]_{\sqrt{2}}^{\sqrt{1+e^2}}$$

$$= \sqrt{1 + e^2} - \sqrt{2} + \frac{1}{2}\ln\frac{\sqrt{1 + e^2} - 1}{\sqrt{1 + e^2} + 1} - \frac{1}{2}\ln\frac{\sqrt{2} - 1}{\sqrt{2} + 1}$$

$$= \sqrt{1 + e^2} - \sqrt{2} + \ln\left(\sqrt{1 + e^2} - 1\right) - 1 - \ln\left(\sqrt{2} - 1\right)$$

Or: Use Formula 23 for $\int \left(\sqrt{1 + u^2}/u\right) du$, or substitute $u = \tan\theta$.

19. $y = x^3 \Rightarrow y' = 3x^2 \Rightarrow 1 + (y')^2 = 1 + 9x^4$. So $L = \int_0^1 \sqrt{1 + 9x^4}\, dx$.

21. $y = e^x \cos x \Rightarrow y' = e^x(\cos x - \sin x) \Rightarrow$

$$1 + (y')^2 = 1 + e^{2x}\left(\cos^2 x - 2\cos x \sin x + \sin^2 x\right) = 1 + e^{2x}(1 - \sin 2x)$$

So $L = \int_0^{\pi/2} \sqrt{1 + e^{2x}(1 - \sin 2x)}\, dx$.

23. $y = x^3 \Rightarrow 1 + (y')^2 = 1 + (3x^2)^2 = 1 + 9x^4 \Rightarrow L = \int_0^1 \sqrt{1 + 9x^4}\, dx$.

Let $f(x) = \sqrt{1 + 9x^4}$. Then by Simpson's Rule with $n = 10$,

$L \approx \frac{1/10}{3}[f(0) + 4f(0.1) + 2f(0.2) + 4f(0.3) + \cdots + 2f(0.8) + 4f(0.9) + f(1)] \approx 1.548$.

25. $y = \sin x$, $1 + (dy/dx)^2 = 1 + \cos^2 x$, $L = \int_0^\pi \sqrt{1 + \cos^2 x}\, dx$. Let $g(x) = \sqrt{1 + \cos^2 x}$. Then

$$L \approx \tfrac{\pi/10}{3}\left[g(0) + 4g\left(\tfrac{\pi}{10}\right) + 2g\left(\tfrac{\pi}{5}\right) + 4g\left(\tfrac{3\pi}{10}\right) + 2g\left(\tfrac{2\pi}{5}\right) + 4g\left(\tfrac{\pi}{2}\right)\right.$$

$$\left. + 2g\left(\tfrac{3\pi}{5}\right) + 4g\left(\tfrac{7\pi}{10}\right) + 2g\left(\tfrac{4\pi}{5}\right) + 4g\left(\tfrac{9\pi}{10}\right) + g(\pi)\right] \approx 3.820$$

27. (a)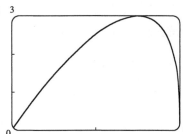

(b)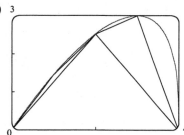

Let $f(x) = y = x\sqrt[3]{4-x}$. The polygon with one side is just the line segment joining the points $(0, f(0)) = (0, 0)$ and $(4, f(4)) = (4, 0)$, and its length is 4. The polygon with two sides joins the points $(0, 0)$, $(2, f(2)) = \left(2, 2\sqrt[3]{2}\right)$ and $(4, 0)$. Its length is

$$\sqrt{(2-0)^2 + \left(2\sqrt[3]{2}-0\right)^2} + \sqrt{(4-2)^2 + \left(0 - 2\sqrt[3]{2}\right)^2} = 2\sqrt{4 + 2^{8/3}} \approx 6.43$$

Similarly, the inscribed polygon with four sides joins the points $(0, 0)$, $\left(1, \sqrt[3]{3}\right)$, $\left(2, 2\sqrt[3]{2}\right)$, $(3, 3)$, and $(4, 0)$, so its length is

$$\sqrt{1 + \left(\sqrt[3]{3}\right)^2} + \sqrt{1 + \left(2\sqrt[3]{2} - \sqrt[3]{3}\right)^2} + \sqrt{1 + \left(3 - 2\sqrt[3]{2}\right)^2} + \sqrt{1 + 9} \approx 7.50$$

(c) Using the arc length formula with $\dfrac{dy}{dx} = x\left[\frac{1}{3}(4-x)^{-2/3}(-1)\right] + \sqrt[3]{4-x} = \dfrac{12 - 4x}{3(4-x)^{2/3}}$, the length of the

curve is $L = \displaystyle\int_0^4 \sqrt{1 + \left(\frac{dy}{dx}\right)^2}\, dx = \int_0^4 \sqrt{1 + \left[\frac{12-4x}{3(4-x)^{2/3}}\right]^2}\, dx.$

(d) According to a CAS, the length of the curve is $L \approx 7.7988$. The actual value is larger than any of the approximations in part (b). This is always true, since any approximating straight line between two points on the curve is shorter than the length of the curve between the two points.

29. $x = \ln\left(1 - y^2\right) \;\Rightarrow\; \dfrac{dx}{dy} = \dfrac{-2y}{1 - y^2} \;\Rightarrow\; 1 + \left(\dfrac{dx}{dy}\right)^2 = 1 + \dfrac{4y^2}{\left(1 - y^2\right)^2} = \dfrac{\left(1 + y^2\right)^2}{\left(1 - y^2\right)^2}.$ So

$$L = \int_0^{1/2} \sqrt{\frac{\left(1 + y^2\right)^2}{\left(1 - y^2\right)^2}}\, dy = \int_0^{1/2} \frac{1 + y^2}{1 - y^2}\, dy = \ln 3 - \tfrac{1}{2} \text{ [from a CAS] } \approx 0.599$$

31. $y^{2/3} = 1 - x^{2/3} \;\Rightarrow\; y = \left(1 - x^{2/3}\right)^{3/2} \;\Rightarrow$

$\dfrac{dy}{dx} = \tfrac{3}{2}\left(1 - x^{2/3}\right)^{1/2}\left(-\tfrac{2}{3}x^{-1/3}\right) = -x^{-1/3}\left(1 - x^{2/3}\right)^{1/2} \;\Rightarrow$

$\left(\dfrac{dy}{dx}\right)^2 = x^{-2/3}\left(1 - x^{2/3}\right) = x^{-2/3} - 1.$ Thus

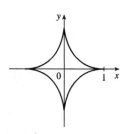

$$L = 4\int_0^1 \sqrt{1 + \left(x^{-2/3} - 1\right)}\, dx = 4\int_0^1 x^{-1/3}\, dx = 4 \lim_{t \to 0^+} \left[\tfrac{3}{2}x^{2/3}\right]_t^1 = 6$$

33. $y = 2x^{3/2} \;\Rightarrow\; y' = 3x^{1/2} \;\Rightarrow\; 1 + \left(y'\right)^2 = 1 + 9x.$ The arc length function with starting point $P_0\,(1, 2)$ is

$$s(x) = \int_1^x \sqrt{1 + 9t}\, dt = \left[\tfrac{2}{27}(1 + 9t)^{3/2}\right]_1^x = \tfrac{2}{27}\left[(1 + 9x)^{3/2} - 10\sqrt{10}\right]$$

35. The prey hits the ground when $y = 0$ $\Leftrightarrow$ $180 - \frac{1}{45}x^2 = 0$ $\Leftrightarrow$ $x^2 = 45 \cdot 180$ $\Rightarrow$ $x = \sqrt{8100} = 90$, since

x must be positive. $y' = -\frac{2}{45}x$ $\Rightarrow$ $1 + (y')^2 = 1 + \frac{4}{45^2}x^2$, so the distance traveled by the prey is

$$L = \int_0^{90} \sqrt{1 + \frac{4}{45^2}x^2}\, dx = \int_0^4 \sqrt{1 + u^2}\left(\frac{45}{2}\, du\right) \quad [u = \frac{2}{45}x,\ du = \frac{2}{45}\, dx]$$

$$\overset{21}{=} \frac{45}{2}\left[\frac{1}{2}u\sqrt{1+u^2} + \frac{1}{2}\ln\left(u + \sqrt{1+u^2}\right)\right]_0^4 = \frac{45}{2}\left[2\sqrt{17} + \frac{1}{2}\ln\left(4 + \sqrt{17}\right)\right]$$

$$= 45\sqrt{17} + \frac{45}{4}\ln\left(4 + \sqrt{17}\right) \approx 209.1 \text{ m}$$

37. The sine wave has amplitude 1 and period 14, since it goes through two periods in a distance of 28 in., so its

equation is $y = 1\sin\left(\frac{2\pi}{14}x\right) = \sin\left(\frac{\pi}{7}x\right)$. The width w of the flat metal sheet needed to make the panel is the arc

length of the sine curve from $x = 0$ to $x = 28$. We set up the integral to evaluate w using the arc length formula

with $\frac{dy}{dx} = \frac{\pi}{7}\cos\left(\frac{\pi}{7}x\right)$: $L = \int_0^{28} \sqrt{1 + \left[\frac{\pi}{7}\cos\left(\frac{\pi}{7}x\right)\right]^2}\, dx = 2\int_0^{14} \sqrt{1 + \left[\frac{\pi}{7}\cos\left(\frac{\pi}{7}x\right)\right]^2}\, dx$. This integral would be

very difficult to evaluate exactly, so we use a CAS, and find that $L \approx 29.36$ inches.

39. $y = \int_1^x \sqrt{t^3 - 1}\, dt$ $\Rightarrow$ $\frac{dy}{dx} = \sqrt{x^3 - 1}$ (by FTC1) $\Rightarrow$ $1 + \left(\frac{dy}{dx}\right)^2 = 1 + \left(\sqrt{x^3 - 1}\right)^2 = x^3$ $\Rightarrow$

$$L = \int_1^4 \sqrt{x^3}\, dx = \int_1^4 x^{3/2}\, dx = \frac{2}{5}\left[x^{5/2}\right]_1^4 = \frac{2}{5}(32 - 1) = \frac{62}{5} = 12.4$$

8.2 Area of a Surface of Revolution

1. $y = \ln x$ $\Rightarrow$ $ds = \sqrt{1 + (dy/dx)^2}\, dx = \sqrt{1 + (1/x)^2}\, dx$ $\Rightarrow$ $S = \int_1^3 2\pi\, (\ln x)\sqrt{1 + (1/x)^2}\, dx$ [by (7)]

3. $y = \sec x$ $\Rightarrow$ $ds = \sqrt{1 + (dy/dx)^2}\, dx = \sqrt{1 + (\sec x \tan x)^2}\, dx$ $\Rightarrow$

$S = \int_0^{\pi/4} 2\pi x\sqrt{1 + (\sec x \tan x)^2}\, dx$ [by (8)]

5. $y = x^3$ $\Rightarrow$ $y' = 3x^2$. So

$$S = \int_0^2 2\pi y\sqrt{1 + (y')^2}\, dx = 2\pi \int_0^2 x^3\sqrt{1 + 9x^4}\, dx \quad (u = 1 + 9x^4,\ du = 36x^3\, dx)$$

$$= \frac{2\pi}{36}\int_1^{145} \sqrt{u}\, du = \frac{\pi}{18}\left[\frac{2}{3}u^{3/2}\right]_1^{145} = \frac{\pi}{27}\left(145\sqrt{145} - 1\right)$$

7. $y = \sqrt{x}$ $\Rightarrow$ $1 + (dy/dx)^2 = 1 + \left[1/(2\sqrt{x})\right]^2 = 1 + 1/(4x)$. So

$$S = \int_4^9 2\pi y\sqrt{1 + \left(\frac{dy}{dx}\right)^2}\, dx = \int_4^9 2\pi\sqrt{x}\sqrt{1 + \frac{1}{4x}}\, dx = 2\pi \int_4^9 \sqrt{x + \frac{1}{4}}\, dx$$

$$= 2\pi\left[\frac{2}{3}\left(x + \frac{1}{4}\right)^{3/2}\right]_4^9 = \frac{4\pi}{3}\left[\frac{1}{8}(4x + 1)^{3/2}\right]_4^9 = \frac{\pi}{6}\left(37\sqrt{37} - 17\sqrt{17}\right)$$

9. $y = \sin x$ $\Rightarrow$ $1 + (dy/dx)^2 = 1 + \cos^2 x$. So

$$S = 2\pi \int_0^\pi \sin x\sqrt{1 + \cos^2 x}\, dx = 2\pi \int_{-1}^1 \sqrt{1 + u^2}\, du \quad (u = -\cos x,\ du = \sin x\, dx)$$

$$= 4\pi \int_0^1 \sqrt{1 + u^2}\, du = 4\pi \int_0^{\pi/4} \sec^3\theta\, d\theta \quad (u = \tan\theta,\ du = \sec^2\theta\, d\theta)$$

$$= 2\pi\left[\sec\theta \tan\theta + \ln|\sec\theta + \tan\theta|\right]_0^{\pi/4} = 2\pi\left[\sqrt{2} + \ln\left(\sqrt{2} + 1\right)\right]$$

11. $y = \cosh x \Rightarrow 1 + (dy/dx)^2 = 1 + \sinh^2 x = \cosh^2 x$. So

$$S = 2\pi \int_0^1 \cosh x \cosh x \, dx = 2\pi \int_0^1 \tfrac{1}{2} (1 + \cosh 2x) \, dx = \pi \left[x + \tfrac{1}{2} \sinh 2x \right]_0^1$$

$$= \pi \left(1 + \tfrac{1}{2} \sinh 2 \right) \text{ or } \pi \left[1 + \tfrac{1}{4} \left(e^2 - e^{-2} \right) \right]$$

13. $x = \tfrac{1}{3} (y^2 + 2)^{3/2} \Rightarrow dx/dy = \tfrac{1}{2} (y^2 + 2)^{1/2} (2y) = y\sqrt{y^2 + 2} \Rightarrow$
$1 + (dx/dy)^2 = 1 + y^2 (y^2 + 2) = (y^2 + 1)^2$. So

$$S = 2\pi \int_1^2 y (y^2 + 1) \, dy = 2\pi \left[\tfrac{1}{4} y^4 + \tfrac{1}{2} y^2 \right]_1^2 = 2\pi \left(4 + 2 - \tfrac{1}{4} - \tfrac{1}{2} \right) = \tfrac{21\pi}{2}$$

15. $y = \sqrt[3]{x} \Rightarrow x = y^3 \Rightarrow 1 + (dx/dy)^2 = 1 + 9y^4$. So

$$S = 2\pi \int_1^2 x \sqrt{1 + (dx/dy)^2} \, dy = 2\pi \int_1^2 y^3 \sqrt{1 + 9y^4} \, dy = \tfrac{2\pi}{36} \int_1^2 \sqrt{1 + 9y^4} \, 36y^3 \, dy$$

$$= \tfrac{\pi}{18} \left[\tfrac{2}{3} \left(1 + 9y^4 \right)^{3/2} \right]_1^2 = \tfrac{\pi}{27} \left(145\sqrt{145} - 10\sqrt{10} \right)$$

17. $x = e^{2y} \Rightarrow 1 + (dx/dy)^2 = 1 + 4e^{4y}$. So

$$S = 2\pi \int_0^{1/2} e^{2y} \sqrt{1 + (2e^{2y})^2} \, dy = 2\pi \int_2^{2e} \sqrt{1 + u^2} \, \tfrac{1}{4} \, du \quad (u = 2e^{2y}, \, du = 4e^{2y} \, dy)$$

$$= \tfrac{\pi}{2} \int_2^{2e} \sqrt{1 + u^2} \, du = \tfrac{\pi}{2} \left[\tfrac{1}{2} u\sqrt{1 + u^2} + \tfrac{1}{2} \ln \left| u + \sqrt{1 + u^2} \right| \right]_2^{2e} \quad (u = \tan\theta \text{ or use Formula 21})$$

$$= \tfrac{\pi}{2} \left[e\sqrt{1 + 4e^2} + \tfrac{1}{2} \ln \left(2e + \sqrt{1 + 4e^2} \right) - \sqrt{5} - \tfrac{1}{2} \ln \left(2 + \sqrt{5} \right) \right]$$

$$= \tfrac{\pi}{4} \left[2e\sqrt{1 + 4e^2} - 2\sqrt{5} + \ln \left(\frac{2e + \sqrt{1 + 4e^2}}{2 + \sqrt{5}} \right) \right]$$

19. $x = \dfrac{1}{2\sqrt{2}} (y^2 - \ln y) \Rightarrow \dfrac{dx}{dy} = \dfrac{1}{2\sqrt{2}} \left(2y - \dfrac{1}{y} \right) \Rightarrow$

$$1 + \left(\frac{dx}{dy} \right)^2 = 1 + \frac{1}{8} \left(2y - \frac{1}{y} \right)^2 = 1 + \frac{1}{8} \left(4y^2 - 4 + \frac{1}{y^2} \right) = \frac{1}{8} \left(4y^2 + 4 + \frac{1}{y^2} \right) = \left[\frac{1}{2\sqrt{2}} \left(2y + \frac{1}{y} \right) \right]^2$$

So

$$S = 2\pi \int_1^2 \frac{1}{2\sqrt{2}} (y^2 - \ln y) \frac{1}{2\sqrt{2}} \left(2y + \frac{1}{y} \right) dy = \frac{\pi}{4} \int_1^2 \left(2y^3 + y - 2y \ln y - \frac{\ln y}{y} \right) dy$$

$$= \tfrac{\pi}{4} \left[\tfrac{1}{2} y^4 + \tfrac{1}{2} y^2 - y^2 \ln y + \tfrac{1}{2} y^2 - \tfrac{1}{2} (\ln y)^2 \right]_1^2 = \tfrac{\pi}{8} \left[y^4 + 2y^2 - 2y^2 \ln y - (\ln y)^2 \right]_1^2$$

$$= \tfrac{\pi}{8} \left[16 + 8 - 8 \ln 2 - (\ln 2)^2 - 1 - 2 \right] = \tfrac{\pi}{8} \left[21 - 8 \ln 2 - (\ln 2)^2 \right]$$

21. With $f(x) = x^4 \sqrt{16x^6 + 1}$,

$$S = 2\pi \int_0^1 x^4 \sqrt{1 + (4x^3)^2} \, dx = 2\pi \int_0^1 x^4 \sqrt{16x^6 + 1} \, dx$$

$$\approx 2\pi \tfrac{1/10}{3} \left[f(0) + 4f(0.1) + 2f(0.2) + 4f(0.3) + 2f(0.4) \right.$$

$$\left. + 4f(0.5) + 2f(0.6) + 4f(0.7) + 2f(0.8) + 4f(0.9) + f(1) \right] \approx 3.44$$

23. $y = 1/x \implies ds = \sqrt{1 + (dy/dx)^2}\, dx = \sqrt{1 + (-1/x^2)^2}\, dx = \sqrt{1 + 1/x^4}\, dx \implies$

$$S = \int_1^2 2\pi \cdot \frac{1}{x} \sqrt{1 + \frac{1}{x^4}}\, dx = 2\pi \int_1^2 \frac{\sqrt{x^4 + 1}}{x^3}\, dx = 2\pi \int_1^4 \frac{\sqrt{u^2 + 1}}{u^2}\left(\tfrac{1}{2}\, du\right) \quad [u = x^2, du = 2x\, dx]$$

$$= \pi \int_1^4 \frac{\sqrt{1 + u^2}}{u^2}\, du \overset{24}{=} \pi \left[-\frac{\sqrt{1 + u^2}}{u} + \ln\left(u + \sqrt{1 + u^2}\right) \right]_1^4$$

$$= \pi\left[-\frac{\sqrt{17}}{4} + \ln\left(4 + \sqrt{17}\right) + \frac{\sqrt{2}}{1} - \ln\left(1 + \sqrt{2}\right) \right] = \pi\left[\sqrt{2} - \frac{\sqrt{17}}{4} + \ln\left(\frac{4 + \sqrt{17}}{1 + \sqrt{2}}\right) \right]$$

25. $y = x^3$ and $0 \le y \le 1 \implies y' = 3x^2$ and $0 \le x \le 1$.

$$S = \int_0^1 2\pi x \sqrt{1 + (3x^2)^2}\, dx = 2\pi \int_0^3 \sqrt{1 + u^2}\, \tfrac{1}{6}\, du \quad [u = 3x^2, du = 6x\, dx]$$

$$= \frac{\pi}{3} \int_0^3 \sqrt{1 + u^2}\, du \overset{21}{=} \frac{\pi}{3}\left[\tfrac{1}{2} u\sqrt{1 + u^2} + \tfrac{1}{2} \ln\left(u + \sqrt{1 + u^2}\right) \right]_0^3$$

$$= \frac{\pi}{3}\left[\tfrac{3}{2}\sqrt{10} + \tfrac{1}{2} \ln\left(3 + \sqrt{10}\right) \right] = \frac{\pi}{6}\left[3\sqrt{10} + \ln\left(3 + \sqrt{10}\right) \right]$$

27. $S = 2\pi \int_1^\infty y \sqrt{1 + \left(\dfrac{dy}{dx}\right)^2}\, dx = 2\pi \int_1^\infty \dfrac{1}{x}\sqrt{1 + \dfrac{1}{x^4}}\, dx = 2\pi \int_1^\infty \dfrac{\sqrt{x^4 + 1}}{x^3}\, dx$. Rather than trying to evaluate

this integral, note that $\sqrt{x^4 + 1} > \sqrt{x^4} = x^2$ for $x > 0$. Thus, if the area is finite,

$$S = 2\pi \int_1^\infty \frac{\sqrt{x^4 + 1}}{x^3}\, dx > 2\pi \int_1^\infty \frac{x^2}{x^3}\, dx = 2\pi \int_1^\infty \frac{1}{x}\, dx$$

But we know that this integral diverges, so the area S is infinite.

29. The curve $8y^2 = x^2\left(1 - x^2\right)$ actually consists of two loops in the region described by the inequalities $|x| \le 1$,

$|y| \le \frac{\sqrt{2}}{8}$. (The maximum value of $|y|$ is attained when $|x| = \frac{1}{\sqrt{2}}$.) If we consider the loop in the region $x \ge 0$, the

surface area S it generates when rotated about the x-axis is calculated as follows: $16y\dfrac{dy}{dx} = 2x - 4x^3$, so

$$\left(\frac{dy}{dx}\right)^2 = \left(\frac{x - 2x^3}{8y}\right)^2 = \frac{x^2\left(1 - 2x^2\right)^2}{64y^2} = \frac{x^2\left(1 - 2x^2\right)^2}{8x^2\left(1 - x^2\right)} = \frac{\left(1 - 2x^2\right)^2}{8\left(1 - x^2\right)} \text{ for } x \ne 0, \pm 1. \text{ The formula also holds}$$

for $x = 0$ by continuity. $1 + \left(\dfrac{dy}{dx}\right)^2 = 1 + \dfrac{\left(1 - 2x^2\right)^2}{8\left(1 - x^2\right)} = \dfrac{9 - 12x^2 + 4x^4}{8\left(1 - x^2\right)} = \dfrac{\left(3 - 2x^2\right)^2}{8\left(1 - x^2\right)}$. So

$$S = 2\pi \int_0^1 \frac{\sqrt{x^2\left(1 - x^2\right)}}{2\sqrt{2}} \cdot \frac{3 - 2x^2}{2\sqrt{2}\sqrt{1 - x^2}}\, dx$$

$$= \frac{\pi}{4} \int_0^1 x\left(3 - 2x^2\right) dx = \frac{\pi}{4}\left[\tfrac{3}{2}x^2 - \tfrac{1}{2}x^4 \right]_0^1 = \frac{\pi}{4}\left(\tfrac{3}{2} - \tfrac{1}{2} \right) = \frac{\pi}{4}$$

31. $\dfrac{x^2}{a^2} + \dfrac{y^2}{b^2} = 1 \implies \dfrac{y\,(dy/dx)}{b^2} = -\dfrac{x}{a^2} \implies \dfrac{dy}{dx} = -\dfrac{b^2 x}{a^2 y} \implies$

$$1 + \left(\frac{dy}{dx}\right)^2 = 1 + \frac{b^4 x^2}{a^4 y^2} = \frac{b^4 x^2 + a^4 y^2}{a^4 y^2} = \frac{b^4 x^2 + a^4 b^2\left(1 - x^2/a^2\right)}{a^4 b^2\left(1 - x^2/a^2\right)} = \frac{a^4 b^2 + b^4 x^2 - a^2 b^2 x^2}{a^4 b^2 - a^2 b^2 x^2}$$

$$= \frac{a^4 + b^2 x^2 - a^2 x^2}{a^4 - a^2 x^2} = \frac{a^4 - \left(a^2 - b^2\right)x^2}{a^2\left(a^2 - x^2\right)}$$

The ellipsoid's surface area is twice the area generated by rotating the first quadrant portion of the ellipse about the x-axis. Thus,

$$S = 2\int_0^a 2\pi y \sqrt{1 + \left(\frac{dy}{dx}\right)^2}\, dx = 4\pi \int_0^a \frac{b}{a}\sqrt{a^2 - x^2}\, \frac{\sqrt{a^4 - (a^2 - b^2)\,x^2}}{a\sqrt{a^2 - x^2}}\, dx$$

$$= \frac{4\pi b}{a^2}\int_0^a \sqrt{a^4 - (a^2 - b^2)\,x^2}\, dx = \frac{4\pi b}{a^2}\int_0^{a\sqrt{a^2 - b^2}} \sqrt{a^4 - u^2}\, \frac{du}{\sqrt{a^2 - b^2}} \quad (u = \sqrt{a^2 - b^2}\, x)$$

$$\overset{30}{=} \frac{4\pi b}{a^2\sqrt{a^2 - b^2}}\left[\frac{u}{2}\sqrt{a^4 - u^2} + \frac{a^4}{2}\sin^{-1}\frac{u}{a^2}\right]_0^{a\sqrt{a^2 - b^2}}$$

$$= \frac{4\pi b}{a^2\sqrt{a^2 - b^2}}\left[\frac{a\sqrt{a^2 - b^2}}{2}\sqrt{a^4 - a^2(a^2 - b^2)} + \frac{a^4}{2}\sin^{-1}\frac{\sqrt{a^2 - b^2}}{a}\right] = 2\pi\left[b^2 + \frac{a^2 b\sin^{-1}\frac{\sqrt{a^2 - b^2}}{a}}{\sqrt{a^2 - b^2}}\right]$$

33. The analogue of $f(x_i^*)$ in the derivation of (4) is now $c - f(x_i^*)$, so

$$S = \lim_{n\to\infty}\sum_{i=1}^n 2\pi\,[c - f(x_i^*)]\sqrt{1 + [f'(x_i^*)]^2}\,\Delta x = \int_a^b 2\pi\,[c - f(x)]\sqrt{1 + [f'(x)]^2}\, dx$$

35. For the upper semicircle, $f(x) = \sqrt{r^2 - x^2}$, $f'(x) = -x/\sqrt{r^2 - x^2}$. The surface area generated is

$$S_1 = \int_{-r}^r 2\pi\left(r - \sqrt{r^2 - x^2}\right)\sqrt{1 + \frac{x^2}{r^2 - x^2}}\, dx = 4\pi\int_0^r\left(r - \sqrt{r^2 - x^2}\right)\frac{r}{\sqrt{r^2 - x^2}}\, dx$$

$$= 4\pi\int_0^r\left(\frac{r^2}{\sqrt{r^2 - x^2}} - r\right)dx$$

For the lower semicircle, $f(x) = -\sqrt{r^2 - x^2}$ and $f'(x) = \dfrac{x}{\sqrt{r^2 - x^2}}$, so $S_2 = 4\pi\int_0^r\left(\dfrac{r^2}{\sqrt{r^2 - x^2}} + r\right)dx$.

Thus, the total area is $S = S_1 + S_2 = 8\pi\int_0^r\left(\dfrac{r^2}{\sqrt{r^2 - x^2}}\right)dx = 8\pi\left[r^2\sin^{-1}\left(\dfrac{x}{r}\right)\right]_0^r = 8\pi r^2\left(\dfrac{\pi}{2}\right) = 4\pi^2 r^2$.

37. In the derivation of (4), we computed a typical contribution to the surface area to be $2\pi\dfrac{y_{i-1} + y_i}{2}\,|P_{i-1}P_i|$, the area of a frustum of a cone. When $f(x)$ is not necessarily positive, the approximations $y_i = f(x_i) \approx f(x_i^*)$ and $y_{i-1} = f(x_{i-1}) \approx f(x_i^*)$ must be replaced by $y_i = |f(x_i)| \approx |f(x_i^*)|$ and $y_{i-1} = |f(x_{i-1})| \approx |f(x_i^*)|$. Thus, $2\pi\dfrac{y_{i-1} + y_i}{2}\,|P_{i-1}P_i| \approx 2\pi\,|f(x_i^*)|\sqrt{1 + [f'(x_i^*)]^2}\,\Delta x$. Continuing with the rest of the derivation as before, we obtain $S = \int_a^b 2\pi\,|f(x)|\sqrt{1 + [f'(x)]^2}\, dx$.

8.3 Applications to Physics and Engineering

1. The weight density of water is $\delta = 62.5$ lb/ft^3.

(a) $P = \delta d \approx (62.5 \text{ lb/ft}^3)\,(3 \text{ ft}) = 187.5$ lb/ft^2

(b) $F = PA \approx (187.5 \text{ lb/ft}^2)\,(5 \text{ ft})\,(2 \text{ ft}) = 1875$ lb. (*A* is the area of the bottom of the tank.)

(c) As in Example 1, the area of the ith strip is $2\,(\Delta x)$ and the pressure is $\delta d = \delta x_i$. Thus,

$$F = \int_0^3 \delta x \cdot 2\,dx \approx (62.5)\,(2) \int_0^3 x\,dx = 125 \left[\tfrac{1}{2}x^2 \right]_0^3 = 125 \left(\tfrac{9}{2} \right) = 562.5 \text{ lb}$$

In Exercises 3–9, n is the number of subintervals of length Δx and x_i^* is a sample point in the ith subinterval $[x_{i-1}, x_i]$.

3. In the middle of the figure in the text, draw a vertical x-axis that increases in the downward direction. The area of the ith rectangular strip is $2\sqrt{100 - (x_i^*)^2}\,\Delta x$ and the pressure on the strip is $\rho g x_i^*$ [$\rho = 1000$ kg/m^3 and $g = 9.8$ m/s^2]. Thus, the hydrostatic force on the ith strip is the product $\rho g x_i^* 2\sqrt{100 - (x_i^*)^2}\,\Delta x$.

$$F = \lim_{n \to \infty} \sum_{i=1}^n \rho g x_i^* 2\sqrt{100 - (x_i^*)^2}\,\Delta x = \int_0^{10} \rho g x \cdot 2\sqrt{100 - x^2}\,dx = 9.8 \times 10^3 \int_0^{10} \sqrt{100 - x^2}\,2x\,dx$$

$$= 9.8 \times 10^3 \int_{100}^0 u^{1/2}\,(-du) \quad (u = 100 - x^2) \quad = 9.8 \times 10^3 \int_0^{100} u^{1/2}\,du$$

$$= 9.8 \times 10^3 \left[\tfrac{2}{3}u^{3/2} \right]_0^{100} = \tfrac{2}{3} \cdot 9.8 \times 10^6 \approx 6.5 \times 10^6 \text{ N}$$

5. Place an x-axis as in Exercise 3. Using similar triangles, $\dfrac{4 \text{ ft wide}}{6 \text{ ft high}} = \dfrac{w \text{ ft wide}}{x_i^* \text{ ft high}}$, so $w = \tfrac{4}{6}x_i^*$ and the area of the ith rectangular strip is $\tfrac{4}{6}x_i^*\,\Delta x$. The pressure on the strip is $\delta\,(x_i^* - 2)$ [$\delta = \rho g = 62.5$ lb/ft^3] and the hydrostatic force is $\delta\,(x_i^* - 2)\,\tfrac{4}{6}x_i^*\,\Delta x$.

$$F = \lim_{n \to \infty} \sum_{i=1}^n \delta\,(x_i^* - 2)\,\tfrac{4}{6}x_i^*\,\Delta x = \int_2^6 \delta\,(x - 2)\,\tfrac{2}{3}x\,dx = \tfrac{2}{3}\delta \int_2^6 (x^2 - 2x)\,dx = \tfrac{2}{3}\delta \left[\tfrac{1}{3}x^3 - x^2 \right]_2^6$$

$$= \tfrac{2}{3}\delta \left[36 - \left(-\tfrac{4}{3} \right) \right] = \tfrac{224}{9}\delta \approx 1.56 \times 10^3 \text{ lb}$$

7. Using similar triangles, $\dfrac{4 \text{ ft wide}}{8 \text{ ft high}} = \dfrac{a \text{ ft wide}}{x_i^* \text{ ft high}}$, so $a = \tfrac{1}{2}x_i^*$ and the width of the ith rectangular strip is $12 + 2a = 12 + x_i^*$. The area of the strip is $(12 + x_i^*)\,\Delta x$. The pressure on the strip is δx_i^*.

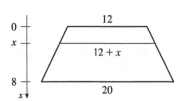

$$F = \lim_{n \to \infty} \sum_{i=1}^n \delta x_i^*\,(12 + x_i^*)\,\Delta x = \int_0^8 \delta x \cdot (12 + x)\,dx$$

$$= \delta \int_0^8 (12x + x^2)\,dx = \delta \left[6x^2 + \tfrac{x^3}{3} \right]_0^8 = \delta \left(384 + \tfrac{512}{3} \right)$$

$$= (62.5)\,\tfrac{1664}{3} \approx 3.47 \times 10^4 \text{ lb}$$

9. From the figure, the area of the ith rectangular strip is $2\sqrt{r^2 - \left(x_i^*\right)^2}\,\Delta x$

and the pressure on it is $\rho g\left(x_i^* + r\right)$.

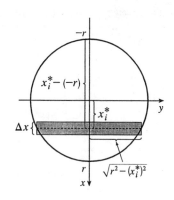

$$F = \lim_{n \to \infty} \sum_{i=1}^n \rho g\left(x_i^* + r\right) 2\sqrt{r^2 - \left(x_i^*\right)^2}\,\Delta x$$

$$= \int_{-r}^r \rho g\left(x + r\right) \cdot 2\sqrt{r^2 - x^2}\,dx$$

$$= \rho g \int_{-r}^r \sqrt{r^2 - x^2}\,2x\,dx + 2\rho g r \int_{-r}^r \sqrt{r^2 - x^2}\,dx$$

The first integral is 0 because the integrand is an odd function. The second
integral can be interpreted as the area of a semicircular disk with radius r,
or we could make the trigonometric substitution $x = r \sin \theta$. Continuing:

$$F = \rho g \cdot 0 + 2\rho g r \cdot \tfrac{1}{2}\pi r^2 = \rho g \pi r^3 = 1000 g \pi r^3 \text{ N (SI units assumed)}.$$

11. $F = \displaystyle\int_0^{4\sqrt{3}} \rho g\left(4\sqrt{3} - x\right) \frac{2x}{\sqrt{3}}\,dx = 8\rho g \int_0^{4\sqrt{3}} x\,dx - \frac{2\rho g}{\sqrt{3}} \int_0^{4\sqrt{3}} x^2\,dx$

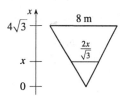

$$= 4\rho g \left[x^2\right]_0^{4\sqrt{3}} - \frac{2\rho g}{3\sqrt{3}}\left[x^3\right]_0^{4\sqrt{3}} = 192\rho g - \frac{2\rho g}{3\sqrt{3}}64 \cdot 3\sqrt{3}$$

$$= 192\rho g - 128\rho g = 64\rho g \approx 64\,(840)\,(9.8) \approx 5.27 \times 10^5 \text{ N}$$

13. (a) $F = \rho g d A \approx (1000)\,(9.8)\,(0.8)\,(0.2)^2 \approx 314 \text{ N}$

(b) $F = \int_{0.8}^1 \rho g x\,(0.2)\,dx = 0.2\rho g \left[\tfrac{1}{2}x^2\right]_{0.8}^1 = (0.2\rho g)\,(0.18) = 0.036\rho g \approx 353 \text{ N}$

15. Assume that the pool is filled with water.

(a) $F = \int_0^3 \delta x 20\,dx = 20\delta\left[\tfrac{1}{2}x^2\right]_0^3 = 20\delta \cdot \tfrac{9}{2} = 90\delta \approx 5625 \text{ lb} \approx 5.63 \times 10^3 \text{ lb}$

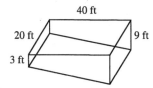

40 ft

20 ft   9 ft

3 ft

(b) $F = \int_0^9 \delta x 20\,dx = 20\delta\left[\tfrac{1}{2}x^2\right]_0^9 = 810\delta \approx 50{,}625 \text{ lb} \approx 5.06 \times 10^4 \text{ lb.}$

(c) For the first 3 ft, the length of the side is constant at 40 ft. For $3 < x \le 9$, we can use similar triangles to find
the length a: $\dfrac{a}{40} = \dfrac{9 - x}{6} \quad \Rightarrow \quad a = 40 \cdot \dfrac{9 - x}{6}$.

$$F = \int_0^3 \delta x 40\,dx + \int_3^9 \delta x\,(40)\,\tfrac{9-x}{6}\,dx = 40\delta\left[\tfrac{1}{2}x^2\right]_0^3 + \tfrac{20}{3}\delta \int_3^9 \left(9x - x^2\right)dx$$

$$= 180\delta + \tfrac{20}{3}\delta\left[\tfrac{9}{2}x^2 - \tfrac{1}{3}x^3\right]_3^9 = 180\delta + \tfrac{20}{3}\delta\left[\left(\tfrac{729}{2} - 243\right) - \left(\tfrac{81}{2} - 9\right)\right]$$

$$= 780\delta \approx 48{,}750 \text{ lb} \approx 4.88 \times 10^4 \text{ lb}$$

(d) For any right triangle with hypotenuse on the bottom,

$$\csc\theta = \frac{\Delta x}{\text{hypotenuse}} \quad\Rightarrow$$

$$\text{hypotenuse} = \Delta x \csc\theta = \Delta x \frac{\sqrt{40^2 + 6^2}}{6} = \frac{\sqrt{409}}{3}\Delta x.$$

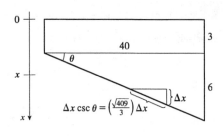

$$F = \int_3^9 \delta x 20\frac{\sqrt{409}}{3}dx = \frac{1}{3}\left(20\sqrt{409}\right)\delta\left[\frac{1}{2}x^2\right]_3^9$$

$$= \frac{1}{3}\cdot 10\sqrt{409}\delta\,(81-9)$$

$$\approx 303,356\text{ lb} \approx 3.03\times 10^5\text{ lb}$$

17. $\bar{x} = A^{-1}\int_a^b xw\,(x)\,dx$ (Equation 8) $\Rightarrow$ $A\bar{x} = \int_a^b xw\,(x)\,dx$ $\Rightarrow$ $(\rho g\bar{x})\,A = \int_a^b \rho gxw\,(x)\,dx = F$ by Exercise 16.

19. $m_1 = 4$, $m_2 = 8$; $P_1\,(-1, 2)$, $P_2\,(2, 4)$. $m = \sum\limits_{i=1}^{2} m_i = m_1 + m_2 = 12$. $M_x = \sum\limits_{i=1}^{2} m_i y_i = 4\cdot 2 + 8\cdot 4 = 40$;

$M_y = \sum\limits_{i=1}^{2} m_i x_i = 4\cdot(-1) + 8\cdot 2 = 12$; $\bar{x} = M_y/m = 1$ and $\bar{y} = M_x/m = \frac{10}{3}$, so the center of mass of the system

is $(\bar{x}, \bar{y}) = \left(1, \frac{10}{3}\right)$.

21. $A = \int_0^2 x^2\,dx = \left[\frac{1}{3}x^3\right]_0^2 = \frac{8}{3}$,

$\bar{x} = A^{-1}\int_0^2 x\cdot x^2\,dx = \frac{3}{8}\left[\frac{1}{4}x^4\right]_0^2 = \frac{3}{8}\cdot 4 = \frac{3}{2}$,

$\bar{y} = A^{-1}\int_0^2 \frac{1}{2}\left(x^2\right)^2\,dx = \frac{3}{8}\cdot\frac{1}{2}\left[\frac{1}{5}x^5\right]_0^2 = \frac{3}{16}\cdot\frac{32}{5} = \frac{6}{5}$.

Centroid $(\bar{x}, \bar{y}) = \left(\frac{3}{2}, \frac{6}{5}\right) = (1.5, 1.2)$.

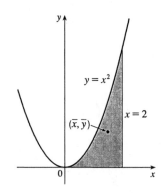

23. $A = \int_0^1 e^x\,dx = [e^x]_0^1 = e - 1$,

$\bar{x} = \frac{1}{A}\int_0^1 xe^x\,dx = \frac{1}{e-1}[xe^x - e^x]_0^1$ (by parts)

$= \frac{1}{e-1}[0 - (-1)] = \frac{1}{e-1}$,

$\bar{y} = \frac{1}{A}\int_0^1 \frac{1}{2}\left(e^x\right)^2\,dx = \frac{1}{e-1}\cdot\frac{1}{4}\left[e^{2x}\right]_0^1 = \frac{1}{4(e-1)}\left(e^2 - 1\right) = \frac{e+1}{4}$.

Centroid $(\bar{x}, \bar{y}) = \left(\frac{1}{e-1}, \frac{e+1}{4}\right) \approx (0.58, 0.93)$.

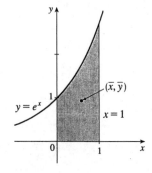

25. From the figure we see that $\bar{x} = \frac{\pi}{4}$ (halfway from $x = 0$ to $\frac{\pi}{2}$). Now

$A = \int_0^{\pi/2} \sin 2x \, dx = -\frac{1}{2} \left[\cos 2x\right]_0^{\pi/2} = -\frac{1}{2}(-1-1) = 1$, so

$\bar{y} = \frac{1}{A} \int_0^{\pi/2} \frac{1}{2} \sin^2 2x \, dx = \frac{1}{1} \int_0^{\pi/2} \frac{1}{2} \cdot \frac{1}{2} (1 - \cos 4x) \, dx$

$\quad = \frac{1}{4} \left[x - \frac{1}{4} \sin 4x \right]_0^{\pi/2} = \frac{\pi}{8}$.

Centroid $(\bar{x}, \bar{y}) = \left(\frac{\pi}{4}, \frac{\pi}{8} \right)$.

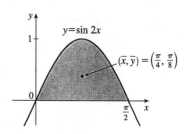

27. $A = \int_0^{\pi/4} (\cos x - \sin x) \, dx = [\sin x + \cos x]_0^{\pi/4} = \sqrt{2} - 1$,

$\bar{x} = A^{-1} \int_0^{\pi/4} x (\cos x - \sin x) \, dx = A^{-1} [x (\sin x + \cos x) + \cos x - \sin x]_0^{\pi/4}$ [integration by parts]

$\quad = A^{-1} \left(\frac{\pi}{4} \sqrt{2} - 1 \right) = \frac{\frac{1}{4} \pi \sqrt{2} - 1}{\sqrt{2} - 1}$

$\bar{y} = A^{-1} \int_0^{\pi/4} \frac{1}{2} (\cos^2 x - \sin^2 x) \, dx = \frac{1}{2A} \int_0^{\pi/4} \cos 2x \, dx = \frac{1}{4A} [\sin 2x]_0^{\pi/4} = \frac{1}{4A} = \frac{1}{4(\sqrt{2}-1)}$.

$(\bar{x}, \bar{y}) = \left(\dfrac{\pi \sqrt{2} - 4}{4 \left(\sqrt{2} - 1 \right)}, \dfrac{1}{4 \left(\sqrt{2} - 1 \right)} \right)$.

29. From the figure we see that $\bar{y} = 0$. Now

$A = \int_0^5 2\sqrt{5-x} \, dx = 2 \left[-\frac{2}{3} (5-x)^{3/2} \right]_0^5$

$\quad = 2 \left(0 + \frac{2}{3} \cdot 5^{3/2} \right) = \frac{20}{3} \sqrt{5}$

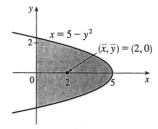

so

$\bar{x} = \frac{1}{A} \int_0^5 x \left[\sqrt{5-x} - (-\sqrt{5-x}) \right] dx = \frac{1}{A} \int_0^5 2x\sqrt{5-x} \, dx$

$\quad = \frac{1}{A} \int_{\sqrt{5}}^0 2 (5 - u^2) u (-2u) \, du \quad (u = \sqrt{5-x}, \, x = 5 - u^2, \, u^2 = 5 - x, \, dx = -2u \, du)$

$\quad = \frac{4}{A} \int_0^{\sqrt{5}} u^2 (5 - u^2) \, du = \frac{4}{A} \left[\frac{5}{3} u^3 - \frac{1}{5} u^5 \right]_0^{\sqrt{5}} = \frac{3}{5\sqrt{5}} \left(\frac{25}{3} \sqrt{5} - 5\sqrt{5} \right) = 5 - 3 = 2$

Thus, the centroid is $(\bar{x}, \bar{y}) = (2, 0)$.

31. By symmetry, $M_y = 0$ and $\bar{x} = 0$. $A = \frac{1}{2} bh = \frac{1}{2} \cdot 2 \cdot 2 = 2$.

$M_x = 2\rho \int_0^1 \frac{1}{2} (2 - 2x)^2 \, dx = \left(2 \cdot 1 \cdot \frac{1}{2} \cdot 2^2 \right) \int_0^1 (1-x)^2 \, dx$

$\quad = 4 \left[-\frac{1}{3} (1-x)^3 \right]_0^1 = 4 \cdot \frac{1}{3} = \frac{4}{3}$

$\bar{y} = \frac{1}{\rho A} M_x = \frac{1}{1 \cdot 2} \cdot \frac{4}{3} = \frac{2}{3}$. $(\bar{x}, \bar{y}) = \left(0, \frac{2}{3} \right)$.

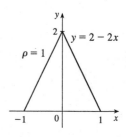

33.

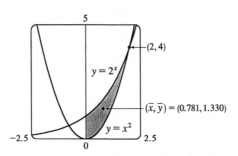

$$A = \int_0^2 \left(2^x - x^2\right) dx = \left[\frac{2^x}{\ln 2} - \frac{x^3}{3}\right]_0^2 = \left(\frac{4}{\ln 2} - \frac{8}{3}\right) - \frac{1}{\ln 2} = \frac{3}{\ln 2} - \frac{8}{3} \approx 1.661418.$$

$$\bar{x} = \frac{1}{A} \int_0^2 x \left(2^x - x^2\right) dx = \frac{1}{A} \int_0^2 \left(x2^x - x^3\right) dx = \frac{1}{A} \int_0^2 \left[x e^{(\ln 2)x} - x^3\right] dx$$

$$\overset{96}{=} \frac{1}{A} \left[\frac{1}{(\ln 2)^2} (x \ln 2 - 1) e^{(\ln 2)x} - \frac{1}{4}x^4\right]_0^2 \quad \text{(or use parts)}$$

$$= \frac{1}{A} \left[\left(\frac{x}{\ln 2} - \frac{1}{(\ln 2)^2}\right) 2^x - \frac{1}{4}x^4\right]_0^2 = \frac{1}{A} \left[\frac{x2^x}{\ln 2} - \frac{2^x}{(\ln 2)^2} - \frac{x^4}{4}\right]_0^2$$

$$= \frac{1}{A} \left[\frac{8}{\ln 2} - \frac{4}{(\ln 2)^2} - 4 + \frac{1}{(\ln 2)^2}\right] = \frac{1}{A} \left[\frac{8}{\ln 2} - \frac{3}{(\ln 2)^2} - 4\right]$$

$$\approx \frac{1}{A} (1.297453) \approx 0.781$$

$$\bar{y} = \frac{1}{A} \int_0^2 \frac{1}{2} \left[(2^x)^2 - (x^2)^2\right] dx = \frac{1}{A} \int_0^2 \frac{1}{2} \left(2^{2x} - x^4\right) dx = \frac{1}{A} \cdot \frac{1}{2} \left[\frac{2^{2x}}{2\ln 2} - \frac{x^5}{5}\right]_0^2$$

$$= \frac{1}{A} \cdot \frac{1}{2} \left(\frac{16}{2\ln 2} - \frac{32}{5} - \frac{1}{2\ln 2}\right) = \frac{1}{A} \left(\frac{15}{4\ln 2} - \frac{16}{5}\right) \approx \frac{1}{A} (2.210106) \approx 1.330$$

35. Choose x- and y-axes so that the base (one side of the triangle) lies along the x-axis with the other vertex along the positive y-axis as shown. From geometry, we know the medians intersect at a point $\frac{2}{3}$ of the way from each vertex (along the median) to the opposite side. The median from B goes to the midpoint $\left(\frac{1}{2}(a+c), 0\right)$ of side AC, so the point of intersection of the

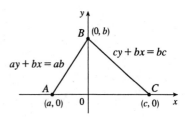

medians is $\left(\frac{2}{3} \cdot \frac{1}{2}(a+c), \frac{1}{3}b\right) = \left(\frac{1}{3}(a+c), \frac{1}{3}b\right)$. This can also be verified by finding the equations of two medians, and solving them simultaneously to find their point of intersection. Now let us compute the location of the centroid of the triangle. The area is $A = \frac{1}{2}(c-a)b$.

$$\bar{x} = \frac{1}{A} \left[\int_a^0 x \cdot \frac{b}{a}(a-x) dx + \int_0^c x \cdot \frac{b}{c}(c-x) dx\right] = \frac{1}{A} \left[\frac{b}{a} \int_a^0 (ax - x^2) dx + \frac{b}{c} \int_0^c (cx - x^2) dx\right]$$

$$= \frac{b}{Aa} \left[\frac{1}{2}ax^2 - \frac{1}{3}x^3\right]_a^0 + \frac{b}{Ac} \left[\frac{1}{2}cx^2 - \frac{1}{3}x^3\right]_0^c = \frac{b}{Aa} \left[-\frac{1}{2}a^3 + \frac{1}{3}a^3\right] + \frac{b}{Ac} \left[\frac{1}{2}c^3 - \frac{1}{3}c^3\right]$$

$$= \frac{2}{a(c-a)} \cdot \frac{-a^3}{6} + \frac{2}{c(c-a)} \cdot \frac{c^3}{6} = \frac{1}{3(c-a)} \left(c^2 - a^2\right) = \frac{a+c}{3}$$

and

$$\bar{y} = \frac{1}{A}\left[\int_a^0 \frac{1}{2}\left(\frac{b}{a}(a-x)\right)^2 dx + \int_0^c \frac{1}{2}\left(\frac{b}{c}(c-x)\right)^2 dx\right]$$

$$= \frac{1}{A}\left[\frac{b^2}{2a^2}\int_a^0 (a^2 - 2ax + x^2)\,dx + \frac{b^2}{2c^2}\int_0^c (c^2 - 2cx + x^2)\,dx\right]$$

$$= \frac{1}{A}\left[\frac{b^2}{2a^2}\left[a^2x - ax^2 + \frac{1}{3}x^3\right]_a^0 + \frac{b^2}{2c^2}\left[c^2x - cx^2 + \frac{1}{3}x^3\right]_0^c\right]$$

$$= \frac{1}{A}\left[\frac{b^2}{2a^2}\left(-a^3 + a^3 - \frac{1}{3}a^3\right) + \frac{b^2}{2c^2}\left(c^3 - c^3 + \frac{1}{3}c^3\right)\right] = \frac{1}{A}\left[\frac{b^2}{6}(-a+c)\right] = \frac{2}{(c-a)b}\cdot\frac{(c-a)b^2}{6} = \frac{b}{3}$$

Thus, $(\bar{x}, \bar{y}) = \left(\dfrac{a+c}{3}, \dfrac{b}{3}\right)$ as claimed.

Remarks: Actually the computation of $\bar{y}$ is all that is needed. By considering each side of the triangle in turn to be the base, we see that the centroid is $\frac{1}{3}$ of the way from each side to the opposite vertex and must therefore be the intersection of the medians.

The computation of $\bar{y}$ in this problem (and many others) can be simplified by using horizontal rather than vertical approximating rectangles. If the length of a thin rectangle at coordinate y is $\ell(y)$, then its area is $\ell(y)\,\Delta y$, its mass is $\rho\ell(y)\,\Delta y$, and its moment about the x-axis is $\Delta M_x = \rho y\ell(y)\,\Delta y$. Thus,

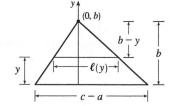

$$M_x = \int \rho y\ell(y)\,dy \quad\text{and}\quad \bar{y} = \frac{\int \rho y\ell(y)\,dy}{\rho A} = \frac{1}{A}\int y\ell(y)\,dy$$

In this problem, $\ell(y) = \dfrac{c-a}{b}(b-y)$ by similar triangles, so

$$\bar{y} = \frac{1}{A}\int_0^b \frac{c-a}{b}y(b-y)\,dy = \frac{2}{b^2}\int_0^b (by - y^2)\,dy = \frac{2}{b^2}\left[\frac{1}{2}by^2 - \frac{1}{3}y^3\right]_0^b = \frac{2}{b^2}\cdot\frac{b^3}{6} = \frac{b}{3}$$

Notice that only one integral is needed when this method is used.

Since the position of a centroid is independent of density when the density is constant, we will assume for convenience that $\rho = 1$ in Exercise 37.

37. Divide the lamina into two triangles and one rectangle with respective masses of 2, 2 and 4, so that the total mass is 8. Using the result of Exercise 35, the triangles have centroids $\left(-1, \frac{2}{3}\right)$ and $\left(1, \frac{2}{3}\right)$. The centroid of the rectangle (its center) is $\left(0, -\frac{1}{2}\right)$. So, using Formulas 5 and 7, we have $\bar{y} = \dfrac{\sum m_i y_i}{m} = \frac{2}{8}\left(\frac{2}{3}\right) + \frac{2}{8}\left(\frac{2}{3}\right) + \frac{4}{8}\left(-\frac{1}{2}\right) = \frac{1}{12}$, and $\bar{x} = 0$, since the lamina is symmetric about the line $x = 0$. Therefore $(\bar{x}, \bar{y}) = \left(0, \frac{1}{12}\right)$.

39. A cone of height h and radius r can be generated by rotating a right triangle about one of its legs as shown. By Exercise 35, $\bar{x} = \frac{1}{3}r$, so by the Theorem of Pappus, the volume of the cone is

$$V = Ad = \frac{1}{2}rh \cdot 2\pi\left(\frac{1}{3}r\right) = \frac{1}{3}\pi r^2 h.$$

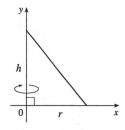

41. Suppose the region lies between two curves $y = f(x)$ and $y = g(x)$ where $f(x) \geq g(x)$, as illustrated in Figure 13. Choose points x_i with $a = x_0 < x_1 < \cdots < x_n = b$ and choose x_i^* to be the midpoint of the ith subinterval; that is, $x_i^* = \bar{x}_i = \frac{1}{2}(x_{i-1} + x_i)$. Then the centroid of the ith approximating rectangle R_i is its center $C_i = \left(\bar{x}_i, \frac{1}{2}[f(\bar{x}_i) + g(\bar{x}_i)]\right)$. Its area is $[f(\bar{x}_i) - g(\bar{x}_i)] \Delta x$, so its mass is $\rho[f(\bar{x}_i) - g(\bar{x}_i)] \Delta x$.

Thus, $M_y(R_i) = \rho[f(\bar{x}_i) - g(\bar{x}_i)] \Delta x \cdot \bar{x}_i = \rho \bar{x}_i [f(\bar{x}_i) - g(\bar{x}_i)] \Delta x$ and

$M_x(R_i) = \rho[f(\bar{x}_i) - g(\bar{x}_i)] \Delta x \cdot \frac{1}{2}[f(\bar{x}_i) + g(\bar{x}_i)] = \rho \cdot \frac{1}{2}[f(\bar{x}_i)^2 - g(\bar{x}_i)^2] \Delta x$. Summing over i and taking the limit as $n \to \infty$, we get $M_y = \lim_{n \to \infty} \sum_i \rho \bar{x}_i [f(\bar{x}_i) - g(\bar{x}_i)] \Delta x = \rho \int_a^b x[f(x) - g(x)] dx$ and

$M_x = \lim_{n \to \infty} \sum_i \rho \cdot \frac{1}{2}[f(\bar{x}_i)^2 - g(\bar{x}_i)^2] \Delta x = \rho \int_a^b \frac{1}{2}[f(x)^2 - g(x)^2] dx$. Thus,

$$\bar{x} = \frac{M_y}{m} = \frac{M_y}{\rho A} = \frac{1}{A} \int_a^b x[f(x) - g(x)] dx \text{ and } \bar{y} = \frac{M_x}{m} = \frac{M_x}{\rho A} = \frac{1}{A} \int_a^b \frac{1}{2}\left[f(x)^2 - g(x)^2\right] dx$$

8.4 Applications to Economics and Biology

1. $C(2000) = C(0) + \int_0^{2000} C'(x) dx = 1{,}500{,}000 + \int_0^{2000} (0.006x^2 - 1.5x + 8) dx$

$= 1{,}500{,}000 + \left[0.002x^3 - 0.75x^2 + 8x\right]_0^{2000} = \$14{,}516{,}000$

3. $C'(x) = 74 + 1.1x - 0.002x^2 + 0.00004x^3$, so the increase in cost is

$$C(1600) - C(1200) = \int_{1200}^{1600} (74 + 1.1x - 0.002x^2 + 0.00004x^3) dx$$

$$= \left[74x + 0.55x^2 - \frac{0.002}{3}x^3 + 0.00001x^4\right]_{1200}^{1600}$$

$$= 64{,}331{,}733.33 - 20{,}464{,}800 = \$43{,}866{,}933.33$$

5. $p(x) = 10 = \dfrac{450}{x + 8} \quad \Rightarrow \quad x + 8 = 45 \quad \Rightarrow \quad x = 37.$

Consumer surplus $= \displaystyle\int_0^{37} [p(x) - 10] dx = \int_0^{37} \left(\frac{450}{x + 8} - 10\right) dx$

$= [450 \ln(x + 8) - 10x]_0^{37}$

$= 450 \ln\left(\frac{45}{8}\right) - 370 \approx \$407.25$

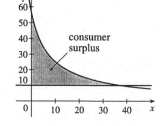

7. $P = p_S(x) = 10 = 5 + \frac{1}{10}\sqrt{x} \quad \Rightarrow \quad 50 = \sqrt{x} \quad \Rightarrow \quad x = 2500.$

Producer surplus $= \int_0^{2500} [P - p_S(x)] dx$

$= \int_0^{2500} \left(10 - 5 - \frac{1}{10}\sqrt{x}\right) dx$

$= \left[5x - \frac{1}{15}x^{3/2}\right]_0^{2500} \approx \$4166.67$

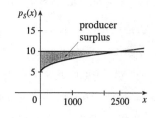

9. $p(x) = \dfrac{800{,}000e^{-x/5000}}{x + 20{,}000} = 16 \Rightarrow x = x_1 \approx 3727.04.$

Consumer surplus $= \int_0^{x_1} [p(x) - 16]\, dx \approx \$37{,}753.01$

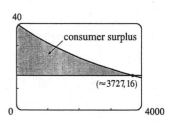

11. $f(8) - f(4) = \int_4^8 f'(t)\, dt = \int_4^8 \sqrt{t}\, dt = \left[\frac{2}{3} t^{3/2}\right]_4^8 = \frac{2}{3}\left(16\sqrt{2} - 8\right) \approx \$9.75$ million

13. $F = \dfrac{\pi P R^4}{8\eta\ell} = \dfrac{\pi (4000)(0.008)^4}{8(0.027)(2)} \approx 1.19 \times 10^{-4}\ \text{cm}^3/\text{s}$

15. $\int_0^{12} c(t)\, dt = \int_0^{12} \frac{1}{4} t(12 - t)\, dt = \left[\frac{3}{2} t^2 - \frac{1}{12} t^3\right]_0^{12} = (216 - 144) = 72\ \text{mg} \cdot \text{s/L}.$ Therefore,

$F = A/72 = \frac{8}{72} = \frac{1}{9}\ \text{L/s} = \frac{60}{9}\ \text{L/min}.$

8.5 Probability

1. (a) $\int_{100}^{200} f(t)\, dt$ is the probability that a randomly chosen battery will have a lifetime of between 100 and 200 hours.

(b) $\int_{200}^{\infty} f(t)\, dt$ is the probability that a randomly chosen battery will have a lifetime of at least 200 hours.

3. (a) In general, we must satisfy the two conditions that are mentioned before Example 1 — namely, (1) $f(x) \geq 0$ for all x, and (2) $\int_{-\infty}^{\infty} f(x)\, dx = 1$. Clearly, condition (1) is satisfied. For condition (2), we see that

$\int_{-\infty}^{\infty} f(x)\, dx = \int_0^{10} 0.1\, dx = \left[\frac{1}{10} x\right]_0^{10} = 1.$ Thus, $f(x)$ is a probability density function.

(b) Since all the numbers between 0 and 10 are equally likely to be selected, we expect the mean to be halfway between the endpoints of the interval; that is, $x = 5$.

$\mu = \int_{-\infty}^{\infty} xf(x)\, dx = \int_0^{10} x(0.1)\, dx = \left[\frac{1}{20} x^2\right]_0^{10} = \frac{100}{20} = 5,$ as expected.

5. We need to find m so that $\int_m^{\infty} f(t)\, dt = \frac{1}{2} \Rightarrow \lim\limits_{x \to \infty} \int_m^x \frac{1}{5} e^{-t/5} dt = \frac{1}{2} \Rightarrow \lim\limits_{x \to \infty} \left[\frac{1}{5}(-5)e^{-t/5}\right]_m^x = \frac{1}{2} \Rightarrow$

$(-1)\left(0 - e^{-m/5}\right) = \frac{1}{2} \Rightarrow e^{-m/5} = \frac{1}{2} \Rightarrow -m/5 = \ln\frac{1}{2} \Rightarrow m = -5\ln\frac{1}{2} = 5\ln 2 \approx 3.47$ min.

7. We use an exponential density function with $\mu = 2.5$ min.

(a) $P(X > 4) = \int_4^{\infty} f(t)\, dt = \lim\limits_{x \to \infty} \int_4^x \frac{1}{2.5} e^{-t/2.5}\, dt = \lim\limits_{x \to \infty} \left[-e^{-t/2.5}\right]_4^x = 0 + e^{-4/2.5} \approx 0.202$

(b) $P(0 \leq X \leq 2) = \int_0^2 f(t)\, dt = \left[-e^{-t/2.5}\right]_0^2 = -e^{-2/2.5} + 1 \approx 0.551$

(c) We need to find a value a so that $P(X \geq a) = 0.02$, or, equivalently, $P(0 \leq X \leq a) = 0.98 \Leftrightarrow$

$\int_0^a f(t)\, dt = 0.98 \Leftrightarrow \left[-e^{-t/2.5}\right]_0^a = 0.98 \Leftrightarrow -e^{-a/2.5} + 1 = 0.98 \Leftrightarrow e^{-a/2.5} = 0.02 \Leftrightarrow$

$-a/2.5 = \ln 0.02 \Leftrightarrow a = -2.5\ln\frac{1}{50} = 2.5\ln 50 \approx 9.78$ min ≈ 10 min. The ad should say that if you aren't served within 10 minutes, you get a free hamburger.

9. $P(X \geq 10) = \int_{10}^{\infty} \frac{1}{4.2\sqrt{2\pi}} \exp\left(-\frac{(x - 9.4)^2}{2 \cdot 4.2^2}\right) dx$. To avoid the improper integral we approximate it by the

integral from 10 to 100. Thus, $P(X \geq 10) \approx \int_{10}^{100} \frac{1}{4.2\sqrt{2\pi}} \exp\left(-\frac{(x - 9.4)^2}{2 \cdot 4.2^2}\right) dx \approx 0.443$ (using a calculator

or computer to estimate the integral), so 44.3% of the households throw out at least 10 lb of paper a week.

11. $P(\mu - 2\sigma \leq X \leq \mu + 2\sigma) = \int_{\mu - 2\sigma}^{\mu + 2\sigma} \frac{1}{\sigma\sqrt{2\pi}} \exp\left(-\frac{(x - \mu)^2}{2\sigma^2}\right) dx$. Substituting $t = \frac{x - \mu}{\sigma}$ and $dt = \frac{1}{\sigma} dx$

gives us

$$\int_{-2}^{2} \frac{1}{\sigma\sqrt{2\pi}} e^{-t^2/2} (\sigma \, dt) = \frac{1}{\sqrt{2\pi}} \int_{-2}^{2} e^{-t^2/2} \, dt \approx 0.9545$$

13. (a) First $p(r) = \frac{4}{a_0^3} r^2 e^{-2r/a_0} \geq 0$ for $r \geq 0$. Next,

$$\int_{-\infty}^{\infty} p(r) \, dr = \int_{0}^{\infty} \frac{4}{a_0^3} r^2 e^{-2r/a_0} \, dr = \frac{4}{a_0^3} \lim_{t \to \infty} \int_{0}^{t} r^2 e^{-2r/a_0} \, dr$$

As in Exercise 12, we use (2) from that solution (with $b = -2/a_0$) and l'Hospital's Rule to get

$\frac{4}{a_0^3} \left[\frac{a_0^3}{-8} (-2)\right] = 1$. This satisfies the second condition for a function to be a probability density function.

(b) Using l'Hospital's Rule, $\frac{4}{a_0^3} \lim_{r \to \infty} \frac{r^2}{e^{2r/a_0}} = \frac{4}{a_0^3} \lim_{r \to \infty} \frac{2r}{(2/a_0) e^{2r/a_0}} = \frac{2}{a_0^2} \lim_{r \to \infty} \frac{2}{(2/a_0) e^{2r/a_0}} = 0$.

To find the maximum of p, we differentiate:

$$p'(r) = \frac{4}{a_0^3} \left[r^2 e^{-2r/a_0} \left(-\frac{2}{a_0}\right) + e^{-2r/a_0} (2r)\right] = \frac{4}{a_0^3} e^{-2r/a_0} (2r) \left(-\frac{r}{a_0} + 1\right)$$

$p'(r) = 0 \iff r = 0$ or $1 = \frac{r}{a_0} \iff r = a_0$. $p'(r)$ changes from positive to negative at $r = a_0$, so $p(r)$

has its maximum value at $r = a_0$.

(c) It is fairly difficult to find a viewing rectangle, but knowing the maximum value from part (b) helps.

$$p(a_0) = \frac{4}{a_0^3} a_0^2 e^{-2a_0/a_0} = \frac{4}{a_0} e^{-2} \approx 9{,}684{,}098{,}979$$

With a maximum of nearly 10 billion and a total area under the curve of 1, we know that the "hump" in the graph must be extremely narrow.

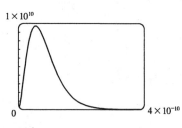

(d) $P(r) = \int_{0}^{r} \frac{4}{a_0^3} s^2 e^{-2s/a_0} \, ds \implies P(4a_0) = \int_{0}^{4a_0} \frac{4}{a_0^3} s^2 e^{-2s/a_0} \, ds$. Using (2) from the solution to

Exercise 12 (with $b = -2/a_0$),

$$P(4a_0) = \frac{4}{a_0^3} \left[\frac{e^{-2s/a_0}}{-8/a_0^3} \left(\frac{4}{a_0^2} s^2 + \frac{4}{a_0} s + 2\right)\right]_{0}^{4a_0} = \frac{4}{a_0^3} \left(\frac{a_0^3}{-8}\right) \left[e^{-8} (64 + 16 + 2) - 1(2)\right]$$

$$= -\frac{1}{2} \left(82 e^{-8} - 2\right) = 1 - 41 e^{-8} \approx 0.986$$

(e) $\mu = \int_{-\infty}^{\infty} r p\,(r)\,dr = \dfrac{4}{a_0^3} \lim_{t\to\infty} \int_0^t r^3 e^{-2r/a_0}\,dr$. Integrating by parts three times or using a CAS, we find that

$\int x^3 e^{bx}\,dx = \dfrac{e^{bx}}{b^4}\left(b^3 x^3 - 3b^2 x^2 + 6bx - 6\right)$. So with $b = -\dfrac{2}{a_0}$, we use l'Hospital's Rule, and get

$\mu = \dfrac{4}{a_0^3}\left[\dfrac{a_0^4}{16}(-6)\right] = \tfrac{3}{2}a_0$.

8 Review

CONCEPT CHECK

1. (a) The length of a curve is defined to be the limit of the lengths of the inscribed polygons, as described near Figure 3 in Section 8.1.

(b) See Equation 8.1.2.

(c) See Equation 8.1.4.

2. (a) $S = \int_a^b 2\pi f\,(x)\sqrt{1 + [f'\,(x)]^2}\,dx$

(b) If $x = g\,(y)$, $c \le y \le d$, then $S = \int_c^d 2\pi y\sqrt{1 + [g'\,(y)]^2}\,dy$.

(c) $S = \int_a^b 2\pi x\sqrt{1 + [f'\,(x)]^2}\,dx$ or $S = \int_c^d 2\pi g\,(y)\sqrt{1 + [g'\,(y)]^2}\,dy$

3. Let $c\,(x)$ be the cross-sectional length of the wall (measured parallel to the surface of the fluid) at depth x. Then the hydrostatic force against the wall is given by $F = \int_a^b \delta x c\,(x)\,dx$, where a and b are the lower and upper limits for x at points of the wall and δ is the weight density of the fluid.

4. (a) The center of mass is the point at which the plate balances horizontally.

(b) See Equations 8.3.8.

5. If a plane region $\mathcal{R}$ that lies entirely on one side of a line ℓ in its plane is rotated about ℓ, then the volume of the resulting solid is the product of the area of $\mathcal{R}$ and the distance traveled by the centroid of $\mathcal{R}$.

6. See Figure 3 in Section 8.4, and the discussion which precedes it.

7. (a) See the definition before Figure 6 in Section 8.4.

(b) See the discussion after Figure 6 in Section 8.4.

8. A probability density function f is a function on the domain of a continuous random variable X such that $\int_a^b f\,(x)\,dx$ measures the probability that X lies between a and b. Such a function f has nonnegative values and satisfies the relation $\int_D f\,(x)\,dx = 1$, where D is the domain of the corresponding random variable X. If $D = \mathbb{R}$, or if we define $f\,(x) = 0$ for real numbers $x \notin D$, then $\int_{-\infty}^{\infty} f\,(x) = 1$. (Of course, to work with f in this way, we must assume that the integrals of f exist.)

9. (a) $\int_0^{100} f\,(x)\,dx$ represents the probability that the weight of a randomly chosen female college student is less than 100 pounds.

(b) $\mu = \int_{-\infty}^{\infty} xf\,(x)\,dx = \int_0^{\infty} xf\,(x)\,dx$

(c) The median of f is the number m such that $\int_m^{\infty} f\,(x)\,dx = \tfrac{1}{2}$.

EXERCISES

1. $3x = 2(y-1)^{3/2}$, $2 \le y \le 5$. $x = \frac{2}{3}(y-1)^{3/2}$, so $dx/dy = (y-1)^{1/2}$ and the arc length formula gives

$$L = \int_2^5 \sqrt{1 + (dx/dy)^2}\, dy = \int_2^5 \sqrt{1 + (y-1)}\, dy = \int_2^5 \sqrt{y}\, dy = \left[\frac{2}{3}y^{3/2}\right]_2^5 = \frac{2}{3}\left(5\sqrt{5} - 2\sqrt{2}\right)$$

3. (a) $y = \frac{1}{6}x^3 + \frac{1}{2x}$, $1 \le x \le 2$ $\Rightarrow$ $y' = \frac{1}{2}\left(x^2 - \frac{1}{x^2}\right)$ $\Rightarrow$ $(y')^2 = \frac{1}{4}\left(x^4 - 2 + \frac{1}{x^4}\right)$ $\Rightarrow$

$$1 + (y')^2 = \frac{1}{4}\left(x^4 + 2 + \frac{1}{x^4}\right) = \frac{1}{4}\left(x^2 + \frac{1}{x^2}\right)^2 \Rightarrow$$

$$L = \int_1^2 \sqrt{1 + (y')^2}\, dy = \frac{1}{2}\int_1^2 \left(x^2 + \frac{1}{x^2}\right) dx = \frac{1}{2}\left[\frac{x^3}{3} - \frac{1}{x}\right]_1^2 = \frac{1}{2}\left(\frac{17}{6}\right) = \frac{17}{12}$$

(b) $S = \int_1^2 2\pi y \sqrt{1 + \left(\frac{dy}{dx}\right)^2}\, dx = 2\pi \int_1^2 \left(\frac{x^3}{6} + \frac{1}{2x}\right)\frac{1}{2}\left(x^2 + \frac{1}{x^2}\right) dx = \pi \int_1^2 \left(\frac{1}{6}x^5 + \frac{2}{3}x + \frac{1}{2}x^{-3}\right) dx$

$$= \pi \left[\frac{1}{36}x^6 + \frac{1}{3}x^2 - \frac{1}{4}x^{-2}\right]_1^2 = \pi \left[\left(\frac{64}{36} + \frac{4}{3} - \frac{1}{16}\right) - \left(\frac{1}{36} + \frac{1}{3} - \frac{1}{4}\right)\right] = \frac{47\pi}{16}$$

5. $y = \sqrt[3]{x} = x^{1/3}$ $\Rightarrow$ $dy/dx = \frac{1}{3}x^{-2/3}$ $\Rightarrow$ $\sqrt{1 + (dy/dx)^2} = \sqrt{1 + 1/(9x^{4/3})}$. Call this $f(x)$. Then

$$L = \int_1^8 f(x)\, dx$$

$$\approx S_{14} = \frac{8-1}{14 \cdot 3}\left[f(1) + 4f(1.5) + 2f(2) + 4f(2.5) + \cdots + 2f(7) + 4f(7.5) + f(8)\right] \approx 7.082581$$

7. The loop lies between $x = 0$ and $x = 3a$ and is symmetric about the x-axis. We can assume without loss of generality that $a > 0$, since if $a = 0$, the graph is the parallel lines $x = 0$ and $x = 3a$, so there is no loop. The upper half of the loop is given by $y = \frac{1}{3\sqrt{a}}\sqrt{x}(3a - x) = \sqrt{a}x^{1/2} - \frac{x^{3/2}}{3\sqrt{a}}$, $0 \le x \le 3a$. The desired surface area

is twice the area generated by the upper half of the loop, that is, $S = 2(2\pi)\int_0^{3a} x\sqrt{1 + \left(\frac{dy}{dx}\right)^2}\, dx$.

$\frac{dy}{dx} = \frac{\sqrt{a}}{2}x^{-1/2} - \frac{x^{1/2}}{2\sqrt{a}}$ $\Rightarrow$ $1 + \left(\frac{dy}{dx}\right)^2 = \frac{a}{4x} + \frac{1}{2} + \frac{x}{4a}$. Therefore

$$S = 2(2\pi)\int_0^{3a} x\left(\frac{\sqrt{a}}{2}x^{-1/2} + \frac{x^{1/2}}{2\sqrt{a}}\right) dx = 2\pi \int_0^{3a} \left(\sqrt{a}x^{1/2} + \frac{x^{3/2}}{\sqrt{a}}\right) dx$$

$$= 2\pi \left[\frac{2\sqrt{a}}{3}x^{3/2} + \frac{2}{5\sqrt{a}}x^{5/2}\right]_0^{3a} = 2\pi \left[\frac{2\sqrt{a}}{3}3a\sqrt{3a} + \frac{2}{5\sqrt{a}}9a^2\sqrt{3a}\right] = \frac{56\sqrt{3}\pi a^2}{5}$$

9. As in Example 1 of Section 8.3,

$$F = \int_0^2 \rho gx(5-x)\, dx = \rho g\left[\frac{5}{2}x^2 - \frac{1}{3}x^3\right]_0^2 = \rho g\left(10 - \frac{8}{3}\right) = \frac{22}{3}\delta \approx \frac{22}{3} \cdot 62.5 \approx 458.3 \text{ lb}$$

11. $A = \int_{-2}^1 \left[(4 - x^2) - (x + 2)\right] dx = \int_{-2}^1 (2 - x - x^2)\, dx = \left[2x - \frac{1}{2}x^2 - \frac{1}{3}x^3\right]_{-2}^1$

$$= \left(2 - \frac{1}{2} - \frac{1}{3}\right) - \left(-4 - 2 + \frac{8}{3}\right) = \frac{9}{2} \Rightarrow$$

$$\bar{x} = A^{-1}\int_{-2}^1 x(2 - x - x^2)\, dx = \frac{2}{9}\int_{-2}^1 (2x - x^2 - x^3)\, dx = \frac{2}{9}\left[x^2 - \frac{1}{3}x^3 - \frac{1}{4}x^4\right]_{-2}^1$$

$$= \frac{2}{9}\left[\left(1 - \frac{1}{3} - \frac{1}{4}\right) - \left(4 + \frac{8}{3} - 4\right)\right] = -\frac{1}{2}$$

and

$$\overline{y} = A^{-1} \int_{-2}^{1} \tfrac{1}{2} \left[(4-x^2)^2 - (x+2)^2 \right] dx = \tfrac{1}{9} \int_{-2}^{1} (x^4 - 9x^2 - 4x + 12)\, dx$$

$$= \tfrac{1}{9} \left[\tfrac{1}{5}x^5 - 3x^3 - 2x^2 + 12x \right]_{-2}^{1} = \tfrac{1}{9} \left[\left(\tfrac{1}{5} - 3 - 2 + 12 \right) - \left(-\tfrac{32}{5} + 24 - 8 - 24 \right) \right] = \tfrac{12}{5}$$

So $(\overline{x}, \overline{y}) = \left(-\tfrac{1}{2}, \tfrac{12}{5} \right)$.

13. The equation of the line passing through $(0,0)$ and $(3,2)$ is $y = \tfrac{2}{3}x$. $A = \tfrac{1}{2} \cdot 3 \cdot 2 = 3$. Therefore, using

Equations 8.3.8, $\overline{x} = \tfrac{1}{3} \int_0^3 x \left(\tfrac{2}{3}x \right) dx = \tfrac{2}{27} \left[x^3 \right]_0^3 = 2$, and $\overline{y} = \tfrac{1}{3} \int_0^3 \tfrac{1}{2} \left(\tfrac{2}{3}x \right)^2 dx = \tfrac{2}{81} \left[x^3 \right]_0^3 = \tfrac{2}{3}$. Thus,

$(\overline{x}, \overline{y}) = \left(2, \tfrac{2}{3} \right)$.

15. The centroid of this circle, $(1,0)$, travels a distance $2\pi(1)$ when the lamina is rotated about the y-axis. The area of the circle is $\pi(1)^2$. So by the Theorem of Pappus, $V = A2\pi\overline{x} = \pi(1)^2 \, 2\pi(1) = 2\pi^2$.

17. $x = 100 \quad \Rightarrow \quad P = 2000 - 0.1(100) - 0.01(100)^2 = 1890$

$$\text{Consumer surplus} = \int_0^{100} [p(x) - P]\, dx = \int_0^{100} (2000 - 0.1x - 0.01x^2 - 1890)\, dx$$

$$= \left[110x - 0.05x^2 - \tfrac{0.01}{3}x^3 \right]_0^{100} = 11{,}000 - 500 - \tfrac{10{,}000}{3} \approx \$7166.67$$

19. $f(x) = \begin{cases} \tfrac{\pi}{20} \sin \left(\tfrac{\pi}{10}x \right) & \text{if } 0 \le x \le 10 \\ 0 & \text{if } x < 0 \text{ or } x > 10 \end{cases}$

(a) $f(x) \ge 0$ for all real numbers x and

$$\int_{-\infty}^{\infty} f(x)\, dx = \int_0^{10} \tfrac{\pi}{20} \sin \left(\tfrac{\pi}{10}x \right) dx = \tfrac{\pi}{20} \cdot \tfrac{10}{\pi} \left[-\cos \left(\tfrac{\pi}{10}x \right) \right]_0^{10} = \tfrac{1}{2}(-\cos \pi + \cos 0) = \tfrac{1}{2}(1+1) = 1.$$

(b) $P(X < 4) = \int_{-\infty}^{4} f(x)\, dx = \int_0^4 \tfrac{\pi}{20} \sin \left(\tfrac{\pi}{10}x \right) dx = \tfrac{1}{2} \left[-\cos \left(\tfrac{\pi}{10}x \right) \right]_0^4 = \tfrac{1}{2} \left(-\cos \tfrac{2\pi}{5} + \cos 0 \right)$

$\approx \tfrac{1}{2}(-0.309017 + 1) \approx 0.345492$

(c) $\mu = \int_{-\infty}^{\infty} xf(x)\, dx = \int_0^{10} \tfrac{\pi}{20}x \sin \left(\tfrac{\pi}{10}x \right) dx$

$= \int_0^\pi \tfrac{\pi}{20} \cdot \tfrac{10}{\pi}u (\sin u) \left(\tfrac{10}{\pi} \right) du \quad [u = \tfrac{\pi}{10}x,\ du = \tfrac{\pi}{10}\, dx]$

$= \tfrac{5}{\pi} \int_0^\pi u \sin u\, du \overset{82}{=} \tfrac{5}{\pi} [\sin u - u \cos u]_0^\pi = \tfrac{5}{\pi}[0 - \pi(-1)] = 5$

This answer is expected because the graph of f is symmetric about the line $x = 5$.

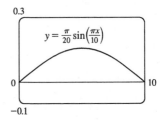

21. (a) The probability density function is $f(t) = \begin{cases} 0 & \text{if } t < 0 \\ \tfrac{1}{8}e^{-t/8} & \text{if } t \ge 0 \end{cases}$

$P(0 \le X \le 3) = \int_0^3 \tfrac{1}{8}e^{-t/8}\, dt = \left[-e^{-t/8} \right]_0^3 = -e^{-3/8} + 1 \approx 0.3127$

(b) $P(X > 10) = \int_{10}^\infty \tfrac{1}{8}e^{-t/8}\, dt = \lim_{x \to \infty} \left[-e^{-t/8} \right]_{10}^x = \lim_{x \to \infty} \left(-e^{-x/8} + e^{-10/8} \right) = 0 + e^{-5/4} \approx 0.2865$

(c) We need to find m such that $P(X \ge m) = \tfrac{1}{2} \quad \Rightarrow \quad \int_m^\infty \tfrac{1}{8}e^{-t/8}\, dt = \tfrac{1}{2} \quad \Rightarrow \quad \lim_{x \to \infty} \left[-e^{-t/8} \right]_m^x = \tfrac{1}{2} \quad \Rightarrow$

$\lim_{x \to \infty} \left(-e^{-x/8} + e^{-m/8} \right) = \tfrac{1}{2} \quad \Rightarrow \quad e^{-m/8} = \tfrac{1}{2} \quad \Rightarrow \quad -m/8 = \ln \tfrac{1}{2} \quad \Rightarrow$

$m = -8 \ln \tfrac{1}{2} = 8 \ln 2 \approx 5.55$ minutes.

Problems Plus

1. $x^2 + y^2 \le 4y \iff x^2 + (y-2)^2 \le 4$, so S is part of a circle, as shown in the diagram. The area of S is

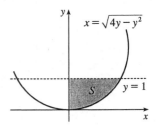

$$\int_0^1 \sqrt{4y - y^2}\,dy = \int_{-2}^{-1} \sqrt{4 - v^2}\,dv \quad (v = y - 2)$$

$$\overset{30}{=} \left[\tfrac{1}{2}v\sqrt{4 - v^2} + \tfrac{1}{2}(4)\sin^{-1}\left(\tfrac{1}{2}v\right) \right]_{-2}^{-1} = \tfrac{2\pi}{3} - \tfrac{\sqrt{3}}{2}$$

Another Method (without calculus): Note that $\theta = \angle ABC = \tfrac{\pi}{3}$, so the area is

$$(\text{area of sector } AOC) - (\text{area of } \triangle ABC) = \tfrac{1}{2}\left(2^2\right)\tfrac{\pi}{3} - \tfrac{1}{2}(1)\sqrt{3} = \tfrac{2\pi}{3} - \tfrac{\sqrt{3}}{2}$$

3. (a) The two spherical zones, whose surface areas we will call S_1 and S_2, are generated by rotation about the y-axis of circular arcs, as indicated in the figure. The arcs are the upper and lower portions of the circle $x^2 + y^2 = r^2$ that are obtained when the circle is cut with the line $y = d$. The portion of the upper arc in the first quadrant is sufficient to generate the upper spherical zone. That portion of the arc can be described

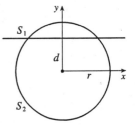

by the relation $x = \sqrt{r^2 - y^2}$ for $d \le y \le r$. Thus, $dy/dx = -y \left/ \sqrt{r^2 - y^2} \right.$ and

$$ds = \sqrt{1 + \left(\frac{dx}{dy}\right)}\,dy = \sqrt{1 + \frac{y^2}{r^2 - y^2}}\,dy = \sqrt{\frac{r^2}{r^2 - y^2}}\,dy = \frac{r\,dy}{\sqrt{r^2 - y^2}}$$

From Formula 8.2.8 we have

$$S_1 = \int_d^r 2\pi x \sqrt{1 + \left(\frac{dx}{dy}\right)^2}\,dy = \int_d^r 2\pi \sqrt{r^2 - y^2}\,\frac{r\,dy}{\sqrt{r^2 - y^2}} = \int_d^r 2\pi r\,dy = 2\pi r\,(r - d)$$

Similarly, we can compute $S_2 = \int_{-r}^d 2\pi x \sqrt{1 + (dx/dy)^2}\,dy = \int_{-r}^d 2\pi r\,dy = 2\pi r\,(r + d)$. Note that $S_1 + S_2 = 4\pi r^2$, the surface area of the entire sphere.

(b) $r = 3960$ mi and $d = r\,(\sin 75°) \approx 3825$ mi, so the surface area of the Arctic Ocean is about $2\pi r\,(r - d) \approx 2\pi\,(3960)\,(135) \approx 3.36 \times 10^6$ mi^2.

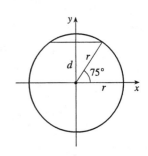

(c) The area on the sphere lies between planes $y = y_1$ and $y = y_2$, where $y_2 - y_1 = h$. Thus, we compute the

surface area on the sphere to be $S = \int_{y_1}^{y_2} 2\pi x \sqrt{1 + \left(\dfrac{dx}{dy}\right)^2}\, dy = \int_{y_1}^{y_2} 2\pi r\, dy = 2\pi r\,(y_2 - y_1) = 2\pi rh$.

This equals the lateral area of a cylinder of radius r and
height h, since such a cylinder is obtained by rotating the line
$x = r$ about the y-axis, so the surface area of the cylinder
between the planes $y = y_1$ and $y = y_2$ is

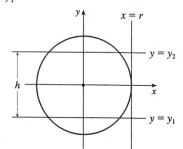

$$A = \int_{y_1}^{y_2} 2\pi x \sqrt{1 + \left(\frac{dx}{dy}\right)^2}\, dy = \int_{y_1}^{y_2} 2\pi r \sqrt{1 + 0^2}\, dy$$

$$= 2\pi ry|_{y=y_1}^{y_2} = 2\pi r\,(y_2 - y_1) = 2\pi rh$$

(d) $h = 2r \sin 23.45° \approx 3152$ mi, so the surface area of the
Torrid Zone is $2\pi rh \approx 2\pi\,(3960)\,(3152) \approx 7.84 \times 10^7$ mi^2.

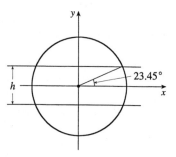

5. (a) Choose a vertical x-axis pointing downward with its origin at the surface. In order to calculate the pressure at
depth z, consider n subintervals of the interval $[0, z]$ by points x_i and choose a point $x_i^* \in [x_{i-1}, x_i]$ for each i.
The thin layer of water lying between depth x_{i-1} and depth x_i has a density of approximately $\rho\,(x_i^*)$, so the
weight of a piece of that layer with unit cross-sectional area is $\rho\,(x_i^*)\,g\Delta x$. The total weight of a column of
water extending from the surface to depth z (with unit cross-sectional area) would be approximately

$\sum_{i=1}^{n} \rho\,(x_i^*)\,g\Delta x$. The estimate becomes exact if we take the limit as $n \to \infty$; weight (or force) per unit area at

depth z is $W = \lim\limits_{n \to \infty} \sum_{i=1}^{n} \rho\,(x_i^*)\,g\Delta x$. In other words, $P\,(z) = \int_0^z \rho\,(x)\,g\,dx$. More generally, if we make no

assumptions about the location of the origin, then $P\,(z) = P_0 + \int_0^z \rho\,(x)\,g\,dx$, where P_0 is the pressure at
$x = 0$. Differentiating, we get $dP/dz = \rho\,(z)\,g$.

(b)

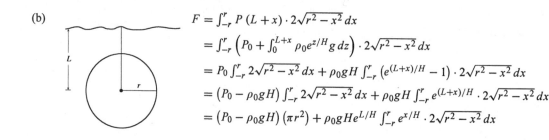

$$F = \int_{-r}^{r} P\,(L + x) \cdot 2\sqrt{r^2 - x^2}\, dx$$

$$= \int_{-r}^{r} \left(P_0 + \int_0^{L+x} \rho_0 e^{z/H} g\, dz\right) \cdot 2\sqrt{r^2 - x^2}\, dx$$

$$= P_0 \int_{-r}^{r} 2\sqrt{r^2 - x^2}\, dx + \rho_0 gH \int_{-r}^{r} \left(e^{(L+x)/H} - 1\right) \cdot 2\sqrt{r^2 - x^2}\, dx$$

$$= (P_0 - \rho_0 gH) \int_{-r}^{r} 2\sqrt{r^2 - x^2}\, dx + \rho_0 gH \int_{-r}^{r} e^{(L+x)/H} \cdot 2\sqrt{r^2 - x^2}\, dx$$

$$= (P_0 - \rho_0 gH)\,(\pi r^2) + \rho_0 gH e^{L/H} \int_{-r}^{r} e^{x/H} \cdot 2\sqrt{r^2 - x^2}\, dx$$

7. To find the height of the pyramid, we use similar triangles. The first figure shows a cross-section of the pyramid passing through the top and through two opposite corners of the square base. Now $|BD| = b$, since it is a radius of the sphere, which has diameter $2b$ since it is tangent to the opposite sides of the square base. Also, $|AD| = b$ since $\triangle ADB$ is isosceles. So the height is $|AB| = \sqrt{b^2 + b^2} = \sqrt{2}b$.

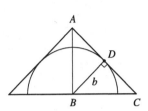

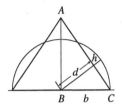

We observe that the shared volume is equal to half the volume of the sphere, minus the sum of the four equal volumes (caps of the sphere) cut off by the triangular faces of the pyramid. See Exercise 6.2.47 for a derivation of the formula for the volume of a cap of a sphere. To use the formula, we need to find the perpendicular distance h of each triangular face from the surface of the sphere. We first find the distance d from the center of the sphere to one of the triangular faces. The third figure shows a cross-section of the pyramid through the top and through the midpoints of opposite sides of the square base. From similar triangles we find that $\dfrac{d}{b} = \dfrac{|AB|}{|AC|} = \dfrac{\sqrt{2}b}{\sqrt{b^2 + \left(\sqrt{2}b\right)^2}}$

$\Rightarrow \quad d = \dfrac{\sqrt{2}b^2}{\sqrt{3b^2}} = \dfrac{\sqrt{6}}{3}b$. So $h = b - \dfrac{\sqrt{6}}{3}b = \dfrac{3-\sqrt{6}}{3}b$. So, using the formula

from Exercise 6.2.47 with $r = b$, we find that the volume of each of the caps is

$\pi \left(\dfrac{3-\sqrt{6}}{3}b\right)^2 \left(b - \dfrac{3-\sqrt{6}}{3\cdot3}b\right) = \dfrac{15-6\sqrt{6}}{9} \cdot \dfrac{6+\sqrt{6}}{9}\pi b^3 = \left(\dfrac{2}{3} - \dfrac{7}{27}\sqrt{6}\right)\pi b^3$. So, using our first observation, the

shared volume is $V = \dfrac{1}{2}\left(\dfrac{4}{3}\pi b^3\right) - 4\left(\dfrac{2}{3} - \dfrac{7}{27}\sqrt{6}\right)\pi b^3 = \left(\dfrac{28}{27}\sqrt{6} - 2\right)\pi b^3$.

9. If $h = L$, then

$$P = \dfrac{\text{area under } y = L\sin\theta}{\text{area of rectangle}} = \dfrac{\int_0^\pi L\sin\theta\, d\theta}{\pi L} = \dfrac{[-\cos\theta]_0^\pi}{\pi} = \dfrac{-(-1)+1}{\pi} = \dfrac{2}{\pi}$$

If $h = L/2$, then

$$P = \dfrac{\text{area under } y = \frac{1}{2}L\sin\theta}{\text{area of rectangle}} = \dfrac{\int_0^\pi \frac{1}{2}L\sin\theta\, d\theta}{\pi L} = \dfrac{[-\cos\theta]_0^\pi}{2\pi} = \dfrac{2}{2\pi} = \dfrac{1}{\pi}$$

9 Differential Equations

9.1 Modeling with Differential Equations

1. $y = 2 + e^{-x^3} \implies y' = -3x^2 e^{-x^3}$.

$$\text{LHS} = y' + 3x^2 y = -3x^2 e^{-x^3} + 3x^2 \left(2 + e^{-x^3}\right) = -3x^2 e^{-x^3} + 6x^2 + 3x^2 e^{-x^3} = 6x^2 = \text{RHS}$$

3. (a) $y = \sin kt \implies y' = k \cos kt \implies y'' = -k^2 \sin kt.$ $y'' + 9y = 0 \implies -k^2 \sin kt + 9 \sin kt = 0 \implies$
$(9 - k^2) \sin kt = 0$ (for all t) $\implies k = \pm 3$

(b) $y = A \sin kt + B \cos kt \implies y' = Ak \cos kt - Bk \sin kt \implies y'' = -Ak^2 \sin kt - Bk^2 \cos kt \implies$
$y'' + 9y = -Ak^2 \sin kt - Bk^2 \cos kt + 9 (A \sin kt + B \cos kt) = (9 - k^2) A \sin kt + (9 - k^2) B \cos kt = 0.$
The last equation is true for all values of A and B if $k = \pm 3$.

5. (a) $y = e^t \implies y' = e^t \implies y'' = e^t.$ LHS $= y'' + 2y' + y = e^t + 2e^t + e^t = 4e^t \neq 0$, so $y = e^t$ is not a solution of the differential equation.

(b) $y = e^{-t} \implies y' = -e^{-t} \implies y'' = e^{-t}.$ LHS $= y'' + 2y' + y = e^{-t} - 2e^{-t} + e^{-t} = 0 = \text{RHS}$, so $y = e^{-t}$ is a solution.

(c) $y = te^{-t} \implies y' = e^{-t} (1 - t) \implies y'' = e^{-t} (t - 2).$

$$\text{LHS} = y'' + 2y' + y = e^{-t} (t - 2) + 2e^{-t} (1 - t) + te^{-t} = e^{-t} [(t - 2) + 2 (1 - t) + t]$$
$$= e^{-t} (0) = 0 = \text{RHS}$$

so $y = te^{-t}$ is a solution.

(d) $y = t^2 e^{-t} \implies y' = te^{-t} (2 - t) \implies y'' = e^{-t} (t^2 - 4t + 2).$

$$\text{LHS} = y'' + 2y' + y = e^{-t} \left(t^2 - 4t + 2\right) + 2te^{-t} (2 - t) + t^2 e^{-t}$$
$$= e^{-t} \left[\left(t^2 - 4t + 2\right) + 2t (2 - t) + t^2\right] = e^{-t} (2) \neq 0$$

so $y = t^2 e^{-t}$ is not a solution.

7. (a) Since the derivative y' is always negative (or 0), the function y must be decreasing (or have a horizontal tangent) on any interval on which it is defined.

(b) $y = \dfrac{1}{x + C} \implies y' = -\dfrac{1}{(x + C)^2}.$ LHS $= y' = -\dfrac{1}{(x + C)^2} = -\left(\dfrac{1}{x + C}\right)^2 = -y^2 = \text{RHS}$

(c) $y = 0$ is a solution of $y' = -y^2$.

(d) $y (0) = \dfrac{1}{0 + C} = \dfrac{1}{C}$ and $y (0) = 0.5 \implies C = 2$, so $y = \dfrac{1}{x + 2}.$

9. (a) $\dfrac{dP}{dt} = 1.2P\left(1 - \dfrac{P}{4200}\right)$. $\dfrac{dP}{dt} > 0 \;\Rightarrow\; 1 - \dfrac{P}{4200} > 0 \;\Rightarrow\; P < 4200 \;\Rightarrow\;$ the population is increasing
for $P < 4200$ (assuming that $P \geq 0$).

(b) $\dfrac{dP}{dt} < 0 \;\Rightarrow\; P > 4200$

(c) $\dfrac{dP}{dt} = 0 \;\Rightarrow\; P = 4200$ or $P = 0$

11. (a) P increases most rapidly at the beginning, since there are usually
many simple, easily-learned sub-skills associated with learning a
skill. As t increases, we would expect dP/dt to remain positive, but
decrease. This is because as time progresses, the only points left to
learn are the more difficult ones.

(b) dP/dt is always positive, so the level of performance is increasing.
As P gets close to M, dP/dt gets close to 0, that is, the performance
levels off, as explained in part (a).

(c)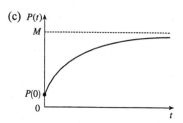

9.2 Direction Fields and Euler's Method

1. (a) 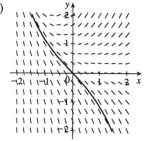   **(b)** 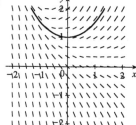   **(c)**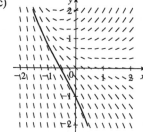

3. $y' = y - 1$. The slopes at each point are independent of x, so the slopes are the same along each line parallel to the
x-axis. Thus, IV is the direction field for this equation. Note that for $y = 1$, $y' = 0$.

5. $y' = y^2 - x^2 = 0 \;\Rightarrow\; y = \pm x$. There are horizontal tangents on these lines only in graph III, so this equation
corresponds to direction field III.

7. (a) 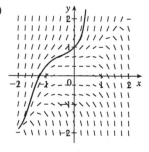   **(b)** 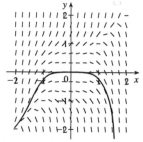   **(c)**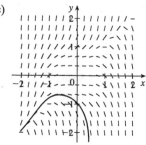

9. $y' = x - y$

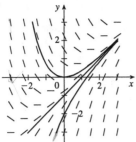

11.

x	y	$y' = y^2$
0	0	0
0	1	1
0	−1	1
1	0	0
−1	0	0
1	−1	1
1	1	1
1	2	4
1	−2	4
−1	2	4
−1	−2	4

The solution curve
through (0, 1)

13.

x	y	$y' = x^2 + y^2$
0	0	0
0	1	1
1	0	1
1	1	2
−1	1	2
0	2	4
2	0	4
2	2	8
2	1	5
−2	−1	5
1	2	5

The solution curve through (0, 0)

15. In Maple, we can use either `directionfield` (in Maple's share library) or `plots[fieldplot]` to plot the direction field. To plot the solution, we can either use the initial-value option in `directionfield`, or actually solve the equation. In Mathematica, we use `PlotVectorField` for the direction field, and the `Plot[Evaluate[...]]` construction to plot the solution, which is $y = e^{(1-\cos 2x)/2}$.

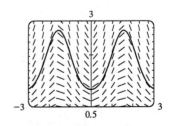

In Derive, use `Direction_Field` (in utility file ODE_APPR) to plot the direction field. Then use `DSOLVE1(-y*SIN(2*x),1,x,y,0,1)` (in utility file ODE1) to solve the equation. Simplify each result.

17.

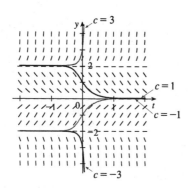

$L = \lim\limits_{t \to \infty} y(t)$ exists for $-2 \le c \le 2$; $L = \pm 2$ for $c = \pm 2$ and

$L = 0$ for $-2 < c < 2$. For other values of c, L does not exist.

19. (a) $y' = F(x, y) = y$ and $y(0) = 1 \Rightarrow x_0 = 0, y_0 = 1$.

(i) $h = 0.4$ and $y_1 = y_0 + hF(x_0, y_0) \Rightarrow y_1 = 1 + 0.4 \cdot 1 = 1.4$. $x_1 = x_0 + h = 0 + 0.4 = 0.4$, so
$y_1 = y(0.4) = 1.4$.

(ii) $h = 0.2 \Rightarrow x_1 = 0.2$ and $x_2 = 0.4$, so we need to find y_2.
$y_1 = y_0 + hF(x_0, y_0) = 1 + 0.2y_0 = 1 + 0.2 \cdot 1 = 1.2$,
$y_2 = y_1 + hF(x_1, y_1) = 1.2 + 0.2y_1 = 1.2 + 0.2 \cdot 1.2 = 1.44$.

(iii) $h = 0.1 \Rightarrow x_4 = 0.4$, so we need to find y_4. $y_1 = y_0 + hF(x_0, y_0) = 1 + 0.1y_0 = 1 + 0.1 \cdot 1 = 1.1$,
$y_2 = y_1 + hF(x_1, y_1) = 1.1 + 0.1y_1 = 1.1 + 0.1 \cdot 1.1 = 1.21$,
$y_3 = y_2 + hF(x_2, y_2) = 1.21 + 0.1y_2 = 1.21 + 0.1 \cdot 1.21 = 1.331$,
$y_4 = y_3 + hF(x_3, y_3) = 1.331 + 0.1y_3 = 1.331 + 0.1 \cdot 1.331 = 1.4641$.

(b)

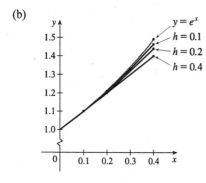

We see that the estimates are
underestimates since they are all below
the graph of $y = e^x$.

(c) (i) For $h = 0.4$:

(exact value) − (approximate value)

$= e^{0.4} - 1.4 \approx 0.0918$

(ii) For $h = 0.2$:

(exact value) − (approximate value)

$= e^{0.4} - 1.44 \approx 0.0518$

(iii) For $h = 0.1$:

(exact value) − (approximate value)

$= e^{0.4} - 1.4641 \approx 0.0277$

Each time the step size is halved, the error estimate also
appears to be halved (approximately).

21. $h = 0.5$, $x_0 = 1$, $y_0 = 2$, and $F(x, y) = 1 + 3x - 2y$. So
$y_n = y_{n-1} + hF(x_{n-1}, y_{n-1}) = y_{n-1} + 0.5(1 + 3x_{n-1} - 2y_{n-1}) = 0.5 + 1.5x_{n-1}$. Thus, $y_1 = 0.5 + 1.5 \cdot 1 = 2$,
$y_2 = 0.5 + 1.5 \cdot 1.5 = 2.75$, $y_3 = 0.5 + 1.5 \cdot 2 = 3.5$, $y_4 = 0.5 + 1.5 \cdot 2.5 = 4.25$.

23. $h = 0.1$, $x_0 = 0$, $y_0 = 1$, and $F(x, y) = x^2 + y^2$. We need to find y_5, because $x_5 = 0.5$. So
$y_n = y_{n-1} + 0.1(x_{n-1}^2 + y_{n-1}^2)$. $y_1 = 1 + 0.1(0^2 + 1^2) = 1.1$, $y_2 = 1.1 + 0.1(0.1^2 + 1.1^2) = 1.222$,
$y_3 = 1.222 + 0.1(0.2^2 + 1.222^2) \approx 1.37533$, $y_4 = 1.37533 + 0.1(0.3^2 + 1.37533^2) \approx 1.57348$,
$y_5 = 1.57348 + 0.1(0.4^2 + 1.57348^2) \approx 1.8371 \approx y(0.5)$.

25. (a) $dy/dx + 3x^2y = 6x^2 \implies y' = 6x^2 - 3x^2y$. Store this expression in Y_1 and use the following simple program to evaluate y (1) for each part, using $H = h = 1$ and $N = 1$ for part (i), $H = 0.1$ and $N = 10$ for part (ii), and so forth.

$$h \to H: 0 \to X: 3 \to Y:$$
$$\text{For}(I, 1, N): Y + HY_1 \to Y: X + H \to X:$$
$$\text{End(loop)}:$$
$$\text{Display } Y.$$

(i) $H = 1, N = 1 \implies y(1) = 3$ (ii) $H = 0.1, N = 10 \implies y(1) \approx 2.3928$

(iii) $H = 0.01, N = 100 \implies y(1) \approx 2.3701$ (iv) $H = 0.001, N = 1000 \implies y(1) \approx 2.3681$

(b) $y = 2 + e^{-x^3} \implies y' = -3x^2e^{-x^3}$

$$\text{LHS} = y' + 3x^2y = -3x^2e^{-x^3} + 3x^2\left(2 + e^{-x^3}\right) = -3x^2e^{-x^3} + 6x^2 + 3x^2e^{-x^3} = 6x^2 = \text{RHS}$$

$$y(0) = 2 + e^{-0} = 2 + 1 = 3$$

(c) (i) For $h = 1$: (exact value) − (approximate value) $= 2 + e^{-1} - 3 \approx -0.6321$

(ii) For $h = 0.1$: (exact value) − (approximate value) $= 2 + e^{-1} - 2.3928 \approx -0.0249$

(iii) For $h = 0.01$: (exact value) − (approximate value) $= 2 + e^{-1} - 2.3701 \approx -0.0022$

(iv) For $h = 0.001$: (exact value) − (approximate value) $= 2 + e^{-1} - 2.3681 \approx -0.0002$

In (ii)–(iv), it seems that when the step size is divided by 10, the error estimate is also divided by 10 (approximately).

27. (a) $R\dfrac{dQ}{dt} + \dfrac{1}{C}Q = E(t)$ becomes

$$5Q' + \tfrac{1}{0.05}Q = 60 \text{ or } Q' + 4Q = 12.$$

(b) From the graph, it appears that the limiting value of the charge Q is about 3.

(c) If $Q' = 0$, then $4Q = 12 \implies Q = 3$ is an equilibrium solution.

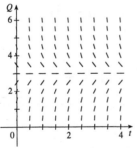

(d)

(e) $Q' + 4Q = 12 \implies Q' = 12 - 4Q$. $Q(0) = 0$, so $t_0 = 0$ and $Q_0 = 0$.

$$Q_1 = Q_0 + hF(t_0, Q_0) = 0 + 0.1(12 - 4 \cdot 0) = 1.2$$
$$Q_2 = Q_1 + hF(t_1, Q_1) = 1.2 + 0.1(12 - 4 \cdot 1.2) = 1.92$$
$$Q_3 = Q_2 + hF(t_2, Q_2) = 1.92 + 0.1(12 - 4 \cdot 1.92) = 2.352$$
$$Q_4 = Q_3 + hF(t_3, Q_3) = 2.352 + 0.1(12 - 4 \cdot 2.352) = 2.6112$$
$$Q_5 = Q_4 + hF(t_4, Q_4) = 2.6112 + 0.1(12 - 4 \cdot 2.6112) = 2.76672$$

Thus, $Q_5 = Q(0.5) \approx 2.77$ C.

9.3 Separable Equations

1. $\dfrac{dy}{dx} = y^2 \;\Rightarrow\; \dfrac{dy}{y^2} = dx \;(y \neq 0) \;\Rightarrow\; \displaystyle\int \dfrac{dy}{y^2} = \int dx \;\Rightarrow\; -\dfrac{1}{y} = x + C \;\Rightarrow\; -y = \dfrac{1}{x+C} \;\Rightarrow$

$y = \dfrac{-1}{x+C}$, and $y = 0$ is also a solution.

3. $yy' = x \;\Rightarrow\; \int y\,dy = \int x\,dx \;\Rightarrow\; \frac{1}{2}y^2 = \frac{1}{2}x^2 + C_1 \;\Rightarrow\; y^2 = x^2 + 2C_1 \;\Rightarrow$
$x^2 - y^2 = C$ (where $C = -2C_1$). This represents a family of hyperbolas.

5. $\dfrac{dy}{dt} = \dfrac{te^t}{y\sqrt{1+y^2}} \;\Rightarrow\; y\sqrt{1+y^2}\,dy = te^t\,dt \;\Rightarrow\; \int y\sqrt{1+y^2}\,dy = \int te^t\,dt \;\Rightarrow$

$\frac{1}{3}\left(1+y^2\right)^{3/2} = te^t - e^t + C$ (where the first integral is evaluated by substitution and the second by parts) $\;\Rightarrow$

$1 + y^2 = \left[3\left(te^t - e^t + C\right)\right]^{2/3} \;\Rightarrow\; y = \pm\sqrt{\left[3\left(te^t - e^t + C\right)\right]^{2/3} - 1}$

7. $\dfrac{du}{dt} = 2 + 2u + t + tu \;\Rightarrow\; \dfrac{du}{dt} = (1+u)(2+t) \;\Rightarrow\; \displaystyle\int \dfrac{du}{1+u} = \int (2+t)\,dt \;(u \neq -1) \;\Rightarrow$

$\ln|1+u| = \frac{1}{2}t^2 + 2t + C \;\Rightarrow\; |1+u| = e^{t^2/2 + 2t + C} = Ke^{t^2/2+2t}$, where $K = e^C \;\Rightarrow\; 1+u = \pm Ke^{t^2/2+2t}$

$\Rightarrow\; u = -1 \pm Ke^{t^2/2+2t}$ where $K > 0$. $u = -1$ is also a solution, so $u = -1 + ke^{t^2/2+2t}$, where k is an arbitrary constant.

9. $\dfrac{dy}{dx} = y^2 + 1, \; y(1) = 0.$ $\displaystyle\int \dfrac{dy}{y^2+1} = \int dx \;\Leftrightarrow\; \tan^{-1} y = x + C.$ $y = 0$ when $x = 1$, so $1 + C = \tan^{-1} 0 = 0$
$\Rightarrow\; C = -1.$ Thus, $\tan^{-1} y = x - 1$ and $y = \tan(x-1).$

11. $xe^{-t}\dfrac{dx}{dt} = t, \; x(0) = 1.$ $\int x\,dx = \int te^t\,dt \;\Rightarrow\; \frac{1}{2}x^2 = (t-1)e^t + C.$ $x(0) = 1$, so $\frac{1}{2} = (0-1)e^0 + C$ and
$C = \frac{3}{2}.$ Thus, $x^2 = 2(t-1)e^t + 3 \;\Rightarrow\; x = \sqrt{2(t-1)e^t + 3}.$

13. $\dfrac{du}{dt} = \dfrac{2t + \sec^2 t}{2u}, \; u(0) = -5.$ $\int 2u\,du = \int \left(2t + \sec^2 t\right)dt \;\Rightarrow\; u^2 = t^2 + \tan t + C$, where
$[u(0)]^2 = 0^2 + \tan 0 + C \;\Rightarrow\; C = 25.$ Therefore, $u^2 = t^2 + \tan t + 25$, so $u = \pm\sqrt{t^2 + \tan t + 25}.$ Since
$u(0) = -5$, we must have $u = -\sqrt{t^2 + \tan t + 25}.$

15. $\dfrac{dy}{dx} = 4x^3 y, \; y(0) = 7.$ $\dfrac{dy}{y} = 4x^3\,dx$ (if $y \neq 0$) $\;\Rightarrow\; \displaystyle\int \dfrac{dy}{y} = \int 4x^3\,dx \;\Rightarrow\; \ln|y| = x^4 + C \;\Rightarrow$

$e^{\ln|y|} = e^{x^4+C} \;\Rightarrow\; |y| = e^{x^4}e^C \;\Rightarrow\; y = Ae^{x^4}; y(0) = 7 \;\Rightarrow\; A = 7 \;\Rightarrow\; y = 7e^{x^4}.$

17. $y' = y \sin x, \; y(0) = 1.$ $\displaystyle\int \dfrac{dy}{y} = \int \sin x\,dx \;\Leftrightarrow\; \ln|y| = -\cos x + C$
$\Rightarrow\; |y| = e^{-\cos x + C} \;\Rightarrow\; y(x) = Ae^{-\cos x}.$ $y(0) = Ae^{-1} = 1 \;\Leftrightarrow$
$A = e^1$, so $y = e \cdot e^{-\cos x} = e^{1-\cos x}.$

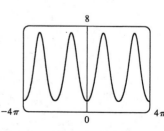

19. $\dfrac{dy}{dx} = \dfrac{\sin x}{\sin y}$, $y\,(0) = \frac{\pi}{2}$. So $\int \sin y\,dy = \int \sin x\,dx$ ⟺

$-\cos y = -\cos x + C$ ⟺ $\cos y = \cos x - C$. From the initial

condition, we need $\cos\frac{\pi}{2} = \cos 0 - C$ ⟹ $0 = 1 - C$ ⟹ $C = 1$, so

the solution is $\cos y = \cos x - 1$. Note that we cannot take $\cos^{-1}$ of both

sides, since that would unnecessarily restrict the solution to the case where

$-1 \le \cos x - 1$ ⟺ $0 \le \cos x$, as $\cos^{-1}$ is defined only on $[-1, 1]$.

Instead we plot the graph using Maple's plots [implicitplot] or

Mathematica's Plot [Evaluate[···]].

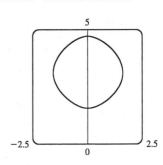

21. (a)

x	y	$y' = 1/y$
0	0.5	2
0	−0.5	−2
0	1	1
0	−1	−1
0	2	0.5
0	−2	−0.5
0	4	0.25
0	3	$0.\overline{3}$
0	0.25	4
0	$0.\overline{3}$	3

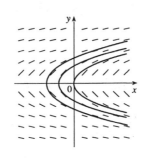

(b) $y\,dy = dx$, so $\frac{1}{2}y^2 = x + c$ or

$y = \pm\sqrt{2\,(x + c)}$.

(c)

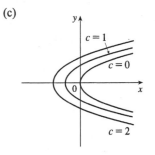

23. The curves $y = kx^2$ form a family of parabolas with axis the y-axis.

Differentiating gives $y' = 2kx$, but $k = y/x^2$, so $y' = 2y/x$. Thus, the

slope of the tangent line at any point (x, y) on one of the parabolas is

$y' = 2y/x$, so the orthogonal trajectories must satisfy $y' = -x\,/(2y)$ ⟺

$2y\,dy = -x\,dx$ ⟺ $y^2 = -x^2/2 + c_1$ ⟺ $x^2 + 2y^2 = c$. This is a

family of ellipses.

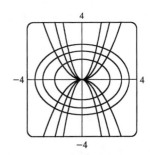

25. Differentiating $y = (x + k)^{-1}$ gives $y' = -\dfrac{1}{(x + k)^2}$, but $k = \dfrac{1}{y} - x$, so

$y' = -\dfrac{1}{(1/y)^2} = -y^2$. Thus, the orthogonal trajectories must satisfy

$y' = -\dfrac{1}{-y^2} = \dfrac{1}{y^2}$ ⟺ $y^2\,dy = dx$ ⟺ $\dfrac{y^3}{3} = x + c$ or

$y = [3\,(x + c)]^{1/3}$

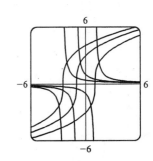

27. From Exercise 9.2.27, $\dfrac{dQ}{dt} = 12 - 4Q$ $\Leftrightarrow$ $\displaystyle\int \dfrac{dQ}{12 - 4Q} = \int dt$ $\Leftrightarrow$ $-\tfrac{1}{4}\ln|12 - 4Q| = t + C$ $\Leftrightarrow$

$\ln|12 - 4Q| = -4t - 4C$ $\Leftrightarrow$ $|12 - 4Q| = e^{-4t-4C}$ $\Leftrightarrow$ $12 - 4Q = Ke^{-4t}$ $(K = \pm e^{-4C})$ $\Leftrightarrow$

$4Q = 12 - Ke^{-4t}$ $\Leftrightarrow$ $Q = 3 - Ae^{-4t}$ $(A = K/3)$. $Q(0) = 0$ $\Leftrightarrow$ $0 = 3 - A$ $\Leftrightarrow$ $A = 3$ $\Leftrightarrow$

$Q(t) = 3 - 3e^{-4t}$. As $t \to \infty$, $Q(t) \to 3 - 0 = 3$ (the limiting value).

29. $\dfrac{dP}{dt} = k(M - P)$ $\Leftrightarrow$ $\displaystyle\int \dfrac{dP}{P - M} = \int (-k)\,dt$ $\Leftrightarrow$ $\ln|P - M| = -kt + C$ $\Leftrightarrow$ $|P - M| = e^{-kt+C}$ $\Leftrightarrow$

$P - M = Ae^{-kt}$ $(A = \pm e^{C})$ $\Leftrightarrow$ $P = M + Ae^{-kt}$. If we assume that performance is at level 0 when $t = 0$,

then $P(0) = 0$ $\Leftrightarrow$ $0 = M + A$ $\Leftrightarrow$ $A = -M$ $\Leftrightarrow$ $P(t) = M - Me^{-kt}$. $\lim\limits_{t \to \infty} P(t) = M - M \cdot 0 = M$.

31. (a) $\dfrac{dC}{dt} = r - kC$ $\Rightarrow$ $\dfrac{dC}{dt} = -(kC - r)$ $\Rightarrow$ $\displaystyle\int \dfrac{dC}{kC - r} = \int -dt$ $\Rightarrow$ $(1/k)\ln|kC - r| = -t + M_1$

$\Rightarrow$ $\ln|kC - r| = -kt + M_2$ $\Rightarrow$ $|kC - r| = e^{-kt+M_2}$ $\Rightarrow$ $kC - r = M_3 e^{-kt}$ $\Rightarrow$ $kC = M_3 e^{-kt} + r$

$\Rightarrow$ $C(t) = M_4 e^{-kt} + r/k$. $C(0) = C_0$ $\Rightarrow$ $C_0 = M_4 + r/k$ $\Rightarrow$ $M_4 = C_0 - r/k$ $\Rightarrow$

$C(t) = (C_0 - r/k)e^{-kt} + r/k$.

(b) If $C_0 < r/k$, then $C_0 - r/k < 0$ and the formula for $C(t)$ shows that $C(t)$ increases and $\lim\limits_{t \to \infty} C(t) = r/k$. As

t increases, the formula for $C(t)$ shows how the role of C_0 steadily diminishes as that of r/k increases.

33. (a) Let $y(t)$ be the amount of salt (in kg) after t minutes. Then $y(0) = 15$. The amount of liquid in the tank is

1000 L at all times, so the concentration at time t (in minutes) is $y(t)/1000$ kg/L and

$\dfrac{dy}{dt} = -\left[\dfrac{y(t)}{1000}\dfrac{\text{kg}}{\text{L}}\right]\left(10 \dfrac{\text{L}}{\text{min}}\right) = -\dfrac{y(t)}{100}\dfrac{\text{kg}}{\text{min}}$. $\displaystyle\int \dfrac{dy}{y} = -\dfrac{1}{100}\int dt$ $\Rightarrow$ $\ln y = -\dfrac{t}{100} + C$, and

$y(0) = 15$ $\Rightarrow$ $\ln 15 = C$, so $\ln y = \ln 15 - \dfrac{t}{100}$. It follows that $\ln\left(\dfrac{y}{15}\right) = -\dfrac{t}{100}$ and $\dfrac{y}{15} = e^{-t/100}$, so

$y = 15e^{-t/100}$ kg.

(b) After 20 minutes, $y = 15e^{-20/100} = 15e^{-0.2} \approx 12.3$ kg.

35. Assume that the raindrop begins at rest, so that $v(0) = 0$. $dm/dt = km$ and $(mv)' = gm$ $\Rightarrow$ $m'v + mv' = gm$

$\Rightarrow$ $(km)v + mv' = gm$ $\Rightarrow$ $v' = g - kv$ $\Rightarrow$ $\displaystyle\int \dfrac{dv}{g - kv} = \int dt$ $\Rightarrow$ $-(1/k)\ln|g - kv| = t + C$ $\Rightarrow$

$g - kv = Ae^{-kt}$. $v(0) = 0$ $\Rightarrow$ $A = g$. So $v = (g/k)(1 - e^{-kt})$. Since $k > 0$, as $t \to \infty$, $e^{-kt} \to 0$ and

therefore, $\lim\limits_{t \to \infty} v(t) = g/k$.

37. (a) The rate of growth of the area is jointly proportional to $\sqrt{A(t)}$ and $M - A(t)$; that is, the rate is proportional to

the product of those two quantities. So for some constant k, $dA/dt = k\sqrt{A}(M - A)$. We are interested in the

maximum of the function dA/dt (when the tissue grows the fastest), so we differentiate, using the Chain Rule

and then substituting for dA/dt from the differential equation:

$$\dfrac{d}{dt}\left(\dfrac{dA}{dt}\right) = k\left[\tfrac{1}{2}A^{-1/2}(M - A)\dfrac{dA}{dt} + \sqrt{A}(-1)\dfrac{dA}{dt}\right] = \tfrac{1}{2}kA^{-1/2}\dfrac{dA}{dt}[(M - A) - 2A]$$

$$= \tfrac{1}{2}k^2(M - A)(M - 3A)$$

This is 0 when $M - A = 0$ [this situation never actually occurs, since the graph of $A(t)$ is asymptotic to the

line $y = M$, as in the logistic model] and when $M - 3A = 0$ $\Leftrightarrow$ $A(t) = M/3$. This represents a maximum

by the First Derivative Test, since $\dfrac{d}{dt}\left(\dfrac{dA}{dt}\right)$ goes from positive to negative when $A(t) = M/3$.

(b) From the CAS, we get $A(t) = M\left(\dfrac{Ce^{\sqrt{Mk}t}-1}{Ce^{\sqrt{Mk}t}+1}\right)^2$. To get C in terms of the initial area A_0 and the maximum

area M, we substitute $t=0$ and $A=A_0$: $A_0 = M\left(\dfrac{C-1}{C+1}\right)^2 \;\Leftrightarrow\; (C+1)\sqrt{A_0}=(C-1)\sqrt{M} \;\Leftrightarrow\;$

$C\sqrt{A_0}+\sqrt{A_0}=C\sqrt{M}-\sqrt{M} \;\Leftrightarrow\; \sqrt{A_0}+\sqrt{M}=C\sqrt{M}-C\sqrt{A_0} \;\Leftrightarrow\; C=\dfrac{\sqrt{M}+\sqrt{A_0}}{\sqrt{M}-\sqrt{A_0}}$. (Notice that

if $A_0=0$, then $C=1$.)

39. (a) We have $V(t)=\pi r^2 y(t) \;\Rightarrow\; \dfrac{dV}{dy}=\pi r^2 = 4\pi$ where $\dfrac{dV}{dt}=\dfrac{dV}{dy}\dfrac{dy}{dt}$. Thus, $\dfrac{dV}{dt}=-a\sqrt{2gy} \;\Rightarrow\;$

$\dfrac{dV}{dy}\dfrac{dy}{dt}=-\pi\left(\tfrac{1}{12}\right)^2\sqrt{2\cdot 32y} \;\Rightarrow\; 4\pi\dfrac{dy}{dt}=-\pi\tfrac{8}{144}\sqrt{y} \;\Rightarrow\; \dfrac{dy}{dt}=-\tfrac{1}{72}\sqrt{y}$.

(b) $\dfrac{dy}{dt}=-\tfrac{1}{72}\sqrt{y} \;\Rightarrow\; y^{-1/2}\,dy=-\tfrac{1}{72}\,dt \;\Rightarrow\; 2\sqrt{y}=-\tfrac{1}{72}t+C.\; y(0)=6 \;\Rightarrow\; 2\sqrt{6}=0+C \;\Rightarrow\;$

$C=2\sqrt{6} \;\Rightarrow\; y=\left(-\tfrac{1}{144}t+\sqrt{6}\right)^2$.

(c) We want to find t when $y=0$, so we set $y=0=\left(-\tfrac{1}{144}t+\sqrt{6}\right)^2 \;\Rightarrow\; t=144\sqrt{6}\approx 5$ min 53 s.

9.4 Exponential Growth and Decay

1. The relative growth rate is $\dfrac{1}{P}\dfrac{dP}{dt}=0.7944$, so $\dfrac{dP}{dt}=0.7944P$ and, by Theorem 2,

$P(t)=P(0)e^{0.7944t}=2e^{0.7944t}$. Thus, $P(6)=2e^{0.7944(6)}\approx 234.99$ or about 235 members.

3. (a) By Theorem 2, $y(t)=y(0)e^{kt}=500e^{kt}.\; y(3)=500e^{3k}=8000 \;\Rightarrow\; e^{3k}=16 \;\Rightarrow\; 3k=\ln 16 \;\Rightarrow\;$

$k=(\ln 16)/3$. So $y(t)=500e^{(\ln 16)t/3}=500\cdot 16^{t/3}$

(b) $y(4)=500\cdot 16^{4/3}\approx 20{,}159$

(c) $dy/dt=ky \;\Rightarrow\; y'(4)=ky(4)=\tfrac{1}{3}\ln 16\left(500\cdot 16^{4/3}\right)$ [from part (a)] $\approx 18{,}631$ cells/h

(d) $y(t)=500\cdot 16^{t/3}=30{,}000 \;\Rightarrow\; 16^{t/3}=60 \;\Rightarrow\; \tfrac{1}{3}t\ln 16=\ln 60 \;\Rightarrow\; t=3(\ln 60)/(\ln 16)\approx 4.4$ h

5. (a) Let the population (in millions) in the year t be $P(t)$. Since the initial time is the year 1750, we substitute
$t-1750$ for t in Theorem 2, so the exponential model gives $P(t)=P(1750)e^{k(t-1750)}$. Then
$P(1800)=906=728e^{k(1800-1750)} \;\Rightarrow\; \ln\tfrac{906}{728}=k(50) \;\Rightarrow\; k=\tfrac{1}{50}\ln\tfrac{906}{728}\approx 0.0043748$. So
with this model, we have $P(1900)=728e^{150(0.0043748)}\approx 1403$ million, and
$P(1950)\approx 728e^{200(0.0043748)}\approx 1746$ million. Both of these estimates are much too low.

(b) In this case, the exponential model gives $P(t)=P(1850)e^{k(t-1850)} \;\Rightarrow\;$
$P(1900)=1608=1171e^{k(1900-1850)} \;\Rightarrow\; \ln\tfrac{1608}{1171}=k(50) \;\Rightarrow\; k=\tfrac{1}{50}\ln\tfrac{1608}{1171}\approx 0.006343$. So with this
model, we estimate $P(1950)\approx 1171e^{100(0.006343)}\approx 2208$ million. This is still too low, but closer than the
estimate of $P(1950)$ in part (a).

(c) The exponential model gives $P(t) = P(1900) e^{k(t-1900)}$ $\Rightarrow$ $P(1950) = 2517 = 1608 e^{k(1950-1900)}$ $\Rightarrow$ $\ln \frac{2517}{1608} = k(50)$ $\Rightarrow$ $k = \frac{1}{50} \ln \frac{2517}{1608} \approx 0.008962$. With this model, we estimate $P(1992) \approx 1608 e^{0.008962(1992-1900)} \approx 3667$ million. This is much too low. The discrepancy is explained by the fact that the world birth rate (average yearly number of births per person) is about the same as always, whereas the mortality rate (especially the infant mortality rate) is much lower, owing mostly to advances in medical science and to the wars in the first part of the twentieth century. The exponential model assumes, among other things, that the birth and mortality rates will remain constant.

7. (a) If $y = [N_2O_5]$ then by Theorem 2, $\dfrac{dy}{dt} = -0.0005y$ $\Rightarrow$ $y(t) = y(0) e^{-0.0005t} = Ce^{-0.0005t}$.

(b) $y(t) = Ce^{-0.0005t} = 0.9C$ $\Rightarrow$ $e^{-0.0005t} = 0.9$ $\Rightarrow$ $-0.0005t = \ln 0.9$ $\Rightarrow$ $t = -2000 \ln 0.9 \approx 211$ s

9. (a) If $y(t)$ is the mass remaining after t days, then $y(t) = y(0) e^{kt} = 50 e^{kt}$. $y(0.00014) = 50 e^{0.00014k} = 25$ $\Rightarrow$ $e^{0.00014k} = \frac{1}{2}$ $\Rightarrow$ $k = -(\ln 2)/0.00014$ $\Rightarrow$ $y(t) = 50 e^{-(\ln 2)t/0.00014} = 50 \cdot 2^{-t/0.00014}$

(b) $y(0.01) = 50 \cdot 2^{-0.01/0.00014} \approx 1.57 \times 10^{-20}$ mg

(c) $50 e^{-(\ln 2)t/0.00014} = 40$ $\Rightarrow$ $-(\ln 2)t/0.00014 = \ln 0.8$ $\Rightarrow$ $t = -0.00014 \dfrac{\ln 0.8}{\ln 2} \approx 4.5 \times 10^{-5}$ s

11. Let $y(t)$ be the level of radioactivity. Thus, $y(t) = y(0) e^{-kt}$ and k is determined by using the half-life:

$y(5730) = \frac{1}{2} y(0)$ $\Rightarrow$ $\frac{1}{2} = e^{-5730k}$ $\Rightarrow$ $k = -\dfrac{\ln \frac{1}{2}}{5730} = \dfrac{\ln 2}{5730}$. If 74% of the ^{14}C remains, then we know

that $y(t) = 0.74 y(0)$ $\Rightarrow$ $0.74 = e^{-t(\ln 2)/5730}$ $\Rightarrow$ $\ln 0.74 = -\dfrac{t \ln 2}{5730}$ $\Rightarrow$

$t = -\dfrac{5730 (\ln 0.74)}{\ln 2} \approx 2489 \approx 2500$ years.

13. (a) If $y = u - 75$, $u(0) = 185$ $\Rightarrow$ $y(0) = 185 - 75 = 110$, and the initial-value problem is $dy/dt = ky$ with $y(0) = 110$. So the solution is $y(t) = 110 e^{kt}$.

(b) $y(30) = 110 e^{30k} = 150 - 75$ $\Rightarrow$ $e^{30k} = \frac{75}{110} = \frac{15}{22}$ $\Rightarrow$ $k = \frac{1}{30} \ln \frac{15}{22}$, so $y(t) = 110 e^{\frac{1}{30} t \ln\left(\frac{15}{22}\right)}$ and $y(45) = 110 e^{\frac{45}{30} \ln\left(\frac{15}{22}\right)} \approx 62°F$. Thus, $u(45) \approx 62 + 75 = 137°F$.

(c) $u(t) = 100$ $\Rightarrow$ $y(t) = 25$. $y(t) = 110 e^{\frac{1}{30} t \ln\left(\frac{15}{22}\right)} = 25$ $\Rightarrow$ $e^{\frac{1}{30} t \ln\left(\frac{15}{22}\right)} = \frac{25}{110}$ $\Rightarrow$ $\frac{1}{30} t \ln \frac{15}{22} = \ln \frac{25}{110}$

$\Rightarrow$ $t = \dfrac{30 \ln \frac{25}{110}}{\ln \frac{15}{22}} \approx 116$ min.

15. (a) Let $P(h)$ be the pressure at altitude h. Then $dP/dh = kP$ $\Rightarrow$ $P(h) = P(0) e^{kh} = 101.3 e^{kh}$.

$P(1000) = 101.3 e^{1000k} = 87.14$ $\Rightarrow$ $1000k = \ln\left(\frac{87.14}{101.3}\right)$ $\Rightarrow$ $P(h) = 101.3 e^{\frac{1}{1000} h \ln\left(\frac{87.14}{101.3}\right)}$, so

$P(3000) = 101.3 e^{3 \ln\left(\frac{87.14}{101.3}\right)} \approx 64.5$ kPa.

(b) $P(6187) = 101.3 e^{\frac{6187}{1000} \ln\left(\frac{87.14}{101.3}\right)} \approx 39.9$ kPa

17. (a) Using $A = A_0 \left(1 + \dfrac{r}{n}\right)^{nt}$ with $A_0 = 3000$, $r = 0.05$, and $t = 5$,

(b) $dA/dt = 0.05A$ and $A(0) = 3000$.

we have:

(i) Annually: $n = 1$; $\quad A = 3000\,(1.05)^5 = \$3828.84$

(ii) Semiannually: $n = 2$; $\quad A = 3000 \left(1 + \frac{0.05}{2}\right)^{10} = \$3840.25$

(iii) Monthly: $n = 12$; $\quad A = 3000 \left(1 + \frac{0.05}{12}\right)^{60} = \$3850.08$

(iv) Weekly: $n = 52$; $\quad A = 3000 \left(1 + \frac{0.05}{52}\right)^{5 \cdot 52} = \$3851.61$

(v) Daily: $n = 365$; $\quad A = 3000 \left(1 + \frac{0.05}{365}\right)^{5 \cdot 365} = \$3852.01$

(vi) Continuously: $\quad A = 3000e^{(0.05)5} = \$3852.08$

19. (a) $\dfrac{dP}{dt} = kP - m = k\left(P - \dfrac{m}{k}\right)$. Let $y = P - \dfrac{m}{k}$, so the equation becomes $\dfrac{dy}{dt} = ky$. The solution is $y = y_0 e^{kt}$

$\Rightarrow \quad P - \dfrac{m}{k} = \left(P_0 - \dfrac{m}{k}\right) e^{kt} \quad \Rightarrow \quad P(t) = \dfrac{m}{k} + \left(P_0 - \dfrac{m}{k}\right) e^{kt}$.

(b) There will be an exponential expansion $\Leftrightarrow \quad P_0 - \dfrac{m}{k} > 0 \quad \Leftrightarrow \quad m < kP_0$.

(c) The population will be constant if $P_0 - \dfrac{m}{k} = 0 \quad \Leftrightarrow \quad m = kP_0$. It will decline if $P_0 - \dfrac{m}{k} < 0 \quad \Leftrightarrow$

$m > kP_0$.

(d) $P_0 = 8{,}000{,}000$, $k = \alpha - \beta = 0.016$, $m = 210{,}000 \quad \Rightarrow \quad m > kP_0 \, (= 128{,}000)$, so by part (c), the population was declining.

9.5 The Logistic Equation

1. (a) $dP/dt = 0.05P - 0.0005P^2 = 0.05P\,(1 - 0.01P) = 0.05P\,(1 - P/100)$. Comparing to Equation 1, $dP/dt = kP\,(1 - P/K)$, we see that the carrying capacity is $K = 100$ and the value of k is 0.05.

(b) The slopes close to 0 occur where P is near 100. The largest slopes appear to be on the line $P = 50$. The solutions are increasing for $0 < P_0 < 100$ and decreasing for $P_0 > 100$.

(c)

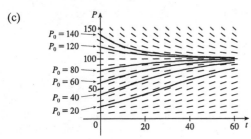

All of the solutions approach $P = 100$ as t increases. As in part (b), the solutions differ since for $0 < P_0 < 100$ they are increasing, and for $P_0 > 100$ they are decreasing. Also, some have an IP and some don't. It appears that the solutions which have $P_0 = 20$ and $P_0 = 40$ have inflection points at $P = 50$.

(d) The equilibrium solutions are $P = 0$ (trivial solution) and $P = 100$. The increasing solutions move away from $P = 0$ and all nonzero solutions approach $P = 100$ as $t \to \infty$.

3. (a) $\dfrac{dy}{dt} = ky\left(1 - \dfrac{y}{K}\right)$ ⟹ $y(t) = \dfrac{K}{1 + Ae^{-kt}}$ with $A = \dfrac{K - y(0)}{y(0)}$. With $K = 8 \times 10^7$, $k = 0.71$, and

$y(0) = 2 \times 10^7$, we get the model $y(t) = \dfrac{8 \times 10^7}{1 + 3e^{-0.71t}}$, so $y(1) = \dfrac{8 \times 10^7}{1 + 3e^{-0.71}} \approx 3.23 \times 10^7$ kg.

(b) $y(t) = 4 \times 10^7$ ⟹ $\dfrac{8 \times 10^7}{1 + 3e^{-0.71t}} = 4 \times 10^7$ ⟹ $2 = 1 + 3e^{-0.71t}$ ⟹ $e^{-0.71t} = \frac{1}{3}$ ⟹

$-0.71t = \ln\frac{1}{3}$ ⟹ $t = \dfrac{\ln 3}{0.71} \approx 1.55$ years

5. (a) We will assume that the difference in the birth and death rates is 20 million/y. Let $t = 0$ correspond to the year
1990 and use a unit of 1 million for all calculations. $k \approx \dfrac{1}{P}\dfrac{dP}{dt} = \frac{1}{5300}(20) = \frac{1}{265}$, so

$$\dfrac{dP}{dt} = kP\left(1 - \dfrac{P}{K}\right) = \dfrac{1}{265}P\left(1 - \dfrac{P}{100,000}\right)$$

(b) $A = \dfrac{K - P_0}{P_0} = \dfrac{100,000 - 5300}{5300} = \dfrac{947}{53} \approx 17.8679$. $P(t) = \dfrac{K}{1 + Ae^{-kt}} = \dfrac{100,000}{1 + \frac{947}{53}e^{-(1/265)t}}$, so

$P(10) \approx 5492.6$ (or 5.5 billion), $P(110) \approx 7813.8$, and $P(510) \approx 27{,}718.3$.

(c) If $K = 50{,}000$, then $P(t) = \dfrac{50{,}000}{1 + \frac{447}{53}e^{-(1/265)t}}$. So $P(10) \approx 5481.5$, $P(110) \approx 7611.8$, and

$P(510) \approx 22{,}412.6$

7. (a) Our assumption is that $\dfrac{dy}{dt} = ky(1 - y)$, where y is the fraction of the population that has heard the rumor.

(b) Using the logistic equation (1), $\dfrac{dP}{dt} = kP\left(1 - \dfrac{P}{K}\right)$, we substitute $y = \dfrac{P}{K}$, $P = Ky$, and $\dfrac{dP}{dt} = K\dfrac{dy}{dt}$, to

obtain $K\dfrac{dy}{dt} = k(Ky)(1 - y)$ ⟺ $\dfrac{dy}{dt} = ky(1 - y)$, our equation in part (a). Now the solution to (1) is

$P(t) = \dfrac{K}{1 + Ae^{-kt}}$, where $A = \dfrac{K - P_0}{P_0}$. We use the same substitution to obtain $Ky = \dfrac{K}{1 + \dfrac{K - Ky_0}{Ky_0}e^{-kt}}$

⟹ $y = \dfrac{y_0}{y_0 + (1 - y_0)e^{-kt}}$.

Alternatively, we could use the same steps as outlined in "The Analytic Solution", following Example 2.

(c) Let t be the number of hours since 8 A.M. Then $y_0 = y(0) = \frac{80}{1000} = 0.08$ and $y(4) = \frac{1}{2}$, so

$\dfrac{1}{2} = y(4) = \dfrac{0.08}{0.08 + 0.92e^{-4k}}$. Thus, $0.08 + 0.92e^{-4k} = 0.16$, $e^{-4k} = \frac{0.08}{0.92} = \frac{2}{23}$, and $e^{-k} = \left(\frac{2}{23}\right)^{1/4}$, so

$y = \dfrac{0.08}{0.08 + 0.92(2/23)^{t/4}} = \dfrac{2}{2 + 23(2/23)^{t/4}}$. Solving this equation for t, we get

$2y + 23y\left(\dfrac{2}{23}\right)^{1/4} = \dfrac{2 - 2y}{23y}$ ⟹ $\left(\dfrac{2}{23}\right)^{1/4} = \dfrac{2}{23}\cdot\dfrac{1 - y}{y}$ ⟹ $\left(\dfrac{2}{23}\right)^{t/4 - 1} = \dfrac{1 - y}{y}$. It follows that

$\dfrac{t}{4} - 1 = \dfrac{\ln\left[(1 - y)/y\right]}{\ln\frac{2}{23}}$, so $t = 4\left[1 + \dfrac{\ln\left((1 - y)/y\right)}{\ln\frac{2}{23}}\right]$. When $y = 0.9$, $\dfrac{1 - y}{y} = \frac{1}{9}$, so

$t = 4\left(1 - \dfrac{\ln 9}{\ln\frac{2}{23}}\right) \approx 7.6$ h or 7 h 36 min. Thus, 90% of the population will have heard the rumor by 3:36 P.M.

9. (a) $\dfrac{dP}{dt} = k(P)\left(1 - \dfrac{P}{K}\right) \Rightarrow$

$$\dfrac{d^2P}{dt^2} = k\left[P\left(-\dfrac{1}{K}\dfrac{dP}{dt}\right) + \left(1 - \dfrac{P}{K}\right)\dfrac{dP}{dt}\right] = k\dfrac{dP}{dt}\left(-\dfrac{P}{K} + 1 - \dfrac{P}{K}\right)$$

$$= k\left[kP\left(1 - \dfrac{P}{K}\right)\right]\left[1 - \dfrac{2P}{K}\right] = k^2 P\left(1 - \dfrac{P}{K}\right)\left(1 - \dfrac{2P}{K}\right)$$

(b) P grows fastest when P' has a maximum, that is, when $P'' = 0$. From part (a), $P'' = 0 \iff P = 0, P = K$, or $P = K/2$. Since $0 < P < K$, we see that $P'' = 0 \iff P = K/2$.

11. (a) The term -15 represents a harvesting of fish at a constant rate — in this case, 15 fish/week. This is the rate at which fish are caught.

(b)

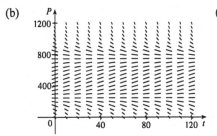

(c) From the graph in part (b), it appears that $P(t) = 250$ and $P(t) = 750$ are the equilibrium solutions. We confirm this analytically by solving the equation $dP/dt = 0$ as follows:

$0.08P(1 - P/1000) - 15 = 0 \Rightarrow$

$0.08P - 0.00008P^2 - 15 = 0 \Rightarrow$

$-0.00008(P^2 - 1000P + 187,500) = 0 \Rightarrow$

$(P - 250)(P - 750) = 0 \Rightarrow P = 250$ or 750.

(d)

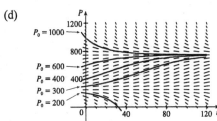

For $0 < P_0 < 250$, $P(t)$ decreases to 0. For $P_0 = 250$, $P(t)$ remains constant. For $250 < P_0 < 750$, $P(t)$ increases and approaches 750. For $P_0 = 750$, $P(t)$ remains constant. For $P_0 > 750$, $P(t)$ decreases and approaches 750.

(e) $\dfrac{dP}{dt} = 0.08P\left(1 - \dfrac{P}{1000}\right) - 15 \iff -\dfrac{100,000}{8}\cdot\dfrac{dP}{dt} = (0.08P - 0.00008P^2 - 15)\cdot\left(-\dfrac{100,000}{8}\right) \iff$

$-12,500\dfrac{dP}{dt} = P^2 - 1000P + 187,500 \iff \dfrac{dP}{(P - 250)(P - 750)} = -\dfrac{1}{12,500}\,dt \iff$

$\displaystyle\int\left(\dfrac{-1/500}{P - 250} + \dfrac{1/500}{P - 750}\right)dP = -\dfrac{1}{12,500}\,dt \iff \int\left(\dfrac{1}{P - 250} - \dfrac{1}{P - 750}\right)dP = \dfrac{1}{25}\,dt \iff$

$\ln|P - 250| - \ln|P - 750| = \dfrac{1}{25}t + C \iff \ln\left|\dfrac{P - 250}{P - 750}\right| = \dfrac{1}{25}t + C \iff \left|\dfrac{P - 250}{P - 750}\right| = e^{t/25+C} = ke^{t/25}$

$\iff \dfrac{P - 250}{P - 750} = ke^{t/25} \iff P - 250 = Pke^{t/25} - 750ke^{t/25} \iff P - Pke^{t/25} = 250 - 750ke^{t/25} \iff$

$P(t) = \dfrac{250 - 750ke^{t/25}}{1 - ke^{t/25}}$. If $t = 0$ and $P = 200$, then $200 = \dfrac{250 - 750k}{1 - k}$

$\iff 200 - 200k = 250 - 750k \iff 550k = 50 \iff k = \dfrac{1}{11}$.

Similarly, if $t = 0$ and $P = 300$, then $k = -\dfrac{1}{9}$. Simplifying P with these two values of k gives us

$P(t) = \dfrac{250\left(3e^{t/25} - 11\right)}{e^{t/25} - 11}$ and $P(t) = \dfrac{750\left(e^{t/25} + 3\right)}{e^{t/25} + 9}$.

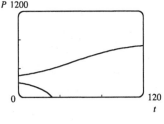

13. (a) $\dfrac{dP}{dt} = (kP)\left(1 - \dfrac{P}{K}\right)\left(1 - \dfrac{m}{P}\right)$. If $m < P < K$, then $dP/dt = (+)\,(+)\,(+) = +\ \Rightarrow\ P$ is increasing. If

$0 < P < m$, then $dP/dt = (+)\,(+)\,(-) = -\ \Rightarrow\ P$ is decreasing.

(b)

$k = 0.08$, $K = 1000$, and $m = 200\ \Rightarrow$

$$\frac{dP}{dt} = 0.08P\left(1 - \frac{P}{1000}\right)\left(1 - \frac{200}{P}\right)$$

For $0 < P_0 < 200$, the population dies out. For $P_0 = 200$, the population is steady. For $200 < P_0 < 1000$, the population increases and approaches 1000. For $P_0 > 1000$, the population decreases and approaches 1000.

The equilibrium solutions are $P\,(t) = 200$ and $P\,(t) = 1000$.

(c) $\dfrac{dP}{dt} = kP\left(1 - \dfrac{P}{K}\right)\left(1 - \dfrac{m}{P}\right) = kP\left(\dfrac{K-P}{K}\right)\left(\dfrac{P-m}{P}\right) = \dfrac{k}{K}(K-P)(P-m)\ \Leftrightarrow$

$\displaystyle\int \frac{dP}{(K-P)(P-m)} = \int \frac{k}{K}\,dt.$

By partial fractions, $\dfrac{1}{(K-P)(P-m)} = \dfrac{A}{K-P} + \dfrac{B}{P-m}$, so $A\,(P-m) + B\,(K-P) = 1$.

If $P = m$, $B = \dfrac{1}{K-m}$; if $P = K$, $A = \dfrac{1}{K-m}$, so $\dfrac{1}{K-m}\displaystyle\int\left(\dfrac{1}{K-P} + \dfrac{1}{P-m}\right)dP = \int \dfrac{k}{K}\,dt\ \Rightarrow$

$\dfrac{1}{K-m}\left(-\ln|K-P| + \ln|P-m|\right) = \dfrac{k}{K}t + M.$

But $m < P < K$, so $\dfrac{1}{K-m}\ln\dfrac{P-m}{K-P} = \dfrac{k}{K}t + M\ \Rightarrow\ \ln\dfrac{P-m}{K-P} = (K-m)\dfrac{k}{K}t + M_1\ \Leftrightarrow$

$\dfrac{P-m}{K-P} = De^{(K-m)(k/K)t}\quad (D = e^{M_1})$. Let $t = 0$: $\dfrac{P_0-m}{K-P_0} = D$. So $\dfrac{P-m}{K-P} = \dfrac{P_0-m}{K-P_0}e^{(K-m)(k/K)t}$.

Solving for P, we get $P\,(t) = \dfrac{m\,(K-P_0) + K\,(P_0-m)\,e^{(K-m)(k/K)t}}{K - P_0 + (P_0-m)\,e^{(K-m)(k/K)t}}.$

(d) If $P_0 < m$, then $P_0 - m < 0$. Let $N\,(t)$ be the numerator of the expression for $P\,(t)$ in part (c). Then

$N\,(0) = P_0\,(K-m) > 0$, and $P_0 - m < 0\ \Leftrightarrow\ \lim\limits_{t\to\infty} K\,(P_0-m)\,e^{(K-m)(k/K)t} = -\infty\ \Rightarrow$

$\lim\limits_{t\to\infty} N\,(t) = -\infty$. Since N is continuous, there is a number t such that $N\,(t) = 0$ and thus $P\,(t) = 0$. So the species will become extinct.

15. (a) $dP/dt = kP \cos(rt - \phi)$ $\Rightarrow$ $(dP)/P = k \cos(rt - \phi) \, dt$ $\Rightarrow$ $\int (dP)/P = k \int \cos(rt - \phi) \, dt$ $\Rightarrow$
$\ln P = (k/r) \sin(rt - \phi) + C$. (Since this is a growth model, $P > 0$ and we can write $\ln P$ instead of $\ln|P|$.)
Since $P(0) = P_0$, we obtain $\ln P_0 = (k/r) \sin(-\phi) + C = -(k/r) \sin \phi + C$ $\Rightarrow$ $C = \ln P_0 + (k/r) \sin \phi$.
Thus, $\ln P = (k/r) \sin(rt - \phi) + \ln P_0 + (k/r) \sin \phi$, which we can rewrite as
$\ln (P/P_0) = (k/r) [\sin(rt - \phi) + \sin \phi]$ or, after exponentiation, $P(t) = P_0 e^{(k/r)[\sin(rt-\phi)+\sin\phi]}$.

(b) As k increases, the amplitude
increases, but the minimum
value stays the same.

As r increases, the amplitude
and the period decrease.

A change in ϕ produces slight
adjustments in the phase shift
and amplitude.

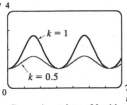

Comparing values of k with
$P_0 = 1$, $r = 2$, and $\phi = \pi/2$

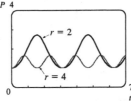

Comparing values of r with
$P_0 = 1$, $k = 1$, and $\phi = \pi/2$

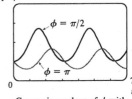

Comparing values of ϕ with
$P_0 = 1$, $k = 1$, and $r = 2$

$P(t)$ oscillates between $P_0 e^{(k/r)(1+\sin\phi)}$ and $P_0 e^{(k/r)(-1+\sin\phi)}$ (the extreme values are attained when $rt - \phi$ is
an odd multiple of $\frac{\pi}{2}$), so $\lim\limits_{t\to\infty} P(t)$ does not exist.

9.6 Linear Equations

1. $y' + e^x y = x^2 y^2$ is not linear since it cannot be put into the standard linear form (1).

3. $xy' + \ln x - x^2 y = 0$ $\Rightarrow$ $xy' - x^2 y = -\ln x$ $\Rightarrow$ $y' + (-x)y = -\dfrac{\ln x}{x}$, which is in the standard linear
form (1), so this equation is linear.

5. Comparing the given equation, $y' + 2y = 2e^x$, with the general form, $y' + P(x)y = Q(x)$, we see that $P(x) = 2$
and the integrating factor is $I(x) = e^{\int P(x)dx} = e^{\int 2\,dx} = e^{2x}$. Multiplying the differential equation by $I(x)$ gives
$e^{2x} y' + 2e^{2x} y = 2e^{3x}$ $\Rightarrow$ $(e^{2x} y)' = 2e^{3x}$ $\Rightarrow$ $e^{2x} y = \int 2e^{3x}\, dx$ $\Rightarrow$ $e^{2x} y = \frac{2}{3} e^{3x} + C$ $\Rightarrow$
$y = \frac{2}{3} e^x + C e^{-2x}$.

7. $I(x) = e^{\int (-2x)dx} = e^{-x^2}$. Multiplying the differential equation by $I(x)$ gives $e^{-x^2} (y' - 2xy) = xe^{-x^2}$ $\Rightarrow$
$\left(e^{-x^2} y\right)' = xe^{-x^2}$ $\Rightarrow$ $y = e^{x^2} \left(\int xe^{-x^2}\, dx + C\right) = Ce^{x^2} - \frac{1}{2}$.

9. $y' - y \tan x = \dfrac{\sin 2x}{\cos x} = 2\sin x$, so $I(x) = e^{\int -\tan x\,dx} = e^{\ln|\cos x|} = \cos x$ (since $-\frac{\pi}{2} < x < \frac{\pi}{2}$). Multiplying the
differential equation by $I(x)$ gives $(y' - y\tan x) \cos x = \dfrac{\sin 2x}{\cos x} \cos x$ $\Rightarrow$ $(y \cos x)' = \sin 2x$ $\Rightarrow$
$y \cos x = \int \sin 2x\, dx + C = -\frac{1}{2} \cos 2x + C = \frac{1}{2} - \cos^2 x + C$ $\Rightarrow$ $y = \dfrac{\sec x}{2} - \cos x + C \sec x$.

11. $I(x) = e^{\int 2x\,dx} = e^{x^2}$. Multiplying the differential equation by $I(x)$
gives $e^{x^2} y' + 2xe^{x^2} y = x^2 e^{x^2}$ $\Rightarrow$ $\left(e^{x^2} y\right)' = x^2 e^{x^2}$. Thus
$y = e^{-x^2} \left[\int x^2 e^{x^2}\, dx + C\right] = e^{-x^2} \left[\frac{1}{2} xe^{x^2} - \int \frac{1}{2} e^{x^2}\, dx + C\right] = \frac{1}{2} x + Ce^{-x^2} - e^{-x^2} \int \frac{1}{2} e^{x^2}\, dx$.

13. $(1+t)\dfrac{du}{dt} + u = 1 + t, t > 0 \quad \Rightarrow \quad \dfrac{d}{dt}[(1+t)u] = 1 + t \quad \Rightarrow \quad (1+t)u = \int (1+t)\,dt = t + \frac{1}{2}t^2 + C \quad \Rightarrow$

$u = \dfrac{t + \frac{1}{2}t^2 + C}{1+t}$ or $u = \dfrac{t^2 + 2t + 2C}{2(t+1)}$.

15. $I(x) = e^{\int dx} = e^x$. Multiplying the differential equation by $I(x)$ gives $e^x y' + e^x y = e^x (x + e^x) \quad \Rightarrow$

$(e^x y)' = e^x (x + e^x)$. Thus $y = e^{-x}\left[\int e^x (x + e^x)\,dx + C\right] = e^{-x}\left[xe^x - e^x + \dfrac{e^{2x}}{2} + C\right] = x - 1 + \dfrac{e^x}{2} + \dfrac{C}{e^x}$.

But $0 = y(0) = -1 + \frac{1}{2} + C$, so $C = \frac{1}{2}$, and the solution to the initial-value problem is

$y = x - 1 + \frac{1}{2}e^x + \frac{1}{2}e^{-x} = x - 1 + \cosh x$.

17. $\dfrac{dv}{dt} - 2tv = 3t^2 e^{t^2}, v(0) = 5$. $I(t) = e^{\int(-2t)dt} = e^{-t^2}$. Multiply the differential equation by $I(t)$ to get

$e^{-t^2}\dfrac{dv}{dt} - 2te^{-t^2}v = 3t^2 \quad \Rightarrow \quad \left(e^{-t^2}v\right)' = 3t^2 \quad \Rightarrow \quad e^{-t^2}v = \int 3t^2\,dt = t^3 + C \quad \Rightarrow \quad v = t^3 e^{t^2} + Ce^{t^2}$.

$5 = v(0) = 0 \cdot 1 + C \cdot 1 = C$, so $v = t^3 e^{t^2} + 5e^{t^2}$.

19. $y' + 2\dfrac{y}{x} = \dfrac{\cos x}{x^2}$ $(x \neq 0)$, so $I(x) = e^{\int(2/x)dx} = x^2$. Multiplying the differential equation by $I(x)$ gives

$x^2 y' + 2xy = \cos x \quad \Rightarrow \quad (x^2 y)' = \cos x \quad \Rightarrow \quad y = x^{-2}\left[\int \cos x\,dx + C\right] = x^{-2}(\sin x + C)$ $(x \neq 0)$. But

$0 = y(\pi) = C$, so the solution to the initial-value problem is $y = (\sin x)/x^2$.

21. $y' + \dfrac{1}{x}y = \cos x$ $(x \neq 0)$, so $I(x) = e^{\int(1/x)dx} = e^{\ln|x|} = x$ (for

$x > 0$). Multiplying the differential equation by $I(x)$ gives

$xy' + y = x\cos x \quad \Rightarrow \quad (xy)' = x\cos x$. Thus,

$y = \dfrac{1}{x}\left[\int x\cos x\,dx + C\right] = \dfrac{1}{x}[x\sin x + \cos x + C]$

$= \sin x + \dfrac{\cos x}{x} + \dfrac{C}{x}$

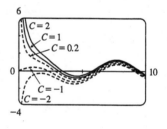

The solutions are asymptotic to the y-axis (except for $C = -1$). In fact, for $C > -1$, $y \to \infty$ as $x \to 0^+$, whereas for $C < -1$, $y \to -\infty$ as $x \to 0^+$. As x gets larger, the solutions approximate $y = \sin x$ more closely. The graphs for larger C lie above those for smaller C. The distance between the graphs lessens as x increases.

23. Setting $u = y^{1-n}$, $\dfrac{du}{dx} = (1-n)y^{-n}\dfrac{dy}{dx}$ or $\dfrac{dy}{dx} = \dfrac{y^n}{1-n}\dfrac{du}{dx} = \dfrac{u^{n/(1-n)}}{1-n}\dfrac{du}{dx}$. Then the Bernoulli differential

equation becomes $\dfrac{u^{n/(1-n)}}{1-n}\dfrac{du}{dx} + P(x)u^{1/(1-n)} = Q(x)u^{n/(1-n)}$ or $\dfrac{du}{dx} + (1-n)P(x)u = Q(x)(1-n)$.

25. Here $n = 3$, $P(x) = \dfrac{2}{x}$, $Q(x) = \dfrac{1}{x^2}$ and setting $u = y^{-2}$, u satisfies $u' - \dfrac{4u}{x} = -\dfrac{2}{x^2}$. Then

$I(x) = e^{\int(-4/x)dx} = x^{-4}$ and $u = x^4\left(\int -\dfrac{2}{x^6}\,dx + C\right) = x^4\left(\dfrac{2}{5x^5} + C\right) = Cx^4 + \dfrac{2}{5x}$. Thus,

$y = \pm\left(Cx^4 + \dfrac{2}{5x}\right)^{-1/2}$.

27. (a) $2\dfrac{dI}{dt} + 10I = 40$ or $\dfrac{dI}{dt} + 5I = 20$. Then the integrating factor is $e^{\int 5\,dt} = e^{5t}$. Multiplying the differential

equation by the integrating factor gives $e^{5t}\dfrac{dI}{dt} + 5Ie^{5t} = 20e^{5t} \;\Rightarrow\; (e^{5t}I)' = 20e^{5t} \;\Rightarrow\;$

$I(t) = e^{-5t}\left[\int 20e^{5t}\,dt + C\right] = 4 + Ce^{-5t}$. But $0 = I(0) = 4 + C$, so $I(t) = 4 - 4e^{-5t}$.

(b) $I(0.1) = 4 - 4e^{-0.5} \approx 1.57$ A

29. $5\dfrac{dQ}{dt} + 20Q = 60$ with $Q(0) = 0C$. Then the integrating factor is $e^{\int 4\,dt} = e^{4t}$, and multiplying the differential

equation by the integrating factor gives $e^{4t}\dfrac{dQ}{dt} + 4e^{4t}Q = 12e^{4t} \;\Rightarrow\; (e^{4t}Q)' = 12e^{4t} \;\Rightarrow\;$

$Q(t) = e^{-4t}\left[\int 12e^{4t}\,dt + C\right] = 3 + Ce^{-4t}$. But $0 = Q(0) = 3 + C$ so $Q(t) = 3\left(1 - e^{-4t}\right)$ is the charge at time

t and $I = dQ/dt = 12e^{-4t}$ is the current at time t.

31. $\dfrac{dP}{dt} + kP = kM$, so $I(t) = e^{\int k\,dt} = e^{kt}$. Multiplying the differential

equation by $I(t)$ gives $e^{kt}\dfrac{dP}{dt} + kPe^{kt} = kMe^{kt} \;\Rightarrow\; (e^{kt}P)' = kMe^{kt}$

$\Rightarrow\; P(t) = e^{-kt}\left(\int kMe^{kt}\,dt + C\right) = M + Ce^{-kt},\, k > 0$. Furthermore,

it is reasonable to assume that $0 \le P(0) \le M$, so $-M \le C \le 0$.

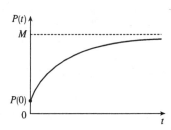

33. $y(0) = 0$ kg. Salt is added at a rate of $\left(0.4\,\dfrac{\text{kg}}{\text{L}}\right)\left(5\,\dfrac{\text{L}}{\text{min}}\right) = 2\,\dfrac{\text{kg}}{\text{min}}$. Since solution is drained from the tank at a

rate of 3 L/min, but salt solution is added at a rate of 5 L/min, the tank, which starts out with 100 L of water,

contains $(100 + 2t)$ L of liquid after t min. Thus, the salt concentration at time t is $\dfrac{y(t)}{100 + 2t}\,\dfrac{\text{kg}}{\text{L}}$. Salt therefore

leaves the tank at a rate of $\left(\dfrac{y(t)}{100 + 2t}\,\dfrac{\text{kg}}{\text{L}}\right)\left(3\,\dfrac{\text{L}}{\text{min}}\right) = \dfrac{3y}{100 + 2t}\,\dfrac{\text{kg}}{\text{min}}$. Combining the rates at which salt enters

and leaves the tank, we get $\dfrac{dy}{dt} = 2 - \dfrac{3y}{100 + 2t}$. Rewriting this equation as $\dfrac{dy}{dt} + \left(\dfrac{3}{100 + 2t}\right)y = 2$, we see that

it is linear. $I(t) = \exp\left(\displaystyle\int \dfrac{3\,dt}{100 + 2t}\right) = \exp\left(\tfrac{3}{2}\ln(100 + 2t)\right) = (100 + 2t)^{3/2}$. Multiplying the differential

equation by $I(t)$ gives $(100 + 2t)^{3/2}\dfrac{dy}{dt} + 3(100 + 2t)^{1/2}y = 2(100 + 2t)^{3/2} \;\Rightarrow\;$

$\left[(100 + 2t)^{3/2}y\right]' = 2(100 + 2t)^{3/2} \;\Rightarrow\; (100 + 2t)^{3/2}y = \tfrac{2}{5}(100 + 2t)^{5/2} + C \;\Rightarrow\;$

$y = \tfrac{2}{5}(100 + 2t) + C(100 + 2t)^{-3/2}$. Now $0 = y(0) = \tfrac{2}{5}(100) + C \cdot 100^{-3/2} = 40 + \tfrac{1}{1000}C \;\Rightarrow\;$

$C = -40{,}000$, so $y = \left[\tfrac{2}{5}(100 + 2t) - 40{,}000(100 + 2t)^{-3/2}\right]$ kg. From this solution (no pun intended), we

calculate the salt concentration at time t to be $C(t) = \dfrac{y(t)}{100 + 2t} = \left[\dfrac{-40{,}000}{(100 + 2t)^{5/2}} + \dfrac{2}{5}\right]\dfrac{\text{kg}}{\text{L}}$. In particular,

$C(20) = \dfrac{-40{,}000}{140^{5/2}} + \dfrac{2}{5} \approx 0.2275\,\dfrac{\text{kg}}{\text{L}}$ and $y(20) = \tfrac{2}{5}(140) - 40{,}000(140)^{-3/2} \approx 31.85$ kg.

35. (a) $\dfrac{dv}{dt} + \dfrac{c}{m}v = g$ and $I\,(t) = e^{\int (c/m)\,dt} = e^{(c/m)t}$, and multiplying the differential equation by $I\,(t)$ gives

$$e^{(c/m)t}\dfrac{dv}{dt} + \dfrac{vce^{(c/m)t}}{m} = ge^{(c/m)t} \quad\Rightarrow\quad \left[e^{(c/m)t}v\right]' = ge^{(c/m)t}. \text{ Hence,}$$

$v\,(t) = e^{-(c/m)t}\left[\int ge^{(c/m)t}\,dt + K\right] = mg/c + Ke^{-(c/m)t}$. But the object is dropped from rest, so $v\,(0) = 0$

and $K = -mg/c$. Thus, the velocity at time t is $v\,(t) = (mg/c)\left[1 - e^{-(c/m)t}\right]$.

(b) $\lim\limits_{t\to\infty} v\,(t) = mg/c$.

(c) $s\,(t) = \int v\,(t)\,dt = (mg/c)\left[t + (m/c)\,e^{-(c/m)t}\right] + c_1$ where $c_1 = s\,(0) - m^2 g/c^2$, $s\,(0)$ is the initial position,

so $s\,(0) = 0$ and $s\,(t) = (mg/c)\left[t + (m/c)\,e^{-(c/m)t}\right] - m^2 g/c^2$.

9.7 Predator-Prey Systems

1. (a) $dx/dt = -0.05x + 0.0001xy$. If $y = 0$, we have $dx/dt = -0.05x$, which indicates that in the absence of y, x declines at a rate proportional to itself. So x represents the predator population and y represents the prey population. The growth of the prey population, $0.1y$ (from $dy/dt = 0.1y - 0.005xy$), is restricted only by encounters with predators (the term $-0.005xy$). The predator population increases only through the term $0.0001xy$, that is, by encounters with the prey and not through additional food sources.

(b) $dy/dt = -0.015y + 0.00008xy$. If $x = 0$, we have $dy/dt = -0.015y$, which indicates that in the absence of x, y would decline at a rate proportional to itself. So y represents the predator population and x represents the prey population. The growth of the prey population, $0.2x$ (from $dx/dt = 0.2x - 0.0002x^2 - 0.006xy = 0.2x\,(1 - 0.001x) - 0.006xy$), is restricted by a carrying capacity of 1000 [from the term $1 - 0.001x = 1 - x/1000$] and by encounters with predators (the term $-0.006xy$). The predator population increases only through the term $0.00008xy$, that is, by encounters with the prey and not through additional food sources.

3. (a) At $t = 0$, there are about 300 rabbits and 100 foxes. At $t = t_1$, the number of foxes reaches a minimum of about 20 while the number of rabbits is about 1000. At $t = t_2$, the number of rabbits reaches a maximum of about 2400, while the number of foxes rebounds to 100. At $t = t_3$, the number of rabbits decreases to about 1000 and the number of foxes reaches a maximum of about 315. As t increases, the number of foxes decreases greatly to 100, and the number of rabbits decreases to 300 (the initial populations), and the cycle starts again.

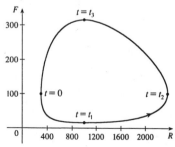

(b)

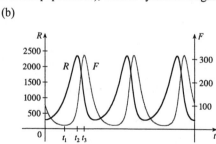

5.

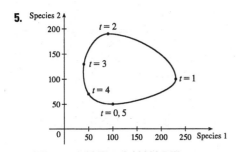

7. $\dfrac{dW}{dR} = \dfrac{-0.02W + 0.00002RW}{0.08R - 0.001RW} \quad \Leftrightarrow \quad (0.08 - 0.001W)\,R\,dW = (-0.02 + 0.00002R)\,W\,dR \quad \Leftrightarrow$

$\dfrac{0.08 - 0.001W}{W}\,dW = \dfrac{-0.02 + 0.00002R}{R}\,dR \quad \Leftrightarrow \quad \displaystyle\int \left(\dfrac{0.08}{W} - 0.001\right) dW = \int \left(-\dfrac{0.02}{R} + 0.00002\right) dR$

$\Leftrightarrow \quad 0.08\ln|W| - 0.001W = -0.02\ln|R| + 0.00002R + K \quad \Leftrightarrow$

$0.08\ln W + 0.02\ln R = 0.001W + 0.00002R + K \quad \Leftrightarrow \quad \ln\left(W^{0.08}R^{0.02}\right) = 0.00002R + 0.001W + K \quad \Leftrightarrow$

$W^{0.08}R^{0.02} = e^{0.00002R + 0.001W + K} \quad \Leftrightarrow \quad R^{0.02}W^{0.08} = Ce^{0.00002R}e^{0.001W} \quad \Leftrightarrow \quad \dfrac{R^{0.02}W^{0.08}}{e^{0.00002R}e^{0.001W}} = C.$

In general, if $\dfrac{dy}{dx} = \dfrac{-ry + bxy}{kx - axy}$, then $C = \dfrac{x^r y^k}{e^{bx}e^{ay}}$.

9. (a) Letting $W = 0$ gives us $dR/dt = 0.08R\,(1 - 0.0002R)$. $dR/dt = 0 \quad \Leftrightarrow \quad R = 0$ or 5000. Since $dR/dt > 0$
for $0 < R < 5000$, we would expect the rabbit population to *increase* to 5000 for these values of R. Since
$dR/dt < 0$ for $R > 5000$, we would expect the rabbit population to *decrease* to 5000 for these values of R.
Hence, in the absence of wolves, we would expect the rabbit population to stabilize at 5000.

(b) R and W are constant $\Rightarrow R' = 0$ and $W' = 0 \Rightarrow \begin{cases} 0 = 0.08R\,(1 - 0.0002R) - 0.001RW \\ 0 = -0.02W + 0.00002RW \end{cases} \Rightarrow$

$\begin{cases} 0 = R\,[0.08\,(1 - 0.0002R) - 0.001W] \\ 0 = W\,(-0.02 + 0.00002R) \end{cases}$

The second equation is true if $W = 0$ or $R = \dfrac{0.02}{0.00002} = 1000$. If $W = 0$ in the first equation, then either $R = 0$

or $R = \dfrac{1}{0.0002} = 5000$ [as in part (a)]. If $R = 1000$, then $0 = 1000\,[0.08\,(1 - 0.0002 \cdot 1000) - 0.001W] \quad \Leftrightarrow$

$0 = 80\,(1 - 0.2) - W \quad \Leftrightarrow \quad W = 64$.

Case (i): $W = 0$, $R = 0$: both populations are zero

Case (ii): $W = 0$, $R = 5000$: see part (a)

Case (iii): $R = 1000$, $W = 64$: the predator/prey interaction balances and the populations are stable.

(c) The populations of wolves and rabbits fluctuate around 64 and 1000, respectively, and eventually stabilize at
those values.

(d)

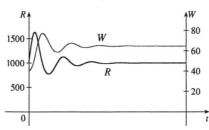

9 Review

1. (a) A differential equation is an equation that contains an unknown function and one or more of its derivatives.

(b) The order of a differential equation is the order of the highest derivative that occurs in the equation.

(c) An initial condition is a condition of the form $y(t_0) = y_0$.

2. $y' = x^2 + y^2 \geq 0$ for all x and y. $y' = 0$ only at the origin, so there is a horizontal tangent at $(0, 0)$, but nowhere else. The graph of the solution is increasing on every interval.

3. See the paragraph preceding Example 1 in Section 9.2.

4. See the paragraph after Figure 14 in Section 9.2.

5. A separable equation is a first-order differential equation in which the expression for dy/dx can be factored as a function of x times a function of y, that is, $dy/dx = g(x) f(y)$. We can solve the equation by integrating both sides of the equation $dy/f(y) = g(x) dx$ and solving for y.

6. A first-order linear differential equation is a differential equation that can be put in the form $\dfrac{dy}{dx} + P(x)y = Q(x)$, where P and Q are continuous functions on a given interval. To solve such an equation, multiply it by the integrating factor $I(x) = e^{\int P(x)dx}$ to put it in the form $[I(x)y]' = I(x) Q(x)$ and then integrate both sides to get $I(x) y = \int I(x) Q(x) dx$, that is, $e^{\int P(x)dx} y = \int e^{\int P(x)dx} Q(x) dx$. Solving for y gives us $y = e^{-\int P(x)dx} \int e^{\int P(x)dx} Q(x) dx$.

7. (a) $dy/dt = ky$

(b) The equation in part (a) is an appropriate model for population growth, assuming that there is enough room and nutrition to support the growth.

(c) If $y(0) = y_0$, then the solution is $y(t) = y_0 e^{kt}$.

8. (a) $dP/dt = kP(1 - P/K)$, where K is the carrying capacity.

(b) The equation in part (a) is an appropriate model for population growth, assuming that the population grows at a rate proportional to the size of the population in the beginning, but eventually levels off and approaches its carrying capacity because of limited resources.

9. (a) $dF/dt = kF - aFS$ and $dS/dt = -rS + bFS$.

(b) In the absence of sharks, an ample food supply would support exponential growth of the fish population, that is, $dF/dt = kF$, where k is a positive constant. In the absence of fish, we assume that the shark population would decline at a rate proportional to itself, that is, $dS/dt = -rS$, where r is a positive constant.

1. False. $y = 0$ is a solution of $y' = -y^4$, but $y = 0$ is not a decreasing function. (All non-trivial solutions are decreasing, however.)

3. False. $x + y$ cannot be written in the form $g(x) f(y)$.

5. True. $e^x y' = y \;\Rightarrow\; y' = e^{-x} y \;\Rightarrow\; y' + (-e^{-x})y = 0$, which is of the form $y' + P(x)y = Q(x)$.

7. True. By comparing $\dfrac{dy}{dt} = 2y\left(1 - \dfrac{y}{5}\right)$ with the logistic differential equation (9.5.1), we see that the carrying capacity is 5, that is, $\displaystyle\lim_{t \to \infty} y = 5$.

EXERCISES

1. (a)

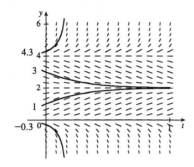

(b) $\lim_{t \to \infty} y(t)$ appears to be finite for $0 \le c \le 4$. In fact $\lim_{t \to \infty} y(t) = 4$ for $c = 4$, $\lim_{t \to \infty} y(t) = 2$ for $0 < c < 4$, and $\lim_{t \to \infty} y(t) = 0$ for $c = 0$. The equilibrium solutions are $y(t) = 0$, $y(t) = 2$, and $y(t) = 4$.

3. (a)

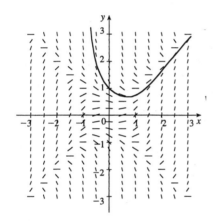

We estimate that when $x = 0.3$, $y = 0.8$, so $y(0.3) \approx 0.8$.

(b) $h = 0.1$, $x_0 = 0$, $y_0 = 1$ and $F(x, y) = x^2 - y^2$. So $y_n = y_{n-1} + 0.1 (x_{n-1}^2 - y_{n-1}^2)$. Thus,
$y_1 = 1 + 0.1 (0^2 - 1^2) = 0.9$,
$y_2 = 0.9 + 0.1 (0.1^2 - 0.9^2) = 0.82$, $y_3 = 0.82 + 0.1 (0.2^2 - 0.82^2) = 0.75676 \approx y(0.3)$.

(c) The centers of the horizontal line segments of the direction field are located on the lines $y = x$ and $y = -x$. When a solution curve crosses one of these lines, it has a local maximum or minimum.

5. $y^2 \dfrac{dy}{dx} = x + \sin x$ $\Rightarrow$ $\int y^2 \, dy = \int (x + \sin x) \, dx$ $\Rightarrow$ $\dfrac{y^3}{3} = \dfrac{x^2}{2} - \cos x + C$ $\Rightarrow$
$y^3 = \frac{3}{2}x^2 - 3 \cos x + K$ (where $K = 3C$) $\Rightarrow$ $y = \sqrt[3]{\frac{3}{2}x^2 - 3 \cos x + K}$

7. Since it's linear, $I(x) = e^{\int -2/x \, dx} = x^{-2}$ and multiplying by $I(x)$ gives $x^{-2} y' - 2y^{-3} = 1$ $\Rightarrow$ $(x^{-2}y)' = 1$
$\Rightarrow$ $y(x) = x^2 (\int 1 \, dx + C) = x^2 [x + C] = Cx^2 + x^3$.

9. $xyy' = \ln x$ $\Rightarrow$ $y \, dy = \dfrac{\ln x}{x} \, dx$ $\Rightarrow$ $\displaystyle\int y \, dy = \int \dfrac{\ln x}{x} \, dx$ (Make the substitution $u = \ln x$; then $du = dx/x$.)
So $\int y \, dy = \int u \, du$ $\Rightarrow$ $\frac{1}{2}y^2 = \frac{1}{2}u^2 + C$ $\Rightarrow \frac{1}{2}y^2 = \frac{1}{2} (\ln x)^2 + C$. $y(1) = 2$ $\Rightarrow$ $\frac{1}{2}2^2 = \frac{1}{2} (\ln 1)^2 + C = C$
$\Leftrightarrow$ $C = 2$. Therefore, $\frac{1}{2}y^2 = \frac{1}{2} (\ln x)^2 + 2$, or $y = \sqrt{(\ln x)^2 + 4}$. The negative square root is inadmissible, since $y(1) > 0$.

11. Since the equation is linear, let $I(x) = e^{\int dx} = e^x$. Then multiplying by $I(x)$ gives $e^x y' + e^x y = \sqrt{x}$ $\Rightarrow$
$(e^x y)' = \sqrt{x}$ $\Rightarrow$ $y(x) = e^{-x} (\int \sqrt{x} \, dx + c) = e^{-x} (\frac{2}{3}x^{3/2} + c)$. But $3 = y(0) = c$, so the solution to the initial-value problem is $y(x) = e^{-x} (\frac{2}{3}x^{3/2} + 3)$.

13. The curves $kx^2 + y^2 = 1$ form a family of ellipses for $k > 0$, a family of hyperbolas for $k < 0$, and two parallel

lines $y = \pm 1$ for $k = 0$. Solving $kx^2 + y^2 = 1$ for k gives $k = \dfrac{1 - y^2}{x^2}$. Differentiating gives $2kx + 2yy' = 0 \Leftrightarrow$

$y' = -\dfrac{kx}{y} = -(1 - y^2)\dfrac{x}{yx^2} = \dfrac{y^2 - 1}{xy}$. Thus, for $k \neq 0$ the orthogonal trajectories must satisfy $y' = -\dfrac{xy}{y^2 - 1}$

$\Rightarrow \dfrac{y^2 - 1}{y}\, dy = -x\, dx \Rightarrow \dfrac{y^2}{2} - \ln|y| = \dfrac{-x^2}{2} + c$. For $k = 0$, the orthogonal trajectories are given by

$x = c_1$ for c_1 an arbitrary constant.

15. (a) $y(t) = y(0)e^{kt} = 1000e^{kt} \Rightarrow y(2) = 1000e^{2k} = 9000 \Rightarrow e^{2k} = 9 \Rightarrow 2k = \ln 9 \Rightarrow$
   $k = \frac{1}{2}\ln 9 = \ln 3 \Rightarrow y(t) = 1000e^{(\ln 3)t} = 1000 \cdot 3^t$

(b) $y(3) = 1000 \cdot 3^3 = 27{,}000$

(c) $dy/dt = 1000 \cdot 3^t \cdot \ln 3;\ t = 3 \Rightarrow dy/dt = 27{,}000\ln 3 \approx 29{,}663$

(d) $1000 \cdot 3^t = 2000 \Rightarrow 3^t = 2 \Rightarrow t\ln 3 = \ln 2 \Rightarrow t = (\ln 2)/(\ln 3) \approx 0.63$ h

17. (a) $C'(t) = -kC(t) \Rightarrow C(t) = C(0)e^{-kt}$ by Theorem 9.4.2. But $C(0) = C_0$, so $C(t) = C_0 e^{-kt}$.

(b) $C(30) = \frac{1}{2}C_0$ since the concentration is reduced by half. Thus, $\frac{1}{2}C_0 = C_0 e^{-30k} \Rightarrow \ln\frac{1}{2} = -30k \Rightarrow$
   $k = -\frac{1}{30}\ln\frac{1}{2} = \frac{1}{30}\ln 2$. Since 10% of the original concentration remains if 90% is eliminated, we want the
   value of t such that $C(t) = \frac{1}{10}C_0$. Therefore, $\frac{1}{10}C_0 = C_0 e^{-t(\ln 2)/30} \Rightarrow \ln 0.1 = -t(\ln 2)/30 \Rightarrow$
   $t = -\frac{30}{\ln 2}\ln 0.1 \approx 100$ h.

19. (a) $\dfrac{dL}{dt} \propto L_\infty - L \Rightarrow \dfrac{dL}{dt} = k(L_\infty - L) \Rightarrow \displaystyle\int \dfrac{dL}{L_\infty - L} = \int k\, dt \Rightarrow -\ln|L_\infty - L| = kt + C \Rightarrow$

$\ln|L_\infty - L| = -kt - C \Rightarrow |L_\infty - L| = e^{-kt - C} \Rightarrow L_\infty - L = Ae^{-kt} \Rightarrow L = L_\infty - Ae^{-kt}$. At

$t = 0,\ L = L(0) = L_\infty - A \Rightarrow A = L_\infty - L(0) \Rightarrow L(t) = L_\infty - [L_\infty - L(0)]e^{-kt}$

(b) $L_\infty = 53$ cm, $L(0) = 10$ cm, and $k = 0.2 \Rightarrow L(t) = 53 - (53 - 10)e^{-0.2t} = 53 - 43e^{-0.2t}$.

21. Let P be the population and I be the number of infected people. The rate of spread dI/dt is jointly proportional to

I and to $P - I$, so for some constant k, $dI/dt = kI(P - I) \Rightarrow I = \dfrac{I_0 P}{I_0 + (P - I_0)e^{-kPt}}$ (from the

discussion of logistic growth in Section 9.5).

Now, measuring t in days, we substitute $t = 7$, $P = 5000$, $I_0 = 160$ and $I(7) = 1200$ to find k:

$1200 = \dfrac{160 \cdot 5000}{160 + (5000 - 160)e^{-5000 \cdot 7 \cdot k}} \Leftrightarrow k \approx 0.00006448$. So, putting $I = 5000 \times 80\% = 4000$, we solve

for t: $4000 = \dfrac{160 \cdot 5000}{160 + (5000 - 160)e^{-0.00006448 \cdot 5000 \cdot t}} \Leftrightarrow 160 + 4840e^{-0.3224t} = 200 \Leftrightarrow$

$-0.3224t = \ln\frac{40}{4840} \Leftrightarrow t \approx 14.9$. So it takes about 15 days for 80% of the population to be infected.

23. $\dfrac{dh}{dt} = -\dfrac{R}{V}\left(\dfrac{h}{k + h}\right) \Rightarrow \displaystyle\int \dfrac{k + h}{h}\, dh = \int\left(-\dfrac{R}{V}\right) dt \Rightarrow \int\left(1 + \dfrac{k}{h}\right) dh = -\dfrac{R}{V}\int 1\, dt \Rightarrow$

$h + k\ln h = -\dfrac{R}{V}t + C$. This equation gives a relationship between h and t, but it is not possible to isolate h and
express it in terms of t.

25. (a) $dx/dt = 0.4x\,(1 - 0.000005x) - 0.002xy$, $dy/dt = -0.2y + 0.000008xy$. If $y = 0$, then
$dx/dt = 0.4x\,(1 - 0.000005x)$, so $dx/dt = 0$ $\Leftrightarrow$ $x = 0$ or $x = 200{,}000$, which shows that the insect
population increases logistically with a carrying capacity of $200{,}000$. Since $dx/dt > 0$ for $0 < x < 200{,}000$
and $dx/dt < 0$ for $x > 200{,}000$, we expect the insect population to stabilize at $200{,}000$.

(b) x and y are constant $\Rightarrow$ $x' = 0$ and $y' = 0$ $\Rightarrow$

$$\begin{cases} 0 = 0.4x\,(1 - 0.000005x) - 0.002xy \\ 0 = -0.2y + 0.000008xy \end{cases} \Rightarrow \begin{cases} 0 = 0.4x\,[(1 - 0.000005x) - 0.005y] \\ 0 = y\,(-0.2 + 0.000008x) \end{cases}$$

The second equation is true if $y = 0$ or $x = \frac{0.2}{0.000008} = 25{,}000$. If $y = 0$ in the first equation, then either $x = 0$
or $x = \frac{1}{0.000005} = 200{,}000$. If $x = 25{,}000$, then $0 = 0.4\,(25{,}000)\,[(1 - 0.000005 \cdot 25{,}000) - 0.005y]$ $\Rightarrow$
$0 = 10{,}000\,[(1 - 0.125) - 0.005y]$ $\Rightarrow$ $0 = 8750 - 50y$ $\Rightarrow$ $y = 175$.

Case (i): $y = 0$, $x = 0$: zero populations

Case (ii): $y = 0$, $x = 200{,}000$: in the absence of birds, the insect population is constantly $200{,}000$.

Case (iii): $x = 25{,}000$, $y = 175$: the predator/prey interaction balances and the populations are stable.

(c) The populations of the birds and insects fluctuate
around 175 and 25,000, respectively, and
eventually stabilize at those values.

(d)

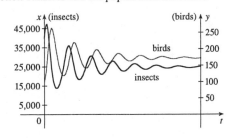

Problems Plus

1. We use the Fundamental Theorem of Calculus to differentiate the given equation:

$$[f(x)]^2 = 100 + \int_0^x \left\{[f(t)]^2 + [f'(t)]^2\right\} dt \quad \Rightarrow \quad 2f(x)f'(x) = [f(x)]^2 + [f'(x)]^2 \quad \Rightarrow$$

$$[f(x)]^2 + [f'(x)]^2 - 2f(x)f'(x) = 0 \quad \Rightarrow \quad [f(x) - f'(x)]^2 = 0 \quad \Leftrightarrow \quad f(x) = f'(x).$$ We can solve this as

a separable equation, or else use Theorem 9.4.2 with $k = 1$, which says that the solutions are $f(x) = Ce^x$. Now

$[f(0)]^2 = 100$, so $f(0) = C = \pm 10$, and hence $f(x) = \pm 10e^x$ are the only functions satisfying the given

equation.

3. $f'(x) = \lim\limits_{h \to 0} \dfrac{f(x+h) - f(x)}{h} = \lim\limits_{h \to 0} \dfrac{f(x)[f(h) - 1]}{h}$ [since $f(x+h) = f(x)f(h)$]

$\qquad = f(x) \lim\limits_{h \to 0} \dfrac{f(h) - 1}{h} = f(x) \lim\limits_{h \to 0} \dfrac{f(h) - f(0)}{h - 0} = f(x)f'(0) = f(x)$

Therefore, $f'(x) = f(x)$ for all x and from Theorem 9.4.2 we get $f(x) = Ae^x$. Now $f(0) = 1 \quad \Rightarrow \quad A = 1$

$\Rightarrow \quad f(x) = e^x$.

5. (a) We are given that $V = \frac{1}{3}\pi r^2 h$, $dV/dt = 60,000\pi$ ft^3/h, and $r = 1.5h = \frac{3}{2}h$. So $V = \frac{1}{3}\pi \left(\frac{3}{2}h\right)^2 h = \frac{3}{4}\pi h^3$

$$\Rightarrow \quad \frac{dV}{dt} = \frac{3}{4}\pi \cdot 3h^2 \frac{dh}{dt} = \frac{9}{4}\pi h^2 \frac{dh}{dt}. \text{ Therefore, } \frac{dh}{dt} = \frac{4(dV/dt)}{9\pi h^2} = \frac{240,000\pi}{9\pi h^2} = \frac{80,000}{3h^2} \; (\bigstar) \quad \Rightarrow$$

$\int 3h^2 dh = \int 80,000 dt \quad \Rightarrow \quad h^3 = 80,000t + C$. When $t = 0$, $h = 60$. Thus, $C = 60^3 = 216,000$, so

$h^3 = 80,000t + 216,000$. Let $h = 100$. Then $100^3 = 1,000,000 = 80,000t + 216,000 \quad \Rightarrow$

$80,000t = 784,000 \quad \Rightarrow \quad t = 9.8$, so the time required is 9.8 hours.

(b) The floor area of the silo is $F = \pi \cdot 200^2 = 40,000\pi$ ft^2, and the area of the base of the pile is

$A = \pi r^2 = \pi \left(\frac{3}{2}h\right)^2 = \frac{9\pi}{4}h^2$. So the area of the floor which is not covered when $h = 60$ is

$F - A = 40,000\pi - 8100\pi = 31,900\pi \approx 100,217$ ft^2. Now $A = \frac{9\pi}{4}h^2 \quad \Rightarrow \quad dA/dt = \frac{9\pi}{4} \cdot 2h \, (dh/dt)$, and

from $(\bigstar)$ in part (a) we know that when $h = 60$, $dh/dt = \frac{80,000}{3(60)^2} = \frac{200}{27}$ ft/h. Therefore,

$dA/dt = \frac{9\pi}{4}(2)(60)\left(\frac{200}{27}\right) = 2000\pi \approx 6283$ ft^2/h.

(c) At $h = 90$ ft, $dV/dt = 60,000\pi - 20,000\pi = 40,000\pi$ ft^3/h. From $(\bigstar)$ in part (a),

$\dfrac{dh}{dt} = \dfrac{4(dV/dt)}{9\pi h^2} = \dfrac{4(40,000\pi)}{9\pi h^2} = \dfrac{160,000}{9h^2} \quad \Rightarrow \quad \int 9h^2 dh = \int 160,000 dt \quad \Rightarrow \quad 3h^3 = 160,000t + C.$

When $t = 0$, $h = 90$; therefore, $C = 3 \cdot 729,000 = 2,187,000$. So $3h^3 = 160,000t + 2,187,000$. At the top,

$h = 100 \Rightarrow 3(100)^3 = 160,000t + 2,187,000 \Rightarrow t = \frac{813,000}{160,000} \approx 5.1$. The pile reaches the top after about 5.1 h.

7. (a) While running from $(L, 0)$ to (x, y), the dog travels a distance

$$s = \int_x^L \sqrt{1 + (dy/dx)^2}\, dx = -\int_L^x \sqrt{1 + (dy/dx)^2}\, dx, \text{ so}$$

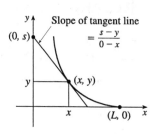

Slope of tangent line
$$= \frac{s - y}{0 - x}$$

$\dfrac{ds}{dx} = -\sqrt{1 + (dy/dx)^2}$. The dog and rabbit run at the same speed,

so the rabbit's position when the dog has traveled a distance s is

$(0, s)$. Since the dog runs straight for the rabbit, $\dfrac{dy}{dx} = \dfrac{s - y}{0 - x}$ (see the

figure).

Thus, $s = y - x\dfrac{dy}{dx}$ $\Rightarrow$ $\dfrac{ds}{dx} = \dfrac{dy}{dx} - \left(x\dfrac{d^2y}{dx^2} + 1\dfrac{dy}{dx}\right) = -x\dfrac{d^2y}{dx^2}$. Equating the two expressions for $\dfrac{ds}{dx}$

gives us $x\dfrac{d^2y}{dx^2} = \sqrt{1 + \left(\dfrac{dy}{dx}\right)^2}$, as claimed.

(b) Letting $z = \dfrac{dy}{dx}$, we obtain the differential equation $x\dfrac{dz}{dx} = \sqrt{1 + z^2}$, or $\dfrac{dz}{\sqrt{1 + z^2}} = \dfrac{dx}{x}$. Integrating:

$\ln x = \displaystyle\int \dfrac{dz}{\sqrt{1 + z^2}} \overset{25}{=} \ln\left|z + \sqrt{1 + z^2}\right| + C$. When $x = L$, $z = dy/dx = 0$, so $\ln L = \ln 1 + C$. Therefore,

$C = \ln L$, so $\ln x = \ln\left(\sqrt{1 + z^2} + z\right) + \ln L = \ln\left(L\left(\sqrt{1 + z^2} + z\right)\right)$ $\Rightarrow$ $x = L\left(\sqrt{1 + z^2} + z\right)$ $\Rightarrow$

$\sqrt{1 + z^2} = \dfrac{x}{L} - z$ $\Rightarrow$ $1 + z^2 = \left(\dfrac{x}{L}\right)^2 - \dfrac{2xz}{L} + z^2$ $\Rightarrow$ $\left(\dfrac{x}{L}\right)^2 - 2z\left(\dfrac{x}{L}\right) - 1 = 0$ $\Rightarrow$

$z = \dfrac{(x/L)^2 - 1}{2(x/L)} = \dfrac{x^2 - L^2}{2Lx} = \dfrac{x}{2L} - \dfrac{L}{2}\dfrac{1}{x}$ (for $x > 0$). Since $z = \dfrac{dy}{dx}$, $y = \dfrac{x^2}{4L} - \dfrac{L}{2}\ln x + C_1$. Since

$y = 0$ when $x = L$, $0 = \dfrac{L}{4} - \dfrac{L}{2}\ln L + C_1$ $\Rightarrow$ $C_1 = \dfrac{L}{2}\ln L - \dfrac{L}{4}$. Thus,

$y = \dfrac{x^2}{4L} - \dfrac{L}{2}\ln x + \dfrac{L}{2}\ln L - \dfrac{L}{4} = \dfrac{x^2 - L^2}{4L} - \dfrac{L}{2}\ln\left(\dfrac{x}{L}\right)$.

(c) As $x \to 0^+$, $y \to \infty$, so the dog never catches the rabbit.

9. (a) $\dfrac{d^2y}{dx^2} = k\sqrt{1 + \left(\dfrac{dy}{dx}\right)^2}$. Setting $z = \dfrac{dy}{dx}$, we get $\dfrac{dz}{dx} = k\sqrt{1 + z^2}$ $\Rightarrow$ $\dfrac{dz}{\sqrt{1 + z^2}} = k\, dx$. Using Formula 25

gives $\ln\left(z + \sqrt{1 + z^2}\right) = kx + c$ $\Rightarrow$ $z + \sqrt{1 + z^2} = Ce^{kx}$ (where $C = e^c$) $\Rightarrow$ $\sqrt{1 + z^2} = Ce^{kx} - z$

$\Rightarrow$ $1 + z^2 = C^2 e^{2kx} - 2Ce^{kx}z + z^2$ $\Rightarrow$ $2Ce^{kx}z = C^2 e^{2kx} - 1$ $\Rightarrow$ $z = \dfrac{C}{2}e^{kx} - \dfrac{1}{2C}e^{-kx}$. Now

$\dfrac{dy}{dx} = \dfrac{C}{2}e^{kx} - \dfrac{1}{2C}e^{-kx}$ $\Rightarrow$ $y = \dfrac{C}{2k}e^{kx} + \dfrac{1}{2Ck}e^{-kx} + C'$. From the diagram in the text, we see that

$y(0) = a$ and $y(\pm b) = h$. $a = y(0) = \dfrac{C}{2k} + \dfrac{1}{2Ck} + C'$ $\Rightarrow$ $C' = a - \dfrac{C}{2k} - \dfrac{1}{2Ck}$

$\Rightarrow$ $y = \dfrac{C}{2k}\left(e^{kx} - 1\right) + \dfrac{1}{2Ck}\left(e^{-kx} - 1\right) + a$. From $h = y(\pm b)$, we find

$h = \dfrac{C}{2k}\left(e^{kb} - 1\right) + \dfrac{1}{2Ck}\left(e^{-kb} - 1\right) + a$ and $h = \dfrac{C}{2k}\left(e^{-kb} - 1\right) + \dfrac{1}{2Ck}\left(e^{kb} - 1\right) + a$. Subtracting the

second equation from the first, we get $0 = \dfrac{C}{k}\dfrac{e^{kb} - e^{-kb}}{2} - \dfrac{1}{Ck}\dfrac{e^{kb} - e^{-kb}}{2} = \dfrac{1}{k}\left(C - \dfrac{1}{C}\right)\sinh(kb)$. Now

$k > 0$ and $b > 0$, so $\sinh(kb) > 0$ and $C = \pm 1$.

If $C = 1$, then $y = \dfrac{1}{2k}\left(e^{kx} - 1\right) + \dfrac{1}{2k}\left(e^{-kx} - 1\right) + a = \dfrac{1}{k}\dfrac{e^{kx} + e^{-kx}}{2} - \dfrac{1}{k} + a = a + \dfrac{1}{k}(\cosh kx - 1)$. If

$C = -1$, then $y = -\dfrac{1}{2k}\left(e^{kx} - 1\right) - \dfrac{1}{2k}\left(e^{-kx} - 1\right) + a = \dfrac{-1}{k}\dfrac{e^{kx} + e^{-kx}}{2} + \dfrac{1}{k} + a = a - \dfrac{1}{k}(\cosh kx - 1)$.

Since $k > 0$, $\cosh kx \geq 1$, and $y \geq a$, we conclude that $C = 1$ and $y = a + \dfrac{1}{k}(\cosh kx - 1)$, where

$h = y(b) = a + \dfrac{1}{k}(\cosh kb - 1)$. Since $\cosh(kb) = \cosh(-kb)$, there is no further information to extract

from the condition that $y(b) = y(-b)$. However, we could replace a with the expression $h - \dfrac{1}{k}(\cosh kb - 1)$,

obtaining $y = h + \dfrac{1}{k}(\cosh kx - \cosh kb)$. It would be better still to keep a in the expression for y, and use the

expression for h to solve for k in terms of a, b, and h. That would enable us to express y in terms of x and the

given parameters a, b, and h. Sadly, it is not possible to solve for k in closed form. That would have to be done

by numerical methods when specific parameter values are given.

(b) The length of the cable is

$$L = \int_{-b}^{b} \sqrt{1 + (dy/dx)^2}\, dx = \int_{-b}^{b} \sqrt{1 + \sinh^2 kx}\, dx = \int_{-b}^{b} \cosh kx\, dx = \left[\tfrac{1}{k}\sinh kx\right]_{-b}^{b}$$

$$= (1/k)\left[\sinh(kb) - \sinh(-kb)\right] = (2/k)\sinh(kb)$$

Parametric Equations and Polar Coordinates

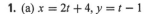 **Curves Defined by Parametric Equations**

1. (a) $x = 2t + 4, y = t - 1$

t	-3	-2	-1	0	1	2
x	-2	0	2	4	6	8
y	-4	-3	-2	-1	0	1

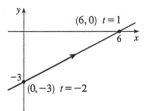

(b) $x = 2t + 4, y = t - 1 \implies x = 2(y+1) + 4 = 2y + 6$ or

$y = \frac{1}{2}x - 3$

3. (a) $x = 1 - 2t, y = t^2 + 4, 0 \le t \le 3$

t	0	1	2	3
x	1	-1	-3	-5
y	4	5	8	13

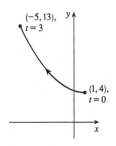

(b) $x = 1 - 2t \implies 2t = 1 - x \implies t = \dfrac{1 - x}{2} \implies$

$y = t^2 + 4 = \left(\dfrac{1-x}{2}\right)^2 + 4 = \frac{1}{4}(x-1)^2 + 4$ or

$y = \frac{1}{4}x^2 - \frac{1}{2}x + \frac{17}{4}$

5. (a) $x = \sqrt{t}, y = 1 - t$

t	0	1	2	3	4
x	0	1	1.414	1.732	2
y	1	0	-1	-2	-3

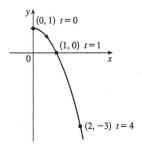

(b) $x = \sqrt{t} \implies t = x^2$. $y = 1 - t = 1 - x^2$. Since $t \ge 0, x \ge 0$.

351

7. (a) $x = \sin\theta$, $y = \cos\theta$, $0 \le \theta \le \pi$.

$x^2 + y^2 = \sin^2\theta + \cos^2\theta = 1$, $0 \le x \le 1$.

(b)

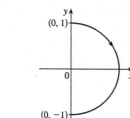

9. (a) $x = \sin^2\theta$, $y = \cos^2\theta$.

$x + y = \sin^2\theta + \cos^2\theta = 1$, $0 \le x \le 1$.

Note that the curve is at $(0, 1)$ whenever $\theta = \pi n$ and is at $(1, 0)$ whenever $\theta = \frac{\pi}{2}n$ for every integer n.

(b)

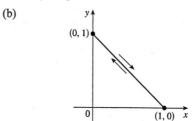

11. (a) $x = e^t$, $y = e^{-t}$, $y = 1/x$, $x > 0$

(b)

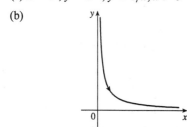

13. (a) $x = \tan\theta + \sec\theta$, $y = \tan\theta - \sec\theta$,

$-\frac{\pi}{2} < \theta < \frac{\pi}{2}$. $xy = \tan^2\theta - \sec^2\theta = -1$

$\Rightarrow \quad y = -1/x$, $x > 0$.

(b)

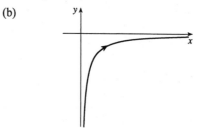

15. (a) $x = \cosh t$, $y = \sinh t$, $x^2 - y^2 = \cosh^2 t - \sinh^2 t = 1$, $x \ge 1$

(b)

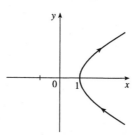

17. $x^2 + y^2 = \cos^2 \pi t + \sin^2 \pi t = 1$, $1 \le t \le 2$, so the particle moves counterclockwise along the circle $x^2 + y^2 = 1$ from $(-1, 0)$ to $(1, 0)$, along the lower half of the circle.

19. $\left(\frac{1}{2}x\right)^2 + \left(\frac{1}{3}y\right)^2 = \sin^2 t + \cos^2 t = 1$, so the particle moves once clockwise along the ellipse $\frac{1}{4}x^2 + \frac{1}{9}y^2 = 1$, starting and ending at $(0, 3)$.

21. $x = \tan t$, $y = \cot t$, $\frac{\pi}{6} \le t \le \frac{\pi}{3}$. $y = 1/x$ for $\frac{1}{\sqrt{3}} \le x \le \sqrt{3}$. The particle moves along the first quadrant branch of the hyperbola $y = 1/x$ from $\left(\frac{1}{\sqrt{3}}, \sqrt{3}\right)$ to $\left(\sqrt{3}, \frac{1}{\sqrt{3}}\right)$.

23. From the graphs, it seems that as $t \to -\infty$, $x \to \infty$ and $y \to -\infty$. So the point $(x(t), y(t))$ will move from far out in the fourth quadrant as t increases. At $t = -\sqrt{3}$, both x and y are 0, so the graph passes through the origin. After that the graph passes through the second quadrant (x is negative, y is positive), then intersects the x-axis at $x = -9$ when $t = 0$. After this, the graph passes through the third quadrant, going through the origin again at $t = \sqrt{3}$, and then as $t \to \infty$, $x \to \infty$ and $y \to \infty$. Note that for every point $(x(t), y(t)) = (3(t^2 - 3), t^3 - 3t)$, we can substitute $-t$ to get the corresponding point $(x(-t), y(-t)) = (3[(-t)^2 - 3], (-t)^3 - 3(-t)) = (x(t), -y(t))$, and so the graph is symmetric about the x-axis. The first figure was obtained using $x_1 = t$, $y_1 = 3(t^2 - 3)$; $x_2 = t$, $y_2 = t^3 - 3t$; and $-2\pi \le t \le 2\pi$.

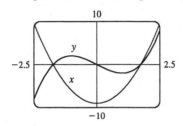

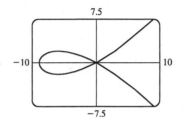

25. As $t \to -\infty$, $x \to \infty$ and $y \to -\infty$. The graph passes through the origin at $t = -1$, and then goes through the second quadrant (x negative, y positive), passing through the point $(-1, 1)$ at $t = 0$. As t increases, the graph passes through the point $(0, 2)$ at $t = 1$, and then as $t \to \infty$, both x and y approach ∞. The first figure was obtained using $x_1 = t$, $y_1 = t^4 - 1$; $x_2 = t$, $y_2 = t^3 + 1$; and $-2\pi \le t \le 2\pi$.

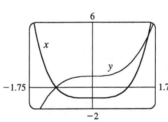

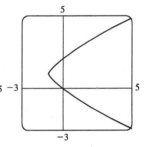

27. As in Example 4, we let $y = t$ and $x = t - 3t^3 + t^5$ and use a t-interval of $[-2\pi, 2\pi]$.

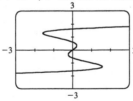

29. The circle $x^2 + y^2 = 4$ can be represented parametrically by $x = 2\cos t$, $y = 2\sin t$; $0 \le t \le 2\pi$. The circle $x^2 + (y - 1)^2 = 4$ can be represented by $x = 2\cos t$, $y = 1 + 2\sin t$; $0 \le t \le 2\pi$. This representation gives us the circle with a counterclockwise orientation starting at $(2, 1)$.

(a) To get a clockwise orientation, we could change the equations to $x = 2\cos t$, $y = 1 - 2\sin t$.

(b) To get three times around in the counterclockwise direction, we use the original equations $x = 2\cos t$, $y = 1 + 2\sin t$ with the domain expanded to $0 \le t \le 6\pi$.

(c) To start at $(0, 3)$ using the original equations, we must have $x_1 = 0$; that is, $2\cos t = 0$. Hence, $t = \frac{\pi}{2}$. So we use $x = 2\cos t$, $y = 1 + 2\sin t$; $\frac{\pi}{2} \le t \le \frac{3\pi}{2}$.

Alternatively, if we want t to start at 0, we could change the equations of the curve. For example, we could use $x = -2\sin t$, $y = 1 + 2\cos t$, $0 \le t \le \pi$.

31. (a) Let $x^2/a^2 = \sin^2 t$ and $y^2/b^2 = \cos^2 t$ to obtain

$x = a\sin t$ and $y = b\cos t$ with $0 \le t \le 2\pi$ as possible

parametric equations for the ellipse $x^2/a^2 + y^2/b^2 = 1$.

(c) As b increases, the ellipse is stretched vertically.

(b) The equations are $x = 3\sin t$ and

$y = b\cos t$ for $b \in \{1, 2, 4, 8\}$.

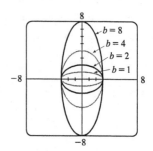

33. The case $\frac{\pi}{2} < \theta < \pi$ is illustrated. C has coordinates $(r\theta, r)$ as before, and Q

has coordinates $(r\theta, r + r\cos(\pi - \theta)) = (r\theta, r(1 - \cos\theta))$ [since

$\cos(\pi - \alpha) = \cos\pi\cos\alpha + \sin\pi\sin\alpha = -\cos\alpha$], so P has coordinates

$(r\theta - r\sin(\pi - \theta), r(1 - \cos\theta)) = (r(\theta - \sin\theta), r(1 - \cos\theta))$ [since

$\sin(\pi - \alpha) = \sin\pi\cos\alpha - \cos\pi\sin\alpha = \sin\alpha$]. Again we have the

parametric equations $x = r(\theta - \sin\theta)$, $y = r(1 - \cos\theta)$.

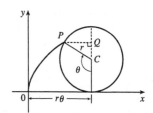

35. It is apparent that $x = |OQ|$ and $y = |QP| = |ST|$. From the

diagram, $x = |OQ| = a\cos\theta$ and $y = |ST| = b\sin\theta$. Thus, the

parametric equations are $x = a\cos\theta$ and $y = b\sin\theta$. To eliminate θ

we rearrange: $\sin\theta = y/b \Rightarrow \sin^2\theta = (y/b)^2$ and $\cos\theta = x/a$

$\Rightarrow \cos^2\theta = (x/a)^2$. Adding the two equations:

$\sin^2\theta + \cos^2\theta = 1 = x^2/a^2 + y^2/b^2$. Thus, we have an ellipse.

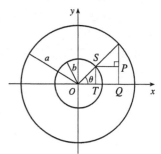

37. $C = (2a\cot\theta, 2a)$, so the x-coordinate of P is $x = 2a\cot\theta$. Let

$B = (0, 2a)$. Then $\angle OAB$ is a right angle and $\angle OBA = \theta$, so

$|OA| = 2a\sin\theta$ and $A = (2a\sin\theta\cos\theta, 2a\sin^2\theta)$. Thus, the

y-coordinate of P is $y = 2a\sin^2\theta$.

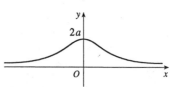

39. (a)

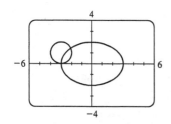

There are 2 points of intersection:

$(-3, 0)$ and approximately $(-2.1, 1.4)$.

(b) As an aid in finding collision points, set your graphing utility

to graph both curves simultaneously and closely observe the

drawing of the graphs. In this case, we have one collision

point: both particles are at $(-3, 0)$ when $t = \frac{3\pi}{2}$. [Notice that

the first curve passes through $(-2.1, 1.4)$ when $t \approx 5.5$, but

the second curve passes through $(-2.1, 1.4)$ when $t \approx 0.4$.]

(c) The circle is centered at $(3, 1)$ instead of $(-3, 1)$. There are

still 2 intersection points: $(3, 0)$ and $(2.1, 1.4)$, but there are

no collision points.

41. $x = t^2$, $y = t^3 - ct$. We use a graphing device to produce the graphs for various values of c with $-\pi \le t \le \pi$. Note that all the members of the family are symmetric about the x-axis. For $c < 0$, the graph does not cross itself, but for $c = 0$ it has a cusp at $(0, 0)$ and for $c > 0$ the graph crosses itself at $x = c$, so the loop grows larger as c increases.

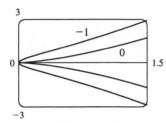
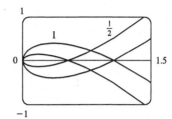

43. Note that all the Lissajous figures are symmetric about the x-axis. The parameters a and b simply stretch the graph in the x- and y-directions respectively. For $a = b = n = 1$ the graph is simply a circle with radius 1. For $n = 2$ the graph crosses itself at the origin and there are loops above and below the x-axis. In general, the figures have $n - 1$ points of intersection, all of which are on the y-axis, and a total of n closed loops.

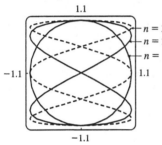

$a = b = 1$

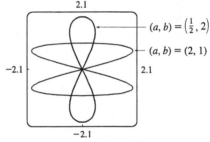

$(a, b) = \left(\frac{1}{2}, 2\right)$

$(a, b) = (2, 1)$

$n = 2$

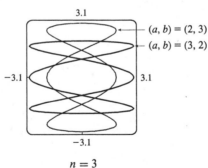

$(a, b) = (2, 3)$

$(a, b) = (3, 2)$

$n = 3$

10.2 Tangents and Areas

1. $x = t - t^3$, $y = 2 - 5t$ $\Rightarrow$ $\dfrac{dy}{dt} = -5$, $\dfrac{dx}{dt} = 1 - 3t^2$, and $\dfrac{dy}{dx} = \dfrac{dy/dt}{dx/dt} = \dfrac{-5}{1 - 3t^2}$ or $\dfrac{5}{3t^2 - 1}$.

3. $x = t \ln t$, $y = \sin^2 t$ $\Rightarrow$ $\dfrac{dy}{dt} = 2 \sin t \cos t$, $\dfrac{dx}{dt} = t\left(\dfrac{1}{t}\right) + (\ln t) \cdot 1 = 1 + \ln t$, and

$$\dfrac{dy}{dx} = \dfrac{dy/dt}{dx/dt} = \dfrac{2 \sin t \cos t}{1 + \ln t}$$

5. $x = t^2 + t$, $y = t^2 - t$; $t = 0$. $\dfrac{dy}{dt} = 2t - 1$, $\dfrac{dx}{dt} = 2t + 1$, so $\dfrac{dy}{dx} = \dfrac{dy/dt}{dx/dt} = \dfrac{2t - 1}{2t + 1}$. When $t = 0$, $x = y = 0$

and $\dfrac{dy}{dx} = -1$. An equation of the tangent is $y - 0 = (-1)(x - 0)$ or $y = -x$.

7. $x = e^{\sqrt{t}}$, $y = t - \ln t^2$; $t = 1$. $\dfrac{dy}{dt} = 1 - \dfrac{2t}{t^2} = 1 - \dfrac{2}{t}$, $\dfrac{dx}{dt} = \dfrac{e^{\sqrt{t}}}{2\sqrt{t}}$, and

$\dfrac{dy}{dx} = \dfrac{dy/dt}{dx/dt} = \dfrac{1 - 2/t}{e^{\sqrt{t}}/(2\sqrt{t})} \cdot \dfrac{2t}{2t} = \dfrac{2t - 4}{\sqrt{t}e^{\sqrt{t}}}$. When $t = 1$, $(x, y) = (e, 1)$ and $\dfrac{dy}{dx} = -\dfrac{2}{e}$, so an equation of the

tangent line is $y - 1 = -\frac{2}{e}(x - e)$ or $y = -\frac{2}{e}x + 3$.

9. (a) $x = e^t$, $y = (t - 1)^2$; $(1, 1)$. $\dfrac{dy}{dt} = 2(t - 1)$, $\dfrac{dx}{dt} = e^t$, and $\dfrac{dy}{dx} = \dfrac{dy/dt}{dx/dt} = \dfrac{2(t - 1)}{e^t}$.

At $(1, 1)$, $t = 0$ and $\dfrac{dy}{dx} = -2$, so an equation of the tangent is $y - 1 = -2(x - 1)$ or $y = -2x + 3$.

(b) $x = e^t \Rightarrow t = \ln x$, so $y = (t - 1)^2 = (\ln x - 1)^2$ and $\dfrac{dy}{dx} = 2(\ln x - 1)\left(\dfrac{1}{x}\right)$. When $x = 1$,

$\dfrac{dy}{dx} = 2(-1)(1) = -2$, so an equation of the tangent is $y = -2x + 3$, as in part (a).

11. $x = 2\sin 2t$, $y = 2\sin t$; $\left(\sqrt{3}, 1\right)$.

$\dfrac{dy}{dx} = \dfrac{dy/dt}{dx/dt} = \dfrac{2\cos t}{2 \cdot 2\cos 2t} = \dfrac{\cos t}{2\cos 2t}$. The point $\left(\sqrt{3}, 1\right)$ corresponds

to $t = \frac{\pi}{6}$, so the slope of the tangent at that point is

$\dfrac{\cos \frac{\pi}{6}}{2\cos \frac{\pi}{3}} = \dfrac{\frac{\sqrt{3}}{2}}{2 \cdot \frac{1}{2}} = \dfrac{\sqrt{3}}{2}$. An equation of the tangent is therefore

$(y - 1) = \frac{\sqrt{3}}{2}\left(x - \sqrt{3}\right)$ or $y = \frac{\sqrt{3}}{2}x - \frac{1}{2}$.

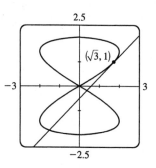

13. $x = t^4 - 1$, $y = t - t^2$ $\Rightarrow$ $\dfrac{dy}{dt} = 1 - 2t$, $\dfrac{dx}{dt} = 4t^3$, $\dfrac{dy}{dx} = \dfrac{dy/dt}{dx/dt} = \dfrac{1 - 2t}{4t^3} = \frac{1}{4}t^{-3} - \frac{1}{2}t^{-2}$;

$\dfrac{d}{dt}\left(\dfrac{dy}{dx}\right) = -\frac{3}{4}t^{-4} + t^{-3}$, $\dfrac{d^2y}{dx^2} = \dfrac{d(dy/dx)/dt}{dx/dt} = \dfrac{-\frac{3}{4}t^{-4} + t^{-3}}{4t^3} \cdot \dfrac{4t^4}{4t^4} = \dfrac{-3 + 4t}{16t^7}$.

15. $x = \sin \pi t$, $y = \cos \pi t$. $\dfrac{dy}{dx} = \dfrac{dy/dt}{dx/dt} = \dfrac{-\pi\sin \pi t}{\pi\cos \pi t} = -\tan \pi t$;

$\dfrac{d^2y}{dx^2} = \dfrac{d}{dx}\left(\dfrac{dy}{dx}\right) = \dfrac{d(dy/dx)/dt}{dx/dt} = \dfrac{-\pi\sec^2 \pi t}{\pi\cos \pi t} = -\sec^3 \pi t$.

17. $x = e^{-t}$, $y = te^{2t}$. $\dfrac{dy}{dx} = \dfrac{dy/dt}{dx/dt} = \dfrac{(2t + 1)e^{2t}}{-e^{-t}} = -(2t + 1)e^{3t}$;

$\dfrac{d}{dt}\left(\dfrac{dy}{dx}\right) = -3(2t + 1)e^{3t} - 2e^{3t} = -(6t + 5)e^{3t}$;

$\dfrac{d^2y}{dx^2} = \dfrac{d}{dx}\left(\dfrac{dy}{dx}\right) = \dfrac{d(dy/dx)/dt}{dx/dt} = \dfrac{-(6t + 5)e^{3t}}{-e^{-t}} = (6t + 5)e^{4t}$.

19. $x = t(t^2 - 3) = t^3 - 3t$, $y = 3(t^2 - 3)$. $\dfrac{dx}{dt} = 3t^2 - 3 = 3(t-1)(t+1)$; $\dfrac{dy}{dt} = 6t$. $\dfrac{dy}{dt} = 0 \iff t = 0 \iff$

$(x, y) = (0, -9)$. $\dfrac{dx}{dt} = 0 \iff t = \pm 1 \iff (x, y) = (-2, -6)$ or $(2, -6)$. So there is a horizontal tangent at

$(0, -9)$ and there are vertical tangents at $(-2, -6)$ and $(2, -6)$.

	$t < -1$	$-1 < t < 0$	$0 < t < 1$	$t > 1$
dx/dt	$+$	$-$	$-$	$+$
dy/dt	$-$	$-$	$+$	$+$
x	$\rightarrow$	$\leftarrow$	$\leftarrow$	$\rightarrow$
y	$\downarrow$	$\downarrow$	$\uparrow$	$\uparrow$
curve	$\searrow$	$\swarrow$	$\nwarrow$	$\nearrow$

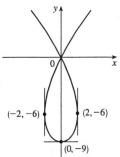

21. $x = \dfrac{3t}{1+t^3}$, $y = \dfrac{3t^2}{1+t^3}$. $\dfrac{dx}{dt} = \dfrac{(1+t^3)3 - 3t(3t^2)}{(1+t^3)^2} = \dfrac{3 - 6t^3}{(1+t^3)^2}$,

$\dfrac{dy}{dt} = \dfrac{(1+t^3)(6t) - 3t^2(3t^2)}{(1+t^3)^2} = \dfrac{6t - 3t^4}{(1+t^3)^2} = \dfrac{3t(2-t^3)}{(1+t^3)^2}$. $\dfrac{dy}{dt} = 0 \iff t = 0$ or $\sqrt[3]{2} \iff$

$(x, y) = (0, 0)$ or $\left(\sqrt[3]{2}, \sqrt[3]{4}\right)$. $\dfrac{dx}{dt} = 0 \iff t^3 = \tfrac{1}{2} \iff t = 2^{-1/3}$

$\iff (x, y) = \left(\sqrt[3]{4}, \sqrt[3]{2}\right)$. There are horizontal tangents at $(0, 0)$ and

$\left(\sqrt[3]{2}, \sqrt[3]{4}\right)$, and there are vertical tangents at $\left(\sqrt[3]{4}, \sqrt[3]{2}\right)$ and $(0, 0)$. [The

vertical tangent at $(0, 0)$ is undetectable by the methods of this section

because that tangent corresponds to the limiting position of the point (x, y)

as $t \to \pm\infty$.] In the following table, $\alpha = \sqrt[3]{2}$.

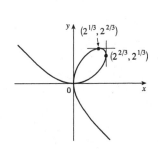

	$t < -1$	$-1 < t < 0$	$0 < t < 1/\alpha$	$1/\alpha < t < \alpha$	$t > \alpha$
dx/dt	$+$	$+$	$+$	$-$	$-$
dy/dt	$-$	$-$	$+$	$+$	$-$
x	$\rightarrow$	$\rightarrow$	$\rightarrow$	$\leftarrow$	$\leftarrow$
y	$\downarrow$	$\downarrow$	$\uparrow$	$\uparrow$	$\downarrow$
curve	$\searrow$	$\searrow$	$\nearrow$	$\nwarrow$	$\swarrow$

23. From the graph, it appears that the leftmost point on the curve $x = t^4 - t^2$,

$y = t + \ln t$ is about $(-0.25, 0.36)$. To find the exact coordinates, we find

the value of t for which the graph has a vertical tangent, that is,

$0 = dx/dt = 4t^3 - 2t \iff 2t(2t^2 - 1) = 0 \iff$

$2t\left(\sqrt{2}t + 1\right)\left(\sqrt{2}t - 1\right) = 0 \iff t = 0$ or $\pm\dfrac{1}{\sqrt{2}}$. The negative and

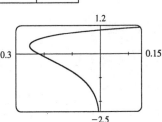

0 roots are inadmissible since $y(t)$ is only defined for $t > 0$, so the leftmost point must be

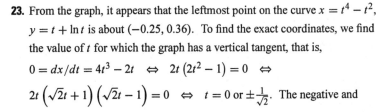

$$\left(x\left(\tfrac{1}{\sqrt{2}}\right), y\left(\tfrac{1}{\sqrt{2}}\right)\right) = \left(\left(\tfrac{1}{\sqrt{2}}\right)^4 - \left(\tfrac{1}{\sqrt{2}}\right)^2, \tfrac{1}{\sqrt{2}} + \ln\tfrac{1}{\sqrt{2}}\right) = \left(-\tfrac{1}{4}, \tfrac{1}{\sqrt{2}} - \tfrac{1}{2}\ln 2\right)$$

25. We graph the curve $x = t^4 - 2t^3 - 2t^2$,

$y = t^3 - t$ in the viewing rectangle
$[-2, 1.1]$ by $[-0.5, 0.5]$. This rectangle
corresponds approximately to
$t \in [-1, 0.8]$. We estimate that the curve
has horizontal tangents at about

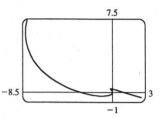

$(-1, -0.4)$ and $(-0.17, 0.39)$ and vertical tangents at about $(0, 0)$ and $(-0.19, 0.37)$. We calculate

$\dfrac{dy}{dx} = \dfrac{dy/dt}{dx/dt} = \dfrac{3t^2 - 1}{4t^3 - 6t^2 - 4t}$. The horizontal tangents occur when $dy/dt = 3t^2 - 1 = 0$ $\Leftrightarrow$ $t = \pm\dfrac{1}{\sqrt{3}}$, so

both horizontal tangents are shown in our graph. The vertical tangents occur when $dx/dt = 2t\left(2t^2 - 3t - 2\right) = 0$

$\Leftrightarrow$ $2t(2t + 1)(t - 2) = 0$ $\Leftrightarrow$ $t = 0, -\frac{1}{2}$ or 2. It seems that we have missed one vertical tangent, and indeed
if we plot the curve on the t-interval $[-1.2, 2.2]$ we see that there is another vertical tangent at $(-8, 6)$.

27. $x = \cos t$, $y = \sin t \cos t$. $\frac{dx}{dt} = -\sin t$, $\frac{dy}{dt} = -\sin^2 t + \cos^2 t = \cos 2t$.

$(x, y) = (0, 0)$ $\Leftrightarrow$ $\cos t = 0$ $\Leftrightarrow$ t is an odd multiple of $\frac{\pi}{2}$. When

$t = \frac{\pi}{2}$, $\frac{dx}{dt} = -1$ and $\frac{dy}{dt} = -1$, so $\frac{dy}{dx} = 1$. When $t = \frac{3\pi}{2}$, $\frac{dx}{dt} = 1$ and

$\frac{dy}{dt} = -1$. So $\frac{dy}{dx} = -1$. Thus, $y = x$ and $y = -x$ are both tangent to the

curve at $(0, 0)$.

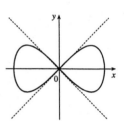

29. (a) $x = r\theta - d\sin\theta$, $y = r - d\cos\theta$; $\dfrac{dx}{d\theta} = r - d\cos\theta$, $\dfrac{dy}{d\theta} = d\sin\theta$. So $\dfrac{dy}{dx} = \dfrac{d\sin\theta}{r - d\cos\theta}$.

(b) If $0 < d < r$, then $|d\cos\theta| \le d < r$, so $r - d\cos\theta \ge r - d > 0$. This shows that $dx/d\theta$ never vanishes, so
the trochoid can have no vertical tangent if $d < r$.

31. The line with parametric equations $x = -7t$, $y = 12t - 5$ is $y = 12\left(-\frac{1}{7}x\right) - 5$, which has slope $-\frac{12}{7}$. The curve

$x = t^3 + 4t$, $y = 6t^2$ has slope $\dfrac{dy}{dx} = \dfrac{dy/dt}{dx/dt} = \dfrac{12t}{3t^2 + 4}$. This equals $-\frac{12}{7}$ $\Leftrightarrow$ $3t^2 + 4 = -7t$ $\Leftrightarrow$

$(3t + 4)(t + 1) = 0$ $\Leftrightarrow$ $t = -1$ or $t = -\frac{4}{3}$ $\Leftrightarrow$ $(x, y) = (-5, 6)$ or $\left(-\frac{208}{27}, \frac{32}{3}\right)$.

33. $A = \int_0^1 (y - 1)\, dx = \int_{\pi/2}^0 \left(e^t - 1\right)(-\sin t)\, dt = \int_0^{\pi/2} \left(e^t \sin t - \sin t\right) dt \overset{98}{=} \left[\frac{1}{2}e^t (\sin t - \cos t) + \cos t\right]_0^{\pi/2}$

$= \frac{1}{2}\left(e^{\pi/2} - 1\right)$

35. By symmetry of the ellipse about the x- and y-axes,

$$A = 4\int_0^a y\, dx = 4\int_{\pi/2}^0 b\sin\theta\,(-a\sin\theta)\, d\theta = 4ab\int_0^{\pi/2} \sin^2\theta\, d\theta = 4ab\int_0^{\pi/2} \tfrac{1}{2}(1 - \cos 2\theta)\, d\theta$$

$$= 2ab\left[\theta - \tfrac{1}{2}\sin 2\theta\right]_0^{\pi/2} = 2ab\left(\tfrac{\pi}{2}\right) = \pi ab$$

37. $A = \int_0^{2\pi r} y\, dx = \int_0^{2\pi} (r - d\cos\theta)(r - d\cos\theta)\, d\theta = \int_0^{2\pi} \left(r^2 - 2dr\cos\theta + d^2\cos^2\theta\right) d\theta$

$= \left[r^2\theta - 2dr\sin\theta + \tfrac{1}{2}d^2\left(\theta + \tfrac{1}{2}\sin 2\theta\right)\right]_0^{2\pi} = 2\pi r^2 + \pi d^2$

39. The graph of $x = \sin t - 2\cos t$, $y = 1 + \sin t \cos t$ is symmetric
about the y-axis. The graph intersects the y-axis when $x = 0 \Rightarrow$
$\sin t - 2\cos t = 0 \Rightarrow \sin t = 2\cos t \Rightarrow \tan t = 2 \Rightarrow$
$t = \tan^{-1} 2 + n\pi$. The left loop is traced in a clockwise direction
from $t = \tan^{-1} 2 - \pi$ to $t = \tan^{-1} 2$, so the area of the loop is given
(as in Example 4) by

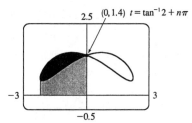

$$A = \int_{\tan^{-1} 2 - \pi}^{\tan^{-1} 2} y \, dx \approx \int_{-2.0344}^{1.1071} (1 + \sin t \cos t)(\cos t + 2\sin t) \, dt \approx 0.8944$$

This integral can be evaluated exactly; its value is $\frac{2}{3}\sqrt{5}$.

41. The coordinates of T are $(r\cos\theta, r\sin\theta)$. Since TP was unwound from
arc TA, TP has length $r\theta$. Also $\angle PTQ = \angle PTR - \angle QTR = \frac{1}{2}\pi - \theta$,
so P has coordinates $x = r\cos\theta + r\theta\cos\left(\frac{1}{2}\pi - \theta\right) = r\left(\cos\theta + \theta\sin\theta\right)$,
$y = r\sin\theta - r\theta\sin\left(\frac{1}{2}\pi - \theta\right) = r\left(\sin\theta - \theta\cos\theta\right)$.

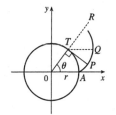

10.3 Arc Length and Surface Area

1. $x = t - t^2$, $y = \frac{4}{3}t^{3/2}$, $1 \le t \le 2$. $dx/dt = 1 - 2t$ and $dy/dt = 2t^{1/2}$, so
$(dx/dt)^2 + (dy/dt)^2 = (1 - 2t)^2 + \left(2t^{1/2}\right)^2 = 1 - 4t + 4t^2 + 4t = 1 + 4t^2$ and
$L = \int_\alpha^\beta \sqrt{(dx/dt)^2 + (dy/dt)^2} \, dt = \int_1^2 \sqrt{1 + 4t^2} \, dt$.

3. $x = t\sin t$, $y = t\cos t$, $0 \le t \le \frac{\pi}{2}$. $dx/dt = t\cos t + \sin t$ and $dy/dt = t(-\sin t) + \cos t$, so

$$(dx/dt)^2 + (dy/dt)^2 = (t\cos t + \sin t)^2 + (\cos t - t\sin t)^2$$
$$= t^2 \cos^2 t + 2t\sin t \cos t + \sin^2 t + \cos^2 t - 2t\sin t \cos t + t^2 \sin^2 t$$
$$= t^2 \left(\cos^2 t + \sin^2 t\right) + \sin^2 t + \cos^2 t = t^2 + 1$$

and $L = \int_0^{\pi/2} \sqrt{t^2 + 1} \, dt$.

5. $x = t^3$, $y = t^2$, $0 \le t \le 4$. $(dx/dt)^2 + (dy/dt)^2 = \left(3t^2\right)^2 + (2t)^2 = 9t^4 + 4t^2$.
$L = \int_0^4 \sqrt{(dx/dt)^2 + (dy/dt)^2} \, dt = \int_0^4 \sqrt{9t^4 + 4t^2} \, dt = \int_0^4 t\sqrt{9t^2 + 4} \, dt = \frac{1}{18} \int_4^{148} \sqrt{u} \, du$ (where $u = 9t^2 + 4$).
So $L = \frac{1}{18} \left(\frac{2}{3}\right) \left[u^{3/2}\right]_4^{148} = \frac{1}{27} \left(148^{3/2} - 4^{3/2}\right) = \frac{8}{27} \left(37^{3/2} - 1\right)$.

7. $x = \dfrac{t}{1+t}$, $y = \ln(1+t)$, $0 \le t \le 2$. $\dfrac{dx}{dt} = \dfrac{(1+t) \cdot 1 - t \cdot 1}{(1+t)^2} = \dfrac{1}{(1+t)^2}$ and $\dfrac{dy}{dt} = \dfrac{1}{1+t}$, so

$$\left(\frac{dx}{dt}\right)^2 + \left(\frac{dy}{dt}\right)^2 = \frac{1}{(1+t)^4} + \frac{1}{(1+t)^2} = \frac{1}{(1+t)^4}\left[1 + (1+t)^2\right] = \frac{t^2 + 2t + 2}{(1+t)^4} \text{ and}$$

$$L = \int_0^2 \frac{\sqrt{t^2 + 2t + 2}}{(1+t)^2}\, dt = \int_1^3 \frac{\sqrt{u^2 + 1}}{u^2}\, du \ [u = t+1, \ du = dt] \overset{24}{=} \left[-\frac{\sqrt{u^2+1}}{u} + \ln\left(u + \sqrt{u^2+1}\right)\right]_1^3$$

$$= -\frac{\sqrt{10}}{3} + \ln\left(3 + \sqrt{10}\right) + \sqrt{2} - \ln\left(1 + \sqrt{2}\right)$$

9. $x = e^t \cos t$, $y = e^t \sin t$, $0 \le t \le \pi$.

$$\left(\frac{dx}{dt}\right)^2 + \left(\frac{dy}{dt}\right)^2 = \left[e^t (\cos t - \sin t)\right]^2 + \left[e^t (\sin t + \cos t)\right]^2$$

$$= e^{2t}\left(2\cos^2 t + 2\sin^2 t\right) = 2e^{2t}$$

$$\Rightarrow \quad L = \int_0^\pi \sqrt{2}\, e^t\, dt = \sqrt{2}\left(e^\pi - 1\right)$$

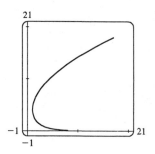

11. $x = e^t - t$, $y = 4e^{t/2}$, $-8 \le t \le 3$.

$$(dx/dt)^2 + (dy/dt)^2 = (e^t - 1)^2 + \left(2e^{t/2}\right)^2 = e^{2t} - 2e^t + 1 + 4e^t$$

$$= e^{2t} + 2e^t + 1 = (e^t + 1)^2$$

$$L = \int_{-8}^3 \sqrt{(e^t + 1)^2}\, dt = \int_{-8}^3 (e^t + 1)\, dt = [e^t + t]_{-8}^3$$

$$= \left(e^3 + 3\right) - \left(e^{-8} - 8\right) = e^3 - e^{-8} + 11$$

13. $x = \ln t$ and $y = e^{-t}$ $\Rightarrow$ $\dfrac{dx}{dt} = \dfrac{1}{t}$ and $\dfrac{dy}{dt} = -e^{-t}$ $\Rightarrow$ $L = \int_1^2 \sqrt{t^{-2} + e^{-2t}}\, dt$. Using Simpson's

Rule with $n = 10$, $\Delta x = (2-1)/10 = 0.1$ and $f(t) = \sqrt{t^{-2} + e^{-2t}}$ we get

$L \approx \frac{0.1}{3}\left[f(1.0) + 4f(1.1) + 2f(1.2) + \cdots + 2f(1.8) + 4f(1.9) + f(2.0)\right] \approx 0.7314$.

15. $x = \sin^2 \theta$, $y = \cos^2 \theta$, $0 \le \theta \le 3\pi$.

$(dx/d\theta)^2 + (dy/d\theta)^2 = (2\sin\theta\cos\theta)^2 + (-2\cos\theta\sin\theta)^2 = 8\sin^2\theta\cos^2\theta = 2\sin^2 2\theta$ $\Rightarrow$

$$\text{Distance} = \int_0^{3\pi} \sqrt{2}\,|\sin 2\theta|\, d\theta = 6\sqrt{2} \int_0^{\pi/2} \sin 2\theta\, d\theta \ \text{(by symmetry)} = \left[-3\sqrt{2}\cos 2\theta\right]_0^{\pi/2}$$

$$= -3\sqrt{2}\,(-1 - 1) = 6\sqrt{2}$$

The full curve is traversed as θ goes from 0 to $\frac{\pi}{2}$, because the curve is the segment of $x + y = 1$ that lies in the first quadrant (since x, $y \ge 0$), and this segment is completely traversed as θ goes from 0 to $\frac{\pi}{2}$.

Thus $L = \int_0^{\pi/2} \sin 2\theta\, d\theta = \sqrt{2}$, as above.

17. $x = a \sin\theta$, $y = b \cos\theta$, $0 \le \theta \le 2\pi$.

$$\left(\tfrac{dx}{d\theta}\right)^2 + \left(\tfrac{dy}{d\theta}\right)^2 = (a \cos\theta)^2 + (-b \sin\theta)^2 = a^2 \cos^2\theta + b^2 \sin^2\theta = a^2 \left(1 - \sin^2\theta\right) + b^2 \sin^2\theta$$

$$= a^2 - \left(a^2 - b^2\right) \sin^2\theta = a^2 - c^2 \sin^2\theta = a^2 \left(1 - \frac{c^2}{a^2}\sin^2\theta\right) = a^2 \left(1 - e^2 \sin^2\theta\right)$$

So $L = 4 \int_0^{\pi/2} \sqrt{a^2 \left(1 - e^2 \sin^2\theta\right)}\, d\theta$ (by symmetry) $= 4a \int_0^{\pi/2} \sqrt{1 - e^2 \sin^2\theta}\, d\theta$

19. (a) Notice that $0 \le t \le 2\pi$ does not give the complete curve because $x(0) \ne x(2\pi)$. In fact, we must take $t \in [0, 4\pi]$ in order to obtain the complete curve, since the first term in each of the parametric equations has period 2π and the second has period $\frac{4\pi}{11}$, and the least common integer multiple of these two numbers is 4π.

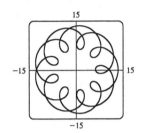

(b) We use the CAS to find the derivatives dx/dt and dy/dt, and then use Theorem 4 to find the arc length. Recent versions of Maple express the integral $\int_0^{4\pi} \sqrt{(dx/dt)^2 + (dy/dt)^2}\, dt$ as $88E\left(2\sqrt{2}i\right)$, where $E(x)$ is the elliptic integral $\int_0^1 \frac{\sqrt{1 - x^2 t^2}}{\sqrt{1 - t^2}}\, dt$ and i is the imaginary number $\sqrt{-1}$. Some earlier versions of Maple (as well as Mathematica) cannot do the integral exactly, so we use the command `evalf(Int(sqrt(diff(x,t)^2+diff(y,t)^2),t=0..4*Pi));` to estimate the length, and find that the arc length is approximately 294.03.

21. $x = t^3$ and $y = t^4 \implies dx/dt = 3t^2$ and $dy/dt = 4t^3$. So
$$S = \int_0^1 2\pi t^4 \sqrt{9t^4 + 16t^6}\, dt = \int_0^1 2\pi t^6 \sqrt{9 + 16t^2}\, dt.$$

23. $x = t^3$, $y = t^2$, $0 \le t \le 1$. $\left(\dfrac{dx}{dt}\right)^2 + \left(\dfrac{dy}{dt}\right)^2 = (3t^2)^2 + (2t)^2 = 9t^4 + 4t^2$.

$$S = \int_0^1 2\pi y \sqrt{\left(\frac{dx}{dt}\right)^2 + \left(\frac{dy}{dt}\right)^2}\, dt = \int_0^1 2\pi t^2 \sqrt{9t^4 + 4t^2}\, dt$$

$$= 2\pi \int_4^{13} \frac{u - 4}{9}\sqrt{u}\left(\tfrac{1}{18}du\right) \text{ (where } u = 9t^2 + 4) = \tfrac{\pi}{81}\left[\tfrac{2}{5}u^{5/2} - \tfrac{8}{3}u^{3/2}\right]_4^{13} = \tfrac{2\pi}{1215}\left(247\sqrt{13} + 64\right)$$

25. $x = a \cos^3\theta$, $y = a \sin^3\theta$, $0 \le \theta \le \frac{\pi}{2}$.

$$\left(\frac{dx}{d\theta}\right)^2 + \left(\frac{dy}{d\theta}\right)^2 = \left(-3a \cos^2\theta \sin\theta\right)^2 + \left(3a \sin^2\theta \cos\theta\right)^2 = 9a^2 \sin^2\theta \cos^2\theta.$$

$$S = \int_0^{\pi/2} 2\pi a \sin^3\theta\, 3a \sin\theta \cos\theta\, d\theta = 6\pi a^2 \int_0^{\pi/2} \sin^4\theta \cos\theta\, d\theta = \tfrac{6}{5}\pi a^2 \left[\sin^5\theta\right]_0^{\pi/2} = \tfrac{6}{5}\pi a^2$$

27. $x = t + t^3$, $y = t - \dfrac{1}{t^2}$, $1 \le t \le 2$. $\dfrac{dx}{dt} = 1 + 3t^2$ and $\dfrac{dy}{dt} = 1 + \dfrac{2}{t^3}$, so

$$\left(\frac{dx}{dt}\right)^2 + \left(\frac{dy}{dt}\right)^2 = (1 + 3t^2)^2 + \left(1 + \frac{2}{t^3}\right)^2 \text{ and}$$

$$S = \int 2\pi y\, ds = \int_1^2 2\pi \left(t - \frac{1}{t^2}\right)\sqrt{(1 + 3t^2)^2 + \left(1 + \frac{2}{t^3}\right)^2}\, dt \approx 59.101.$$

29. $\left(\dfrac{dx}{dt}\right)^2 + \left(\dfrac{dy}{dt}\right)^2 = (6t)^2 + \left(6t^2\right)^2 = 36t^2\left(1+t^2\right) \quad \Rightarrow$

$S = \int_0^5 2\pi x \sqrt{(dx/dt)^2 + (dy/dt)^2}\, dt = \int_0^5 2\pi\left(3t^2\right)6t\sqrt{1+t^2}\, dt = 18\pi \int_0^5 t^2\sqrt{1+t^2}\,2t\, dt$

$= 18\pi \int_1^{26} (u-1)\sqrt{u}\, du \ \ \text{(where } u = 1+t^2) \ = 18\pi \int_1^{26} \left(u^{3/2} - u^{1/2}\right) du = 18\pi \left[\tfrac{2}{5}u^{5/2} - \tfrac{2}{3}u^{3/2}\right]_1^{26}$

$= 18\pi \left[\left(\tfrac{2}{5}\cdot 676\sqrt{26} - \tfrac{2}{3}\cdot 26\sqrt{26}\right) - \left(\tfrac{2}{5} - \tfrac{2}{3}\right)\right] = \tfrac{24}{5}\pi\left(949\sqrt{26} + 1\right)$

31. $x = a\cos\theta,\, y = b\sin\theta,\, 0 \le \theta \le 2\pi$.

$(dx/d\theta)^2 + (dy/d\theta)^2 = (-a\sin\theta)^2 + (b\cos\theta)^2 = a^2\sin^2\theta + b^2\cos^2\theta = a^2\left(1 - \cos^2\theta\right) + b^2\cos^2\theta$

$$= a^2 - \left(a^2 - b^2\right)\cos^2\theta = a^2 - c^2\cos^2\theta = a^2\left(1 - \frac{c^2}{a^2}\cos^2\theta\right) = a^2\left(1 - e^2\cos^2\theta\right)$$

(a) $S = \int_0^\pi 2\pi b\sin\theta\, a\sqrt{1 - e^2\cos^2\theta}\, d\theta = 2\pi ab \int_{-e}^e \sqrt{1 - u^2}\left(\tfrac{1}{e}\right) du \ \text{(where } u = -e\cos\theta,\, du = e\sin\theta\, d\theta)$

$= \frac{4\pi ab}{e}\int_0^e \left(1 - u^2\right)^{1/2} du = \frac{4\pi ab}{e}\int_0^{\sin^{-1} e}\cos^2 v\, dv \ \text{(where } u = \sin v) = \frac{2\pi ab}{e}\int_0^{\sin^{-1} e}\left(1 + \cos 2v\right) dv$

$= \frac{2\pi ab}{e}\left[v + \tfrac{1}{2}\sin 2v\right]_0^{\sin^{-1} e} = \frac{2\pi ab}{e}\left[v + \sin v\cos v\right]_0^{\sin^{-1} e} = \frac{2\pi ab}{e}\left(\sin^{-1} e + e\sqrt{1 - e^2}\right)$

But $\sqrt{1 - e^2} = \sqrt{1 - \dfrac{c^2}{a^2}} = \sqrt{\dfrac{a^2 - c^2}{a^2}} = \sqrt{\dfrac{b^2}{a^2}} = \dfrac{b}{a}$, so $S = \dfrac{2\pi ab}{e}\sin^{-1} e + 2\pi b^2$.

(b) $S = \int_{-\pi/2}^{\pi/2} 2\pi a\cos\theta\, a\sqrt{1 - e^2\cos^2\theta}\, d\theta = 4\pi a^2 \int_0^{\pi/2}\cos\theta\sqrt{\left(1 - e^2\right) + e^2\sin^2\theta}\, d\theta$

$= \frac{4\pi a^2\left(1 - e^2\right)}{e}\int_0^{\pi/2}\frac{e}{\sqrt{1 - e^2}}\cos\theta\sqrt{1 + \left(\frac{e\sin\theta}{\sqrt{1 - e^2}}\right)^2}\, d\theta$

$= \frac{4\pi a^2\left(1 - e^2\right)}{e}\int_0^{e/\sqrt{1 - e^2}}\sqrt{1 + u^2}\, du \left(\text{where } u = \frac{e\sin\theta}{\sqrt{1 - e^2}}\right)$

$= \frac{4\pi a^2\left(1 - e^2\right)}{e}\int_0^{\sin^{-1} e}\sec^3 v\, dv \ \ \text{(where } u = \tan v,\, du = \sec^2 v\, dv)$

$= \frac{2\pi a^2\left(1 - e^2\right)}{e}\left[\sec v\tan v + \ln|\sec v + \tan v|\right]_0^{\sin^{-1} e}$

$= \frac{2\pi a^2\left(1 - e^2\right)}{e}\left[\frac{1}{\sqrt{1 - e^2}}\frac{e}{\sqrt{1 - e^2}} + \ln\left|\frac{1}{\sqrt{1 - e^2}} + \frac{e}{\sqrt{1 - e^2}}\right|\right]$

$= 2\pi a^2 + \frac{2\pi a^2\left(1 - e^2\right)}{e}\ln\sqrt{\frac{1 + e}{1 - e}} = 2\pi a^2 + \frac{2\pi b^2}{e}\frac{1}{2}\ln\left(\frac{1 + e}{1 - e}\right) \ \left(\text{since } 1 - e^2 = \frac{b^2}{a^2}\right)$

$= 2\pi\left[a^2 + \frac{b^2}{2e}\ln\frac{1 + e}{1 - e}\right]$

33. (a) $\phi = \tan^{-1}\left(\dfrac{dy}{dx}\right) \Rightarrow \dfrac{d\phi}{dt} = \dfrac{d}{dt}\tan^{-1}\left(\dfrac{dy}{dx}\right) = \dfrac{1}{1+(dy/dx)^2}\left[\dfrac{d}{dt}\left(\dfrac{dy}{dx}\right)\right]$. But $\dfrac{dy}{dx} = \dfrac{dy/dt}{dx/dt} = \dfrac{\dot{y}}{\dot{x}} \Rightarrow$

$\dfrac{d}{dt}\left(\dfrac{dy}{dx}\right) = \dfrac{d}{dt}\left(\dfrac{\dot{y}}{\dot{x}}\right) = \dfrac{\ddot{y}\dot{x}-\ddot{x}\dot{y}}{\dot{x}^2} \Rightarrow \dfrac{d\phi}{dt} = \dfrac{1}{1+(\dot{y}/\dot{x})^2}\left(\dfrac{\ddot{y}\dot{x}-\ddot{x}\dot{y}}{\dot{x}^2}\right) = \dfrac{\dot{x}\ddot{y}-\ddot{x}\dot{y}}{\dot{x}^2+\dot{y}^2}$.

Using the Chain Rule, and the fact that $s = \displaystyle\int_0^t \sqrt{\left(\dfrac{dx}{dt}\right)^2 + \left(\dfrac{dy}{dt}\right)^2}\, dt \Rightarrow$

$\dfrac{ds}{dt} = \sqrt{\left(\dfrac{dx}{dt}\right)^2 + \left(\dfrac{dy}{dt}\right)^2} = (\dot{x}^2+\dot{y}^2)^{1/2}$, we have that

$\dfrac{d\phi}{ds} = \dfrac{d\phi/dt}{ds/dt} = \left(\dfrac{\dot{x}\ddot{y}-\ddot{x}\dot{y}}{\dot{x}^2+\dot{y}^2}\right)\dfrac{1}{(\dot{x}^2+\dot{y}^2)^{1/2}} = \dfrac{\dot{x}\ddot{y}-\ddot{x}\dot{y}}{(\dot{x}^2+\dot{y}^2)^{3/2}}$. So

$\kappa = \left|\dfrac{d\phi}{ds}\right| = \left|\dfrac{\dot{x}\ddot{y}-\ddot{x}\dot{y}}{(\dot{x}^2+\dot{y}^2)^{3/2}}\right| = \dfrac{|\dot{x}\ddot{y}-\ddot{x}\dot{y}|}{(\dot{x}^2+\dot{y}^2)^{3/2}}$.

(b) $x = x$ and $y = f(x) \Rightarrow \dot{x} = 1, \ddot{x} = 0$ and $\dot{y} = \dfrac{dy}{dx}, \ddot{y} = \dfrac{d^2y}{dx^2}$.

So $\kappa = \dfrac{\left|1\cdot(d^2y/dx^2) - 0\cdot(dy/dx)\right|}{\left[1+(dy/dx)^2\right]^{3/2}} = \dfrac{|d^2y/dx^2|}{\left[1+(dy/dx)^2\right]^{3/2}}$.

35. $x = \theta - \sin\theta \Rightarrow \dot{x} = 1 - \cos\theta \Rightarrow \ddot{x} = \sin\theta$, and $y = 1 - \cos\theta \Rightarrow \dot{y} = \sin\theta \Rightarrow \ddot{y} = \cos\theta$.

Therefore, $\kappa = \dfrac{|\cos\theta - \cos^2\theta - \sin^2\theta|}{\left[(1-\cos\theta)^2 + \sin^2\theta\right]^{3/2}} = \dfrac{|\cos\theta - (\cos^2\theta + \sin^2\theta)|}{(1 - 2\cos\theta + \cos^2\theta + \sin^2\theta)^{3/2}} = \dfrac{|\cos\theta - 1|}{(2 - 2\cos\theta)^{3/2}}$. The top of

the arch is characterized by a horizontal tangent, and from Example 1 in Section 10.2, the tangent is horizontal

when $\theta = (2n-1)\pi$, so take $n = 1$ and substitute $\theta = \pi$ into the expression for κ:

$\kappa = \dfrac{|\cos\pi - 1|}{(2 - 2\cos\pi)^{3/2}} = \dfrac{|-1-1|}{[2 - 2(-1)]^{3/2}} = \dfrac{1}{4}$.

10.4 Polar Coordinates

1. (a) By adding 2π to $\dfrac{\pi}{2}$, we obtain the   (b) $\left(-2, \dfrac{\pi}{4}\right)$ (c) $(3, 2)$

point $\left(1, \dfrac{5\pi}{2}\right)$. The direction

opposite $\dfrac{\pi}{2}$ is $\dfrac{3\pi}{2}$, so $\left(-1, \dfrac{3\pi}{2}\right)$ is a

point that satisfies the $r < 0$

requirement.

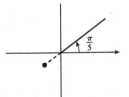

$\left(2, \dfrac{5\pi}{4}\right), \left(-2, \dfrac{9\pi}{4}\right)$

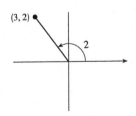

$(3, 2+2\pi), (-3, 2+\pi)$

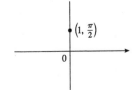

3. (a)

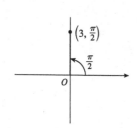

$x = 3\cos\frac{\pi}{2} = 3\,(0) = 0$ and

$y = 3\sin\frac{\pi}{2} = 3\,(1) = 3$ give us

$(0, 3)$.

(b)

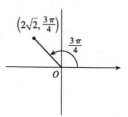

$x = 2\sqrt{2}\cos\frac{3\pi}{4}$

$= 2\sqrt{2}\left(-\frac{1}{\sqrt{2}}\right) = -2$ and

$y = 2\sqrt{2}\sin\frac{3\pi}{4} = 2\sqrt{2}\left(\frac{1}{\sqrt{2}}\right) = 2$

give us $(-2, 2)$.

(c)

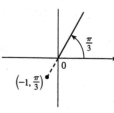

$x = -1\cos\frac{\pi}{3} = -\frac{1}{2}$ and
$y = -1\sin\frac{\pi}{3} = -\frac{\sqrt{3}}{2}$ give
us $\left(-\frac{1}{2}, -\frac{\sqrt{3}}{2}\right)$.

5. (a) $x = 1$ and $y = 1 \;\Rightarrow\; r = \sqrt{1^2 + 1^2} = \sqrt{2}$ and $\theta = \tan^{-1}\left(\frac{1}{1}\right) = \frac{\pi}{4}$. Since $(1, 1)$ is in the first quadrant, the

polar coordinates are (i) $\left(\sqrt{2}, \frac{\pi}{4}\right)$ and (ii) $\left(-\sqrt{2}, \frac{5\pi}{4}\right)$.

(b) $x = 2\sqrt{3}$ and $y = -2 \;\Rightarrow\; r = \sqrt{\left(2\sqrt{3}\right)^2 + (-2)^2} = 4$ and $\theta = \tan^{-1}\left(-\frac{2}{2\sqrt{3}}\right) = -\frac{\pi}{6}$. Since $\left(2\sqrt{3}, -2\right)$

is in the fourth quadrant and $0 \le \theta \le 2\pi$, the polar coordinates are (i) $\left(4, \frac{11\pi}{6}\right)$ and (ii) $\left(-4, \frac{5\pi}{6}\right)$.

7. $r > 1$

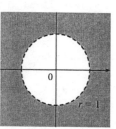

9. $0 \le r \le 2,\ \frac{\pi}{2} \le \theta \le \pi$

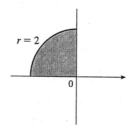

11. $2 < r < 3,\ \frac{5\pi}{3} \le \theta \le \frac{7\pi}{3}$

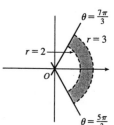

13. $\left(1, \frac{\pi}{6}\right)$ is $\left(\frac{\sqrt{3}}{2}, \frac{1}{2}\right)$ Cartesian and $\left(3, \frac{3\pi}{4}\right)$ is $\left(-\frac{3}{\sqrt{2}}, \frac{3}{\sqrt{2}}\right)$ Cartesian. The square of the distance between them is

$\left(\frac{\sqrt{3}}{2} + \frac{\sqrt{3}}{2}\right)^2 + \left(\frac{1}{2} - \frac{3}{\sqrt{2}}\right)^2 = \frac{1}{4}\left(40 + 6\sqrt{6} - 6\sqrt{2}\right)$, so the distance is $\frac{1}{2}\sqrt{40 + 6\sqrt{6} - 6\sqrt{2}}$.

15. $r = 2 \;\Leftrightarrow\; \sqrt{x^2 + y^2} = 2 \;\Leftrightarrow\; x^2 + y^2 = 4$, a circle of radius 2 centered at the origin.

17. $r = 3\sin\theta \;\Rightarrow\; r^2 = 3r\sin\theta \;\Leftrightarrow\; x^2 + y^2 = 3y \;\Leftrightarrow\; x^2 + \left(y - \frac{3}{2}\right)^2 = \left(\frac{3}{2}\right)^2$, a circle of radius $\frac{3}{2}$ centered

at $\left(0, \frac{3}{2}\right)$. The first two equations are actually equivalent since $r^2 = 3r\sin\theta \;\Rightarrow\; r\,(r - 3\sin\theta) = 0 \;\Rightarrow$

$r = 0$ or $r = 3\sin\theta$. But $r = 3\sin\theta$ gives the point $r = 0$ (the pole) when $\theta = 0$. Thus, the single equation

$r = 3\sin\theta$ is equivalent to the compound condition $(r = 0$ or $r = 3\sin\theta)$.

19. $r^2 = \sin 2\theta = 2\sin\theta\cos\theta \;\Leftrightarrow\; r^4 = 2r\sin\theta r\cos\theta \;\Leftrightarrow\; \left(x^2 + y^2\right)^2 = 2yx$

21. $y = 5 \;\Leftrightarrow\; r\sin\theta = 5$

23. $x^2 + y^2 = 25 \;\Leftrightarrow\; r^2 = 25 \;\Rightarrow\; r = 5$

25. $2xy = 1 \iff 2r\cos\theta\, r\sin\theta = 1 \iff r^2 \sin 2\theta = 1 \iff r^2 = \csc 2\theta$

27. (a) The description leads immediately to the polar equation $\theta = \frac{\pi}{6}$, and the Cartesian equation
$y = \tan\left(\frac{\pi}{6}\right) x = \frac{1}{\sqrt{3}} x$ is slightly more difficult to derive.

(b) The easier description here is the Cartesian equation $x = 3$.

29. $r = -2\sin\theta \iff r^2 = -2r\sin\theta$ (since the
possibility $r = 0$ is covered by the equation
$r = -2\sin\theta$) $\iff x^2 + y^2 = -2y \iff$
$x^2 + y^2 + 2y + 1 = 1 \iff x^2 + (y+1)^2 = 1.$

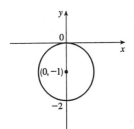

31. $r = \csc\theta = \dfrac{1}{\sin\theta} \iff r\sin\theta = 1.$ (The
right-hand equation implies that $\sin\theta \neq 0$, so we
can divide by $\sin\theta$ to get the left-hand equation)
$\iff y = 1.$

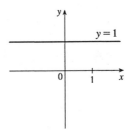

33. As in Example 4, $r = 5$ represents the circle with
center O and radius 5.

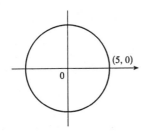

35. $r = \sin\theta \iff r^2 = r\sin\theta \iff x^2 + y^2 = y$
$\iff x^2 + \left(y - \frac{1}{2}\right)^2 = \left(\frac{1}{2}\right)^2.$ The reasoning here
is the same as in Exercise 29. This is a circle of
radius $\frac{1}{2}$ centered at $\left(0, \frac{1}{2}\right).$

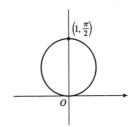

37. $r = 2(1 - \sin\theta)$. This curve is a cardioid.

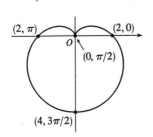

39. $r = \theta, \theta \geq 0$

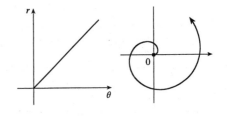

41. $r = 1/\theta$

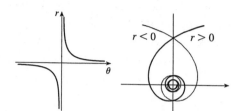

43. $r = \sin 2\theta$

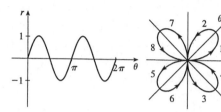

45. $r = 2\cos 4\theta$

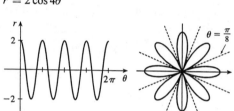
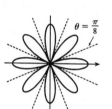

47. $r^2 = 4\cos 2\theta$

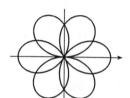
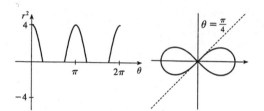

49. $r = 2\cos\left(\frac{3}{2}\theta\right)$

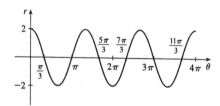
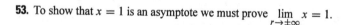

51. $x = (r)\cos\theta = (4 + 2\sec\theta)\cos\theta = 4\cos\theta + 2$. Now, $r \to \infty$ $\Rightarrow$

$(4 + 2\sec\theta) \to \infty$ $\Rightarrow$ $\theta \to \left(\frac{\pi}{2}\right)^-$ or $\theta \to \left(\frac{3\pi}{2}\right)^+$ (since we need only

consider $0 \le \theta < 2\pi$), so $\lim\limits_{r\to\infty} x = \lim\limits_{\theta\to\pi/2^-} (4\cos\theta + 2) = 2$. Also, $r \to -\infty$

$\Rightarrow$ $(4 + 2\sec\theta) \to -\infty$ $\Rightarrow$ $\theta \to \left(\frac{\pi}{2}\right)^+$ or $\theta \to \left(\frac{3\pi}{2}\right)^-$, so

$\lim\limits_{r\to-\infty} x = \lim\limits_{\theta\to\pi/2^+} (4\cos\theta + 2) = 2$. Therefore, $\lim\limits_{r\to\pm\infty} x = 2$ $\Rightarrow$ $x = 2$ is

a vertical asymptote.

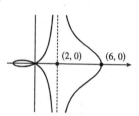

53. To show that $x = 1$ is an asymptote we must prove $\lim\limits_{r\to\pm\infty} x = 1$.

$x = (r)\cos\theta = (\sin\theta\tan\theta)\cos\theta = \sin^2\theta$. Now, $r \to \infty$ $\Rightarrow$

$\sin\theta\tan\theta \to \infty$ $\Rightarrow$ $\theta \to \left(\frac{\pi}{2}\right)^-$, so $\lim\limits_{r\to\infty} x = \lim\limits_{\theta\to\pi/2^-} \sin^2\theta = 1$. Also,

$r \to -\infty$ $\Rightarrow$ $\sin\theta\tan\theta \to -\infty$ $\Rightarrow$ $\theta \to \left(\frac{\pi}{2}\right)^+$, so

$\lim\limits_{r\to-\infty} x = \lim\limits_{\theta\to\pi/2^+} \sin^2\theta = 1$.

Therefore, $\lim\limits_{r\to\pm\infty} x = 1$ $\Rightarrow$ $x = 1$ is a vertical asymptote. Also notice that $x = \sin^2\theta \ge 0$ for all θ, and

$x = \sin^2\theta \le 1$ for all θ. And $x \ne 1$, since the curve is not defined at odd multiples of $\frac{\pi}{2}$. Therefore, the curve lies

entirely within the vertical strip $0 \le x < 1$.

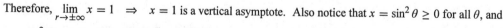

55. (a) We see that the curve crosses itself at the origin, where $r = 0$ (in fact the inner loop corresponds to negative r-values,) so we solve the equation of the limaçon for $r = 0 \Leftrightarrow c \sin\theta = -1 \Leftrightarrow \sin\theta = -1/c$. Now if $|c| < 1$, then this equation has no solution and hence there is no inner loop. But if $c < -1$, then on the interval $(0, 2\pi)$ the equation has the two solutions $\theta = \sin^{-1}(-1/c)$ and $\theta = \pi - \sin^{-1}(-1/c)$, and if $c > 1$, the solutions are $\theta = \pi + \sin^{-1}(1/c)$ and $\theta = 2\pi - \sin^{-1}(1/c)$. In each case, $r < 0$ for θ between the two solutions, indicating a loop.

(b) For $0 < c < 1$, the dimple (if it exists) is characterized by the fact that y has a local maximum at $\theta = \frac{3\pi}{2}$. So we determine for what c-values $\dfrac{d^2y}{d\theta^2}$ is negative at $\theta = \frac{3\pi}{2}$, since by the Second Derivative Test this indicates a

maximum: $y = r\sin\theta = \sin\theta + c\sin^2\theta \;\Rightarrow\; \dfrac{dy}{d\theta} = \cos\theta + 2c\sin\theta\cos\theta = \cos\theta + c\sin 2\theta \;\Rightarrow$

$\dfrac{d^2y}{d\theta^2} = -\sin\theta + 2c\cos 2\theta$. At $\theta = \frac{3\pi}{2}$, this is equal to $-(-1) + 2c(-1) = 1 - 2c$, which is negative only for $c > \frac{1}{2}$. A similar argument shows that for $-1 < c < 0$, y only has a local minimum at $\theta = \frac{\pi}{2}$ (indicating a dimple) for $c < -\frac{1}{2}$.

57. Using Equation 3 with $r = 3\cos\theta$, we have

$$\dfrac{dy}{dx} = \dfrac{dy/d\theta}{dx/d\theta} = \dfrac{(dr/d\theta)(\sin\theta) + r\cos\theta}{(dr/d\theta)(\cos\theta) - r\sin\theta} = \dfrac{-3\sin\theta\sin\theta + 3\cos\theta\cos\theta}{-3\sin\theta\cos\theta - 3\cos\theta\sin\theta} = \dfrac{3(\cos^2\theta - \sin^2\theta)}{-3(2\sin\theta\cos\theta)}$$

$$= -\dfrac{\cos 2\theta}{\sin 2\theta} = -\cot 2\theta = \dfrac{1}{\sqrt{3}} \text{ when } \theta = \dfrac{\pi}{3}$$

Another Solution: $r = 3\cos\theta \;\Rightarrow\; x = r\cos\theta = 3\cos^2\theta, \; y = r\sin\theta = 3\sin\theta\cos\theta \;\Rightarrow$

$$\dfrac{dy}{dx} = \dfrac{dy/d\theta}{dx/d\theta} = \dfrac{-3\sin^2\theta + 3\cos^2\theta}{-6\cos\theta\sin\theta} = \dfrac{\cos 2\theta}{-\sin 2\theta} = -\cot 2\theta = \dfrac{1}{\sqrt{3}} \text{ when } \theta = \dfrac{\pi}{3}$$

59. $r = 1/\theta \;\Rightarrow\; x = r\cos\theta = (\cos\theta)/\theta, \; y = r\sin\theta = (\sin\theta)/\theta \;\Rightarrow$

$$\dfrac{dy}{dx} = \dfrac{dy/d\theta}{dx/d\theta} = \dfrac{\sin\theta(-1/\theta^2) + (1/\theta)\cos\theta}{\cos\theta(-1/\theta^2) - (1/\theta)\sin\theta} \cdot \dfrac{\theta^2}{\theta^2} = \dfrac{-\sin\theta + \theta\cos\theta}{-\cos\theta - \theta\sin\theta} = -\pi \text{ when } \theta = \pi$$

61. $r = 1 + \cos\theta \;\Rightarrow\; x = r\cos\theta = \cos\theta + \cos^2\theta, \; y = r\sin\theta = \sin\theta + \sin\theta\cos\theta \;\Rightarrow$

$$\dfrac{dy}{dx} = \dfrac{dy/d\theta}{dx/d\theta} = \dfrac{\cos\theta + \cos^2\theta - \sin^2\theta}{-\sin\theta - 2\cos\theta\sin\theta} = \dfrac{\cos\theta + \cos 2\theta}{-\sin\theta - \sin 2\theta} = -1 \text{ when } \theta = \dfrac{\pi}{6}$$

63. $r = 3\cos\theta \;\Rightarrow\; x = r\cos\theta = 3\cos\theta\cos\theta, \; y = r\sin\theta = 3\cos\theta\sin\theta \;\Rightarrow$
$dy/d\theta = -3\sin^2\theta + 3\cos^2\theta = 3\cos 2\theta = 0 \;\Rightarrow\; 2\theta = \frac{\pi}{2}$ or $\frac{3\pi}{2} \Leftrightarrow \theta = \frac{\pi}{4}$ or $\frac{3\pi}{4}$. So the tangent is horizontal at $\left(\frac{3}{\sqrt{2}}, \frac{\pi}{4}\right)$ and $\left(-\frac{3}{\sqrt{2}}, \frac{3\pi}{4}\right)$ $\left[\text{same as } \left(\frac{3}{\sqrt{2}}, -\frac{\pi}{4}\right)\right]$. $dx/d\theta = -6\sin\theta\cos\theta = -3\sin 2\theta = 0 \;\Rightarrow$
$2\theta = 0$ or $\pi \Leftrightarrow \theta = 0$ or $\frac{\pi}{2}$. So the tangent is vertical at $(3, 0)$ and $\left(0, \frac{\pi}{2}\right)$.

65. $r = 1 + \cos\theta \;\Rightarrow\; x = r\cos\theta = \cos\theta\,(1 + \cos\theta),\; y = r\sin\theta = \sin\theta\,(1 + \cos\theta) \;\Rightarrow$
$dy/d\theta = (1 + \cos\theta)\cos\theta - \sin^2\theta = 2\cos^2\theta + \cos\theta - 1 = (2\cos\theta - 1)(\cos\theta + 1) = 0 \;\Rightarrow\; \cos\theta = \frac{1}{2}$ or -1
$\Rightarrow\; \theta = \frac{\pi}{3}, \pi, \text{ or } \frac{5\pi}{3} \;\Rightarrow\;$ horizontal tangent at $\left(\frac{3}{2}, \frac{\pi}{3}\right),\, (0, \pi),\,$ and $\left(\frac{3}{2}, \frac{5\pi}{3}\right)$.
$dx/d\theta = -(1 + \cos\theta)\sin\theta - \cos\theta\sin\theta = -\sin\theta\,(1 + 2\cos\theta) = 0 \;\Rightarrow\; \sin\theta = 0 \text{ or } \cos\theta = -\frac{1}{2} \;\Rightarrow$
$\theta = 0, \pi, \frac{2\pi}{3}, \text{ or } \frac{4\pi}{3} \;\Rightarrow\;$ vertical tangent at $(2, 0),\, \left(\frac{1}{2}, \frac{2\pi}{3}\right),\,$ and $\left(\frac{1}{2}, \frac{4\pi}{3}\right)$. Note that the tangent is horizontal,
not vertical when $\theta = \pi$, since $\displaystyle\lim_{\theta\to\pi} \frac{dy/d\theta}{dx/d\theta} = 0$.

67. $r = \cos 2\theta \;\Rightarrow\; x = r\cos\theta = \cos 2\theta\cos\theta,\; y = r\sin\theta = \cos 2\theta\sin\theta \;\Rightarrow$

$$dy/d\theta = -2\sin 2\theta\sin\theta + \cos 2\theta\cos\theta = -4\sin^2\theta\cos\theta + \left(\cos^3\theta - \sin^2\theta\cos\theta\right)$$

$$= \cos\theta\left(\cos^2\theta - 5\sin^2\theta\right) = \cos\theta\left(1 - 6\sin^2\theta\right) = 0 \;\Rightarrow$$

$\cos\theta = 0 \text{ or } \sin\theta = \pm\frac{1}{\sqrt{6}} \;\Rightarrow\; \theta = \frac{\pi}{2}, \frac{3\pi}{2}, \alpha, \pi - \alpha, \pi + \alpha, \text{ or } 2\pi - \alpha \quad (\text{where } \alpha = \sin^{-1}\frac{1}{\sqrt{6}}).$
So the tangent is horizontal at $\left(-1, \frac{\pi}{2}\right), \left(-1, \frac{3\pi}{2}\right), \left(\frac{2}{3}, \alpha\right), \left(\frac{2}{3}, \pi - \alpha\right), \left(\frac{2}{3}, \pi + \alpha\right),$ and $\left(\frac{2}{3}, 2\pi - \alpha\right)$.

$$dx/d\theta = -2\sin 2\theta\cos\theta - \cos 2\theta\sin\theta = -4\sin\theta\cos^2\theta - \left(2\cos^2\theta - 1\right)\sin\theta$$

$$= \sin\theta\left(1 - 6\cos^2\theta\right) = 0 \;\Rightarrow$$

$\sin\theta = 0 \text{ or } \cos\theta = \pm\frac{1}{\sqrt{6}} \;\Rightarrow\; \theta = 0, \pi, \frac{\pi}{2} - \alpha, \frac{\pi}{2} + \alpha, \frac{3\pi}{2} - \alpha, \text{ or } \frac{3\pi}{2} + \alpha \;(\text{where } \alpha = \cos^{-1}\frac{1}{\sqrt{6}}).$
So the tangent is vertical at $(1, 0), (1, \pi), \left(\frac{2}{3}, \frac{3\pi}{2} - \alpha\right), \left(\frac{2}{3}, \frac{3\pi}{2} + \alpha\right), \left(\frac{2}{3}, \frac{\pi}{2} - \alpha\right),$ and $\left(\frac{2}{3}, \frac{\pi}{2} + \alpha\right)$.

69. $r = a\sin\theta + b\cos\theta \;\Rightarrow\; r^2 = ar\sin\theta + br\cos\theta \;\Rightarrow\; x^2 + y^2 = ay + bx \;\Rightarrow$
$\left(x - \frac{1}{2}b\right)^2 + \left(y - \frac{1}{2}a\right)^2 = \frac{1}{4}(a^2 + b^2),$ and this is a circle with center $\left(\frac{1}{2}b, \frac{1}{2}a\right)$ and radius $\frac{1}{2}\sqrt{a^2 + b^2}$.

Note for Exercises 71–76: Maple is able to plot polar curves using the `polarplot` command, or using the `coords=polar` option in a regular `plot` command. In Mathematica, use `PolarPlot`. In Derive, change to `Polar` under `Options State`. If your graphing device cannot plot polar equations, you must convert to parametric equations. For example, in Exercise 71, $x = r\cos\theta = [1 + 2\sin(\theta/2)]\cos\theta,\, y = r\sin\theta = [1 + 2\sin(\theta/2)]\sin\theta$.

71. $r = 1 + 2\sin(\theta/2)$. The parameter interval is $[0, 4\pi]$.

73. $r = e^{\sin\theta} - 2\cos(4\theta)$. The parameter interval is $[0, 2\pi]$.

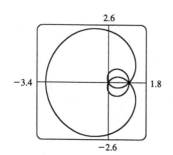

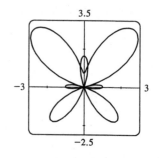

75. $r = \sin(9\theta/4)$. The parameter interval is $[0, 8\pi]$.

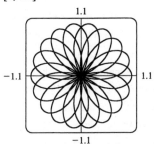

77.

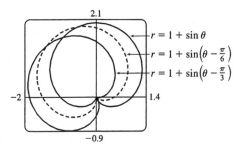

It appears that the graph of $r = 1 + \sin\left(\theta - \frac{\pi}{6}\right)$ is the same shape as the graph of $r = 1 + \sin\theta$, but rotated counterclockwise about the origin by $\frac{\pi}{6}$. Similarly, the graph of $r = 1 + \sin\left(\theta - \frac{\pi}{3}\right)$ is rotated by $\frac{\pi}{3}$. In general, the graph of $r = f(\theta - \alpha)$ is the same shape as that of $r = f(\theta)$, but rotated counterclockwise through α about the origin. That is, for any point (r_0, θ_0) on the curve $r = f(\theta)$, the point $(r_0, \theta_0 + \alpha)$ is on the curve $r = f(\theta - \alpha)$, since $r_0 = f(\theta_0) = f((\theta_0 + \alpha) - \alpha)$.

79. (a) $r = \sin n\theta$. From the graphs, it seems that when n is even, the number of loops in the curve (called a rose) is $2n$, and when n is odd, the number of loops is simply n.

This is because in the case of n odd, every point on the graph is traversed twice, due to the fact that

$$r(\theta + \pi) = \sin[n(\theta + \pi)] = \sin n\theta \cos n\pi + \cos n\theta \sin n\pi = \begin{cases} \sin n\theta & \text{if } n \text{ is even} \\ -\sin n\theta & \text{if } n \text{ is odd} \end{cases}$$

$n = 2$

$n = 3$

$n = 4$

$n = 5$

(b) The graph of $r = |\sin n\theta|$ has $2n$ loops whether n is odd or even, since $r(\theta + \pi) = r(\theta)$.

$n = 2$

$n = 3$

$n = 4$

$n = 5$

81. $r = \dfrac{1 - a\cos\theta}{1 + a\cos\theta}$. We start with $a = 0$, since in this case the curve is simply the circle $r = 1$.

As a increases, the graph moves to the left, and its right side becomes flattened. As a increases through about 0.4, the right side seems to grow a dimple, which upon closer investigation (with narrower θ-ranges) seems to appear at $a \approx 0.42$ (the actual value is $\sqrt{2} - 1$). As $a \to 1$, this dimple becomes more pronounced, and the curve begins to stretch out horizontally, until at $a = 1$ the denominator vanishes at $\theta = \pi$, and the dimple becomes an actual cusp. For $a > 1$ we must choose our parameter interval carefully, since $r \to \infty$ as $1 + a\cos\theta \to 0$ ⟺ $\theta \to \pm\cos^{-1}(-1/a)$. As a increases from 1, the curve splits into two parts. The left part has a loop, which grows larger as a increases, and the right part grows broader vertically, and its left tip develops a dimple when $a \approx 2.42$ (actually, $\sqrt{2} + 1$). As a increases, the dimple grows more and more pronounced.

If $a < 0$, we get the same graph as we do for the corresponding positive a-value, but with a rotation through π about the pole, as happened when c was replaced with $-c$ in Exercise 80.

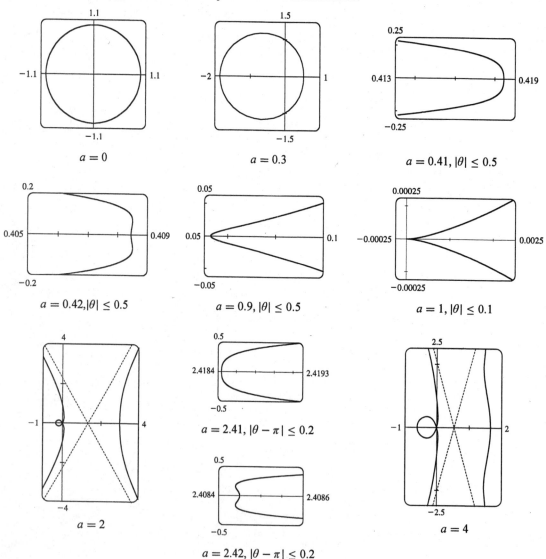

$a = 0$

$a = 0.3$

$a = 0.41, |\theta| \le 0.5$

$a = 0.42, |\theta| \le 0.5$

$a = 0.9, |\theta| \le 0.5$

$a = 1, |\theta| \le 0.1$

$a = 2$

$a = 2.41, |\theta - \pi| \le 0.2$

$a = 2.42, |\theta - \pi| \le 0.2$

$a = 4$

83. $\tan \psi = \tan(\phi - \theta) = \dfrac{\tan \phi - \tan \theta}{1 + \tan \phi \tan \theta} = \dfrac{\dfrac{dy}{dx} - \tan \theta}{1 + \dfrac{dy}{dx}\tan \theta} = \dfrac{\dfrac{dy/d\theta}{dx/d\theta} - \tan \theta}{1 + \dfrac{dy/d\theta}{dx/d\theta}\tan \theta}$

$= \dfrac{\dfrac{dy}{d\theta} - \dfrac{dx}{d\theta}\tan \theta}{\dfrac{dx}{d\theta} + \dfrac{dy}{d\theta}\tan \theta} = \dfrac{\left(\dfrac{dr}{d\theta}\sin\theta + r\cos\theta\right) - \tan\theta\left(\dfrac{dr}{d\theta}\cos\theta - r\sin\theta\right)}{\left(\dfrac{dr}{d\theta}\cos\theta - r\sin\theta\right) + \tan\theta\left(\dfrac{dr}{d\theta}\sin\theta + r\cos\theta\right)}$

$= \dfrac{r\cos\theta + r\cdot\dfrac{\sin^2\theta}{\cos\theta}}{\dfrac{dr}{d\theta}\cos\theta + \dfrac{dr}{d\theta}\cdot\dfrac{\sin^2\theta}{\cos\theta}} = \dfrac{r\cos^2\theta + r\sin^2\theta}{\dfrac{dr}{d\theta}\cos^2\theta + \dfrac{dr}{d\theta}\sin^2\theta} = \dfrac{r}{dr/d\theta}$

10.5 Areas and Lengths in Polar Coordinates

1. $r = \sqrt{\theta}, 0 \le \theta \le \frac{\pi}{4}$. $A = \int_0^{\pi/4} \frac{1}{2}r^2\,d\theta = \int_0^{\pi/4}\frac{1}{2}\left(\sqrt{\theta}\right)^2 d\theta = \int_0^{\pi/4}\frac{1}{2}\theta\,d\theta = \left[\frac{1}{4}\theta^2\right]_0^{\pi/4} = \frac{1}{64}\pi^2$

3. $r = \sin\theta, \frac{\pi}{3} \le \theta \le \frac{2\pi}{3}$.

$A = \int_{\pi/3}^{2\pi/3}\frac{1}{2}\sin^2\theta\,d\theta = \frac{1}{4}\int_{\pi/3}^{2\pi/3}(1 - \cos 2\theta)\,d\theta = \frac{1}{4}\left[\theta - \frac{1}{2}\sin 2\theta\right]_{\pi/3}^{2\pi/3}$

$= \frac{1}{4}\left[\frac{2\pi}{3} - \frac{1}{2}\sin\frac{4\pi}{3} - \frac{\pi}{3} + \frac{1}{2}\sin\frac{2\pi}{3}\right] = \frac{1}{4}\left[\frac{2\pi}{3} - \frac{1}{2}\left(-\frac{\sqrt{3}}{2}\right) - \frac{\pi}{3} + \frac{1}{2}\left(\frac{\sqrt{3}}{2}\right)\right] = \frac{1}{4}\left(\frac{\pi}{3} + \frac{\sqrt{3}}{2}\right) = \frac{\pi}{12} + \frac{\sqrt{3}}{8}$

5. $r = \theta, 0 \le \theta \le \pi$. $A = \int_0^{\pi}\frac{1}{2}\theta^2\,d\theta = \left[\frac{1}{6}\theta^3\right]_0^{\pi} = \frac{1}{6}\pi^3$

7. $r = 4 + 3\sin\theta, -\frac{\pi}{2} \le \theta \le \frac{\pi}{2}$.

$A = \int_{-\pi/2}^{\pi/2}\frac{1}{2}(4 + 3\sin\theta)^2\,d\theta = \frac{1}{2}\int_{-\pi/2}^{\pi/2}(16 + 24\sin\theta + 9\sin^2\theta)\,d\theta$

$= \frac{1}{2}\int_{-\pi/2}^{\pi/2}(16 + 9\sin^2\theta)\,d\theta$ [by Theorem 5.5.7(b)]

$= \frac{1}{2}\cdot 2\int_0^{\pi/2}\left[16 + 9\cdot\frac{1}{2}(1 - \cos 2\theta)\right]d\theta$ [by Theorem 5.5.7(a)]

$= \int_0^{\pi/2}\left(\frac{41}{2} - \frac{9}{2}\cos 2\theta\right)d\theta = \left[\frac{41}{2}\theta - \frac{9}{4}\sin 2\theta\right]_0^{\pi/2} = \left(\frac{41\pi}{4} - 0\right) - (0 - 0) = \frac{41\pi}{4}$

9. $A = \int_0^{\pi}\frac{1}{2}(5\sin\theta)^2\,d\theta$

$= \frac{25}{4}\int_0^{\pi}(1 - \cos 2\theta)\,d\theta$

$= \frac{25}{4}\left[\theta - \frac{1}{2}\sin 2\theta\right]_0^{\pi} = \frac{25}{4}\pi$

11. $A = 4\int_0^{\pi/4}\frac{1}{2}r^2\,d\theta = 2\int_0^{\pi/4}(4\cos 2\theta)\,d\theta$

$= 8\int_0^{\pi/4}\cos 2\theta\,d\theta = 4[\sin 2\theta]_0^{\pi/4} = 4$

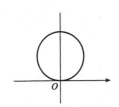

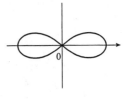

13. $A = 2 \int_{-\pi/2}^{\pi/2} \frac{1}{2} (4 - \sin\theta)^2 \, d\theta = \int_{-\pi/2}^{\pi/2} (16 - 8\sin\theta + \sin^2\theta) \, d\theta$

$= \int_{-\pi/2}^{\pi/2} (16 + \sin^2\theta) \, d\theta$ [by Theorem 5.5.7(b)]

$= 2 \int_{0}^{\pi/2} (16 + \sin^2\theta) \, d\theta$ [by Theorem 5.5.7(a)]

$= 2 \int_{0}^{\pi/2} \left[16 + \frac{1}{2} (1 - \cos 2\theta) \right] d\theta = 2 \left[\frac{33}{2}\theta - \frac{1}{4} \sin 2\theta \right]_{0}^{\pi/2}$

$= \frac{33\pi}{2}$

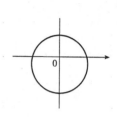

15. By symmetry, the total area is twice the area enclosed above the polar axis, so

$A = 2 \int_{0}^{\pi} \frac{1}{2} r^2 d\theta = \int_{0}^{\pi} [2 + \cos 6\theta]^2 \, d\theta = \int_{0}^{\pi} (4 + 4\cos 6\theta + \cos^2 6\theta) \, d\theta$

$= \left[4\theta + 4 \left(\frac{1}{6} \sin 6\theta \right) + \left(\frac{1}{24} \sin 12\theta + \frac{1}{2}\theta \right) \right]_{0}^{\pi} = 4\pi + \frac{\pi}{2} = \frac{9\pi}{2}$

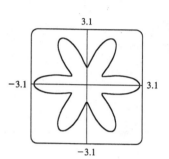

3.1

−3.1

3.1

−3.1

17. $A = \int_{0}^{\pi/2} \frac{1}{2} \sin^2 2\theta \, d\theta = \int_{0}^{\pi/2} \frac{1}{4} (1 - \cos 4\theta) \, d\theta = \frac{1}{4} \left[\theta - \frac{1}{4} \sin 4\theta \right]_{0}^{\pi/2} = \frac{\pi}{8}$

19. $r = 0 \;\Rightarrow\; 3\cos 5\theta = 0 \;\Rightarrow\; 5\theta = \frac{\pi}{2} \;\Rightarrow\; \theta = \frac{\pi}{10}.$

$A = \int_{-\pi/10}^{\pi/10} \frac{1}{2} (3\cos 5\theta)^2 \, d\theta = \int_{0}^{\pi/10} 9\cos^2 5\theta \, d\theta = \frac{9}{2} \int_{0}^{\pi/10} (1 + \cos 10\theta) \, d\theta = \frac{9}{2} \left[\theta + \frac{1}{10} \sin 10\theta \right]_{0}^{\pi/10} = \frac{9\pi}{20}$

21.

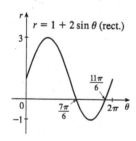

$r = 1 + 2\sin\theta$ (rect.)

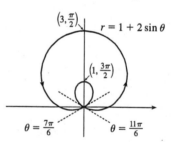

$\left(3, \frac{\pi}{2} \right)$ $r = 1 + 2\sin\theta$

$\left(1, \frac{3\pi}{2} \right)$

$\theta = \frac{7\pi}{6}$ $\theta = \frac{11\pi}{6}$

This is a limaçon, with inner loop traced out between $\theta = \frac{7\pi}{6}$ and $\frac{11\pi}{6}$ [found by solving $r = 0$].

$A = 2 \int_{7\pi/6}^{3\pi/2} \frac{1}{2} (1 + 2\sin\theta)^2 \, d\theta = \int_{7\pi/6}^{3\pi/2} (1 + 4\sin\theta + 4\sin^2\theta) \, d\theta = [\theta - 4\cos\theta + 2\theta - \sin 2\theta]_{7\pi/6}^{3\pi/2}$

$= \left(\frac{9\pi}{2} \right) - \left(\frac{7\pi}{2} + 2\sqrt{3} - \frac{\sqrt{3}}{2} \right) = \pi - \frac{3\sqrt{3}}{2}$

23. $1 - \cos\theta = \frac{3}{2} \;\Rightarrow\; \cos\theta = -\frac{1}{2} \;\Rightarrow\; \theta = \frac{2\pi}{3}$ or $\frac{4\pi}{3} \;\Rightarrow\;$

$A = 2 \int_{2\pi/3}^{\pi} \frac{1}{2} \left[(1 - \cos\theta)^2 - \left(\frac{3}{2} \right)^2 \right] d\theta = \int_{2\pi/3}^{\pi} \left(-\frac{5}{4} - 2\cos\theta + \cos^2\theta \right) d\theta$

$= \left[-\frac{5}{4}\theta - 2\sin\theta \right]_{2\pi/3}^{\pi} + \frac{1}{2} \int_{2\pi/3}^{\pi} (1 + \cos 2\theta) \, d\theta$

$= -\frac{5}{12}\pi + \sqrt{3} + \frac{1}{2} \left[\theta + \frac{1}{2} \sin 2\theta \right]_{2\pi/3}^{\pi}$

$= -\frac{5}{12}\pi + \sqrt{3} + \frac{1}{6}\pi + \frac{\sqrt{3}}{8} = \frac{9\sqrt{3}}{8} - \frac{1}{4}\pi$

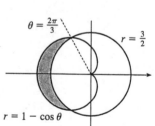

$\theta = \frac{2\pi}{3}$ $r = \frac{3}{2}$

$r = 1 - \cos\theta$

25. $4 \sin \theta = 2 \iff \sin \theta = \frac{1}{2} \iff \theta = \frac{\pi}{6}$ or $\frac{5\pi}{6} \iff$

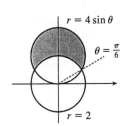

$r = 4 \sin \theta$

$\theta = \frac{\pi}{6}$

$r = 2$

$$A = 2 \int_{\pi/6}^{\pi/2} \frac{1}{2} \left[(4 \sin \theta)^2 - 2^2 \right] d\theta = \int_{\pi/6}^{\pi/2} (16 \sin^2 \theta - 4) \, d\theta$$

$$= \int_{\pi/6}^{\pi/2} [8 (1 - \cos 2\theta) - 4] \, d\theta = [4\theta - 4 \sin 2\theta]_{\pi/6}^{\pi/2}$$

$$= \frac{4}{3}\pi + 2\sqrt{3}$$

27. $3 \cos \theta = 1 + \cos \theta \iff \cos \theta = \frac{1}{2} \implies \theta = \frac{\pi}{3}$ or $-\frac{\pi}{3}$.

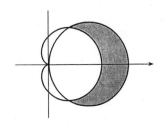

$$A = 2 \int_0^{\pi/3} \frac{1}{2} \left[(3 \cos \theta)^2 - (1 + \cos \theta)^2 \right] d\theta$$

$$= \int_0^{\pi/3} (8 \cos^2 \theta - 2 \cos \theta - 1) \, d\theta = \int_0^{\pi/3} [4 (1 + \cos 2\theta) - 2 \cos \theta - 1] \, d\theta$$

$$= [3\theta + 2 \sin 2\theta - 2 \sin \theta]_0^{\pi/3} = \pi + \sqrt{3} - \sqrt{3} = \pi$$

29. $A = 2 \int_0^{\pi/4} \frac{1}{2} \sin^2 \theta d\theta = \int_0^{\pi/4} \frac{1}{2} (1 - \cos 2\theta) \, d\theta$

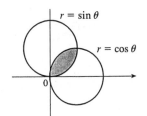

$r = \sin \theta$

$r = \cos \theta$

0

$$= \left[\frac{1}{2}\theta - \frac{1}{4} \sin 2\theta \right]_0^{\pi/4}$$

$$= \frac{1}{8}\pi - \frac{1}{4}$$

31. $\sin 2\theta = \cos 2\theta \implies \tan 2\theta = 1 \implies 2\theta = \frac{\pi}{4} \implies \theta = \frac{\pi}{8} \implies$

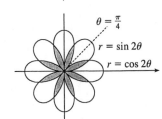

$\theta = \frac{\pi}{4}$

$r = \sin 2\theta$

$r = \cos 2\theta$

$$A = 16 \int_0^{\pi/8} \frac{1}{2} \sin^2 2\theta d\theta = 4 \int_0^{\pi/8} (1 - \cos 4\theta) \, d\theta$$

$$= 4 \left[\theta - \frac{1}{4} \sin 4\theta \right]_0^{\pi/8} = \frac{1}{2}\pi - 1$$

33. $A = 2 \left[\int_{-\pi/2}^{-\pi/6} \frac{1}{2} (3 + 2 \sin \theta)^2 \, d\theta + \int_{-\pi/6}^{\pi/2} \frac{1}{2} 2^2 \, d\theta \right]$

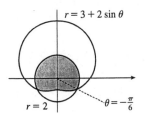

$r = 3 + 2 \sin \theta$

$r = 2$

$\theta = -\frac{\pi}{6}$

$$= \int_{-\pi/2}^{-\pi/6} (9 + 12 \sin \theta + 4 \sin^2 \theta) \, d\theta + [4\theta]_{-\pi/6}^{\pi/2}$$

$$= [9\theta - 12 \cos \theta + 2\theta - \sin 2\theta]_{-\pi/2}^{-\pi/6} + \frac{8\pi}{3} = \frac{19\pi}{3} - \frac{11\sqrt{3}}{2}$$

35. $A = 2 \left[\int_0^{2\pi/3} \frac{1}{2} \left(\frac{1}{2} + \cos \theta \right)^2 \, d\theta - \int_{2\pi/3}^{\pi} \frac{1}{2} \left(\frac{1}{2} + \cos \theta \right)^2 \, d\theta \right]$

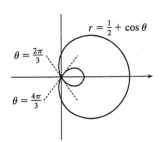

$r = \frac{1}{2} + \cos \theta$

$\theta = \frac{2\pi}{3}$

$\theta = \frac{4\pi}{3}$

$$= \int_0^{2\pi/3} \left(\frac{1}{4} + \cos \theta + \cos^2 \theta \right) d\theta - \int_{2\pi/3}^{\pi} \left(\frac{1}{4} + \cos \theta + \cos^2 \theta \right) d\theta$$

$$= \left[\frac{\theta}{4} + \sin \theta + \frac{\theta}{2} + \frac{\sin 2\theta}{4} \right]_0^{2\pi/3} - \left[\frac{\theta}{4} + \sin \theta + \frac{\theta}{2} + \frac{\sin 2\theta}{4} \right]_{2\pi/3}^{\pi}$$

$$= \left(\frac{\pi}{2} + \frac{3\sqrt{3}}{8} \right) - \left(\frac{3\pi}{4} \right) + \left(\frac{\pi}{2} + \frac{3\sqrt{3}}{8} \right) = \frac{1}{4} \left(\pi + 3\sqrt{3} \right)$$

37. The two circles intersect at the pole since $(0, 0)$ satisfies the first equation

and $\left(0, \frac{\pi}{2}\right)$ the second. The other intersection point $\left(\frac{1}{\sqrt{2}}, \frac{\pi}{4}\right)$ occurs

where $\sin\theta = \cos\theta$.

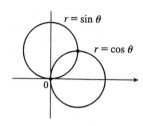

39. The curves intersect at the pole since $\left(0, \frac{\pi}{2}\right)$ satisfies $r = \cos\theta$ and

$(0, 0)$ satisfies $r = 1 - \cos\theta$. $\cos\theta = 1 - \cos\theta \Longrightarrow \cos\theta = \frac{1}{2} \Longrightarrow$

$\theta = \frac{\pi}{3}$ or $\frac{5\pi}{3} \Longrightarrow$ the other intersection points are $\left(\frac{1}{2}, \frac{\pi}{3}\right)$ and

$\left(\frac{1}{2}, \frac{5\pi}{3}\right)$.

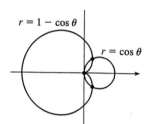

41. The pole is a point of intersection. $\sin\theta = \sin 2\theta = 2\sin\theta\cos\theta \quad \Leftrightarrow$

$\sin\theta(1 - 2\cos\theta) = 0 \quad \Leftrightarrow \quad \sin\theta = 0$ or $\cos\theta = \frac{1}{2} \quad \Rightarrow \quad \theta = 0, \pi,$

$\frac{\pi}{3}, -\frac{\pi}{3} \quad \Rightarrow \quad \left(\frac{\sqrt{3}}{2}, \frac{\pi}{3}\right)$ and $\left(\frac{\sqrt{3}}{2}, \frac{2\pi}{3}\right)$ (by symmetry) are the other

intersection points.

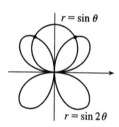

43.

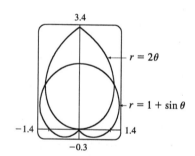

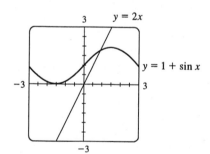

From the first graph, we see that the pole is one point of intersection. By zooming in or using the cursor, we
estimate the θ-values of the intersection points to be about 0.89 and $\pi - 0.89 \approx 2.25$. (The first of these values
may be more easily estimated by plotting $y = 1 + \sin x$ and $y = 2x$ in rectangular coordinates; see the second
graph.)

By symmetry, the total area contained is twice the area contained in the first quadrant, that is,

$$A \approx 2\int_0^{0.89} \tfrac{1}{2}(2\theta)^2\, d\theta + 2\int_{0.89}^{\pi/2} \tfrac{1}{2}(1 + \sin\theta)^2\, d\theta$$

$$= \left[\tfrac{4}{3}\theta^3\right]_0^{0.89} + \left[\theta - 2\cos\theta + \left(\tfrac{1}{2}\theta - \tfrac{1}{4}\sin 2\theta\right)\right]_{0.89}^{\pi/2} \approx 3.46$$

45. $L = \int_a^b \sqrt{r^2 + (dr/d\theta)^2}\, d\theta = \int_0^{3\pi/4} \sqrt{(5\cos\theta)^2 + (-5\sin\theta)^2}\, d\theta = 5\int_0^{3\pi/4} \sqrt{\cos^2\theta + \sin^2\theta}\, d\theta$

$= 5\int_0^{3\pi/4} d\theta = \tfrac{15}{4}\pi$

47. $L = \int_a^b \sqrt{r^2 + (dr/d\theta)^2}\, d\theta = \int_0^{2\pi} \sqrt{(2^\theta)^2 + [(\ln 2)\, 2^\theta]^2}\, d\theta = \int_0^{2\pi} 2^\theta \sqrt{1 + \ln^2 2}\, d\theta$

$= \left[\sqrt{1 + \ln^2 2}\, \left(\dfrac{2^\theta}{\ln 2} \right) \right]_0^{2\pi} = \dfrac{\sqrt{1 + \ln^2 2}\, (2^{2\pi} - 1)}{\ln 2}$

49. $L = \int_0^{2\pi} \sqrt{(\theta^2)^2 + (2\theta)^2}\, d\theta = \int_0^{2\pi} \theta \sqrt{\theta^2 + 4}\, d\theta = \frac{1}{2} \cdot \frac{2}{3} \left[(\theta^2 + 4)^{3/2} \right]_0^{2\pi} = \frac{8}{3} \left[(\pi^2 + 1)^{3/2} - 1 \right]$

51. From Figure 4 in Example 1,

$$L = \int_{-\pi/4}^{\pi/4} \sqrt{r^2 + (r')^2}\, d\theta = 2 \int_0^{\pi/4} \sqrt{\cos^2 2\theta + 4 \sin^2 2\theta}\, d\theta \approx 2\,(1.211056) \approx 2.4221$$

53. $L = 2 \int_0^{2\pi} \sqrt{\cos^8 (\theta/4) + \cos^6 (\theta/4) \sin^2 (\theta/4)}\, d\theta$

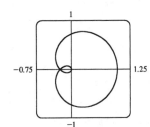

$= 2 \int_0^{2\pi} |\cos^3 (\theta/4)| \sqrt{\cos^2 (\theta/4) + \sin^2 (\theta/4)}\, d\theta$

$= 2 \int_0^{2\pi} |\cos^3 (\theta/4)|\, d\theta = 8 \int_0^{\pi/2} \cos^3 u\, du$ (where $u = \frac{1}{4}\theta$)

$= 8 \left[\sin u - \frac{1}{3} \sin^3 u \right]_0^{\pi/2} = \frac{16}{3}$

Note that the curve is retraced after every interval of length 4π.

55. (a) From (10.3.5),

$S = \int_a^b 2\pi y \sqrt{(dx/d\theta)^2 + (dy/d\theta)^2}\, d\theta$

$= \int_a^b 2\pi y \sqrt{r^2 + (dr/d\theta)^2}\, d\theta$ (see the derivation of Equation 5) $= \int_a^b 2\pi r \sin\theta \sqrt{r^2 + (dr/d\theta)^2}\, d\theta$

(b) $r^2 = \cos 2\theta \;\Rightarrow\; 2r \dfrac{dr}{d\theta} = -2 \sin 2\theta \;\Rightarrow\; \left(\dfrac{dr}{d\theta} \right)^2 = \dfrac{\sin^2 2\theta}{r^2} = \dfrac{\sin^2 2\theta}{\cos 2\theta}$.

$S = 2 \int_0^{\pi/4} 2\pi \sqrt{\cos 2\theta}\, \sin\theta \sqrt{\cos 2\theta + (\sin^2 2\theta)/\cos 2\theta}\, d\theta = 4\pi \int_0^{\pi/4} \sin\theta\, d\theta$

$= [-4\pi \cos\theta]_0^{\pi/4} = -4\pi \left(\tfrac{1}{\sqrt{2}} - 1 \right) = 2\pi \left(2 - \sqrt{2} \right)$

10.6 Conic Sections

1. $x = 2y^2 \;\Rightarrow\; y^2 = \frac{1}{2}x$. $4p = \frac{1}{2}$, so $p = \frac{1}{8}$. The

vertex is $(0, 0)$, the focus is $\left(\frac{1}{8}, 0 \right)$, and the

directrix is $x = -\frac{1}{8}$.

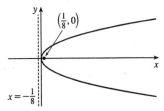

3. $4x^2 = -y \;\Rightarrow\; x^2 = -\frac{1}{4}y$. $4p = -\frac{1}{4}$, so

$p = -\frac{1}{16}$. The vertex is $(0, 0)$, the focus is

$\left(0, -\frac{1}{16} \right)$, and the directrix is $y = \frac{1}{16}$.

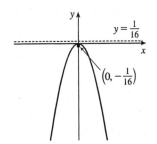

5. $(x + 2)^2 = 8(y - 3)$. $4p = 8$, so $p = 2$. The vertex is $(-2, 3)$, the focus is $(-2, 5)$, and the directrix is $y = 1$.

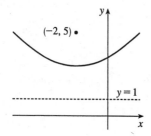

7. $2x + y^2 - 8y + 12 = 0 \Rightarrow$
$(y - 4)^2 = -2(x - 2) \Rightarrow p = -\frac{1}{2} \Rightarrow$
vertex $(2, 4)$, focus $\left(\frac{3}{2}, 4\right)$, directrix $x = \frac{5}{2}$

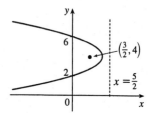

9. The equation has the form $y^2 = 4px$, where $p < 0$. Since the parabola passes through $(-1, 1)$, we have $1^2 = 4p(-1)$, so $4p = -1$ and an equation is $y^2 = -x$ or $x = -y^2$. $4p = -1$, so $p = -\frac{1}{4}$ and the focus is $\left(-\frac{1}{4}, 0\right)$ while the directrix is $x = \frac{1}{4}$.

11. $x^2/16 + y^2/4 = 1 \Rightarrow a = 4, b = 2$,
$c = \sqrt{16 - 4} = 2\sqrt{3} \Rightarrow$ center $(0, 0)$, vertices $(\pm 4, 0)$, foci $\left(\pm 2\sqrt{3}, 0\right)$

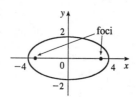

13. $25x^2 + 9y^2 = 225 \Leftrightarrow \frac{1}{9}x^2 + \frac{1}{25}y^2 = 1 \Rightarrow$
$a = 5, b = 3, c = 4 \Rightarrow$ center $(0, 0)$, vertices $(0, \pm 5)$, foci $(0, \pm 4)$

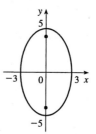

15. $9x^2 - 18x + 4y^2 = 27 \Leftrightarrow \frac{(x - 1)^2}{4} + \frac{y^2}{9} = 1$
$\Rightarrow a = 3, b = 2, c = \sqrt{5} \Rightarrow$ center $(1, 0)$,
vertices $(1, \pm 3)$, foci $\left(1, \pm\sqrt{5}\right)$

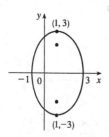

17. The center is $(0, 0)$, $a = 3$, and $b = 2$, so an equation is $\frac{x^2}{4} + \frac{y^2}{9} = 1$. $c = \sqrt{a^2 - b^2} = \sqrt{5}$, so the foci are $\left(0, \pm\sqrt{5}\right)$.

19. $\dfrac{x^2}{144} - \dfrac{y^2}{25} = 1 \;\Rightarrow\; a = 12, b = 5,$

$c = \sqrt{144 + 25} = 13 \;\Rightarrow\;$ center $(0, 0)$, vertices

$(\pm 12, 0)$, foci $(\pm 13, 0)$, asymptotes $y = \pm\frac{5}{12}x$

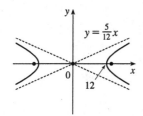

21. $9y^2 - x^2 = 9 \;\Rightarrow\; y^2 - \frac{1}{9}x^2 = 1 \;\Rightarrow\; a = 1,$

$b = 3, c = \sqrt{10} \;\Rightarrow\;$ center $(0, 0)$, vertices

$(0, \pm 1)$, foci $\left(0, \pm\sqrt{10}\right)$, asymptotes $y = \pm\frac{1}{3}x$

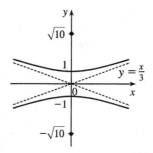

23. $2y^2 - 4y - 3x^2 + 12x = -8 \;\Leftrightarrow\; \dfrac{(x-2)^2}{6} - \dfrac{(y-1)^2}{9} = 1 \;\Rightarrow\; a = \sqrt{6}, b = 3, c = \sqrt{15} \;\Rightarrow\;$ center

$(2, 1)$, vertices $\left(2 \pm \sqrt{6}, 1\right)$, foci $\left(2 \pm \sqrt{15}, 1\right)$, asymptotes $y - 1 = \pm\frac{3}{\sqrt{6}}(x - 2)$

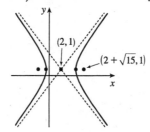

25. The parabola with vertex $(0, 0)$ and focus $(0, -2)$ opens downward and has $p = -2$, so its equation is
$x^2 = 4py = -8y$.

27. Vertex at $(2, 0)$, $p = 1$, opens to right $\;\Rightarrow\; y^2 = 4p(x - 2) = 4(x - 2)$

29. The parabola must have equation $y^2 = 4px$, so $(-4)^2 = 4p(1) \;\Rightarrow\; p = 4 \;\Rightarrow\; y^2 = 16x$.

31. The ellipse with foci $(\pm 2, 0)$ and vertices $(\pm 5, 0)$ has center $(0, 0)$ and a horizontal major axis, with $a = 5$ and

$c = 2$, so $b = \sqrt{a^2 - c^2} = \sqrt{21}$. An equation is $\dfrac{x^2}{25} + \dfrac{y^2}{21} = 1$.

33. Center $(3, 0)$, $c = 1$, $a = 3 \;\Rightarrow\; b = \sqrt{8} = 2\sqrt{2} \;\Rightarrow\; \frac{1}{8}(x - 3)^2 + \frac{1}{9}y^2 = 1$

35. Center $(2, 2)$, $c = 2$, $a = 3 \;\Rightarrow\; b = \sqrt{5} \;\Rightarrow\; \frac{1}{9}(x - 2)^2 + \frac{1}{5}(y - 2)^2 = 1$

37. Center $(0, 0)$, vertical axis, $c = 3$, $a = 1 \;\Rightarrow\; b = \sqrt{8} = 2\sqrt{2} \;\Rightarrow\; y^2 - \frac{1}{8}x^2 = 1$

39. Center $(4, 3)$, horizontal axis, $c = 3$, $a = 2 \;\Rightarrow\; b = \sqrt{5} \;\Rightarrow\; \frac{1}{4}(x - 4)^2 - \frac{1}{5}(y - 3)^2 = 1$

41. Center $(0, 0)$, horizontal axis, $a = 3$, $\frac{b}{a} = 2 \;\Rightarrow\; b = 6 \;\Rightarrow\; \frac{1}{9}x^2 - \frac{1}{36}y^2 = 1$

43. In Figure 8, we see that the point on the ellipse closest to a focus is the closer vertex (which is a distance $a - c$ from
it) while the farthest point is the other vertex (at a distance of $a + c$). So for this lunar orbit,
$(a - c) + (a + c) = 2a = (1728 + 110) + (1728 + 314)$, or $a = 1940$; and $(a + c) - (a - c) = 2c = 314 - 110$,

or $c = 102$. Thus, $b^2 = a^2 - c^2 = 3{,}753{,}196$, and the equation is $\dfrac{x^2}{3{,}763{,}600} + \dfrac{y^2}{3{,}753{,}196} = 1$.

45. (a) Set up the coordinate system so that A is $(-200, 0)$ and B is $(200, 0)$.

$|PA| - |PB| = (1200)(980) = 1{,}176{,}000 \text{ ft} = \frac{2450}{11} \text{ mi} = 2a \implies a = \frac{1225}{11}$, and $c = 200$ so

$b^2 = c^2 - a^2 = \frac{3{,}339{,}375}{121} \implies \dfrac{121x^2}{1{,}500{,}625} - \dfrac{121y^2}{3{,}339{,}375} = 1.$

(b) Due north of $B \implies x = 200 \implies \dfrac{(121)(200)^2}{1{,}500{,}625} - \dfrac{121y^2}{3{,}339{,}375} = 1 \implies y = \dfrac{133{,}575}{539} \approx 248 \text{ mi}$

47. The function whose graph is the upper branch of this hyperbola is concave upward. The

function is $y = f(x) = a\sqrt{1 + \dfrac{x^2}{b^2}} = \dfrac{a}{b}\sqrt{b^2 + x^2}$, so $y' = \dfrac{a}{b}x\left(b^2 + x^2\right)^{-1/2}$ and

$y'' = \dfrac{a}{b}\left[\left(b^2 + x^2\right)^{-1/2} - x^2\left(b^2 + x^2\right)^{-3/2}\right] = ab\left(b^2 + x^2\right)^{-3/2} > 0$ for all x, and so f is concave upward.

49. (a) ellipse

(b) hyperbola

(c) empty graph (no curve)

(d) In case (a), $a^2 = k$, $b^2 = k - 16$, and $c^2 = a^2 - b^2 = 16$, so the foci are at $(\pm 4, 0)$. In case (b), $k - 16 < 0$, so $a^2 = k$, $b^2 = 16 - k$, and $c^2 = a^2 + b^2 = 16$, and so again the foci are at $(\pm 4, 0)$.

51. Use the parametrization $x = 2\cos t$, $y = \sin t$, $0 \le t \le 2\pi$ to get

$$L = 4\int_0^{\pi/2}\sqrt{(dx/dt)^2 + (dy/dt)^2}\,dt = 4\int_0^{\pi/2}\sqrt{4\sin^2 t + \cos^2 t}\,dt = 4\int_0^{\pi/2}\sqrt{3\sin^2 t + 1}\,dt$$

Using Simpson's Rule with $n = 10$,

$$L \approx \tfrac{4}{3}\left(\tfrac{\pi}{20}\right)\left[f(0) + 4f\left(\tfrac{\pi}{20}\right) + 2f\left(\tfrac{\pi}{10}\right) + \cdots + 2f\left(\tfrac{2\pi}{5}\right) + 4f\left(\tfrac{9\pi}{20}\right) + f\left(\tfrac{\pi}{2}\right)\right]$$

with $f(t) = \sqrt{3\sin^2 t + 1}$, so $L \approx 9.69$.

53. $\dfrac{x^2}{a^2} + \dfrac{y^2}{b^2} = 1 \implies \dfrac{2x}{a^2} + \dfrac{2yy'}{b^2} = 0 \implies y' = -\dfrac{b^2 x}{a^2 y}$ $(y \ne 0)$. Thus, the slope of the tangent line at P is

$-\dfrac{b^2 x_1}{a^2 y_1}$. The slope of F_1P is $\dfrac{y_1}{x_1 + c}$ and of F_2P is $\dfrac{y_1}{x_1 - c}$. By the formula from Problems Plus, we have

$$\tan\alpha = \dfrac{\dfrac{y_1}{x_1 + c} + \dfrac{b^2 x_1}{a^2 y_1}}{1 - \dfrac{b^2 x_1 y_1}{a^2 y_1(x_1 + c)}} = \dfrac{a^2 y_1^2 + b^2 x_1(x_1 + c)}{a^2 y_1(x_1 + c) - b^2 x_1 y_1} = \dfrac{a^2 b^2 + b^2 c x_1}{c^2 x_1 y_1 + a^2 c y_1} \quad \left(\begin{array}{l}\text{using } b^2 x_1^2 + a^2 y_1^2 = a^2 b^2 \\ \text{and } a^2 - b^2 = c^2\end{array}\right)$$

$$= \dfrac{b^2(c x_1 + a^2)}{c y_1(c x_1 + a^2)} = \dfrac{b^2}{c y_1}$$

and

$$\tan\beta = \dfrac{-\dfrac{y_1}{x_1 - c} - \dfrac{b^2 x_1}{a^2 y_1}}{1 - \dfrac{b^2 x_1 y_1}{a^2 y_1(x_1 - c)}} = \dfrac{-a^2 y_1^2 - b^2 x_1(x_1 - c)}{a^2 y_1(x_1 - c) - b^2 x_1 y_1} = \dfrac{-a^2 b^2 + b^2 c x_1}{c^2 x_1 y_1 - a^2 c y_1} = \dfrac{b^2(c x_1 - a^2)}{c y_1(c x_1 - a^2)} = \dfrac{b^2}{c y_1}$$

So $\alpha = \beta$.

10.7 Conic Sections in Polar Coordinates

1. The directrix $y = 6$ is above the focus at the origin, so we use the form with " $+ e \sin \theta$ " in the denominator.

$$r = \frac{ed}{1 + e \sin \theta} = \frac{\frac{7}{4} \cdot 6}{1 + \frac{7}{4} \sin \theta} = \frac{42}{4 + 7 \sin \theta}$$

3. The directrix $x = -5$ is to the left of the focus at the origin, so we use the form with " $- e \cos \theta$ " in the

denominator. $r = \dfrac{ed}{1 - e \cos \theta} = \dfrac{\frac{3}{4} \cdot 5}{1 - \frac{3}{4} \cos \theta} = \dfrac{15}{4 - 3 \cos \theta}$

5. $r = 5 \sec \theta \iff x = r \cos \theta = 5$, so $r = \dfrac{ed}{1 + e \cos \theta} = \dfrac{4 \cdot 5}{1 + 4 \cos \theta} = \dfrac{20}{1 + 4 \cos \theta}$

7. Focus $(0, 0)$, vertex $\left(5, \frac{\pi}{2}\right)$ $\Rightarrow$ directrix $y = 10$ $\Rightarrow$ $r = \dfrac{ed}{1 + e \sin \theta} = \dfrac{10}{1 + \sin \theta}$

9. $r = \dfrac{4}{1 + 3 \cos \theta}$

(a) $e = 3$

(b) Since $e = 3 > 1$, the conic is a hyperbola.

(c) $ed = 4$ $\Rightarrow$ $d = \frac{4}{3}$ $\Rightarrow$ directrix $x = \frac{4}{3}$

(d) The vertices are $(1, 0)$ and $(-2, \pi) = (2, 0)$; the center is $\left(\frac{3}{2}, 0\right)$; the

asymptotes are parallel to $\theta = \pm \cos^{-1}\left(-\frac{1}{3}\right)$

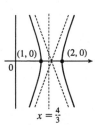

11. $r = \dfrac{2}{1 - \cos \theta}$

(a) $e = 1$

(b) Parabola

(c) $ed = 2$ $\Rightarrow$ $d = 2$ $\Rightarrow$ directrix $x = -2$

(d) Vertex $(-1, 0) = (1, \pi)$

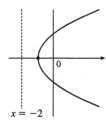

13. $r = \dfrac{3}{1 + \frac{1}{2} \sin \theta}$

(a) $e = \frac{1}{2}$

(b) Ellipse

(c) $ed = 3$ $\Rightarrow$ $d = 6$ $\Rightarrow$ directrix $y = 6$

(d) Vertices $\left(2, \frac{\pi}{2}\right)$ and $\left(6, \frac{3\pi}{2}\right)$; center $\left(2, \frac{3\pi}{2}\right)$

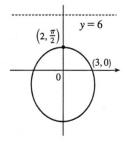

15. $r = \dfrac{7/2}{1 - \frac{5}{2}\sin\theta}$

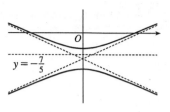

(a) $e = \frac{5}{2}$

(b) Hyperbola

(c) $ed = \frac{7}{2} \Rightarrow d = \frac{7}{5} \Rightarrow$ directrix $y = -\frac{7}{5}$

$y = -\frac{7}{5}$

(d) Center $\left(\frac{5}{3}, \frac{3\pi}{2}\right)$; vertices $\left(-\frac{7}{3}, \frac{\pi}{2}\right) = \left(\frac{7}{3}, \frac{3\pi}{2}\right)$ and $\left(1, \frac{3\pi}{2}\right)$

17. (a) The equation is $r = \dfrac{1}{4 - 3\cos\theta} = \dfrac{1/4}{1 - \frac{3}{4}\cos\theta}$, so $e = \frac{3}{4}$ and

$ed = \frac{1}{4} \Rightarrow d = \frac{1}{3}$. The conic is an ellipse, and the equation of its

directrix is $x = r\cos\theta = -\frac{1}{3} \Rightarrow r = -\dfrac{1}{3\cos\theta}$. We must be

careful in our choice of parameter values in this equation

$(-1 \le \theta \le 1$ works well).

(b) The equation is obtained by
replacing θ with $\theta - \frac{\pi}{3}$ in the
equation of the original conic
(see Example 4), so

$$r = \dfrac{1}{4 - 3\cos\left(\theta - \frac{\pi}{3}\right)}.$$

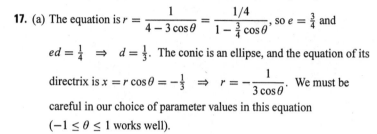

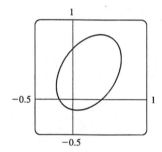

19. For $e < 1$ the curve is an ellipse. It is nearly circular when e is close to 0.
As e increases, the graph is stretched out to the right, and grows larger
(that is, its right-hand focus moves to the right while its left-hand focus
remains at the origin.) At $e = 1$, the curve becomes a parabola with focus
at the origin.

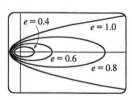

21. $|PF| = e\,|Pl| \Rightarrow r = e\,[d - r\cos(\pi - \theta)] = e\,(d + r\cos\theta) \Rightarrow$

$r\,(1 - e\cos\theta) = ed \Rightarrow r = \dfrac{ed}{1 - e\cos\theta}$

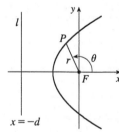

23. $|PF| = e\,|Pl| \Rightarrow r = e\,[d - r\sin(\theta - \pi)] = e\,(d + r\sin\theta) \Rightarrow$

$r\,(1 - e\sin\theta) = ed \Rightarrow r = \dfrac{ed}{1 - e\sin\theta}$

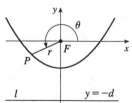

25. (a) If the directrix is $x = -d$, then $r = \dfrac{ed}{1 - e \cos\theta}$ [see Figure 2(b)], and, from (4), $a^2 = \dfrac{e^2 d^2}{(1 - e^2)^2}$ $\Rightarrow$

$ed = a(1 - e^2)$. Therefore, $r = \dfrac{a(1 - e^2)}{1 - e \cos\theta}$.

(b) $e = 0.017$ and the major axis $= 2a = 2.99 \times 10^8$ $\Rightarrow$ $a = 1.495 \times 10^8$.

Therefore $r = \dfrac{1.495 \times 10^8 \left[1 - (0.017)^2\right]}{1 - 0.017 \cos\theta} \approx \dfrac{1.49 \times 10^8}{1 - 0.017 \cos\theta}$.

27. Here $2a = $ length of major axis $= 36.18$ AU $\Rightarrow$ $a = 18.09$ AU and $e = 0.97$. By Exercise 25(a), the equation

of the orbit is $r = \dfrac{18.09\left[1 - (0.97)^2\right]}{1 - 0.97 \cos\theta} \approx \dfrac{1.07}{1 - 0.97 \cos\theta}$. By Exercise 26(a), the maximum distance from the

comet to the sun is $18.09\,(1 + 0.97) \approx 35.64$ AU or about 3.314 billion miles.

29. The minimum distance is at perihelion where $4.6 \times 10^7 = r = a(1 - e) = a(1 - 0.206) = a(0.794)$

$\Rightarrow$ $a = 4.6 \times 10^7 / 0.794$. So the maximum distance, which is at aphelion, is

$r = a(1 + e) = \left(4.6 \times 10^7 / 0.794\right) \times 10^7\,(1.206) \approx 7.0 \times 10^7$ km.

31. From Exercise 29, we have $e = 0.206$ and $a(1 - e) = 4.6 \times 10^7$ km. Thus, $a = 4.6 \times 10^7 / 0.794$. From

Exercise 25, we can write the equation of Mercury's orbit as $r = a\dfrac{1 - e^2}{1 - e \cos\theta}$. So since

$\dfrac{dr}{d\theta} = \dfrac{-a(1 - e^2)\, e \sin\theta}{(1 - e \cos\theta)^2}$ $\Rightarrow$

$r^2 + \left(\dfrac{dr}{d\theta}\right)^2 = \dfrac{a^2(1 - e^2)^2}{(1 - e \cos\theta)^2} + \dfrac{a^2(1 - e^2)^2 e^2 \sin^2\theta}{(1 - e \cos\theta)^4} = \dfrac{a^2(1 - e^2)^2}{(1 - e \cos\theta)^4}\left(1 - 2e \cos\theta + e^2\right)$

the length of the orbit is

$$L = \int_0^{2\pi} \sqrt{r^2 + (dr/d\theta)^2}\; d\theta = a\left(1 - e^2\right)\int_0^{2\pi} \dfrac{\sqrt{1 + e^2 - 2e \cos\theta}}{(1 - e \cos\theta)^2}\; d\theta \approx 3.6 \times 10^8 \text{ km}$$

This seems reasonable, since Mercury's orbit is nearly circular, and the circumference of a circle of radius a is $2\pi a \approx 3.6 \times 10^8$ km.

10 Review

CONCEPT CHECK

1. (a) A parametric curve is a set of points of the form $(x, y) = (f(t), g(t))$, where f and g are continuous functions of a variable t.

(b) Sketching a parametric curve, like sketching the graph of a function, is difficult to do in general. We can plot points on the curve by finding $f(t)$ and $g(t)$ for various values of t, either by hand or with a calculator or computer. Sometimes, when f and g are given by formulas, we can eliminate t from the equations $x = f(t)$ and $y = g(t)$ to get a Cartesian equation relating x and y. It may be easier to graph that equation than to work with the original formulas for x and y in terms of t.

2. (a) You can find $\dfrac{dy}{dx}$ as a function of t by calculating $\dfrac{dy}{dx} = \dfrac{dy/dt}{dx/dt}$ (if $dx/dt \neq 0$).

(b) Calculate the area as $\int_a^b y\, dx = \int_\alpha^\beta g(t)\, f'(t)\, dt$ [or $\int_\beta^\alpha g(t)\, f'(t)\, dt$ if the leftmost point is $(f(\beta), g(\beta))$ rather than $(f(\alpha), g(\alpha))$].

3. (a) $L = \int_\alpha^\beta \sqrt{(dx/dt)^2 + (dy/dt)^2}\, dt = \int_\alpha^\beta \sqrt{[f'(t)]^2 + [g'(t)]^2}\, dt$

(b) $S = \int_\alpha^\beta 2\pi y \sqrt{(dx/dt)^2 + (dy/dt)^2}\, dt = \int_\alpha^\beta 2\pi g(t) \sqrt{[f'(t)]^2 + [g'(t)]^2}\, dt$

4. (a) See Figure 5 in Section 10.4.

(b) $x = r\cos\theta$, $y = r\sin\theta$

(c) To find a polar representation (r, θ) with $r \geq 0$ and $0 \leq \theta < 2\pi$, first calculate $r = \sqrt{x^2 + y^2}$. Then θ is specified by $\cos\theta = x/r$ and $\sin\theta = y/r$.

5. (a) Calculate $\dfrac{dy}{dx} = \dfrac{dy/d\theta}{dx/d\theta} = \dfrac{(d/d\theta)(r\sin\theta)}{(d/d\theta)(r\cos\theta)} = \dfrac{(dr/d\theta)\sin\theta + r\cos\theta}{(dr/d\theta)\cos\theta - r\sin\theta}$, where $r = f(\theta)$.

(b) Calculate $A = \int_a^b \frac{1}{2}r^2\, d\theta = \int_a^b \frac{1}{2}[f(\theta)]^2\, d\theta$

(c) $L = \int_a^b \sqrt{(dx/d\theta)^2 + (dy/d\theta)^2}\, d\theta = \int_a^b \sqrt{r^2 + (dr/d\theta)^2}\, d\theta = \int_a^b \sqrt{[f(\theta)]^2 + [f'(\theta)]^2}\, d\theta$

6. (a) A parabola is a set of points in a plane whose distances from a fixed point F (the focus) and a fixed line ℓ (the directrix) are equal.

(b) $x^2 = 4py$; $y^2 = 4px$

7. (a) An ellipse is a set of points in a plane the sum of whose distances from two fixed points (the foci) is a constant.

(b) $\dfrac{x^2}{a^2} + \dfrac{y^2}{a^2 - c^2} = 1$.

8. (a) A hyperbola is a set of points in a plane the difference of whose distances from two fixed points (the foci) is a constant.

(b) $\dfrac{x^2}{a^2} - \dfrac{y^2}{c^2 - a^2} = 1$

(c) $y = \pm\dfrac{\sqrt{c^2 - a^2}}{a}x$

9. (a) If a conic section has focus F and corresponding directrix ℓ, then the eccentricity e is the fixed ratio $|PF|/|P\ell|$ for points P of the conic section.

(b) $e < 1$ for an ellipse; $e > 1$ for a hyperbola; $e = 1$ for a parabola.

(c) $x = d: r = \dfrac{ed}{1 + e\cos\theta}$. $x = -d: r = \dfrac{ed}{1 - e\cos\theta}$. $y = d: r = \dfrac{ed}{1 + e\sin\theta}$. $y = -d: r = \dfrac{ed}{1 - e\sin\theta}$.

EXERCISES

1. $x = t^2 + 4t$, $y = 2 - t$, $-4 \leq t \leq 1$. $t = 2 - y$, so
$x = (2 - y)^2 + 4(2 - y) = 4 - 4y + y^2 + 8 - 4y = y^2 - 8y + 12$. This
is part of a parabola with vertex $(-4, 4)$, opening to the right.

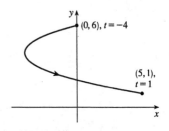

3. $x = \tan\theta$, $y = \cot\theta$. $y = 1/\tan\theta = 1/x$. The whole curve is traced out as
θ ranges over the open interval $\left(-\frac{\pi}{2}, \frac{\pi}{2}\right)$ [or any open interval of the form
$\left(-\frac{\pi}{2} + n\pi, \frac{\pi}{2} + n\pi\right)$, where n is an integer].

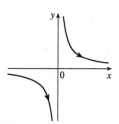

5. $r = 1 + 3\cos\theta$

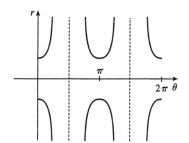
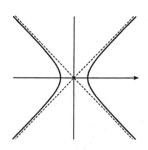

7. $r^2 = \sec 2\theta \;\Rightarrow\; r^2\cos 2\theta = 1$
$\Rightarrow\; r^2\left(\cos^2\theta - \sin^2\theta\right) = 1$
$\Rightarrow\; r^2\cos^2\theta - r^2\sin^2\theta = 1$
$\Rightarrow\; x^2 - y^2 = 1$, a hyperbola

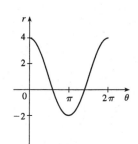
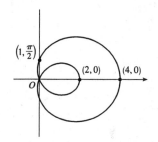

9. $r = 2\cos^2(\theta/2) = 1 + \cos\theta$

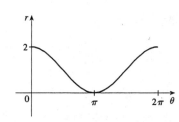
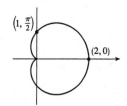

11. $r = \dfrac{1}{1 + \cos\theta} \;\Rightarrow\; e = 1 \;\Rightarrow\;$ parabola; $d = 1 \;\Rightarrow\;$ directrix $x = 1$
and vertex $\left(\frac{1}{2}, 0\right)$; y-intercepts are $\left(1, \frac{\pi}{2}\right)$ and $\left(1, \frac{3\pi}{2}\right)$.

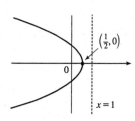

13. $x + y = 2 \;\Leftrightarrow\; r\cos\theta + r\sin\theta = 2 \;\Leftrightarrow\; r(\cos\theta + \sin\theta) = 2 \;\Leftrightarrow\; r = \dfrac{2}{\cos\theta + \sin\theta}$

15. $r = (\sin\theta)/\theta$. As $\theta \to \pm\infty$, $r \to 0$.

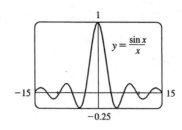

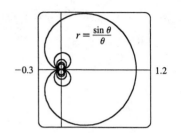

17. $x = \ln t$, $y = 1 + t^2$; $t = 1$. $\dfrac{dy}{dt} = 2t$ and $\dfrac{dx}{dt} = \dfrac{1}{t}$, so $\dfrac{dy}{dx} = \dfrac{dy/dt}{dx/dt} = \dfrac{2t}{1/t} = 2t^2$. When $t = 1$, $\dfrac{dy}{dx} = 2$.

19. $\dfrac{dy}{dx} = \dfrac{(dr/d\theta)\sin\theta + r\cos\theta}{(dr/d\theta)\cos\theta - r\sin\theta} = \dfrac{\sin\theta + \theta\cos\theta}{\cos\theta - \theta\sin\theta} = \dfrac{\frac{1}{\sqrt{2}} + \frac{\pi}{4}\cdot\frac{1}{\sqrt{2}}}{\frac{1}{\sqrt{2}} - \frac{\pi}{4}\cdot\frac{1}{\sqrt{2}}} = \dfrac{4 + \pi}{4 - \pi}$ when $\theta = \frac{\pi}{4}$.

21. $\dfrac{dy}{dx} = \dfrac{dy/dt}{dx/dt} = \dfrac{t\cos t + \sin t}{-t\sin t + \cos t}$. $\dfrac{d^2y}{dx^2} = \dfrac{\frac{d}{dt}\left(\frac{dy}{dx}\right)}{dx/dt}$, where

$\dfrac{d}{dt}\left(\dfrac{dy}{dx}\right) = \dfrac{(-t\sin t + \cos t)(-t\sin t + 2\cos t) - (t\cos t + \sin t)(-t\cos t - 2\sin t)}{(-t\sin t + \cos t)^2} = \dfrac{t^2 + 2}{(-t\sin t + \cos t)^2}$

$\Rightarrow \dfrac{d^2y}{dx^2} = \dfrac{t^2 + 2}{(-t\sin t + \cos t)^3}$.

23. We graph the curve for $-2.2 \le t \le 1.2$. By zooming in or using a cursor, we find that the lowest point is about $(1.4, 0.75)$. To find the exact values, we find the t-value at which $dy/dt = 2t + 1 = 0$ $\Leftrightarrow$ $t = -\frac{1}{2}$ $\Leftrightarrow$

$(x, y) = \left(\frac{11}{8}, \frac{3}{4}\right)$.

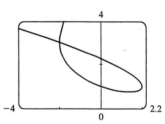

25. $\dfrac{dx}{dt} = -2a\sin t + 2a\sin 2t = 2a\sin t\,(2\cos t - 1) = 0$ $\Leftrightarrow$ $\sin t = 0$ or $\cos t = \frac{1}{2}$ $\Rightarrow$ $t = 0, \frac{\pi}{3}, \pi,$ or $\frac{5\pi}{3}$.

$\dfrac{dy}{dt} = 2a\cos t - 2a\cos 2t = 2a\left(1 + \cos t - 2\cos^2 t\right) = 2a\,(1 - \cos t)\,(1 + 2\cos t) = 0$ $\Rightarrow$ $t = 0, \frac{2\pi}{3},$ or $\frac{4\pi}{3}$.

Thus the graph has vertical tangents where

$t = \frac{\pi}{3}, \pi$ and $\frac{5\pi}{3}$, and horizontal tangents where

$t = \frac{2\pi}{3}$ and $\frac{4\pi}{3}$. To determine what the slope is

where $t = 0$, we use l'Hospital's Rule to evaluate

$\displaystyle\lim_{t \to 0}\dfrac{dy/dt}{dx/dt} = 0$, so there is a horizontal tangent

there.

t	x	y
0	a	0
$\frac{\pi}{3}$	$\frac{3}{2}a$	$\frac{\sqrt{3}}{2}a$
$\frac{2\pi}{3}$	$-\frac{1}{2}a$	$\frac{3\sqrt{3}}{2}a$
π	$-3a$	0
$\frac{4\pi}{3}$	$-\frac{1}{2}a$	$-\frac{3\sqrt{3}}{2}a$
$\frac{5\pi}{3}$	$\frac{3}{2}a$	$-\frac{\sqrt{3}}{2}a$

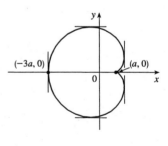

27. This curve has 10 "petals". For instance, for $-\frac{\pi}{10} \le \theta \le \frac{\pi}{10}$, there are two petals, one with $r > 0$ and one with $r < 0$.

$$A = 10\int_{-\pi/10}^{\pi/10}\tfrac{1}{2}r^2\,d\theta = 5\int_{-\pi/10}^{\pi/10}9\cos 5\theta\,d\theta = 90\int_{0}^{\pi/10}\cos 5\theta\,d\theta = [18\sin 5\theta]_0^{\pi/10} = 18$$

29. The curves intersect where $4\cos\theta = 2$; that is, at $\left(2, \frac{\pi}{3}\right)$ and $\left(2, -\frac{\pi}{3}\right)$.

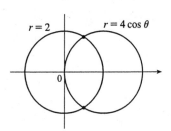

31. The curves intersect where $2\sin\theta = \sin\theta + \cos\theta \Rightarrow$
$\sin\theta = \cos\theta \Rightarrow \theta = \frac{\pi}{4}$, and also at the origin (at which $\theta = \frac{3\pi}{4}$ on the second curve).

$$A = \int_0^{\pi/4} \frac{1}{2}(2\sin\theta)^2\,d\theta + \int_{\pi/4}^{3\pi/4} \frac{1}{2}(\sin\theta + \cos\theta)^2\,d\theta$$

$$= \int_0^{\pi/4}(1 - \cos 2\theta)\,d\theta + \frac{1}{2}\int_{\pi/4}^{3\pi/4}(1 + \sin 2\theta)\,d\theta$$

$$= \left[\theta - \frac{1}{2}\sin 2\theta\right]_0^{\pi/4} + \left[\frac{1}{2}\theta - \frac{1}{4}\cos 2\theta\right]_{\pi/4}^{3\pi/4} = \frac{1}{2}(\pi - 1)$$

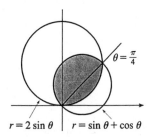

33. $x = 3t^2,\ y = 2t^3$.

$$L = \int_0^2 \sqrt{(dx/dt)^2 + (dy/dt)^2}\,dt = \int_0^2 \sqrt{(6t)^2 + (6t^2)^2}\,dt = 6\int_0^2 t\sqrt{1+t^2}\,dt$$

$$= \left[2\left(1+t^2\right)^{3/2}\right]_0^2 = 2\left(5\sqrt{5} - 1\right)$$

35. $L = \int_\pi^{2\pi} \sqrt{r^2 + (dr/d\theta)^2}\,d\theta = \int_\pi^{2\pi} \sqrt{(1/\theta)^2 + (-1/\theta^2)^2}\,d\theta$

$$\overset{24}{=} \int_\pi^{2\pi} \frac{\sqrt{\theta^2 + 1}}{\theta^2}\,d\theta = \left[-\frac{\sqrt{\theta^2 + 1}}{\theta} + \ln\left|\theta + \sqrt{\theta^2 + 1}\right|\right]_\pi^{2\pi}$$

$$= \frac{\sqrt{\pi^2 + 1}}{\pi} - \frac{\sqrt{4\pi^2 + 1}}{2\pi} + \ln\left|\frac{2\pi + \sqrt{4\pi^2 + 1}}{\pi + \sqrt{\pi^2 + 1}}\right|$$

37. $S = \int_1^4 2\pi y\sqrt{(dx/dt)^2 + (dy/dt)^2}\,dt = \int_1^4 2\pi \left(\frac{1}{3}t^3 + \frac{1}{2}t^{-2}\right)\sqrt{\left(2/\sqrt{t}\right)^2 + (t^2 - t^{-3})^2}\,dt$

$$= 2\pi \int_1^4 \left(\frac{1}{3}t^3 + \frac{1}{2}t^{-2}\right)\sqrt{(t^2 + t^{-3})^2}\,dt = 2\pi \int_1^4 \left(\frac{1}{3}t^5 + \frac{5}{6} + \frac{1}{2}t^{-5}\right)dt = 2\pi \left[\frac{1}{18}t^6 + \frac{5}{6}t - \frac{1}{8}t^{-4}\right]_1^4 = \frac{471{,}295}{1024}\pi$$

39. For all c except -1, the curve is asymptotic to the line $x = 1$. For $c < -1$, the curve bulges to the right near $y = 0$. As c increases, the bulge becomes smaller, until at $c = -1$ the curve is the straight line $x = 1$. As c continues to increase, the curve bulges to the left, until at $c = 0$ there is a cusp at the origin. For $c > 0$, there is a loop to the left of the origin, whose size and roundness increase as c increases. Note that the x-intercept of the curve is always $-c$.

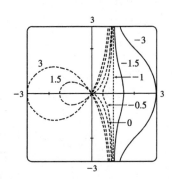

41. Ellipse, center $(0,0)$, $a = 3$, $b = 2\sqrt{2}$, $c = 1$ $\Rightarrow$
foci $(\pm 1, 0)$, vertices $(\pm 3, 0)$.

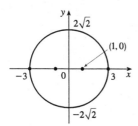

43. $6\left(y^2 - 6y + 9\right) = -(x+1)$ $\Leftrightarrow$
$(y-3)^2 = -\frac{1}{6}(x+1)$, a parabola with vertex
$(-1,3)$, opening to the left, $p = -\frac{1}{24}$ $\Rightarrow$ focus
$\left(-\frac{25}{24}, 3\right)$ and directrix $x = -\frac{23}{24}$.

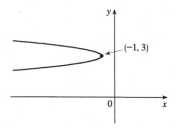

45. The parabola opens upward with vertex $(0,4)$ and $p = 2$, so its equation is $(x-0)^2 = 4 \cdot 2\,(y-4)$ $\Leftrightarrow$
$x^2 = 8\,(y-4)$.

47. The hyperbola has center $(0,0)$ and foci on the x-axis. $c = 3$ and $b/a = \frac{1}{2}$ (from the asymptotes)
$\Rightarrow$ $9 = c^2 = a^2 + b^2 = (2b)^2 + b^2 = 5b^2$ $\Rightarrow$ $b = \frac{3}{\sqrt{5}}$ $\Rightarrow$ $a = \frac{6}{\sqrt{5}}$ $\Rightarrow$ the equation is
$\dfrac{x^2}{36/5} - \dfrac{y^2}{9/5} = 1$ $\Leftrightarrow$ $5x^2 - 20y^2 = 36$.

49. $x^2 = -y + 100$ has its vertex at $(0,100)$, so one of the vertices of the ellipse is $(0,100)$. Another form of the
equation of a parabola is $x^2 = 4p\,(y-100)$ so $4p\,(y-100) = -y+100$ $\Rightarrow$ $4py - 4p\,(100) = 100 - y$ $\Rightarrow$
$4p = \dfrac{100 - y}{y - 100}$ $\Rightarrow$ $p = -\frac{1}{4}$. Therefore the shared focus is found at $\left(0, \frac{399}{4}\right)$ so $2c = \frac{399}{4} - 0$ $\Rightarrow$ $c = \frac{399}{8}$
and the center of the ellipse is $\left(0, \frac{399}{8}\right)$. So $a = 100 - \frac{399}{8} = \frac{401}{8}$ and $b^2 = a^2 - c^2 = \frac{401^2 - 399^2}{8^2} = 25$. So the
equation of the ellipse is $\dfrac{x^2}{b^2} + \dfrac{\left(y - \frac{399}{8}\right)^2}{a^2} = 1$ $\Rightarrow$ $\dfrac{x^2}{25} + \dfrac{\left(y - \frac{399}{8}\right)^2}{\left(\frac{401}{8}\right)^2} = 1$ or $\dfrac{x^2}{25} + \dfrac{(8y - 399)^2}{160{,}801} = 1$.

51. Directrix $x = 4$ $\Rightarrow$ $d = 4$, so $e = \frac{1}{3}$ $\Rightarrow$ $r = \dfrac{ed}{1 + e\cos\theta} = \dfrac{4}{3 + \cos\theta}$.

53. In polar coordinates, an equation for the circle is $r = 2a\sin\theta$. Thus, the coordinates of Q are
$x = r\cos\theta = 2a\sin\theta\cos\theta$ and $y = r\sin\theta = 2a\sin^2\theta$. The coordinates of R are $x = 2a\cot\theta$ and $y = 2a$.
Since P is the midpoint of QR, we use the midpoint formula to get $x = a\,(\sin\theta\cos\theta + \cot\theta)$ and
$y = a\left(1 + \sin^2\theta\right)$.

Problems Plus

1. $x = \int_1^t \frac{\cos u}{u}\, du$, $y = \int_1^t \frac{\sin u}{u}\, du$, so by FTC1, we have $\dfrac{dx}{dt} = \dfrac{\cos t}{t}$ and $\dfrac{dy}{dt} = \dfrac{\sin t}{t}$. Vertical tangent lines

occur when $\dfrac{dx}{dt} = 0$ $\Leftrightarrow$ $\cos t = 0$. The parameter value corresponding to $(x, y) = (0, 0)$ is $t = 1$, so the

nearest vertical tangent occurs when $t = \frac{\pi}{2}$. Therefore, the arc length between these points is

$$L = \int_1^{\pi/2} \sqrt{\left(\frac{dx}{dt}\right)^2 + \left(\frac{dy}{dt}\right)^2}\, dt = \int_1^{\pi/2} \sqrt{\frac{\cos^2 t}{t^2} + \frac{\sin^2 t}{t^2}}\, dt = \int_1^{\pi/2} \frac{dt}{t} = [\ln t]_1^{\pi/2} = \ln \frac{\pi}{2}$$

3. (a) If $\tan\theta = \sqrt{\dfrac{y}{C - y}}$, then $\tan^2\theta = \dfrac{y}{C - y}$, so $C\tan^2\theta - y\tan^2\theta = y$ and

$$y = \frac{C\tan^2\theta}{1 + \tan^2\theta} = \frac{C\tan^2\theta}{\sec^2\theta} = C\tan^2\theta\cos^2\theta = C\sin^2\theta = \frac{C}{2}(1 - \cos 2\theta). \text{ Now}$$

$$dx = \sqrt{\frac{y}{C - y}}\, dy = \tan\theta \cdot \frac{C}{2} \cdot 2\sin 2\theta\, d\theta = C\tan\theta \cdot 2\sin\theta\cos\theta\, d\theta = 2C\sin^2\theta\, d\theta = C\,(1 - \cos 2\theta)\, d\theta$$

Thus, $x = C\left(\theta - \frac{1}{2}\sin 2\theta\right) + K$ for some constant K. When $\theta = 0$, we have $y = 0$. We require that $x = 0$

when $\theta = 0$ so that the curve passes through the origin when $\theta = 0$. This yields $K = 0$. We now have

$x = \frac{1}{2}C\,(2\theta - \sin 2\theta)$, $y = \frac{1}{2}C\,(1 - \cos 2\theta)$.

(b) Setting $\phi = 2\theta$ and $r = \frac{1}{2}C$, we get $x = r\,(\phi - \sin\phi)$, $y = r\,(1 - \cos\phi)$. Comparison with Equations 10.1.1 shows that the curve is a cycloid.

5. (a) If (a, b) lies on the curve, then there is some parameter value t_1 such that $\dfrac{3t_1}{1 + t_1^3} = a$ and $\dfrac{3t_1^2}{1 + t_1^3} = b$. If

$t_1 = 0$, the point is $(0, 0)$, which lies on the line $y = x$. If $t_1 \neq 0$, then the point corresponding to $t = \dfrac{1}{t_1}$ is

given by $x = \dfrac{3\,(1/t_1)}{1 + (1/t_1)^3} = \dfrac{3t_1^2}{t_1^3 + 1} = b$, $y = \dfrac{3\,(1/t_1)^2}{1 + (1/t_1)^3} = \dfrac{3t_1}{t_1^3 + 1} = a$. So (b, a) also lies on the curve.

[Another way to see this is to do part (e) first; the result is immediate.] The curve intersects the line $y = x$

when $\dfrac{3t}{1 + t^3} = \dfrac{3t^2}{1 + t^3}$ $\Rightarrow$ $t = t^2$ $\Rightarrow$ $t = 0$ or 1, so the points are $(0, 0)$ and $\left(\frac{3}{2}, \frac{3}{2}\right)$.

(b) $\dfrac{dy}{dt} = \dfrac{(1 + t^3)\,(6t) - 3t^2\,(3t^2)}{(1 + t^3)^2} = \dfrac{6t - 3t^4}{(1 + t^3)^2} = 0$ when $6t - 3t^4 = 3t\,(2 - t^3) = 0$ $\Rightarrow$ $t = 0$ or $t = \sqrt[3]{2}$, so

there are horizontal tangents at $(0, 0)$ and $\left(\sqrt[3]{2}, \sqrt[3]{4}\right)$. Using the symmetry from part (a), we see that there are

vertical tangents at $(0, 0)$ and $\left(\sqrt[3]{4}, \sqrt[3]{2}\right)$.

(c) Notice that as $t \to -1^+$, we have $x \to -\infty$ and $y \to \infty$. As $t \to -1^-$, we have $x \to \infty$ and $y \to -\infty$.

Also $y - (-x - 1) = y + x + 1 = \dfrac{3t + 3t^2 + \left(1 + t^3\right)}{1 + t^3} = \dfrac{(t + 1)^3}{1 + t^3} = \dfrac{(t + 1)^2}{t^2 - t + 1} \to 0$ as $t \to -1$. So

$y = -x - 1$ is a slant asymptote.

(d) $\dfrac{dx}{dt} = \dfrac{(1+t^3)\,(3) - 3t\,(3t^2)}{(1+t^3)^2} = \dfrac{3 - 6t^3}{(1+t^3)^2}$ and from part (b) we have $\dfrac{dy}{dt} = \dfrac{6t - 3t^4}{(1+t^3)^2}$. So

$\dfrac{dy}{dx} = \dfrac{dy/dt}{dx/dt} = \dfrac{t\,(2 - t^3)}{1 - 2t^3}$. Also $\dfrac{d^2y}{dx^2} = \dfrac{\frac{d}{dt}\left(\frac{dy}{dx}\right)}{dx/dt} = \dfrac{2\,(1+t^3)^4}{3\,(1 - 2t^3)^3} > 0 \;\Leftrightarrow\; t < \frac{1}{\sqrt[3]{2}}$. So the curve is

concave upward there and has a minimum point at $(0, 0)$ and a maximum point at $\left(\sqrt[3]{2}, \sqrt[3]{4}\right)$. Using this together with the information from parts (a), (b), and (c), we sketch the curve.

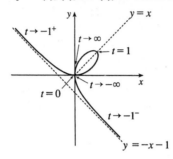

(e) $x^3 + y^3 = \left(\dfrac{3t}{1+t^3}\right)^3 + \left(\dfrac{3t^2}{1+t^3}\right)^3 = \dfrac{27t^3 + 27t^6}{(1+t^3)^3} = \dfrac{27t^3\,(1+t^3)}{(1+t^3)^3} = \dfrac{27t^3}{(1+t^3)^2}$ and

$3xy = 3\left(\dfrac{3t}{1+t^3}\right)\left(\dfrac{3t^2}{1+t^3}\right) = \dfrac{27t^3}{(1+t^3)^2}$, so $x^3 + y^3 = 3xy$.

(f) We start with the equation from part (e) and substitute $x = r\cos\theta$, $y = r\sin\theta$. Then $x^3 + y^3 = 3xy \;\Rightarrow\;$
$r^3\cos^3\theta + r^3\sin^3\theta = 3r^2\cos\theta\sin\theta$. For $r \neq 0$, this gives $r = \dfrac{3\cos\theta\sin\theta}{\cos^3\theta + \sin^3\theta}$. Dividing numerator and

denominator by $\cos^3\theta$, we obtain $r = \dfrac{3\left(\dfrac{1}{\cos\theta}\right)\dfrac{\sin\theta}{\cos\theta}}{1 + \dfrac{\sin^3\theta}{\cos^3\theta}} = \dfrac{3\sec\theta\tan\theta}{1 + \tan^3\theta}$.

(g) The loop corresponds to $\theta \in \left(0, \frac{\pi}{2}\right)$, so its area is

$$A = \int_0^{\pi/2}\dfrac{r^2}{2}\,d\theta = \dfrac{1}{2}\int_0^{\pi/2}\left(\dfrac{3\sec\theta\tan\theta}{1+\tan^3\theta}\right)^2 d\theta = \dfrac{9}{2}\int_0^{\pi/2}\dfrac{\sec^2\theta\tan^2\theta}{(1+\tan^3\theta)^2}\,d\theta$$

$$= \dfrac{9}{2}\int_0^{\infty}\dfrac{u^2\,du}{(1+u^3)^2}\ \ (\text{put }u = \tan\theta) = \lim_{b\to\infty}\dfrac{9}{2}\left[-\dfrac{1}{3}\left(1+u^3\right)^{-1}\right]_0^b = \dfrac{3}{2}$$

(h) By symmetry, the area between the folium and the line $y = -x - 1$ is equal to the enclosed area in the third quadrant, plus twice the enclosed area in the fourth quadrant. The area in the third quadrant is $\frac{1}{2}$, and since

$y = -x - 1 \;\Rightarrow\; r\sin\theta = -r\cos\theta - 1 \;\Rightarrow\; r = -\dfrac{1}{\sin\theta + \cos\theta}$, the area in the fourth quadrant is

$\dfrac{1}{2}\displaystyle\int_{-\pi/2}^{-\pi/4}\left[\left(-\dfrac{1}{\sin\theta + \cos\theta}\right)^2 - \left(\dfrac{3\sec\theta\tan\theta}{1 + \tan^3\theta}\right)^2\right]d\theta = \dfrac{1}{2}$ (using a CAS). Therefore, the total area is

$\frac{1}{2} + 2\left(\frac{1}{2}\right) = \frac{3}{2}$.

Infinite Sequences and Series

11.1 Sequences

1. (a) A sequence is an ordered list of numbers. It can also be defined as a function whose domain is the set of positive integers.

(b) The terms a_n approach 8 as n becomes large. In fact, we can make a_n as close to 8 as we like by taking n sufficiently large.

(c) The terms a_n become large as n becomes large.

3. $a_n = 1 - (0.2)^n$, so the sequence is $\{0.8, 0.96, 0.992, 0.9984, 0.99968, \ldots\}$.

5. $a_n = \dfrac{3(-1)^n}{n!}$, so the sequence is $\left\{\dfrac{-3}{1}, \dfrac{3}{2}, \dfrac{-3}{6}, \dfrac{3}{24}, \dfrac{-3}{120}, \ldots\right\} = \left\{-3, \dfrac{3}{2}, -\dfrac{1}{2}, \dfrac{1}{8}, -\dfrac{1}{40}, \ldots\right\}$.

7. $a_n = \sin\dfrac{n\pi}{2}$, so the sequence is $\{1, 0, -1, 0, 1, \ldots\}$.

9. The numerators are all 1 and the denominators are powers of 2, so $a_n = \dfrac{1}{2^n}$.

11. $\{2, 7, 12, 17, \ldots\}$. Each term is larger than the preceding one by 5, so
$a_n = a_1 + d(n-1) = 2 + 5(n-1) = 5n - 3$.

13. $\left\{1, -\frac{2}{3}, \frac{4}{9}, -\frac{8}{27}, \ldots\right\}$. Each term is $-\frac{2}{3}$ times the preceding one, so $a_n = \left(-\frac{2}{3}\right)^{n-1}$.

15. $a_n = n(n-1)$. $a_n \to \infty$ as $n \to \infty$, so the sequence diverges.

17. $a_n = \dfrac{3 + 5n^2}{n + n^2} = \dfrac{5 + 3/n^2}{1 + 1/n}$, so $a_n \to \dfrac{5+0}{1+0} = 5$ as $n \to \infty$. Converges

19. $a_n = \dfrac{2^n}{3^{n+1}} = \dfrac{1}{3}\left(\dfrac{2}{3}\right)^n$, so $\lim\limits_{n\to\infty} a_n = \frac{1}{3} \lim\limits_{n\to\infty}\left(\frac{2}{3}\right)^n = \frac{1}{3} \cdot 0 = 0$ by (7) with $r = \frac{2}{3}$. Converges

21. $a_n = \dfrac{(-1)^{n-1} n}{n^2 + 1} = \dfrac{(-1)^{n-1}}{n + 1/n}$, so $0 \le |a_n| = \dfrac{1}{n + 1/n} \le \dfrac{1}{n} \to 0$ as $n \to \infty$, so $a_n \to 0$ by the Squeeze Theorem and Theorem 5. Converges

23. $a_n = 2 + \cos n\pi$, so
$\{a_n\} = \{2 + \cos\pi, 2 + \cos 2\pi, 2 + \cos 3\pi, 2 + \cos 4\pi, \ldots\} = \{2 - 1, 2 + 1, 2 - 1, 2 + 1, \ldots\}$
$= \{1, 3, 1, 3, \ldots\}$
This sequence oscillates between 1 and 3, so it diverges.

25. $0 < \dfrac{3 + (-1)^n}{n^2} \le \dfrac{4}{n^2}$ and $\lim\limits_{n\to\infty} \dfrac{4}{n^2} = 0$, so $\left\{\dfrac{3 + (-1)^n}{n^2}\right\}$ converges to 0 by the Squeeze Theorem.

27. $\lim\limits_{x\to\infty} \dfrac{\ln(x^2)}{x} = \lim\limits_{x\to\infty} \dfrac{2\ln x}{x} \overset{\text{H}}{=} \lim\limits_{x\to\infty} \dfrac{2/x}{1} = 0$, so by Theorem 2, $\left\{\dfrac{\ln(n^2)}{n}\right\}$ converges to 0.

29. $b_n = \sqrt{n+2} - \sqrt{n} = (\sqrt{n+2} - \sqrt{n}) \dfrac{\sqrt{n+2}+\sqrt{n}}{\sqrt{n+2}+\sqrt{n}} = \dfrac{2}{\sqrt{n+2}+\sqrt{n}} < \dfrac{2}{2\sqrt{n}} = \dfrac{1}{\sqrt{n}} \to 0$ as $n \to \infty$. So by

the Squeeze Theorem with $a_n = 0$ and $c_n = 1/\sqrt{n}$, $\{\sqrt{n+2} - \sqrt{n}\}$ converges to 0.

31. $\lim\limits_{x \to \infty} \dfrac{x}{2^x} \overset{H}{=} \lim\limits_{x \to \infty} \dfrac{1}{(\ln 2)\,2^x} = 0$, so by Theorem 2, $\{n2^{-n}\}$ converges to 0.

33. $0 \le \dfrac{\cos^2 n}{2^n} \le \dfrac{1}{2^n}$ [since $0 \le \cos^2 n \le 1$], so since $\lim\limits_{n \to \infty} \dfrac{1}{2^n} = 0$, $\left\{\dfrac{\cos^2 n}{2^n}\right\}$ converges to 0 by the Squeeze

Theorem.

35. The series converges, since

$$a_n = \dfrac{1+2+3+\cdots+n}{n^2} = \dfrac{n(n+1)/2}{n^2} \quad \text{[sum of the first } n \text{ positive integers]} \quad = \dfrac{n+1}{2n} = \dfrac{1+1/n}{2} \to \dfrac{1}{2}$$

as $n \to \infty$.

37. $a_n = \dfrac{1}{2} \cdot \dfrac{2}{2} \cdot \dfrac{3}{2} \cdot \ldots \cdot \dfrac{(n-1)}{2} \cdot \dfrac{n}{2} \ge \dfrac{1}{2} \cdot \dfrac{n}{2} = \dfrac{n}{4} \to \infty$ as $n \to \infty$, so $\{a_n\}$ diverges.

39.

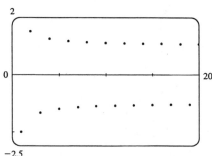

From the graph, we see that the sequence

$\left\{(-1)^n \dfrac{n+1}{n}\right\}$ is divergent, since it oscillates

between 1 and -1 (approximately).

41.

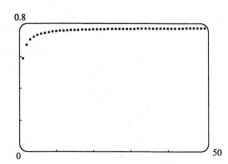

From the graph, it appears that the sequence

converges to about 0.78.

$$\lim\limits_{n \to \infty} \dfrac{2n}{2n+1} = \lim\limits_{n \to \infty} \dfrac{2}{2+1/n} = 1, \text{ so}$$

$$\lim\limits_{n \to \infty} \arctan\left(\dfrac{2n}{2n+1}\right) = \arctan 1 = \dfrac{\pi}{4}.$$

43.

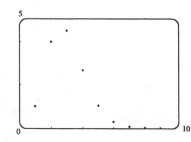

From the graph, it appears that the sequence converges to 0.

$$0 < a_n = \dfrac{n^3}{n!} = \dfrac{n}{n} \cdot \dfrac{n}{(n-1)} \cdot \dfrac{n}{(n-2)} \cdot \dfrac{1}{(n-3)} \cdot \ldots \cdot \dfrac{1}{3} \cdot \dfrac{1}{2} \cdot \dfrac{1}{1}$$

$$\le \dfrac{n^2}{(n-1)(n-2)(n-3)} \quad \text{(for } n \ge 4)$$

$$= \dfrac{1/n}{(1-1/n)(1-2/n)(1-3/n)} \to 0 \text{ as } n \to \infty$$

So by the Squeeze Theorem, $\{n^3/n!\}$ converges to 0.

45.

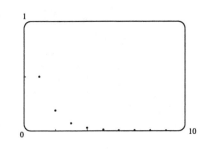

From the graph, it appears that the sequence approaches 0.

$$0 < a_n = \frac{1 \cdot 3 \cdot 5 \cdot \cdots \cdot (2n-1)}{(2n)^n} = \frac{1}{2n} \cdot \frac{3}{2n} \cdot \frac{5}{2n} \cdot \cdots \cdot \frac{2n-1}{2n}$$

$$\leq \frac{1}{2n} \cdot (1) \cdot (1) \cdot \cdots \cdot (1) = \frac{1}{2n} \to 0 \text{ as } n \to \infty$$

So by the Squeeze Theorem, $\left\{ \dfrac{1 \cdot 3 \cdot 5 \cdot \cdots \cdot (2n-1)}{(2n)^n} \right\}$ converges to 0.

47. (a) $a_n = 1000 \, (1.06)^n \Rightarrow a_1 = 1060, a_2 = 1123.60, a_3 = 1191.02, a_4 = 1262.48$, and $a_5 = 1338.23$.

(b) $\lim\limits_{n \to \infty} a_n = 1000 \lim\limits_{n \to \infty} (1.06)^n$, so the sequence diverges by (7) with $r = 1.06 > 1$.

49. If $|r| \geq 1$, then $\{r^n\}$ diverges by (7), so $\{nr^n\}$ diverges also, since $|nr^n| = n \, |r^n| \geq |r^n|$. If $|r| < 1$ then

$$\lim_{x \to \infty} xr^x = \lim_{x \to \infty} \frac{x}{r^{-x}} \overset{H}{=} \lim_{x \to \infty} \frac{1}{(-\ln r) r^{-x}} = \lim_{x \to \infty} \frac{r^x}{-\ln r} = 0, \text{ so } \lim_{n \to \infty} nr^n = 0, \text{ and hence } \{nr^n\} \text{ converges}$$

whenever $|r| < 1$.

51. Since $\{a_n\}$ is a decreasing sequence, $a_n > a_{n+1}$ for all $n \geq 1$. Because all of its terms lie between 5 and 8, $\{a_n\}$ is a bounded sequence. By the Monotonic Sequence Theorem, $\{a_n\}$ is convergent, that is, $\{a_n\}$ has a limit L. L must be less than 8 since $\{a_n\}$ is decreasing, so $5 \leq L < 8$.

53. $a_n = \dfrac{1}{2n+3}$ is decreasing since $a_{n+1} = \dfrac{1}{2(n+1)+3} = \dfrac{1}{2n+5} < \dfrac{1}{2n+3} = a_n$ for each $n \geq 1$. The sequence is bounded since $0 < a_n \leq \frac{1}{5}$ for all $n \geq 1$.

55. $a_n = \cos(n\pi/2)$ is not monotonic. The first few terms are $0, -1, 0, 1, 0, -1, 0, 1, \ldots$. In fact, the sequence consists of the terms $0, -1, 0, 1$ repeated over and over again in that order. The sequence is bounded since $|a_n| \leq 1$ for all $n \geq 1$.

57. $a_n = \dfrac{n}{n^2+1}$ defines a decreasing sequence since for $f(x) = \dfrac{x}{x^2+1}$,

$$f'(x) = \frac{(x^2+1)(1) - x(2x)}{(x^2+1)^2} = \frac{1-x^2}{(x^2+1)^2} \leq 0 \text{ for } x \geq 1. \text{ The sequence is bounded since } 0 < a_n \leq \tfrac{1}{2} \text{ for all}$$

$n \geq 1$.

59. $a_1 = 2^{1/2}, a_2 = 2^{3/4}, a_3 = 2^{7/8}, \ldots$, so $a_n = 2^{(2^n - 1)/2^n} = 2^{1-(1/2^n)}$. $\lim\limits_{n \to \infty} a_n = \lim\limits_{n \to \infty} 2^{1-(1/2^n)} = 2^1 = 2$.

Alternate Solution: Let $L = \lim\limits_{n \to \infty} a_n$. (We could show the limit exists by showing that $\{a_n\}$ is bounded and increasing.) So L must satisfy $L = \sqrt{2 \cdot L} \Rightarrow L^2 = 2L \Rightarrow L(L-2) = 0$ ($L \neq 0$ since the sequence increases), so $L = 2$.

61. We show by induction that $\{a_n\}$ is increasing and bounded above by 3.

Let P_n be the proposition that $a_{n+1} > a_n$ and $0 < a_n < 3$. Clearly P_1 is true. Assume that P_n is true. Then

$$a_{n+1} > a_n \quad \Rightarrow \quad \frac{1}{a_{n+1}} < \frac{1}{a_n} \quad \Rightarrow \quad -\frac{1}{a_{n+1}} > -\frac{1}{a_n}.$$

Now $a_{n+2} = 3 - \dfrac{1}{a_{n+1}} > 3 - \dfrac{1}{a_n} = a_{n+1} \quad \Leftrightarrow \quad P_{n+1}$. This proves that $\{a_n\}$ is increasing and bounded above by

3, so $1 = a_1 < a_n < 3$, that is, $\{a_n\}$ is bounded, and hence convergent by the Monotonic Sequence Theorem. If

$L = \lim\limits_{n\to\infty} a_n$, then $\lim\limits_{n\to\infty} a_{n+1} = L$ also, so L must satisfy $L = 3 - 1/L \quad \Rightarrow \quad L^2 - 3L + 1 = 0 \quad \Rightarrow$

$L = \frac{3\pm\sqrt{5}}{2}$. But $L > 1$, so $L = \frac{3+\sqrt{5}}{2}$.

63. (a) Let a_n be the number of rabbit pairs in the nth month. Clearly $a_1 = 1 = a_2$. In the nth month, each pair that is 2 or more months old (that is, a_{n-2} pairs) will produce a new pair to add to the a_{n-1} pairs already present. Thus, $a_n = a_{n-1} + a_{n-2}$, so that $\{a_n\} = \{f_n\}$, the Fibonacci sequence.

(b) $a_n = \dfrac{f_{n+1}}{f_n} \quad \Rightarrow \quad a_{n-1} = \dfrac{f_n}{f_{n-1}} = \dfrac{f_{n-1} + f_{n-2}}{f_{n-1}} = 1 + \dfrac{f_{n-2}}{f_{n-1}} = 1 + \dfrac{1}{f_{n-1}/f_{n-2}} = 1 + \dfrac{1}{a_{n-2}}$. If

$L = \lim\limits_{n\to\infty} a_n$, then $L = \lim\limits_{n\to\infty} a_{n-1}$ and $L = \lim\limits_{n\to\infty} a_{n-2}$, so L must satisfy $L = 1 + \dfrac{1}{L} \quad \Rightarrow \quad L^2 - L - 1 = 0$

$\Rightarrow \quad L = \frac{1+\sqrt{5}}{2}$ (since L must be positive).

65. (a)

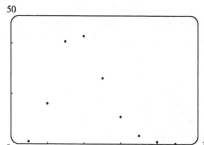

From the graph, it appears that the

sequence $\left\{\dfrac{n^5}{n!}\right\}$ converges to 0, that is,

$$\lim_{n\to\infty} \dfrac{n^5}{n!} = 0.$$

(b)

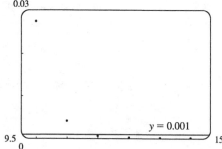

From the first graph, it seems that the smallest possible value of N corresponding to $\varepsilon = 0.1$ is 9, since $n^5/n! < 0.1$ whenever $n \geq 10$, but $9^5/9! > 0.1$. From the second graph, it seems that for $\varepsilon = 0.001$, the smallest possible value for N is 11.

67. If $\lim\limits_{n\to\infty} |a_n| = 0$ then $\lim\limits_{n\to\infty} -|a_n| = 0$, and since $-|a_n| \leq a_n \leq |a_n|$, we have that $\lim\limits_{n\to\infty} a_n = 0$ by the Squeeze Theorem.

69. (a) First we show that $a > a_1 > b_1 > b$.

$a_1 - b_1 = \frac{a+b}{2} - \sqrt{ab} = \frac{1}{2}\left(a - 2\sqrt{ab} + b\right) = \frac{1}{2}\left(\sqrt{a} - \sqrt{b}\right)^2 > 0$ (since $a > b$) $\Rightarrow$ $a_1 > b_1$. Also

$a - a_1 = a - \frac{1}{2}(a + b) = \frac{1}{2}(a - b) > 0$ and $b - b_1 = b - \sqrt{ab} = \sqrt{b}\left(\sqrt{b} - \sqrt{a}\right) < 0$, so $a > a_1 > b_1 > b$.

In the same way we can show that $a_1 > a_2 > b_2 > b_1$ and so the given assertion is true for $n = 1$. Suppose it is true for $n = k$, that is, $a_k > a_{k+1} > b_{k+1} > b_k$. Then

$$a_{k+2} - b_{k+2} = \frac{1}{2}(a_{k+1} + b_{k+1}) - \sqrt{a_{k+1}b_{k+1}} = \frac{1}{2}\left(a_{k+1} - 2\sqrt{a_{k+1}b_{k+1}} + b_{k+1}\right)$$

$$= \frac{1}{2}\left(\sqrt{a_{k+1}} - \sqrt{b_{k+1}}\right)^2 > 0$$

$$a_{k+1} - a_{k+2} = a_{k+1} - \frac{1}{2}(a_{k+1} + b_{k+1}) = \frac{1}{2}(a_{k+1} - b_{k+1}) > 0$$

and $b_{k+1} - b_{k+2} = b_{k+1} - \sqrt{a_{k+1}b_{k+1}} = \sqrt{b_{k+1}}\left(\sqrt{b_{k+1}} - \sqrt{a_{k+1}}\right) < 0$ $\Rightarrow$ $a_{k+1} > a_{k+2} > b_{k+2} > b_{k+1}$,

so the assertion is true for $n = k + 1$. Thus, it is true for all n by mathematical induction.

(b) From part (a) we have $a > a_n > a_{n+1} > b_{n+1} > b_n > b$, which shows that both sequences, $\{a_n\}$ and $\{b_n\}$, are monotonic and bounded. So they are both convergent by the Monotonic Sequence Theorem.

(c) Let $\lim\limits_{n\to\infty} a_n = \alpha$ and $\lim\limits_{n\to\infty} b_n = \beta$. Then $\lim\limits_{n\to\infty} a_{n+1} = \lim\limits_{n\to\infty} \dfrac{a_n + b_n}{2}$ $\Rightarrow$ $\alpha = \dfrac{\alpha + \beta}{2}$ $\Rightarrow$ $2\alpha = \alpha + \beta$

$\Rightarrow$ $\alpha = \beta$.

11.2 Series

1. (a) A sequence is an ordered list of numbers whereas a series is the *sum* of a list of numbers.

(b) A series is convergent if the sequence of partial sums is a convergent sequence. A series is divergent if it is not convergent.

3.

n	s_n
1	3.33333
2	4.44444
3	4.81481
4	4.93827
5	4.97942
6	4.99314
7	4.99771
8	4.99924
9	4.99975
10	4.99992
11	4.99997
12	4.99999

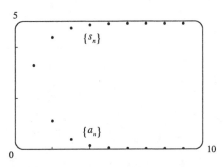

From the graph, it seems that the series converges. In fact, it is a geometric series with $a = \frac{10}{3}$ and $r = \frac{1}{3}$, so its sum is $\displaystyle\sum_{n=1}^{\infty} \frac{10}{3^n} = \frac{10/3}{1 - 1/3} = 5$. Note that the dot corresponding to $n = 1$ is part of both $\{a_n\}$ and $\{s_n\}$.

5.

n	s_n
1	0.50000
2	1.16667
3	1.91667
4	2.71667
5	3.55000
6	4.40714
7	5.28214
8	6.17103
9	7.07103
10	7.98012

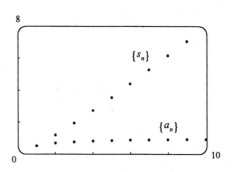

The series diverges, since its terms do not approach 0.

7.

n	s_n
1	0.64645
2	0.80755
3	0.87500
4	0.91056
5	0.93196
6	0.94601
7	0.95581
8	0.96296
9	0.96838
10	0.97259

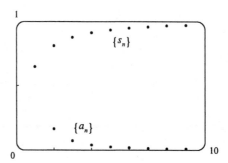

From the graph, it seems that the series converges to 1. To find the sum, we write

$$s_n = \sum_{i=1}^{n} \left(\frac{1}{i^{1.5}} - \frac{1}{(i+1)^{1.5}} \right) = \left(1 - \frac{1}{2^{1.5}} \right) + \left(\frac{1}{2^{1.5}} - \frac{1}{3^{1.5}} \right)$$

$$+ \left(\frac{1}{3^{1.5}} - \frac{1}{4^{1.5}} \right) + \cdots + \left(\frac{1}{n^{1.5}} - \frac{1}{(n+1)^{1.5}} \right) = 1 - \frac{1}{(n+1)^{1.5}}$$

So the sum is $\lim_{n \to \infty} s_n = 1$.

9. (a) $\lim_{n \to \infty} a_n = \lim_{n \to \infty} \frac{2n}{3n+1} = \frac{2}{3}$, so the *sequence* $\{a_n\}$ is convergent by (11.1.1).

(b) Since $\lim_{n \to \infty} a_n = \frac{2}{3} \neq 0$, the *series* $\sum_{n=1}^{\infty} a_n$ is divergent by the Test for Divergence (7).

11. $4 + \frac{8}{5} + \frac{16}{25} + \frac{32}{125} + \cdots$ is a geometric series with $a = 4$ and $r = \frac{2}{5}$. Since $|r| = \frac{2}{5} < 1$, the series converges to $\frac{4}{1-2/5} = \frac{4}{3/5} = \frac{20}{3}$.

13. $-2 + \frac{5}{2} - \frac{25}{8} + \frac{125}{32} - \cdots$ is a geometric series with $a = -2$ and $r = \frac{5/2}{-2} = -\frac{5}{4}$. Since $|r| = \frac{5}{4} > 1$, the series diverges by (4).

15. $\sum_{n=1}^{\infty} 5 \left(\frac{2}{3}\right)^{n-1}$ is a geometric series with $a = 5$ and $r = \frac{2}{3}$. Since $|r| = \frac{2}{3} < 1$, the series converges to

$\frac{a}{1-r} = \frac{5}{1-2/3} = \frac{5}{1/3} = 15$.

17. $\sum_{n=1}^{\infty} \frac{(-3)^{n-1}}{4^n} = \frac{1}{4} \sum_{n=1}^{\infty} \left(-\frac{3}{4}\right)^{n-1}$. The latter series is geometric with $a = 1$ and $r = -\frac{3}{4}$. Since $|r| = \frac{3}{4} < 1$, it

converges to $\frac{1}{1-(-3/4)} = \frac{4}{7}$. Thus, the given series converges to $\left(\frac{1}{4}\right)\left(\frac{4}{7}\right) = \frac{1}{7}$.

19. For $\sum_{n=1}^{\infty} 3^{-n} 8^{n+1} = \sum_{n=1}^{\infty} 8 \left(\frac{8}{3}\right)^n$, $a = \frac{64}{3}$ and $r = \frac{8}{3} > 1$, so the series diverges.

21. $\sum_{n=1}^{\infty} \frac{n}{n+5}$ diverges since $\lim_{n\to\infty} a_n = \lim_{n\to\infty} \frac{n}{n+5} = 1 \neq 0$. [Use (7), the Test for Divergence.]

23. Converges. $s_n = \sum_{i=1}^{n} \frac{1}{i(i+2)} = \sum_{i=1}^{n} \left(\frac{1/2}{i} - \frac{1/2}{i+2}\right)$ (using partial fractions) $= \frac{1}{2} \sum_{i=1}^{n} \left(\frac{1}{i} - \frac{1}{i+2}\right)$. The latter

sum is a telescoping series:

$$\left(1 - \frac{1}{3}\right) + \left(\frac{1}{2} - \frac{1}{4}\right) + \left(\frac{1}{3} - \frac{1}{5}\right) + \cdots + \left(\frac{1}{n-1} - \frac{1}{n+1}\right) + \left(\frac{1}{n} - \frac{1}{n+2}\right) = 1 + \frac{1}{2} - \frac{1}{n+1} - \frac{1}{n+2}$$

Thus, $\sum_{n=1}^{\infty} \frac{1}{n(n+2)} = \frac{1}{2} \lim_{n\to\infty} \left(1 + \frac{1}{2} - \frac{1}{n+1} - \frac{1}{n+2}\right) = \frac{1}{2}\left(1 + \frac{1}{2}\right) = \frac{3}{4}$.

25. $\sum_{n=1}^{\infty} \left[2(0.1)^n + (0.2)^n\right] = 2 \sum_{n=1}^{\infty} (0.1)^n + \sum_{n=1}^{\infty} (0.2)^n$. These are convergent geometric series and so by

Theorem 8, their sum is also convergent. $2\left(\frac{0.1}{1-0.1}\right) + \frac{0.2}{1-0.2} = \frac{2}{9} + \frac{1}{4} = \frac{17}{36}$

27. $\lim_{n\to\infty} a_n = \lim_{n\to\infty} \frac{n}{\sqrt{1+n^2}} = \lim_{n\to\infty} \frac{1}{\sqrt{1+1/n^2}} = 1 \neq 0$, so the series diverges by the Test for Divergence.

29. Converges. $\sum_{n=1}^{\infty} \frac{3^n + 2^n}{6^n} = \sum_{n=1}^{\infty} \left(\frac{3^n}{6^n} + \frac{2^n}{6^n}\right) = \sum_{n=1}^{\infty} \left[\left(\frac{1}{2}\right)^n + \left(\frac{1}{3}\right)^n\right] = \frac{1/2}{1-1/2} + \frac{1/3}{1-1/3} = 1 + \frac{1}{2} = \frac{3}{2}$

31. $\lim_{n\to\infty} a_n = \lim_{n\to\infty} \arctan n = \frac{\pi}{2} \neq 0$, so the series diverges by the Test for Divergence.

33. $s_n = (\ln 1 - \ln 2) + (\ln 2 - \ln 3) + (\ln 3 - \ln 4) + \cdots + [\ln n - \ln(n+1)] = \ln 1 - \ln(n+1) = -\ln(n+1)$

(telescoping series). Thus, $\lim_{n\to\infty} s_n = -\infty$, so the series is divergent.

35. $0.\overline{2} = \frac{2}{10} + \frac{2}{10^2} + \cdots = \frac{2/10}{1-1/10} = \frac{2}{9}$

37. $3.\overline{417} = 3 + \frac{417}{10^3} + \frac{417}{10^6} + \cdots = 3 + \frac{417/10^3}{1-1/10^3} = 3 + \frac{417}{999} = \frac{3414}{999} = \frac{1138}{333}$

39. $0.12\overline{3456} = \frac{123}{1000} + \frac{0.000456}{1-0.001} = \frac{123}{1000} + \frac{456}{999,000} = \frac{123,333}{999,000} = \frac{41,111}{333,000}$

41. $\sum_{n=1}^{\infty} \frac{x^n}{3^n}$ is a geometric series with $r = \frac{x}{3}$, so the series converges $\Leftrightarrow |r| < 1 \Leftrightarrow \frac{|x|}{3} < 1 \Leftrightarrow |x| < 3$. In

that case, the sum of the series is $\frac{x/3}{1-x/3} = \frac{x}{3-x}$.

43. $\sum_{n=0}^{\infty} 4^n x^n = \sum_{n=0}^{\infty} (4x)^n$ is a geometric series with $r = 4x$, so the series converges $\Leftrightarrow |r| < 1 \Leftrightarrow$

$4|x| < 1 \Leftrightarrow |x| < \frac{1}{4}$. In that case, the sum of the series is $\frac{1}{1-4x}$.

45. $\displaystyle\sum_{n=0}^{\infty}\left(\frac{1}{x}\right)^n$ is geometric with $r = \dfrac{1}{x}$, so it converges whenever $\left|\dfrac{1}{x}\right| < 1 \Leftrightarrow |x| > 1 \Leftrightarrow x > 1$ or $x < -1$,

and the sum is $\dfrac{1}{1 - 1/x} = \dfrac{x}{x - 1}$.

47. After defining f, We use `convert(f,parfrac);` in Maple, `Apart` in Mathematica, or `Expand Rational`

and `Simplify` in Derive to find that the general term is $\dfrac{1}{(4n + 1)(4n - 3)} = -\dfrac{1/4}{4n + 1} + \dfrac{1/4}{4n - 3}$. So the nth

partial sum is

$$s_n = \sum_{k=1}^{n}\left(-\frac{1/4}{4k + 1} + \frac{1/4}{4k - 3}\right) = \frac{1}{4}\left(\frac{1}{4k - 3} - \frac{1}{4k + 1}\right)$$

$$= \frac{1}{4}\left[\left(1 - \frac{1}{5}\right) + \left(\frac{1}{5} - \frac{1}{9}\right) + \left(\frac{1}{9} - \frac{1}{13}\right) + \cdots + \left(\frac{1}{4n - 3} - \frac{1}{4n + 1}\right)\right] = \frac{1}{4}\left(1 - \frac{1}{4n + 1}\right)$$

The series converges to $\lim\limits_{n\to\infty} s_n = \frac{1}{4}$. This can be confirmed by directly computing the sum using

`sum(f,1..infinity);` (in Maple), `Sum[f,{n,1,Infinity}]` (in Mathematica), or `Calculus Sum`

(from 1 to ∞) and `Simplify` (in Derive).

49. For $n = 1$, $a_1 = 0$ since $s_1 = 0$. For $n > 1$,

$$a_n = s_n - s_{n-1} = \frac{n - 1}{n + 1} - \frac{(n - 1) - 1}{(n - 1) + 1} = \frac{(n - 1)n - (n + 1)(n - 2)}{(n + 1)n} = \frac{2}{n(n + 1)}$$

Also, $\displaystyle\sum_{n=1}^{\infty} a_n = \lim_{n\to\infty} s_n = \lim_{n\to\infty}\frac{1 - 1/n}{1 + 1/n} = 1$.

51. (a) The first step in the chain occurs when the local government spends D dollars. The people who receive it spend a fraction c of those D dollars, that is, Dc dollars. Those who receive the Dc dollars spend a fraction c of it, that is, Dc^2 dollars. Continuing in this way, we see that the total spending after n transactions is

$$S_n = D + Dc + Dc^2 + \cdots + Dc^{n-1} = \frac{D(1 - c^n)}{1 - c} \text{ by (3)}.$$

(b) $\displaystyle\lim_{n\to\infty} S_n = \lim_{n\to\infty}\frac{D(1 - c^n)}{1 - c} = \frac{D}{1 - c}\lim_{n\to\infty}(1 - c^n) = \frac{D}{1 - c}$ (since $0 < c < 1 \Rightarrow \lim\limits_{n\to\infty} c^n = 0$)

$$= \frac{D}{s} \text{ (since } c + s = 1) = kD \text{ (since } k = 1/s)$$

If $c = 0.8$, then $s = 1 - c = 0.2$ and the multiplier is $k = 1/s = 5$.

53. $\sum_{n=2}^{\infty}(1 + c)^{-n}$ is a geometric series with $a = (1 + c)^{-2}$ and $r = (1 + c)^{-1}$, so the series converges when

$\left|(1 + c)^{-1}\right| < 1 \Leftrightarrow |1 + c| > 1 \Leftrightarrow 1 + c > 1$ or $1 + c < -1 \Leftrightarrow c > 0$ or $c < -2$. We calculate the sum

of the series and set it equal to 2: $\dfrac{(1 + c)^{-2}}{1 - (1 + c)^{-1}} = 2 \Leftrightarrow \left(\dfrac{1}{1 + c}\right)^2 = 2 - 2\left(\dfrac{1}{1 + c}\right) \Leftrightarrow$

$1 = 2(1 + c)^2 - 2(1 + c) = 0 \Leftrightarrow 2c^2 + 2c - 1 = 0 \Leftrightarrow c = \frac{-2 \pm \sqrt{12}}{4} = \frac{\pm\sqrt{3} - 1}{2}$. However, the negative

root is inadmissible because $-2 < \frac{-\sqrt{3} - 1}{2} < 0$. So $c = \frac{\sqrt{3} - 1}{2}$.

55. Let d_n be the diameter of C_n. We draw lines from the centers of the C_i to the center of D (or C), and using the Pythagorean Theorem, we can write $1^2 + \left(1 - \frac{1}{2}d_1\right)^2 = \left(1 + \frac{1}{2}d_1\right)^2 \Leftrightarrow$

$1 = \left(1 + \frac{1}{2}d_1\right)^2 - \left(1 - \frac{1}{2}d_1\right)^2 = 2d_1$ (difference of squares)

$\Rightarrow d_1 = \frac{1}{2}$. Similarly,

$1 = \left(1 + \frac{1}{2}d_2\right)^2 - \left(1 - d_1 - \frac{1}{2}d_2\right)^2 = 2d_2 + 2d_1 - d_1^2 - d_1 d_2$

$= (2 - d_1)(d_1 + d_2) \Leftrightarrow$

$d_2 = \dfrac{1}{2 - d_1} - d_1 = \dfrac{(1 - d_1)^2}{2 - d_1}, 1 = \left(1 + \frac{1}{2}d_3\right)^2 - \left(1 - d_1 - d_2 - \frac{1}{2}d_3\right)^2 \Leftrightarrow d_3 = \dfrac{[1 - (d_1 + d_2)]^2}{2 - (d_1 + d_2)}$, and in

general, $d_{n+1} = \dfrac{\left(1 - \sum_{i=1}^{n} d_i\right)^2}{2 - \sum_{i=1}^{n} d_i}$. If we actually calculate d_2 and d_3 from the formulas above, we find that they are

$\dfrac{1}{6} = \dfrac{1}{2 \cdot 3}$ and $\dfrac{1}{12} = \dfrac{1}{3 \cdot 4}$ respectively, so we suspect that in general, $d_n = \dfrac{1}{n(n+1)}$. To prove this, we

use induction: assume that for all $k \leq n$, $d_k = \dfrac{1}{k(k+1)} = \dfrac{1}{k} - \dfrac{1}{k+1}$. Then

$\displaystyle\sum_{i=1}^{n} d_i = 1 - \dfrac{1}{n+1} = \dfrac{n}{n+1}$ (telescoping sum). Substituting this into our formula for d_{n+1}, we get

$d_{n+1} = \dfrac{\left[1 - \dfrac{n}{n+1}\right]^2}{2 - \left(\dfrac{n}{n+1}\right)} = \dfrac{\dfrac{1}{(n+1)^2}}{\dfrac{n+2}{n+1}} = \dfrac{1}{(n+1)(n+2)}$, and the induction is complete.

Now, we observe that the partial sums $\sum_{i=1}^{n} d_i$ of the diameters of the circles approach 1 as $n \to \infty$; that is,

$\displaystyle\sum_{n=1}^{\infty} a_n = \sum_{n=1}^{\infty} \dfrac{1}{n(n+1)} = 1$, which is what we wanted to prove.

57. The series $1 - 1 + 1 - 1 + 1 - 1 + \cdots$ diverges (geometric series with $r = -1$) so we cannot say that $0 = 1 - 1 + 1 - 1 + 1 - 1 + \cdots$.

59. $\sum_{n=1}^{\infty} ca_n = \lim_{n \to \infty} \sum_{i=1}^{n} ca_i = \lim_{n \to \infty} c \sum_{i=1}^{n} a_i = c \lim_{n \to \infty} \sum_{i=1}^{n} a_i = c \sum_{n=1}^{\infty} a_n$, which exists by hypothesis.

61. Suppose on the contrary that $\sum (a_n + b_n)$ converges. Then by Theorem 8(iii), so would $\sum [(a_n + b_n) - a_n] = \sum b_n$, a contradiction.

63. The partial sums $\{s_n\}$ form an increasing sequence, since $s_n - s_{n-1} = a_n > 0$ for all n. Also, the sequence $\{s_n\}$ is bounded since $s_n \leq 1000$ for all n. So by Theorem 11.1.10, the sequence of partial sums converges, that is, the series $\sum a_n$ is convergent.

65. (a) At the first step, only the interval $\left(\frac{1}{3}, \frac{2}{3}\right)$ (length $\frac{1}{3}$) is removed. At the second step, we remove the intervals

$\left(\frac{1}{9}, \frac{2}{9}\right)$ and $\left(\frac{7}{9}, \frac{8}{9}\right)$, which have a total length of $2 \cdot \left(\frac{1}{3}\right)^2$. At the third step, we remove 2^2 intervals, each of

length $\left(\frac{1}{3}\right)^3$. In general, at the nth step we remove 2^{n-1} intervals, each of length $\left(\frac{1}{3}\right)^n$, for a length of

$2^{n-1} \cdot \left(\frac{1}{3}\right)^n = \frac{1}{3}\left(\frac{2}{3}\right)^{n-1}$. Thus, the total length of all removed intervals is $\sum_{n=1}^{\infty} \frac{1}{3}\left(\frac{2}{3}\right)^{n-1} = \frac{1/3}{1-2/3} = 1$

(geometric series with $a = \frac{1}{3}$ and $r = \frac{2}{3}$). Notice that at the nth step, the leftmost interval that is removed is

$\left(\left(\frac{1}{3}\right)^n, \left(\frac{2}{3}\right)^n\right)$, so we never remove 0, and 0 is in the Cantor set. Also, the rightmost interval removed is

$\left(1 - \left(\frac{2}{3}\right)^n, 1 - \left(\frac{1}{3}\right)^n\right)$, so 1 is never removed. Some other numbers in the Cantor set are $\frac{1}{3}, \frac{2}{3}, \frac{1}{9}, \frac{2}{9}, \frac{7}{9}$, and

$\frac{8}{9}$.

(b) The area removed at the first step is $\frac{1}{9}$; at the second step, $8 \cdot \left(\frac{1}{9}\right)^2$; at the third step, $(8)^2 \cdot \left(\frac{1}{9}\right)^3$. In general, the

area removed at the nth step is $(8)^{n-1}\left(\frac{1}{9}\right)^n = \frac{1}{9}\left(\frac{8}{9}\right)^{n-1}$, so the total area of all removed squares is

$\sum_{n=1}^{\infty} \frac{1}{9}\left(\frac{8}{9}\right)^{n-1} = \frac{1/9}{1-8/9} = 1$.

67. (a) $\sum_{n=1}^{\infty} \frac{n}{(n+1)!} \Rightarrow s_1 = \frac{1}{1 \cdot 2} = \frac{1}{2}, s_2 = \frac{1}{2} + \frac{2}{1 \cdot 2 \cdot 3} = \frac{5}{6}, s_3 = \frac{5}{6} + \frac{3}{1 \cdot 2 \cdot 3 \cdot 4} = \frac{23}{24}$,

$s_4 = \frac{23}{24} + \frac{4}{1 \cdot 2 \cdot 3 \cdot 4 \cdot 5} = \frac{119}{120}$. The denominators are $(n+1)!$, so a guess would be $s_n = \frac{(n+1)! - 1}{(n+1)!}$.

(b) For $n = 1, s_1 = \frac{1}{2} = \frac{2! - 1}{2!}$, so the formula holds for $n = 1$. Assume $s_k = \frac{(k+1)! - 1}{(k+1)!}$. Then

$$s_{k+1} = \frac{(k+1)! - 1}{(k+1)!} + \frac{k+1}{(k+2)!} = \frac{(k+1)! - 1}{(k+1)!} + \frac{k+1}{(k+1)!\,(k+2)}$$

$$= \frac{(k+2)! - (k+2) + k + 1}{(k+2)!} = \frac{(k+2)! - 1}{(k+2)!}$$

Thus, the formula is true for $n = k + 1$. So by induction, the guess is correct.

(c) $\lim_{n \to \infty} s_n = \lim_{n \to \infty} \frac{(n+1)! - 1}{(n+1)!} = \lim_{n \to \infty} \left[1 - \frac{1}{(n+1)!}\right] = 1$ and so $\sum_{n=0}^{\infty} \frac{n}{(n+1)!} = 1$.

11.3 The Integral Test and Estimates of Sums

1. The picture shows that $a_2 = \dfrac{1}{2^{1.3}} < \displaystyle\int_1^2 \dfrac{1}{x^{1.3}}\,dx$,

$a_3 = \dfrac{1}{3^{1.3}} < \displaystyle\int_2^3 \dfrac{1}{x^{1.3}}\,dx$, and so on, so $\displaystyle\sum_{n=2}^{\infty} \dfrac{1}{n^{1.3}} < \int_1^{\infty} \dfrac{1}{x^{1.3}}\,dx$. The

integral converges by (7.8.2) with $p = 1.3 > 1$, so the series converges.

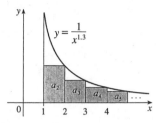

3. The function $f(x) = 1/x^4$ is continuous, positive, and decreasing on $[1, \infty)$, so the Integral Test applies.

$\displaystyle\int_1^{\infty} \dfrac{1}{x^4}\,dx = \lim_{b\to\infty} \int_1^b x^{-4}\,dx = \lim_{b\to\infty} \left[\dfrac{x^{-3}}{-3}\right]_1^b = \lim_{b\to\infty} \left(-\dfrac{1}{3b^3} + \dfrac{1}{3}\right) = \dfrac{1}{3}$, so $\displaystyle\sum_{n=1}^{\infty} \dfrac{1}{n^4}$ converges.

5. The function $f(x) = 1/(3x+1)$ is continuous, positive, and decreasing on $[1, \infty)$, so the Integral Test applies.

$\displaystyle\int_1^{\infty} \dfrac{dx}{3x+1} = \lim_{b\to\infty} \int_1^b \dfrac{dx}{3x+1} = \lim_{b\to\infty}\left[\tfrac{1}{3}\ln(3x+1)\right]_1^b = \lim_{b\to\infty}\left[\tfrac{1}{3}\ln(3b+1) - \tfrac{1}{3}\ln 4\right] = \infty$

so the improper integral diverges, and so does the series $\sum_{n=1}^{\infty} 1/(3n+1)$.

7. $f(x) = xe^{-x}$ is continuous and positive on $[1, \infty)$. $f'(x) = -xe^{-x} + e^{-x} = e^{-x}(1-x) < 0$ for $x > 1$, so f is decreasing on $[1, \infty)$. Thus, the Integral Test applies.

$\displaystyle\int_1^{\infty} xe^{-x}\,dx = \lim_{b\to\infty} \int_1^b xe^{-x}\,dx = \lim_{b\to\infty}\left[-xe^{-x} - e^{-x}\right]_1^b \text{ (by parts)} = \lim_{b\to\infty}\left[-be^{-b} - e^{-b} + e^{-1} + e^{-1}\right] = 2/e$

since $\lim_{b\to\infty} be^{-b} = \lim_{b\to\infty}(b/e^b) \overset{H}{=} \lim_{b\to\infty}(1/e^b) = 0$ and $\lim_{b\to\infty} e^{-b} = 0$. Thus, $\sum_{n=1}^{\infty} ne^{-n}$ converges.

9. $\sum_{n=5}^{\infty} (1/n^{1.0001})$ is a p-series, $p = 1.0001 > 1$, so it converges.

11. $1 + \tfrac{1}{8} + \tfrac{1}{27} + \tfrac{1}{64} + \tfrac{1}{125} + \cdots = \sum_{n=1}^{\infty} (1/n^3)$. This is a p-series with $p = 3 > 1$, so it converges by (1).

13. $\displaystyle\sum_{n=1}^{\infty} \dfrac{5 - 2\sqrt{n}}{n^3} = 5\sum_{n=1}^{\infty} \dfrac{1}{n^3} - 2\sum_{n=1}^{\infty} \dfrac{1}{n^{5/2}}$ by Theorem 11.2.8, since $\sum_{n=1}^{\infty} \dfrac{1}{n^3}$ and $\sum_{n=1}^{\infty} \dfrac{1}{n^{5/2}}$ both converge by (1) (with

$p = 3$ and $p = \tfrac{5}{2}$). Thus, $\displaystyle\sum_{n=1}^{\infty} \dfrac{5 - 2\sqrt{n}}{n^3}$ converges.

15. $f(x) = xe^{-x^2}$ is continuous and positive on $[1, \infty)$, and since $f'(x) = e^{-x^2}(1 - 2x^2) < 0$ for $x > 1$, f is decreasing as well. Thus, we can use the Integral Test.

$\displaystyle\int_1^{\infty} xe^{-x^2}\,dx = \lim_{t\to\infty}\left[-\tfrac{1}{2}e^{-x^2}\right]_1^t = 0 - \left(-\tfrac{1}{2}e^{-1}\right) = 1/(2e)$. Since the integral converges, the series converges.

17. $f(x) = \dfrac{x}{x^2+1}$ is continuous and positive on $[1, \infty)$, and since $f'(x) = \dfrac{1-x^2}{(x^2+1)^2} < 0$ for $x > 1$, f is also

decreasing. Using the Integral Test, $\displaystyle\int_1^{\infty} \dfrac{x}{x^2+1}\,dx = \lim_{t\to\infty}\left[\dfrac{\ln(x^2+1)}{2}\right]_1^t = \infty$, so the series diverges.

19. $f(x) = \dfrac{1}{x\ln x}$ is continuous and positive on $[2, \infty)$, and also decreasing since $f'(x) = -\dfrac{1+\ln x}{x^2(\ln x)^2} < 0$ for

$x > 2$, so we can use the Integral Test. $\displaystyle\int_2^{\infty} \dfrac{1}{x\ln x}\,dx = \lim_{t\to\infty}\left[\ln(\ln x)\right]_2^t = \lim_{t\to\infty}\left[\ln(\ln t) - \ln(\ln 2)\right] = \infty$, so the

series diverges.

21. $f(x) = \dfrac{\arctan x}{1 + x^2}$ is continuous and positive on $[1, \infty)$. $f'(x) = \dfrac{1 - 2x \arctan x}{(1 + x^2)^2} < 0$ for $x > 1$, since

$2x \arctan x \geq \frac{\pi}{2} > 1$ for $x \geq 1$. So f is decreasing and we can use the Integral Test.

$\displaystyle\int_1^\infty \dfrac{\arctan x}{1 + x^2}\, dx = \lim_{t\to\infty} \left[\tfrac{1}{2} \arctan^2 x \right]_1^t = \dfrac{(\pi/2)^2}{2} - \dfrac{(\pi/4)^2}{2} = \dfrac{3\pi^2}{32}$, so the series converges.

23. $f(x) = \dfrac{1}{x^2 + 2x + 2}$ is continuous and positive on $[1, \infty)$, and $f'(x) = -\dfrac{2x + 2}{(x^2 + 2x + 2)^2} < 0$

for $x \geq 1$, so f is decreasing and we can use the Integral Test.

$\displaystyle\int_1^\infty \dfrac{1}{x^2 + 2x + 2}\, dx = \int_1^\infty \dfrac{1}{(x + 1)^2 + 1}\, dx = \lim_{t\to\infty} [\arctan(x + 1)]_1^t = \frac{\pi}{2} - \arctan 2$, so the series converges as
well.

25. We have already shown (in Exercise 19) that when $p = 1$ the series $\displaystyle\sum_{n=2}^{\infty} \dfrac{1}{n\,(\ln n)^p}$ diverges, so assume that $p \neq 1$.

$f(x) = \dfrac{1}{x\,(\ln x)^p}$ is continuous and positive on $[2, \infty)$, and $f'(x) = -\dfrac{p + \ln x}{x^2\,(\ln x)^{p+1}} < 0$ if $x > e^{-p}$, so that f is

eventually decreasing and we can use the Integral Test.

$\displaystyle\int_2^\infty \dfrac{1}{x\,(\ln x)^p}\, dx = \lim_{t\to\infty} \left[\dfrac{(\ln x)^{1-p}}{1 - p} \right]_2^t \quad (\text{for } p \neq 1) = \lim_{t\to\infty} \left[\dfrac{(\ln t)^{1-p}}{1 - p} \right] - \dfrac{(\ln 2)^{1-p}}{1 - p}$

This limit exists whenever $1 - p < 0 \iff p > 1$, so the series converges for $p > 1$.

27. Clearly the series cannot converge if $p \geq -\frac{1}{2}$, because then $\displaystyle\lim_{n\to\infty} n\,(1 + n^2)^p \neq 0$. Also, if $p = -1$ the series

diverges (see Exercise 17). So assume $p < -\frac{1}{2}$, $p \neq -1$. Then $f(x) = x\,(1 + x^2)^p$ is continuous,

positive, and eventually decreasing on $[1, \infty)$, and we can use the Integral Test.

$\displaystyle\int_1^\infty x\,(1 + x^2)^p\, dx = \lim_{t\to\infty} \left[\dfrac{1}{2} \cdot \dfrac{(1 + x^2)^{p+1}}{p + 1} \right]_1^t = \lim_{t\to\infty} \dfrac{1}{2} \cdot \dfrac{(1 + t^2)^{p+1}}{p + 1} - \dfrac{2^p}{p + 1}$. This limit exists and is finite

$\iff p + 1 < 0 \iff p < -1$, so the series converges whenever $p < -1$.

29. Since this is a p-series with $p = x$, $\zeta(x)$ is defined when $x > 1$. Unless specified otherwise, the domain of a
function f is the set of numbers x such that the expression for $f(x)$ makes sense and defines a real number. So, in
the case of a series, it's the set of numbers x such that the series is convergent.

31. (a) $f(x) = \dfrac{1}{x^2}$ is positive and continuous and $f'(x) = -\dfrac{2}{x^3}$ is negative for $x > 1$, and so the Integral

Test applies. $\displaystyle\sum_{n=1}^{\infty} \dfrac{1}{n^2} \approx s_{10} = \dfrac{1}{1^2} + \dfrac{1}{2^2} + \dfrac{1}{3^2} + \cdots + \dfrac{1}{10^2} \approx 1.549768$.

$R_{10} \leq \displaystyle\int_{10}^\infty \dfrac{1}{x^2}\, dx = \lim_{t\to\infty} \left[\dfrac{-1}{x} \right]_{10}^t = \lim_{t\to\infty} \left(-\dfrac{1}{t} + \dfrac{1}{10} \right) = \dfrac{1}{10}$, so the error is at most 0.1.

(b) $s_{10} + \displaystyle\int_{11}^\infty \dfrac{1}{x^2}\, dx \leq s \leq s_{10} + \int_{10}^\infty \dfrac{1}{x^2}\, dx \implies s_{10} + \dfrac{1}{11} \leq s \leq s_{10} + \dfrac{1}{10} \implies$

$1.549768 + 0.090909 = 1.640677 \leq s \leq 1.549768 + 0.1 = 1.649768$, so we get $s \approx 1.64522$ (the average of
1.640677 and 1.649768) with error ≤ 0.005 (the maximum of $1.649768 - 1.64522$ and $1.64522 - 1.640677$,
rounded up).

(c) $R_n \leq \displaystyle\int_n^\infty \dfrac{1}{x^2}\, dx = \dfrac{1}{n}$. So $R_n < 0.001$ if $\dfrac{1}{n} < \dfrac{1}{1000} \iff n > 1000$.

33. $f(x) = x^{-3/2}$ is positive and continuous and $f'(x) = -\frac{3}{2}x^{-5/2}$ is negative for $x > 1$, so the Integral Test applies. From the end of Example 6, we see that the error is at most half the length of the interval. From (3), the interval is $\left(s_n + \int_{n+1}^{\infty} f(x)\,dx, s_n + \int_{n}^{\infty} f(x)\,dx\right)$, so its length is $\int_{n}^{\infty} f(x)\,dx - \int_{n+1}^{\infty} f(x)\,dx$. Thus, we need n such that

$$0.01 > \tfrac{1}{2}\left(\int_{n}^{\infty} x^{-3/2}\,dx - \int_{n+1}^{\infty} x^{-3/2}\,dx\right) = \frac{1}{2}\left(\lim_{t\to\infty}\left[\frac{-2}{\sqrt{x}}\right]_{n}^{t} - \lim_{t\to\infty}\left[\frac{-2}{\sqrt{x}}\right]_{n+1}^{t}\right) = \frac{1}{\sqrt{n}} - \frac{1}{\sqrt{n+1}}$$

⇔ $n > 13.08$. Again from the end of Example 6, we approximate s by the midpoint of this interval. In general, the midpoint is $\frac{1}{2}\left[\left(s_n + \int_{n+1}^{\infty} f(x)\,dx\right) + \left(s_n + \int_{n}^{\infty} f(x)\,dx\right)\right] = s_n + \frac{1}{2}\left(\int_{n+1}^{\infty} f(x)\,dx + \int_{n}^{\infty} f(x)\,dx\right)$. So using $n = 14$, we have $s \approx s_{14} + \frac{1}{2}\left(\int_{14}^{\infty} x^{-3/2}\,dx + \int_{15}^{\infty} x^{-3/2}\,dx\right) = 2.0872 + \frac{1}{\sqrt{14}} + \frac{1}{\sqrt{15}} \approx 2.6127$. Any larger value of n will also work. For instance, $s \approx s_{30} + \frac{1}{\sqrt{30}} + \frac{1}{\sqrt{31}} \approx 2.6124$.

35. (a) From the figure, $a_2 + a_3 + \cdots + a_n \le \int_{1}^{n} f(x)\,dx$, so with

$$f(x) = \frac{1}{x},$$

$$\frac{1}{2} + \frac{1}{3} + \frac{1}{4} + \cdots + \frac{1}{n} \le \int_{1}^{n} \frac{1}{x}\,dx = \ln n.$$

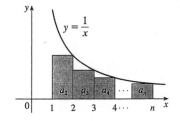

Thus, $s_n = 1 + \frac{1}{2} + \frac{1}{3} + \frac{1}{4} + \cdots + \frac{1}{n} \le 1 + \ln n$.

(b) By part (a), $s_{10^6} \le 1 + \ln 10^6 \approx 14.82 < 15$ and $s_{10^9} \le 1 + \ln 10^9 \approx 21.72 < 22$.

37. $b^{\ln n} = \left(e^{\ln b}\right)^{\ln n} = \left(e^{\ln n}\right)^{\ln b} = n^{\ln b} = \frac{1}{n^{-\ln b}}$. This is a p-series, which converges for all b such that $-\ln b > 1$ ⇔ $\ln b < -1$ ⇔ $b < e^{-1}$ ⇔ $b < 1/e$.

11.4 The Comparison Tests

1. (a) We cannot say anything about $\sum a_n$. If $a_n > b_n$ for all n and $\sum b_n$ is convergent, then $\sum a_n$ could be convergent or divergent. (See Note 2 on page 723.)

(b) If $a_n < b_n$ for all n, then $\sum a_n$ is convergent. [This is part (i) of the Comparison Test.]

3. $\frac{1}{n^2 + n + 1} < \frac{1}{n^2}$ for all $n \ge 1$, so $\sum_{n=1}^{\infty} \frac{1}{n^2 + n + 1}$ converges by comparison with $\sum_{n=1}^{\infty} \frac{1}{n^2}$, which converges because it is a p-series with $p = 2 > 1$.

5. $\frac{5}{2 + 3^n} < \frac{5}{3^n}$ for all $n \ge 1$, so $\sum_{n=1}^{\infty} \frac{5}{2 + 3^n}$ converges by comparison with $\sum_{n=1}^{\infty} \frac{5}{3^n} = 5\sum_{n=1}^{\infty} \frac{1}{3^n}$, which converges because $\sum_{n=1}^{\infty} \frac{1}{3^n}$ is a convergent geometric series with $r = \frac{1}{3}$.

7. $\frac{n+1}{n^2} > \frac{n}{n^2} = \frac{1}{n}$ for all $n \ge 1$, so $\sum_{n=1}^{\infty} \frac{n+1}{n^2}$ diverges by comparison with the harmonic series $\sum_{n=1}^{\infty} \frac{1}{n}$.

9. $\frac{3}{n2^n} \le \frac{3}{2^n} \cdot \sum_{n=1}^{\infty} \frac{3}{2^n}$ is a geometric series with $|r| = \frac{1}{2} < 1$, and hence converges, so $\sum_{n=1}^{\infty} \frac{3}{n2^n}$ converges also, by the Comparison Test.

11. $\dfrac{1}{\sqrt{n(n+1)(n+2)}} < \dfrac{1}{\sqrt{n \cdot n \cdot n}} = \dfrac{1}{n^{3/2}}$ and since $\displaystyle\sum_{n=1}^{\infty} \dfrac{1}{n^{3/2}}$ converges $(p = \frac{3}{2} > 1)$, so does

$\displaystyle\sum_{n=1}^{\infty} \dfrac{1}{\sqrt{n(n+1)(n+2)}}$ by the Comparison Test.

13. If $a_n = \dfrac{n^2+1}{n^3-1}$ and $b_n = \dfrac{1}{n}$, then $\displaystyle\lim_{n\to\infty} \dfrac{a_n}{b_n} = \lim_{n\to\infty} \dfrac{n^3+n}{n^3-1} = \lim_{n\to\infty} \dfrac{1+1/n^2}{1-1/n^3} = 1$, so $\displaystyle\sum_{n=2}^{\infty} \dfrac{n^2+1}{n^3-1}$ diverges by the

Limit Comparison Test with the divergent (partial) harmonic series $\displaystyle\sum_{n=2}^{\infty} \dfrac{1}{n}$.

15. $\dfrac{3+\cos n}{3^n} \le \dfrac{4}{3^n}$ since $\cos n \le 1$. $\displaystyle\sum_{n=1}^{\infty} \dfrac{4}{3^n}$ is a geometric series with $|r| = \frac{1}{3} < 1$ so it converges, and so

$\displaystyle\sum_{n=1}^{\infty} \dfrac{3+\cos n}{3^n}$ converges by the Comparison Test.

17. $\dfrac{n}{\sqrt{n^5+4}} < \dfrac{n}{\sqrt{n^5}} = \dfrac{1}{n^{3/2}}$. $\displaystyle\sum_{n=1}^{\infty} \dfrac{1}{n^{3/2}}$ is a convergent p-series $(p = \frac{3}{2} > 1)$ so $\displaystyle\sum_{n=1}^{\infty} \dfrac{n}{\sqrt{n^5+4}}$ converges by the

Comparison Test.

19. $\dfrac{2^n}{1+3^n} < \dfrac{2^n}{3^n} = \left(\dfrac{2}{3}\right)^n$. $\displaystyle\sum_{n=1}^{\infty} \left(\dfrac{2}{3}\right)^n$ is a convergent geometric series $(|r| = \frac{2}{3} < 1)$, so $\displaystyle\sum_{n=1}^{\infty} \dfrac{2^n}{1+3^n}$ converges by the

Comparison Test.

21. Use the Limit Comparison Test with $a_n = \dfrac{1}{1+\sqrt{n}}$ and $b_n = \dfrac{1}{\sqrt{n}}$: $\displaystyle\lim_{n\to\infty} \dfrac{a_n}{b_n} = \lim_{n\to\infty} \dfrac{\sqrt{n}}{1+\sqrt{n}} = 1 > 0$. Since

$\displaystyle\sum_{n=1}^{\infty} \dfrac{1}{\sqrt{n}}$ is a divergent p-series $(p = \frac{1}{2} \le 1)$, $\displaystyle\sum_{n=1}^{\infty} \dfrac{1}{1+\sqrt{n}}$ also diverges.

23. Let $a_n = \dfrac{n^2+1}{n^4+1}$ and $b_n = \dfrac{1}{n^2}$. Then $\displaystyle\lim_{n\to\infty} \dfrac{a_n}{b_n} = \lim_{n\to\infty} \dfrac{n^4+n^2}{n^4+1} = 1$. Since $\displaystyle\sum_{n=1}^{\infty} \dfrac{1}{n^2}$ is a convergent p-series

$(p = 2 > 1)$, so is $\displaystyle\sum_{n=1}^{\infty} \dfrac{n^2+1}{n^4+1}$ by the Limit Comparison Test.

25. If $a_n = \dfrac{1+n+n^2}{\sqrt{1+n^2+n^6}}$ and $b_n = \dfrac{1}{n}$, then $\displaystyle\lim_{n\to\infty} \dfrac{a_n}{b_n} = \lim_{n\to\infty} \dfrac{n+n^2+n^3}{\sqrt{1+n^2+n^6}} = \lim_{n\to\infty} \dfrac{1/n^2+1/n+1}{\sqrt{1/n^6+1/n^4+1}} = 1$, so

$\displaystyle\sum_{n=1}^{\infty} \dfrac{1+n+n^2}{\sqrt{1+n^2+n^6}}$ diverges by the Limit Comparison Test with the divergent harmonic series $\displaystyle\sum_{n=1}^{\infty} \dfrac{1}{n}$.

27. Let $a_n = \dfrac{n+1}{n2^n}$ and $b_n = \dfrac{1}{2^n}$. Then $\displaystyle\lim_{n\to\infty} \dfrac{a_n}{b_n} = \lim_{n\to\infty} \dfrac{n+1}{n} = 1$. Since $\displaystyle\sum_{n=1}^{\infty} \dfrac{1}{2^n}$ is a convergent geometric series

$(|r| = \frac{1}{2} < 1)$, $\displaystyle\sum_{n=1}^{\infty} \dfrac{n+1}{n2^n}$ converges by the Limit Comparison Test.

29. Clearly $n! = n(n-1)(n-2)\cdots(3)(2) \ge 2 \cdot 2 \cdot 2 \cdots \cdots 2 \cdot 2 = 2^{n-1}$, so $\dfrac{1}{n!} \le \dfrac{1}{2^{n-1}}$. $\displaystyle\sum_{n=1}^{\infty} \dfrac{1}{2^{n-1}}$ is a convergent

geometric series $(|r| = \frac{1}{2} < 1)$ so $\displaystyle\sum_{n=1}^{\infty} \dfrac{1}{n!}$ converges by the Comparison Test.

31. Use the Limit Comparison Test with $a_n = \sin\left(\dfrac{1}{n}\right)$ and $b_n = \dfrac{1}{n}$: $\displaystyle\lim_{n\to\infty} \dfrac{a_n}{b_n} = \lim_{n\to\infty} \dfrac{\sin(1/n)}{1/n} = \lim_{\theta\to 0} \dfrac{\sin\theta}{\theta} = 1 > 0$.

Since $\sum_{n=1}^{\infty} b_n$ is the divergent harmonic series, $\sum_{n=1}^{\infty} \sin(1/n)$ also diverges.

33. $\displaystyle\sum_{n=1}^{10} \frac{1}{n^4+n^2} = \frac{1}{2}+\frac{1}{20}+\frac{1}{90}+\cdots+\frac{1}{10,100} \approx 0.567975.$ Now $\dfrac{1}{n^4+n^2} < \dfrac{1}{n^4}$, so using the reasoning and

notation of Example 5, the error is $R_{10} \le T_{10} = \displaystyle\sum_{n=11}^{\infty} \frac{1}{n^4} \le \int_{10}^{\infty} \frac{dx}{x^4} = \lim_{t\to\infty} \left[-\frac{x^{-3}}{3}\right]_{10}^{t} = \frac{1}{3000} = 0.000\overline{3}.$

35. $\displaystyle\sum_{n=1}^{10} \frac{1}{1+2^n} = \frac{1}{3}+\frac{1}{5}+\frac{1}{9}+\cdots+\frac{1}{1025} \approx 0.76352.$ Now $\dfrac{1}{1+2^n} < \dfrac{1}{2^n}$, so the error is

$R_{10} \le T_{10} = \displaystyle\sum_{n=11}^{\infty} \frac{1}{2^n} = \frac{1/2^{11}}{1-1/2}$ (geometric series) $\approx 0.00098.$

37. Since $\dfrac{d_n}{10^n} \le \dfrac{9}{10^n}$ for each n, and since $\displaystyle\sum_{n=1}^{\infty} \frac{9}{10^n}$ is a convergent geometric series $(|r| = \frac{1}{10} < 1)$,

$0.d_1 d_2 d_3 \ldots = \displaystyle\sum_{n=1}^{\infty} \frac{d_n}{10^n}$ will always converge by the Comparison Test.

39. Since $\sum a_n$ converges, $\displaystyle\lim_{n\to\infty} a_n = 0$, so there exists N such that $|a_n - 0| < 1$ for all $n > N$ $\Rightarrow$ $0 \le a_n < 1$ for

all $n > N$ $\Rightarrow$ $0 \le a_n^2 \le a_n$. Since $\sum a_n$ converges, so does $\sum a_n^2$ by the Comparison Test.

41. (a) We wish to prove that if $\displaystyle\lim_{n\to\infty} \frac{a_n}{b_n} = \infty$ and $\sum b_n$ diverges, then so does $\sum a_n$. So suppose on the contrary that

$\sum a_n$ converges. Since $\displaystyle\lim_{n\to\infty} \frac{a_n}{b_n} = \infty$, we have that $\displaystyle\lim_{n\to\infty} \frac{b_n}{a_n} = 0$, so by the extension of the Limit Comparison

Test proved in Exercise 40(a), if $\sum a_n$ converges, so must $\sum b_n$. But this contradicts our hypothesis, so $\sum a_n$

must diverge.

(b) If $a_n = \dfrac{1}{\ln n}$ and $b_n = \dfrac{1}{n}$ for $n \ge 2$, then $\displaystyle\lim_{n\to\infty} \frac{a_n}{b_n} = \lim_{n\to\infty} \frac{n}{\ln n} = \lim_{x\to\infty} \frac{x}{\ln x} \stackrel{H}{=} \lim_{x\to\infty} \frac{1}{1/x} = \lim_{x\to\infty} x = \infty$, so

by part (a), $\displaystyle\sum_{n=2}^{\infty} \frac{1}{\ln n}$ is divergent.

43. $\displaystyle\lim_{n\to\infty} na_n = \lim_{n\to\infty} \frac{a_n}{1/n}$, so we apply the Limit Comparison Test with $b_n = \dfrac{1}{n}$. Since $\displaystyle\lim_{n\to\infty} na_n > 0$ we know that

either both series converge or both series diverge, and we also know that $\displaystyle\sum_{n=0}^{\infty} \frac{1}{n}$ diverges (p-series with $p = 1$).

Therefore, $\sum a_n$ must be divergent.

45. Yes. Since $\sum a_n$ converges, its terms approach 0 as $n \to \infty$, so $\displaystyle\lim_{n\to\infty} \frac{\sin a_n}{a_n} = 1$ by Theorem 3.4.2. Thus,

$\sum \sin a_n$ converges by the Limit Comparison Test.

11.5 Alternating Series

1. (a) An alternating series is a series whose terms are alternately positive and negative.

(b) An alternating series $\sum_{n=1}^{\infty} (-1)^{n-1} b_n$ converges if $0 < b_{n+1} \le b_n$ for all n and $\displaystyle\lim_{n\to\infty} b_n = 0$. (This is the
Alternating Series Test.)

(c) The error involved in using the partial sum s_n as an approximation to the total sum s is the remainder
$R_n = s - s_n$ and the size of the error is smaller than b_{n+1}, that is, $|R_n| \le b_{n+1}$. (This is the Alternating Series
Estimation Theorem.)

3. $\dfrac{4}{7} - \dfrac{4}{8} + \dfrac{4}{9} - \dfrac{4}{10} + \dfrac{4}{11} - \cdots = \sum_{n=1}^{\infty} (-1)^{n-1} \dfrac{4}{n+6}$. $b_n = \dfrac{4}{n+6} > 0$, $\{b_n\}$ is decreasing, and $\lim_{n\to\infty} b_n = 0$, so the
series converges by the Alternating Series Test.

5. $b_n = \dfrac{1}{\sqrt{n}} > 0$, $\{b_n\}$ is decreasing, and $\lim_{n\to\infty} b_n = 0$, so $\sum_{n=1}^{\infty} \dfrac{(-1)^{n-1}}{\sqrt{n}}$ converges by the Alternating Series Test.

7. $a_n = (-1)^n \dfrac{2n}{4n+1}$, so $|a_n| = \dfrac{2n}{4n+1} \to \dfrac{1}{2}$ as $n \to \infty$. Therefore, $\lim_{n\to\infty} a_n \neq 0$ (in fact the limit does not exist)
and the series $\sum_{n=1}^{\infty} (-1)^n \dfrac{2n}{4n+1}$ diverges by the Test for Divergence.

9. $b_n = \dfrac{1}{4n^2+1} > 0$, $\{b_n\}$ is decreasing, and $\lim_{n\to\infty} b_n = 0$, so the series $\sum_{n=1}^{\infty} \dfrac{(-1)^{n+1}}{4n^2+1}$ converges by the Alternating
Series Test.

11. $\sum_{n=1}^{\infty} (-1)^{n-1} \dfrac{\sqrt{n}}{n+4}$. $b_n = \dfrac{\sqrt{n}}{n+4} > 0$ for all n. Let $f(x) = \dfrac{\sqrt{x}}{x+4}$. Then $f'(x) = \dfrac{4-x}{2\sqrt{x}\,(x+4)^2} < 0$ if $x > 4$, so

$\{b_n\}$ is decreasing after $n = 4$. $\lim_{n\to\infty} \dfrac{\sqrt{n}}{n+4} = \lim_{n\to\infty} \dfrac{1}{\sqrt{n}+4/\sqrt{n}} = 0$. So the series converges by the Alternating

Series Test.

13. $\sum_{n=2}^{\infty} (-1)^n \dfrac{n}{\ln n}$. $\lim_{n\to\infty} \dfrac{n}{\ln n} \overset{\text{H}}{=} \lim_{n\to\infty} \dfrac{1}{1/n} = \infty$, so the series diverges by the Test for Divergence.

15. $\sum_{n=1}^{\infty} \dfrac{\cos n\pi}{n^{3/4}} = \sum_{n=1}^{\infty} \dfrac{(-1)^n}{n^{3/4}}$. $b_n = \dfrac{1}{n^{3/4}}$ is decreasing and positive and $\lim_{n\to\infty} \dfrac{1}{n^{3/4}} = 0$, so the series converges by the
Alternating Series Test.

17. $\sum_{n=1}^{\infty} (-1)^n \sin \dfrac{\pi}{n}$. $b_n = \sin \dfrac{\pi}{n} > 0$ for $n \geq 2$ and $\sin \dfrac{\pi}{n} \geq \sin \dfrac{\pi}{n+1}$, and $\lim_{n\to\infty} \sin \dfrac{\pi}{n} = \sin 0 = 0$, so the series
converges by the Alternating Series Test.

19. $\dfrac{n^n}{n!} = \dfrac{n \cdot n \cdots \cdots n}{1 \cdot 2 \cdots \cdots n} \geq n \;\Rightarrow\; \lim_{n\to\infty} \dfrac{n^n}{n!} = \infty \;\Rightarrow\; \lim_{n\to\infty} \dfrac{(-1)^n n^n}{n!}$ does not exist. So the series diverges by the
Test for Divergence.

21.

n	a_n	S_n
1	1	1
2	-0.35355	0.64645
3	0.19245	0.83890
4	-0.125	0.71390
5	0.08944	0.80334
6	-0.06804	0.73530
7	0.05399	0.78929
8	-0.04419	0.74510
9	0.03704	0.78214
10	-0.03162	0.75051

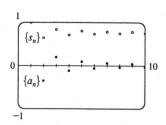

By the Alternating Series Estimation Theorem, the error in the

approximation $\sum_{n=1}^{\infty} \dfrac{(-1)^{n-1}}{n^{3/2}} \approx 0.75051$ is

$|s - s_{10}| \leq b_{11} = 1/(11)^{3/2} \approx 0.0275$ (to four decimal places,
rounded up).

23. With $b_n = 1/n^2$, $b_{10} = 1/10^2 = 0.01$ and $b_{11} = 1/11^2 = 1/121 \approx 0.008 < 0.01$, so by the Alternating Series
Estimation Theorem, $n = 10$.

25. $b_7 = 2^7/7! \approx 0.025 > 0.01$ and $b_8 = 2^8/8! \approx 0.006 < 0.01$, so by the Alternating Series Estimation Theorem, $n = 7$. (That is, since the 8th term is less than the desired error, we need to add the first 7 terms to get the sum to the desired accuracy.)

27. $\sum\limits_{n=1}^{\infty} \dfrac{(-1)^{n-1}}{(2n-1)!}$. $b_5 = \dfrac{1}{(2\cdot5-1)!} = \dfrac{1}{362,880} < 0.00001$, so $\sum\limits_{n=1}^{\infty} \dfrac{(-1)^{n-1}}{(2n-1)!} \approx \sum\limits_{n=1}^{4} \dfrac{(-1)^{n-1}}{(2n-1)!} \approx 0.8415$.

29. $b_6 = \dfrac{1}{2^6 6!} = \dfrac{1}{46,080} \approx 0.000022 < 0.00001$, so $\sum\limits_{n=0}^{\infty} \dfrac{(-1)^n}{2^n n!} \approx \sum\limits_{n=0}^{5} \dfrac{(-1)^n}{2^n n!} \approx 0.6065$.

31. $\sum\limits_{n=1}^{\infty} \dfrac{(-1)^{n-1}}{n} = 1 - \dfrac{1}{2} + \dfrac{1}{3} - \dfrac{1}{4} + \cdots + \dfrac{1}{49} - \dfrac{1}{50} + \dfrac{1}{51} - \dfrac{1}{52} + \cdots$. The 50th partial sum of this series is an

underestimate, since $\sum\limits_{n=1}^{\infty} \dfrac{(-1)^{n-1}}{n} = s_{50} + \left(\dfrac{1}{51} - \dfrac{1}{52}\right) + \left(\dfrac{1}{53} - \dfrac{1}{54}\right) + \cdots$, and the terms in parentheses are all positive. The result can be seen geometrically in Figure 1.

33. Clearly $b_n = \dfrac{1}{n+p}$ is decreasing and eventually positive and $\lim\limits_{n\to\infty} b_n = 0$ for any p. So the series converges (by the Alternating Series Test) for any p for which every b_n is defined, that is, $n + p \neq 0$ for $n \geq 1$, or p is not a negative integer.

35. $\sum b_{2n} = \sum 1/(2n)^2$ clearly converges (by comparison with the p-series for $p = 2$). So suppose that $\sum(-1)^{n-1} b_n$ converges. Then by Theorem 11.2.8(ii), so does

$$\sum\left[(-1)^{n-1} b_n + b_n\right] = 2\left(1 + \tfrac{1}{3} + \tfrac{1}{5} + \cdots\right) = 2\sum \dfrac{1}{2n-1}.$$ But this diverges by comparison with the harmonic

series, a contradiction. Therefore, $\sum(-1)^{n-1} b_n$ must diverge. The Alternating Series Test does not apply since $\{b_n\}$ is not decreasing.

11.6 Absolute Convergence and the Ratio and Root Tests

1. (a) Since $\lim\limits_{n\to\infty} \left|\dfrac{a_{n+1}}{a_n}\right| = 8 > 1$, part (ii) of the Ratio Test tells us that $\sum a_n$ is divergent.

(b) Since $\lim\limits_{n\to\infty} \left|\dfrac{a_{n+1}}{a_n}\right| = 0.8 < 1$, part (i) of the Ratio Test tells us that $\sum a_n$ is convergent.

(c) Since $\lim\limits_{n\to\infty} \left|\dfrac{a_{n+1}}{a_n}\right| = 1$, the Ratio Test fails and $\sum a_n$ might converge or it might diverge.

3. $\sum\limits_{n=1}^{\infty} \dfrac{1}{n\sqrt{n}} = \sum\limits_{n=1}^{\infty} \dfrac{1}{n^{3/2}}$ is a convergent p-series ($p = \tfrac{3}{2} > 1$), so the given series is absolutely convergent.

5. Using the Ratio Test, $\lim\limits_{n\to\infty} \left|\dfrac{a_{n+1}}{a_n}\right| = \lim\limits_{n\to\infty} \left|\dfrac{(-3)^{n+1}/(n+1)^3}{(-3)^n/n^3}\right| = 3 \lim\limits_{n\to\infty} \left(\dfrac{n}{n+1}\right)^3 = 3 > 1$, so the series

diverges.

7. $\sum\limits_{n=1}^{\infty} \dfrac{(-1)^n}{5+n}$ converges by the Alternating Series Test, but $\sum\limits_{n=1}^{\infty} \dfrac{1}{5+n}$ diverges by the Limit Comparison Test with the

harmonic series $\sum\limits_{n=1}^{\infty} \dfrac{1}{n}$, so the given series is conditionally convergent.

9. $\lim\limits_{n\to\infty} |a_n| = \lim\limits_{n\to\infty} \dfrac{n}{5+n} = \lim\limits_{n\to\infty} \dfrac{1}{5/n+1} = 1$, so $\lim\limits_{n\to\infty} a_n \neq 0$. Thus, the given series is divergent by the Test for Divergence.

11. $\lim\limits_{n\to\infty} \left| \dfrac{a_{n+1}}{a_n} \right| = \lim\limits_{n\to\infty} \dfrac{1/(2n+2)!}{1/(2n)!} = \lim\limits_{n\to\infty} \dfrac{(2n)!}{(2n+2)!} = \lim\limits_{n\to\infty} \dfrac{(2n)!}{(2n+2)(2n+1)(2n)!} =$

$\lim\limits_{n\to\infty} \dfrac{1}{(2n+2)(2n+1)} = 0 < 1$, so the series is absolutely convergent by the Ratio Test. Of course, absolute convergence is the same as convergence for this series, since all of its terms are positive.

13. $\left| \dfrac{\sin 2n}{n^2} \right| \leq \dfrac{1}{n^2}$ and $\sum\limits_{n=1}^{\infty} \dfrac{1}{n^2}$ converges (p-series, $p = 2 > 1$), so $\sum\limits_{n=1}^{\infty} \dfrac{\sin 2n}{n^2}$ converges absolutely by the Comparison Test.

15. $\lim\limits_{n\to\infty} \left| \dfrac{a_{n+1}}{a_n} \right| = \lim\limits_{n\to\infty} \left[\dfrac{(n+1)3^{n+1}}{4^n} \cdot \dfrac{4^{n-1}}{n \cdot 3^n} \right] = \lim\limits_{n\to\infty} \left(\dfrac{3}{4} \cdot \dfrac{n+1}{n} \right) = \dfrac{3}{4} < 1$, so the series is absolutely convergent by the Ratio Test.

17. $\lim\limits_{n\to\infty} \left| \dfrac{a_{n+1}}{a_n} \right| = \lim\limits_{n\to\infty} \left[\dfrac{10^{n+1}}{(n+2)4^{2n+3}} \cdot \dfrac{(n+1)4^{2n+1}}{10^n} \right] = \lim\limits_{n\to\infty} \left(\dfrac{10}{4^2} \cdot \dfrac{n+1}{n+2} \right) = \dfrac{5}{8} < 1$, so the series is absolutely convergent by the Ratio Test. Since the terms of this series are positive, absolute convergence is the same as convergence.

19. $\lim\limits_{n\to\infty} \left| \dfrac{a_{n+1}}{a_n} \right| = \lim\limits_{n\to\infty} \dfrac{(n+1)!/10^{n+1}}{n!/10^n} = \lim\limits_{n\to\infty} \dfrac{n+1}{10} = \infty$, so the series diverges by the Ratio Test.

21. $\dfrac{|\cos(n\pi/3)|}{n!} \leq \dfrac{1}{n!}$ and $\sum\limits_{n=1}^{\infty} \dfrac{1}{n!}$ converges (Exercise 11.4.29), so the given series converges absolutely by the Comparison Test.

23. $\lim\limits_{n\to\infty} \sqrt[n]{|a_n|} = \lim\limits_{n\to\infty} \left(\dfrac{n^n}{3^{1+3n}} \right)^{1/n} = \lim\limits_{n\to\infty} \dfrac{n}{\sqrt[n]{3} \cdot 3^3} = \infty$, so the series is divergent by the Root Test.

Or: $\lim\limits_{n\to\infty} \left| \dfrac{a_{n+1}}{a_n} \right| = \lim\limits_{n\to\infty} \left[\dfrac{(n+1)^{n+1}}{3^{4+3n}} \cdot \dfrac{3^{1+3n}}{n^n} \right] = \lim\limits_{n\to\infty} \left[\dfrac{1}{3^3} \cdot \left(\dfrac{n+1}{n} \right)^n (n+1) \right]$

$= \dfrac{1}{27} \lim\limits_{n\to\infty} \left(1 + \dfrac{1}{n} \right)^n \lim\limits_{n\to\infty} (n+1) = \dfrac{1}{27} e \lim\limits_{n\to\infty} (n+1) = \infty$

so the series is divergent by the Ratio Test.

25. $\lim\limits_{n\to\infty} \sqrt[n]{|a_n|} = \lim\limits_{n\to\infty} \dfrac{n^2+1}{2n^2+1} = \lim\limits_{n\to\infty} \dfrac{1+1/n^2}{2+1/n^2} = \dfrac{1}{2} < 1$, so the series is absolutely convergent by the Root Test.

27. $\lim\limits_{n\to\infty} \left| \dfrac{a_{n+1}}{a_n} \right| = \lim\limits_{n\to\infty} \dfrac{(n+1)!/[1 \cdot 3 \cdot 5 \cdots (2n+1)]}{n!/[1 \cdot 3 \cdot 5 \cdots (2n-1)]} = \lim\limits_{n\to\infty} \dfrac{n+1}{2n+1} = \dfrac{1}{2} < 1$, so the series converges absolutely by the Ratio Test.

29. $\sum\limits_{n=1}^{\infty} \dfrac{2 \cdot 4 \cdot 6 \cdots (2n)}{n!} = \sum\limits_{n=1}^{\infty} \dfrac{2^n n!}{n!} = \sum\limits_{n=1}^{\infty} 2^n$ which diverges by the Test for Divergence since $\lim\limits_{n\to\infty} 2^n = \infty$.

31. By the recursive definition, $\lim\limits_{n\to\infty} \left| \dfrac{a_{n+1}}{a_n} \right| = \lim\limits_{n\to\infty} \left| \dfrac{5n+1}{4n+3} \right| = \dfrac{5}{4} > 1$, so the series diverges by the Ratio Test.

33. (a) $\lim\limits_{n\to\infty}\left|\dfrac{1/(n+1)^3}{1/n^3}\right| = \lim\limits_{n\to\infty}\dfrac{n^3}{(n+1)^3} = \lim\limits_{n\to\infty}\dfrac{1}{(1+1/n)^3} = 1$. Inconclusive.

(b) $\lim\limits_{n\to\infty}\left|\dfrac{(n+1)}{2^{n+1}}\cdot\dfrac{2^n}{n}\right| = \lim\limits_{n\to\infty}\dfrac{n+1}{2n} = \lim\limits_{n\to\infty}\left(\dfrac{1}{2}+\dfrac{1}{2n}\right) = \dfrac{1}{2}$. Conclusive (convergent).

(c) $\lim\limits_{n\to\infty}\left|\dfrac{(-3)^n}{\sqrt{n+1}}\cdot\dfrac{\sqrt{n}}{(-3)^{n-1}}\right| = 3\lim\limits_{n\to\infty}\sqrt{\dfrac{n}{n+1}} = 3\lim\limits_{n\to\infty}\sqrt{\dfrac{1}{1+1/n}} = 3$. Conclusive (divergent).

(d) $\lim\limits_{n\to\infty}\left|\dfrac{\sqrt{n+1}}{1+(n+1)^2}\cdot\dfrac{1+n^2}{\sqrt{n}}\right| = \lim\limits_{n\to\infty}\left[\sqrt{1+\dfrac{1}{n}}\cdot\dfrac{1/n^2+1}{1/n^2+(1+1/n)^2}\right] = 1$. Inconclusive.

35. (a) $\lim\limits_{n\to\infty}\left|\dfrac{a_{n+1}}{a_n}\right| = \lim\limits_{n\to\infty}\dfrac{|x|^{n+1}/(n+1)!}{|x|^n/n!} = |x|\lim\limits_{n\to\infty}\dfrac{1}{n+1} = 0$, so by the Ratio Test the series converges for all

x.

(b) Since the series of part (a) always converges, we must have $\lim\limits_{n\to\infty}\dfrac{x^n}{n!} = 0$ by Theorem 11.2.6.

37. (a) $s_5 = \sum\limits_{n=1}^{5}\dfrac{1}{n2^n} = \dfrac{1}{2}+\dfrac{1}{8}+\dfrac{1}{24}+\dfrac{1}{64}+\dfrac{1}{160} = \dfrac{661}{960} \approx 0.68854$. Now the ratios

$r_n = \dfrac{a_{n+1}}{a_n} = \dfrac{n2^n}{(n+1)\,2^{n+1}} = \dfrac{n}{2(n+1)}$ form an increasing sequence, since

$r_{n+1} - r_n = \dfrac{n+1}{2(n+2)} - \dfrac{n}{2(n+1)} = \dfrac{(n+1)^2 - n(n+2)}{2(n+1)(n+2)} = \dfrac{1}{2(n+1)(n+2)} > 0$. So by Exercise 36(b),

the error is less than $\dfrac{a_6}{1-\lim\limits_{n\to\infty} r_n} = \dfrac{1/(6\cdot 2^6)}{1-1/2} = \dfrac{1}{192} \approx 0.00521$.

(b) The error in using s_n as an approximation to the sum is $R_n = \dfrac{a_{n+1}}{1-\frac{1}{2}} = \dfrac{2}{(n+1)\,2^{n+1}}$. We want $R_n < 0.00005$

$\Leftrightarrow\quad \dfrac{1}{(n+1)\,2^n} < 0.00005 \quad\Leftrightarrow\quad (n+1)\,2^n > 20{,}000$. To find such an n we can use trial and error or a

graph. We calculate $(11+1)\,2^{11} = 24{,}576$, so $s_{11} = \sum\limits_{n=1}^{11}\dfrac{1}{n2^n} \approx 0.693109$ is within 0.00005 of the actual

sum.

39. Summing the inequalities $-|a_i| \le a_i \le |a_i|$ for $i = 1, 2, \ldots, n$, we get $-\sum_{i=1}^{n}|a_i| \le \sum_{i=1}^{n}a_i \le \sum_{i=1}^{n}|a_i| \Rightarrow$
$-\lim\limits_{n\to\infty}\sum_{i=1}^{n}|a_i| \le \lim\limits_{n\to\infty}\sum_{i=1}^{n}a_i \le \lim\limits_{n\to\infty}\sum_{i=1}^{n}|a_i| \Rightarrow -\sum_{n=1}^{\infty}|a_n| \le \sum_{n=1}^{\infty}a_n \le \sum_{n=1}^{\infty}|a_n| \Rightarrow$
$\left|\sum_{n=1}^{\infty}a_n\right| \le \sum_{n=1}^{\infty}|a_n|$.

41. (a) Since $\sum a_n$ is absolutely convergent, and since $|a_n^+| \le |a_n|$ and $|a_n^-| \le |a_n|$ (because a_n^+ and a_n^- each equal
either a_n or 0), we conclude by the Comparison Test that both $\sum a_n^+$ and $\sum a_n^-$ must be absolutely convergent.
(Or use Theorem 11.2.8.)

(b) We will show by contradiction that both $\sum a_n^+$ and $\sum a_n^-$ must diverge. For suppose that

$\sum a_n^+$ converged. Then so would $\sum\left(a_n^+ - \frac{1}{2}a_n\right)$ by Theorem 11.2.8. But

$\sum\left(a_n^+ - \frac{1}{2}a_n\right) = \sum\left[\frac{1}{2}(a_n + |a_n|) - \frac{1}{2}a_n\right] = \frac{1}{2}\sum|a_n|$, which diverges because $\sum a_n$ is only conditionally

convergent. Hence, $\sum a_n^+$ can't converge. Similarly, neither can $\sum a_n^-$.

11.7 Strategy for Testing Series

1. $\lim\limits_{n\to\infty} a_n = \lim\limits_{n\to\infty} \dfrac{n^2-1}{n^2+1} = \lim\limits_{n\to\infty} \dfrac{1-1/n^2}{1+1/n} = 1 \neq 0$, so the series diverges by the Test for Divergence.

3. $\dfrac{1}{n^2+n} < \dfrac{1}{n^2}$ for all $n \geq 1$, so $\sum\limits_{n=1}^{\infty} \dfrac{1}{n^2+n}$ converges by the Comparison Test with $\sum\limits_{n=1}^{\infty} \dfrac{1}{n^2}$, a p-series that converges because $p = 2 > 1$.

5. $\lim\limits_{n\to\infty} \left| \dfrac{a_{n+1}}{a_n} \right| = \lim\limits_{n\to\infty} \left(\dfrac{3^{n+2}}{2^{3n+3}} \cdot \dfrac{2^{3n}}{3^{n+1}} \right) = \lim\limits_{n\to\infty} \dfrac{3}{2^3} = \dfrac{3}{8} < 1$, so the series is absolutely convergent by the Ratio Test.

7. $\sum\limits_{k=1}^{\infty} k^{-1.7} = \sum\limits_{k=1}^{\infty} \dfrac{1}{k^{1.7}}$ is a convergent p-series ($p = 1.7 > 1$).

9. $\lim\limits_{n\to\infty} \left| \dfrac{a_{n+1}}{a_n} \right| = \lim\limits_{n\to\infty} \dfrac{(n+1)/e^{n+1}}{n/e^n} = \dfrac{1}{e} \lim\limits_{n\to\infty} \dfrac{n+1}{n} = \dfrac{1}{e} < 1$, so the series converges by the Ratio Test.

11. $b_n = \dfrac{1}{n \ln n} > 0$ for $n \geq 2$, $\{b_n\}$ is decreasing, and $\lim\limits_{n\to\infty} b_n = 0$, so $\sum\limits_{n=2}^{\infty} \dfrac{(-1)^{n+1}}{n \ln n}$ converges by the Alternating

Series Test. The series is conditionally convergent since $\sum\limits_{n=2}^{\infty} \dfrac{1}{n \ln n}$ diverges by Exercise 11.3.19.

13. Let $f(x) = \dfrac{2}{x (\ln x)^3}$. $f(x)$ is clearly positive and decreasing for $x \geq 2$, so we apply the Integral Test.

$\displaystyle\int_2^{\infty} \dfrac{2}{x (\ln x)^3} \, dx = \lim\limits_{t\to\infty} \left[\dfrac{-1}{(\ln x)^2} \right]_2^t = 0 - \dfrac{-1}{(\ln 2)^2}$, which is finite, so $\sum\limits_{n=2}^{\infty} \dfrac{2}{n (\ln n)^3}$ converges.

15. $\lim\limits_{n\to\infty} \left| \dfrac{a_{n+1}}{a_n} \right| = \lim\limits_{n\to\infty} \dfrac{3^{n+1} (n+1)^2 / (n+1)!}{3^n n^2 / n!} = 3 \lim\limits_{n\to\infty} \dfrac{n+1}{n^2} = 0$, so the series converges by the Ratio Test.

17. $\dfrac{3^n}{5^n+n} \leq \dfrac{3^n}{5^n} = \left(\dfrac{3}{5} \right)^n$. Since $\sum\limits_{n=1}^{\infty} \left(\dfrac{3}{5} \right)^n$ is a convergent geometric series ($|r| = \dfrac{3}{5} < 1$), $\sum\limits_{n=1}^{\infty} \dfrac{3^n}{5^n+n}$ converges by the Comparison Test.

19. $\lim\limits_{n\to\infty} \left| \dfrac{a_{n+1}}{a_n} \right| = \lim\limits_{n\to\infty} \dfrac{\dfrac{(n+1)!}{2\cdot 5\cdot 8\cdots\cdots(3n+5)}}{\dfrac{n!}{2\cdot 5\cdot 8\cdots\cdots(3n+2)}} = \lim\limits_{n\to\infty} \dfrac{n+1}{3n+5} = \dfrac{1}{3} < 1$, so the series converges by the Ratio Test.

21. Use the Limit Comparison Test with $a_i = \dfrac{1}{\sqrt{i}\,(i+1)}$ and $b_i = \dfrac{1}{i}$.

$\lim\limits_{i\to\infty} \dfrac{a_i}{b_i} = \lim\limits_{i\to\infty} \dfrac{i}{\sqrt{i}\,(i+1)} = \lim\limits_{i\to\infty} \dfrac{1}{\sqrt{1+1/i}} = 1$. Since $\sum\limits_{i=1}^{\infty} b_i$ diverges (harmonic series) so does

$\sum\limits_{i=1}^{\infty} \dfrac{1}{\sqrt{i}\,(i+1)}$.

23. $\lim\limits_{n\to\infty} 2^{1/n} = 2^0 = 1$, so $\lim\limits_{n\to\infty} (-1)^n 2^{1/n}$ does not exist and the series diverges by the Test for Divergence.

25. Let $f(x) = \dfrac{\ln x}{\sqrt{x}}$. Then $f'(x) = \dfrac{2 - \ln x}{2x^{3/2}} < 0$ when $\ln x > 2$ or $x > e^2$, so $\dfrac{\ln n}{\sqrt{n}}$ is decreasing for $n > e^2$.

By l'Hospital's Rule, $\displaystyle\lim_{n\to\infty} \frac{\ln n}{\sqrt{n}} = \lim_{n\to\infty} \frac{1/n}{1/(2\sqrt{n})} = \lim_{n\to\infty} \frac{2}{\sqrt{n}} = 0$, so the series converges by the Alternating

Series Test.

27. $\displaystyle\sum_{n=1}^{\infty} \frac{(-2)^{2n}}{n^n} = \sum_{n=1}^{\infty} \left(\frac{4}{n}\right)^n$. $\displaystyle\lim_{n\to\infty} \sqrt[n]{|a_n|} = \lim_{n\to\infty} \frac{4}{n} = 0$, so the series converges by the Root Test.

29. $\displaystyle\int_2^{\infty} \frac{\ln x}{x^2}\,dx = \lim_{t\to\infty} \left[-\frac{\ln x}{x} - \frac{1}{x}\right]_1^t$ (using integration by parts) $\overset{\text{H}}{=} 1$. So $\displaystyle\sum_{n=1}^{\infty} \frac{\ln n}{n^2}$ converges by the Integral Test,

and since $\dfrac{k \ln k}{(k+1)^3} < \dfrac{k \ln k}{k^3} = \dfrac{\ln k}{k^2}$, the given series converges by the Comparison Test.

31. $\displaystyle\lim_{n\to\infty} \left|\frac{a_{n+1}}{a_n}\right| = \lim_{n\to\infty} \frac{2^{n+1}/(2n+3)!}{2^n/(2n+1)!} = 2 \lim_{n\to\infty} \frac{1}{(2n+3)(2n+2)} = 0$, so the series converges by the Ratio Test.

33. $0 < \dfrac{\tan^{-1} n}{n^{3/2}} < \dfrac{\pi/2}{n^{3/2}}$. $\displaystyle\sum_{n=1}^{\infty} \frac{\pi/2}{n^{3/2}} = \frac{\pi}{2} \sum_{n=1}^{\infty} \frac{1}{n^{3/2}}$ which is a convergent p-series ($p = \frac{3}{2} > 1$), so

$\displaystyle\sum_{n=1}^{\infty} \frac{\tan^{-1} n}{n^{3/2}}$ converges by the Comparison Test.

35. $\displaystyle\lim_{n\to\infty} \sqrt[n]{|a_n|} = \lim_{n\to\infty} \left(\frac{n}{n+1}\right)^{n^2/n} = \lim_{n\to\infty} \frac{1}{[(n+1)/n]^n} = \frac{1}{\displaystyle\lim_{n\to\infty}(1+1/n)^n} = \frac{1}{e} < 1$ (see Equation 3.8.6), so

the series converges by the Root Test.

37. $\displaystyle\lim_{n\to\infty} \sqrt[n]{|a_n|} = \lim_{n\to\infty} (2^{1/n} - 1) = 1 - 1 = 0$, so the series converges by the Root Test.

11.8 Power Series

1. A power series is a series of the form $\sum_{n=0}^{\infty} c_n x^n = c_0 + c_1 x + c_2 x^2 + c_3 x^3 + \cdots$, where x is a variable and the c_n's are constants called the coefficients of the series.
More generally, a series of the form $\sum_{n=0}^{\infty} c_n (x-a)^n = c_0 + c_1(x-a) + c_2(x-a)^2 + \cdots$ is called a power series in $(x-a)$ or a power series centered at a or a power series about a.

3. $\displaystyle\lim_{n\to\infty} \left|\frac{a_{n+1}}{a_n}\right| = \lim_{n\to\infty} \left|\frac{x^{n+1}}{\sqrt{n+1}} \cdot \frac{\sqrt{n}}{x}\right| = \lim_{n\to\infty} \left|\frac{x}{\sqrt{n+1}/\sqrt{n}}\right| = \lim_{n\to\infty} \frac{|x|}{\sqrt{1+1/n}} = |x|$. By the Ratio Test, the

series converges when $|x| < 1$, so the radius of convergence $R = 1$. When $x = 1$, the series $\sum_{n=1}^{\infty} 1/\sqrt{n}$ diverges because it is a p-series with $p = \frac{1}{2} \le 1$, but when $x = -1$, it converges by the Alternating Series Test. So the interval of convergence is $I = [-1, 1)$.

5. If $a_n = nx^n$, then $\displaystyle\lim_{n\to\infty} \left|\frac{a_{n+1}}{a_n}\right| = \lim_{n\to\infty} \left|\frac{(n+1)x^{n+1}}{nx^n}\right| = |x| \lim_{n\to\infty} \frac{n+1}{n} = |x| < 1$ for convergence (by the Ratio

Test). So $R = 1$. When $x = 1$ or -1, $\displaystyle\lim_{n\to\infty} nx^n$ does not exist, so $\sum_{n=0}^{\infty} nx^n$ diverges for $x = \pm 1$. So $I = (-1, 1)$.

7. If $a_n = \dfrac{x^n}{n!}$, then $\displaystyle\lim_{n\to\infty} \left|\frac{a_{n+1}}{a_n}\right| = \lim_{n\to\infty} \left|\frac{x^{n+1}/(n+1)!}{x^n/n!}\right| = |x| \lim_{n\to\infty} \frac{1}{n+1} = 0 < 1$ for all x. So, by the Ratio

Test, $R = \infty$, and $I = (-\infty, \infty)$.

9. $\lim\limits_{n\to\infty}\left|\dfrac{a_{n+1}}{a_n}\right| = \lim\limits_{n\to\infty}\dfrac{(n+1)\,4^{n+1}\,|x|^{n+1}}{n4^n\,|x|^n} = \lim\limits_{n\to\infty}\left(1+\dfrac{1}{n}\right)4\,|x| = 4\,|x|.$ Now $4\,|x| < 1 \iff |x| < \frac{1}{4}$, so by the

Ratio Test, $R = \frac{1}{4}$. When $x = \frac{1}{4}$, we get the divergent series $\sum_{n=1}^{\infty}(-1)^n\,n$, and when $x = -\frac{1}{4}$, we get the

divergent series $\sum_{n=1}^{\infty} n$. Thus, $I = \left(-\frac{1}{4}, \frac{1}{4}\right)$.

11. If $a_n = \dfrac{3^n x^n}{(n+1)^2}$, then $\lim\limits_{n\to\infty}\left|\dfrac{a_{n+1}}{a_n}\right| = \lim\limits_{n\to\infty}\left|\dfrac{3^{n+1}x^{n+1}}{(n+2)^2}\cdot\dfrac{(n+1)^2}{3^n x^n}\right| = 3\,|x|\lim\limits_{n\to\infty}\left(\dfrac{n+1}{n+2}\right)^2 = 3\,|x| < 1$ for

convergence, so $|x| < \frac{1}{3}$ and $R = \frac{1}{3}$. When $x = \frac{1}{3}$, $\sum\limits_{n=0}^{\infty}\dfrac{3^n x^n}{(n+1)^2} = \sum\limits_{n=0}^{\infty}\dfrac{1}{(n+1)^2} = \sum\limits_{n=1}^{\infty}\dfrac{1}{n^2}$ which is a convergent

p-series $(p = 2 > 1)$. When $x = -\frac{1}{3}$, $\sum\limits_{n=0}^{\infty}\dfrac{3^n x^n}{(n+1)^2} = \sum\limits_{n=0}^{\infty}\dfrac{(-1)^n}{(n+1)^2}$ which converges by the Alternating Series

Test, so $I = \left[-\frac{1}{3}, \frac{1}{3}\right]$.

13. If $a_n = \dfrac{x^n}{\ln n}$, then $\lim\limits_{n\to\infty}\left|\dfrac{a_{n+1}}{a_n}\right| = \lim\limits_{n\to\infty}\left|\dfrac{x^{n+1}}{\ln(n+1)}\cdot\dfrac{\ln n}{x^n}\right| = |x|\lim\limits_{n\to\infty}\dfrac{\ln n}{\ln(n+1)} \overset{\text{H}}{=} |x|$, so $R = 1$. When $x = 1$,

$\sum\limits_{n=2}^{\infty}\dfrac{x^n}{\ln n} = \sum\limits_{n=2}^{\infty}\dfrac{1}{\ln n}$ which diverges because $\dfrac{1}{\ln n} > \dfrac{1}{n}$ and $\sum\limits_{n=2}^{\infty}\dfrac{1}{n}$ is the divergent harmonic series. When $x = -1$,

$\sum\limits_{n=2}^{\infty}\dfrac{x^n}{\ln n} = \sum\limits_{n=2}^{\infty}\dfrac{(-1)^n}{\ln n}$ which converges by the Alternating Series Test. So $I = [-1, 1)$.

15. $\lim\limits_{n\to\infty}\left|\dfrac{a_{n+1}}{a_n}\right| = \lim\limits_{n\to\infty}\dfrac{\sqrt{n+1}\,|x-1|^{n+1}}{\sqrt{n}\,|x-1|^n} = \lim\limits_{n\to\infty}\sqrt{1+\dfrac{1}{n}}\,|x-1| = |x-1|$, so by the Ratio Test, the series

converges when $|x-1| < 1 \iff -1 < x-1 < 1 \iff 0 < x < 2.$ $R = 1$. When $x = 2$, the series becomes

$\sum_{n=0}^{\infty}\sqrt{n}$, which diverges by the Test for Divergence. When $x = 0$, the series becomes $\sum_{n=0}^{\infty}(-1)^n\sqrt{n}$, which

also diverges by the Test for Divergence. Thus, $I = (0, 2)$.

17. $\lim\limits_{n\to\infty}\left|\dfrac{a_{n+1}}{a_n}\right| = \lim\limits_{n\to\infty}\left[\dfrac{|x+2|^{n+1}}{(n+1)\,2^{n+1}}\cdot\dfrac{n2^n}{|x+2|^n}\right] = \lim\limits_{n\to\infty}\dfrac{n}{n+1}\cdot\dfrac{|x+2|}{2} = \dfrac{|x+2|}{2}$, so by the Ratio Test, the

series converges when $\dfrac{|x+2|}{2} < 1 \iff |x+2| < 2 \iff -2 < x+2 < 2 \iff -4 < x < 0.$ $R = 2$. When

$x = -4$, the series becomes $\sum\limits_{n=1}^{\infty}(-1)^n\dfrac{(-2)^n}{n2^n} = \sum\limits_{n=1}^{\infty}\dfrac{2^n}{n2^n} = \sum\limits_{n=1}^{\infty}\dfrac{1}{n}$, which is the divergent harmonic series. When

$x = 0$, the series is $\sum\limits_{n=1}^{\infty}\dfrac{(-1)^n}{n}$, the alternating harmonic series, which converges by the Alternating Series Test.

Thus, $I = (-4, 0]$.

19. If $a_n = \dfrac{(x-2)^n}{n^n}$, then $\lim\limits_{n\to\infty}\sqrt[n]{|a_n|} = \lim\limits_{n\to\infty}\dfrac{|x-2|}{n} = 0$, so the series converges for all x (by the Root Test).

$R = \infty$ and $I = (-\infty, \infty)$.

21. If $a_n = \dfrac{2^n (x - 3)^n}{n + 3}$, then

$$\lim_{n \to \infty} \left| \frac{a_{n+1}}{a_n} \right| = \lim_{n \to \infty} \left| \frac{2^{n+1} (x - 3)^{n+1}}{n + 4} \cdot \frac{n + 3}{2^n (x - 3)^n} \right| = 2|x - 3| \lim_{n \to \infty} \frac{n + 3}{n + 4} = 2|x - 3| < 1 \text{ for convergence, or}$$

$|x - 3| < \frac{1}{2} \Leftrightarrow \frac{5}{2} < x < \frac{7}{2}$, and $R = \frac{1}{2}$. When $x = \frac{5}{2}$, $\displaystyle\sum_{n=0}^{\infty} \frac{2^n (x - 3)^n}{n + 3} = \sum_{n=0}^{\infty} \frac{(-1)^n}{n + 3}$ which converges by the

Alternating Series Test. When $x = \frac{7}{2}$, $\displaystyle\sum_{n=0}^{\infty} \frac{2^n (x - 3)^n}{n + 3} = \sum_{n=0}^{\infty} \frac{1}{n + 3}$, similar to the harmonic series, which

diverges. So $I = \left[\frac{5}{2}, \frac{7}{2} \right)$.

23. If $a_n = n! (2x - 1)^n$, then $\displaystyle\lim_{n \to \infty} \left| \frac{a_{n+1}}{a_n} \right| = \lim_{n \to \infty} \left| \frac{(n + 1)! (2x - 1)^{n+1}}{n! (2x - 1)^n} \right| = \lim_{n \to \infty} (n + 1) |2x - 1| \to \infty$ as $n \to \infty$

for all $x \neq \frac{1}{2}$. Since the series diverges for all $x \neq \frac{1}{2}$, $R = 0$ and $I = \left\{ \frac{1}{2} \right\}$.

25. $\displaystyle\lim_{n \to \infty} \left| \frac{a_{n+1}}{a_n} \right| = \lim_{n \to \infty} \left[\frac{|4x + 1|^{n+1}}{(n + 1)^2} \cdot \frac{n^2}{|4x + 1|^n} \right] = \lim_{n \to \infty} \frac{|4x + 1|}{(1 + 1/n)^2} = |4x + 1|$, so by the Ratio Test, the series

converges when $|4x + 1| < 1 \Leftrightarrow -1 < 4x + 1 < 1 \Leftrightarrow -2 < 4x < 0 \Leftrightarrow -\frac{1}{2} < x < 0$, so $R = \frac{1}{4}$. When

$x = -\frac{1}{2}$, the series becomes $\displaystyle\sum_{n=1}^{\infty} \frac{(-1)^n}{n^2}$, which converges by the Alternating Series Test. When $x = 0$, the series

becomes $\displaystyle\sum_{n=1}^{\infty} \frac{1}{n^2}$, a convergent p-series ($p = 2 > 1$). $I = \left[-\frac{1}{2}, 0 \right]$.

27. If $a_n = \dfrac{x^n}{(\ln n)^n}$ then $\displaystyle\lim_{n \to \infty} \sqrt[n]{|a_n|} = \lim_{n \to \infty} \frac{|x|}{\ln n} = 0 < 1$ for all x, so $R = \infty$ and $I = (-\infty, \infty)$ by the Root Test.

29. (a) We are given that the power series $\sum_{n=0}^{\infty} c_n x^n$ is convergent for $x = 4$. So by Theorem 3, it must converge for
    at least $-4 < x \leq 4$. In particular, it converges when $x = -2$, that is, $\sum_{n=0}^{\infty} c_n (-2)^n$ is convergent.

   (b) It does not follow that $\sum_{n=0}^{\infty} c_n (-4)^n$ is necessarily convergent. [See the comments after Theorem 3 about
    convergence at the endpoint of an interval. An example is $c_n = (-1)^n / (n4^n)$.]

31. If $a_n = \dfrac{(n!)^k}{(kn)!} x^n$, then

$$\lim_{n \to \infty} \left| \frac{a_{n+1}}{a_n} \right| = \lim_{n \to \infty} \frac{[(n + 1)!]^k (kn)!}{(n!)^k [k (n + 1)]!} |x| = \lim_{n \to \infty} \frac{(n + 1)^k}{(kn + k) (kn + k - 1) \cdots (kn + 2) (kn + 1)} |x|$$

$$= \lim_{n \to \infty} \left[\frac{(n + 1)}{(kn + 1)} \frac{(n + 1)}{(kn + 2)} \cdots \frac{(n + 1)}{(kn + k)} \right] |x|$$

$$= \lim_{n \to \infty} \left[\frac{n + 1}{kn + 1} \right] \lim_{n \to \infty} \left[\frac{n + 1}{kn + 2} \right] \cdots \lim_{n \to \infty} \left[\frac{n + 1}{kn + k} \right] |x| = \left(\frac{1}{k} \right)^k |x| < 1 \quad \Leftrightarrow$$

$|x| < k^k$ for convergence, and the radius of convergence is $R = k^k$.

33. (a) If $a_n = \dfrac{(-1)^n x^{2n+1}}{n!\,(n+1)!\,2^{2n+1}}$, then $\lim\limits_{n\to\infty}\left|\dfrac{a_{n+1}}{a_n}\right| = \left(\dfrac{x}{2}\right)^2 \lim\limits_{n\to\infty}\dfrac{1}{(n+1)\,(n+2)} = 0$ for all x. So $J_1(x)$ converges

for all x; the domain is $(-\infty, \infty)$.

(b), (c) The initial terms of $J_1(x)$ up to $n = 5$ are

$$a_0 = \frac{x}{2},\ a_1 = -\frac{x^3}{16},\ a_2 = \frac{x^5}{384},\ a_3 = -\frac{x^7}{18{,}432},$$

$$a_4 = \frac{x^9}{1{,}474{,}560},\ \text{and}\ a_5 = -\frac{x^{11}}{176{,}947{,}200}.\ \text{The}$$

partial sums seem to approximate $J_1(x)$ well
near the origin, but as $|x|$ increases, we need to
take a large number of terms to get a good
approximation.

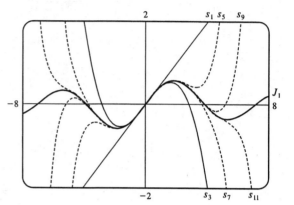

35. $s_{2n-1} = 1 + 2x + x^2 + 2x^3 + \cdots + x^{2n-2} + 2x^{2n-1} = (1+2x)\left(1 + x^2 + x^4 + \cdots + x^{2n-2}\right)$

$$= (1+2x)\,\frac{1 - x^{2n}}{1 - x^2}\ \text{[by (11.2.3) with}\ r = x^2] \to \frac{1+2x}{1-x^2}\ \text{as}\ n \to \infty\ \text{[by (11.2.4)]},$$

when $|x| < 1$. Also $s_{2n} = s_{2n-1} + x^{2n} \to \dfrac{1+2x}{1-x^2}$ since $x^{2n} \to 0$ for $|x| < 1$. Therefore, $s_n \to \dfrac{1+2x}{1-x^2}$ since s_{2n}

and s_{2n-1} both approach $\dfrac{1+2x}{1-x^2}$ as $n \to \infty$. Thus, the interval of convergence is $(-1, 1)$ and $f(x) = \dfrac{1+2x}{1-x^2}$.

37. We use the Root Test on the series $\sum c_n x^n$. $\lim\limits_{n\to\infty}\sqrt[n]{|c_n x^n|} = |x|\lim\limits_{n\to\infty}\sqrt[n]{|c_n|} = c\,|x| < 1$ for convergence, or

$|x| < 1/c$, so $R = 1/c$.

39. For $2 < x < 3$, $\sum c_n x^n$ diverges and $\sum d_n x^n$ converges. By Exercise 11.2.61, $\sum (c_n + d_n)\,x^n$ diverges. Since both
series converge for $|x| < 2$, the radius of convergence of $\sum (c_n + d_n)\,x^n$ is 2.

11.9 Representations of Functions as Power Series

1. If $f(x) = \sum\limits_{n=0}^{\infty} c_n x^n$ has radius of convergence 10, then $f'(x) = \sum\limits_{n=1}^{\infty} n c_n x^{n-1}$ also has radius of convergence 10 by

Theorem 2.

3. $f(x) = \dfrac{1}{1+x} = \dfrac{1}{1-(-x)} = \sum\limits_{n=0}^{\infty}(-x)^n = \sum\limits_{n=0}^{\infty}(-1)^n x^n$ with $|-x| < 1 \iff |x| < 1$, so $R = 1$ and

$I = (-1, 1)$.

5. Replacing x with x^3 in (1) gives $f(x) = \dfrac{1}{1-x^3} = \sum\limits_{n=0}^{\infty}(x^3)^n = \sum\limits_{n=0}^{\infty}x^{3n}$. The series converges when $|x^3| < 1$; that

is, when $|x| < 1$, so $I = (-1, 1)$.

7. $f(x) = \dfrac{1}{4+x^2} = \dfrac{1}{4}\left(\dfrac{1}{1+x^2/4}\right) = \dfrac{1}{4}\sum\limits_{n=0}^{\infty}(-1)^n\left(\dfrac{x^2}{4}\right)^n$ (using Exercise 3) $= \sum\limits_{n=0}^{\infty}\dfrac{(-1)^n x^{2n}}{4^{n+1}}$, with $\left|\dfrac{x^2}{4}\right| < 1$

$\iff x^2 < 4 \iff |x| < 2$, so $R = 2$ and $I = (-2, 2)$.

9. $f(x) = \dfrac{1}{x-5} = -\dfrac{1}{5}\left(\dfrac{1}{1-x/5}\right) = -\dfrac{1}{5}\sum_{n=0}^{\infty}\left(\dfrac{x}{5}\right)^n$. The series converges when $\left|\dfrac{x}{5}\right| < 1$; that is, when $|x| < 5$, so

$I = (-5, 5)$.

11. $f(x) = \dfrac{3}{x^2+x-2} = \dfrac{3}{(x+2)(x-1)} = \dfrac{A}{x+2} + \dfrac{B}{x-1} \quad\Rightarrow\quad 3 = A(x-1) + B(x+2)$. Taking $x = -2$, we

get $A = -1$. Taking $x = 1$, we get $B = 1$. Thus,

$$\dfrac{3}{x^2+x-2} = \dfrac{1}{x-1} - \dfrac{1}{x+2} = -\dfrac{1}{1-x} - \dfrac{1}{2}\dfrac{1}{1+x/2} = -\sum_{n=0}^{\infty}x^n - \dfrac{1}{2}\sum_{n=0}^{\infty}\left(-\dfrac{x}{2}\right)^n$$

$$= \sum_{n=0}^{\infty}\left[-1 - \dfrac{1}{2}\left(-\dfrac{1}{2}\right)^n\right]x^n = \sum_{n=0}^{\infty}\left[-1 + \left(-\dfrac{1}{2}\right)^{n+1}\right]x^n = \sum_{n=0}^{\infty}\left[\dfrac{(-1)^{n+1}}{2^{n+1}} - 1\right]x^n$$

We represented the given function as the sum of two geometric series; the first converges for $x \in (-1, 1)$ and the second converges for $x \in (-2, 2)$. Thus, the sum converges for $x \in (-1, 1) = I$.

13. $f(x) = \dfrac{1}{(1+x)^2} = -\dfrac{d}{dx}\left(\dfrac{1}{1+x}\right) = -\dfrac{d}{dx}\left[\sum_{n=0}^{\infty}(-1)^n x^n\right]$ (from Exercise 3)

$= \sum_{n=1}^{\infty}(-1)^{n+1} nx^{n-1}$ [from Theorem 2(a)] $= \sum_{n=0}^{\infty}(-1)^n (n+1)x^n$ with $R = 1$.

15. $f(x) = \dfrac{1}{(1+x)^3} = -\dfrac{1}{2}\dfrac{d}{dx}\left[\dfrac{1}{(1+x)^2}\right] = -\dfrac{1}{2}\dfrac{d}{dx}\left[\sum_{n=0}^{\infty}(-1)^n (n+1)x^n\right]$ (from Exercise 13)

$= -\dfrac{1}{2}\sum_{n=1}^{\infty}(-1)^n (n+1)nx^{n-1} = \dfrac{1}{2}\sum_{n=0}^{\infty}(-1)^n (n+2)(n+1)x^n$ with $R = 1$.

17. $f(x) = \ln(5-x) = -\displaystyle\int\dfrac{dx}{5-x} = -\dfrac{1}{5}\displaystyle\int\dfrac{dx}{1-x/5}$

$= -\dfrac{1}{5}\displaystyle\int\left[\sum_{n=0}^{\infty}\left(\dfrac{x}{5}\right)^n\right]dx = C - \dfrac{1}{5}\sum_{n=0}^{\infty}\dfrac{x^{n+1}}{5^n (n+1)} = C - \sum_{n=1}^{\infty}\dfrac{x^n}{n5^n}$

Putting $x = 0$, we get $C = \ln 5$. The series converges for $|x/5| < 1 \iff |x| < 5$, so $R = 5$.

19. $\dfrac{1}{2-x} = \dfrac{1}{2(1-x/2)} = \dfrac{1}{2}\sum_{n=0}^{\infty}\left(\dfrac{x}{2}\right)^n = \sum_{n=0}^{\infty}\dfrac{1}{2^{n+1}}x^n$ for $\left|\dfrac{x}{2}\right| < 1 \iff |x| < 2$. Now

$\dfrac{1}{(x-2)^2} = \left(\dfrac{1}{2-x}\right)' = \left(\sum_{n=0}^{\infty}\dfrac{1}{2^{n+1}}x^n\right)' = \sum_{n=0}^{\infty}\left(\dfrac{1}{2^{n+1}}x^n\right)' = \sum_{n=1}^{\infty}\dfrac{n}{2^{n+1}}x^{n-1} = \sum_{n=0}^{\infty}\dfrac{n+1}{2^{n+2}}x^n$. So

$f(x) = \dfrac{x^3}{(x-2)^2} = x^3\sum_{n=0}^{\infty}\dfrac{n+1}{2^{n+2}}x^n = \sum_{n=0}^{\infty}\dfrac{n+1}{2^{n+2}}x^{n+3}$ or $\sum_{n=3}^{\infty}\dfrac{n-2}{2^{n-1}}x^n$ for $|x| < 2$. Thus, $R = 2$ and $I = (-2, 2)$.

21. $f(x) = \ln(3+x) = \displaystyle\int \frac{dx}{3+x} = \frac{1}{3}\int \frac{dx}{1+x/3} = \frac{1}{3}\int \sum_{n=0}^{\infty}(-1)^n \left(\frac{x}{3}\right)^n dx$ (from Exercise 3)

$= C + \dfrac{1}{3}\displaystyle\sum_{n=0}^{\infty} \dfrac{(-1/3)^n}{n+1}x^{n+1} = \ln 3 + \dfrac{1}{3}\sum_{n=1}^{\infty}\dfrac{(-1/3)^{n-1}}{n}x^n$ $[C = f(0) = \ln 3]$

$= \ln 3 + \displaystyle\sum_{n=1}^{\infty}\dfrac{(-1)^{n-1}}{n3^n}x^n$ with $R = 3$.

The terms of the series are $a_0 = \ln 3$, $a_1 = \dfrac{x}{3}$, $a_2 = -\dfrac{x^2}{18}$, $a_3 = \dfrac{x^3}{81}$, $a_4 = -\dfrac{x^4}{324}$, $a_5 = \dfrac{x^5}{1215}$, ...

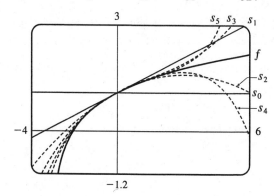

As n increases, $s_n(x)$ approximates f better on the interval of convergence, which is $(-3, 3)$.

23. $f(x) = \ln\left(\dfrac{1+x}{1-x}\right) = \ln(1+x) - \ln(1-x) = \displaystyle\int \dfrac{dx}{1+x} + \int \dfrac{dx}{1-x}$

$= \displaystyle\int\left[\sum_{n=0}^{\infty}(-1)^n x^n + \sum_{n=0}^{\infty}x^n\right]dx = \int \sum_{n=0}^{\infty}2x^{2n}\,dx = \sum_{n=0}^{\infty}\dfrac{2x^{2n+1}}{2n+1} + C$

But $f(0) = \ln\frac{1}{1} = 0$, so $C = 0$ and we have $f(x) = \displaystyle\sum_{n=0}^{\infty}\dfrac{2x^{2n+1}}{2n+1}$ with $R = 1$. If $x = \pm 1$, then

$f(x) = \pm 2\displaystyle\sum_{n=0}^{\infty}\dfrac{1}{2n+1}$, which both diverge by the Limit Comparison Test with $b_n = \dfrac{1}{n}$.

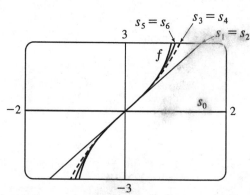

As n increases, $s_n(x)$ approximates f better on the interval of convergence, which is $(-1, 1)$.

25. $\displaystyle\int \dfrac{dx}{1+x^4} = \int \sum_{n=0}^{\infty}(-1)^n x^{4n}\,dx = C + \sum_{n=0}^{\infty}\dfrac{(-1)^n x^{4n+1}}{4n+1}$ with $R = 1$.

27. By Example 7, $\arctan x = \sum_{n=0}^{\infty} (-1)^n \frac{x^{2n+1}}{2n+1}$, so

$$\int \frac{\arctan x}{x} \, dx = \int \sum_{n=0}^{\infty} (-1)^n \frac{x^{2n}}{2n+1} \, dx = C + \sum_{n=0}^{\infty} (-1)^n \frac{x^{2n+1}}{(2n+1)^2} \text{ with } R = 1.$$

29. We use the representation $\int \frac{dx}{1+x^4} = C + \sum_{n=0}^{\infty} \frac{(-1)^n x^{4n+1}}{4n+1}$ from Exercise 25 with $C = 0$. So

$$\int_0^{0.2} \frac{dx}{1+x^4} = \left[x - \frac{x^5}{5} + \frac{x^9}{9} - \frac{x^{13}}{13} + \cdots \right]_0^{0.2} = 0.2 - \frac{0.2^5}{5} + \frac{0.2^9}{9} - \frac{0.2^{13}}{13} + \cdots$$

Since the series is alternating, the error in the nth-order approximation is less than the first neglected term, by The Alternating Series Estimation Theorem. If we use only the first two terms of the series, then the error is at most $0.2^9/9 \approx 5.7 \times 10^{-8}$. So, to six decimal places, $\int_0^{0.2} \frac{dx}{1+x^4} \approx 0.2 - \frac{0.2^5}{5} = 0.199936$.

31. We substitute x^4 for x in Example 7, and find that

$$\int x^2 \tan^{-1} \left(x^4 \right) dx = \int x^2 \sum_{n=0}^{\infty} (-1)^n \frac{\left(x^4 \right)^{2n+1}}{2n+1} \, dx$$

$$= \int \sum_{n=0}^{\infty} (-1)^n \frac{x^{8n+6}}{2n+1} \, dx = C + \sum_{n=0}^{\infty} (-1)^n \frac{x^{8n+7}}{(2n+1)(8n+7)}$$

So $\int_0^{1/3} x^2 \tan^{-1} \left(x^4 \right) dx = \left[\frac{x^7}{7} - \frac{x^{15}}{45} + \cdots \right]_0^{1/3} = \frac{1}{7 \cdot 3^7} - \frac{1}{45 \cdot 3^{15}} + \cdots$. The series is alternating, so if we use only one term, the error is at most $1/ \left(45 \cdot 3^{15} \right) \approx 1.5 \times 10^{-9}$. So $\int_0^{1/3} x^2 \tan^{-1} \left(x^4 \right) dx \approx 1/ \left(7 \cdot 3^7 \right) \approx 0.000065$ to six decimal places.

33. Using the result of Example 6, $\ln(1-x) = -\sum_{n=1}^{\infty} x^n/n$, with $x = -0.1$, we have

$\ln 1.1 = \ln[1 - (-0.1)] = 0.1 - \frac{0.01}{2} + \frac{0.001}{3} - \frac{0.0001}{4} + \frac{0.00001}{5} - \cdots$. The series is alternating, so if we use only the first four terms, the error is at most $\frac{0.00001}{5} = 0.000002$. So $\ln 1.1 \approx 0.1 - \frac{0.01}{2} + \frac{0.001}{3} - \frac{0.0001}{4} \approx 0.09531$.

35. (a) $J_0(x) = \sum_{n=0}^{\infty} \frac{(-1)^n x^{2n}}{2^{2n} (n!)^2}$, $J_0'(x) = \sum_{n=1}^{\infty} \frac{(-1)^n 2nx^{2n-1}}{2^{2n} (n!)^2}$, and $J_0''(x) = \sum_{n=1}^{\infty} \frac{(-1)^n 2n(2n-1) x^{2n-2}}{2^{2n} (n!)^2}$, so

$$x^2 J_0''(x) + x J_0'(x) + x^2 J_0(x) = \sum_{n=1}^{\infty} \frac{(-1)^n 2n(2n-1) x^{2n}}{2^{2n} (n!)^2} + \sum_{n=1}^{\infty} \frac{(-1)^n 2nx^{2n}}{2^{2n} (n!)^2} + \sum_{n=0}^{\infty} \frac{(-1)^n x^{2n+2}}{2^{2n} (n!)^2}$$

$$= \sum_{n=1}^{\infty} \frac{(-1)^n 2n(2n-1) x^{2n}}{2^{2n} (n!)^2} + \sum_{n=1}^{\infty} \frac{(-1)^n 2nx^{2n}}{2^{2n} (n!)^2} + \sum_{n=1}^{\infty} \frac{(-1)^{n-1} x^{2n}}{2^{2n-2} [(n-1)!]^2}$$

$$= \sum_{n=1}^{\infty} (-1)^n \left[\frac{2n(2n-1) + 2n - 2^2 n^2}{2^{2n} (n!)^2} \right] x^{2n} = \sum_{n=1}^{\infty} (-1)^n \left[\frac{4n^2 - 2n + 2n - 4n^2}{2^{2n} (n!)^2} \right] x^{2n} = 0$$

(b) $\int_0^1 J_0(x) \, dx = \int_0^1 \left[\sum_{n=0}^{\infty} \frac{(-1)^n x^{2n}}{2^{2n} (n!)^2} \right] dx = \int_0^1 \left(1 - \frac{x^2}{4} + \frac{x^4}{64} - \frac{x^6}{2304} + \cdots \right) dx$

$$= \left[x - \frac{x^3}{3 \cdot 4} + \frac{x^5}{5 \cdot 64} - \frac{x^7}{7 \cdot 2304} + \cdots \right]_0^1 = 1 - \frac{1}{12} + \frac{1}{320} - \frac{1}{16,128} + \cdots$$

Since $\frac{1}{16,128} \approx 0.000062$, it follows from The Alternating Series Estimation Theorem that, correct to three decimal places, $\int_0^1 J_0(x) \, dx \approx 1 - \frac{1}{12} + \frac{1}{320} \approx 0.920$.

37. (a) $f(x) = \sum\limits_{n=0}^{\infty} \dfrac{x^n}{n!} \Rightarrow f'(x) = \sum\limits_{n=1}^{\infty} \dfrac{nx^{n-1}}{n!} = \sum\limits_{n=1}^{\infty} \dfrac{x^{n-1}}{(n-1)!} = \sum\limits_{n=0}^{\infty} \dfrac{x^n}{n!} = f(x)$

(b) By Theorem 9.4.2, the only solution to the differential equation $df(x)/dx = f(x)$ is $f(x) = Ke^x$, but $f(0) = 1$, so $K = 1$ and $f(x) = e^x$.

Or: We could solve the equation $df(x)/dx = f(x)$ as a separable differential equation.

39. If $a_n = \dfrac{x^n}{n^2}$, then by the Ratio Test, $\lim\limits_{n\to\infty}\left|\dfrac{a_{n+1}}{a_n}\right| = |x|\lim\limits_{n\to\infty}\left(\dfrac{n}{n+1}\right)^2 = |x| < 1$ for convergence, so $R = 1$.

When $x = \pm 1$, $\sum\limits_{n=1}^{\infty}\left|\dfrac{x^n}{n^2}\right| = \sum\limits_{n=1}^{\infty}\dfrac{1}{n^2}$ which is a convergent p-series ($p = 2 > 1$), so the interval of convergence for f is $[-1, 1]$. By Theorem 2, the radii of convergence of f' and f'' are both 1, so we need only check the

endpoints. $f(x) = \sum\limits_{n=1}^{\infty}\dfrac{x^n}{n^2} \Rightarrow f'(x) = \sum\limits_{n=1}^{\infty}\dfrac{nx^{n-1}}{n^2} = \sum\limits_{n=0}^{\infty}\dfrac{x^n}{n+1}$, and this series diverges for $x = 1$ (harmonic

series) and converges for $x = -1$ (Alternating Series Test), so the interval of convergence is $[-1, 1)$.

$f''(x) = \sum\limits_{n=1}^{\infty}\dfrac{nx^{n-1}}{n+1}$ diverges at both 1 and -1 (Test for Divergence) since $\lim\limits_{n\to\infty}\dfrac{n}{n+1} = 1 \neq 0$, so its interval of

convergence is $(-1, 1)$.

11.10 Taylor and Maclaurin Series

1. Using Theorem 5 with $\sum\limits_{n=0}^{\infty} b_n (x-5)^n$, $b_n = \dfrac{f^{(n)}(a)}{n!}$, so $b_8 = \dfrac{f^{(8)}(5)}{8!}$.

3.

n	$f^{(n)}(x)$	$f^{(n)}(0)$
0	$\cos x$	1
1	$-\sin x$	0
2	$-\cos x$	-1
3	$\sin x$	0
4	$\cos x$	1
...	...	...

$\cos x = f(0) + f'(0)x + \dfrac{f''(0)}{2!}x^2 + \dfrac{f^{(3)}(0)}{3!}x^3 + \dfrac{f^{(4)}(0)}{4!}x^4 + \cdots$

$= 1 - \dfrac{x^2}{2!} + \dfrac{x^4}{4!} - \cdots = \sum\limits_{n=0}^{\infty}\dfrac{(-1)^n x^{2n}}{(2n)!}$

If $a_n = \dfrac{(-1)^n x^{2n}}{(2n)!}$, then

$\lim\limits_{n\to\infty}\left|\dfrac{a_{n+1}}{a_n}\right| = x^2 \lim\limits_{n\to\infty}\dfrac{1}{(2n+2)(2n+1)} = 0 < 1$ for all x. So

$R = \infty$ (Ratio Test).

5.

n	$f^{(n)}(x)$	$f^{(n)}(0)$
0	$(1+x)^{-3}$	1
1	$-3(1+x)^{-4}$	-3
2	$12(1+x)^{-5}$	12
3	$-60(1+x)^{-6}$	-60
4	$360(1+x)^{-7}$	360
...	...	...

$(1+x)^{-3} = f(0) + f'(0)x + \dfrac{f''(0)}{2!}x^2 + \dfrac{f'''(0)}{3!}x^3$

$\qquad + \dfrac{f^{(4)}(0)}{4!}x^4 + \cdots$

$= 1 - 3x + \tfrac{12}{2}x^2 - \tfrac{60}{6}x^3 + \tfrac{360}{24}x^4 + \cdots$

$= \sum\limits_{n=0}^{\infty}\dfrac{(-1)^n (n+2)!\, x^n}{2(n!)} = \sum\limits_{n=0}^{\infty}\dfrac{(-1)^n (n+2)(n+1)x^n}{2}$

$\lim\limits_{n\to\infty}\left|\dfrac{a_{n+1}}{a_n}\right| = \lim\limits_{n\to\infty}\dfrac{(n+3)(n+2)|x|^{n+1}}{(n+2)(n+1)|x|} = |x| < 1$ for convergence,

so $R = 1$.

7.

n	$f^{(n)}(x)$	$f^{(n)}(0)$
0	$\sinh x$	0
1	$\cosh x$	1
2	$\sinh x$	0
3	$\cosh x$	1
4	$\sinh x$	0
...	...	...

So $f^{(n)}(0) = \begin{cases} 0 & \text{if } n \text{ is even} \\ 1 & \text{if } n \text{ is odd} \end{cases}$ and $\sinh x = \sum\limits_{n=0}^{\infty} \dfrac{x^{2n+1}}{(2n+1)!}$. If

$a_n = \dfrac{x^{2n+1}}{(2n+1)!}$ then

$\lim\limits_{n\to\infty} \left| \dfrac{a_{n+1}}{a_n} \right| = x^2 \lim\limits_{n\to\infty} \dfrac{1}{(2n+3)(2n+2)} = 0 < 1$ for all x, so

$R = \infty$.

9.

n	$f^{(n)}(x)$	$f^{(n)}(2)$
0	$1 + x + x^2$	7
1	$1 + 2x$	5
2	2	2
3	0	0
4	0	0
...	...	...

$f(x) = 7 + 5(x-2) + \dfrac{2}{2!}(x-2)^2 + \sum\limits_{n=3}^{\infty} \dfrac{0}{n!}(x-2)^n$

$= 7 + 5(x-2) + (x-2)^2$

Since $a_n = 0$ for large n, $R = \infty$.

11. Clearly, $f^{(n)}(x) = e^x$, so $f^{(n)}(3) = e^3$ and $e^x = \sum\limits_{n=0}^{\infty} \dfrac{e^3}{n!}(x-3)^n$. If $a_n = \dfrac{e^3}{n!}(x-3)^n$, then

$\lim\limits_{n\to\infty} \left| \dfrac{a_{n+1}}{a_n} \right| = \lim\limits_{n\to\infty} \dfrac{|x-3|}{n+1} = 0$ for all x, so $R = \infty$.

13.

n	$f^{(n)}(x)$	$f^{(n)}(1)$
0	x^{-1}	1
1	$-x^{-2}$	-1
2	$2x^{-3}$	2
3	$-3 \cdot 2x^{-4}$	$-3 \cdot 2$
4	$4 \cdot 3 \cdot 2x^{-5}$	$4 \cdot 3 \cdot 2$
...	...	...

So $f^{(n)}(1) = (-1)^n n!$, and $\dfrac{1}{x} = \sum\limits_{n=0}^{\infty} \dfrac{(-1)^n n!}{n!}(x-1)^n = \sum\limits_{n=0}^{\infty} (-1)^n (x-1)^n$. If $a_n = (-1)^n (x-1)^n$ then

$\lim\limits_{n\to\infty} \left| \dfrac{a_{n+1}}{a_n} \right| = |x-1| < 1$ for convergence, so $0 < x < 2$ and $R = 1$.

15.

n	$f^{(n)}(x)$	$f^{(n)}\left(\frac{\pi}{4}\right)$
0	$\sin x$	$\sqrt{2}/2$
1	$\cos x$	$\sqrt{2}/2$
2	$-\sin x$	$-\sqrt{2}/2$
3	$-\cos x$	$-\sqrt{2}/2$
4	$\sin x$	$\sqrt{2}/2$
...	...	...

$$\sin x = f\left(\tfrac{\pi}{4}\right) + f'\left(\tfrac{\pi}{4}\right)\left(x - \tfrac{\pi}{4}\right) + \frac{f''\left(\tfrac{\pi}{4}\right)}{2!}\left(x - \tfrac{\pi}{4}\right)^2$$

$$+ \frac{f^{(3)}\left(\tfrac{\pi}{4}\right)}{3!}\left(x - \tfrac{\pi}{4}\right)^3 + \frac{f^{(4)}\left(\tfrac{\pi}{4}\right)}{4!}\left(x - \tfrac{\pi}{4}\right)^4 + \cdots$$

$$= \tfrac{\sqrt{2}}{2}\left[1 + \left(x - \tfrac{\pi}{4}\right) - \tfrac{1}{2!}\left(x - \tfrac{\pi}{4}\right)^2\right.$$

$$\left. - \tfrac{1}{3!}\left(x - \tfrac{\pi}{4}\right)^3 + \tfrac{1}{4!}\left(x - \tfrac{\pi}{4}\right)^4 + \cdots\right]$$

$$= \tfrac{\sqrt{2}}{2}\left[1 - \tfrac{1}{2!}\left(x - \tfrac{\pi}{4}\right)^2 + \tfrac{1}{4!}\left(x - \tfrac{\pi}{4}\right)^4 - \cdots\right]$$

$$+ \tfrac{\sqrt{2}}{2}\left[\left(x - \tfrac{\pi}{4}\right) - \tfrac{1}{3!}\left(x - \tfrac{\pi}{4}\right)^3 + \cdots\right]$$

$$= \tfrac{\sqrt{2}}{2}\sum_{n=0}^{\infty}(-1)^n\left[\tfrac{1}{(2n)!}\left(x - \tfrac{\pi}{4}\right)^{2n} + \tfrac{1}{(2n+1)!}\left(x - \tfrac{\pi}{4}\right)^{2n+1}\right]$$

The series can also be written in the more elegant form $\sin x = \dfrac{\sqrt{2}}{2}\sum_{n=0}^{\infty}\dfrac{(-1)^{n(n-1)/2}\left(x - \frac{\pi}{4}\right)^n}{n!}$. If

$a_n = \dfrac{(-1)^{n(n-1)/2}\left(x - \frac{\pi}{4}\right)^n}{n!}$, then $\lim\limits_{n\to\infty}\left|\dfrac{a_{n+1}}{a_n}\right| = \lim\limits_{n\to\infty}\dfrac{\left|x - \frac{\pi}{4}\right|}{n+1} = 0 < 1$ for all x, so $R = \infty$.

17. If $f(x) = \cos x$, then by Formula 9 with $a = 0$, $|R_n(x)| \le \dfrac{\left|f^{(n+1)}(x)\right|}{(n+1)!}|x|^{n+1}$. But $f^{(n+1)}(x) = \pm\sin x$ or

$\pm\cos x$. In each case, $\left|f^{(n+1)}(x)\right| \le 1$, so $|R_n(x)| \le \dfrac{1}{(n+1)!}|x|^{n+1} \to 0$ as $n \to \infty$ by Equation 10. So

$\lim\limits_{n\to\infty} R_n(x) = 0$ and, by Theorem 8, the series in Exercise 3 represents $\cos x$ for all x.

19. If $f(x) = \sinh x$, then $R_n(x) = \dfrac{f^{(n+1)}(z)}{(n+1)!}x^{n+1}$, where $0 < |z| < |x|$. But for all n,

$\left|f^{(n+1)}(z)\right| \le \cosh z \le \cosh x$ (since all derivatives are either sinh or cosh, $|\sinh z| < |\cosh z|$ for all z, and

$|z| < |x| \implies \cosh z < \cosh x$), so $|R_n(z)| \le \dfrac{\cosh x}{(n+1)!}x^{n+1} \to 0$ as $n \to \infty$ (by Equation 10). So by

Theorem 8, the series represents $\sinh x$ for all x.

21. $\cos x = \sum\limits_{n=0}^{\infty}(-1)^n\dfrac{x^{2n}}{(2n)!} \implies f(x) = \cos(\pi x) = \sum\limits_{n=0}^{\infty}\dfrac{(-1)^n(\pi x)^{2n}}{(2n)!} = \sum\limits_{n=0}^{\infty}\dfrac{(-1)^n\pi^{2n}x^{2n}}{(2n)!}$, $R = \infty$

23. $\tan^{-1}x = \sum\limits_{n=0}^{\infty}(-1)^n\dfrac{x^{2n+1}}{2n+1} \implies f(x) = x\tan^{-1}x = x\sum\limits_{n=0}^{\infty}(-1)^n\dfrac{x^{2n+1}}{2n+1} = \sum\limits_{n=0}^{\infty}(-1)^n\dfrac{x^{2n+2}}{2n+1}$, $R = 1$

25. $e^x = \sum\limits_{n=0}^{\infty}\dfrac{x^n}{n!} \implies f(x) = x^2e^{-x} = x^2\sum\limits_{n=0}^{\infty}\dfrac{(-x)^n}{n!} = \sum\limits_{n=0}^{\infty}\dfrac{(-1)^n x^{n+2}}{n!}$, $R = \infty$

27. $\sin^2 x = \tfrac{1}{2}[1 - \cos 2x] = \dfrac{1}{2}\left[1 - \sum\limits_{n=0}^{\infty}\dfrac{(-1)^n(2x)^{2n}}{(2n)!}\right] = 2^{-1}\left[1 - 1 - \sum\limits_{n=1}^{\infty}\dfrac{(-1)^n(2x)^{2n}}{(2n)!}\right]$

$$= \sum\limits_{n=1}^{\infty}\dfrac{(-1)^{n+1}2^{2n-1}x^{2n}}{(2n)!}, \quad R = \infty$$

29. $\dfrac{\sin x}{x} = \dfrac{1}{x}\sum\limits_{n=0}^{\infty}\dfrac{(-1)^n x^{2n+1}}{(2n+1)!} = \sum\limits_{n=0}^{\infty}\dfrac{(-1)^n x^{2n}}{(2n+1)!}$ and this series also gives the required value at $x = 0$ (namely 1), so

$R = \infty$.

31.

n	$f^{(n)}(x)$	$f^{(n)}(0)$
0	$(1+x)^{1/2}$	1
1	$\frac{1}{2}(1+x)^{-1/2}$	$\frac{1}{2}$
2	$-\frac{1}{4}(1+x)^{-3/2}$	$-\frac{1}{4}$
3	$\frac{3}{8}(1+x)^{-5/2}$	$\frac{3}{8}$
4	$-\frac{15}{16}(1+x)^{-7/2}$	$-\frac{15}{16}$
...	...	...

So $f^{(n)}(0) = \dfrac{(-1)^{n-1}\,1\cdot 3\cdot 5\cdot\ \cdots\ \cdot(2n-3)}{2^n}$ for $n \geq 2$,

and

$$\sqrt{1+x} = 1 + \frac{x}{2} + \sum_{n=2}^{\infty} \frac{(-1)^{n-1}\,1\cdot 3\cdot 5\cdot\ \cdots\ \cdot(2n-3)}{2^n n!} x^n.$$

If $a_n = \dfrac{(-1)^{n-1}\,1\cdot 3\cdot 5\cdot\ \cdots\ \cdot(2n-3)}{2^n n!} x^n$, then

$$\lim_{n\to\infty}\left|\frac{a_{n+1}}{a_n}\right| = \frac{|x|}{2}\lim_{n\to\infty}\frac{2n-1}{n+1} = |x| < 1 \text{ for convergence,}$$

so $R = 1$.

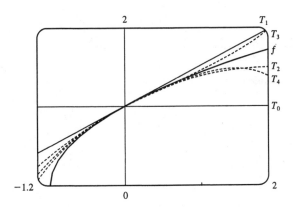

33. $\cos x = \displaystyle\sum_{n=0}^{\infty}(-1)^n \frac{x^{2n}}{(2n)!} \quad\Rightarrow\quad f(x) = \cos(x^2) = \sum_{n=0}^{\infty}\frac{(-1)^n (x^2)^{2n}}{(2n)!} = \sum_{n=0}^{\infty}\frac{(-1)^n x^{4n}}{(2n)!},\ R=\infty$

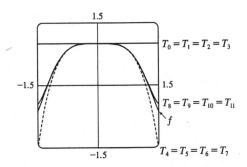

35. $\ln(1+x) = \displaystyle\int \frac{dx}{1+x} = \int \sum_{n=0}^{\infty}(-1)^n x^n\,dx = C + \sum_{n=0}^{\infty}(-1)^n \frac{x^{n+1}}{n+1} = \sum_{n=1}^{\infty}\frac{(-1)^{n-1} x^n}{n}$ with $C=0$ and $R=1$,

so $\ln(1.1) = \displaystyle\sum_{n=1}^{\infty}\frac{(-1)^{n-1}(0.1)^n}{n}$. This is an alternating series with $b_5 = \dfrac{(0.1)^5}{5} = 0.000002$, so to five decimal

places, $\ln(1.1) \approx \displaystyle\sum_{n=1}^{4}\frac{(-1)^{n-1}(0.1)^n}{n} \approx 0.09531$.

37. $\displaystyle\int \sin(x^2)\,dx = \int \sum_{n=0}^{\infty}(-1)^n \frac{(x^2)^{2n+1}}{(2n+1)!}\,dx = \int \sum_{n=0}^{\infty}\frac{(-1)^n x^{4n+2}}{(2n+1)!}\,dx = C + \sum_{n=0}^{\infty}\frac{(-1)^n x^{4n+3}}{(4n+3)(2n+1)!}$

39. Using the series from Exercise 31 and substituting x^3 for x, we get

$$\int \sqrt{x^3+1}\, dx = \int \left[1 + \frac{x^3}{2} + \sum_{n=2}^{\infty} \frac{(-1)^{n-1}\, 1 \cdot 3 \cdot 5 \cdots (2n-3)}{2^n n!} x^{3n} \right] dx$$

$$= C + x + \frac{x^4}{8} + \sum_{n=2}^{\infty} \frac{(-1)^{n-1}\, 1 \cdot 3 \cdot 5 \cdots (2n-3)}{2^n n!\, (3n+1)} x^{3n+1}$$

41. Using our series from Exercise 37, we get $\displaystyle \int_0^1 \sin\left(x^2\right) dx = \sum_{n=0}^{\infty} \left[\frac{(-1)^n x^{4n+3}}{(4n+3)(2n+1)!} \right]_0^1 = \sum_{n=0}^{\infty} \frac{(-1)^n}{(4n+3)(2n+1)!}$

and $|c_3| = \dfrac{1}{75,600} < 0.000014$, so by the Alternating Series Estimation Theorem, we have

$$\sum_{n=0}^{2} \frac{(-1)^n}{(4n+3)(2n+1)!} = \frac{1}{3} - \frac{1}{42} + \frac{1}{1320} \approx 0.310 \text{ (correct to three decimal places)}.$$

43. We first find a series representation for $f(x) = (1+x)^{-1/2}$, and then substitute.

n	$f^{(n)}(x)$	$f^{(n)}(0)$
0	$(1+x)^{-1/2}$	1
1	$-\frac{1}{2}(1+x)^{-3/2}$	$-\frac{1}{2}$
2	$\frac{3}{4}(1+x)^{-5/2}$	$\frac{3}{4}$
3	$-\frac{15}{8}(1+x)^{-7/2}$	$-\frac{15}{8}$
...	...	...

$$\frac{1}{\sqrt{1+x}} = 1 - \frac{x}{2} + \frac{3}{4}\left(\frac{x^2}{2!}\right) - \frac{15}{8}\left(\frac{x^3}{3!}\right) + \cdots \quad \Rightarrow \quad \frac{1}{\sqrt{1+x^3}} = 1 - \frac{1}{2}x^3 + \frac{3}{8}x^6 - \frac{5}{16}x^9 + \cdots \quad \Rightarrow$$

$$\int_0^{0.1} \frac{dx}{\sqrt{1+x^3}} = \left[x - \frac{1}{8}x^4 + \frac{3}{56}x^7 - \frac{1}{32}x^{10} + \cdots \right]_0^{0.1} \approx (0.1) - \frac{1}{8}(0.1)^4 \text{, by the Alternating Series Estimation}$$

Theorem, since $\frac{3}{56}(0.1)^7 \approx 0.0000000054 < 10^{-8}$, which is the maximum desired error. Therefore,

$$\int_0^{0.1} \frac{dx}{\sqrt{1+x^3}} \approx 0.09998750.$$

45. $\displaystyle \lim_{x \to 0} \frac{x - \tan^{-1} x}{x^3} = \lim_{x \to 0} \frac{x - \left(x - \frac{1}{3}x^3 + \frac{1}{5}x^5 - \frac{1}{7}x^7 + \cdots \right)}{x^3} = \lim_{x \to 0} \frac{\frac{1}{3}x^3 - \frac{1}{5}x^5 + \frac{1}{7}x^7 - \cdots}{x^3}$

$$= \lim_{x \to 0} \left(\frac{1}{3} - \frac{1}{5}x^2 + \frac{1}{7}x^4 - \cdots \right) = \frac{1}{3}$$

since power series are continuous functions.

47. $\displaystyle \lim_{x \to 0} \frac{\sin x - x + \frac{1}{6}x^3}{x^5} = \lim_{x \to 0} \frac{\left(x - \frac{1}{3!}x^3 + \frac{1}{5!}x^5 - \frac{1}{7!}x^7 + \cdots \right) - x + \frac{1}{6}x^3}{x^5}$

$$= \lim_{x \to 0} \frac{\frac{1}{5!}x^5 - \frac{1}{7!}x^7 + \cdots}{x^5} = \lim_{x \to 0} \left(\frac{1}{5!} - \frac{x^2}{7!} + \frac{x^4}{9!} - \cdots \right) = \frac{1}{5!} = \frac{1}{120}$$

since power series are continuous functions.

49. As in Example 8(a), we have $e^{-x^2} = 1 - \frac{x^2}{1!} + \frac{x^4}{2!} - \frac{x^6}{3!} + \cdots$ and we know that $\cos x = 1 - \frac{x^2}{2!} + \frac{x^4}{4!} - \cdots$ from

Equation 16. Therefore, $e^{-x^2} \cos x = \left(1 - x^2 + \frac{1}{2}x^4 - \cdots\right)\left(1 - \frac{1}{2}x^2 + \frac{1}{24}x^4 - \cdots\right)$. Writing only the terms

with degree ≤ 4, we get $e^{-x^2} \cos x = 1 - \frac{1}{2}x^2 + \frac{1}{24}x^4 - x^2 + \frac{1}{2}x^4 + \frac{1}{2}x^4 + \cdots = 1 - \frac{3}{2}x^2 + \frac{25}{24}x^4 + \cdots$.

51.

$$
\begin{array}{r}
-x + \frac{1}{2}x^2 - \frac{1}{3}x^3 + \cdots \\
1 + x + \frac{1}{2}x^2 + \frac{1}{6}x^3 + \cdots \overline{\big)\, -x - \frac{1}{2}x^2 - \frac{1}{3}x^3 - \cdots} \\
-x - x^2 - \frac{1}{2}x^3 - \cdots \\
\hline
\frac{1}{2}x^2 + \frac{1}{6}x^3 - \cdots \\
\frac{1}{2}x^2 + \frac{1}{2}x^3 + \cdots \\
\hline
-\frac{1}{3}x^3 + \cdots \\
-\frac{1}{3}x^3 + \cdots \\
\hline
\cdots
\end{array}
$$

From Example 6 in Section 11.9, we have $\ln(1-x) = -x - \frac{1}{2}x^2 - \frac{1}{3}x^3 - \cdots$, $|x| < 1$. Therefore,

$y = \dfrac{\ln(1-x)}{e^x} = \dfrac{-x - \frac{1}{2}x^2 - \frac{1}{3}x^3 - \cdots}{1 + x + \frac{1}{2}x^2 + \frac{1}{6}x^3 + \cdots}$. So by the long division above, $\dfrac{\ln(1-x)}{e^x} = -x + \dfrac{x^2}{2} - \dfrac{x^3}{3} + \cdots$,

$|x| < 1$.

53. $\displaystyle\sum_{n=0}^{\infty} (-1)^n \frac{x^{4n}}{n!} = \sum_{n=0}^{\infty} \frac{(-x^4)^n}{n!} = e^{-x^4}$, by (11).

55. $\displaystyle\sum_{n=0}^{\infty} \frac{(-1)^n \pi^{2n+1}}{4^{2n+1}(2n+1)!} = \sum_{n=0}^{\infty} \frac{(-1)^n \left(\frac{\pi}{4}\right)^{2n+1}}{(2n+1)!} = \sin \frac{\pi}{4} = \frac{1}{\sqrt{2}}$, by (15).

57. $3 + \dfrac{9}{2!} + \dfrac{27}{3!} + \dfrac{81}{4!} + \cdots = \dfrac{3^1}{1!} + \dfrac{3^2}{2!} + \dfrac{3^3}{3!} + \dfrac{3^4}{4!} + \cdots = \displaystyle\sum_{n=1}^{\infty} \frac{3^n}{n!} = \sum_{n=0}^{\infty} \frac{3^n}{n!} - 1 = e^3 - 1$, by (11).

59. Assume that $|f'''(x)| \leq M$, so $f'''(x) \leq M$ for $a \leq x \leq d$. Now $\int_a^x f'''(t)\,dt \leq \int_a^x M\,dt$

$\Rightarrow \quad f''(x) - f''(a) \leq M(x-a) \quad \Rightarrow \quad f''(x) \leq f''(a) + M(x-a)$. Thus,

$\int_a^x f''(t)\,dt \leq \int_a^x [f''(a) + M(t-a)]\,dt \quad \Rightarrow \quad f'(x) - f'(a) \leq f''(a)(x-a) + \frac{1}{2}M(x-a)^2$

$\Rightarrow \quad f'(x) \leq f'(a) + f''(a)(x-a) + \frac{1}{2}M(x-a)^2 \quad \Rightarrow$

$\int_a^x f'(t)\,dt \leq \int_a^x \left[f'(a) + f''(a)(t-a) + \frac{1}{2}M(t-a)^2\right]dt$

$\Rightarrow \quad f(x) - f(a) \leq f'(a)(x-a) + \frac{1}{2}f''(a)(x-a)^2 + \frac{1}{6}M(x-a)^3$. So

$f(x) - f(a) - f'(a)(x-a) - \frac{1}{2}f''(a)(x-a)^2 \leq \frac{1}{6}M(x-a)^3$. But

$R_2(x) = f(x) - T_2(x) = f(x) - f(a) - f'(a)(x-a) - \frac{1}{2}f''(a)(x-a)^2$, so $R_2(x) \leq \frac{1}{6}M(x-a)^3$. A

similar argument using $f'''(x) \geq -M$ shows that $R_2(x) \geq -\frac{1}{6}M(x-a)^3$. So $|R_2(x_2)| \leq \frac{1}{6}M|x-a|^3$.

Although we have assumed that $x > a$, a similar calculation shows that this inequality is also true if $x < a$.

11.11 The Binomial Series

1. The general binomial series in (2) is

$$(1+x)^k = \sum_{n=0}^{\infty} \binom{k}{n} x^n = 1 + kx + \frac{k(k-1)}{2!} x^2 + \frac{k(k-1)(k-2)}{3!} x^3 + \cdots.$$

$$(1+x)^{1/2} = \sum_{n=0}^{\infty} \binom{\frac{1}{2}}{n} x^n = 1 + \left(\tfrac{1}{2}\right) x + \frac{\left(\frac{1}{2}\right)\left(-\frac{1}{2}\right)}{2!} x^2 + \frac{\left(\frac{1}{2}\right)\left(-\frac{1}{2}\right)\left(-\frac{3}{2}\right)}{3!} x^3 + \cdots$$

$$= 1 + \frac{x}{2} - \frac{x^2}{2^2 \cdot 2!} + \frac{1 \cdot 3 \cdot x^3}{2^3 \cdot 3!} - \frac{1 \cdot 3 \cdot 5 \cdot x^4}{2^4 \cdot 4!} + \cdots$$

$$= 1 + \frac{x}{2} + \sum_{n=2}^{\infty} \frac{(-1)^{n-1} 1 \cdot 3 \cdot 5 \cdots (2n-3) x^n}{2^n \cdot n!}, \quad R = 1$$

3. $\dfrac{1}{(2+x)^3} = \dfrac{1}{[2(1+x/2)]^3} = \dfrac{1}{8}\left(1 + \dfrac{x}{2}\right)^{-3} = \dfrac{1}{8} \displaystyle\sum_{n=0}^{\infty} \binom{-3}{n} \left(\dfrac{x}{2}\right)^n$. The binomial coefficient is

$$\binom{-3}{n} = \frac{(-3)(-4)(-5) \cdots (-3-n+1)}{n!} = \frac{(-1)^n \cdot 2 \cdot 3 \cdot 4 \cdot 5 \cdots (n+1)(n+2)}{2 \cdot n!}$$

$$= \frac{(-1)^n (n+1)(n+2)}{2}$$

so $\dfrac{1}{(2+x)^3} = \dfrac{1}{8}\displaystyle\sum_{n=0}^{\infty} \dfrac{(-1)^n (n+1)(n+2)}{2} \dfrac{x^n}{2^n} = \displaystyle\sum_{n=0}^{\infty} \dfrac{(-1)^n (n+1)(n+2) x^n}{2^{n+4}}$ for $\left|\dfrac{x}{2}\right| < 1 \Leftrightarrow |x| < 2$, so $R = 2$.

5. $\sqrt[4]{1-8x} = (1-8x)^{1/4} = \displaystyle\sum_{n=0}^{\infty} \binom{\frac{1}{4}}{n} (-8x)^n$

$$= 1 + \tfrac{1}{4}(-8x) + \frac{\frac{1}{4}\left(-\frac{3}{4}\right)}{2!}(-8x)^2 + \frac{\left(\frac{1}{4}\right)\left(-\frac{3}{4}\right)\left(-\frac{7}{4}\right)}{3!}(-8x)^3 + \cdots$$

$$= 1 - 2x + \sum_{n=2}^{\infty} \frac{(-1)^n (-1)^{n-1} \cdot 3 \cdot 7 \cdots (4n-5) 8^n}{4^n \cdot n!} x^n$$

$$= 1 - 2x - \sum_{n=2}^{\infty} \frac{3 \cdot 7 \cdots (4n-5) 2^n}{n!} x^n$$

and $|-8x| < 1 \Leftrightarrow |x| < \tfrac{1}{8}$, so $R = \tfrac{1}{8}$.

7. $\dfrac{x}{\sqrt{4+x^2}} = \dfrac{x}{2\sqrt{1+x^2/4}} = \dfrac{x}{2}\left(1 + \dfrac{x^2}{4}\right)^{-1/2} = \dfrac{x}{2}\displaystyle\sum_{n=0}^{\infty} \binom{-\frac{1}{2}}{n}\left(\dfrac{x^2}{4}\right)^n$

$$= \frac{x}{2}\left[1 + \left(-\tfrac{1}{2}\right)\frac{x^2}{4} + \frac{\left(-\frac{1}{2}\right)\left(-\frac{3}{2}\right)}{2!}\left(\frac{x^2}{4}\right)^2 + \frac{\left(-\frac{1}{2}\right)\left(-\frac{3}{2}\right)\left(-\frac{5}{2}\right)}{3!}\left(\frac{x^2}{4}\right)^3 + \cdots\right]$$

$$= \frac{x}{2} + \sum_{n=1}^{\infty} (-1)^n \frac{1 \cdot 3 \cdot 5 \cdots (2n-1)}{n! \, 2^{3n+1}} x^{2n+1} \text{ and } \frac{x^2}{4} < 1 \Leftrightarrow \frac{|x|}{2} < 1 \Leftrightarrow |x| < 2, \text{ so } R = 2.$$

9. $\dfrac{1}{\sqrt[3]{8+x}} = (8+x)^{-1/3} = 8^{-1/3}\left(1+\dfrac{x}{8}\right)^{-1/3} = \dfrac{1}{2}\left(1+\dfrac{x}{8}\right)^{-1/3}$

$= \dfrac{1}{2}\left[1+\left(-\dfrac{1}{3}\right)\left(\dfrac{x}{8}\right)+\dfrac{\left(-\frac{1}{3}\right)\left(-\frac{4}{3}\right)}{2!}\left(\dfrac{x}{8}\right)^2+\cdots\right]$

$= \dfrac{1}{2}\left[1+\displaystyle\sum_{n=1}^{\infty}\dfrac{(-1)^n\,1\cdot 4\cdot 7\cdots(3n-2)}{3^n\cdot n!\,8^n}x^n\right]$ and $\left|\dfrac{x}{8}\right|<1 \iff |x|<8$, so $R=8$.

The three Taylor polynomials are $T_1(x) = \dfrac{1}{2}-\dfrac{1}{48}x$, $T_2(x) = \dfrac{1}{2}-\dfrac{1}{48}x+\dfrac{1}{576}x^2$, and

$T_3(x) = \dfrac{1}{2}-\dfrac{1}{48}x+\dfrac{1}{576}x^2-\dfrac{4\cdot 7}{2\cdot 27\cdot 6\cdot 512}x^3 = \dfrac{1}{2}-\dfrac{1}{48}x+\dfrac{1}{576}x^2-\dfrac{7}{41{,}472}x^3$.

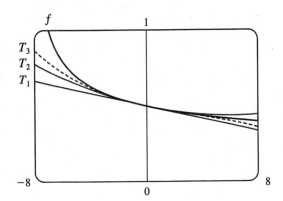

11. (a) $\left[1+(-x^2)\right]^{-1/2} = 1+\left(-\dfrac{1}{2}\right)(-x^2)+\dfrac{\left(-\frac{1}{2}\right)\left(-\frac{3}{2}\right)}{2!}(-x^2)^2+\dfrac{\left(-\frac{1}{2}\right)\left(-\frac{3}{2}\right)\left(-\frac{5}{2}\right)}{3!}(-x^2)^3+\cdots$

$= 1+\displaystyle\sum_{n=1}^{\infty}\dfrac{1\cdot 3\cdot 5\cdots(2n-1)}{2^n\cdot n!}x^{2n}$

(b) $\sin^{-1}x = \displaystyle\int\dfrac{1}{\sqrt{1-x^2}}\,dx = C+x+\displaystyle\sum_{n=1}^{\infty}\dfrac{1\cdot 3\cdot 5\cdots(2n-1)}{(2n+1)\,2^n\cdot n!}x^{2n+1}$

$= x+\displaystyle\sum_{n=1}^{\infty}\dfrac{1\cdot 3\cdot 5\cdots(2n-1)}{(2n+1)\,2^n\cdot n!}x^{2n+1}$ since $0 = \sin^{-1}0 = C$.

13. (a) $(1+x)^{-1/2} = 1+\left(-\dfrac{1}{2}\right)x+\dfrac{\left(-\frac{1}{2}\right)\left(-\frac{3}{2}\right)}{2!}x^2+\dfrac{\left(-\frac{1}{2}\right)\left(-\frac{3}{2}\right)\left(-\frac{5}{2}\right)}{3!}x^3+\cdots$

$= 1+\displaystyle\sum_{n=1}^{\infty}\dfrac{(-1)^n\,1\cdot 3\cdot 5\cdots(2n-1)}{2^n\cdot n!}x^n$

(b) Take $x=0.1$ in the above series. $\dfrac{1\cdot 3\cdot 5\cdot 7}{2^4 4!}(0.1)^4 < 0.00003$, so

$\dfrac{1}{\sqrt{1.1}} \approx 1-\dfrac{0.1}{2}+\dfrac{1\cdot 3}{2^2\cdot 2!}(0.1)^2-\dfrac{1\cdot 3\cdot 5}{2^3\cdot 3!}(0.1)^3 \approx 0.953$.

15. (a) $[1 + (-x)]^{-2} = 1 + (-2)(-x) + \dfrac{(-2)(-3)}{2!}(-x)^2 + \dfrac{(-2)(-3)(-4)}{3!}(-x)^3 + \cdots$

$$= 1 + 2x + 3x^2 + 4x^3 + \cdots = \sum_{n=0}^{\infty} (n+1)x^n,$$

so $\dfrac{x}{(1-x)^2} = \sum_{n=0}^{\infty} (n+1)x^{n+1} = \sum_{n=1}^{\infty} nx^n.$

(b) With $x = \frac{1}{2}$ in part (a), we have $\sum_{n=1}^{\infty} \dfrac{n}{2^n} = \dfrac{\frac{1}{2}}{\left(1 - \frac{1}{2}\right)^2} = 2.$

17. (a) $(1 + x^2)^{1/2} = 1 + \left(\frac{1}{2}\right)x^2 + \dfrac{\left(\frac{1}{2}\right)\left(-\frac{1}{2}\right)}{2!}(x^2)^2 + \dfrac{\left(\frac{1}{2}\right)\left(-\frac{1}{2}\right)\left(-\frac{3}{2}\right)}{3!}(x^2)^3 + \cdots$

$$= 1 + \dfrac{x^2}{2} + \sum_{n=2}^{\infty} \dfrac{(-1)^{n-1} 1 \cdot 3 \cdot 5 \cdot \cdots \cdot (2n-3)}{2^n \cdot n!} x^{2n}$$

(b) The coefficient of x^{10} (corresponding to $n = 5$) in the above Maclaurin series is $\dfrac{f^{(10)}(0)}{10!}$, so

$$\dfrac{f^{(10)}(0)}{10!} = \dfrac{(-1)^4 \cdot 1 \cdot 3 \cdot 5 \cdot 7}{2^5 \cdot 5!} \quad \Rightarrow \quad f^{(10)}(0) = 10!\left(\dfrac{1 \cdot 3 \cdot 5 \cdot 7}{2^5 \cdot 5!}\right) = 99{,}225.$$

19. (a) $g(x) = \sum_{n=0}^{\infty} \binom{k}{n} x^n \quad \Rightarrow \quad g'(x) = \sum_{n=1}^{\infty} \binom{k}{n} nx^{n-1}$, so

$$(1+x)g'(x) = (1+x) \sum_{n=1}^{\infty} \binom{k}{n} nx^{n-1} = \sum_{n=1}^{\infty} \binom{k}{n} nx^{n-1} + \sum_{n=1}^{\infty} \binom{k}{n} nx^n$$

$$= \sum_{n=0}^{\infty} \binom{k}{n+1}(n+1)x^n + \sum_{n=0}^{\infty} \binom{k}{n} nx^n \qquad \begin{bmatrix} \text{Replace } n \text{ with } n+1 \\ \text{in the first series} \end{bmatrix}$$

$$= \sum_{n=0}^{\infty} (n+1)\dfrac{k(k-1)(k-2)\cdots(k-n+1)(k-n)}{(n+1)!}x^n$$

$$\quad + \sum_{n=0}^{\infty} \left[(n)\dfrac{k(k-1)(k-2)\cdots(k-n+1)}{n!}\right]x^n$$

$$= \sum_{n=0}^{\infty} \dfrac{(n+1)k(k-1)(k-2)\cdots(k-n+1)}{(n+1)!}[(k-n)+n]x^n$$

$$= k \sum_{n=0}^{\infty} \dfrac{k(k-1)(k-2)\cdots(k-n+1)}{n!}x^n = k \sum_{n=0}^{\infty} \binom{k}{n} x^n = kg(x)$$

Thus, $g'(x) = \dfrac{kg(x)}{1+x}.$

(b) $h(x) = (1+x)^{-k} g(x) \quad \Rightarrow$

$$h'(x) = -k(1+x)^{-k-1} g(x) + (1+x)^{-k} g'(x) \quad \text{[Product Rule]}$$

$$= -k(1+x)^{-k-1} g(x) + (1+x)^{-k} \dfrac{kg(x)}{1+x} \quad \text{[from part (a)]}$$

$$= -k(1+x)^{-k-1} g(x) + k(1+x)^{-k-1} g(x) = 0$$

(c) From part (b) we see that $h(x)$ must be constant for $x \in (-1, 1)$, so $h(x) = h(0) = 1$ for $x \in (-1, 1)$. Thus, $h(x) = 1 = (1+x)^{-k} g(x) \quad \Leftrightarrow \quad g(x) = (1+x)^k$ for $x \in (-1, 1)$.

11.12 Applications of Taylor Polynomials

1. (a)

n	$f^{(n)}(x)$	$f^{(n)}(0)$	$T_n(x)$
0	$\cos x$	1	1
1	$-\sin x$	0	1
2	$-\cos x$	-1	$1 - \frac{1}{2}x^2$
3	$\sin x$	0	$1 - \frac{1}{2}x^2$
4	$\cos x$	1	$1 - \frac{1}{2}x^2 + \frac{1}{24}x^4$
5	$-\sin x$	0	$1 - \frac{1}{2}x^2 + \frac{1}{24}x^4$
6	$-\cos x$	-1	$1 - \frac{1}{2}x^2 + \frac{1}{24}x^4 - \frac{1}{720}x^6$

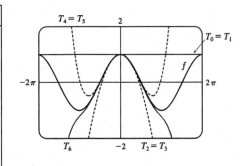

(b)

x	f	$T_0 = T_1$	$T_2 = T_3$	$T_4 = T_5$	T_6
$\frac{\pi}{4}$	0.7071	1	0.6916	0.7074	0.7071
$\frac{\pi}{2}$	0	1	-0.2337	0.0200	-0.0009
π	-1	1	-3.9348	0.1239	-1.2114

(c) As n increases, $T_n(x)$ is a good approximation to $f(x)$ on a larger and larger interval.

3.

n	$f^{(n)}(x)$	$f^{(n)}(1)$
0	$\ln x$	0
1	$1/x$	1
2	$-1/x^2$	-1
3	$2/x^3$	2
4	$-6/x^4$	-6

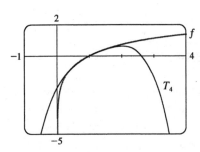

$$T_4(x) = \sum_{n=0}^{4} \frac{f^{(n)}(1)}{n!}(x-1)^n = 0 + (x-1) - \frac{1}{2}(x-1)^2 + \frac{1}{3}(x-1)^3 - \frac{1}{4}(x-1)^4$$

5.

n	$f^{(n)}(x)$	$f^{(n)}\left(\frac{\pi}{6}\right)$
0	$\sin x$	$\frac{1}{2}$
1	$\cos x$	$\frac{\sqrt{3}}{2}$
2	$-\sin x$	$-\frac{1}{2}$
3	$-\cos x$	$-\frac{\sqrt{3}}{2}$

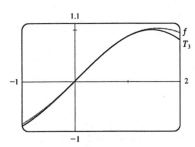

$$T_3(x) = \sum_{n=0}^{3} \frac{f^{(n)}\left(\frac{\pi}{6}\right)}{n!}\left(x-\frac{\pi}{6}\right)^n = \frac{1}{2} + \frac{\sqrt{3}}{2}\left(x-\frac{\pi}{6}\right) - \frac{1}{4}\left(x-\frac{\pi}{6}\right)^2 - \frac{\sqrt{3}}{12}\left(x-\frac{\pi}{6}\right)^3$$

7.

n	$f^{(n)}(x)$	$f^{(n)}(0)$
0	$\tan x$	0
1	$\sec^2 x$	1
2	$2\sec^2 x \tan x$	0
3	$4\sec^2 x \tan^2 x + 2\sec^4 x$	2
4	$8\sec^2 x \tan^3 x + 16\sec^4 x \tan x$	0

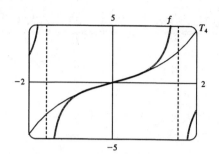

$$T_4(x) = \sum_{n=0}^{4} \frac{f^{(n)}(0)}{n!} x^n = x + \frac{2x^3}{3!} = x + \frac{x^3}{3}$$

9.

n	$f^{(n)}(x)$	$f^{(n)}(0)$
0	$e^x \sin x$	0
1	$e^x (\sin x + \cos x)$	1
2	$2e^x \cos x$	2
3	$2e^x (\cos x - \sin x)$	2

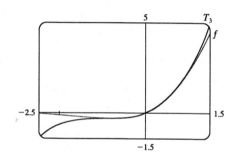

$$T_3(x) = \sum_{n=0}^{3} \frac{f^{(n)}(0)}{n!} x^n = x + x^2 + \tfrac{1}{3}x^3$$

11. In Maple, we can find the Taylor polynomials by the following method: first define `f:=sec(x);` and then set
`T2:=convert(taylor(f,x=0,3),polynom);`, `T4:=convert(taylor(f,x=0,5),polynom);`,
etc. (The third argument in the `taylor` function is one more than the degree of the desired polynomial). We must
convert to the type `polynom` because the output of the
`taylor` function contains an error term which we do not
want. In Mathematica, we use
`Tn:=Normal[Series[f,{x,0,n}]]`, with n=2, 4,
etc. Note that in Mathematica, the "degree" argument is the
same as the degree of the desired polynomial. In Derive,
author $\sec x$, then enter `Calculus,Taylor,8,0`; and
then simplify the expression. The eighth Taylor polynomial is
$T_8(x) = 1 + \tfrac{1}{2}x^2 + \tfrac{5}{24}x^4 + \tfrac{61}{720}x^6 + \tfrac{277}{8064}x^8$.

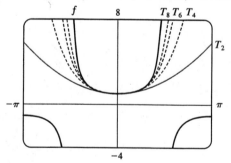

13.

$$f(x) = \sqrt{x} \qquad\qquad f(4) = 2$$
$$f'(x) = \tfrac{1}{2}x^{-1/2} \qquad\qquad f'(4) = \tfrac{1}{4}$$
$$f''(x) = -\tfrac{1}{4}x^{-3/2} \qquad\qquad f''(4) = -\tfrac{1}{32}$$
$$f'''(x) = \tfrac{3}{8}x^{-5/2}$$

(a) $\sqrt{x} \approx T_2(x) = 2 + \frac{1}{4}(x-4) - \frac{1/32}{2!}(x-4)^2$

$= 2 + \frac{1}{4}(x-4) - \frac{1}{64}(x-4)^2$

(b) $|R_2(x)| \le \dfrac{M}{3!}|x-4|^3$, where $|f'''(x)| \le M$. Now

$4 \le x \le 4.2 \implies |x-4| \le 0.2 \implies |x-4|^3 \le 0.008.$

Since $f'''(x)$ is decreasing on $[4, 4.2]$, we can take

$M = |f'''(4)| = \frac{3}{8}4^{-5/2} = \frac{3}{256}$, so

$|R_2(x)| \le \frac{3/256}{6}(0.008) = \frac{0.008}{512} = 0.000015625.$

(c)

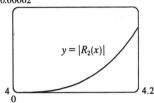

0.00002

$y = |R_2(x)|$

4   4.2
0

From the graph of

$|R_2(x)| = |\sqrt{x} - T_2(x)|$, it seems that

the error is less than 1.52×10^{-5} on

$[4, 4.2]$.

15.

$f(x) = \sin x$ $f\left(\frac{\pi}{4}\right) = \frac{\sqrt{2}}{2}$ $f^{(4)}(x) = \sin x$ $f^{(4)}\left(\frac{\pi}{4}\right) = \frac{\sqrt{2}}{2}$

$f'(x) = \cos x$ $f'\left(\frac{\pi}{4}\right) = \frac{\sqrt{2}}{2}$ $f^{(5)}(x) = \cos x$ $f^{(5)}\left(\frac{\pi}{4}\right) = \frac{\sqrt{2}}{2}$

$f''(x) = -\sin x$ $f''\left(\frac{\pi}{4}\right) = -\frac{\sqrt{2}}{2}$ $f^{(6)}(x) = -\sin x$

$f'''(x) = -\cos x$ $f'''\left(\frac{\pi}{4}\right) = -\frac{\sqrt{2}}{2}$

(a) $\sin x \approx T_5(x)$

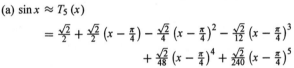

$= \frac{\sqrt{2}}{2} + \frac{\sqrt{2}}{2}\left(x - \frac{\pi}{4}\right) - \frac{\sqrt{2}}{4}\left(x - \frac{\pi}{4}\right)^2 - \frac{\sqrt{2}}{12}\left(x - \frac{\pi}{4}\right)^3$

$+ \frac{\sqrt{2}}{48}\left(x - \frac{\pi}{4}\right)^4 + \frac{\sqrt{2}}{240}\left(x - \frac{\pi}{4}\right)^5$

(b) $|R_5(x)| \le \dfrac{M}{6!}\left|x - \frac{\pi}{4}\right|^6$, where $|f^{(6)}(x)| \le M$. Now

$0 \le x \le \frac{\pi}{2} \implies \left(x - \frac{\pi}{4}\right)^6 \le \left(\frac{\pi}{4}\right)^6$, and letting $x = \frac{\pi}{2}$ gives

$M = 1$, so $|R_5(x)| \le \frac{1}{6!}\left(\frac{\pi}{4}\right)^6 = \frac{1}{720}\left(\frac{\pi}{4}\right)^6 \approx 0.00033.$

(c)

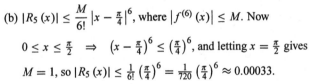

0.0003

0            $\frac{\pi}{2}$

From the graph of

$|R_5(x)| = |\sin x - T_5(x)|$, it seems that

the error is less than 0.00026 on $\left[0, \frac{\pi}{2}\right]$.

17.

$f(x) = \tan x$ $f(0) = 0$ $f'''(x) = 4\sec^2 x \tan^2 x + 2\sec^4 x$ $f'''(0) = 2$

$f'(x) = \sec^2 x$ $f'(0) = 1$ $f^{(4)}(x) = 8\sec^2 x \tan^3 x + 16\sec^4 x \tan x$

$f''(x) = 2\sec^2 x \tan x$ $f''(0) = 0$

(a) $\tan x \approx T_3(x) = x + \frac{1}{3}x^3$

(b) $|R_3(x)| \le \dfrac{M}{4!}|x|^4$, where $|f^{(4)}(x)| \le M$. Now $0 \le x \le \frac{\pi}{6}$

$\implies x^4 \le \left(\frac{\pi}{6}\right)^4$, and letting $x = \frac{\pi}{6}$ gives

$|R_3(x)| \le \dfrac{8\left(\frac{2}{\sqrt{3}}\right)^2 \left(\frac{1}{\sqrt{3}}\right)^3 + 16\left(\frac{2}{\sqrt{3}}\right)^4 \left(\frac{1}{\sqrt{3}}\right)}{4!}\left(\frac{\pi}{6}\right)^4$

$= \frac{4\sqrt{3}}{9}\left(\frac{\pi}{6}\right)^4 \approx 0.057859$

(c)

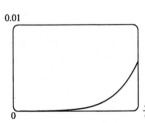

0.01

0            $\frac{\pi}{6}$

From the graph, it seems that the error is

less than 0.006 on $[0, \pi]$.

19. $f(x) = e^{x^2}$ $f(0) = 1$ $f'''(x) = e^{x^2}(12x + 8x^3)$ $f'''(0) = 0$

$f'(x) = e^{x^2}(2x)$ $f'(0) = 0$ $f^{(4)}(x) = e^{x^2}(12 + 48x^2 + 16x^4)$

$f''(x) = e^{x^2}(2 + 4x^2)$ $f''(0) = 2$

(a) $e^{x^2} \approx T_3(x) = 1 + \frac{2}{2!}x^2 = 1 + x^2$

(b) $|R_3(x)| \le \dfrac{M}{4!}|x|^4$, where $|f^{(4)}(x)| \le M$. Now $0 \le x \le 0.1$

     $\Rightarrow$ $x^4 \le (0.1)^4$, and letting $x = 0.1$ gives

$|R_3(x)| \le \dfrac{e^{0.01}(12 + 0.48 + 0.0016)}{24}(0.1)^4 \approx 0.00006.$

(c)

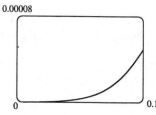

0.00008

0   0.1

From the graph of

$|R_3(x)| = \left| e^{x^2} - (1 + x^2) \right|$, it appears

that the error is less than 0.000051 on

$[0, 0.1]$.

21. $f(x) = x^{3/4}$ $f(16) = 8$ $f'''(x) = \frac{15}{64}x^{-9/4}$ $f'''(16) = \frac{15}{32,768}$

$f'(x) = \frac{3}{4}x^{-1/4}$ $f'(16) = \frac{3}{8}$ $f^{(4)}(x) = -\frac{135}{256}x^{-13/4}$

$f''(x) = -\frac{3}{16}x^{-5/4}$ $f''(16) = -\frac{3}{512}$

(a) $x^{3/4} \approx T_3(x) = 8 + \frac{3}{8}(x - 16) - \frac{3}{1024}(x - 16)^2$

                 $+ \frac{5}{65,536}(x - 16)^3$

(b) $|R_3(x)| \le \dfrac{M}{4!}|x - 16|^4$, where $|f^{(4)}(x)| \le M$. Now

$15 \le x \le 17$ $\Rightarrow$ $|x - 16|^4 \le 1^4 = 1$, and letting $x = 15$

to minimize the denominator of $f^{(4)}(x)$ gives

$|R_3(x)| \le \dfrac{135/\left[256(15)^{13/4}\right]}{4!}(1) \approx 0.0000033.$

(c)

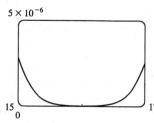

5×10^{-6}

15   17

0

It appears that the error is less than

3×10^{-6} on $(15, 17)$.

23. From Exercise 5, $\sin x = \frac{1}{2} + \frac{\sqrt{3}}{2}\left(x - \frac{\pi}{6}\right) - \frac{1}{4}\left(x - \frac{\pi}{6}\right)^2 - \frac{\sqrt{3}}{12}\left(x - \frac{\pi}{6}\right)^3 + R_3(x)$, where $R_3(x) \le \dfrac{M}{4!}\left|x - \frac{\pi}{6}\right|^4$

with $|f^{(4)}(x)| = |\sin x| \le M = 1$. Now $35° = \left(\frac{\pi}{6} + \frac{\pi}{36}\right)$ radians, so the error is $\left|R_3\left(\frac{\pi}{36}\right)\right| \le \dfrac{\left(\frac{\pi}{36}\right)^4}{4!} < 0.000003.$

Therefore, to five decimal places, $\sin 35° \approx \frac{1}{2} + \frac{\sqrt{3}}{2}\left(\frac{\pi}{36}\right) - \frac{1}{4}\left(\frac{\pi}{36}\right)^2 - \frac{\sqrt{3}}{12}\left(\frac{\pi}{36}\right)^3 \approx 0.57358.$

25. All derivatives of e^x are e^x, so $|R_n(x)| \le \dfrac{e^x}{(n+1)!}|x|^{n+1}$, where $0 < x < 0.1$. Letting $x = 0.1$,

$R_n(0.1) \le \dfrac{e^{0.1}}{(n+1)!}(0.1)^{n+1} < 0.00001$, and by trial and error we find that $n = 3$ satisfies this inequality since

$R_3(0.1) < 0.0000046$. Thus, by adding the three terms of the Maclaurin series for e^x corresponding to $n = 0, 1$,

and 2, we can estimate $e^{0.1}$ to within 0.00001.

27. $\sin x = x - \frac{1}{3!}x^3 + \frac{1}{5!}x^5 - \cdots$. By the Alternating

Series Estimation Theorem, the error in the

approximation $\sin x = x - \frac{1}{3!}x^3$ is less than

$\left|\frac{1}{5!}x^5\right| < 0.01 \iff |x^5| < 120\,(0.01) \iff$

$|x| < (1.2)^{1/5} \approx 1.037$. The curves intersect at

$x \approx 1.043$, so the graph confirms our estimate. Since

both the sine function and the given approximation are

odd functions, we need to check the estimate only for

$x > 0$.

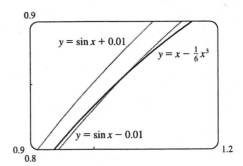

29. Let $s\,(t)$ be the position function of the car, and for convenience set $s\,(0) = 0$. The velocity of the car is

$v\,(t) = s'\,(t)$ and the acceleration is $a\,(t) = s''\,(t)$, so the second degree Taylor polynomial is

$T_2\,(t) = s\,(0) + v\,(0)\,t + \dfrac{a\,(0)}{2}t^2 = 20t + t^2$. We estimate the distance travelled during the next second to be

$s\,(1) \approx T_2\,(1) = 20 + 1 = 21$ m. The function $T_2\,(t)$ would not be accurate over a full minute, since the car could

not possibly maintain an acceleration of 2 m/s^2 for that long (if it did, its final speed would be

140 m/s ≈ 315 mi/h!)

31. $E = \dfrac{q}{D^2} - \dfrac{q}{(D+d)^2} = \dfrac{q}{D^2} - \dfrac{q}{D^2\,(1+d/D)^2} = \dfrac{q}{D^2}\left[1 - \left(1 + \dfrac{d}{D}\right)^{-2}\right]$.

We use the Binomial Series to expand $(1 + d/D)^{-2}$:

$$E = \frac{q}{D^2}\left[1 - \left(1 - 2\left(\frac{d}{D}\right) + \frac{2\cdot 3}{2!}\left(\frac{d}{D}\right)^2 - \frac{2\cdot 3\cdot 4}{3!}\left(\frac{d}{D}\right)^3 + \cdots\right)\right]$$

$$= \frac{q}{D^2}\left[2\left(\frac{d}{D}\right) - 3\left(\frac{d}{D}\right)^2 + 4\left(\frac{d}{D}\right)^3 - \cdots\right] \approx 2qd \cdot \frac{1}{D^3}$$

when D is much larger than d, that is, when P is far away from the dipole.

33. (a) If the water is deep, then $2\pi d/L$ is large, and we know that $\tanh x \to 1$ as $x \to \infty$. So we can approximate

$\tanh\,(2\pi d/L) \approx 1$, and so $v^2 \approx gL/\,(2\pi) \iff v \approx \sqrt{gL/\,(2\pi)}$.

(b) From the calculations at right, the first term in the 　　　　$f\,(x) = \tanh x$ 　　　　　　　　$f\,(0) = 0$

Maclaurin series of $\tanh x$ is x, so if the water is 　　　　$f'\,(x) = \operatorname{sech}^2 x$ 　　　　　　$f'\,(0) = 1$

shallow, we can approximate $\tanh\dfrac{2\pi d}{L} \approx \dfrac{2\pi d}{L}$, 　$f''\,(x) = -2\operatorname{sech}^2 x \tanh x$ 　$f''\,(0) = 0$

and so $v^2 \approx \dfrac{gL}{2\pi} \cdot \dfrac{2\pi d}{L} \iff v \approx \sqrt{gd}$. 　　$f'''\,(x) = 2\operatorname{sech}^2 x \,(3\tanh^2 x - 1)$ 　$f'''\,(0) = -2$

(c) Since $\tanh x$ is an odd function, its Maclaurin series is alternating, so the error in the approximation

$\tanh\dfrac{2\pi d}{L} \approx \dfrac{2\pi d}{L}$ is less than the first neglected term, which is $\dfrac{|f'''\,(0)|}{3!}\left(\dfrac{2\pi d}{L}\right)^3 = \dfrac{1}{3}\left(\dfrac{2\pi d}{L}\right)^3$. If

$L > 10d$, then $\dfrac{1}{3}\left(\dfrac{2\pi d}{L}\right)^3 < \dfrac{1}{3}\left(2\pi \cdot \dfrac{1}{10}\right)^3 = \dfrac{\pi^3}{375}$, so the error in the approximation $v^2 = gd$ is less than

$\dfrac{gL}{2\pi} \cdot \dfrac{\pi^3}{375} \approx 0.0132gL$.

35. Using $f(x) = T_n(x) + R_n(x)$ with $n = 1$ and $x = r$, we have $f(r) = T_1(r) + R_1(r)$, where T_1 is the first-degree Taylor polynomial of f at a. Because $a = x_n$, $f(r) = f(x_n) + f'(x_n)(r - x_n) + R_1(r)$. But r is a root of f, so $f(r) = 0$ and we have $0 = f(x_n) + f'(x_n)(r - x_n) + R_1(r)$. Taking the first two terms to the left side and dividing by $f'(x_n)$, we have $f'(x_n)(x_n - r) - f(x_n) = R_1(r)$ $\Rightarrow$ $x_n - r - \dfrac{f(x_n)}{f'(x_n)} = \dfrac{R_1(r)}{f'(x_n)}$. By the formula for Newton's method, the left side of the preceding equation is $x_{n+1} - r$, so $|x_{n+1} - r| = \left| \dfrac{R_1(r)}{f'(x_n)} \right|$.

Taylor's Inequality gives us $|R_1(r)| \leq \dfrac{|f''(r)|}{2!} |r - x_n|^2$. Combining this inequality with the facts $|f''(x)| \leq M$ and $|f'(x)| \geq K$ gives us $|x_{n+1} - r| \leq \dfrac{M}{2K} |x_n - r|^2$.

11 Review

CONCEPT CHECK

1. (a) See Definition 11.1.1.

(b) See Definition 11.2.2.

(c) The terms of the sequence $\{a_n\}$ approach 3 as n becomes large.

(d) By adding sufficiently many terms of the series, we can make the partial sums as close to 3 as we like.

2. (a) See Definition 11.1.9.

(b) A sequence is monotonic if it is either increasing or decreasing.

(c) By Theorem 11.1.10, every bounded, monotonic sequence is convergent.

3. (a) See (4) in Section 11.2.

(b) See (1) in Section 11.3.

4. If $\sum a_n = 3$, then $\lim\limits_{n \to \infty} a_n = 0$ and $\lim\limits_{n \to \infty} s_n = 3$.

5. (a) See the Test for Divergence (11.2.7).

(b) See the Integral Test on page 715.

(c) See the Comparison Test on page 722.

(d) See the Limit Comparison Test on page 723.

(e) See the Alternating Series Test on page 727.

(f) See the Ratio Test on page 733.

(g) See the Root Test on page 735.

6. (a) See Definition 11.6.1.

(b) By (11.6.3), it is convergent.

(c) See Definition 11.6.2.

7. (a) Use either (2) or (3) in Section 11.3.

(b) See Example 5 in Section 11.4.

(c) By adding terms until you reach the desired accuracy given by the Alternating Series Estimation Theorem on page 729.

8. (a) $\sum_{n=0}^{\infty} c_n (x-a)^n$

(b) Given the power series $\sum_{n=0}^{\infty} c_n (x-a)^n$, the radius of convergence is:

(i) 0 if the series converges only when $x = a$

(ii) ∞ if the series converges for all x, or

(iii) a positive number R such that the series converges if $|x-a| < R$ and diverges if $|x-a| > R$.

(c) The interval of convergence of a power series is the interval that consists of all values of x for which the series converges. Corresponding to the cases in part (b), the interval of convergence is: (i) the single point $\{a\}$, (ii) all real numbers, that is, the real number line $(-\infty, \infty)$, or (iii) an interval with endpoints $a - R$ and $a + R$ which can contain neither, either, or both of the endpoints. In this case, we must test the series for convergence at each endpoint to determine the interval of convergence.

9. (a), (b) See Theorem 11.9.2.

10. (a) $T_n (x) = \sum_{i=0}^{n} \dfrac{f^{(i)} (a)}{i!} (x-a)^i$

(b) $\sum_{n=0}^{\infty} \dfrac{f^{(n)} (a)}{n!} (x-a)^n$

(c) $\sum_{n=0}^{\infty} \dfrac{f^{(n)} (0)}{n!} x^n$ [$a = 0$ in part (b)]

(d) See Theorem 11.10.8.

(e) See Taylor's Inequality (11.10.9).

11. (a) – (e) See the table on page 758.

12. See the Binomial Series (11.11.2) for the expansion. The radius of convergence for the binomial series is 1.

TRUE-FALSE QUIZ

1. False. See Note 2 after Theorem 11.2.6.

3. False. For example, take $c_n = (-1)^n / (n6^n)$.

5. False, since $\lim\limits_{n \to \infty} \left| \dfrac{a_{n+1}}{a_n} \right| = \lim\limits_{n \to \infty} \left| \dfrac{n^3}{(n+1)^3} \right| = \lim\limits_{n \to \infty} \dfrac{1}{(1+1/n)^3} = 1$.

7. False. See the note after Example 2 in Section 11.4.

9. True. See (7) in Section 11.1.

11. True. By Theorem 11.10.5 the coefficient of x^3 is $\dfrac{f'''(0)}{3!} = \dfrac{1}{3} \Rightarrow f'''(0) = 2$.

Or: Use Theorem 11.9.2 to differentiate f three times.

13. False. For example, let $a_n = b_n = (-1)^n$. Then $\{a_n\}$ and $\{b_n\}$ are divergent, but $a_n b_n = 1$, so $\{a_n b_n\}$ is convergent.

15. True by Theorem 11.6.3. [$\sum (-1)^n a_n$ is absolutely convergent and hence convergent.]

EXERCISES

1. $\left\{ \dfrac{2+n^3}{1+2n^3} \right\}$ converges since $\lim\limits_{n\to\infty} \dfrac{2+n^3}{1+2n^3} = \lim\limits_{n\to\infty} \dfrac{2/n^3+1}{1/n^3+2} = \dfrac{1}{2}$.

3. $\lim\limits_{n\to\infty} a_n = \lim\limits_{n\to\infty} \dfrac{n^3}{1+n^2} = \lim\limits_{n\to\infty} \dfrac{n}{1/n^2+1} = \infty$, so the sequence diverges.

5. $\{\sin n\}$ is divergent since $\lim\limits_{n\to\infty} \sin n$ does not exist.

7. $\left\{ \left(1+\dfrac{3}{n}\right)^{4n} \right\}$ is convergent. Let $y = \left(1+\dfrac{3}{x}\right)^{4x}$. Then

$$\lim_{x\to\infty} \ln y = \lim_{x\to\infty} 4x \ln(1+3/x) = \lim_{x\to\infty} \dfrac{\ln(1+3/x)}{1/(4x)} \overset{\mathrm{H}}{=} \lim_{x\to\infty} \dfrac{\dfrac{1}{1+3/x}\left(-\dfrac{3}{x^2}\right)}{-1/(4x^2)} = \lim_{x\to\infty} \dfrac{12}{1+3/x} = 12$$

so $\lim\limits_{x\to\infty} y = \lim\limits_{n\to\infty}\left(1+\dfrac{3}{n}\right)^{4n} = e^{12}$.

9. We use induction, hypothesizing that $a_{n-1} < a_n < 2$. Note first that $1 < a_2 = \frac{1}{3}(1+5) = \frac{5}{3} < 2$, so the hypothesis holds for $n=2$. Now assume that $a_{k-1} < a_k < 2$. Then $a_k = \frac{1}{3}(a_{k-1}+4) < \frac{1}{3}(a_k+4) < \frac{1}{3}(2+4) = 2$. So $a_k < a_{k+1} < 2$, and the induction is complete. To find the limit of the sequence, we note that $L = \lim\limits_{n\to\infty} a_n = \lim\limits_{n\to\infty} a_{n+1} \;\Rightarrow\; L = \frac{1}{3}(L+4) \;\Rightarrow\; L = 2$.

11. $\dfrac{n}{n^3+1} < \dfrac{n}{n^3} = \dfrac{1}{n^2}$, so $\sum\limits_{n=1}^{\infty} \dfrac{n}{n^3+1}$ converges by the Comparison Test with the convergent p-series $\sum\limits_{n=1}^{\infty} \dfrac{1}{n^2}$ $(p=2>1)$.

13. $\lim\limits_{n\to\infty} \left| \dfrac{a_{n+1}}{a_n} \right| = \lim\limits_{n\to\infty} \left[\dfrac{(n+1)^3}{5^{n+1}} \cdot \dfrac{5^n}{n^3} \right] = \lim\limits_{n\to\infty}\left(1+\dfrac{1}{n}\right)^3 \cdot \dfrac{1}{5} = \dfrac{1}{5} < 1$, so $\sum\limits_{n=1}^{\infty} \dfrac{n^3}{5^n}$ converges by the Ratio Test.

15. $\lim\limits_{n\to\infty} \sqrt[n]{|a_n|} = \lim\limits_{n\to\infty} \dfrac{n}{3n+1} = \dfrac{1}{3} < 1$, so the series converges by the Root Test.

17. $\left| \dfrac{\sin n}{1+n^2} \right| \le \dfrac{1}{1+n^2} < \dfrac{1}{n^2}$ and since $\sum\limits_{n=1}^{\infty} \dfrac{1}{n^2}$ converges (p-series with $p=2>1$), so does $\sum\limits_{n=1}^{\infty} \left| \dfrac{\sin n}{1+n^2} \right|$ by the Comparison Test, and so does $\sum\limits_{n=1}^{\infty} \dfrac{\sin n}{1+n^2}$ by Theorem 11.6.3.

19. $\lim\limits_{n\to\infty} \left| \dfrac{a_{n+1}}{a_n} \right| = \lim\limits_{n\to\infty} \dfrac{1\cdot3\cdot5\cdots\cdots(2n-1)(2n+1)}{5^{n+1}(n+1)!} \cdot \dfrac{5^n n!}{1\cdot3\cdot5\cdots\cdots(2n-1)} = \lim\limits_{n\to\infty} \dfrac{2n+1}{5(n+1)} = \dfrac{2}{5} < 1$, so the series converges by the Ratio Test.

21. Let $b_n = \dfrac{\sqrt{n}}{n+1} > 0$. Then $0 \le \lim\limits_{n\to\infty} b_n = \lim\limits_{n\to\infty} \dfrac{\sqrt{n}}{n+1} \le \lim\limits_{n\to\infty} \dfrac{\sqrt{n}}{n} = \lim\limits_{n\to\infty} \dfrac{1}{\sqrt{n}} = 0$, so $\lim\limits_{n\to\infty} b_n = 0$. If

$$f(x) = \dfrac{\sqrt{x}}{x+1} \text{ for } x>0, \text{ then } f'(x) = \dfrac{(x+1)\cdot\frac{1}{2\sqrt{x}} - \sqrt{x}\cdot1}{(x+1)^2} = \dfrac{(x+1)-2x}{2\sqrt{x}(x+1)^2} = \dfrac{1-x}{2\sqrt{x}(x+1)^2}, \text{ so } f'(x) < 0$$

for $x>1$. It follows that $f(1) > f(2) > f(3) > \cdots$; that is, $b_n > b_{n+1}$ for all n. Thus, $\sum\limits_{n=1}^{\infty} (-1)^{n-1} \dfrac{\sqrt{n}}{n+1}$ converges by the Alternating Series Test.

23. Consider the series of absolute values: $\sum_{n=1}^{\infty} n^{-1/3}$ is a p-series with $p = \frac{1}{3} < 1$ and is therefore divergent. But if we apply the Alternating Series Test we see that $a_{n+1} < a_n$ and $\lim\limits_{n\to\infty} n^{-1/3} = 0$. Therefore $\sum_{n=1}^{\infty} (-1)^{n-1} n^{-1/3}$ is conditionally convergent.

25. $\left| \dfrac{a_{n+1}}{a_n} \right| = \left| \dfrac{(-1)^{n+1}(n+2)3^{n+1}}{2^{2n+3}} \cdot \dfrac{2^{2n+1}}{(-1)^n(n+1)3^n} \right| = \dfrac{n+2}{n+1} \cdot \dfrac{3}{4} = \dfrac{1+(2/n)}{1+(1/n)} \cdot \dfrac{3}{4} \to \dfrac{3}{4} < 1$ as $n \to \infty$, so by

the Ratio Test, $\displaystyle\sum_{n=1}^{\infty} \dfrac{(-1)^n(n+1)3^n}{2^{2n+1}}$ is absolutely convergent.

27. Convergent geometric series. $\displaystyle\sum_{n=1}^{\infty} \dfrac{2^{2n+1}}{5^n} = \sum_{n=1}^{\infty} \dfrac{(2^2)^n \cdot 2^1}{5^n} = 2 \sum_{n=1}^{\infty} \dfrac{4^n}{5^n} = 2 \left(\dfrac{\frac{4}{5}}{1 - \frac{4}{5}} \right) = 8.$

29. $\displaystyle\sum_{n=1}^{\infty} \left[\tan^{-1}(n+1) - \tan^{-1} n \right] = \lim_{n\to\infty} \left[(\tan^{-1} 2 - \tan^{-1} 1) + (\tan^{-1} 3 - \tan^{-1} 2) + \cdots \right.$
$$\left. + (\tan^{-1}(n+1) - \tan^{-1} n) \right]$$
$$= \lim_{n\to\infty} \left[\tan^{-1}(n+1) - \tan^{-1} 1 \right] = \tfrac{\pi}{2} - \tfrac{\pi}{4} = \tfrac{\pi}{4}$$

31. $1.2 + 0.0\overline{345} = \dfrac{12}{10} + \dfrac{345/10,000}{1 - 1/1000} = \dfrac{12}{10} + \dfrac{345}{9990} = \dfrac{4111}{3330}$

33. $\displaystyle\sum_{n=1}^{\infty} \dfrac{(-1)^{n+1}}{n^5} = 1 - \dfrac{1}{32} + \dfrac{1}{243} - \dfrac{1}{1024} + \dfrac{1}{3125} - \dfrac{1}{7776} + \dfrac{1}{16,807} - \dfrac{1}{32,768} + \cdots.$ Since $\dfrac{1}{32,768} < 0.000031$,

$\displaystyle\sum_{n=1}^{\infty} \dfrac{(-1)^{n+1}}{n^5} \approx \sum_{n=1}^{7} \dfrac{(-1)^{n+1}}{n^5} \approx 0.9721.$

35. $\displaystyle\sum_{n=1}^{\infty} \dfrac{1}{2 + 5^n} \approx \sum_{n=1}^{8} \dfrac{1}{2 + 5^n} \approx 0.18976224.$ To estimate the error, note that $\dfrac{1}{2 + 5^n} < \dfrac{1}{5^n}$, so the remainder term is

$R_8 = \displaystyle\sum_{n=9}^{\infty} \dfrac{1}{2 + 5^n} < \sum_{n=9}^{\infty} \dfrac{1}{5^n} = \dfrac{1/5^9}{1 - 1/5} = 6.4 \times 10^{-7}$ (geometric series with $a = \frac{1}{5^9}$ and $r = \frac{1}{5}$).

37. Use the Limit Comparison Test. $\lim\limits_{n\to\infty} \left| \dfrac{\left(\frac{n+1}{n}\right) a_n}{a_n} \right| = \lim\limits_{n\to\infty} \dfrac{n+1}{n} = \lim\limits_{n\to\infty} \left(1 + \dfrac{1}{n}\right) = 1 > 0.$ Since $\sum |a_n|$ is

convergent, so is $\displaystyle\sum \left| \left(\dfrac{n+1}{n} \right) a_n \right|$, by the Limit Comparison Test.

39. $\lim\limits_{n\to\infty} \left| \dfrac{a_{n+1}}{a_n} \right| = \lim\limits_{n\to\infty} \left[\dfrac{|x+2|^{n+1}}{(n+1)4^{n+1}} \cdot \dfrac{n4^n}{|x+2|^n} \right] = \lim\limits_{n\to\infty} \left[\dfrac{n}{n+1} \dfrac{|x+2|}{4} \right] = \dfrac{|x+2|}{4} < 1 \iff |x+2| < 4$, so

$R = 4.$ $|x + 2| < 4 \iff -4 < x + 2 < 4 \iff -6 < x < 2.$ If $x = -6$, then the series becomes

$\displaystyle\sum_{n=1}^{\infty} \dfrac{(-4)^n}{n4^n} = \sum_{n=1}^{\infty} \dfrac{(-1)^n}{n}$, the alternating harmonic series, which converges by the Alternating Series Test. When

$x = 2$, the series becomes the harmonic series $\displaystyle\sum_{n=1}^{\infty} \dfrac{1}{n}$, which diverges. Thus, $I = [-6, 2)$.

41. $\lim\limits_{n\to\infty} \left| \dfrac{a_{n+1}}{a_n} \right| = \lim\limits_{n\to\infty} \left| \dfrac{2^{n+1}(x-3)^{n+1}}{\sqrt{n+4}} \cdot \dfrac{\sqrt{n+3}}{2^n(x-3)^n} \right| = 2|x-3| \lim\limits_{n\to\infty} \sqrt{\dfrac{n+3}{n+4}} = 2|x-3| < 1 \quad \Leftrightarrow$

$|x-3| < \frac{1}{2}$, so $R = \frac{1}{2}$. For $x = \frac{7}{2}$, the series becomes $\sum\limits_{n=0}^{\infty} \dfrac{1}{\sqrt{n+3}} = \sum\limits_{n=3}^{\infty} \dfrac{1}{n^{1/2}}$, which diverges ($p = \frac{1}{2} \le 1$), but

for $x = \frac{5}{2}$, we get $\sum\limits_{n=0}^{\infty} \dfrac{(-1)^n}{\sqrt{n+3}}$, which is a convergent alternating series, so $I = \left[\frac{5}{2}, \frac{7}{2} \right)$.

43.

$f(x) = \sin x$	$f\left(\frac{\pi}{6}\right) = \frac{1}{2}$	$f'''(x) = -\cos x$	$f'''\left(\frac{\pi}{6}\right) = -\frac{\sqrt{3}}{2}$
$f'(x) = \cos x$	$f'\left(\frac{\pi}{6}\right) = \frac{\sqrt{3}}{2}$	$f^{(4)}(x) = \sin x$	$f^{(4)}\left(\frac{\pi}{6}\right) = \frac{1}{2}$
$f''(x) = -\sin x$	$f''\left(\frac{\pi}{6}\right) = -\frac{1}{2}$	$\cdots$	$\cdots$

$f^{(2n)}\left(\frac{\pi}{6}\right) = (-1)^n \cdot \frac{1}{2}$ and $f^{(2n+1)}\left(\frac{\pi}{6}\right) = (-1)^n \cdot \frac{\sqrt{3}}{2}$.

$$\sin x = \sum_{n=0}^{\infty} \frac{f^{(n)}\left(\frac{\pi}{6}\right)}{n!} \left(x - \frac{\pi}{6}\right)^n = \sum_{n=0}^{\infty} \frac{(-1)^n}{2(2n)!} \left(x - \frac{\pi}{6}\right)^{2n} + \sum_{n=0}^{\infty} \frac{(-1)^n \sqrt{3}}{2(2n+1)!} \left(x - \frac{\pi}{6}\right)^{2n+1}$$

45. $\dfrac{1}{1+x} = \dfrac{1}{1-(-x)} = \sum\limits_{n=0}^{\infty} (-1)^n x^n$ for $|x| < 1 \quad \Rightarrow \quad \dfrac{x^2}{1+x} = \sum\limits_{n=0}^{\infty} (-1)^n x^{n+2}$ with $R = 1$.

47. $\dfrac{1}{1-x} = \sum\limits_{n=0}^{\infty} x^n$ for $|x| < 1 \quad \Rightarrow \quad \ln(1-x) = -\displaystyle\int \dfrac{dx}{1-x} = -\displaystyle\int \sum_{n=0}^{\infty} x^n \, dx = C - \sum\limits_{n=0}^{\infty} \dfrac{x^{n+1}}{n+1}$.

$\ln(1-0) = C - 0 \quad \Rightarrow \quad C = 0 \quad \Rightarrow \quad \ln(1-x) = -\sum\limits_{n=0}^{\infty} \dfrac{x^{n+1}}{n+1} = \sum\limits_{n=1}^{\infty} \dfrac{-x^n}{n}$ with $R = 1$.

49. $\sin x = \sum\limits_{n=0}^{\infty} \dfrac{(-1)^n x^{2n+1}}{(2n+1)!} \quad \Rightarrow \quad \sin(x^4) = \sum\limits_{n=0}^{\infty} \dfrac{(-1)^n (x^4)^{2n+1}}{(2n+1)!} = \sum\limits_{n=0}^{\infty} \dfrac{(-1)^n x^{8n+4}}{(2n+1)!}$ for all x, so the radius of

convergence is ∞.

51. $f(x) = 1/\sqrt[4]{16-x} = (16-x)^{-1/4} = \frac{1}{2}\left(1 - \frac{1}{16}x\right)^{-1/4}$

$$= \frac{1}{2}\left[1 + \left(-\frac{1}{4}\right)\left(-\frac{x}{16}\right) + \frac{\left(-\frac{1}{4}\right)\left(-\frac{5}{4}\right)}{2!}\left(-\frac{x}{16}\right)^2 + \cdots \right]$$

$$= \frac{1}{2} + \sum_{n=1}^{\infty} \frac{1 \cdot 5 \cdot 9 \cdots (4n-3)}{2 \cdot 4^n \cdot n! \cdot 16^n} x^n = \frac{1}{2} + \sum_{n=1}^{\infty} \frac{1 \cdot 5 \cdot 9 \cdots (4n-3)}{2^{6n+1} n!} x^n$$

for $\left| -\dfrac{x}{16} \right| < 1 \quad \Rightarrow \quad R = 16$.

53. $e^x = \sum\limits_{n=0}^{\infty} \dfrac{x^n}{n!}$ so $\dfrac{e^x}{x} = \dfrac{1}{x} + \sum\limits_{n=1}^{\infty} \dfrac{x^{n-1}}{n!}$ and $\displaystyle\int \dfrac{e^x}{x} dx = C + \ln|x| + \sum\limits_{n=1}^{\infty} \dfrac{x^n}{n \cdot n!}$.

55. (a)

$$f(x) = x^{1/2} \qquad\qquad f(1) = 1 \qquad\qquad f'''(x) = \tfrac{3}{8}x^{-5/2} \qquad f'''(1) = \tfrac{3}{8}$$

$$f'(x) = \tfrac{1}{2}x^{-1/2} \qquad\quad f'(1) = \tfrac{1}{2} \qquad\quad f^{(4)}(x) = -\tfrac{15}{16}x^{-7/2}$$

$$f''(x) = -\tfrac{1}{4}x^{-3/2} \qquad f''(1) = -\tfrac{1}{4}$$

$$\sqrt{x} \approx T_3(x) = 1 + \frac{1/2}{1!}(x-1) - \frac{1/4}{2!}(x-1)^2 + \frac{3/8}{3!}(x-1)^3$$

$$= 1 + \tfrac{1}{2}(x-1) - \tfrac{1}{8}(x-1)^2 + \tfrac{1}{16}(x-1)^3$$

(b)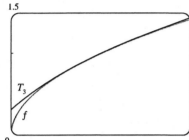

(c) $|R_3(x)| \le \dfrac{M}{4!}|x-1|^4$, where $\left|f^{(4)}(x)\right| \le M$ with

$f^{(4)}(x) = -\tfrac{15}{16}x^{-7/2}$. Now $0.9 \le x \le 1.1 \;\Rightarrow$

$(x-1)^4 \le (0.1)^4$, and letting $x = 0.9$ gives

$$M = \frac{15}{16\,(0.9)^{7/2}}, \text{ so}$$

$$|R_3(x)| \le \frac{15}{16\,(0.9)^{7/2}\,4!}(0.1)^4 \approx 0.000005648.$$

(d)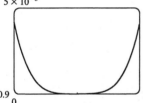

From the graph of $|R_3(x)| = \left|\sqrt{x} - T_3(x)\right|$, it appears that the error is less than 5×10^{-6} on $[0.9, 1.1]$.

57. $\sin x = \displaystyle\sum_{n=0}^{\infty}(-1)^n \frac{x^{2n+1}}{(2n+1)!} = x - \frac{x^3}{3!} + \frac{x^5}{5!} - \frac{x^7}{7!} + \cdots$, so $\sin x - x = -\dfrac{x^3}{3!} + \dfrac{x^5}{5!} - \dfrac{x^7}{7!} + \cdots$ and

$\dfrac{\sin x - x}{x^3} = -\dfrac{1}{3!} + \dfrac{x^2}{5!} - \dfrac{x^4}{7!} + \cdots$ and $\displaystyle\lim_{x\to 0}\frac{\sin x - x}{x^3} = \lim_{x\to 0}\left(-\frac{1}{6} + \frac{x^2}{120} - \frac{x^4}{5040} + \cdots\right) = -\frac{1}{6}$.

59. $f(x) = \sum_{n=0}^{\infty} c_n x^n \;\Rightarrow\; f(-x) = \sum_{n=0}^{\infty} c_n(-x)^n = \sum_{n=0}^{\infty}(-1)^n c_n x^n$

(a) If f is an odd function, then $f(-x) = -f(x) \;\Rightarrow\; \sum_{n=0}^{\infty}(-1)^n c_n x^n = \sum_{n=0}^{\infty} -c_n x^n$. The coefficients of any power series are uniquely determined (by Theorem 11.10.5), so $(-1)^n c_n = -c_n$. If n is even, then $(-1)^n = 1$, so $c_n = -c_n \;\Rightarrow\; 2c_n = 0 \;\Rightarrow\; c_n = 0$. Thus, all even coefficients are 0.

(b) If f is even, then $f(-x) = f(x) \;\Rightarrow\; \sum_{n=0}^{\infty}(-1)^n c_n x^n = \sum_{n=0}^{\infty} c_n x^n \;\Rightarrow\; (-1)^n c_n = c_n$. If n is odd, then $(-1)^n = -1$, so $-c_n = c_n \;\Rightarrow\; 2c_n = 0 \;\Rightarrow\; c_n = 0$. Thus, all odd coefficients are 0.

Problems Plus

1. It would be far too much work to compute 15 derivatives of f. The key idea is to remember that $f^{(n)}(0)$ occurs in the coefficient of x^n in the Maclaurin series of f. We start with the Maclaurin series for sin:

$$\sin x = x - \frac{x^3}{3!} + \frac{x^5}{5!} - \cdots. \quad \text{Then } \sin\left(x^3\right) = x^3 - \frac{x^9}{3!} + \frac{x^{15}}{5!} - \cdots \text{ and so the coefficient of } x^{15} \text{ is}$$

$$\frac{f^{(15)}(0)}{15!} = \frac{1}{5!}. \quad \text{Therefore, } f^{(15)}(0) = \frac{15!}{5!} = 6 \cdot 7 \cdot 8 \cdot 9 \cdot 10 \cdot 11 \cdot 12 \cdot 13 \cdot 14 \cdot 15 = 10{,}897{,}286{,}400.$$

3. (a) From Formula 14a in Appendix D, with $x = y = \theta$, we get $\tan 2\theta = \dfrac{2\tan\theta}{1 - \tan^2\theta}$, so $\cot 2\theta = \dfrac{1 - \tan^2\theta}{2\tan\theta}$ $\Rightarrow$

$$2\cot 2\theta = \frac{1 - \tan^2\theta}{\tan\theta} = \cot\theta - \tan\theta. \text{ Replacing } \theta \text{ by } \tfrac{1}{2}x, \text{ we get } 2\cot x = \cot\tfrac{1}{2}x - \tan\tfrac{1}{2}x,$$

or $\tan\tfrac{1}{2}x = \cot\tfrac{1}{2}x - 2\cot x$.

(b) From part (a), $\tan\dfrac{x}{2^n} = \cot\dfrac{x}{2^n} - 2\cot\dfrac{x}{2^{n-1}}$, so the nth partial sum of $\displaystyle\sum_{n=1}^{\infty}\dfrac{1}{2^n}\tan\dfrac{x}{2^n}$ is

$$s_n = \frac{\tan(x/2)}{2} + \frac{\tan(x/4)}{4} + \frac{\tan(x/8)}{8} + \cdots + \frac{\tan(x/2^n)}{2^n}$$

$$= \left[\frac{\cot(x/2)}{2} - \cot x\right] + \left[\frac{\cot(x/4)}{4} - \frac{\cot(x/2)}{2}\right] + \left[\frac{\cot(x/8)}{8} - \frac{\cot(x/4)}{4}\right] + \cdots$$

$$+ \left[\frac{\cot(x/2^n)}{2^n} - \frac{\cot\left(x/2^{n-1}\right)}{2^{n-1}}\right] = -\cot x + \frac{\cot(x/2^n)}{2^n} \quad \text{(telescoping sum)}$$

Now $\dfrac{\cot(x/2^n)}{2^n} = \dfrac{\cos(x/2^n)}{2^n\sin(x/2^n)} = \dfrac{\cos(x/2^n)}{x} \cdot \dfrac{x/2^n}{\sin(x/2^n)} \to \dfrac{1}{x} \cdot 1 = \dfrac{1}{x}$ as $n \to \infty$ since $x/2^n \to 0$ for

$x \neq 0$. Therefore, if $x \neq 0$ and $x \neq n\pi$, then $\displaystyle\sum_{n=1}^{\infty}\dfrac{1}{2^n}\tan\dfrac{x}{2^n} = \lim_{n\to\infty}\left(-\cot x + \dfrac{1}{2^n}\cot\dfrac{x}{2^n}\right) = -\cot x + \dfrac{1}{x}$. If

$x = 0$, then all terms in the series are 0, so the sum is 0.

5. (a) At each stage, each side is replaced by four shorter sides, each of length $\tfrac{1}{3}$ of the side length at the preceding stage. Writing s_0 and ℓ_0 for the number of sides and the length of the side of the initial triangle, we generate the table at right. In general, we have $s_n = 3 \cdot 4^n$ and $\ell_n = \left(\tfrac{1}{3}\right)^n$, so the length of the perimeter at the nth stage of

$s_0 = 3$	$\ell_0 = 1$
$s_1 = 3 \cdot 4$	$\ell_1 = 1/3$
$s_2 = 3 \cdot 4^2$	$\ell_2 = 1/3^2$
$s_3 = 3 \cdot 4^3$	$\ell_3 = 1/3^3$
$\cdots$	$\cdots$

construction is $p_n = s_n\ell_n = 3 \cdot 4^n \cdot \left(\tfrac{1}{3}\right)^n = 3 \cdot \left(\tfrac{4}{3}\right)^n$.

(b) $p_n = \dfrac{4^n}{3^{n-1}} = 4\left(\dfrac{4}{3}\right)^{n-1}$. Since $\tfrac{4}{3} > 1$, $p_n \to \infty$ as $n \to \infty$.

(c) The area of each of the small triangles added at a given stage is one-ninth of the area of the triangle added at the preceding stage. Let a be the area of the original triangle. Then the area a_n of each of the small triangles added at stage n is $a_n = a \cdot \dfrac{1}{9^n} = \dfrac{a}{9^n}$. Since a small triangle is added to each side at every stage, it follows that the total area A_n added to the figure at the nth stage is

$$A_n = s_{n-1} \cdot a_n = 3 \cdot 4^{n-1} \cdot \frac{a}{9^n} = a \cdot \frac{4^{n-1}}{3^{2n-1}}. \quad \text{Then the total area enclosed by the snowflake curve is}$$

$$A = a + A_1 + A_2 + A_3 + \cdots = a + a \cdot \frac{1}{3} + a \cdot \frac{4}{3^3} + a \cdot \frac{4^2}{3^5} + a \cdot \frac{4^3}{3^7} + \cdots. \quad \text{After the first term, this is a}$$

geometric series with common ratio $\frac{4}{9}$, so $A = a + \dfrac{a/3}{1 - \frac{4}{9}} = a + \dfrac{a}{3} \cdot \dfrac{9}{5} = \dfrac{8a}{5}$. But the area of the original

equilateral triangle with side 1 is $a = \frac{1}{2} \cdot 1 \cdot \sin \frac{\pi}{3} = \frac{\sqrt{3}}{4}$. So the area enclosed by the snowflake curve is $\frac{8}{5} \cdot \frac{\sqrt{3}}{4} = \frac{2\sqrt{3}}{5}$.

7. (a) Let $a = \arctan x$ and $b = \arctan y$. Then, from Formula 14b in Appendix D,

$$\tan (a - b) = \frac{\tan a - \tan b}{1 + \tan a \tan b} = \frac{\tan (\arctan x) - \tan (\arctan y)}{1 + \tan (\arctan x) \tan (\arctan y)} = \frac{x - y}{1 + xy} \quad \Rightarrow$$

$$\arctan x - \arctan y = a - b = \arctan \frac{x - y}{1 + xy} \quad \text{since } -\frac{\pi}{2} < \arctan x - \arctan y < \frac{\pi}{2}$$

(b) From part (a) we have

$$\arctan \tfrac{120}{119} - \arctan \tfrac{1}{239} = \arctan \frac{\frac{120}{119} - \frac{1}{239}}{1 + \frac{120}{119} \cdot \frac{1}{239}} = \arctan \frac{\frac{28,561}{28,441}}{\frac{28,561}{28,441}} = \arctan 1 = \tfrac{\pi}{4}$$

(c) Replacing y by $-y$ in the formula of part (a), we get $\arctan x + \arctan y = \arctan \dfrac{x + y}{1 - xy}$. So

$$4 \arctan \tfrac{1}{5} = 2 \left(\arctan \tfrac{1}{5} + \arctan \tfrac{1}{5} \right) = 2 \arctan \frac{\frac{1}{5} + \frac{1}{5}}{1 - \frac{1}{5} \cdot \frac{1}{5}} = 2 \arctan \tfrac{5}{12} = \arctan \tfrac{5}{12} + \arctan \tfrac{5}{12}$$

$$= \arctan \frac{\frac{5}{12} + \frac{5}{12}}{1 - \frac{5}{12} \cdot \frac{5}{12}} = \arctan \tfrac{120}{119}$$

Thus, from part (b), we have $4 \arctan \frac{1}{5} - \arctan \frac{1}{239} = \arctan \frac{120}{119} - \arctan \frac{1}{239} = \frac{\pi}{4}$.

(d) From Example 7 in Section 11.9 we have $\arctan x = x - \dfrac{x^3}{3} + \dfrac{x^5}{5} - \dfrac{x^7}{7} + \dfrac{x^9}{9} - \dfrac{x^{11}}{11} + \cdots$, so

$$\arctan \frac{1}{5} = \frac{1}{5} - \frac{1}{3 \cdot 5^3} + \frac{1}{5 \cdot 5^5} - \frac{1}{7 \cdot 5^7} + \frac{1}{9 \cdot 5^9} - \frac{1}{11 \cdot 5^{11}} + \cdots$$

This is an alternating series and the size of the terms decreases to 0, so by the Alternating Series Estimation Theorem, the sum lies between s_5 and s_6, that is, $0.197395560 < \arctan \frac{1}{5} < 0.197395562$.

(e) From the series in part (d) we get $\arctan \dfrac{1}{239} = \dfrac{1}{239} - \dfrac{1}{3 \cdot 239^3} + \dfrac{1}{5 \cdot 239^5} - \cdots$. The third term is less than 2.6×10^{-13}, so by the Alternating Series Estimation Theorem, we have, to nine decimal places, $\arctan \frac{1}{239} \approx s_2 \approx 0.004184076$. Thus, $0.004184075 < \arctan \frac{1}{239} < 0.004184077$.

(f) From part (c) we have $\pi = 16\arctan\frac{1}{5} - 4\arctan\frac{1}{239}$, so from parts (d) and (e) we have

$16(0.197395560) - 4(0.004184077) < \pi < 16(0.197395562) - 4(0.004184075) \Rightarrow$

$3.141592652 < \pi < 3.141592692$. So, to 7 decimal places, $\pi \approx 3.1415927$.

9. We start with the geometric series $\sum\limits_{n=0}^{\infty} x^n = \dfrac{1}{1-x}$, $|x| < 1$, and differentiate:

$\sum\limits_{n=1}^{\infty} nx^{n-1} = \dfrac{d}{dx}\left(\sum\limits_{n=0}^{\infty} x^n\right) = \dfrac{d}{dx}\left(\dfrac{1}{1-x}\right) = \dfrac{1}{(1-x)^2}$ for $|x| < 1 \Rightarrow \sum\limits_{n=1}^{\infty} nx^n = x\sum\limits_{n=1}^{\infty} nx^{n-1} = \dfrac{x}{(1-x)^2}$ for

$|x| < 1$. Differentiate again:

$\sum\limits_{n=1}^{\infty} n^2 x^{n-1} = \dfrac{d}{dx}\dfrac{x}{(1-x)^2} = \dfrac{(1-x)^2 - x\cdot 2(1-x)(-1)}{(1-x)^4} = \dfrac{x+1}{(1-x)^3} \Rightarrow \sum\limits_{n=1}^{\infty} n^2 x^n = \dfrac{x^2+x}{(1-x)^3} \Rightarrow$

$\sum\limits_{n=1}^{\infty} n^3 x^{n-1} = \dfrac{d}{dx}\dfrac{x^2+x}{(1-x)^3} = \dfrac{(1-x)^3(2x+1) - (x^2+x)3(1-x)^2(-1)}{(1-x)^6} = \dfrac{x^2+4x+1}{(1-x)^4} \Rightarrow$

$\sum\limits_{n=1}^{\infty} n^3 x^n = \dfrac{x^3+4x^2+x}{(1-x)^4}$, $|x| < 1$. The radius of convergence is 1 because that is the radius of convergence for

the geometric series we started with. If $x = \pm 1$, the series is $\sum n^3(\pm 1)^n$, which diverges by the Test For

Divergence, so the interval of convergence is $(-1, 1)$.

11. $a_{n+1} = \dfrac{a_n + b_n}{2}$, $b_{n+1} = \sqrt{b_n a_{n+1}}$. So $a_1 = \cos\theta$, $b_1 = 1 \Rightarrow a_2 = \dfrac{1+\cos\theta}{2} = \cos^2\frac{\theta}{2}$,

$b_2 = \sqrt{b_1 a_2} = \sqrt{\cos^2\frac{\theta}{2}} = \cos\frac{\theta}{2}$ since $-\frac{\pi}{2} \le \theta \le \frac{\pi}{2}$. Then

$a_3 = \frac{1}{2}\left(\cos\frac{\theta}{2} + \cos^2\frac{\theta}{2}\right) = \cos\frac{\theta}{2}\cdot\frac{1}{2}\left(1+\cos\frac{\theta}{2}\right) = \cos\frac{\theta}{2}\cos^2\frac{\theta}{4} \Rightarrow$

$b_3 = \sqrt{b_2 a_3} = \sqrt{\cos\frac{\theta}{2}\cos\frac{\theta}{2}\cos^2\frac{\theta}{4}} = \cos\frac{\theta}{2}\cos\frac{\theta}{4} \Rightarrow$

$a_4 = \frac{1}{2}\left(\cos\frac{\theta}{2}\cos^2\frac{\theta}{4} + \cos\frac{\theta}{2}\cos\frac{\theta}{4}\right) = \cos\frac{\theta}{2}\cos\frac{\theta}{4}\cdot\frac{1}{2}\left(1+\cos\frac{\theta}{4}\right) = \cos\frac{\theta}{2}\cos\frac{\theta}{4}\cos^2\frac{\theta}{8} \Rightarrow$

$b_4 = \sqrt{\cos\frac{\theta}{2}\cos\frac{\theta}{4}\cos\frac{\theta}{2}\cos\frac{\theta}{4}\cos^2\frac{\theta}{8}} = \cos\frac{\theta}{2}\cos\frac{\theta}{4}\cos\frac{\theta}{8}$. By now we see the pattern:

$b_n = \cos\frac{\theta}{2}\cos\frac{\theta}{2^2}\cos\frac{\theta}{2^3}\cdots\cdots\cos\frac{\theta}{2^{n-1}}$ and $a_n = b_n\cos\frac{\theta}{2^{n-1}}$. (This could be proved by mathematical

induction.) By Exercise 10(a), $\sin\theta = 2^{n-1}\sin\frac{\theta}{2^{n-1}}\cos\frac{\theta}{2}\cos\frac{\theta}{4}\cdots\cos\frac{\theta}{2^{n-1}}$. So

$b_n = \cos\frac{\theta}{2}\cos\frac{\theta}{2^2}\cos\frac{\theta}{2^3}\cdots\cos\frac{\theta}{2^{n-1}} \to \frac{\sin\theta}{\theta}$ as $n \to \infty$ by Exercise 10(b), and

$a_n = b_n\cos\frac{\theta}{2^{n-1}} \to \frac{\sin\theta}{\theta}\cdot 1 = \frac{\sin\theta}{\theta}$ as $n \to \infty$. So $\lim\limits_{n\to\infty} a_n = \lim\limits_{n\to\infty} b_n = \frac{\sin\theta}{\theta}$.

13. Let $f(x) = \sum\limits_{m=0}^{\infty} c_m x^m$ and $g(x) = e^{f(x)} = \sum\limits_{n=0}^{\infty} d_n x^n$. Then $g'(x) = \sum\limits_{n=0}^{\infty} nd_n x^{n-1}$, so nd_n occurs as the

coefficient of x^{n-1}. But also

$g'(x) = e^{f(x)}f'(x) = \left(\sum\limits_{n=0}^{\infty} d_n x^n\right)\left(\sum\limits_{m=1}^{\infty} mc_m x^{m-1}\right)$

$= \left(d_0 + d_1 x + d_2 x^2 + \cdots + d_{n-1}x^{n-1} + \cdots\right)\left(c_1 + 2c_2 x + 3c_3 x^2 + \cdots + nc_n x^{n-1} + \cdots\right)$

so the coefficient of x^{n-1} is $c_1 d_{n-1} + 2c_2 d_{n-2} + 3c_3 d_{n-3} + \cdots + nc_n d_0 = \sum\limits_{i=1}^{n} ic_i d_{n-i}$. Therefore,

$nd_n = \sum\limits_{i=1}^{n} ic_i d_{n-i}$.

15. $u = 1 + \dfrac{x^3}{3!} + \dfrac{x^6}{6!} + \dfrac{x^9}{9!} + \cdots, v = x + \dfrac{x^4}{4!} + \dfrac{x^7}{7!} + \dfrac{x^{10}}{10!} + \cdots, w = \dfrac{x^2}{2!} + \dfrac{x^5}{5!} + \dfrac{x^8}{8!} + \cdots$. The key idea is to

differentiate: $\dfrac{du}{dx} = \dfrac{3x^2}{3!} + \dfrac{6x^5}{6!} + \dfrac{9x^8}{9!} + \cdots = \dfrac{x^2}{2!} + \dfrac{x^5}{5!} + \dfrac{x^8}{8!} + \cdots = w$. Similarly,

$\dfrac{dv}{dx} = 1 + \dfrac{x^3}{3!} + \dfrac{x^6}{6!} + \dfrac{x^9}{9!} + \cdots = u$, and $\dfrac{dw}{dx} = x + \dfrac{x^4}{4!} + \dfrac{x^7}{7!} + \dfrac{x^{10}}{10!} + \cdots = v$. So $u' = w$, $v' = u$, and $w' = v$.

Now differentiate the left hand side of the desired equation:

$$\dfrac{d}{dx}\left(u^3 + v^3 + w^3 - 3uvw\right) = 3u^2u' + 3v^2v' + 3w^2w' - 3\left(u'vw + uv'w + uvw'\right)$$

$$= 3u^2w + 3v^2u + 3w^2v - 3\left(vw^2 + u^2w + uv^2\right) = 0 \quad \Rightarrow$$

$u^3 + v^3 + w^3 - 3uvw = C$. To find the value of the constant C, we put $x = 0$ in the last equation and get

$1^3 + 0^3 + 0^3 - 3\left(1 \cdot 0 \cdot 0\right) = C \quad \Rightarrow \quad C = 1$, so $u^3 + v^3 + w^3 - 3uvw = 1$.

17. If L is the length of a side of the equilateral triangle, then the area is $A = \frac{1}{2}L \cdot \frac{\sqrt{3}}{2}L = \frac{\sqrt{3}}{4}L^2$ and so $L^2 = \frac{4}{\sqrt{3}}A$.

Let r be the radius of one of the circles. When there are n rows of circles, the figure shows that

$L = \sqrt{3}r + r + (n-2)(2r) + r + \sqrt{3}r = r\left(2n - 2 + 2\sqrt{3}\right)$, so $r = \dfrac{L}{2\left(n + \sqrt{3} - 1\right)}$. The number of circles is

$1 + 2 + \cdots + n = \dfrac{n(n+1)}{2}$ and so the total area of the circles is

$A_n = \dfrac{n(n+1)}{2}\pi r^2 = \dfrac{n(n+1)}{2}\pi \dfrac{L^2}{4\left(n + \sqrt{3} - 1\right)^2} = \dfrac{n(n+1)}{2}\pi \dfrac{4A/\sqrt{3}}{4\left(n + \sqrt{3} - 1\right)^2}$

$= \dfrac{n(n+1)}{\left(n + \sqrt{3} - 1\right)^2}\dfrac{\pi A}{2\sqrt{3}} \quad \Rightarrow$

$\dfrac{A_n}{A} = \dfrac{n(n+1)}{\left(n + \sqrt{3} - 1\right)^2}\dfrac{\pi}{2\sqrt{3}}$

$= \dfrac{1 + 1/n}{\left[1 + \left(\sqrt{3} - 1\right)/n\right]^2}\dfrac{\pi}{2\sqrt{3}} \rightarrow \dfrac{\pi}{2\sqrt{3}}$ as $n \rightarrow \infty$

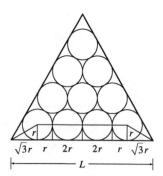

$$\sqrt{3}r \quad r \quad 2r \quad 2r \quad r \quad \sqrt{3}r$$
$$\vdash\!\!\!-\!\!\!-\!\!\!- L -\!\!\!-\!\!\!-\!\!\!\dashv$$

19. Call the series S. We group the terms according to the number of digits in their denominators:

$$S = \underbrace{\left(1 + \tfrac{1}{2} + \cdots + \tfrac{1}{8} + \tfrac{1}{9}\right)}_{g_1} + \underbrace{\left(\tfrac{1}{11} + \cdots + \tfrac{1}{99}\right)}_{g_2} + \underbrace{\left(\tfrac{1}{111} + \cdots + \tfrac{1}{999}\right)}_{g_3} + \cdots$$

Now in the group g_n, there are 9^n terms, since we have 9 choices for each of the n digits in the denominator.

Furthermore, each term in g_n is less than $\frac{1}{10^{n-1}}$. So $g_n < 9^n \cdot \frac{1}{10^{n-1}} = 9\left(\frac{9}{10}\right)^{n-1}$. Now $\sum_{n=1}^{\infty} 9\left(\frac{9}{10}\right)^{n-1}$ is a

geometric series with $a = 9$ and $r = \frac{9}{10} < 1$. Therefore, by the Comparison Test,

$S = \sum_{n=1}^{\infty} g_n < \sum_{n=1}^{\infty} 9\left(\frac{9}{10}\right)^{n-1} = \frac{9}{1 - 9/10} = 90$.

Appendixes

Intervals, Inequalities, and Absolute Values

1. $|5 - 23| = |-18| = 18$

3. $|-\pi| = \pi$ because $\pi > 0$.

5. $\left|\sqrt{5} - 5\right| = -\left(\sqrt{5} - 5\right) = 5 - \sqrt{5}$ because $\sqrt{5} - 5 < 0$.

7. For $x < 2$, $x - 2 < 0$, so $|x - 2| = -(x - 2) = 2 - x$.

9. $|x + 1| = \begin{cases} x + 1 & \text{for } x + 1 \geq 0 \iff x \geq -1 \\ -(x + 1) & \text{for } x + 1 < 0 \iff x < -1 \end{cases}$

11. $\left|x^2 + 1\right| = x^2 + 1$ (since $x^2 + 1 \geq 0$ for all x).

13. $2x + 7 > 3 \iff 2x > -4 \iff x > -2$, so $x \in (-2, \infty)$.

15. $1 - x \leq 2 \iff -x \leq 1 \iff x \geq -1$, so $x \in [-1, \infty)$.

17. $2x + 1 < 5x - 8 \iff 9 < 3x \iff 3 < x$, so $x \in (3, \infty)$.

19. $-1 < 2x - 5 < 7 \iff 4 < 2x < 12 \iff 2 < x < 6$, so $x \in (2, 6)$.

21. $0 \leq 1 - x < 1 \iff -1 \leq -x < 0 \iff 1 \geq x > 0$, so $x \in (0, 1]$.

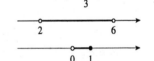

23. $4x < 2x + 1 \leq 3x + 2$. So $4x < 2x + 1 \iff 2x < 1 \iff x < \frac{1}{2}$, and $2x + 1 \leq 3x + 2 \iff -1 \leq x$. Thus, $x \in \left[-1, \frac{1}{2}\right)$.

25. $(x - 1)(x - 2) > 0$. *Case 1:* $x - 1 > 0 \iff x > 1$, and $x - 2 > 0 \iff x > 2$, so $x \in (2, \infty)$. *Case 2:* $x - 1 < 0 \iff x < 1$, and $x - 2 < 0 \iff x < 2$, so $x \in (-\infty, 1)$. Thus, the solution set is $(-\infty, 1) \cup (2, \infty)$.

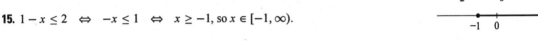

27. $2x^2 + x \leq 1 \iff 2x^2 + x - 1 \leq 0 \iff (2x - 1)(x + 1) \leq 0$. *Case 1:* $2x - 1 \geq 0 \iff x \geq \frac{1}{2}$, and $x + 1 \leq 0 \iff x \leq -1$, which is impossible. *Case 2:* $2x - 1 \leq 0 \iff x \leq \frac{1}{2}$, and $x + 1 \geq 0 \iff x \geq -1$, so $x \in \left[-1, \frac{1}{2}\right]$. Thus, the solution set is $\left[-1, \frac{1}{2}\right]$.

29. $x^2 + x + 1 > 0 \iff x^2 + x + \frac{1}{4} + \frac{3}{4} > 0 \iff \left(x + \frac{1}{2}\right)^2 + \frac{3}{4} > 0.$ But since $\left(x + \frac{1}{2}\right)^2 \geq 0$ for every real x, the original inequality will be true for all real x as well. Thus, the solution set is $(-\infty, \infty)$.

31. $x^2 < 3 \iff x^2 - 3 < 0 \iff \left(x - \sqrt{3}\right)\left(x + \sqrt{3}\right) < 0.$ *Case 1:* $x > \sqrt{3}$ and $x < -\sqrt{3}$, which is impossible.
Case 2: $x < \sqrt{3}$ and $x > -\sqrt{3}$. Thus, the solution set is $\left(-\sqrt{3}, \sqrt{3}\right)$.
Another Method: $x^2 < 3 \iff |x| < \sqrt{3} \iff -\sqrt{3} < x < \sqrt{3}.$

$$-\sqrt{3} \quad 0 \quad \sqrt{3}$$

33. $x^3 - x^2 \leq 0 \iff x^2(x - 1) \leq 0.$ Since $x^2 \geq 0$ for all x, the inequality is satisfied when $x - 1 \leq 0 \iff x \leq 1$. Thus, the solution set is $(-\infty, 1]$.

$$0 \quad 1$$

35. $x^3 > x \iff x^3 - x > 0 \iff x(x^2 - 1) > 0 \iff x(x - 1)(x + 1) > 0.$ Constructing a table:

Interval	x	$x - 1$	$x + 1$	$x(x-1)(x+1)$
$x < -1$	$-$	$-$	$-$	$-$
$-1 < x < 0$	$-$	$-$	$+$	$+$
$0 < x < 1$	$+$	$-$	$+$	$-$
$x > 1$	$+$	$+$	$+$	$+$

Since $x^3 > x$ when the last column is positive, the solution set is $(-1, 0) \cup (1, \infty)$.

$$-1 \quad 0 \quad 1$$

37. $1/x < 4.$ This is clearly true for $x < 0$. So suppose $x > 0$. then $1/x < 4 \iff 1 < 4x \iff \frac{1}{4} < x$. Thus, the solution set is $(-\infty, 0) \cup \left(\frac{1}{4}, \infty\right)$.

$$0 \quad \tfrac{1}{4}$$

39. $C = \frac{5}{9}(F - 32) \implies F = \frac{9}{5}C + 32.$ So $50 \leq F \leq 95 \implies 50 \leq \frac{9}{5}C + 32 \leq 95 \implies 18 \leq \frac{9}{5}C \leq 63 \implies 10 \leq C \leq 35.$ So the interval is $[10, 35]$.

41. (a) Let T represent the temperature in degrees Celsius and h the height in km. $T = 20$ when $h = 0$ and T decreases by $10\,°C$ for every km. Thus, $T = 20 - 10h$ when $0 \leq h \leq 12$.

(b) From (a), $T = 20 - 10h \implies h = 2 - T/10.$ So $0 \leq h \leq 5 \implies 0 \leq 2 - T/10 \leq 5 \implies -2 \leq -T/10 \leq 3 \implies -20 \leq -T \leq 30 \implies 20 \geq T \geq -30 \implies -30 \leq T \leq 20.$ Thus, the range of temperatures (in $°C$) to be expected is $[-30, 20]$.

43. $|2x| = 3 \iff$ either $2x = 3$ or $2x = -3 \iff x = \frac{3}{2}$ or $x = -\frac{3}{2}$.

45. $|x + 3| = |2x + 1| \iff$ either $x + 3 = 2x + 1$ or $x + 3 = -(2x + 1).$ In the first case, $x = 2$, and in the second case, $3x = -4 \iff x = -\frac{4}{3}$.

47. By Property 5 of absolute values, $|x| < 3 \iff -3 < x < 3$, so $x \in (-3, 3)$.

49. $|x - 4| < 1 \quad \Leftrightarrow \quad -1 < x - 4 < 1 \quad \Leftrightarrow \quad 3 < x < 5$, so $x \in (3, 5)$.

51. $|x + 5| \geq 2 \quad \Leftrightarrow \quad x + 5 \geq 2$ or $x + 5 \leq -2 \quad \Leftrightarrow \quad x \geq -3$ or $x \leq -7$, so $x \in (-\infty, -7] \cup [-3, \infty)$.

53. $|2x - 3| \leq 0.4 \quad \Leftrightarrow \quad -0.4 \leq 2x - 3 \leq 0.4 \quad \Leftrightarrow \quad 2.6 \leq 2x \leq 3.4 \quad \Leftrightarrow \quad 1.3 \leq x \leq 1.7$, so $x \in [1.3, 1.7]$.

55. $1 \leq |x| \leq 4$. So either $1 \leq x \leq 4$ or $1 \leq -x \leq 4 \quad \Leftrightarrow \quad -1 \geq x \geq -4$. Thus, $x \in [-4, -1] \cup [1, 4]$.

57. $a(bx - c) \geq bc \quad \Leftrightarrow \quad bx - c \geq \dfrac{bc}{a} \quad \Leftrightarrow \quad bx \geq \dfrac{bc}{a} + c = \dfrac{bc + ac}{a} \quad \Leftrightarrow \quad x \geq \dfrac{bc + ac}{ab}$

59. $ax + b < c \quad \Leftrightarrow \quad ax < c - b \quad \Leftrightarrow \quad x > \dfrac{c - b}{a}$ (since $a < 0$)

61. $|(x + y) - 5| = |(x - 2) + (y - 3)| \leq |x - 2| + |y - 3| < 0.01 + 0.04 = 0.05$

63. If $a < b$ then $a + a < a + b$ and $a + b < b + b$. So $2a < a + b < 2b$. Dividing by 2, $a < \frac{1}{2}(a + b) < b$.

65. $|ab| = \sqrt{(ab)^2} = \sqrt{a^2 b^2} = \sqrt{a^2}\sqrt{b^2} = |a|\,|b|$

67. If $0 < a < b$, then $a \cdot a < a \cdot b$ and $a \cdot b < b \cdot b$ [using Rule 3 of Inequalities]. So $a^2 < ab < b^2$ and hence $a^2 < b^2$.

69. Observe that the sum, difference and product of two integers is always an integer. Let the rational numbers be represented by $r = m/n$ and $s = p/q$ (where m, n, p and q are integers with $n \neq 0$, $q \neq 0$). Now
$r + s = \dfrac{m}{n} + \dfrac{p}{q} = \dfrac{mq + pn}{nq}$, but $mq + pn$ and nq are both integers, so $\dfrac{mq + pn}{nq} = r + s$ is a rational number by definition. Similarly, $r - s = \dfrac{m}{n} - \dfrac{p}{q} = \dfrac{mq - pn}{nq}$ is a rational number. Finally, $r \cdot s = \dfrac{m}{n} \cdot \dfrac{p}{q} = \dfrac{mp}{nq}$ but mp and nq are both integers, so $\dfrac{mp}{nq} = r \cdot s$ is a rational number by definition.

B Coordinate Geometry and Lines

1. From the Distance Formula with $x_1 = 1$, $x_2 = 4$, $y_1 = 1$, $y_2 = 5$, we find the distance from $(1, 1)$ to $(4, 5)$ to be $\sqrt{(4 - 1)^2 + (5 - 1)^2} = \sqrt{3^2 + 4^2} = \sqrt{25} = 5$.

3. $\sqrt{(-1 - 6)^2 + [3 - (-2)]^2} = \sqrt{(-7)^2 + 5^2} = \sqrt{74}$

5. $\sqrt{(4 - 2)^2 + (-7 - 5)^2} = \sqrt{2^2 + (-12)^2} = \sqrt{148} = 2\sqrt{37}$

7. From (2), the slope is $\dfrac{11 - 5}{4 - 1} = \dfrac{6}{3} = 2$.

9. With $P(-3, 3)$ and $Q(-1, -6)$, the slope m of the line through P and Q is $m = \dfrac{-6 - 3}{-1 - (-3)} = -\dfrac{9}{2}$.

11. Since $|AC| = \sqrt{(-4 - 0)^2 + (3 - 2)^2} = \sqrt{(-4)^2 + 1^2} = \sqrt{17}$ and $|BC| = \sqrt{[-4 - (-3)]^2 + [3 - (-1)]^2} = \sqrt{(-1)^2 + 4^2} = \sqrt{17}$, the triangle has two sides of equal length, and so is isosceles.

13. Using $A\,(-2, 9)$, $B\,(4, 6)$, $C\,(1, 0)$, and $D\,(-5, 3)$, we have

$|AB| = \sqrt{[4-(-2)]^2 + (6-9)^2} = \sqrt{6^2 + (-3)^2} = 3\sqrt{5},$

$|BC| = \sqrt{(1-4)^2 + (0-6)^2} = \sqrt{(-3)^2 + (-6)^2} = 3\sqrt{5},$

$|CD| = \sqrt{(-5-1)^2 + (3-0)^2} = \sqrt{(-6)^2 + 3^2} = 3\sqrt{5},$ and

$|DA| = \sqrt{[-2-(-5)]^2 + (9-3)^2} = \sqrt{3^2 + 6^2} = 3\sqrt{5}.$ So all sides are of equal length. Moreover,

$m_{AB} = \dfrac{6-9}{4-(-2)} = -\dfrac{1}{2}$, $m_{BC} = \dfrac{0-6}{1-4} = 2$, $m_{CD} = \dfrac{3-0}{-5-1} = -\dfrac{1}{2}$, and $m_{DA} = \dfrac{9-3}{-2-(-5)} = 2$, so the sides

are perpendicular. Thus, it is a square.

15. The slope of the line segment AB is $\dfrac{4-1}{7-1} = \dfrac{1}{2}$, the slope of CD is $\dfrac{7-10}{-1-5} = \dfrac{1}{2}$, the slope of BC is

$\dfrac{10-4}{5-7} = -3$, and the slope of DA is $\dfrac{1-7}{1-(-1)} = -3$. So AB is parallel to CD and BC is parallel to DA. Hence

$ABCD$ is a parallelogram.

17. $x = 3$

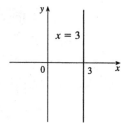

19. $xy = 0 \iff x = 0$ or $y = 0$

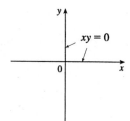

21. By the point-slope form of the equation of a line, an equation of the line through $(2, -3)$ with slope 6 is

$y - (-3) = 6\,(x - 2)$ or $y = 6x - 15$.

23. $y - 7 = \frac{2}{3}\,(x - 1)$ or $2x - 3y + 19 = 0$

25. The slope of the line through $(2, 1)$ and $(1, 6)$ is $m = \dfrac{6-1}{1-2} = -5$, so an equation of the line is

$y - 1 = -5\,(x - 2)$ or $y = -5x + 11$.

27. By the slope-intercept form of the equation of a line, an equation of the line is $y = 3x - 2$.

29. Since the line passes through $(1, 0)$ and $(0, -3)$, its slope is $m = \dfrac{-3-0}{0-1} = 3$, so an equation is $y = 3x - 3$.

31. Since $m = 0$, $y - 5 = 0\,(x - 4)$ or $y = 5$.

33. Putting the line $x + 2y = 6$ into its slope-intercept form $y = -\frac{1}{2}x + 3$, we see that this line has slope $-\frac{1}{2}$. So we
want the line of slope $-\frac{1}{2}$ that passes through the point $(1, -6)$: $y - (-6) = -\frac{1}{2}\,(x - 1) \iff y = -\frac{1}{2}x - \frac{11}{2}$.

35. $2x + 5y + 8 = 0 \iff y = -\frac{2}{5}x - \frac{8}{5}$. Since this line has slope $-\frac{2}{5}$, a line perpendicular to it would have slope $\frac{5}{2}$,
so the required line is $y - (-2) = \frac{5}{2}\,[x - (-1)] \iff y = \frac{5}{2}x + \frac{1}{2}$.

37. $x + 3y = 0 \iff y = -\frac{1}{3}x$, so the slope is $-\frac{1}{3}$ and the y-intercept is 0.

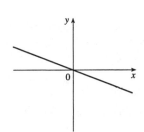

39. $y = -2$ is a horizontal line with slope 0 and y-intercept -2.

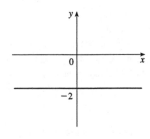

41. $3x - 4y = 12 \iff$ $y = \frac{3}{4}x - 3$, so the slope is $\frac{3}{4}$ and the y-intercept is -3.

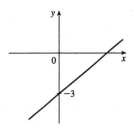

43. $\{(x, y) \mid x < 0\}$

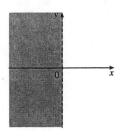

45. $\{(x, y) \mid xy < 0\} =$ $\{(x, y) \mid x < 0 \text{ and } y > 0\}$ $\cup \{(x, y) \mid x > 0 \text{ and } y < 0\}$

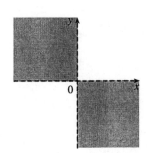

47. $\{(x, y) \mid |x| \le 2\} =$ $\{(x, y) \mid -2 \le x \le 2\}$

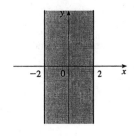

49. $\{(x, y) \mid 0 \le y \le 4, x \le 2\}$

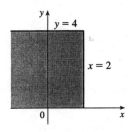

51. $\{(x, y) \mid 1 + x \le y \le 1 - 2x\}$

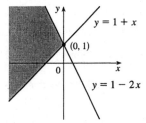

53. Let $P\,(0, y)$ be a point on the y-axis. The distance from P to $(5, -5)$ is

$\sqrt{(5 - 0)^2 + (-5 - y)^2} = \sqrt{5^2 + (y + 5)^2}$. The distance from P to $(1, 1)$ is

$\sqrt{(1 - 0)^2 + (1 - y)^2} = \sqrt{1^2 + (y - 1)^2}$. We want these distances to be equal:

$\sqrt{5^2 + (y + 5)^2} = \sqrt{1^2 + (y - 1)^2} \iff 5^2 + (y + 5)^2 = 1^2 + (y - 1)^2 \iff$

$25 + (y^2 + 10y + 25) = 1 + (y^2 - 2y + 1) \iff 12y = -48 \iff y = -4$. So the desired point is $(0, -4)$.

55. (a) Using the midpoint formula from Exercise 54 with $(1, 3)$ and $(7, 15)$, we get $\left(\frac{1+7}{2}, \frac{3+15}{2}\right) = (4, 9)$.

(b) Using the same formula, we get $\left(\frac{-1+8}{2}, \frac{6-12}{2}\right) = \left(\frac{7}{2}, -3\right)$.

57. $2x - y = 4 \Leftrightarrow y = 2x - 4 \Rightarrow m_1 = 2$ and $6x - 2y = 10 \Leftrightarrow 2y = 6x - 10 \Leftrightarrow y = 3x - 5 \Rightarrow$
$m_2 = 3$. Since $m_1 \neq m_2$, the two lines are not parallel. To find the point of intersection: $2x - 4 = 3x - 5 \Leftrightarrow$
$x = 1 \Rightarrow y = -2$. Thus, the point of intersection is $(1, -2)$.

59. With $A(1, 4)$ and $B(7, -2)$, the slope of segment AB is $\frac{-2-4}{7-1} = -1$, so its perpendicular bisector has slope 1.
The midpoint of AB is $\left(\frac{1+7}{2}, \frac{4+(-2)}{2}\right) = (4, 1)$, so an equation of the perpendicular bisector is $y - 1 = 1(x - 4)$
or $y = x - 3$.

61. (a) Since the x-intercept is a, the point $(a, 0)$ is on the line, and similarly since the y-intercept is b, $(0, b)$ is on the
line. Hence, the slope of the line is $m = \dfrac{b - 0}{0 - a} = -\dfrac{b}{a}$. Substituting into $y = mx + b$ gives $y = -\dfrac{b}{a}x + b \Leftrightarrow$
$\dfrac{b}{a}x + y = b \Leftrightarrow \dfrac{x}{a} + \dfrac{y}{b} = 1$.

(b) Letting $a = 6$ and $b = -8$ gives $\dfrac{x}{6} + \dfrac{y}{-8} = 1 \Leftrightarrow -8x + 6y = -48 \Leftrightarrow 6y = 8x - 48 \Leftrightarrow$
$y = \frac{4}{3}x - 8$.

C Graphs of Second-Degree Equations

1. From (1), the equation is $(x - 3)^2 + (y + 1)^2 = 25$.

3. The equation has the form $x^2 + y^2 = r^2$. Since $(4, 7)$ lies on the circle, we have $4^2 + 7^2 = r^2 \Rightarrow r^2 = 65$. So
the required equation is $x^2 + y^2 = 65$.

5. $x^2 + y^2 - 4x + 10y + 13 = 0 \Leftrightarrow x^2 - 4x + y^2 + 10y = -13 \Leftrightarrow$
$(x^2 - 4x + 4) + (y^2 + 10y + 25) = -13 + 4 + 25 = 16 \Leftrightarrow (x - 2)^2 + (y + 5)^2 = 4^2$. Thus, we have a circle
with center $(2, -5)$ and radius 4.

7. $x^2 + y^2 + x = 0 \Leftrightarrow \left(x^2 + x + \frac{1}{4}\right) + y^2 = \frac{1}{4} \Leftrightarrow \left(x + \frac{1}{2}\right)^2 + y^2 = \left(\frac{1}{2}\right)^2$. Thus, we have a circle with
center $\left(-\frac{1}{2}, 0\right)$ and radius $\frac{1}{2}$.

9. $2x^2 + 2y^2 - x + y = 1 \Leftrightarrow 2\left(x^2 - \frac{1}{2}x + \frac{1}{16}\right) + 2\left(y^2 + \frac{1}{2}y + \frac{1}{16}\right) = 1 + \frac{1}{8} + \frac{1}{8} \Leftrightarrow$
$2\left(x - \frac{1}{4}\right)^2 + 2\left(y + \frac{1}{4}\right)^2 = \frac{5}{4} \Leftrightarrow \left(x - \frac{1}{4}\right)^2 + \left(y + \frac{1}{4}\right)^2 = \frac{5}{8}$. Thus, we have a circle with center $\left(\frac{1}{4}, -\frac{1}{4}\right)$ and
radius $\frac{\sqrt{5}}{2\sqrt{2}} = \frac{\sqrt{10}}{4}$.

11. $y = -x^2$. Parabola

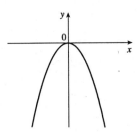

13. $x^2 + 4y^2 = 16$ ⟺ $\dfrac{x^2}{16} + \dfrac{y^2}{4} = 1$. Ellipse

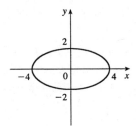

15. $16x^2 - 25y^2 = 400$ ⟺ $\dfrac{x^2}{25} - \dfrac{y^2}{16} = 1$.

   Hyperbola

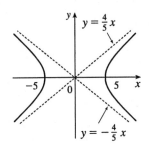

17. $4x^2 + y^2 = 1$ ⟺ $\dfrac{x^2}{1/4} + y^2 = 1$. Ellipse

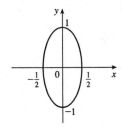

19. $x = y^2 - 1$. Parabola with vertex at $(-1, 0)$

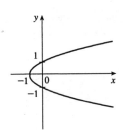

21. $9y^2 - x^2 = 9$ ⟺ $y^2 - \dfrac{x^2}{9} = 1$. Hyperbola

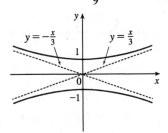

23. $xy = 4$. Hyperbola

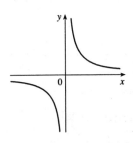

25. $9(x - 1)^2 + 4(y - 2)^2 = 36$ ⟺

   $\dfrac{(x - 1)^2}{4} + \dfrac{(y - 2)^2}{9} = 1$. Ellipse centered at

   $(1, 2)$

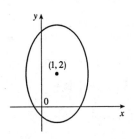

27. $y = x^2 - 6x + 13 = (x^2 - 6x + 9) + 4 = (x - 3)^2 + 4$. Parabola with vertex at $(3, 4)$

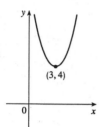

29. $x = -y^2 + 4$. Parabola with vertex at $(4, 0)$

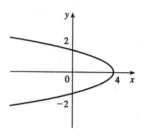

31. $x^2 + 4y^2 - 6x + 5 = 0 \Leftrightarrow$
$(x^2 - 6x + 9) + 4y^2 = -5 + 9 = 4 \Leftrightarrow$
$\dfrac{(x - 3)^2}{4} + y^2 = 1$. Ellipse centered at $(3, 0)$

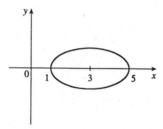

33. $y = 3x$ and $y = x^2$ intersect where $3x = x^2 \Leftrightarrow$
$0 = x^2 - 3x = x(x - 3)$, that is, at $(0, 0)$ and $(3, 9)$.

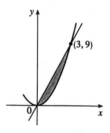

35. The parabola must have an equation of the form $y = a(x - 1)^2 - 1$. Substituting $x = 3$ and $y = 3$ into the equation gives $3 = a(3 - 1)^2 - 1$, so $a = 1$, and the equation is $y = (x - 1)^2 - 1 = x^2 - 2x$.
Note that using the other point $(-1, 3)$ would have given the same value for a, and hence the same equation.

37. $\{(x, y) \mid x^2 + y^2 \le 1\}$

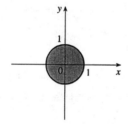

39. $\{(x, y) \mid y \ge x^2 - 1\}$

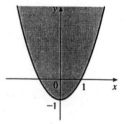

D Trigonometry

1. $210° = 210\left(\frac{\pi}{180}\right) = \frac{7\pi}{6}$ rad

3. $9° = 9\left(\frac{\pi}{180}\right) = \frac{\pi}{20}$ rad

5. $900° = 900\left(\frac{\pi}{180}\right) = 5\pi$ rad

7. 4π rad $= 4\pi\left(\frac{180}{\pi}\right) = 720°$

9. $\frac{5\pi}{12}$ rad $= \frac{5\pi}{12}\left(\frac{180}{\pi}\right) = 75°$

11. $-\frac{3\pi}{8}$ rad $= -\frac{3\pi}{8}\left(\frac{180}{\pi}\right) = -67.5°$

13. Using Formula 3, $a = r\theta = 36 \cdot \frac{\pi}{12} = 3\pi$ cm.

15. Using Formula 3, $\theta = a/r = \frac{1}{1.5} = \frac{2}{3}$ rad $= \frac{2}{3}\left(\frac{180}{\pi}\right) = \left(\frac{120}{\pi}\right)° \approx 38.2°$.

17.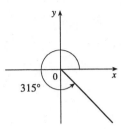

19.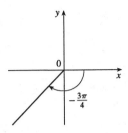

21.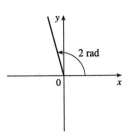

23.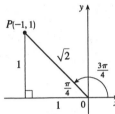

From the diagram we see that a point on the terminal side is $P(-1, 1)$. Therefore, taking $x = -1$, $y = 1$, $r = \sqrt{2}$ in the definitions of the trigonometric ratios, we have $\sin\frac{3\pi}{4} = \frac{1}{\sqrt{2}}$, $\cos\frac{3\pi}{4} = -\frac{1}{\sqrt{2}}$, $\tan\frac{3\pi}{4} = -1$, $\csc\frac{3\pi}{4} = \sqrt{2}$, $\sec\frac{3\pi}{4} = -\sqrt{2}$, and $\cot\frac{3\pi}{4} = -1$.

25.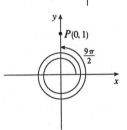

From the diagram we see that a point on the terminal line is $P(0, 1)$. Therefore taking $x = 0$, $y = 1$, $r = 1$ in the definitions of the trigonometric ratios, we have $\sin\frac{9\pi}{2} = 1$, $\cos\frac{9\pi}{2} = 0$, $\tan\frac{9\pi}{2} = y/x$ is undefined since $x = 0$, $\csc\frac{9\pi}{2} = 1$, $\sec\frac{9\pi}{2} = r/x$ is undefined since $x = 0$, and $\cot\frac{9\pi}{2} = 0$.

27.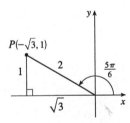

Using Figure 8 we see that a point on the terminal line is $P\left(-\sqrt{3}, 1\right)$. Therefore taking $x = -\sqrt{3}$, $y = 1$, $r = 2$ in the definitions of the trigonometric ratios, we have $\sin\frac{5\pi}{6} = \frac{1}{2}$, $\cos\frac{5\pi}{6} = -\frac{\sqrt{3}}{2}$, $\tan\frac{5\pi}{6} = -\frac{1}{\sqrt{3}}$, $\csc\frac{5\pi}{6} = 2$, $\sec\frac{5\pi}{6} = -\frac{2}{\sqrt{3}}$, and $\cot\frac{5\pi}{6} = -\sqrt{3}$.

29. $\sin\theta = y/r = \frac{3}{5} \implies y = 3$, $r = 5$, and $x = \sqrt{r^2 - y^2} = 4$ (since $0 < \theta < \frac{\pi}{2}$). Therefore taking $x = 4$, $y = 3$, $r = 5$ in the definitions of the trigonometric ratios, we have $\cos\theta = \frac{4}{5}$, $\tan\theta = \frac{3}{4}$, $\csc\theta = \frac{5}{3}$, $\sec\theta = \frac{5}{4}$, and $\cot\theta = \frac{4}{3}$.

31. $\frac{\pi}{2} < \phi < \pi \implies \phi$ is in the second quadrant, where x is negative and y is positive. Therefore $\sec\phi = r/x = -1.5 = -\frac{3}{2} \implies r = 3$, $x = -2$, and $y = \sqrt{r^2 - x^2} = \sqrt{5}$. Taking $x = -2$, $y = \sqrt{5}$, and $r = 3$ in the definitions of the trigonometric ratios, we have $\sin\phi = \frac{\sqrt{5}}{3}$, $\cos\phi = -\frac{2}{3}$, $\tan\phi = -\frac{\sqrt{5}}{2}$, $\csc\phi = \frac{3}{\sqrt{5}}$, and $\cot\theta = -\frac{2}{\sqrt{5}}$.

33. $\pi < \beta < 2\pi$ means that β is in the third or fourth quadrant where y is negative. Also since $\cot\beta = x/y = 3$ which is positive, x must also be negative. Therefore $\cot\beta = x/y = \frac{3}{1} \implies x = -3$, $y = -1$, and $r = \sqrt{x^2 + y^2} = \sqrt{10}$. Taking $x = -3$, $y = -1$ and $r = \sqrt{10}$ in the definitions of the trigonometric ratios, we have $\sin\beta = -\frac{1}{\sqrt{10}}$, $\cos\beta = -\frac{3}{\sqrt{10}}$, $\tan\beta = \frac{1}{3}$, $\csc\beta = -\sqrt{10}$, and $\sec\beta = -\frac{\sqrt{10}}{3}$.

35. $\sin 35° = \dfrac{x}{10} \;\Rightarrow\; x = 10 \sin 35° \approx 5.73576$ cm

37. $\tan \frac{2\pi}{5} = \dfrac{x}{8} \;\Rightarrow\; x = 8 \tan \frac{2\pi}{5} \approx 24.62147$ cm

39.

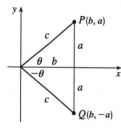

(a) From the diagram we see that $\sin \theta = \dfrac{y}{r} = \dfrac{a}{c}$, and
$$\sin(-\theta) = \dfrac{-a}{c} = -\dfrac{a}{c} = -\sin\theta.$$

(b) Again from the diagram we see that $\cos\theta = \dfrac{x}{r} = \dfrac{b}{c} = \cos(-\theta).$

41. (a) Using (12a) and (13a), we have
$$\tfrac{1}{2}\left[\sin(x+y)+\sin(x-y)\right] = \tfrac{1}{2}\left[\sin x \cos y + \cos x \sin y + \sin x \cos y - \cos x \sin y\right]$$
$$= \tfrac{1}{2}\,(2\sin x \cos y) = \sin x \cos y$$

(b) This time, using (12b) and (13b), we have
$$\tfrac{1}{2}\left[\cos(x+y)+\cos(x-y)\right] = \tfrac{1}{2}\left[\cos x \cos y - \sin x \sin y + \cos x \cos y + \sin x \sin y\right]$$
$$= \tfrac{1}{2}\,(2\cos x \cos y) = \cos x \cos y$$

(c) Again using (12b) and (13b), we have
$$\tfrac{1}{2}\left[\cos(x-y)-\cos(x+y)\right] = \tfrac{1}{2}\left[\cos x \cos y + \sin x \sin y - \cos x \cos y + \sin x \sin y\right]$$
$$= \tfrac{1}{2}\,(2\sin x \sin y) = \sin x \sin y$$

43. Using (12a), $\sin\left(\frac{\pi}{2}+x\right) = \sin\frac{\pi}{2}\cos x + \cos\frac{\pi}{2}\sin x = 1\cdot\cos x + 0\cdot\sin x = \cos x.$

45. Using (6), $\sin\theta\cot\theta = \sin\theta\cdot\dfrac{\cos\theta}{\sin\theta} = \cos\theta.$

47. $\sec y - \cos y = \dfrac{1}{\cos y} - \cos y$ [by (6)] $= \dfrac{1-\cos^2 y}{\cos y} = \dfrac{\sin^2 y}{\cos y}$ [by (7)] $= \dfrac{\sin y}{\cos y}\sin y = \tan y \sin y$ [by (6)]

49. $\cot^2\theta + \sec^2\theta = \dfrac{\cos^2\theta}{\sin^2\theta} + \dfrac{1}{\cos^2\theta}$ [by (6)] $= \dfrac{\cos^2\theta\cos^2\theta + \sin^2\theta}{\sin^2\theta\cos^2\theta}$
$$= \dfrac{(1-\sin^2\theta)(1-\sin^2\theta)+\sin^2\theta}{\sin^2\theta\cos^2\theta}\;\text{[by (7)]}\;= \dfrac{1-\sin^2\theta+\sin^4\theta}{\sin^2\theta\cos^2\theta}$$
$$= \dfrac{\cos^2\theta + \sin^4\theta}{\sin^2\theta\cos^2\theta}\;\text{[by (7)]}\;= \dfrac{1}{\sin^2\theta} + \dfrac{\sin^2\theta}{\cos^2\theta} = \csc^2\theta + \tan^2\theta\;\text{[by (6)]}$$

51. Using (14a), we have $\tan 2\theta = \tan(\theta+\theta) = \dfrac{\tan\theta+\tan\theta}{1-\tan\theta\tan\theta} = \dfrac{2\tan\theta}{1-\tan^2\theta}.$

53. Using (15a) and (16a),
$$\sin x \sin 2x + \cos x \cos 2x = \sin x\,(2\sin x \cos x) + \cos x\left(2\cos^2 x - 1\right) = 2\sin^2 x \cos x + 2\cos^3 x - \cos x$$
$$= 2\left(1-\cos^2 x\right)\cos x + 2\cos^3 x - \cos x \;\text{[by (7)]}$$
$$= 2\cos x - 2\cos^3 x + 2\cos^3 x - \cos x = \cos x$$

55. $\dfrac{\sin \phi}{1 - \cos \phi} = \dfrac{\sin \phi}{1 - \cos \phi} \cdot \dfrac{1 + \cos \phi}{1 + \cos \phi} = \dfrac{\sin \phi (1 + \cos \phi)}{1 - \cos^2 \phi} = \dfrac{\sin \phi (1 + \cos \phi)}{\sin^2 \phi}$ [by (7)]

$= \dfrac{1 + \cos \phi}{\sin \phi} = \dfrac{1}{\sin \phi} + \dfrac{\cos \phi}{\sin \phi} = \csc \phi + \cot \phi$ [by (6)]

57. Using (12a),

$\sin 3\theta + \sin \theta = \sin (2\theta + \theta) + \sin \theta = \sin 2\theta \cos \theta + \cos 2\theta \sin \theta + \sin \theta$

$= \sin 2\theta \cos \theta + \left(2 \cos^2 \theta - 1\right) \sin \theta + \sin \theta$ [by (16a)]

$= \sin 2\theta \cos \theta + 2 \cos^2 \theta \sin \theta - \sin \theta + \sin \theta = \sin 2\theta \cos \theta + \sin 2\theta \cos \theta$ [by (15a)]

$= 2 \sin 2\theta \cos \theta$

59. Since $\sin x = \frac{1}{3}$ we can label the opposite side as having
length 1, the hypotenuse as having length 3, and use the
Pythagorean Theorem to get that the adjacent side has
length $\sqrt{8}$. Then, from the diagram, $\cos x = \frac{\sqrt{8}}{3}$. Similarly
we have that $\sin y = \frac{3}{5}$. Now use (12a):

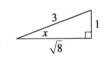
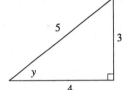

$\sin (x + y) = \sin x \cos y + \cos x \sin y = \frac{1}{3} \cdot \frac{4}{5} + \frac{\sqrt{8}}{3} \cdot \frac{3}{5}$

$= \frac{4}{15} + \frac{3\sqrt{8}}{15} = \frac{4 + 6\sqrt{2}}{15}$

61. Using (13b) and the values for $\cos x$ and $\sin y$ obtained in Exercise 59, we have

$\cos (x - y) = \cos x \cos y + \sin x \sin y = \frac{\sqrt{8}}{3} \cdot \frac{4}{5} + \frac{1}{3} \cdot \frac{3}{5} = \frac{8\sqrt{2} + 3}{15}$

63. Using (15a) and the value for $\sin y$ obtained in Exercise 59, we have

$\sin 2y = 2 \sin y \cos y = \dfrac{2 \sin y}{\sec y} = 2 \left(\frac{3}{5}\right) \left(\frac{4}{5}\right) = \frac{24}{25}$

65. $2 \cos x - 1 = 0 \iff \cos x = \frac{1}{2} \implies x = \frac{\pi}{3}, \frac{5\pi}{3}$

67. $2 \sin^2 x = 1 \iff \sin^2 x = \frac{1}{2} \iff \sin x = \pm \frac{1}{\sqrt{2}} \implies x = \frac{\pi}{4}, \frac{3\pi}{4}, \frac{5\pi}{4}, \frac{7\pi}{4}$.

69. Using (15a), $\sin 2x = \cos x \implies 2 \sin x \cos x - \cos x = 0 \iff \cos x (2 \sin x - 1) = 0 \iff \cos x = 0$ or
$2 \sin x - 1 = 0 \implies x = \frac{\pi}{2}, \frac{3\pi}{2}$ or $\sin x = \frac{1}{2} \implies x = \frac{\pi}{6}$ or $\frac{5\pi}{6}$. Therefore, the solutions are $x = \frac{\pi}{6}, \frac{\pi}{2}, \frac{5\pi}{6}$,
$\frac{3\pi}{2}$.

71. $\sin x = \tan x \iff \sin x - \tan x = 0 \iff \sin x - \dfrac{\sin x}{\cos x} = 0 \iff \sin x \left(1 - \dfrac{1}{\cos x}\right) = 0 \iff \sin x = 0$ or
$1 - \dfrac{1}{\cos x} = 0 \implies x = 0, \pi, 2\pi$ or $1 = \dfrac{1}{\cos x} \implies \cos x = 1 \implies x = 0, 2\pi$. Therefore the solutions are
$x = 0, \pi, 2\pi$.

73. We know that $\sin x = \frac{1}{2}$ when $x = \frac{\pi}{6}$ or $\frac{5\pi}{6}$, and from Figure 13(a), we see that $\sin x \leq \frac{1}{2} \implies 0 \leq x \leq \frac{\pi}{6}$ or
$\frac{5\pi}{6} \leq x \leq 2\pi$.

75. $\tan x = -1$ when $x = \frac{3\pi}{4}, \frac{7\pi}{4}$, and $\tan x = 1$ when $x = \frac{\pi}{4}$ or $\frac{5\pi}{4}$. From Figure 14 we see that $-1 < \tan x < 1 \implies$
$0 \leq x < \frac{\pi}{4}, \frac{3\pi}{4} < x < \frac{5\pi}{4}$, and $\frac{7\pi}{4} < x \leq 2\pi$.

77. $y = \cos\left(x - \frac{\pi}{3}\right)$. We start with the graph of $y = \cos x$ and shift it $\frac{\pi}{3}$ units to the right.

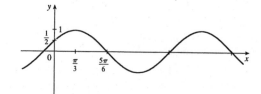

79. $y = \frac{1}{3}\tan\left(x - \frac{\pi}{2}\right)$. We start with the graph of $y = \tan x$, shift it $\frac{\pi}{2}$ units to the right and compress it to $\frac{1}{3}$ of its original vertical size.

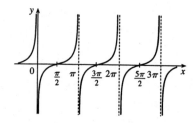

81. $y = |\sin x|$. We start with the graph of $y = \sin x$ and reflect the parts below the x-axis about the x-axis.

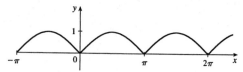

83. (a) $\sin^{-1}\left(\frac{\sqrt{3}}{2}\right) = \frac{\pi}{3}$ since $\sin\frac{\pi}{3} = \frac{\sqrt{3}}{2}$ and $\frac{\pi}{3}$ is in $\left[-\frac{\pi}{2}, \frac{\pi}{2}\right]$.

(b) $\cos^{-1}(-1) = \pi$ since $\cos\pi = -1$ and π is in $[0, \pi]$.

85. (a) $\tan^{-1}\sqrt{3} = \frac{\pi}{3}$ since $\tan\frac{\pi}{3} = \sqrt{3}$ and $\frac{\pi}{3}$ is in $\left(-\frac{\pi}{2}, \frac{\pi}{2}\right)$.

(b) $\arcsin\left(-\frac{1}{\sqrt{2}}\right) = -\frac{\pi}{4}$ since $\sin\left(-\frac{\pi}{4}\right) = -\frac{1}{\sqrt{2}}$ and $-\frac{\pi}{4}$ is in $\left[-\frac{\pi}{2}, \frac{\pi}{2}\right]$.

87. (a) $\sin\left(\sin^{-1}0.7\right) = 0.7$ since 0.7 is in $[-1, 1]$.

(b) $\tan^{-1}\left(\tan\frac{4\pi}{3}\right) = \tan^{-1}\sqrt{3} = \frac{\pi}{3}$ since $\frac{\pi}{3}$ is in $\left[-\frac{\pi}{2}, \frac{\pi}{2}\right]$.

89. Let $\theta = \cos^{-1}\frac{4}{5}$, so $\cos\theta = \frac{4}{5}$. Then $\sin\left(\cos^{-1}\frac{4}{5}\right) = \sin\theta = \sqrt{1 - \left(\frac{4}{5}\right)^2} = \sqrt{\frac{9}{25}} = \frac{3}{5}$.

91. Let $\theta = \sin^{-1}\frac{5}{13}$. Then $\sin\theta = \frac{5}{13}$, so $\cos\left(2\sin^{-1}\frac{5}{13}\right) = \cos 2\theta = 1 - 2\sin^2\theta = 1 - 2\left(\frac{5}{13}\right)^2 = \frac{119}{169}$.

93. Let $y = \sin^{-1}x$. Then $-\frac{\pi}{2} \le y \le \frac{\pi}{2} \Rightarrow \cos y \ge 0$, so $\cos\left(\sin^{-1}x\right) = \cos y = \sqrt{1 - \sin^2 y} = \sqrt{1 - x^2}$

95. Let $y = \tan^{-1}x$. Then $\tan y = x$, so from the triangle we see that
$$\sin\left(\tan^{-1}x\right) = \sin y = \frac{x}{\sqrt{1 + x^2}}.$$

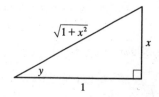

97.

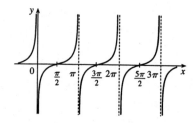

The graph of $\sin^{-1}x$ is the reflection of the graph of $\sin x$ about the line $y = x$.

99. From the figure in the text, we see that $x = b\cos\theta$, $y = b\sin\theta$, and from the distance formula we have that the distance c from (x, y) to $(a, 0)$ is $c = \sqrt{(x - a)^2 + (y - 0)^2} \Rightarrow$

$$c^2 = (b\cos\theta - a)^2 + (b\sin\theta)^2 = b^2\cos^2\theta - 2ab\cos\theta + a^2 + b^2\sin^2\theta$$
$$= a^2 + b^2\left(\cos^2\theta + \sin^2\theta\right) - 2ab\cos\theta = a^2 + b^2 - 2ab\cos\theta \quad [\text{by (7)}]$$

101. Using the Law of Cosines, we have $c^2 = 1^2 + 1^2 - 2(1)(1)\cos(\alpha - \beta) = 2[1 - \cos(\alpha - \beta)]$. Now, using the distance formula, $c^2 = |AB|^2 = (\cos\alpha - \cos\beta)^2 + (\sin\alpha - \sin\beta)^2$. Equating these two expressions for c^2, we get $2[1 - \cos(\alpha - \beta)] = \cos^2\alpha + \sin^2\alpha + \cos^2\beta + \sin^2\beta - 2\cos\alpha\cos\beta - 2\sin\alpha\sin\beta \Rightarrow$ $1 - \cos(\alpha - \beta) = 1 - \cos\alpha\cos\beta - \sin\alpha\sin\beta \Rightarrow \cos(\alpha - \beta) = \cos\alpha\cos\beta + \sin\alpha\sin\beta$.

103. In Exercise 102 we used the subtraction formula for cosine to prove the addition formula for cosine. Using that formula with $x = \frac{\pi}{2} - \alpha$, $y = \beta$, we get $\cos\left[\left(\frac{\pi}{2} - \alpha\right) + \beta\right] = \cos\left(\frac{\pi}{2} - \alpha\right)\cos\beta - \sin\left(\frac{\pi}{2} - \alpha\right)\sin\beta \Rightarrow$ $\cos\left[\frac{\pi}{2} - (\alpha - \beta)\right] = \cos\left(\frac{\pi}{2} - \alpha\right)\cos\beta - \sin\left(\frac{\pi}{2} - \alpha\right)\sin\beta$. Now we use the identities given in the problem, $\cos\left(\frac{\pi}{2} - \theta\right) = \sin\theta$ and $\sin\left(\frac{\pi}{2} - \theta\right) = \cos\theta$, to get $\sin(\alpha - \beta) = \sin\alpha\cos\beta - \cos\alpha\sin\beta$.

105. Using the formula from Exercise 104, the area of the triangle is $\frac{1}{2}(10)(3)\sin 107° \approx 14.34457$.

E Sigma Notation

1. $\sum_{i=1}^{5}\sqrt{i} = \sqrt{1} + \sqrt{2} + \sqrt{3} + \sqrt{4} + \sqrt{5}$

3. $\sum_{i=4}^{6} 3^i = 3^4 + 3^5 + 3^6$

5. $\sum_{k=0}^{4}\frac{2k-1}{2k+1} = -1 + \frac{1}{3} + \frac{3}{5} + \frac{5}{7} + \frac{7}{9}$

7. $\sum_{i=1}^{n} i^{10} = 1^{10} + 2^{10} + 3^{10} + \cdots + n^{10}$

9. $\sum_{j=0}^{n-1}(-1)^j = 1 - 1 + 1 - 1 + \cdots + (-1)^{n-1}$

11. $1 + 2 + 3 + 4 + \cdots + 10 = \sum_{i=1}^{10} i$

13. $\frac{1}{2} + \frac{2}{3} + \frac{3}{4} + \frac{4}{5} + \cdots + \frac{19}{20} = \sum_{i=1}^{19}\frac{i}{i+1}$

15. $2 + 4 + 6 + 8 + \cdots + 2n = \sum_{i=1}^{n} 2i$

17. $1 + 2 + 4 + 8 + 16 + 32 = \sum_{i=0}^{5} 2^i$

19. $x + x^2 + x^3 + \cdots + x^n = \sum_{i=1}^{n} x^i$

21. $\sum_{i=4}^{8}(3i - 2) = 10 + 13 + 16 + 19 + 22 = 80$

23. $\sum_{j=1}^{6} 3^{j+1} = 3^2 + 3^3 + 3^4 + 3^5 + 3^6 + 3^7 = 9 + 27 + 81 + 243 + 729 + 2187 = 3276$
(For a more general method, see Exercise 47.)

25. $\sum_{n=1}^{20}(-1)^n = -1 + 1 - 1 + 1 - 1 + 1 - 1 + 1 - 1 + 1 - 1 + 1 - 1 + 1 - 1 + 1 - 1 + 1 - 1 + 1 = 0$

27. $\sum_{i=0}^{4}(2^i + i^2) = (1 + 0) + (2 + 1) + (4 + 4) + (8 + 9) + (16 + 16) = 61$

29. $\sum_{i=1}^{n} 2i = 2\sum_{i=1}^{n} i = n(n + 1)$

31. $\displaystyle\sum_{i=1}^{n}(i^2 + 3i + 4) = \sum_{i=1}^{n} i^2 + 3\sum_{i=1}^{n} i + \sum_{i=1}^{n} 4 = \frac{n(n + 1)(2n + 1)}{6} + \frac{3n(n + 1)}{2} + 4n$
$$= \frac{1}{6}\left[(2n^3 + 3n^2 + n) + (9n^2 + 9n) + 24n\right] = \frac{1}{6}\left(2n^3 + 12n^2 + 34n\right) = \frac{1}{3}n\left(n^2 + 6n + 17\right)$$

33. $\displaystyle\sum_{i=1}^{n}(i + 1)(i + 2) = \sum_{i=1}^{n}(i^2 + 3i + 2) = \sum_{i=1}^{n} i^2 + 3\sum_{i=1}^{n} i + \sum_{i=1}^{n} 2$
$$= \frac{n(n + 1)(2n + 1)}{6} + \frac{3n(n + 1)}{2} + 2n = \frac{n(n + 1)}{6}\left[(2n + 1) + 9\right] + 2n$$
$$= \frac{n(n + 1)}{3}(n + 5) + 2n = \frac{n}{3}\left[(n + 1)(n + 5) + 6\right] = \frac{n}{3}\left(n^2 + 6n + 11\right)$$

35. $\sum_{i=1}^{n} (i^3 - i - 2) = \sum_{i=1}^{n} i^3 - \sum_{i=1}^{n} i - \sum_{i=1}^{n} 2 = \left[\dfrac{n(n+1)}{2}\right]^2 - \dfrac{n(n+1)}{2} - 2n$

$$= \tfrac{1}{4}n(n+1)[n(n+1) - 2] - 2n = \tfrac{1}{4}n(n+1)(n+2)(n-1) - 2n$$

$$= \tfrac{1}{4}n[(n+1)(n-1)(n+2) - 8] = \tfrac{1}{4}n[(n^2-1)(n+2) - 8] = \tfrac{1}{4}n(n^3 + 2n^2 - n - 10)$$

37. By Theorem 2(a) and Example 3, $\sum_{i=1}^{n} c = c \sum_{i=1}^{n} 1 = cn$.

39. $\sum_{i=1}^{n} [(i+1)^4 - i^4] = (2^4 - 1^4) + (3^4 - 2^4) + (4^4 - 3^4) + \cdots + [(n+1)^4 - n^4]$

$$= (n+1)^4 - 1^4 = n^4 + 4n^3 + 6n^2 + 4n$$

On the other hand,

$$\sum_{i=1}^{n} [(i+1)^4 - i^4] = \sum_{i=1}^{n} (4i^3 + 6i^2 + 4i + 1) = 4\sum_{i=1}^{n} i^3 + 6\sum_{i=1}^{n} i^2 + 4\sum_{i=1}^{n} i + \sum_{i=1}^{n} 1$$

$$= 4S + n(n+1)(2n+1) + 2n(n+1) + n \quad \left(\text{where } S = \sum_{i=1}^{n} i^3\right)$$

$$= 4S + 2n^3 + 3n^2 + n + 2n^2 + 2n + n = 4S + 2n^3 + 5n^2 + 4n$$

Thus, $n^4 + 4n^3 + 6n^2 + 4n = 4S + 2n^3 + 5n^2 + 4n$, from which it follows that

$$4S = n^4 + 2n^3 + n^2 = n^2(n^2 + 2n + 1) = n^2(n+1)^2 \text{ and } S = \left[\dfrac{n(n+1)}{2}\right]^2.$$

41. (a) $\sum_{i=1}^{n} [i^4 - (i-1)^4] = (1^4 - 0^4) + (2^4 - 1^4) + (3^4 - 2^4) + \cdots + [n^4 - (n-1)^4] = n^4 - 0 = n^4$

(b) $\sum_{i=1}^{100} (5^i - 5^{i-1}) = (5^1 - 5^0) + (5^2 - 5^1) + (5^3 - 5^2) + \cdots + (5^{100} - 5^{99}) = 5^{100} - 5^0 = 5^{100} - 1$

(c) $\sum_{i=3}^{99} \left(\dfrac{1}{i} - \dfrac{1}{i+1}\right) = \left(\dfrac{1}{3} - \dfrac{1}{4}\right) + \left(\dfrac{1}{4} - \dfrac{1}{5}\right) + \left(\dfrac{1}{5} - \dfrac{1}{6}\right) + \cdots + \left(\dfrac{1}{99} - \dfrac{1}{100}\right) = \dfrac{1}{3} - \dfrac{1}{100} = \dfrac{97}{300}$

(d) $\sum_{i=1}^{n} (a_i - a_{i-1}) = (a_1 - a_0) + (a_2 - a_1) + (a_3 - a_2) + \cdots + (a_n - a_{n-1}) = a_n - a_0$

43. $\lim_{n\to\infty} \sum_{i=1}^{n} \dfrac{1}{n} \left(\dfrac{i}{n}\right)^2 = \lim_{n\to\infty} \dfrac{1}{n^3} \sum_{i=1}^{n} i^2 = \lim_{n\to\infty} \dfrac{1}{n^3} \dfrac{n(n+1)(2n+1)}{6} = \lim_{n\to\infty} \dfrac{1}{6} \left(1 + \dfrac{1}{n}\right)\left(2 + \dfrac{1}{n}\right) = \tfrac{1}{6}(1)(2) = \tfrac{1}{3}$

45. $\lim_{n\to\infty} \sum_{i=1}^{n} \dfrac{2}{n} \left[\left(\dfrac{2i}{n}\right)^3 + 5\left(\dfrac{2i}{n}\right)\right] = \lim_{n\to\infty} \sum_{i=1}^{n} \left[\dfrac{16}{n^4} i^3 + \dfrac{20}{n^2} i\right] = \lim_{n\to\infty} \left[\dfrac{16}{n^4} \sum_{i=1}^{n} i^3 + \dfrac{20}{n^2} \sum_{i=1}^{n} i\right]$

$$= \lim_{n\to\infty} \left[\dfrac{16}{n^4} \dfrac{n^2(n+1)^2}{4} + \dfrac{20}{n^2} \dfrac{n(n+1)}{2}\right] = \lim_{n\to\infty} \left[\dfrac{4(n+1)^2}{n^2} + \dfrac{10n(n+1)}{n^2}\right]$$

$$= \lim_{n\to\infty} \left[4\left(1 + \dfrac{1}{n}\right)^2 + 10\left(1 + \dfrac{1}{n}\right)\right] = 4 \cdot 1 + 10 \cdot 1 = 14$$

47. Let $S = \sum_{i=1}^{n} ar^{i-1} = a + ar + ar^2 + \cdots + ar^{n-1}$. Then $rS = ar + ar^2 + \cdots + ar^{n-1} + ar^n$. Subtracting the

first equation from the second, we find $(r-1)S = ar^n - a = a(r^n - 1)$, so $S = \dfrac{a(r^n - 1)}{r - 1}$ (since $r \neq 1$).

49. $\sum_{i=1}^{n} (2i + 2^i) = 2\sum_{i=1}^{n} i + \sum_{i=1}^{n} 2 \cdot 2^{i-1} = 2\dfrac{n(n+1)}{2} + \dfrac{2(2^n - 1)}{2 - 1} = 2^{n+1} + n^2 + n - 2.$

For the first sum we have used Theorem 3(c), and for the second, Exercise 47 with $a = r = 2$.

G Complex Numbers

1. $(3 + 2i) + (7 - 3i) = (3 + 7) + (2 - 3)i = 10 - i$

3. $(3 - i)(4 + i) = 12 + 3i - 4i - (-1) = 13 - i$

5. $\overline{12 + 7i} = 12 - 7i$

7. $\dfrac{2 + 3i}{1 - 5i} = \dfrac{2 + 3i}{1 - 5i} \cdot \dfrac{1 + 5i}{1 + 5i} = \dfrac{2 + 10i + 3i + 15(-1)}{1 - 25(-1)} = \dfrac{-13 + 13i}{26} = -\tfrac{1}{2} + \tfrac{1}{2}i$

9. $\dfrac{1}{1 + i} = \dfrac{1}{1 + i} \cdot \dfrac{1 - i}{1 - i} = \dfrac{1 - i}{1 - (-1)} = \dfrac{1 - i}{2} = \tfrac{1}{2} - \tfrac{1}{2}i$

11. $i^3 = i^2 \cdot i = (-1)i = -i$

13. $\sqrt{-25} = \sqrt{25}\,i = 5i$

15. $\overline{3 + 4i} = 3 - 4i,\ |3 + 4i| = \sqrt{3^2 + 4^2} = \sqrt{25} = 5$

17. $\overline{-4i} = \overline{0 - 4i} = 0 + 4i = 4i,\ |-4i| = \sqrt{0^2 + (-4)^2} = 4$

19. $4x^2 + 9 = 0 \iff 4x^2 = -9 \iff x^2 = -\tfrac{9}{4} \iff x = \pm\sqrt{-\tfrac{9}{4}} = \pm\sqrt{\tfrac{9}{4}}\,i = \pm\tfrac{3}{2}i.$

21. By the quadratic formula, $x^2 - 8x + 17 = 0 \iff x = \dfrac{8 \pm \sqrt{(-8)^2 - 4(1)(17)}}{2(1)} = \dfrac{8 \pm \sqrt{-4}}{2} = \dfrac{8 \pm 2i}{2} = 4 \pm i.$

23. By the quadratic formula, $z^2 + z + 2 = 0 \iff z = \dfrac{-1 \pm \sqrt{1^2 - 4(1)(2)}}{2(1)} = \dfrac{-1 \pm \sqrt{-7}}{2} = -\tfrac{1}{2} \pm \tfrac{\sqrt{7}}{2}i.$

25. For $z = -3 + 3i,\ r = \sqrt{(-3)^2 + 3^2} = 3\sqrt{2}$ and $\tan\theta = \tfrac{3}{-3} = -1 \implies \theta = \tfrac{3}{4}\pi$ (since z lies in the second quadrant). Therefore, $-3 + 3i = 3\sqrt{2}\left(\cos\tfrac{3\pi}{4} + i\sin\tfrac{3\pi}{4}\right).$

27. For $z = 3 + 4i,\ r = \sqrt{3^2 + 4^2} = 5$ and $\tan\theta = \tfrac{4}{3} \implies \theta = \tan^{-1}\tfrac{4}{3}$ (since z lies in the second quadrant). Therefore, $3 + 4i = 5\left[\cos\left(\tan^{-1}\tfrac{4}{3}\right) + i\sin\left(\tan^{-1}\tfrac{4}{3}\right)\right].$

29. For $z = \sqrt{3} + i,\ r = \sqrt{\left(\sqrt{3}\right)^2 + 1^2} = 2$ and $\tan\theta = \tfrac{1}{\sqrt{3}} \implies \theta = \tfrac{\pi}{6} \implies z = 2\left(\cos\tfrac{\pi}{6} + i\sin\tfrac{\pi}{6}\right)$. For $w = 1 + \sqrt{3}i,\ r = 2$ and $\tan\theta = \sqrt{3} \implies \theta = \tfrac{\pi}{3} \implies w = 2\left(\cos\tfrac{\pi}{3} + i\sin\tfrac{\pi}{3}\right).$

Therefore, $zw = 2 \cdot 2\left[\cos\left(\tfrac{\pi}{6} + \tfrac{\pi}{3}\right) + i\sin\left(\tfrac{\pi}{6} + \tfrac{\pi}{3}\right)\right] = 4\left(\cos\tfrac{\pi}{2} + i\sin\tfrac{\pi}{2}\right),$

$z/w = \tfrac{2}{2}\left[\cos\left(\tfrac{\pi}{6} - \tfrac{\pi}{3}\right) + i\sin\left(\tfrac{\pi}{6} - \tfrac{\pi}{3}\right)\right] = \cos\left(-\tfrac{\pi}{6}\right) + i\sin\left(-\tfrac{\pi}{6}\right),$ and $1 = 1 + 0i = 1(\cos 0 + i\sin 0) \implies$

$1/z = \tfrac{1}{2}\left[\cos\left(0 - \tfrac{\pi}{6}\right) + i\sin\left(0 - \tfrac{\pi}{6}\right)\right] = \tfrac{1}{2}\left[\cos\left(-\tfrac{\pi}{6}\right) + i\sin\left(-\tfrac{\pi}{6}\right)\right].$ For $1/z$, we could also use the formula that precedes Example 5 to obtain $1/z = \tfrac{1}{8}\left(\cos\tfrac{\pi}{6} - i\sin\tfrac{\pi}{6}\right).$

31. For $z = 2\sqrt{3} - 2i,\ r = \sqrt{\left(2\sqrt{3}\right)^2 + (-2)^2} = 4$ and $\tan\theta = \tfrac{-2}{2\sqrt{3}} = -\tfrac{1}{\sqrt{3}} \implies$

$\theta = -\tfrac{\pi}{6} \implies z = 4\left[\cos\left(-\tfrac{\pi}{6}\right) + i\sin\left(-\tfrac{\pi}{6}\right)\right].$ For $w = -1 + i,\ r = \sqrt{2},$

$\tan\theta = \tfrac{1}{-1} = -1 \implies \theta = \tfrac{3\pi}{4} \implies z = \sqrt{2}\left(\cos\tfrac{3\pi}{4} + i\sin\tfrac{3\pi}{4}\right).$ Therefore,

$zw = 4\sqrt{2}\left[\cos\left(-\tfrac{\pi}{6} + \tfrac{3\pi}{4}\right) + i\sin\left(-\tfrac{\pi}{6} + \tfrac{3\pi}{4}\right)\right] = 4\sqrt{2}\left(\cos\tfrac{7\pi}{12} + i\sin\tfrac{7\pi}{12}\right),$

$z/w = \tfrac{4}{\sqrt{2}}\left[\cos\left(-\tfrac{\pi}{6} - \tfrac{3\pi}{4}\right) + i\sin\left(-\tfrac{\pi}{6} - \tfrac{3\pi}{4}\right)\right] = \tfrac{4}{\sqrt{2}}\left[\cos\left(-\tfrac{11\pi}{12}\right) + i\sin\left(-\tfrac{11\pi}{12}\right)\right]$

$= 2\sqrt{2}\left(\cos\tfrac{13\pi}{12} + i\sin\tfrac{13\pi}{12}\right),$ and

$1/z = \tfrac{1}{4}\left[\cos\left(-\tfrac{\pi}{6}\right) - i\sin\left(-\tfrac{\pi}{6}\right)\right] = \tfrac{1}{4}\left(\cos\tfrac{\pi}{6} + i\sin\tfrac{\pi}{6}\right).$

33. For $z = 1 + i$, $r = \sqrt{2}$ and $\tan\theta = \frac{1}{1} = 1 \;\Rightarrow\; \theta = \frac{\pi}{4} \;\Rightarrow\; z = \sqrt{2}\left(\cos\frac{\pi}{4} + i\sin\frac{\pi}{4}\right)$. So by De Moivre's Theorem,

$$(1 + i)^{20} = \left[\sqrt{2}\left(\cos\frac{\pi}{4} + i\sin\frac{\pi}{4}\right)\right]^{20} = \left(2^{1/2}\right)^{20}\left(\cos\frac{20\cdot\pi}{4} + i\sin\frac{20\cdot\pi}{4}\right)$$

$$= 2^{10}\left(\cos 5\pi + i\sin 5\pi\right) = 2^{10}\left[-1 + i\,(0)\right] = -2^{10} = -1024$$

35. For $z = 2\sqrt{3} + 2i$, $r = 4$ and $\tan\theta = \frac{2}{2\sqrt{3}} = \frac{1}{\sqrt{3}} \;\Rightarrow\; \theta = \frac{\pi}{6} \;\Rightarrow\; z = 4\left(\cos\frac{\pi}{6} + i\sin\frac{\pi}{6}\right)$. So by De Moivre's Theorem,

$$\left(2\sqrt{3} + 2i\right)^{5} = \left[4\left(\cos\frac{\pi}{6} + i\sin\frac{\pi}{6}\right)\right]^{5} = 4^{5}\left(\cos\frac{5\pi}{6} + i\sin\frac{5\pi}{6}\right) = 1024\left[-\frac{\sqrt{3}}{2} + \frac{1}{2}i\right] = -512\sqrt{3} + 512i$$

37. $1 = 1 + 0i = 1\,(\cos 0 + i\sin 0)$. Using Equation 3 with $r = 1$, $n = 8$, and $\theta = 0$, we have

$$w_k = 1^{1/8}\left[\cos\left(\frac{0 + 2k\pi}{8}\right) + i\sin\left(\frac{0 + 2k\pi}{8}\right)\right] = \cos\frac{k\pi}{4} + i\sin\frac{k\pi}{4},\ \text{where } k = 0, 1, 2, \ldots, 7.$$

$w_0 = 1\,(\cos 0 + i\sin 0) = 1$, $w_1 = 1\left(\cos\frac{\pi}{4} + i\sin\frac{\pi}{4}\right) = \frac{1}{\sqrt{2}} + \frac{1}{\sqrt{2}}i$,

$w_2 = 1\left(\cos\frac{\pi}{2} + i\sin\frac{\pi}{2}\right) = i$, $w_3 = 1\left(\cos\frac{3\pi}{4} + i\sin\frac{3\pi}{4}\right) = -\frac{1}{\sqrt{2}} + \frac{1}{\sqrt{2}}i$,

$w_4 = 1\,(\cos\pi + i\sin\pi) = -1$, $w_5 = 1\left(\cos\frac{5\pi}{4} + i\sin\frac{5\pi}{4}\right) = -\frac{1}{\sqrt{2}} - \frac{1}{\sqrt{2}}i$,

$w_6 = 1\left(\cos\frac{3\pi}{2} + i\sin\frac{3\pi}{2}\right) = -i$, $w_7 = 1\left(\cos\frac{7\pi}{4} + i\sin\frac{7\pi}{4}\right) = \frac{1}{\sqrt{2}} - \frac{1}{\sqrt{2}}i$

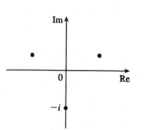

39. $i = 0 + i = 1\left(\cos\frac{\pi}{2} + i\sin\frac{\pi}{2}\right)$. Using Equation 3 with $r = 1$, $n = 3$, and $\theta = \frac{\pi}{2}$, we have

$$w_k = 1^{1/3}\left[\cos\left(\frac{\frac{\pi}{2} + 2k\pi}{3}\right) + i\sin\left(\frac{\frac{\pi}{2} + 2k\pi}{3}\right)\right],\ \text{where } k = 0, 1, 2.$$

$w_0 = \left(\cos\frac{\pi}{6} + i\sin\frac{\pi}{6}\right) = \frac{\sqrt{3}}{2} + \frac{1}{2}i$

$w_1 = \left(\cos\frac{5\pi}{6} + i\sin\frac{5\pi}{6}\right) = -\frac{\sqrt{3}}{2} + \frac{1}{2}i$

$w_2 = \left(\cos\frac{9\pi}{6} + i\sin\frac{9\pi}{6}\right) = -i$

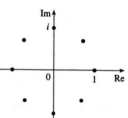

41. Using Euler's formula (6) with $y = \frac{\pi}{2}$, we have $e^{i\pi/2} = \cos\frac{\pi}{2} + i\sin\frac{\pi}{2} = 0 + 1i = i$.

43. Using Euler's formula with $y = \frac{3\pi}{4}$, we have $e^{i3\pi/4} = \cos\frac{3\pi}{4} + i\sin\frac{3\pi}{4} = -\frac{1}{\sqrt{2}} + \frac{1}{\sqrt{2}}i$.

45. Using Equation 7 with $x = 2$ and $y = \pi$, we have $e^{2+i\pi} = e^{2}e^{i\pi} = e^{2}\,(\cos\pi + i\sin\pi) = e^{2}\,(-1 + 0) = -e^{2}$.

47. $F(x) = e^{rx} = e^{(a+bi)x} = e^{ax+bxi} = e^{ax}\,(\cos bx + i\sin bx) = e^{ax}\cos bx + i\,(e^{ax}\sin bx) \;\Rightarrow\;$

$F'(x) = (e^{ax}\cos bx)' + i\,(e^{ax}\sin bx)' = (ae^{ax}\cos bx - be^{ax}\sin bx) + i\,(ae^{ax}\sin bx + be^{ax}\cos bx)$

$= a\left[e^{ax}\,(\cos bx + i\sin bx)\right] + b\left[e^{ax}\,(-\sin bx + i\cos bx)\right] = ae^{rx} + b\left[e^{ax}\,(i^{2}\sin bx + i\cos bx)\right]$

$= ae^{rx} + bi\left[e^{ax}\,(\cos bx + i\sin bx)\right] = ae^{rx} + bie^{rx} = (a + bi)\,e^{rx} = re^{rx}$